Innovations in Army Energy and Power Materials Technologies

Edited by

Dr. Edward C. Shaffer
Chief, Energy and Power Division
Sensors and Electronic Devices Directorate
US Army Research Laboratory, Adelphi, MD 20783

Dr. Tsvetanka S. Zheleva
Associate Chief, Energy and Power Division
Sensors and Electronic Devices Directorate
US Army Research Laboratory, Adelphi, MD 20783

Published under license by **Materials Research Forum LLC**
Millersville, PA 17551, USA

Published as part of the book series
Materials Research Foundations
Volume 36 (2018)
ISSN 2471-8890 (Print)
ISSN 2471-8904 (Online)

Print ISBN 978-1-945291-78-4
ePDF ISBN 978-1-945291-79-1

Distributed worldwide by

Materials Research Forum LLC
105 Springdale Lane
Millersville, PA 17551
USA
http://www.mrforum.com

Manufactured in the United States of America
10 9 8 7 6 5 4 3 2 1

Table of Contents

Preface

Dr. Edward C. Shaffer
Chef, Energy and Power Division
US Army Research Laboratory
Edward.c.shaffer.civ@mail.mil

This compendium captures the considerable experience and technical depth residing within the US Army Research Labratory (ARL) specifically within the area of Energy and Power (E&P) technologies. ARLEnergy and Power Division researchers leverage an extensive partner network within the lab, with academe, with the Army Research Development and Engineering Centers, with other services and agencies, and with industry.

Defense Science and Technology Context

The primary purpose of this work is to document the "why, what and how" of Army research in the E&P area, providing a snapshot of how top-driven Army needs and bottom-up internal discovery and innovation are reconciled in practice by investigator approaches and results. As such, we hope to convey not only why we are researching selected aspects of energy and power areas, but the unique aspects and dynamics of the Army/ defense research milieu that distinguish the work from strictly academic inquiry or commercial product focused efforts.

ARL is part of both the Department of Defense (DoD) as well as the Army Research and Engineering (R&E) enterprise. We work closely in partnership with other labs and activities within this enterprise. DoD's research and engineering focus, led by the Assistant Secretary of Defense for R&E within that enterprise, is critical to the nation's defense. DoD R&E strategic guiding imperatives include:

1. Mitigate current and emerging adversary threats that could degrade U.S. (and allied) capabilities;
2. Affordably enable new or extended capabilities in existing military systems;
3. Create technology surprise through science and engineering applications to military problems.

These imperatives complement the seven DoD S&T priority areas which provide the focus to meet future technological goals. Three of these areas address "Complex Threats" (Electronic Warfare/ Electronic Protection; Cyber Science and Technology; and

Counter Weapons of Mass Destruction); the remaining four enable "Force Multipliers" (Data-to-Decisions; Autonomy; Engineered Resilient Systems; and Human Systems). DoD/joint and service S&T portfolios are organized to address these priorities and are augmented by ten significant underpinning technology areas, each of which has a "Community of Interest" (COI) to advise DoD and service S&T leadership on the health of US defense focused efforts in these areas [1]:

- Advanced Electronics
- Air Platforms Biomedical
- Counter-Improvised Explosive Devices
- Energy & Power Technologies
- Ground & Sea Platforms
- Materials and Manufacturing Processes
- Sensors & Processing
- Space
- Weapons

The technologies addressed within this compendium most closely align to the Energy & Power COI, focused on E&P technologies to enable and enhance operational capabilities and effectiveness.

Both DoD and the Army use a "Technologies Taxonomy" to help organize, coordinate, and analyze S&T efforts in underlying areas across activities as well as to help inform interagency (e.g. Department of Energy, National Aeronautics and Space Administration) S&T engagements. For example, DoD and Army E&P taxonomy includes the following areas:

• **Electromechanical Conversion:** Increase the power density, efficiency, and robustness of motors, generators, and actuators while also reducing their life cycle costs.
• **Energy Storage**: Improve electrical and electrochemical energy storage devices to decrease device size, weight, and cost as well as increase their capabilities in extreme temperatures and operating conditions.

• **Power Control and Distribution**: Develop tactical, deployable power systems using conventional fuels, alternative fuels, and energy harvested from renewable/ambient sources.

• **Power Generation/Energy Conversion:** Enable smart energy networks for platforms, forward operating bases, and facilities using modeling and simulation tools as well as new, greater capability and efficiency components.

• **Thermal Transport and Control:** Efficiently manage heat and enable higher power density systems through advanced thermal science and technology: advanced components, system modeling, and adaptive or hybrid-cycle technologies.

Defense and Army Laboratories Context

DoD laboratories execute mission areas within the DoD Research and Engineering Enterprise as outlined above. These mission areas continue to evolve to address both common DoD as well as service specific needs and priorities. Traditionally, DoD labs, including ARL, have been responsible for conducting basic science, supporting technology transition, and facilitating acquisition of technology for DoD/Army (smart buyer):

> "The Labs' traditional missions are defined as *"high quality science, technology transition, and smart acquisition."* While these missions remain important and relevant, the rapidly changing technology and threat landscapes dictate the Labs should also adapt their missions to continue serving the warfighter; this includes an expanded set of missions to focus on technology leadership, open innovation, and technology defense." [2]

The Defense Science Board Task Force Report [2] notes that DoD labs should embrace open innovation by engagement with non-traditional defense R&T partners, and incorporate: Technology Leadership (supporting DoD through disruptive and emerging major technology shifts); Technology Defense (protecting US technological advantage); and Open Innovation (better tapping and integrating technologies developed outside of DoD).

ARL has several initiatives that are enabling such an open innovation environment: ARL Open Campus and ARL Extended. Open Campus is providing more seamless mechanisms and access for partners to work with ARL researchers while ARL Extended is establishing ARL research innovation hubs in selected geographic areas within the United States. As an early adopter of Open Campus, the ARL Energy & Power Division established the Center for Research in Extreme Batteries (CREB). CREB leverages resources from ARL and funding from industry, academic, and Government partners, to initiate collaborative projects. The consortium is sponsoring novel battery research with a focus on extreme performance parameters and environments. A CREB technical accomplishment (detailed in this compendium) is a recent discovery of "water in salt" aqueous electrolytes which offers potential new directions for energy storage and water science.

ARL Technical Strategy

As the Army's corporate research lab, ARL conducts relevant, transformative research rooted in the tenets of discovery, innovation, and transition to deliver S&T options to meet the complex demands of the future multi-domain battlefield. ARL recently focused its investment strategy to address Chief of Staff of the Army (CSA) priorities and shape how the future Army will fight and win in a complex world. ARL has organized its technical strategy via **S&T Campaigns** which include **Essential Research Areas (ERAs)** that are critical to reducing technology uncertainty, filling knowledge gaps, and identifying technology risk associated with delivering new capabilities.

ARL's S&T Campaigns – "a systematic course of aggressive science and technology activities envisioned to lead to enhanced land power capabilities in the deep future" - form ARL's intellectual framework for planning technical strategy, aligning and synchronizing resources, people, and infrastructure. The ARL S&T Campaigns are:

- *Human Sciences*
- *Information Sciences*
- *Sciences for Maneuver*
- *Sciences for Lethality and Protection*
- *Materials Research*
- *Computational Sciences*
- *Assessment and Analysis*
- *Extramural Basic Research.*

ARL's **Essential Research Areas** are cross-cutting, capability driven focus areas; these ERAs address both enduring and future Army challenges as well as Chief of Staff of the Army (CSA) priorities:

- *Human Agent Teaming*
- *Artificial Intelligence and Machine Learning*
- *Accelerated Learning for a Ready and Responsive Force*
- *Cyber and Electromagnetic Technologies for Complex Environments*
- *Distributed and Cooperative Engagement in Contested Environments*
- *Manipulating Physics of Failure for Robust Performance of Materials*
- *Tactical Unit Energy Independence*
- *Science of Manufacturing at the Point of Need*
- *Discovery*

Each S&T Campaign has key campaign initiatives (KCIs) that specifically address one or more of the ERAs. Several focus on humans collaborating closely with artificially intelligent systems on the future battlefield: *Human-Agent Teaming, Artificial*

Intelligence and Machine Learning, and Accelerated Learning for a Ready and Responsive Force, which address, respectively, enabling highly effective information exchange between Soldiers and intelligent systems in performing warfighting functions as a cohesive team; increasing the intelligence of engineered systems to become the true teammates of Soldiers; and preparing humans to engage with intelligent systems within their team to create an effective fighting force.

The complexity of the virtual domain and its melding with the physical realm for capabilities shapes the following three ERAs: Cyber and Electromagnetic Technologies for Complex Environments, Distributed and Cooperative Engagement in Contested Environments, and Manipulating Physics of Failure for Robust Performance of Materials. The reliance of intelligent systems on wireless communication and networked processes makes them vulnerable to cyber, physical, and electronic attacks. Specific application are needed to enable dispersed entities to deliver overwhelming kinetic and non-kinetic effects while combining scalable speed, radical maneuverability, and extremely efficient payload kill mechanisms. Enabling such systems requires integrating machine intelligence and human-agent teaming concepts. Enhancing the protection properties of friendly assets, as well as the lethality of munitions requires advanced materials that can sense their environment and change their state as a consequence.

Combining foundational advances in the above mentioned ERAs with innovative ways to enhance energy independence and agility and unburden tactical units is addressed by two ERAs: *Tactical Unit Energy Independence and Science of Manufacturing at the Point of Need.* Energy independence is essential to maintaining forces in the field without resupply and the reliance on intelligent systems creates increased energy demand, and an acute need to ensure sufficient reserves of power.

The *Discovery* ERA is focused on identifying, creating, developing, and exploiting innovative yet Army-relevant fundamental science and engineering advances. Discovery is essential to ARL's mission. It helps to ensure the Army's continuing and future technological superiority, and creates future offset against our adversaries while avoiding technological surprise.

Energy and Power S&T Alignment

Concepts for future military operations suggest a move away from large, enduring, fixed (in time and space) forward deployed bases. As such, the need for more agile, flexible, smaller footprint, rapidly reconfigurable E&P capabilities will continue to grow. Current and planned military power and energy hardware provide limited solutions for the rapidly changing dynamics anticipated on the future battlefield. Current Army operating concepts envision units capable of cross-domain maneuver. Independent expeditionary brigades and smaller units will need to be able to fight in a multi-domain environment. Evolving leadership, tactics, techniques, procedures, as well as underlying maneuver, support, fires, and Command, Control, Communications and Intelligence (C3I) must be technically informed. Technologies must constructively support and not overwhelm unit leaders juggling demands in the land, air, maritime, cyber and space domains while maneuvering.

ARL E&P S&T efforts are addressing these dynamic future Army needs within ARL's S&T Campaigns and ERAs. Materials, device and component work are addressed primarily via the *Materials Research Campaign* whereas more much of the integrative, more mature efforts are included within the *Sciences for Maneuver Campaign*. ARL's materials research aims to address emerging requirements and capabilities for all Army platforms, with emphasis on Photonics, Electronics, E&P, Biological and Bio-inspired Materials, Structural Materials, High Strain and Ballistic Materials, and Manufacturing Science. ARL's research in Sciences-for-Maneuver focuses on gaining a greater fundamental understanding of advanced mobility technologies that enable innovative configurations and subsystems architectures – critical to the future Army's movement, sustainment, and maneuverability.

ARL E&P power work in the **Materials Research Campaign** includes materials and devices for more compact, high density, high efficiency Energy Storage; Power Generation and Energy Harvesting; Fuel Cells and Fuel Processing; and Thermal Science solutions for Soldiers, platforms, and systems. ARL's E&P focus in the **Sciences for Maneuver Campaign** is focused on leveraging and integrating underlying materials and components for applications supporting Army mobility and advanced maneuver. Knowledge gained through these research efforts will lead to technologies for the design, fabrication, integration, control, and platforms support that will significantly improve Power Projection Superiority for the Army of 2030. Sciences for Maneuver Campaign E&P efforts encompass Energy and Propulsion; Energy Storage for Mobility; Power/Energy Conversion; Compact Power Generation; and Intelligent Power. E&P efforts within the Sciences for Maneuver Campaign continue to be refined to support Army future platform modernization.

ESSENTIAL RESEARCH AREA: TACTICAL UNIT ENERGY INDEPENDENCE (TUEI)

ARL's ERAs are designed to ensure that researchers are addressing Army user-inspired problems in order to distinguish our work from purely academic efforts. The thinking underlying the TUEI ERA is illustrative. In future peer-contested multi-domain battlefields, Army units will be required to operate at high operational tempos in decentralized, dispersed and semi-independent formations. Reducing overall power consumption and logistics demands will be a key enabler to realize highly mobile extended operating range and duration for extended operations. More efficient, energy dense and energy adaptable systems with increased ability to operate in degraded conditions, enhanced by more autonomous energy management will be essential in this environment.

Increases in Soldier, platform and system energy demand for capabilities have not been matched by innovations in the ability to generate power, convert and store energy, and manage and distribute power. The battlefield of the future will become increasingly digital and network-centric, with new generations of sensing, weaponry, targeting, fire control, communications, and even unmanned air and ground systems adding to the energy demand of tactical units. Additionally, emerging Army robotic and autonomous systems (RAS) will enable human-machine collaboration across all mission areas by extending the area and time over which a force can be effective. RAS capabilities are envisioned to lighten Soldier load and manually and cognitively unburden the Soldier, including for E&P.

Emerging TUEI ERA initiatives include using robotic platforms to deliver and manage small unit energy. For example, Squad Maneuver Equipment Transport (SMET) [4] platforms in development include plans to carry 1000 pounds to lighten the Soldier load, have a range of 96 km while carrying additional energy sources without resupply for up to 72 hours, be able to off-load 3 kW of electrical power, and be capable of recharging all rechargeable batteries in a dismounted Army infantry platoon.

Artificial Intelligence (AI)/ Machine Learning (ML) technology is a key enabler for future intelligent energy systems. A long term goal would be to develop autonomous self-sustaining energy capabilities enabled by AI, autonomous agents and systems, and seamless energy transfer among Soldiers, weapons systems, vehicles, and robotic platforms. This could include a hub, which could reside on a future highly mobile and agile autonomous squad support platform. Today and in the near future, energy management is done at the local and individual Soldier level. Current methods are inflexible in relation to their capability for power distribution as they have no global understanding of the overall available energy sources, no energy balance across the systems and no efficient method to impact that balance.

The TUEI ERA is also leveraging a number of ongoing research efforts to improve and realize a self-sustaining energy balance by: enabling self-sustaining autonomous systems; increasing flexibility in collecting and using locally available fuel/energy sources through development of thermophotovolatics and multi-fuel tolerant hydrocarbon based engines; increasing the energy density of Soldier wearable and conformable batteries; augmenting Soldier and platform energy via alternative and energy scavenged power sources; mitigating high power drain components (such as Soldier radios, sensors, and computing) via low-power electronics; and enhancing mobile energy operations and flexible distribution via near- and far-range wireless power transfer.

ARL anticipates innovations that will someday enable a tactical unit to collect, store, convert and distribute as much energy as needed for tactical missions a "net-zero" unit, capable of creating as much energy as it consumes, operating within an AI/ML enabled seamless energy "cloud". Whether missions last three days, seven days, or indefinitely, the Army will be better equipped to deploy an expeditionary mobile force, under extreme, remote and perhaps unforeseeable conditions.

ORGANIZATION OF THE COMPENDIUM;

This compendium is organized to address specific technical work associated with three technical branches within ARL's Energy and Power Division: *Electrochemistry; Power Sciences; and Power Integration.* Work in the Electrochemistry Branch, currently led by Dr. Cynthia Lundgren, is focused primary within the Energy Storage and Power Generation/Energy Conversion (Novel Energy) technical areas as reflected within the E&P S&T taxonomy areas outlined previously. Within ARL, this work is primary material and component focused, including electrochemical energy storage (batteries and capacitors) and electrochemical energy conversion (fuel cells, photoelectrochemistry, and photochemistry) as presented in this compendium.

Recent Power Sciences (also known as Power Components) Branch work (up to the present) has been led by Dr. Paul Barnes. Technology focus falls primarily under the Power Generation/ Energy Conversion and Thermal Transport and Control taxonomy areas previously presented. Within ARL, power generation / energy conversion work is mostly fundamental research focused on developing compact and alternative sources of power and energy capture/recovery for Soldiers and platforms, including thermophotovoltaics; thermoelectrics; pyroelectrics; flexible photovoltaics; micropower and flexible power components; compact combustion/catalysis for compact, fuel flexible sources; and isotopic and isomeric energy sources. Thermal transport and control efforts include single and multiphase cooling materials and components, multichannel and novel power electronics cooling structures, and thermal packaging.

The research and engineering work within the Power Integration (also known as Power Conditioning) Branch is led by Mr. Bruce Geil and falls mostly under the Power Control and Distribution taxonomy area. It is arranged into three primary interactional areas.

These are: wide bandgap devices (modeling, characterization and analysis for high power/ pulse power applications); power conditioning and thermal management (designs for active (power electronics) and passive (packaging) for high power, pulse power, and military unique applications); and Intelligent energy/ energy networks (architectures, designs, and test beds for advanced control of integrated and distributed energy systems.

The research performed within Power Integration is more applied in nature compared to the work in the other two branches. The focus includes understanding power device, component, and subsystem performance to address specific needs, including Army advanced mobility/propulsion, protection, lethality, and sustainment. This research is documented in ARL technical reports as well as published reports shared primarily within the engineering community.

Prefaces to each section highlight key aspects of the work within Energy and Power Division at ARL.

[1] DoD Research and Engineering Enterprise, May 2014 (https://www.acq.osd.mil/chieftechnologist/publications/docs/ASD(R&E)_Strategic_Guidance_May_2014.pdf)

[2] Defense Science Board Task Force "Defense Research Enterprise Assessment Report", Jan 2017 (https://www.acq.osd.mil/dsb/reports/2010s/Defense_Research_Enterprise_Assessment.pdf)

[3] "Army Research Laboratory S&T Campaign Plans 2015-2035", Sept 2014

[4] Army seeks to lighten Soldier loads using robots, Army News Service, Mar 2016 (https://www.army.mil/article/163385/Army_seeks_to_lighten_Soldier_loads_using_robots/)

Section I

Electrochemistry

Cynthia Lundgren
U.S. Army Research Laboratory
Chief, Electrochemsitry Branch
E-mail: cynthia.a.lundgren2.civ@mail.mil

1. Introduction

The research being conducted in the Electrochemistry Branch is arranged into 2 main areas:

Electrochemical Energy Storage: Batteries and Capacitors

Electrochemical Energy Conversion: Fuel cells, Photoelectrochemistry, Photochemistry

Much of the research performed in the Electrochemistry Branch is materials based, with efforts targeted to specific Army applications. The Army has a real and continuing need for electrochemical energy storage at all scales. Batteries and capacitors are in a wide variety of currently fielded systems such as sensors, radios, night vision goggles, weapon sights, and vehicles, and are being considered for load leveling at forward operating bases (FOB's). Fuel cells have been developed for many of the same Army applications and are currently being fielded for vehicles, unmanned aerial vehicles and squad-level battery chargers with development targeting Soldier applications. Research related to photochemistry and photoelectrochemistry are looking toward an Army future where a significantly reduced logistical footprint extends the fighting efficiency of Soldiers. The intent of this high risk / high-payoff research is to provide a basis for new approaches to warfighting. The research being conducted results in refereed publications, technical reports, and briefings shared broadly within both the scientific and larger DoD community as well as patents for licensing. An introductory overview is given.

2. Overview

2.1 Electrochemical Energy Storage

This research has traditionally been focused on primary and secondary batteries for Soldier applications with the main goal being reduction in weight of the carried battery. This has meant development of lithium (Li) metal primary and lithium-ion rechargeable batteries, which have the highest specific energy due to the light weight and standard electrode potential of Li, with the result that the bulk of the current Army battery inventory is lithium based. The goals have broadened recently to include performance related to safety, calendar life, and amortized cost. This has been the direct result of the larger range of applications that are considering electrochemical energy storage including high power applications from directed energy to battery SWaP, silent watch, and bulk storage for load leveling at FOB's.

2.1.1 Primary Batteries

The Li-sulfur dioxide (Li-SO_2) and Li-manganese dioxide (Li-MnO_2) primary batteries have been part of the Army inventory for several decades. Efforts to improve on the specific energy of primary batteries has focused on new chemistries including Li-carbon monofluoride (Li-CF_x) and Li-air. These chemistries can provide some of the highest specific energies available but also have some of the largest challenges to development for Soldier applications. Li-CF_x batteries, under typical Soldier radio use conditions, generate significant heat from the discharge reaction that can result in unsafe conditions. So far, most attempts to resolve this problem have not been successful. The Li-CF_x chemistry has been hybridized with Li-MnO_2 to address this heat issue with the resulting BA-5790 battery performing at twice the specific energy of the older Li-SO_2 battery. Li-air batteries require the use of a membrane system that is difficult and expensive to manufacture, susceptible to degradation during storage, and that limits cell discharge rate. The chemistry has been demonstrated in prototype cells, but further development is required before it can become commercialized.

2.1.2 Lithium-Metal Rechargeable Batteries

A number of rechargeable lithium metal batteries have been investigated for their higher specific energy including Li-sulfur (Li-S), Li-MnO_2, and Li-iron disulfide (Li-FeS_2). The MnO_2 and FeS_2 chemistries are not fully rechargeable due to phase changes that occur in the crystal structure on discharge. Modifications to the structure using dopants and the formation of composite cathodes have not been successful enough to make these systems viable. The most promising chemistry is Li-S due to its low cost and high specific energy. The Li-S chemistry has several well-known issues that stem from the fact that the discharge products of sulfur are soluble in the electrolyte. This leads to self-discharge being a significant problem, and to the eventual passivation of the Li anode upon extended cycling. Efforts to solve this issue have been undertaken by us and others including modification of the electrolyte to reduce solubility, electrolyte additives to prevent passivation, cell design including permeable and non-permeable membranes to mitigate the solubility issue, and the reaction of sulfur with carbon or infiltration into porous carbon to immobilize it, but to date, no commercial cells are available for use. Li metal batteries also suffer from the problem of poor Li plating that leads to safety concerns after 10s-100s of cycles. Current efforts to resolve this issue have focused on additives and conducting membranes to prevent the growth of dendrites.

2.1.3 Lithium-Ion Rechargeable Batteries

Li-ion battery research has focused on three main areas, development of new electrode materials, development of new electrolytes, and development of a fundamental understanding of the interface reactions and solvation behavior that controls passivation and performance. Current electrode material development has focused on replacing the conventional 4V cathodes with a 5V systems such as $LiNi_{0.5}Mn_{1.5}O_4$, in order to increase specific energy. Our in-house efforts have focused on the development of lithium cobalt phosphate ($LiCoPO_4$) (4.8V), which has 30% higher specific energy than Li-iron phosphate ($LiFePO_4$), and improved safety over Li-cobalt oxide ($LiCoO_2$). Progress has been steady and the scale-up to pilot scale production is underway. This work also requires the development of new electrolytes that can withstand the high voltage without oxidation and has led to the discovery of new materials such as tris(hexafluoro-iso-propyl)phosphate (HFiP) as well as the development of fluorinated electrolytes that show better oxidative stability.

The development of a better understanding of the formation of a passivation layer on graphite anodes in Li-ion batteries has been undertaken with the goal of improving both high temperature storage, low temperature discharge, and cycle life. This understanding has led to the development of new electrolytes that replace carbonate solvents with water and allow for a wide operating window (>3V) and new Li-ion chemistries. This promises to make Li-ion batteries safer and cheaper.

2.1.4 Lithium-Ion Capacitors

Li-ion capacitors are a hybrid technology that incorporate hard carbon Li intercalation anodes and activated carbon electric-double layer capacitor cathodes. This technology is just being explored as a commercially viable technology that can provide high power and long life at moderate energy density (10 Wh/kg). The current focus for Army applications is to develop improved high temperature storage and lifetime through the incorporation of advanced electrolytes. The use of additives to both improve anode passivation (life) and high voltage stability (energy) if successful can lead to a wider adoption of this technology for embedded applications. The development of hybrid cathode systems, where a relatively small percentage of $LiFePO_4$ is incorporated with the activated carbon cathode promised higher specific energy (25 Wh/kg), which has wide applications in the Army.

2.2 Electrochemical Energy Conversion

A number of electrochemical energy conversion technologies have been developed to either convert fuels directly to electrical energy (fuel cells) or to convert other forms of energy (JP8, solar, CO_2) into fuel. This effort has traditionally been focused on fuel cells, but recent focus has shifted to harvesting energy locally to convert to simple fuels for later use.

2.2.1 Fuel Cells

A significant effort has been undertaken by the Army to develop fuel cells for Soldier and Army vehicle applications. The Soldier application targets 10's of watts at specific energies greater than available in traditional batteries. The vehicle efforts, targeting 100's of watts, have been directed towards auxiliary power units (APU) that can be used to power electronics during Silent watch where operation of the engine would be inefficient and unwanted. Several different technologies have been investigated including proton exchange membrane (PEM) fuel cells using reformed methanol or chemical hydrides to generate hydrogen as fuel and direct methanol fuel cells (DMFC). For Soldier carried systems, the challenge has been to provide 20W in a package comparable in size and weight to a conventional battery.

2.2.1.1 Reformed Methanol Fuel Cell

The reformed methanol fuel cell uses a PEM fuel cell stack that has been shown to operate for 1000's of hours without degradation, so the power generating component of the device is not an issue. The challenge has been to provide hydrogen in a compact and lightweight form. Methanol is such a compact source of hydrogen, but it requires a conversion process to generate the usable hydrogen. To accomplish this, there is a fuel processing system (reformer) that takes methanol and water, and through a series of reactions at 300°C generates H_2 and CO_2. The CO_2 is exhausted and the H_2 is then used in the PEM fuel cell to generate power. The reformer, fans, pumps, and other controls all encompass the balance of plant (BOP) that make up the bulk of the device. The PEM fuel cell stack itself is a small component of the overall unit. The fuel cell unit also requires a battery for start-up and load leveling. The result is that the reformed methanol fuel cell is a complex device that is costly and difficult to operate for extended periods in fielded conditions. A second approach to generate hydrogen is through the use of chemical hydrides such as LiH, $NaBH_4$, or AlH_3. These materials are typically 10% by weight H_2 and are therefore dense sources of H_2. There is also a fuel processing step that in the case of LiH and NaBH4 involves reaction with water; in the case of AlH_3, the addition of heat. Control of the reaction is the challenge here along with storage of the H_2

generated. This requires a BOP and therefore the specific energy is reduced. The AlH_3 system is the most promising as control of the H_2 generation is easier and the yield is excellent.

2.2.1.2 Direct Methanol Fuel Cell

Direct methanol fuel cells (DMFC) use similar technology to H_2 fuel cells, but operate at lower temperatures (60°C) than the reformed methanol system. The methanol is used directly, and because it is harder to oxidize than H_2, the resulting energy efficiency is lower. Crossover of the methanol in the cell also results in oxidation of the fuel without useful energy being produced, and this further reduces efficiency. The system is beneficial for carried applications because of its lower operating temperature and overall energy density in the finished fuel cell unit. The DMFC also has a BOP that is used to dilute the methanol, as it cannot be used at 100% concentration, and to recycle water in the system. Soldier carried systems have been developed with the required 20W power, but due to cost and complexity, have not been adopted.

2.2.1.3 Reforming Battlefield Logistics Fuel for H_2

Reforming the fuel used by most Army vehicles to produce H_2 for PEM fuel cells has been an Army specific approach to incorporating fuel cells on the battlefield. Reforming JP8 requires first removing the sulfur compounds using absorbent materials, and then developing reformation catalysts that are tolerant to low (10 ppm) residual levels of sulfur. This effort has been to develop new materials with high capacities to trap organosulfur compounds from the fuel stream, and to re-use the absorbent after a regeneration (heating) step. The efforts have resulted in sulfur absorbent materials that are 5 times better than conventional materials. A second effort in this area has focused on developing a microns thick supported palladium (Pd) membrane for separating H_2 from other byproducts of the JP8 reformation process. Utilizing cleanroom technology, ARL has been successful in developing this Pd membrane that supports the high rates of H_2 diffusion for practically relevant separation conditions. This membrane has a small fraction of the Pd found in conventional Pd membranes with significantly faster H_2 transport.

2.2.1.4 Hybrid Acid-Alkaline Fuel Cells

The benefit of a hybrid acid-alkaline fuel cell utilizing methanol as fuel is significantly reduced BOP. The hybrid acid-alkaline fuel cell efforts have three components, design of the fuel cell system with modeling calculations on the concentrations of reactants and products (CO_2 and H_2O) under different operating configurations and conditions. This effort has led to the conclusion that a bipolar membrane cell should function with less water management required and therefore less BOP than conventional DMFC's resulting in a more compact system. Additional work has been done to develop membranes that can operate for extended periods in alkaline conditions. There has been both internal and external efforts in this area with ARL evaluating materials produced by external partners. There has also been an effort to develop new non-precious metal catalysts that can be used in alkaline conditions which will result in lower cost systems as platinum is now the typical cathode catalyst.

2.2.1.6 Sunlight to H_2

There are several approaches being considered to convert readily available waste streams in the field to usable fuel for generators or fuel cells. Urea and grey water can be remediated and simultaneously, H_2 can be produced to be used as fuel or to facilitate production of small molecule fuels.

2.2.1.7 CO_2 to Fuel

The Army uses significant amounts of JP8 on the battlefield. This generates large volumes of CO_2 gas. This effort is designed to convert waste streams (CO_2, H_2O) from field generators to liquid fuel by harvesting solar energy. In order to accomplish this, new catalysts are being designed and evaluated that convert CO_2 and water into small molecule fuels such as CH_4, CH_3OH, and HCOOH, which can then be used directly or upgraded to longer chain hydrocarbons. The use of membrane cells that take advantage of separating the products of of CO_2 reduction from the counter electrode reaction are also being investigated to improve efficiency and lower recombination in these systems.This effort focuses on development using catalysts and solar absorbers.

3. Summary

An overview of some of the development work being carried out in the Electrochemistry Branch is given. These material and system development efforts are supported by fundamental research in electrolyte interfaces, electrolyte structures, the interaction of light with nanostructures to produce catalytic effects, as well as extensive computational modeling of these same systems from the molecular to the macroscopic device level. The research areas are more comprehensive than given here, but this sampling will serve as an orientation into the research and as an introduction for the next several chapters in this book.

Chapter 1

Recent Trends in Double Layer Capacitors and Dual Intercalation Batteries from Molecular Prospective

Jenel Vatamanu,[1,2,*] Oleg Borodin,[1*] Marco Olguin,[1] Gleb Yushin,[3] Dmitry Bedrov[3]

[1] Electrochemistry Branch, Sensors and Electron Devices Directorate, Army Research Laboratory, 2800 Powder Mill Rd., Adelphi, MD, 20703, USA

[2] School of Materials Science, Georgia Institute of Technology, 771 Ferst Dr. NW, Atlanta, GA, 30332, USA

[3] Materials Science & Engineering Department, University of Utah, 122 South Central Campus Drive, Salt Lake City, UT, 84112, USA

*jenel.vatamanu@utah.edu and *oleg.a.borodin.civ@mail.mil

Abstract

Supercapacitors or electrical double layer (EDL) capacitors store charge *via* rearrangement of ions in electrolyte and their adsorption on electrode surfaces. They are actively researched due to multiple potential applications requiring longer cycling life, broader operational temperature range, and higher power density compared to batteries. Recent developments of nanostructured carbon-based electrodes with high specific surface area have demonstrated the potential to significantly increase supercapacitor energy density. Molecular modeling of electrolytes near charged electrode surfaces has provided key insights into fundamental aspects of charge storage at nanoscale, including an understanding of the mechanisms of ion adsorption and dynamics at flat surfaces and inside nanopores, and the influence of curvature, roughness, and electronic structure of electrode surfaces. Here we review these molecular modeling findings together with available experimental observations and put this analysis into perspective of future developments in this field. Current research trends and future directions are discussed.

Keywords

Battery, Capacitor, Molecular Dynamics Simulations

This work is derived from Vatamanu, J.; Borodin, O.; Olguin M.; Yushin, G.;Bedrov, D., Charge storage at the nanoscale: understanding the trends from the molecular scale perspective. J. Mater. Chem. A 2017, 5 (40), 21049-21076. https://doi.org/10.1039/C7TA05153K

Contents

1. Introduction

The increasing demand for efficient, reliable and affordable energy storage devices stimulated research into the Faradaic and non-Faradaic phenomena occurring at the electrode-electrolyte interfaces. In the Faradaic devices, such as batteries or pseudo-capacitors, the charge is stored due to electrochemical reactions occurring at electrodes. In contrast, capacitors store energy *via* non-Faradaic processes (*i.e.*, involving physical phenomena without chemical reactions). Depending on the type of materials and charge storage mechanisms involved there are three types of capacitive storage devices: dielectric capacitors, electrolytic capacitors and electric double layer capacitors (EDLCs) or supercapacitors.[1, 2] In dielectric capacitors, the space between electrodes is filled with a dielectric material and the energy storage involves reorientation of material local dipoles such that they oppose the externally applied electrostatic field between the plates. These capacitors have the fastest charge-discharge rate, operate at high voltages (kV) and in alternating current. However, the stored energy normalized per mass of dielectric material in this type of capacitors is small, as shown in Figure 1. The energy density of dielectric capacitors can be increased by the utilization of dielectrics with high electric permittivity or ferroelectrics, e.g., the class II ceramic capacitors. [3, 4] The electrolytic capacitors are comprised of two metal electrodes, typically made of aluminum, tantalum or niobium, with ion conducting liquid or solid electrolyte between one of the electrodes and a dielectric material formed by oxidation of the surface of another electrode [5]. These capacitors are commonly utilized in electronics or as power supplies and their current market size exceeds US$6 billions/year.[1, 6, 7] Among the disadvantages of electrolytic capacitors are their relatively short lifetime and large leakage currents.

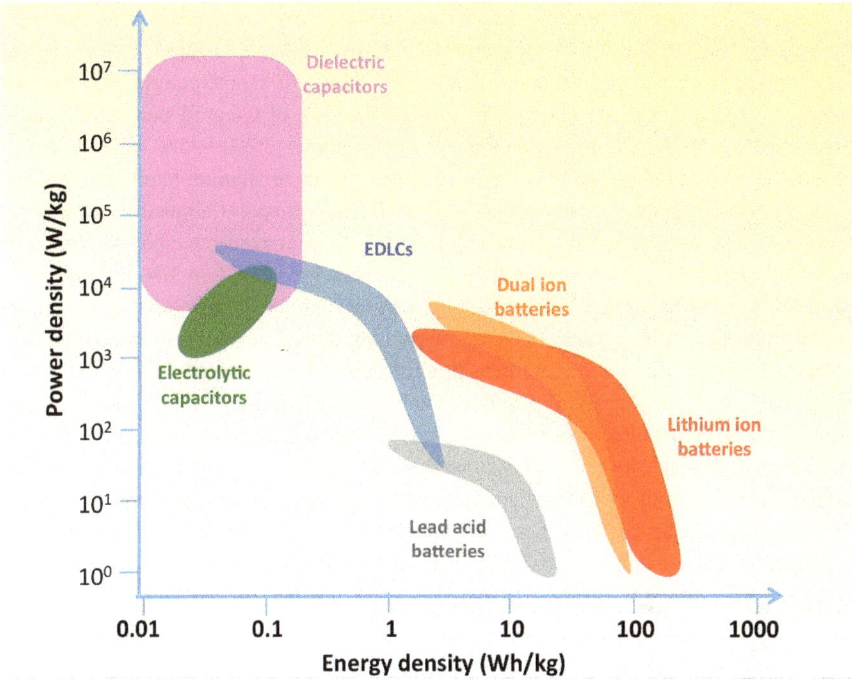

Figure 1. Ragone plot illustrating the power versus energy density for various electricity based energy storage devices. Figure 1 is based on data compiled from several sources.[24, 152-157]

The EDLCs, on the other hand, are usually comprised of porous electrodes and liquid electrolyte. The charge storage is achieved *via* double layer restructuring due to applied electric field causing the ion diffusion and adsorption at the electrode surface. As the energy in EDLCs is stored within this thin interfacial layer, the enhancement of the energy density arises primarily from the increased specific surface area (SSA) of porous electrodes. Activated carbon is one of the most currently used electrode materials made by charring and partially oxidizing the precursor using steam or carbon dioxide to create or enhance nanoscopic pores, often followed by subsequent purification to reduce impurities. While coconut shell activated carbon (SSA around 2000 m^2/g) is a popular, low cost ($4 for commodity activated carbon and $15 per kg for the EDLC-grade carbon in 2013) material that has been considered as the most common for practical applications in EDLCs, other natural precursor derived activated carbons[8, 9] as well as more advanced C-based materials have been widely investigated by academic groups.[8, 10-12] Extensive experimental work focusing on understanding and design of nanostructured materials for energy storage applications allowed an increase in SSA of C-based porous electrodes from a few hundreds m^2/g in powders or activated carbons to 2000-3500 m^2/g in various carbide-derived carbons (CDC), C-onions, zeolite or silica or alumina-templated carbons, Kroll carbons, graphene, polymer-derived carbons, vertically aligned C-nanotubes, aerogels, nanotubes and carbon-fibers.[2, 8, 11, 13-22] Recent research showed that it is technically possible to synthesize porous electrode materials based on 3D-graphenes with SSA as large as 4000 m^2/g.[23] Many of these nanostructured materials, such as nanotubes or 3D-graphenes, unfortunately, have a low packing density resulting in low volumetric energy density.[24] The low volume density electrode materials increase the amount (and hence the weight and cost) of inactive components present in EDLCs such as bulk electrolyte, current collectors, separator, binder, connectors, and packaging therefore making such low-density electrodes much less attractive for assembly of thin electrode cells.[10, 24] Hence, both volumetric and gravimetric energy densities as well as electrode thickness need to be considered as performance metrics in addition to the commonly considered values of capacitance per unit surface area or per mass of electrode material.

In comparison with dielectric and electrolytic capacitors, the EDLCs with highly porous electrodes achieve the highest non-Faradaic capacitance and charge/energy density[25] making them suitable for applications requiring intermediate power density over longer-time discharge, e.g., emergency doors, memory backups, car starters, detonators, pulsed lasers, pacemaker devices, defibrillators, electric forklifts, cranes, and even electric buses and trains.[10] Almost half of today's supercapacitor production is designated to electric vehicle transportation (EVT).[26] In EVT, the addition of supercapacitors to electrochemical batteries or fuel cells enables regenerative braking and faster acceleration, increases the overall energetic efficiency up to 20-25% and improves the life-time of batteries.[27, 28] EVT based exclusively on supercapacitors (or the so called capa vehicles) was already tested as a non-polluting and energy efficient alternative for inner-city transportation that requires multiple stops (where recharging could be done in as little as half a minute) and relatively short distances (few miles) between stops. Such supercapacitor-only based EVT apparently is 20-50% more energy efficient than over-the-line electric vehicles (trolleys, electric trains).[29] EDLCs are also utilized in the bridge power applications, where immediate power availability may be difficult to achieve, such as in Uninterruptable Power Supply (UPS) systems utilizing generators, fuel cells or flywheels as the main power backup. The EDLC burst power applications are still limited to frequencies lower than 100 Hz and often sacrifice the energy density to increase the power density.[30] The supercapacitors could also be ideal for storing relatively small amounts of solar, mechanical, and thermal energy harvested from (relatively small) environmental energy fluxes.[31-33] For example, traffic speed indicators or video cameras could be powered by a combination of solar cells and supercapacitors. The supercapacitors can be scaled to any desired size from nano/micro-meter in electronics to locomotive/wagon size (of tens of tons) in railroad transportation.

Because of the inherent advantages of the reversible charge storage, it is expected that the demand for non-Faradaic energy storage will increase in the near future. Specifically, it is projected that the supercapacitor market will rise from about US$1 billion per year today to several billions in the next decades.[7, 29] However, the larger scale deployment of supercapacitors will depend on how the current and future research addresses their shortcomings which will be discussed in the next sections.

The EDLC electrolytes are usually based on aqueous or organic based solvents, such as propylene carbonate (PC) or acetonitrile (AN), while room temperature ionic liquids (RTILs) are currently investigated as a promising nonflammable alternative for higher temperature, higher voltage applications.[8] Electrolytes with protic solvents at typical salt concentrations (0.1-3M) suffer from narrow electrochemical stability window, but offer higher conductivity and could be advantageous in high rate applications. On the other hand, it was recently demonstrated that super-concentrated aqueous solutions of Li-salts (~20m for single salt and up to 27m for a mixture of two salts) can increase the operational potential window slightly above 3V due to the suppression of the hydrogen/oxygen chemical evolution at electrodes.[34-37] Aprotic solvents have the intermediate stability and rate characteristics between aqueous electrolytes and RTILs, but may operate at extremely low temperatures down to – 70 °C.[38] The RTILs consist of relatively bulky organic ions that are liquid at room temperature. They yield acceptable ionic conductivity and viscosity, low vapor pressure, low toxicity and low flammability, which makes them attractive candidates for EDLCs if their cost can be reduced.[39,40] Also, a wide variety of possible chemical structures of cations and anions suitable for RTILs allows for further chemical tailoring of these liquids to obtain the desired properties. In some cases, addition of an aprotic solvent to RTIL is useful because the solvent can decrease the viscosity, increase the mobility of ions, and extend the lower end of the operating temperatures. Usage of asymmetric (in size and shape) ions or mixtures of ionic liquids also provided a viable strategy for extending the liquid range to lower temperatures beyond the melting point of traditional solvents.[41]

Unlike continuous double layer formation occurring in EDLC with increasing voltage, the anion insertion into graphite occurs at well-defined potentials, typically around 4-5 V vs. Li/Li^+. Such anion intercalation expands the graphite interlayer spacing to dimensions comparable to the smallest pore sizes used in EDLC, giving rise to some similarity between them as we discuss below. Due to a well-defined anion intercalation potential that is analogous in many ways to the lithium intercalation in graphite, devices relying on both anion and cation Faradaic reactions are classified as a dual-ion battery. In addition to dual-ion batteries with all-carbon electrodes, alloyed anodes could replace graphite negative electrodes leading to a further increase in the energy density of dual-ion batteries. Some of the challenges of the dual-ion battery (DIB) are a relatively low volumetric capacity experimentally observed in the graphite cathode and the need to identify electrolytes that are compatible with the relatively high cathode voltages (4-5 V vs Li/Li^+). DIBs exhibit energy densities higher than EDLC capacitors but lower than lithium ion batteries (LIB) as shown in Figure 1. DIB energy density is determined by the capacity of electrodes, voltage between electrodes and ability of electrolyte to provide both lithium cations and anions for intercalation into anode and cathode. Thus, it is important to choose electrolytes with high salt solubility and conductivity over wide concentration range in addition to ensuring that electrolyte in electrochemically compatible with both electrodes. Unlike DIB, current lithium-ion batteries (LIBs) rely on shuttling lithium ions between graphite negative anode and intercalation cathode. LIB energy density is largely determined by electrode capacities and voltage between electrode after accounting for the weight of electrolyte, current collectors, separator and packaging. Current lithium-ion batteries are cathode limited, similar to DIBs, and require electrolytes to be either electrochemically stable at these potentials or form a stable ionically conducting and electronically insulating passivation layer to kinetically protect electrolyte from further redox reactions at the electrodes. A high gravimetric and, especially, high volumetric capacity and voltage of lithium-ion battery cathodes such as lithium cobalt oxide and lithium nickel cobalt manganese oxide yield high energy densities for lithium-ion batteries (see Figure 1) when paired with graphite or alloyed anodes, making lithium-ion batteries the technology of choice for portable electronics, hybrid and electric vehicles, and power tools.[42]

In EDLCs and DIBs devices discussed above, the molecular scale phenomena are key in controlling the performance and design strategies. These phenomena include: i) more efficient packing of ions in EDL or inside pores at lower voltages to increase capacitance; ii) role of quantum capacitance to ensure that it is not a limiting factor; iii) faster transport of ions inside the pores; iv) intercalation kinetics for faradic devices; v) improved electrochemical stability of electrolytes to expand electrochemical window or to control the passivation layer for hybrid devices; vi) design of hybrid micro and macro-pores architectures to optimize transport and capacitance. Below we will discuss these molecular scale phenomena and their influence on the efficiency of energy storage devices.

2. Structural properties of electric double layer

In this section, we discuss recent theoretical and experimental advances in understanding the structure of the electrode-electrolyte interface, the EDL capacitance, and the energy density stored by various electrolytes at different electrode structures. We reiterate that the energy/charge stored by the capacitor can be quantified with the integral capacitance (C_I). The C_I is defined as the ratio between the electrode charge σ (normalized either per unit specific surface area (SSA) or mass of the electrode) and the applied potential ΔU between electrodes, $C_I = \sigma/(\Delta U/2)$. For characterization of the electrode-electrolyte interface and the sensitivity of its properties to the electrode potential, the differential capacitance (C_D) defined as the derivative of the electrode charge with respect to electrode potential, $C_D = d\sigma/dU_{electrode}$, is commonly used both in experiments and modeling. The electrode potential ($U_{electrode}$) is readily available from simulations by integrating the Poisson equation of the ensemble averaged charge distribution from the electrode surface to bulk electrolyte.

The fundamental understanding of C_I and C_D magnitudes and their dependence on voltage, temperature, and the chemical structure of electrode and electrolyte is key to elucidating correlations between the structural changes within the EDL and routes to achieve capacitance enhancement. For example, in light of simple EDL models[43] one would associate the increase of capacitance upon increasing the electrode potential with low concentration of ions near the electrode surface. In contrast, a decrease of capacitance as the electrode is charged can be caused by ion crowding and oversaturating at the surface. Also, sharp peaks in C_D at certain voltages can indicate possible phase transitions within the interfacial layer.

As an electric field (or charge) is applied on the electrode, the ionic electrolyte will restructure such that it will diminish (or screen out) the impact of the applied field. Extensive experimental work based on atomic force microscopy and theoretical work based on MD simulations showed a consistent picture regarding the EDL structure formed at the electrode-electrolyte interface. Specifically, in pure RTIL and concentrated electrolytes, the layers locally rich either in counterions or coions are forming near the surface along the direction perpendicular to the electrode surface. It is quite remarkable that the AFM experiments reached a level of resolution that can accurately pinpoint individual electrolyte layers near the electrode surface.[44] This EDL structure generates space charge oscillations in the electrolyte, as exemplified in Figure 2. For pure RTILs it is often the case that the innermost electrolyte layer near the surface (about 5-6Å width next to electrode) can be over-screened at low voltages, meaning that it can contain more counter-charge than the charge on the electrode. This extra (over-screened) charge in the innermost layer is then compensated by the net co-ion charge in the follow up layers of electrolyte formed near the surface such that the total charge on the electrode is equal and opposite in sign to the total charge in the entire EDL structure. Interestingly, solutions of ions in aprotic solvents also generated multiple structural layers near the electrode qualitatively similar to those of RTILs.[45] These structural oscillations in electrolyte are fairly short-ranged, specifically they extend over a few nanometers from the electrode surface, typically 3-5nm for pure RTILs and 1-2nm for solutions.[45] Low temperature, high solute concentration, and high electrolyte density can extend the range of these oscillations.

The multilayer structure formed near the surface screens out and limits the range in the ability for the electrostatic field to penetrate inside the electrolyte. This aspect can be illustrated in the variation of the screened Poisson potential with respect to the distance from the electrode. As illustrated in Figure 2, within a couple of nanometers from the electrode surface the density oscillations vanished and the Poisson potential in electrolyte becomes almost constant, indicating that the externally applied field is mostly screened out. Note that recent AFM measurements [46, 47] revealed a presence of relatively weak forces between two charged mica surfaces with pure RTIL electrolyte in between. These forces were extending to large surface separations and therefore suggest a relatively large Debye screening length in pure RTILs and the presence of a broad (7-8nm) diffuse layer.

Despite having a multilayer structure, the adsorption/desorption and reorientation of ions in the innermost layer as a function of electrode potential are expected to have the main contribution to the magnitude of the capacitance. Although a quantitative experimental characterization of ion distribution at the surface is challenging, appropriately tuned atomistic simulations can address this issue quite straightforwardly. Figure 3 compares the dependence of ion density in the interfacial layer (defined as 5 Å electrolyte width from the surface) on the electrode potential obtained from simulations of 1-butyl-3-methylimidazolium bis(fluoromethylsulfonyl)imide ($c_4$mim-TFSI) pure RTIL and a solution of tetraethylammonium tetrafluoroborate (TEA-BF_4) in a polar solvent (acetonitrile, AN). As illustrated in Figure 3, the solution of ions behaves differently from pure RTILs. Specifically, in solutions, as the surface becomes weakly charged (or near the potential of zero charge, PZC) the co-ions are almost completely repelled[45, 48] from the surface and the counter-ions accumulate with increasing potential. Somewhat smaller changes in ion adsorption/desorption from the interfacial layer were observed for the smaller Li^+ ions in carbonate solvents.[49] Such behavior for Li^+ at the electrode interface is attributed to the strong coordination of Li^+ by oxygen atoms from carbonate solvent molecules.[50] In contrast to ionic solutions, pure RTILs require electrode potentials as large as 2V (relative to PZC) to repel the co-ions from the electrode surface and about 3V to saturate the innermost layer with counter-ions. Therefore, within the electrode potential window of -2V to +2V the dominant mechanism of charging the interfacial layer in RTIL electrolytes involves a monotonic swapping of co-ions and counter-ions as the potential varies, while for ionic solutions it primarily involves the accumulation of solvated counter-ions next to the charged surface that is relatively depleted of co-ions. The mechanism of monotonic swaps of co-ions and counter-ions in RTILs is consistent with the weakly changing C_D vs potential observed both experimentally[51, 52] and theoretically[53] for these systems.

Considering the difference in the mechanism of ion adsorption/desorption for pure RTILs and ionic solutions observed in Figure 3, one should expect a noticeably different behavior of the C_D on voltage for these two types of electrolyte. For example, in electrolytes with solvated ions the EDL restructuring should be more sensitive to the electrode potential change than in RTILs. As a result, at low voltages the solutions of ions might generate a larger C_D than ionic liquids. Considering steric aspects (*i.e.*, surface not crowded with ions) both the Gouy-Chapman and the mean field theories predict a sharp increase of capacitance versus potential increase for diluted solutions. Models based on the mean field theories[43, 54, 55], however, tend to over-predict the magnitude of capacitance for both solutions and RTILs as shown for example by Breitsprecher et al.[56]. Quantitative predictions of theoretical models for the EDL structure and capacitance can be significantly improved based on classical density functional approaches[57-60] (which are quite popular for EDL studies), mean field[61, 62] or integral equation theories[63, 64]. In contrast to basic EDL theories, various atomistic simulations predict consistent trends and variations of C_D for RTIL electrolytes and solutions of ions as illustrated in Figure 4.[45, 65, 66] Specifically, there is a rather weak dependence of C_D on the applied voltage within the potential range defined by the electrochemical stability window of commonly used electrolytes (roughly between -3V and +3V). Typical values of C_D for these electrolytes range between 4 and 6μF/cm^2. At larger voltages, *i.e.*, above ±3V, the electrode surface becomes crowded with counter-ions and C_D begins to decrease. While such a behavior is expected for RTILs, it is surprising that several solutions of ions show a similar dependence of C_D on the potential, which is not consistent with the Gouy-Chapman theory prediction.[45] This clearly indicates a more complex influence of polar solvent molecules on the delicate balance of electrolyte structuring near the electrode surface. We believe that the observed deviation from the expected Gouy-Chapman behavior can be rationalized by taking into account a strong adsorption of solvent molecules onto the charged electrode surface (as expected for polar solvents) and by a large energy penalty required to de-solvate the ions.

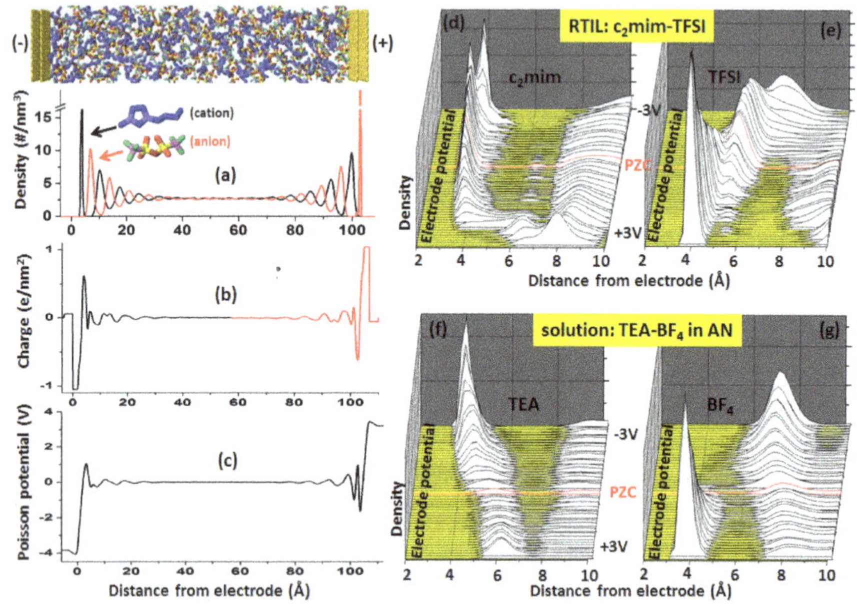

Figure 2: Typical distribution of average ion density (a), cumulative space charge (b) and the Poisson potential (c) observed between for a room temperature ionic liquid between two charged electrodes. The local electrolyte structuring near electrode surfaces consists of several successive layers rich either in co-ions or counterions and generates space charge near electrodes that completely screen out the applied electrostatic field within 2-3 nanometers from the surface. The panels (d) and (e) show the distribution of ions in the interfacial layer for a $c_2$mim-TFSI RTIL and panels (f) and (g) for TEA-BF_4 solution in acetonitrile solvent at various electrode voltages. This figure is based on the simulations presented in refs. [66, 158] *for RTIL ($c_2$mim-TFSI) and ref.* [45] *for the solution of ions (TEA-BF_4 in AN).*

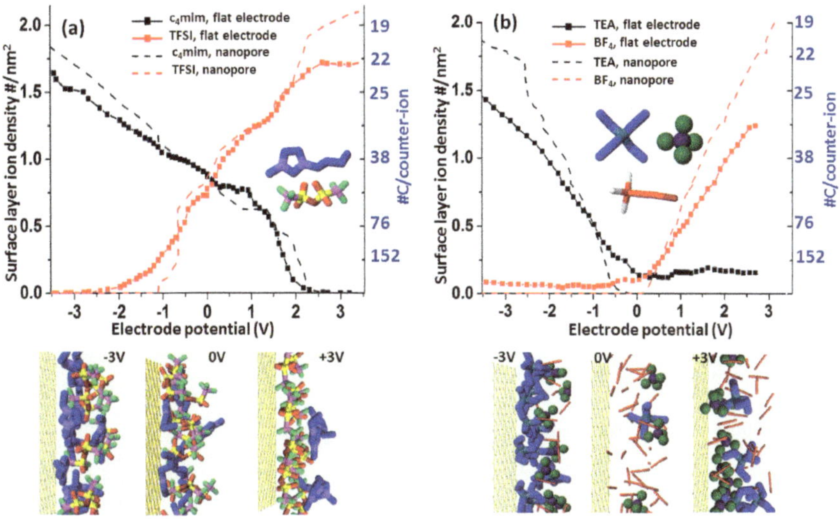

Figure 3. The dependence of surface layer ion density for the innermost adsorbed electrolyte layer near a flat electrode (within 5 Å from electrode surface) and within the same layer inside a 7.5 Å slit pore for (a) $c_4$mim-TFSI RTIL and (b) solution of TEA-BF_4 in AN. The dashed lines represent data for electrolyte inside the 7.5 Å width slit pore and the symbols represent data for electrolyte near a flat electrode surface. The left y-axis shows the total number of ions centers of mass in the layer per 1 nm^2 cross-sectional area. The right y-axis shows the density in units of number of C atoms from one graphene sheet per one counterion in the interfacial layer. The flat electrode and the pore surface were modeled as graphene layers. Other details regarding these simulations can be found in refs. [45, 66, 80].

The atomic scale structure of the electrode surface can also affect the EDL structure dramatically and hence influence the magnitude of the capacitance and its dependence on the applied voltage. Figure 5 compares C_D vs. electrode potential for a typical RTIL and an ionic solution on atomically flat and rough electrode surfaces. For the latter, the roughness patterns have dimensions comparable to the dimensions of electrolyte ions. For ionic solutions the surface roughness does not significantly change the overall capacitance and its dependence on the electrode potential. However, for pure RTILs the surface roughness noticeably affects the capacitance dependence and leads to increased values.[67-70] In contrast to smooth surfaces, the electrodes with high density of atomically rough patterns can generate multiple minima and maxima in C_D with magnitude variations as much as 400% (from 4 to ≈16μF/cm^2)[69] and increase the C_I by almost 100% as compared to that of atomically flat surface.[68] DFT calculation by Zhan et el. [70] also showed an increase of capacitance at zig-zag graphene edges. Such behavior of the C_D on rough electrodes can be explained by the presence of large local electric fields at the edges of the roughness features. These large local fields can facilitate the co-ion/counter-ion separation in the EDL.[71] Simulation studies found that the C_I increase at the rough electrodes is comparable to the C_I observed in narrow nanopores, which will be discussed in the next section. Curved electrodes can also increase the capacitance by as much as 100% (or up to 10-12μF/cm^2) compared to atomically smooth and flat electrodes, as has been shown both by theory and experiments for C-onions and C-nanotubes[72,67] Similar to atomically flat surfaces the electrodes comprised of curved nanoparticles generate almost constant C_D vs. potential in the wide range of potentials.[73]

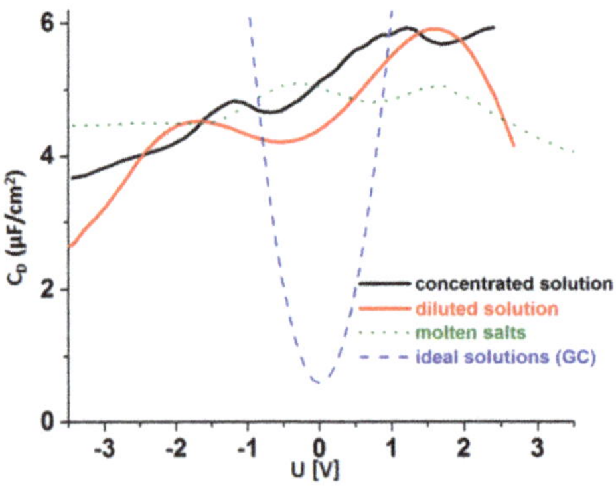

Figure 4. Differential capacitance for ideal Gouy-Chapman solution is compared to the differential capacitance of TEA-BF$_4$ in AN using data from ref. [45] *and c$_2$mim-TFSI RTIL using data from ref.* [158].

Finally, it was found that the temperature has a rather modest impact on the EDL structure and the C_D for electrodes with atomically smooth surfaces[51] but a more pronounced effect for the electrodes with atomically rough surfaces.[53, 73] On the rough electrodes, the increase of temperature smoothens out the magnitude of the oscillations between minima and maxima of C_D while maintaining the overall similar $\sigma = f(U)$ dependence and therefore similar C_I.[53] Both simulations [53, 73] and experiments [51] reported a slight decrease in capacitance as the temperature increases, with the exception of a narrow range of voltages for atomically rough surfaces where the overall broadening of peaks in C_D vs. electrode potential can lead to an increase of C_D with T at the specific $U_{electrode}$.

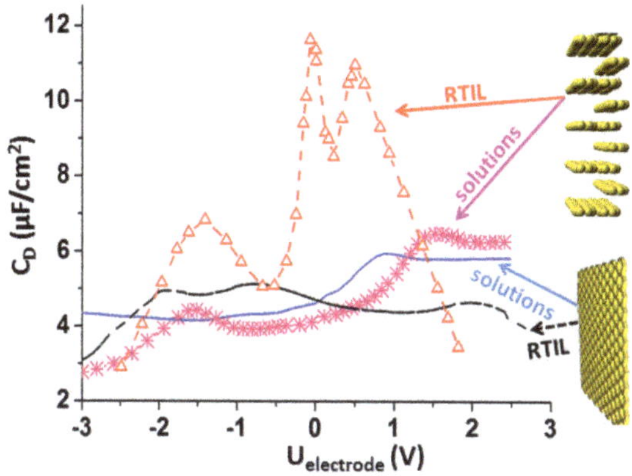

Figure 5. The influence of electrode surface topography on C_D dependence for RTILs and solutions obtained from MD simulations described in refs. [45, 53]. *The pyr$_{13}$-FSI RTIL was simulated as described in ref.* [53]. *The solution of ions consisted of $c_2$mim-BF_4 an AN solvent as described in ref.* [45]

3. Electrolytes in charged nanopores.

An ever-increasing need to increase the electrode surface area in order to achieve a higher capacitance without resorting to the low or ultralow density materials with a low volumetric energy density, led to the consideration of nanoporous materials with pore widths comparable to ions' diameters. Clearly, the ionic packing in such narrow pores might noticeably deviate from the ion packing in EDLs at the "open" electrode surfaces, *i.e.*, electrode surfaces in contact with bulk electrolyte, and hence suggesting that the charge storage mechanism can be different. For example, the carbide derived carbon (CDC) electrodes with sub-nanometer size pores demonstrated an increased area-normalized capacitance up to 13-14μF/cm^2, which is almost double the capacitances of the same electrolyte in mesopores or at open electrodes.[25, 74, 75] As shown by Largeot et al. [75] for the RTIL electrolyte ($c_2$mim-TFSI), a sharp increase in capacitance is possible in sub-nanometer pores with widths comparable to the ions dimensions. This enhancement however, vanishes when the pore widths are modified by as little as 0.5Å. In other words, by carefully controlling the pore widths and their polydispersity (or their lack off),[75] in principle, a doubling of the stored energy density is possible. Note that the experimental measurements of the carbon specific surface area generally rely on the simplified gas sorption models that often use inappropriate (for small nanopores) assumptions and do not consider a realistic shape of the pores. As such, these estimations and thus the area-normalized capacitance values should be considered only qualitatively. Other, more direct studies of ion sorption in pores of various sizes, such as small angle neutron scattering[76, 77] are typically significantly more expensive.

The experimentally observed increase in capacitance of narrow pores has raised an intriguing theoretical question of why the sub-nanometer pores allow supposedly denser countercharge packing and enhanced capacitances. Kornyshev and co-workers elucidated one of the physical origins explaining this phenomenon. They demonstrated that the charge distribution generated by ions on the polarized metallic walls results in enhanced screening of electrostatic interactions between ions inside charged nanopores.[78] Therefore, the electrostatic repulsion between ions with the same charge become short-ranged and as a result of such strong screening, it is then possible to densely pack electrode counterions inside a nanopore generating the so-called "superionic states" [78] that lead to increased capacitance.[75] According to this model and subsequent extensive molecular simulations using simplified coarse-grained models, the C_I enhancement in slit nanopores is originated either by a sharp (vs. potential) expulsion of co-ions from charged nanopores at a certain value of the electrode potential or by the fast swapping of co-ions and counter-ions occurring at low voltages. [79-81]

Atomistic simulations probed this phenomenon for a variety of RTILs and reported mixed results. For nanopores with atomically smooth surfaces, such as slit or nanotubes, typical RTILs generated a smaller capacitance enhancement than reported experimentally. Specifically, depending on electrolyte and the force field utilized in simulations, several simulation works showed about 50% enhancement in C_I for sub-nanometer (slit) pores relative to the mesopores, or C_I with values around 8-9μF/cm^2.[79, 80, 82] Also, multiple simulations[79, 83] and classical DFT calculations[84, 85] showed a complex oscillating dependence of the capacitance versus pore width (Figure 6a). Utilization of asymmetric pores (corresponding to a setup where the positive and negative pores had different available volumes for corresponding ion insertion), were shown to further increase the energy density if the ratio of volumes available for the ions on the positive and negative electrodes were commensurate with the asymmetry of electrolyte anion and cation dimensions [80, 86, 87]. Capacitance enhancements as much as 100% (compared to flat electrodes) were observed in simulations of NaCl aqueous solution in slit pores tighter than 2nm,[88] which is in agreement with classical DFT theory that correlated the dielectric constant of the solvent with the capacitance enhancement.[89]

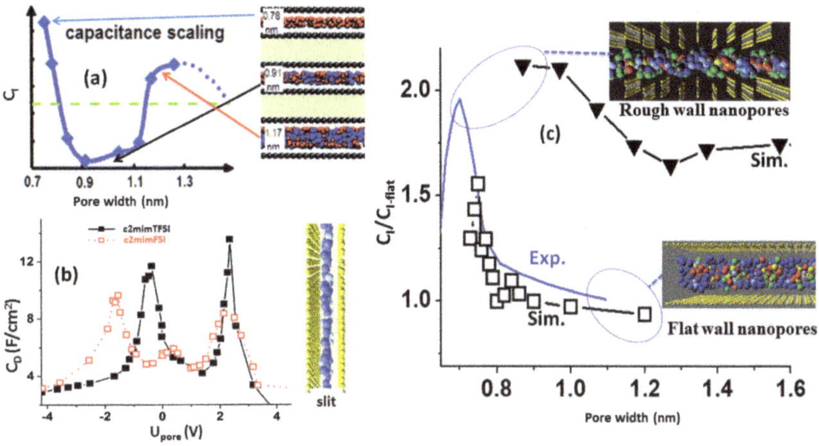

Figure 6. (a) A typical dependence of the integral capacitance (C_I) on the pore width for an RTIL inserted into a slit pore as obtained from MD simulations. An increase of capacitance is observed in narrow pores followed by oscillating dependence at larger pore widths. Reproduced from ref. [83] *with permission of the American Chemical Society. (b) Differential capacitance versus potential for 2D slit pores. Reproduced from ref.* [80] *with permission of the American Chemical Society. (c) The dependence of the pore capacitance (C_I) relative to the capacitance of the same electrolyte on a flat unconfined electrode ($C_{I\text{-flat}}$) on the pore width for RTIL inside slit pores with flat and rough wall surfaces. Pores with atomically rough walls, whose surface topography combines the nanoconfinement effect with surface roughness and curvature, can generate additional capacitance enhancement as compared to pores with smooth surfaces. The symbols represent data from MD simulations while the blue line is based the experimental data from ref.* [75] *Reproduced from ref.* [80] *with permission of the American Chemical Society.*

More complex nanopore geometries were also investigated. A systematic study of the charge density enhancement inside the pores, as a function of pore structure, found that pores having surface roughness patterns can significantly enhance the capacitance.[80] For example, simulations of nanopores with atomically rough surfaces predicted C_I as large as 11-12 μF/cm^2 for typical RTIL electrolytes (see Figure 6c), demonstrating that atomic scale surface roughness of nanopore walls (expected to be present in realistic carbon based subnanometer wide pores) can further increase the stored energy density[80] and is a desirable feature of the electrode structure. The additional C_I enhancement in nanopores with atomically rough walls is related to the same phenomena observed for the open electrodes with atomically rough surfaces discussed above. While the overall enhancement of C_I in nanopores with atomically rough walls is still only 50% compared to the open electrodes with the same surface structure, the absolute values of C_I predicted by atomistic simulations in such nanopores are much closer to experimentally reported values for the same electrolytes and pore dimensions.

The electrolyte structural changes inside charged nanopores can be summarized as follows. At low voltages, the charge is accumulated inside the pore by swapping the co-ions with counter-ions. The observed capacitance enhancement is usually related to a sharp (with respect to voltage change) de-mixing (sometimes considered as a first-order phase transition) of co-ions and counter-ions inside the nanopore at some intermediate voltage (electrode potential around 1-2V for typical RTILs). At this voltage, the co-ions completely leave the pore and only counter-ions are left inside the nanopore. While the total density of ions inside the nanopore at this transition voltage can drop, the stored charge density is increasing due to co-ion and counter-ion separation. In order to maximize the enhancement in C_I, it is desirable that such de-mixing of ions occurs at similar voltages on both the negative and the positive electrodes, leading to peaks in C_D on both electrodes. Both slit[80] and cylindrical[90, 91] pores generated peaks in capacitance (Figure 6b), suggesting that such behavior is a fundamental characteristic of interactions of electrolyte species in confined environments.[92] As the electrode potential is increased further (*i.e.*, high voltages) the charge accumulation occurs *via* counter-ion condensation. However, there is a limit (due to ions' excluded volume) to how much the counter-ions can pack. Several studies showed that the capacitance enhancement inside carbon based nanoporous electrodes is limited to ΔU of 2-3V and it diminishes significantly at potential differences higher than 6V.[80]

To further comprehend the influence of nanoconfinement on the electrolyte charge reorganization, we have compared the variation of the co- and counter-ion charge density in the innermost layer of unconfined electrolyte (*i.e.*, in the EDL at the open structure electrode) as a function of voltage with the behavior of electrolyte inside charged nanopores. Figure 2a-b shows the cumulative charge density inside the nanopore normalized per unit surface area as a function voltage. In both cases the atomic structure of the electrode surface (*i.e.*, the nanopore walls and for the open structure electrode) are the same. A quick glance uncovers a surprising similarity between the ionic response to the applied potential during electrode charging for the nanoconfined and open structures. This similarity suggests that the nanoconfiment effects on the electrolyte layer near the surface are relatively small and the composition of electrolyte inside nanopores can be approximated by the composition of the inner Helmholtz layer at the open electrode. Examination of the differences indicates that a more pronounced variation for the confined system vs. open one in the region of 0-1 V for the $c_4$mim cation leads to higher capacitance in this voltage range. Similarly, the TFSI anion desorbs from the nanopore somewhat more expediently than from the open surface with the potential decreasing from -1 to -2 V. Another observation is that at high voltages (>2 V or < -2 V) the counter-ion density is higher in the nanopore compared to that near the open electrode, with a larger difference observed for the positive electrode. The analysis of realistic materials is complicated by irregular pore shapes that can have broad distributions and have varying electronic properties. In order to start addressing such more realistic representations Merlet et al. [93] simulated CDC pore structures reconstructed using a reverse Monte Carlo technique[94] from experimental data. They categorized the charge enhancement as a function of the local pore structure/geometric pattern and identified steric environments that enhance the capacitances.[93] Finally, the asymmetry in charge storage mechanisms versus electrode polarity, consisting of cation adsorption in the negative pores and anion-cation exchange in the positive pores was observed experimentally with in-situ NMR spectroscopy.[95] Similar asymmetry of charge storage inside positive and negative nanopores were observed for RTIL inside narrow pores in other experimental[95-98] and simulation works.[80]

Similar to RTILs, the solutions of ions can also enhance the C_I in sub-nanometer pores[99], albeit to a smaller extent. [100] In order to obtain a sufficiently high concentration of ions inside nanopores such that the C_I is enhanced compared to the open structure electrodes, the ions should first de-solvate at least partially upon insertion and then displace the solvent from the pore. Snapshots illustrating the ionic composition inside a 7.5Å width slit nanopore at several potentials are shown in Figure 7 both for a typical RTIL and a solution of ions.

4. Dual-ion intercalation into graphite

The ability to control ion insertion into graphite is also of great importance to the development of the DIB , which is a promising energy storage technology due to several environmental, safety, and material cost benefits over the state-of-the-art Li-ion batteries. For example, the mining and processing of rather expensive and toxic cobalt and nickel metals needed for the commercially used nickel and cobalt containing cathodes may cause adverse respiratory, pulmonary, and neurological effects in those exposed. The lack of appropriate personal protection in some of the metal mining facilities in the developing countries is a major concern. As such, the possible use of graphite as a lower-cost, broadly available and safer electrode material for both the anode and the cathode may be an attractive alternative.

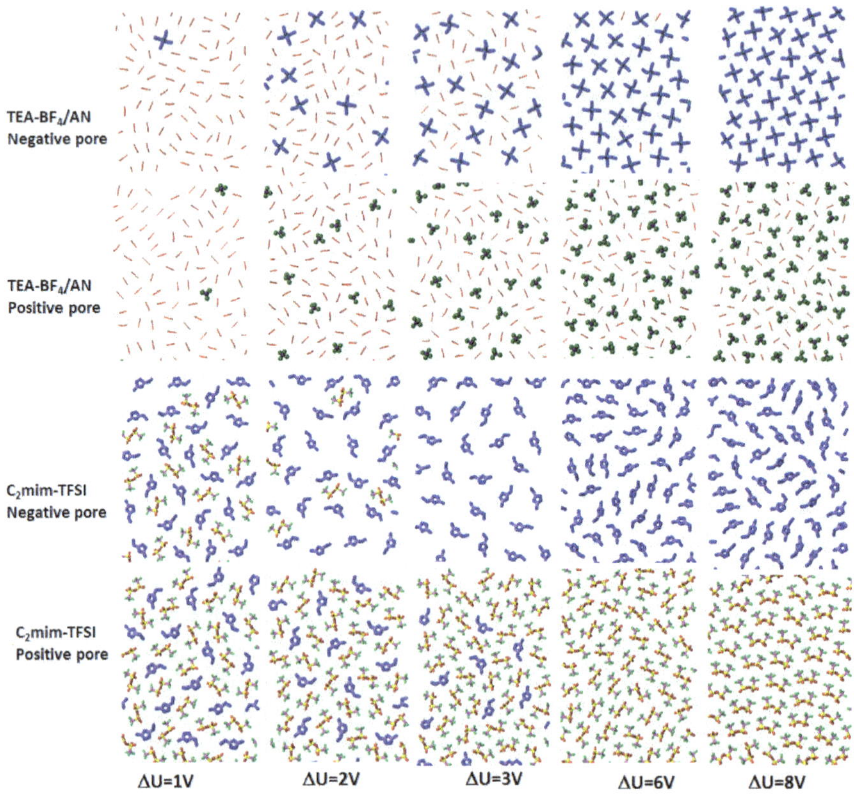

Figure 7. Snapshots showing the distribution of ions inside a 7.5Å width slit pore for electrolytes consisting of solution of ions and RTIL at several different potential differences between electrodes. The TEA and $c_2$mim ions are colored blue, BF_4 is green, and the AN is colored in red/pink. Details regarding simulation conditions based on which these snapshots were generated are given in refs. [45, 80].

The electrochemical energy storage mechanism in this system consists of the simultaneous intercalation of the Li^+, K^+ cation into the graphite negative electrode or usage of Al metal and intercalation of the anion into the graphite positive electrode during charging, with anion intercalation proceeding in well-defined and sequential phases.[101-117] One of the challenges of the dual-ion battery (DIB) is a relatively low volumetric capacity experimentally observed for the graphite cathode. Another challenge is the need to identify suitable electrolyte mixtures that exhibit a high oxidative stability at the graphite cathode and, in addition, form a stable solid electrolyte interphase (SEI) at the graphite anode, permitting stable and highly reversible ion intercalation/de-intercalation into graphite electrodes at operating voltages up to 5 V. A number of electrolytes were examined for use in a dual ion battery including propylene carbonate (PC), ethylene carbonate (EC), diethyl carbonate (DEC), and dimethyl carbonate (DMC), as well as dimethyl sulfoxide (DMSO), dimethylformamide (DMF), ethyl methyl sulfone (EMS) doped with $LiPF_6$, LiTFSI and other salts.[101] The DIB with ionic liquid and IL-solvent such as EMS showed good capacity and efficiency.[114] [103, 104, 106, 108, 114] Recent work by Read et al.[105] demonstrated that electrolytes based on monofluoroethylene carbonate (FEC) and ethylmethyl carbonate (EMC) exhibited high oxidative stability on graphite cathodes and formed stable SEIs on graphite anodes that allowed for full dual-graphite electrochemical cells to be evaluated for energy storage applications. Among various electrolyte mixtures used in dual-graphite batteries to successfully achieve reversible cycling performance, a noteworthy requirement to obtain a sufficiently high energy density is an electrolyte with a high molarity. A recently reported integrated DIB design consisting of an Al film deposited by magnetron sputtering onto one side of the 3D porous glass fiber separator forming a porous anode, and cathode natural graphite directly loaded onto the other side of the separator with Al film deposited on the cathode top, achieved good rate performance up to 120 C with a high capacity of 116.1 mA h g^{-1} (charging/discharging within 30 s) using 4 m $LiPF_6$ EMC -4% vinylene carbonate additive as electrolyte.[115]

Despite the advances made in electrolyte formulations, which, for example, have led to the mitigation of problems related to graphite exfoliation by preventing the co-intercalation of solvent into the graphite anode, the fact that different electrolyte solvents result in a similar stage-I graphite anion intercalated structure would suggest that the solvent does not dictate the structure/distribution of ions inside the graphite and hence may play an auxiliary role in the intercalation process. The structure and arrangement of anions in the graphite will set the limits on critical performance parameters for the dual-graphite cell, where intercalant structure determines the discharge capacity and affects the graphite lattice expansion/contraction properties. To better understand these correlations DFT calculations have been used to investigate the graphite intercalation structure with PF_6^- as anion. In particular, these studies aim to map the anion intercalate composition with the well-defined staging phases de-lineated by the voltage vs. specific capacity profiles, as given in the in-situ work of Read et al.[105, 107] The in-situ XRD and dilatometry results provide evidence for specific compositions being formed at the onset of each of the identified staging phases during the anion intercalation process (*i.e.*, charging). The following sequence of phase formation observed upon charge was proposed:

$C_{24}PF_6(IV) \rightarrow C_{24}PF_6(III) \rightarrow C_{24}PF_6(II) \rightarrow C_{20}PF_6(II) \rightarrow C_{24}PF_6(I) \rightarrow C_{20}PF_6(I)$

Although real-time spectroscopic measurements of stage transitions in Graphite Intercalation Compounds (GICs) have shown the intercalation mechanism to be a complex electrochemical process which propagates in the form of independent concentric sub-fronts interacting with the graphite surface, a conventional and practical interpretation of stage transitions views domains of intercalant as sliding along the graphite galleries to create local arrangements of stage n (n > 0) compositions. To model the nano-domains of intercalant formed in each graphite layer for the various stage n phases within the DFT super cell approach would require a very computationally expensive model due to the large size of the simulation cell. Therefore, the authors have adopted four different model systems corresponding to various concentrations of anion intercalant within the stage-I composition. The model systems employed assume a full graphite gallery intercalation with PF_6^- anions in an ordered configuration as shown in Figure 8. The progression of phase changes shown in Figure 8a to 8d would correspond to the following stage transition scheme:

$C_{48}PF_6(I) \rightarrow C_{24}PF_6(I) \rightarrow C_{20}PF_6(I) \rightarrow C_{12}PF_6(I)$

We expect the calculated capacity values as a function of voltage to be higher than the reported experimental values due to the assumption of full stage-I graphite layer intercalation in this model system. Staging phenomena will be reported later. The intercalation voltage was calculated as follows:

$$\Delta E = (E[GrPF_6] - E[Gr] - n_{anion}*(E[LiPF_6(s)] - E[LiPF_6(g)]) + n_{anion}*E[Li(bulk)]/n_{Li})/n_{anion}$$

where $E[GrPF_6]$ is the electronic energy of the GIC structure, E[Gr] corresponds to the electronic energy of the graphite system, $E[LiPF_6(s)]$ represents the solvation energy for the contact ion pair and was estimated using PBE/6-31+G(d,p) DFT using cluster-continuum approach with $LiPF_6(PC)_3$ treated explicitly and solvent beyond the first solvation shell implicitly using PCM model with PC parameter,[118] $E[LiPF_6(g)]$ is the electronic energy of the isolated (gas-phase) contact ion pair, E[Li(bulk)] corresponds to the electronic energy of bulk Li metal, n_{anion} denotes the number of PF_6^- anions, and n_{Li} designates the number of Li atoms in the Li bulk model system.

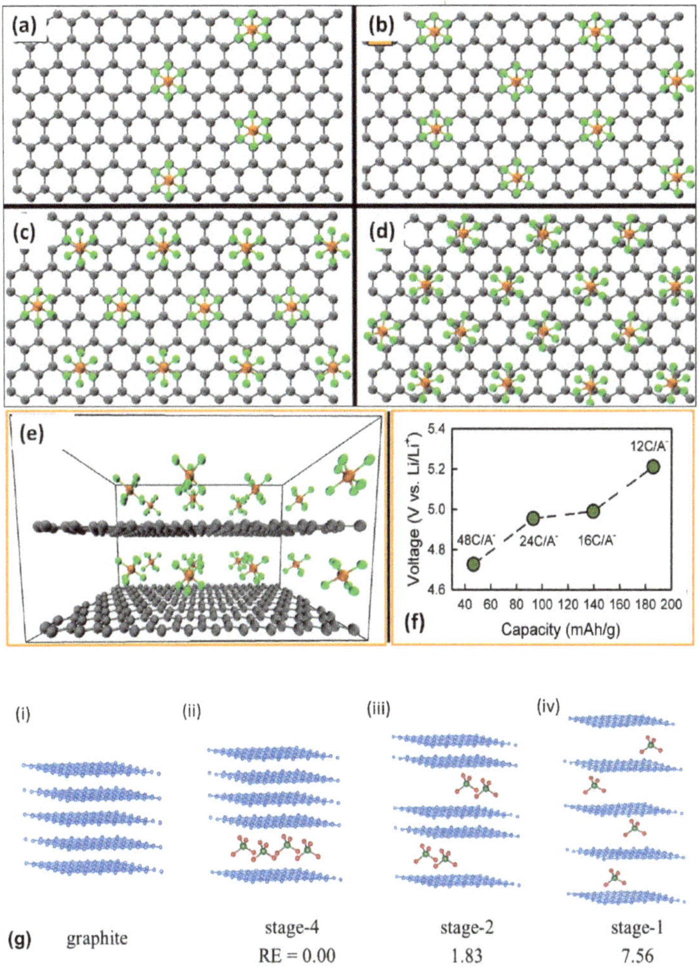

Figure 8. Optimized structures of the four GIC model systems. Color coding: carbon (gray); phosphorous (orange); fluorine (green). (a) $C_{48}PF_6$; (b) $C_{24}PF_6$; (c) $C_{20}PF_6$; (d) $C_{12}PF_6$; (e) orthographic perspective highlighting the Stage (I) model of the supercell employed in DFT calculations; (f) Voltage versus capacity plot for the four stage (I) model systems. (M. Olguin and O. Borodin, unpublished results) (g) Systematic illustration of staging mechanism for (i) graphite, (ii) stage-4, (iii) stage-2, and (iv) stage-1 of the $AlCl_4^-$ anion, RE (in eV) is the relative energetics for same concentrations from Bhauriyal el al.[112] *Reproduced with permission from the Royal Society of Chemistry*

The DFT predictions show that a cathode capacity of 100-140 mAh/g can potentially be achieved using $LiPF_6$-based electrolytes that are stable up to 5 V (Figure 8). A increase in capacity requires a significant increase of the electrode voltage beyond 5 V that is currently not supported by traditional liquid electrolytes. While overestimating the experimentally determined capacity at 5 V[105, 107] by ~30%. DFT calculations accurately predict the plateau around 4.9-5V for PF_6^- intercalation. Since the DFT calculations were performed at 0 K, it is expected that inclusion of entropic effects in MD simulations at room temperature will result in lower PF_6^- intercalation capacities.

A comparison of the PF_6-intercalation in graphite examined by DFT with the anion populations in the smallest nanopores obtained from MD simulations as shown in Figure 3a uncovers intriguing similarities. First, the curves yield a similar number of anions per carbon at 5 V between electrodes: 16 carbon atoms per PF_6^- anion in DFT calculations (Figure 8) and ~20 carbons per larger $TFSI^-$ anions in MD simulations (Figure 3a). Intriguingly, in both cases the slope changes around 5 V (2V vs. PZC in Figure 3a), indicating that the work to densify ions within the hard nano-carbon or graphite pore is significantly higher beyond 5V when the pore is filled primarily with counterions only.

The energetics of staging was examined for the $AlCl_4$ anions in graphite using DFT calculations as shown in Figure 8g. This study demonstrated on the molecular level that intercalation into a single gallery is favored until it reaches its maximum occupancy before starting to fill adjacent empty galleries. This behavior can be understood on the basis of ion–ion Coulomb repulsion and interlayer van der Waals attraction[112] and is in accord with the DFT studies of the lithium staging.[119, 120]

Importantly, the overall electrolyte weight and concentration should also be considered during supercapacitor and dual-ion carbon battery optimization. At a typical 1M PC-based electrolyte composition, the weight of PC solvent per anion is about 60 heavy atoms per anion, which is higher than the weight of carbons at the C_{24}/PF_6 stage of dual ion intercalation compounds or equivalent EDL capacitor. Increasing electrolyte ion concentration or switching to more expensive RTIL electrolytes would result in a higher energy density. It is consistent with the need to utilize highly concentrated electrolytes suggested in the pioneering work by Dahn et al. that focused on the effect of the maximum salt concentration difference in the electrolyte between the fully discharged and fully charged states.[101] In fact, improved energy density was demonstrated by increasing $LiPF_6$ salt concentration essentially up to the solubility limit.[111]

Achieving a fast diffusion coefficient of the PF_6^- or other anions in graphite is important for realizing high charge-discharge rates using large graphite flakes. The PF_6^- anion diffusion coefficient was estimated using galvanostatic intermittent titration technique

(GITT) yielding values between 10^{-11} cm^2 s^{-1} to 10^{-14} cm^2 s^{-1} depending on the electrode voltage.[110] The measured activation energy of 0.366 eV is slightly higher than the 0.23 eV value for the PF_6^- anion hopping barrier obtained from DFT calculations. [110] Similar barriers from 0.308 eV to 0.4 eV were reported for the lithium diffusion in graphite.[121] We find it somewhat unexpected because the estimated PF_6^- chemical diffusion coefficient is significantly lower than the Li^+ cation diffusion coefficient parallel to the graphene plane of 10^{-7} cm^2 s^{-1} reported from DFT calculations.[121] A much higher $TFSI^-$ anion diffusion coefficient was also predicted in MD simulations of the TFSI anion in the 0.75 nm slit pore yielding the $TFSI^-$ self-diffusion coefficient almost as high as bulk ionic liquid (~10^{-5}-10^{-6} cm^2 s^{-1}) as will be discussed below in details.[122]

The anion intercalation in graphite was recently extended to the Al-graphite battery using the IL-based electrolyte achieving an energy density of 68.7 $Whkg^{-1}$ (based on 110 $mAhg^{-1}$ cathode capacity and the masses of active materials in electrodes and electrolyte).[117] A novel potassium ion-based DIB utilized a potassium ion containing, IL-based electrolyte and showed promise for stationary ("grid") energy storage by employing environmentally friendly, abundant and recyclable materials with a capacity retention of 95% after 1500 cycles.[102]

5. Future research and concluding remarks

A diverse set of electric devices requires energy storage with markedly different power, energy and lifetime requirements. EDL and hybrid capacitors and ion intercalation devices will play an important role in future energy storage. For non-Faradaic supercapacitors, it is essential to increase their energy density and reduce the self-discharge rates. It was shown experimentally that the preparation of porous materials with high specific surface area and sub-nanometer pore widths could significantly increase the non-Faradaic capacitance. For pseudocapacitors, which can store more charge than EDLCs, it is important to increase their lifetime and cycling ability while decreasing material cost. The mechanical stability of the pseudocapacitive electrodes might be significantly improved by the utilization of composite materials of conducting polymers or metal oxides with porous carbon. Organometallic frameworks are also promising candidates for high energy density supercapacitors because they can store charge *via* both non-Faradaic and pseudocapacitive processes[123, 124] and have high specific surface areas.

Another possibility to optimize the power and energy densities is the utilization of supercapacitor-battery asymmetric devices, where at least one of the electrodes is Faradaic (or pseudocapacitive). This concept was demonstrated, for example, by Zhang et al.[125] and Zhao et al.[126] with porous carbon (having EDL capacitance) as positive electrode and Fe_3O_4/graphene[125] or nanoconfined $Li_4Ti_5O_{12}$ / porous carbon[126] composites as negative electrodes (having Faradaic capacitance involving Li-insertion into Fe_3O_4 or $Li_4Ti_5O_{12}$ as in Li-ion batteries). The C- $Li_4Ti_5O_{12}$ /C cells can be charged or discharged very fast.[126] Similarly, hybrids with electrodes containing mixture of nitroxide-polymer (which store charge pseudocapacitively) with $LiFePO_4$ (which is a typical Li-battery cathode involving Li insertion in the bulk electrode) improved the power and energy densities because of the initial fast electrochemical response of the pseudocapacitive polymer which later transfers charge to bulk $LiFePO_4$.[127]

The C-based pores (graphene, nanotubes, various porous carbons) can be chemically modified, or functionalized, by either inserting hetero-atoms (N, P, S, B) within the C-framework, or by attaching certain chemical groups (OH, OR, COOR, NR_2, etc.) to C atoms in order to achieve desired properties. [128, 129] Such pore modifications can increase the energy density *via* both non-Faradaic and pseudocapacitive mechanisms as well as *via* increase in quantum capacitance. The functionalized porous C can be tuned for specialized applications in medicine, nanoelectronics, energy fibers, or all-solid[130, 131] supercapacitors for flexible/foldable devices, etc. Gaining a better understanding of which functional groups may lead to slower self-discharge and how it depends on electrolyte composition and applied potential, is important for future optimization of the chemical functionalization strategy.

Of practical interest is the decrease of the production cost of electrolytes and electrode materials. With respect to electrolytes, the replacement of the more expensive Li-salts with cheaper alternatives consisting of Na or Mg solutes could be promising for both supercapacitors and electrochemical batteries. [132, 133] As for electrodes, experiments showed many options to prepare porous materials for energy storage from renewable or waste products[21, 22, 134, 135] such as wood,[136, 137] paper/cellulose,[138] plants such as hemp,[139, 140, 141] recycled cigarette filters,[142] recycled plastic materials,[143] scrap waste tires,[144] and conversion of biomass[140, 145-148] or bio-waste[149-151] therefore potentially significantly reducing the cost and having positive environmental impact.

Computational chemistry and molecular modeling could play an important role in assisting experiments and complex realistic systems, particularly with increasing electrolyte packing in nanoconfined geometries for high power density, understanding supercapacitor self-discharge and degradation mechanisms, elucidating the role of impurities in charge-discharge rates, predicting the device lifetime, and predicting mechanisms of Faradaic processes in batteries, pseudocapacitors, and hybrid devices.

Acknowledgments

Authors gratefully acknowledge the support from project sponsored by the Army Research Laboratory under Cooperative Agreement Number W911NF-12-2-0023. The views and conclusions contained in this document are those of the authors and should not be interpreted as representing the official policies, either expressed or implied, of ARL or the U.S. Government. The U.S. Government is authorized to reproduce and distribute reprints for Government purposes notwithstanding any copyright notation herein. JV acknowledges support through ORAU (contract # W911NF-16-2-0107) to U.S. Army Research Laboratory. MO was supported by an Oak Ridge Associated Universities (ORAU) Postdoctoral Fellowship. GY and OB acknowledge partial support from NASA Minority University Research and Education Project (MUREP) project (NASA grant NNX15AP44A and IAA NND16AA29I). Dr. Jeffrey Read (ARL) is acknowledged for critical reading of the manuscript and suggestions.

This work is derived from Vatamanu, J.; Borodin, O.; Olguin M.; Yushin, G.;Bedrov, D., Charge storage at the nanoscale: understanding the trends from the molecular scale perspective. J. Mater. Chem. A 2017, 5 (40), 21049-21076. https://doi.org/10.1039/C7TA05153K

References

[1] Nishino, A. Capacitors: Operating Principles, Current Market and Technical Trends. *J. Power Sources* 1996, 60, 137-147. https://doi.org/10.1016/S0378-7753(96)80003-6

[2] Simon, P.; Gogotsi, Y. Materials for Electrochemical Capacitors. *Nat. Mater.* 2008, 7, 845-854. https://doi.org/10.1038/nmat2297

[3] Pan, M. J.; Randall, C. A. A Brief Introduction to Ceramic Capacitors. *IEEE Electrical Insulation Magazine* 2010, 26, 44-50. https://doi.org/10.1109/MEI.2010.5482787

[4] González, A.; Goikolea, E.; Barrena, J. A.; Mysyk, R. Review on Supercapacitors: Technologies and Materials. *Renewable and Sustainable Energy Reviews* 2016, 58, 1189-1206. https://doi.org/10.1016/j.rser.2015.12.249

[5] Wei, L.; Yushin, G. Capacitive Electric Storage. In *World Scientific Handbook of Energy*, Crawley, G. M., Ed. 2013; Vol. 3, pp 373-404. https://doi.org/10.1142/9789814343527_0017

[6] Business, H. Electronic Capacitors Industry Report. https://business.highbeam.com/industry-reports/equipment/electronic-capacitors.

[7] Jayalakshmi, M.; Balasubramanian, K. Simple Capacitors to Supercapacitors - an Overview. *Int. J. Electrochem. Sc.* 2008, 3, 1196-1217.

[8] Gu, W.; Yushin, G. Review of Nanostructured Carbon Materials for Electrochemical Capacitor Applications: Advantages and Limitations of Activated Carbon, Carbide-Derived Carbon, Zeolite-Templated Carbon, Carbon Aerogels, Carbon Nanotubes, Onion-Like Carbon, and Graphene. *Wiley Interdisciplinary Reviews-Energy and Environment* 2014, 3, 424-473. https://doi.org/10.1002/wene.102

[9] Wei, L.; Yushin, G. Electrical Double Layer Capacitors with Activated Sucrose-Derived Carbon Electrodes. *Carbon* 2011, 49, 4830-4838. https://doi.org/10.1016/j.carbon.2011.07.003

[10] Weinstein, L.; Dash, R. Supercapacitor Carbons Have Exotic Carbons Failed? *Mater Today* 2013, 16, 356-357. https://doi.org/10.1016/j.mattod.2013.09.005

[11] Wei, L.; Sevilla, M.; Fuertes, A. B.; Mokaya, R.; Yushin, G. Polypyrrole-Derived Activated Carbons for High-Performance Electrical Double-Layer Capacitors with Ionic Liquid Electrolyte. *Adv. Funct. Mater.* 2012, 22, 827-834. https://doi.org/10.1002/adfm.201101866

[12] Choi, N.-S.; Chen, Z.; Freunberger, S. A.; Ji, X.; Sun, Y.-K.; Amine, K.; Yushin, G.; Nazar, L. F.; Cho, J.; Bruce, P. G. Challenges Facing Lithium Batteries and Electrical Double-Layer Capacitors. *Angewandte Chemie-International Edition* 2012, 51, 9994-10024. https://doi.org/10.1002/anie.201201429

[13] Zhong, H.; Xu, F.; Li, Z.; Fu, R.; Wu, D. High-Energy Supercapacitors Based on Hierarchical Porous Carbon with an Ultrahigh Ion-Accessible Surface Area in Ionic Liquid Electrolytes. *Nanoscale* 2013, 5, 4678-4682. https://doi.org/10.1039/c3nr00738c

[14] Li, Y.; Li, Z.; Shen, P. K. Simultaneous Formation of Ultrahigh Surface Area and Three-Dimensional Hierarchical Porous Graphene-Like Networks for Fast and Highly Stable Supercapacitors. *Adv. Mater.* 2013, 25, 2474-2480. https://doi.org/10.1002/adma.201205332

[15] Arcila-Velez, M. R.; Zhu, J.; Childress, A.; Karakaya, M.; Podila, R.; Rao, A. M.; Roberts, M. E. Roll-to-Roll Synthesis of Vertically Aligned Carbon Nanotube Electrodes for Electrical Double Layer Capacitors. *Nano Energy* 2014, 8, 9-16. https://doi.org/10.1016/j.nanoen.2014.05.004

[16] Oschatz, M.; Boukhalfa, S.; Nickel, W.; Hofmann, J. P.; Fischer, C.; Yushin, G.; Kaskel, S. Carbide-Derived Carbon Aerogels with Tunable Pore Structure as Versatile Electrode Material in High Power Supercapacitors. *Carbon* 2017, 113, 283-291. https://doi.org/10.1016/j.carbon.2016.11.050

[17] Moon, J. S.; Kim, H.; Lee, D.-C.; Lee, J. T.; Yushin, G. Increasing Capacitance of Zeolite-Templated Carbons in Electric Double Layer Capacitors. *J. Electrochem. Soc.* 2015, 162, A5070-A5076. https://doi.org/10.1149/2.0131505jes

[18] Oschatz, M.; Boukhalfa, S.; Nickel, W.; Lee, J. T.; Klosz, S.; Borchardt, L.; Eychmueller, A.; Yushin, G.; Kaskel, S. Kroll-Carbons Based on Silica and Alumina Templates as High-Rate Electrode Materials in Electrochemical Double-Layer Capacitors. *Journal of Materials Chemistry A* 2014, 2, 5131-5139. https://doi.org/10.1039/c3ta14815g

[19] Korenblit, Y.; Yushin, G. *Carbide-Derived Carbons*. 2014; ASAP, doi: p 303-329.

[20] Wei, L.; Nitta, N.; Yushin, G. Lithographically Patterned Thin Activated Carbon Films as a New Technology Platform for on-Chip Devices. *Acs Nano* 2013, 7, 6498-6506. https://doi.org/10.1021/nn4028129

[21] Gu, W.; Sevilla, M.; Magasinski, A.; Fuertes, A. F.; Yushin, G. Sulfur-Containing Activated Carbons with Greatly Reduced Content of Bottle Neck Pores for Double-Layer Capacitors: A Case Study for Pseudocapacitance Detection. *Energy Environ. Sci.* 2013, 6, 2465-2476. https://doi.org/10.1039/c3ee41182f

[22] Wei, L.; Sevilla, M.; Fuertesc, A. B.; Mokaya, R.; Yushin, G. Hydrothermal Carbonisation of Abundant Renewable Natural Organic Chemicals for High-Performance Supercapacitor Electrodes. *Advanced Energy Materials* 2011, 1, 356-361. https://doi.org/10.1002/aenm.201100019

[23] Zhang, L.; Zhang, F.; Yang, X.; Long, G.; Wu, Y.; Zhang, T.; Leng, K.; Huang, Y.; Ma, Y.; Yu, A., *et al.* Porous 3d Graphene-Based Bulk Materials with Exceptional High Surface Area and Excellent Conductivity for Supercapacitors. *Sci. Rep.* 2013, 3, 1408. https://doi.org/10.1038/srep01408

[24] Gogotsi, Y.; Simon, P. True Performance Metrics in Electrochemical Energy Storage. *Science* 2011, 334, 917-918. https://doi.org/10.1126/science.1213003

[25] Ji, H.; Zhao, X.; Qiao, Z.; Jung, J.; Zhu, Y.; Lu, Y.; Zhang, L. L.; MacDonald, A. H.; Ruoff, R. S. Capacitance of Carbon-Based Electrical Double-Layer Capacitors. *Nat. Commun.* 2014, 5. https://doi.org/10.1038/ncomms4317

[26] Karden, E.; Ploumen, S.; Fricke, B.; Miller, T.; Snyder, K. Energy Storage Devices for Future Hybrid Electric Vehicles. *J. Power Sources* 2007, 168, 2-11. https://doi.org/10.1016/j.jpowsour.2006.10.090

[27] Zhao, H.; Burke, A. F. Fuel Cell Powered Vehicles Using Supercapacitors–Device Characteristics, Control Strategies, and Simulation Results. *Fuel Cells* 2010, 10, 879-896. https://doi.org/10.1002/fuce.200900214

[28] Thounthong, P.; Raël, S.; Davat, B. Energy Management of Fuel Cell/Battery/Supercapacitor Hybrid Power Source for Vehicle Applications. *J. Power Sources* 2009, 193, 376-385. https://doi.org/10.1016/j.jpowsour.2008.12.120

[29] Winter, M.; Brodd, R. J. What Are Batteries, Fuel Cells, and Supercapacitors? *Chem. Rev.* 2004, 104, 4245-4270. https://doi.org/10.1021/cr020730k

[30] Pech, D.; Brunet, M.; Durou, H.; Huang, P. H.; Mochalin, V.; Gogotsi, Y.; Taberna, P. L.; Simon, P. Ultrahigh-Power Micrometre-Sized Supercapacitors Based on Onion-Like Carbon. *Nat Nanotechnol* 2010, 5, 651-654. https://doi.org/10.1038/nnano.2010.162

[31] Zhao, Z.; Wang, S.; You, C. Piezoelectric Micro-Power Generation to Charge Supercapacitor with Optimized Duty Cycle. *J. Intell. Mater. Syst. Struct.* 2010, 21, 1131-1140. https://doi.org/10.1177/1045389X10376843

[32] Xia, X.; Luo, J.; Zeng, Z.; Guan, C.; Zhang, Y.; Tu, J.; Zhang, H.; Fan, H. J. Integrated Photoelectrochemical Energy Storage: Solar Hydrogen Generation and Supercapacitor. *Scientific Reports* 2012, 2, 981. https://doi.org/10.1038/srep00981

[33] Stavrinidou, E.; Gabrielsson, R.; Nilsson, K. P. R.; Singh, S. K.; Franco-Gonzalez, J. F.; Volkov, A. V.; Jonsson, M. P.; Grimoldi, A.; Elgland, M.; Zozoulenko, I. V., *et al.* In Vivo Polymerization and Manufacturing of Wires and Supercapacitors in Plants. *Proc. Natl. Acad. Sci. U. S. A.* 2017, 114, 2807-2812. https://doi.org/10.1073/pnas.1616456114

[34] Suo, L.; Borodin, O.; Sun, W.; Fan, X.; Yang, C.; Wang, F.; Gao, T.; Ma, Z.; Schroeder, M.; von Cresce, A., *et al.* Advanced High-Voltage Aqueous Lithium-Ion Battery Enabled by "Water-in-Bisalt" Electrolyte. *Angew. Chem. Int. Ed.* 2016, 55, 7136-7141.

[35] Suo, L.; Han, F.; Fan, X.; Liu, H.; Xu, K.; Wang, C. "Water-in-Salt" Electrolytes Enable Green and Safe Li-Ion Batteries for Large Scale Electric Energy Storage

Applications. *J. Mater. Chem. A* 2016, 4, 6639-6644. https://doi.org/10.1039/C6TA00451B

[36] Wang, F.; Lin, Y.; Suo, L.; Fan, X.; Gao, T.; Yang, C.; Han, F.; Qi, Y.; Xu, K.; Wang, C. Stabilizing High Voltage Licoo2 Cathode in Aqueous Electrolyte with Interphase-Forming Additive. *Energy Environ. Sci.* 2016, 9, 3666-3673. https://doi.org/10.1039/C6EE02604D

[37] Yamada, Y.; Usui, K.; Sodeyama, K.; Ko, S.; Tateyama, Y.; Yamada, A. Hydrate-Melt Electrolytes for High-Energy-Density Aqueous Batteries. *Nature Energy* 2016, 1, 16129. https://doi.org/10.1038/nenergy.2016.129

[38] Korenblit, Y.; Kajdos, A.; West, W. C.; Smart, M. C.; Brandon, E. J.; Kvit, A.; Jagiello, J.; Yushin, G. In Situ Studies of Ion Transport in Microporous Supercapacitor Electrodes at Ultralow Temperatures. *Adv. Funct. Mater.* 2012, 22, 1655-1662. https://doi.org/10.1002/adfm.201102573

[39] Fedorov, M. V.; Kornyshev, A. A. Ionic Liquids at Electrified Interfaces. *Chem. Rev.* 2014, 114, 2978-3036. https://doi.org/10.1021/cr400374x

[40] Hayes, R.; Warr, G. G.; Atkin, R. Structure and Nanostructure in Ionic Liquids. *Chem. Rev.* 2015, 115, 6357-6426 https://doi.org/10.1021/cr500411q

[41] Lin, R.; Taberna, P.-L.; Fantini, S.; Presser, V.; Pérez, C. R.; Malbosc, F.; Rupesinghe, N. L.; Teo, K. B. K.; Gogotsi, Y.; Simon, P. Capacitive Energy Storage from −50 to 100 °C Using an Ionic Liquid Electrolyte. *J. Phys. Chem. Lett.* 2011, 2, 2396-2401. https://doi.org/10.1021/jz201065t

[42] Nitta, N.; Wu, F.; Lee, J. T.; Yushin, G. Li-Ion Battery Materials: Present and Future. *Materials Today* 2015, 18, 252-264. https://doi.org/10.1016/j.mattod.2014.10.040

[43] Kornyshev, A. A. Double-Layer in Ionic Liquids: Paradigm Change? *J. Phys. Chem. B* 2007, 111, 5545-5557. https://doi.org/10.1021/jp067857o

[44] Yamamoto, R.; Morisaki, H.; Sakata, O.; Shimotani, H.; Yuan, H.; Iwasa, Y.; Kimura, T.; Wakabayashi, Y. External Electric Field Dependence of the Structure of the Electric Double Layer at an Ionic Liquid/Au Interface. *Appl. Phys. Lett.* 2012, 101, 053122. https://doi.org/10.1063/1.4742920

[45] Vatamanu, J.; Vatamanu, M.; Borodin, O.; Bedrov, D. A Comparative Study of Room Temperature Ionic Liquids and Their Organic Solvent Mixtures near Charged Electrodes. *J. Phys.: Condens. Matter* 2016, 28, 464002. https://doi.org/10.1088/0953-8984/28/46/464002

[46] Gebbie, M. A.; Valtiner, M.; Banquy, X.; Fox, E. T.; Henderson, W. A.; Israelachvili, J. N. Ionic Liquids Behave as Dilute Electrolyte Solutions. *Proc. Natl. Acad. Sci. U. S. A.* 2013, 110, 9674-9679. https://doi.org/10.1073/pnas.1307871110

[47] Lee, A. A.; Vella, D.; Perkin, S.; Goriely, A. Are Room-Temperature Ionic Liquids Dilute Electrolytes? *J. Phys. Chem. Lett.* 2015, 6, 159-163. https://doi.org/10.1021/jz502250z

[48] Merlet, C.; Salanne, M.; Rotenberg, B.; Madden, P. A. Influence of Solvation on the Structural and Capacitive Properties of Electrical Double Layer Capacitors. *Electrochim. Acta* 2013, 101, 262-271. https://doi.org/10.1016/j.electacta.2012.12.107

[49] Vatamanu, J.; Borodin, O.; Smith, G. D. Molecular Dynamics Simulation Studies of the Structure of a Mixed Carbonate/LiPF6 Electrolyte near Graphite Surface as a Function of Electrode Potential. *J. Phys. Chem. C* 2011, 116, 1114-1121. https://doi.org/10.1021/jp2101539

[50] Borodin, O.; Smith, G. D.; Henderson, W. Li+ Cation Environment, Transport, and Mechanical Properties of the Litfsi Doped N-Methyl-N-Alkylpyrrolidinium+TFSI- Ionic Liquids. *J. Phys. Chem. B* 2006, 110, 16879-16886. https://doi.org/10.1021/jp061930t

[51] Druschler, M.; Borisenko, N.; Wallauer, J.; Winter, C.; Huber, B.; Endres, F.; Roling, B. New Insights into the Interface between a Single-Crystalline Metal Electrode and an Extremely Pure Ionic Liquid: Slow Interfacial Processes and the Influence of Temperature on Interfacial Dynamics. *Phys. Chem. Chem. Phys.* 2012, 14, 5090-5099. https://doi.org/10.1039/c2cp40288b

[52] Cannes, C.; Cachet, H.; Debiemme-Chouvy, C.; Deslouis, C.; de Sanoit, J.; Le Naour, C.; Zinovyeva, V. A. Double Layer at [Bumeim][Tf2N] Ionic Liquid–Pt or −C Material Interfaces. *J. Phys. Chem. C* 2013, 117, 22915-22925. https://doi.org/10.1021/jp407665q

[53] Vatamanu, J.; Xing, L.; Li, W.; Bedrov, D. Influence of Temperature on the Capacitance of Ionic Liquid Electrolytes on Charged Surfaces. *Phys. Chem. Chem. Phys.* 2014, 16, 5174-82. https://doi.org/10.1039/c3cp54705a

[54] Oldham, K. B. A Gouy–Chapman–Stern Model of the Double Layer at a (Metal)/(Ionic Liquid) Interface. *J. Electroanal. Chem.* 2008, 613, 131-138. https://doi.org/10.1016/j.jelechem.2007.10.017

[55] Yining, H.; Shanghui, H.; Tianying, Y. A Mean-Field Theory on the Differential Capacitance of Asymmetric Ionic Liquid Electrolytes. *J. Phys.: Condens. Matter* 2014, 26, 284103. https://doi.org/10.1088/0953-8984/26/28/284103

[56] Konrad, B.; Peter, K.; Christian, H. Coarse-Grained Simulations of an Ionic Liquid-Based Capacitor: I. Density, Ion Size, and Valency Effects. *J. Phys.: Condens. Matter* 2014, 26, 284108. https://doi.org/10.1088/0953-8984/26/28/284103

[57] Henderson, D.; Lamperski, S.; Bari Bhuiyan, L.; Wu, J. The Tail Effect on the Shape of an Electrical Double Layer Differential Capacitance Curve. *J. Chem. Phys.* 2013, 138, 144704. https://doi.org/10.1063/1.4799886

[58] Henderson, D.; Jiang, D.-e.; Jin, Z.; Wu, J. Application of Density Functional Theory to Study the Double Layer of an Electrolyte with an Explicit Dimer Model for the Solvent. *J. Phys. Chem. B* 2012, 116, 11356-11361. https://doi.org/10.1021/jp305400z

[59] Jiang, D.-e.; Meng, D.; Wu, J. Density Functional Theory for Differential Capacitance of Planar Electric Double Layers in Ionic Liquids. *Chem. Phys. Lett.* 2011, 504, 153-158. https://doi.org/10.1016/j.cplett.2011.01.072

[60] Forsman, J.; Woodward, C. E.; Trulsson, M. A Classical Density Functional Theory of Ionic Liquids. *J. Phys. Chem. B* 2011, 115, 4606-4612. https://doi.org/10.1021/jp111747w

[61] Reszko-Zygmunt, J.; Sokołowski, S.; Henderson, D.; Boda, D. Temperature Dependence of the Double Layer Capacitance for the Restricted Primitive Model of an Electrolyte Solution from a Density Functional Approach. *J. Chem. Phys.* 2005, 122, 084504. https://doi.org/10.1063/1.1850453

[62] Lamperski, S.; Outhwaite, C. W.; Bhuiyan, L. B. The Electric Double-Layer Differential Capacitance at and near Zero Surface Charge for a Restricted Primitive Model Electrolyte. *J. Phys. Chem. B* 2009, 113, 8925-8929. https://doi.org/10.1021/jp900037h

[63] Sarkisov, L.; Van Tassel, P. R. Theories of Molecular Fluids Confined in Disordered Porous Materials. *J. Phys.: Condens. Matter* 2008, 20. https://doi.org/10.1088/0953-8984/20/33/333101

[64] Woelki, S.; Kohler, H.-H.; Krienke, H. A Singlet-RISM Theory for Solid/Liquid Interfaces Part I: Uncharged Walls. *J. Phys. Chem. B* 2007, 111, 13386-13397. https://doi.org/10.1021/jp068998t

[65] Yang, H.; Yang, J.; Bo, Z.; Chen, X.; Shuai, X.; Kong, J.; Yan, J.; Cen, K. Kinetic-Dominated Charging Mechanism within Representative Aqueous Electrolyte-Based Electric Double-Layer Capacitors. *J. Phys. Chem. Lett.* 2017, ASAP, doi: 10.1021/acs.jpclett.7b01525, 3703-3710. https://doi.org/10.1021/acs.jpclett.7b01525

[66] Hu, Z.; Vatamanu, J.; Borodin, O.; Bedrov, D. A Comparative Study of Alkylimidazolium Room Temperature Ionic Liquids with FSI and TFSI Anions near Charged Electrodes. *Electrochim. Acta* 2014, 145, 40-52. https://doi.org/10.1016/j.electacta.2014.08.072

[67] Yang, H.; Zhang, X.; Yang, J.; Bo, Z.; Hu, M.; Yan, J.; Cen, K. Molecular Origin of Electric Double-Layer Capacitance at Multilayer Graphene Edges. *J. Phys. Chem. Lett.* 2017, 8, 153-160. https://doi.org/10.1021/acs.jpclett.6b02659

[68] Ho, T. A.; Striolo, A. Capacitance Enhancement via Electrode Patterning. *J. Chem. Phys.* 2013, 139, 204708. https://doi.org/10.1063/1.4833316

[69] Xing, L.; Vatamanu, J.; Smith, G. D.; Bedrov, D. Nanopatterning of Electrode Surfaces as a Potential Route to Improve the Energy Density of Electric Double-Layer Capacitors: Insight from Molecular Simulations. *J. Phys. Chem. Lett.* 2012, 3, 1124-1129. https://doi.org/10.1021/jz300253p

[70] Zhan, C.; Zhang, Y.; Cummings, P. T.; Jiang, D.-e. Computational Insight into the Capacitive Performance of Graphene Edge Planes. *Carbon* 2017, 116, 278-285. https://doi.org/10.1016/j.carbon.2017.01.104

[71] Vatamanu, J.; Hu, Z.; Bedrov, D.; Perez, C.; Gogotsi, Y. Increasing Energy Storage in Electrochemical Capacitors with Ionic Liquid Electrolytes and Nanostructured Carbon Electrodes. *J. Phys. Chem. Lett.* 2013, 4, 2829-2837. https://doi.org/10.1021/jz401472c

[72] Feng, G.; Jiang, D.-e.; Cummings, P. T. Curvature Effect on the Capacitance of Electric Double Layers at Ionic Liquid/Onion-Like Carbon Interfaces. *J. Chem. Theory Comput.* 2012, 8, 1058-1063. https://doi.org/10.1021/ct200914j

[73] Feng, G.; Li, S.; Atchison, J. S.; Presser, V.; Cummings, P. T. Molecular Insights into Carbon Nanotube Supercapacitors: Capacitance Independent of Voltage and Temperature. *J. Phys. Chem. C* 2013, 117, 9178-9186. https://doi.org/10.1021/jp403547k

[74] Lin, R.; Taberna, P. L.; Chmiola, J.; Guay, D.; Gogotsi, Y.; Simon, P. Microelectrode Study of Pore Size, Ion Size, and Solvent Effects on the

Charge/Discharge Behavior of Microporous Carbons for Electrical Double-Layer Capacitors. *J. Electrochem. Soc.* 2009, 156, A7-A12. https://doi.org/10.1149/1.3002376

[75] Largeot, C.; Portet, C.; Chmiola, J.; Taberna, P.-L.; Gogotsi, Y.; Simon, P. Relation between the Ion Size and Pore Size for an Electric Double-Layer Capacitor. *J. Am. Chem. Soc.* 2008, 130, 2730-2731. https://doi.org/10.1021/ja7106178

[76] Boukhalfa, S.; Gordon, D.; He, L.; Melnichenko, Y. B.; Nitta, N.; Magasinski, A.; Yushin, G. In Situ Small Angle Neutron Scattering Revealing Ion Sorption in Microporous Carbon Electrical Double Layer Capacitors. *ACS Nano* 2014, 8, 2495-2503. https://doi.org/10.1021/nn406077n

[77] Boukhalfa, S.; He, L.; Melnichenko, Y. B.; Yushin, G. Small-Angle Neutron Scattering for Insitu Probing of Ion Adsorption inside Micropores. *Angewandte Chemie-International Edition* 2013, 52, 4618-4622. https://doi.org/10.1002/anie.201209141

[78] Kondrat, S.; Kornyshev, A. Superionic State in Double-Layer Capacitors with Nanoporous Electrodes. *J. Phys.: Condens. Matter* 2011, 23, 022201. https://doi.org/10.1088/0953-8984/23/2/022201

[79] Feng, G.; Cummings, P. T. Supercapacitor Capacitance Exhibits Oscillatory Behavior as a Function of Nanopore Size. *J. Phys. Chem. Lett.* 2011, 2, 2859-2864. https://doi.org/10.1021/jz201312e

[80] Vatamanu, J.; Vatamanu, M.; Bedrov, D. Non-Faradaic Energy Storage by Room Temperature Ionic Liquids in Nanoporous Electrodes. *ACS Nano* 2015, 9, 5999–6017. https://doi.org/10.1021/acsnano.5b00945

[81] Wu, P.; Huang, J.; Meunier, V.; Sumpter, B. G.; Qiao, R. Voltage Dependent Charge Storage Modes and Capacity in Subnanometer Pores. *J. Phys. Chem. Lett.* 2012, 3, 1732-1737. https://doi.org/10.1021/jz300506j

[82] Wu, Z. S.; Parvez, K.; Feng, X.; Müllen, K. Graphene-Based in-Plane Micro-Supercapacitors with High Power and Energy Densities. *Nat. Commun.* 2013, 4, 2487. https://doi.org/10.1038/ncomms3487

[83] Wu, P.; Huang, J.; Meunier, V.; Sumpter, B. G.; Qiao, R. Complex Capacitance Scaling in Ionic Liquids-Filled Nanopores. *ACS Nano* 2011, 5, 9044-9051. https://doi.org/10.1021/nn203260w

[84] Henderson, D. Oscillations in the Capacitance of a Nanopore Containing an Electrolyte Due to Pore Width and Nonzero Size Ions. *J. Colloid Interface Sci.* 2012, 374, 345-347. https://doi.org/10.1016/j.jcis.2012.01.050

[85] Jiang, D.-e.; Jin, Z.; Henderson, D.; Wu, J. Solvent Effect on the Pore-Size Dependence of an Organic Electrolyte Supercapacitor. *J. Phys. Chem. Lett.* 2012, 3, 1727-1731. https://doi.org/10.1021/jz3004624

[86] Van Aken, K. L.; Beidaghi, M.; Gogotsi, Y. Formulation of Ionic-Liquid Electrolyte to Expand the Voltage Window of Supercapacitors. *Angew. Chem. Int. Ed.* 2015, 54, 4806-4809. https://doi.org/10.1002/anie.201412257

[87] Varanasi, S. R.; Bhatia, S. K. Optimal Electrode Mass Ratio in Nanoporous Carbon Electrochemical Supercapacitors. *J. Phys. Chem. C* 2016, 120, 27925-27933. https://doi.org/10.1021/acs.jpcc.6b07759

[88] Qiu, Y.; Chen, Y. Capacitance Performance of Sub-2 Nm Graphene Nanochannels in Aqueous Electrolyte. *J. Phys. Chem. C* 2015, 119, 23813-23819. https://doi.org/10.1021/acs.jpcc.5b06401

[89] Jiang, D.-e.; Wu, J. Unusual Effects of Solvent Polarity on Capacitance for Organic Electrolytes in a Nanoporous Electrode. *Nanoscale* 2014, 6, 5545-5550. https://doi.org/10.1039/C4NR00046C

[90] Kornyshev, A. A. The Simplest Model of Charge Storage in Single File Metallic Nanopores. *Faraday Discuss.* 2013, 164, 117-133. https://doi.org/10.1039/c3fd00026e

[91] Kondrat, S.; Kornyshev, A. A. Pressing a Spring: What Does It Take to Maximize the Energy Storage in Nanoporous Supercapacitors? *Nanoscale Horizons* 2016, 1, 45-52. https://doi.org/10.1039/C5NH00004A

[92] Mohammadzadeh, L.; Quaino, P.; Schmickler, W. Interactions of Anions and Cations in Carbon Nanotubes. *Faraday Discuss.* 2016, 193, 415-426. https://doi.org/10.1039/C6FD00076B

[93] Merlet, C.; Péan, C.; Rotenberg, B.; Madden, P. A.; Daffos, B.; Taberna, P. L.; Simon, P.; Salanne, M. Highly Confined Ions Store Charge More Efficiently in Supercapacitors. *Nat Commun* 2013, 4. https://doi.org/10.1038/ncomms3701

[94] Jain, S. K.; Pellenq, R. J. M.; Pikunic, J. P.; Gubbins, K. E. Molecular Modeling of Porous Carbons Using the Hybrid Reverse Monte Carlo Method. *Langmuir* 2006, 22, 9942-9948. https://doi.org/10.1021/la053402z

[95] Griffin, J. M.; Forse, A. C.; Tsai, W.-Y.; Taberna, P.-L.; Simon, P.; Grey, C. P. In Situ NMR and Electrochemical Quartz Crystal Microbalance Techniques Reveal the Structure of the Electrical Double Layer in Supercapacitors. *Nat. Mater.* 2015, 14, 812-819. https://doi.org/10.1038/nmat4318

[96] Wang, H.; Köster, T. K. J.; Trease, N. M.; Ségalini, J.; Taberna, P.-L.; Simon, P.; Gogotsi, Y.; Grey, C. P. Real-Time NMR Studies of Electrochemical Double-Layer Capacitors. *J. Am. Chem. Soc.* 2011, 133, 19270-19273. https://doi.org/10.1021/ja2072115

[97] Hu, L.; Guo, D.; Feng, G.; Li, H.; Zhai, T. Asymmetric Behavior of Positive and Negative Electrodes in Carbon/Carbon Supercapacitors and Its Underlying Mechanism. *J. Phys. Chem. C* 2016, 120, 24675-24681. https://doi.org/10.1021/acs.jpcc.6b09898

[98] Prehal, C.; Koczwara, C.; Jäckel, N.; Schreiber, A.; Burian, M.; Amenitsch, H.; Hartmann, M. A.; Presser, V.; Paris, O. Quantification of Ion Confinement and Desolvation in Nanoporous Carbon Supercapacitors with Modelling and in Situ X-Ray Scattering. *Nature Energy* 2017, 2, 16215. https://doi.org/10.1038/nenergy.2016.215

[99] Chmiola, J.; Yushin, G.; Gogotsi, Y.; Portet, C.; Simon, P.; Taberna, P. L. Anomalous Increase in Carbon Capacitance at Pore Sizes Less Than 1 Nanometer. *Science* 2006, 313, 1760-1763. https://doi.org/10.1126/science.1132195

[100] Centeno, T. A.; Sereda, O.; Stoeckli, F. Capacitance in Carbon Pores of 0.7 to 15 Nm: A Regular Pattern. *Phys. Chem. Chem. Phys.* 2011, 13, 12403-12406. https://doi.org/10.1039/c1cp20748b

[101] Dahn, J. R.; Seel, J. A. Energy and Capacity Projections for Practical Dual-Graphite Cells. *J. Electrochem. Soc.* 2000, 147, 899-901. https://doi.org/10.1149/1.1393289

[102] Beltrop, K.; Beuker, S.; Heckmann, A.; Winter, M.; Placke, T. Alternative Electrochemical Energy Storage: Potassium-Based Dual-Graphite Batteries. *Energy & Environ. Sci.* 2017, ASAP. https://doi.org/10.1039/C7EE01535F

[103] Placke, T.; Fromm, O.; Lux, S. F.; Bieker, P.; Rothermel, S.; Meyer, H.-W.; Passerini, S.; Winter, M. Reversible Intercalation of Bis(Trifluoromethanesulfonyl)Imide Anions from an Ionic Liquid Electrolyte into Graphite for High Performance Dual-Ion Cells. *J. Electrochem. Soc.* 2012, 159, A1755-A1765. https://doi.org/10.1149/2.011211jes

[104] Qi, X.; Blizanac, B.; DuPasquier, A.; Meister, P.; Placke, T.; Oljaca, M.; Li, J.; Winter, M. Investigation of PF6- and TFSI- Anion Intercalation into Graphitized Carbon Blacks and Its Influence on High Voltage Lithium Ion Batteries. *Phys. Chem. Chem. Phys.* 2014, 16, 25306-25313. https://doi.org/10.1039/C4CP04113E

[105] Read, J. A.; Cresce, A. V.; Ervin, M. H.; Xu, K. Dual-Graphite Chemistry Enabled by a High Voltage Electrolyte. *Energ. & Env. Sci.* 2014, 7, 617-620. https://doi.org/10.1039/C3EE43333A

[106] Rothermel, S.; Meister, P.; Schmuelling, G.; Fromm, O.; Meyer, H. W.; Nowak, S.; Winter, M.; Placke, T. Dual-Graphite Cells Based on the Reversible Intercalation of Bis(Trifluoromethanesulfonyl)Imide Anions from an Ionic Liquid Electrolyte. *Energ. & Env. Sci.* 2014, 7, 3412-3423. https://doi.org/10.1039/C4EE01873G

[107] Read, J. A. In-Situ Studies on the Electrochemical Intercalation of Hexafluorophosphate Anion in Graphite with Selective Cointercalation of Solvent. *J. Phys. Chem. C* 2015, 119, 8438-8446. https://doi.org/10.1021/jp5115465

[108] Beltrop, K.; Meister, P.; Klein, S.; Heckmann, A.; Grunebaum, M.; Wiemhofer, H. D.; Winter, M.; Placke, T. Does Size Really Matter? New Insights into the Intercalation Behavior of Anions into a Graphite-Based Positive Electrode for Dual-Ion Batteries. *Electrochim Acta* 2016, 209, 44-55. https://doi.org/10.1016/j.electacta.2016.05.012

[109] Fukutsuka, T.; Yamane, F.; Miyazaki, K.; Abe, T. Electrochemical Intercalation of Bis(Fluorosulfonyl)Amide Anion into Graphite. *J. Electrochem. Soc.* 2016, 163, A499-A503. https://doi.org/10.1149/2.0881603jes

[110] Miyoshi, S.; Akbay, T.; Kurihara, T.; Fukuda, T.; Staykov, A. T.; Ida, S.; Ishihara, T. Fast Diffusivity of PF_6^- Anions in Graphitic Carbon for a Dual-Carbon Rechargeable Battery with Superior Rate Property. *J. Phys. Chem. C* 2016, 120, 22887-22894. https://doi.org/10.1021/acs.jpcc.6b07957

[111] Miyoshi, S.; Nagano, H.; Fukuda, T.; Kurihara, T.; Watanabe, M.; Ida, S.; Ishihara, T. Dual-Carbon Battery Using High Concentration LiPF6 in Dimethyl Carbonate (Dmc) Electrolyte. *J. Electrochem. Soc.* 2016, 163, A1206-A1213. https://doi.org/10.1149/2.0381607jes

[112] Bhauriyal, P.; Mahata, A.; Pathak, B. The Staging Mechanism of $Alcl_4$ Intercalation in a Graphite Electrode for an Aluminium-Ion Battery. *Phys. Chem. Chem. Phys.* 2017, 19, 7980-7989. https://doi.org/10.1039/C7CP00453B

[113] Fan, H.; Qi, L.; Wang, H. Hexafluorophosphate Anion Intercalation into Graphite Electrode from Methyl Propionate. *Solid State Ionics* 2017, 300, 169-174. https://doi.org/10.1016/j.ssi.2016.12.032

[114] Fan, J.; Zhang, Z.; Liu, Y.; Wang, A.; Li, L.; Yuan, W. An Excellent Rechargeable Pp14tfsi Ionic Liquid Dual-Ion Battery. *Chemical Communications* 2017, 53, 6891-6894. https://doi.org/10.1039/C7CC02534C

[115] Jiang, C.; Fang, Y.; Lang, J.; Tang, Y. Integrated Configuration Design for Ultrafast Rechargeable Dual-Ion Battery. *Advanced Energy Materials* 2017, ASAP, doi: 10.1002/aenm.201700913, 1700913. https://doi.org/10.1002/aenm.201700913

[116] Rodríguez-Pérez, I. A.; Ji, X. Anion Hosting Cathodes in Dual-Ion Batteries. *Acs Energy Lett* 2017, ASAP, doi: 10.1021/acsenergylett.7b00321, 1762-1770. https://doi.org/10.1021/acsenergylett.7b00321

[117] Wang, D.-Y.; Wei, C.-Y.; Lin, M.-C.; Pan, C.-J.; Chou, H.-L.; Chen, H.-A.; Gong, M.; Wu, Y.; Yuan, C.; Angell, M., *et al.* Advanced Rechargeable Aluminium Ion Battery with a High-Quality Natural Graphite Cathode. *Nature Comm.* 2017, 8, 14283. https://doi.org/10.1038/ncomms14283

[118] Chapman, N.; Borodin, O.; Yoon, T.; Nguyen, C. C.; Lucht, B. L. Spectroscopic and Density Functional Theory Characterization of Common Lithium Salt Solvates in Carbonate Electrolytes for Lithium Batteries. *J. Phys. Chem. C* 2017, 121, 2135-2148. https://doi.org/10.1021/acs.jpcc.6b12234

[119] Okamoto, Y. Density Functional Theory Calculations of Alkali Metal (Li, Na, and K) Graphite Intercalation Compounds. *J. Phys. Chem. C* 2014, 118, 16-19. https://doi.org/10.1021/jp4063753

[120] Persson, K.; Hinuma, Y.; Meng, Y. S.; Van der Ven, A.; Ceder, G. Thermodynamic and Kinetic Properties of the Li-Graphite System from First-Principles Calculations. *Phys. Rev. B* 2010, 82. https://doi.org/10.1103/PhysRevB.82.125416

[121] Persson, K.; Sethuraman, V. A.; Hardwick, L. J.; Hinuma, Y.; Meng, Y. S.; van der Ven, A.; Srinivasan, V.; Kostecki, R.; Ceder, G. Lithium Diffusion in Graphitic Carbon. *J. Phys. Chem. Lett.* 2010, 1, 1176-1180. https://doi.org/10.1021/jz100188d

[122] He, Y.; Qiao, R.; Vatamanu, J.; Borodin, O.; Bedrov, D.; Huang, J.; Sumpter, B. G. Importance of Ion Packing on the Dynamics of Ionic Liquids During Micropore

Charging. *J. Phys. Chem. Lett.* 2016, 7, 36-42. https://doi.org/10.1021/acs.jpclett.5b02378

[123] Choi, K. M.; Jeong, H. M.; Park, J. H.; Zhang, Y.-B.; Kang, J. K.; Yaghi, O. M. Supercapacitors of Nanocrystalline Metal–Organic Frameworks. *ACS Nano* 2014, 8, 7451-7457. https://doi.org/10.1021/nn5027092

[124] Liu, X.; Shi, C.; Zhai, C.; Cheng, M.; Liu, Q.; Wang, G. Cobalt-Based Layered Metal–Organic Framework as an Ultrahigh Capacity Supercapacitor Electrode Material. *ACS Appl. Mater. Inter.* 2016, 8, 4585-4591. https://doi.org/10.1021/acsami.5b10781

[125] Zhang, F.; Zhang, T.; Yang, X.; Zhang, L.; Leng, K.; Huang, Y.; Chen, Y. A High-Performance Supercapacitor-Battery Hybrid Energy Storage Device Based on Graphene-Enhanced Electrode Materials with Ultrahigh Energy Density. *Energy Environ. Sci.* 2013, 6, 1623-1632. https://doi.org/10.1039/c3ee40509e

[126] Zhao, E.; Qin, C.; Jung, H.-R.; Berdichevsky, G.; Nese, A.; Marder, S.; Yushin, G. Lithium Titanate Confined in Carbon Nanopores for Asymmetric Supercapacitors. *Acs Nano* 2016, 10, 3977-3984. https://doi.org/10.1021/acsnano.6b00479

[127] Guo, W.; Xin, S.; Ji, M.; Guo, Y.; Wan, L. Supercapacitor-Battery Hybrid Energy Storage Devices from an Aqueous Nitroxide Radical Active Material. *Chin. Sci. Bull.* 2011, 56, 2433-2436. https://doi.org/10.1007/s11434-011-4575-z

[128] Liu, D.; Fu, C.; Zhang, N.; Zhou, H.; Kuang, Y. Three-Dimensional Porous Nitrogen Doped Graphene Hydrogel for High Energy Density Supercapacitors. *Electrochim. Acta* 2016, 213, 291-297. https://doi.org/10.1016/j.electacta.2016.07.131

[129] Georgakilas, V.; Otyepka, M.; Bourlinos, A. B.; Chandra, V.; Kim, N.; Kemp, K. C.; Hobza, P.; Zboril, R.; Kim, K. S. Functionalization of Graphene: Covalent and Non-Covalent Approaches, Derivatives and Applications. *Chem. Rev.* 2012, 112, 6156-6214. https://doi.org/10.1021/cr3000412

[130] Chen, T.; Tang, Y.; Qiao, Y.; Liu, Z.; Guo, W.; Song, J.; Mu, S.; Yu, S.; Zhao, Y.; Gao, F. All-Solid-State High Performance Asymmetric Supercapacitors Based on Novel Mns Nanocrystal and Activated Carbon Materials. *Scientific Reports* 2016, 6, 23289. https://doi.org/10.1038/srep23289

[131] Liu, Y.; Weng, B.; Razal, J. M.; Xu, Q.; Zhao, C.; Hou, Y.; Seyedin, S.; Jalili, R.; Wallace, G. G.; Chen, J. High-Performance Flexible All-Solid-State

Supercapacitor from Large Free-Standing Graphene-Pedot/Pss Films. *Scientific Reports* 2015, 5, 17045. https://doi.org/10.1038/srep17045

[132] Chen, Z.; Augustyn, V.; Jia, X.; Xiao, Q.; Dunn, B.; Lu, Y. High-Performance Sodium-Ion Pseudocapacitors Based on Hierarchically Porous Nanowire Composites. *ACS Nano* 2012, 6, 4319-4327. https://doi.org/10.1021/nn300920e

[133] Li, H.; Peng, L.; Zhu, Y.; Zhang, X.; Yu, G. Achieving High-Energy–High-Power Density in a Flexible Quasi-Solid-State Sodium Ion Capacitor. *Nano Lett.* 2016, 16, 5938-5943. https://doi.org/10.1021/acs.nanolett.6b02932

[134] Wei, L.; Yushin, G. Nanostructured Activated Carbons from Natural Precursors for Electrical Double Layer Capacitors. *Nano Energy* 2012, 1, 552-565. https://doi.org/10.1016/j.nanoen.2012.05.002

[135] Liu, J.; Deng, Y.; Li, X.; Wang, L. Promising Nitrogen-Rich Porous Carbons Derived from One-Step Calcium Chloride Activation of Biomass-Based Waste for High Performance Supercapacitors. *ACS Sustainable Chemistry & Engineering* 2016, 4, 177-187. https://doi.org/10.1021/acssuschemeng.5b00926

[136] Jiang, J.; Zhang, L.; Wang, X.; Holm, N.; Rajagopalan, K.; Chen, F.; Ma, S. Highly Ordered Macroporous Woody Biochar with Ultra-High Carbon Content as Supercapacitor Electrodes. *Electrochim. Acta* 2013, 113, 481-489. https://doi.org/10.1016/j.electacta.2013.09.121

[137] Jin, H.; Wang, X.; Gu, Z.; Polin, J. Carbon Materials from High Ash Biochar for Supercapacitor and Improvement of Capacitance with Hno3 Surface Oxidation. *J. Power Sources* 2013, 236, 285-292. https://doi.org/10.1016/j.jpowsour.2013.02.088

[138] Cai, J.; Niu, H.; Li, Z.; Du, Y.; Cizek, P.; Xie, Z.; Xiong, H.; Lin, T. High-Performance Supercapacitor Electrode Materials from Cellulose-Derived Carbon Nanofibers. *ACS Appl. Mater. Inter.* 2015, 7, 14946-14953. https://doi.org/10.1021/acsami.5b03757

[139] Zhu, H.; Luo, W.; Ciesielski, P. N.; Fang, Z.; Zhu, J. Y.; Henriksson, G.; Himmel, M. E.; Hu, L. Wood-Derived Materials for Green Electronics, Biological Devices, and Energy Applications. *Chem. Rev.* 2016, 116, 9305-9374. https://doi.org/10.1021/acs.chemrev.6b00225

[140] Sevilla, M.; Gu, W.; Falco, C.; Titirici, M. M.; Fuertes, A. B.; Yushin, G. Hydrothermal Synthesis of Microalgae-Derived Microporous Carbons for

Electrochemical Capacitors. *Journal of Power Sources* 2014, 267, 26-32. https://doi.org/10.1016/j.jpowsour.2014.05.046

[141] Wang, H.; Xu, Z.; Kohandehghan, A.; Li, Z.; Cui, K.; Tan, X.; Stephenson, T. J.; King'ondu, C. K.; Holt, C. M. B.; Olsen, B. C., *et al.* Interconnected Carbon Nanosheets Derived from Hemp for Ultrafast Supercapacitors with High Energy. *ACS Nano* 2013, 7, 5131-5141. https://doi.org/10.1021/nn400731g

[142] Minzae, L.; Gil-Pyo, K.; Hyeon Don, S.; Soomin, P.; Jongheop, Y. Preparation of Energy Storage Material Derived from a Used Cigarette Filter for a Supercapacitor Electrode. *Nanotechnology* 2014, 25, 345601. https://doi.org/10.1088/0957-4484/25/34/345601

[143] Boota, M.; Paranthaman, M. P.; Naskar, A. K.; Li, Y.; Akato, K.; Gogotsi, Y. Waste Tire Derived Carbon–Polymer Composite Paper as Pseudocapacitive Electrode with Long Cycle Life. *ChemSusChem* 2015, 8, 3576-3581. https://doi.org/10.1002/cssc.201500866

[144] Zhi, M.; Yang, F.; Meng, F.; Li, M.; Manivannan, A.; Wu, N. Effects of Pore Structure on Performance of an Activated-Carbon Supercapacitor Electrode Recycled from Scrap Waste Tires. *ACS Sustainable Chemistry & Engineering* 2014, 2, 1592-1598. https://doi.org/10.1021/sc500336h

[145] Jain, A.; Balasubramanian, R.; Srinivasan, M. P. Hydrothermal Conversion of Biomass Waste to Activated Carbon with High Porosity: A Review. *Chem. Eng. J.* 2016, 283, 789-805. https://doi.org/10.1016/j.cej.2015.08.014

[146] Jurewicz, K.; Babeł, K. Efficient Capacitor Materials from Active Carbons Based on Coconut Shell/Melamine Precursors. *Energy & Fuels* 2010, 24, 3429-3435. https://doi.org/10.1021/ef901554j

[147] Wang, J.; Nie, P.; Ding, B.; Dong, S.; Hao, X.; Dou, H.; Zhang, X. Biomass Derived Carbon for Energy Storage Devices. *J. Mater. Chem. A* 2017, 5, 2411-2428. https://doi.org/10.1039/C6TA08742F

[148] Jiang, J. High Temperature Monolithic Biochar Supercapacitor Using Ionic Liquid Electrolyte. *J. Electrochem. Soc.* 2017, 164, H5043-H5048. https://doi.org/10.1149/2.0211708jes

[149] Tian, W.; Gao, Q.; Tan, Y.; Li, Z. Unusual Interconnected Graphitized Carbon Nanosheets as the Electrode of High-Rate Ionic Liquid-Based Supercapacitor. *Carbon* 2017, 119, 287-295. https://doi.org/10.1016/j.carbon.2017.04.050

[150] Sennu, P.; Choi, H.-J.; Baek, S.-G.; Aravindan, V.; Lee, Y.-S. Tube-Like Carbon for Li-Ion Capacitors Derived from the Environmentally Undesirable Plant: Prosopis Juliflora. *Carbon* 2016, 98, 58-66. https://doi.org/10.1016/j.carbon.2015.10.087

[151] Deng, J.; Li, M.; Wang, Y. Biomass-Derived Carbon: Synthesis and Applications in Energy Storage and Conversion. *Green Chem.* 2016, 18, 4824-4854. https://doi.org/10.1039/C6GC01172A

[152] McCloskey, B. D. Expanding the Ragone Plot: Pushing the Limits of Energy Storage. *J. Phys. Chem. Lett.* 2015, 6, 3592-3593. https://doi.org/10.1021/acs.jpclett.5b01813

[153] Christen, T.; Carlen, M. W. Theory of Ragone Plots. *J. Power Sources* 2000, 91, 210-216. https://doi.org/10.1016/S0378-7753(00)00474-2

[154] Zhai, Y.; Dou, Y.; Zhao, D.; Fulvio, P. F.; Mayes, R. T.; Dai, S. Carbon Materials for Chemical Capacitive Energy Storage. *Adv. Mater.* 2011, 23, 4828-4850. https://doi.org/10.1002/adma.201100984

[155] Faraji, S.; Ani, F. N. Microwave-Assisted Synthesis of Metal Oxide/Hydroxide Composite Electrodes for High Power Supercapacitors – a Review. *J. Power Sources* 2014, 263, 338-360. https://doi.org/10.1016/j.jpowsour.2014.03.144

[156] Rui, X.; Tan, H.; Yan, Q. Nanostructured Metal Sulfides for Energy Storage. *Nanoscale* 2014, 6, 9889-9924. https://doi.org/10.1039/C4NR03057E

[157] Pech, D.; Brunet, M.; Durou, H.; Huang, P.; Mochalin, V.; Gogotsi, Y.; Taberna, P.-L.; Simon, P. Ultrahigh-Power Micrometre-Sized Supercapacitors Based on Onion-Like Carbon. *Nat Nano* 2010, 5, 651-654. https://doi.org/10.1038/nnano.2010.162

[158] Vatamanu, J.; Borodin, O.; Bedrov, D.; Smith, G. D. Molecular Dynamics Simulation Study of the Interfacial Structure and Differential Capacitance of Alkylimidazolium Bis(Trifluoromethanesulfonyl)Imide [Cnmim][TFSI] Ionic Liquids at Graphite Electrodes. *J. Phys. Chem. C* 2012, 116, 7940-7951. https://doi.org/10.1021/jp301399b

Chapter 2

Importance of Reduction and Oxidation Stability of High Voltage Electrolytes and Additives

Samuel A. Delp*,[a] Oleg Borodin,[a] Marco Olguin,[a] Claire G. Eisner,[b] Joshua L. Allen,[a] T. Richard Jow[a]

[a]Electrochemistry Branch, US Army Research Laboratory, Adelphi, MD, 20783-1138

[b]Simulation Sciences Branch, US Army Research Laboratory, Aberdeen Proving Ground, MD, 21005-5066

samuel.delp.ctr@mail.mil, oleg.a.borodin.civ@mail.mil

Abstract

The electrolyte is a critical component for rechargeable Li-ion batteries, especially batteries containing high voltage cathodes. A series of electrolyte salts, solvents and additives was investigated via cyclic voltammetry (CV) on glassy carbon (GC) electrodes. Quantum chemistry (QC) calculations were used for prediction of oxidation and reduction stability of electrolyte components such as ethylene carbonate (EC), dimethyl carbonate (DMC), fluoroethylene carbonate (FEC), vinylene carbonate (VC), tris(trimethylsilyl) phosphate (TMSP), 1,3-propane sultone (PS), tris(hexafluoroisopropyl) phosphate (HFiPP) and lithium salts such as $LiPF_6$, $LiBF_4$, lithium difluoro(oxalato)borate (LiDFOB), lithium 4,5-dicyano-2-trifluoromethyl-imidazolide (LiTDI), and lithium bis(trifluoromethanesulfonimide) (LiTFSI). QC calculations predicted that defluorination of $LiPF_6$ and LiTFSI aggregates coupled with electron transfer significantly increased their reduction potential, while H-transfer upon oxidation lowered oxidation potential for many solvents. The composition of the Li^+ cation solvation shell was estimated from the binding energies for the Li^+–solvent complexes using cluster-continuum calculations and was used to provide insight into the experimental data on electrolyte reduction. Full coin cell data was acquired using $LiNi_{0.5}Mn_{1.5}O_4$ (LNMO) cathodes with graphite anodes at 25°C and 55°C. Differential capacity plots (dQ/dE vs. E) and electrochemical impedance spectroscopy (EIS) results. The electrolyte with the most desirable performance was 1M $LiPF_6$ in 3:7 EC:EMC (wt%) with 1wt% TMSP. Oxidative CV experiments show that the TMSP containing

The work has been previously published by Elsevier: Electrochimica Acta Volume 209, 10 August 2016, Pages 498-510 Importance of Reduction and Oxidation Stability of High Voltage Electrolytes and Additives Samuel A. Delp, Oleg Borodin, Marco Olguin, Claire G. Eisner, Joshua L. Allen, T. Richard Jow
https://doi.org/10.1016/j.electacta.2016.05.100

electrolyte has a slightly lower oxidation stability compared to the baseline, which is consistent with the order of oxidation stability of electrolyte components from QC calculations. The EIS measurements showed that the TMSP containing electrolyte had the lowest impedance after cycling and the dQ/dE plots show that the redox reactions retained their peak shape and area more so than an electrolyte without TMSP, indicating a greater capacity retention. Density functional theory calculations of TMSP oxidation on the LNMO surface were performed in order to provide insight into the additive role in improving cell performance.

Keywords

Lithium Battery, Electrolyte, Quantum Chemistry, Density Functional Theory, Reduction, Oxidation

Contents

1. Introduction

One approach for increasing the energy density over that of the state-of-the-art Li-ion batteries is to employ cathodes that operate at higher voltages, such as the spinel structure lithium nickel manganese oxide, {$LiNi_{0.5}Mn_{1.5}O_4$ (LNMO) [1], 4.7V vs. Li/Li^+}, and the olivine structure lithium cobalt phosphate, {$LiCoPO_4$ (LCP) [2], 4.8V vs. Li/Li^+}. With capacities similar to their lower voltage counterparts, $LiMn_2O_4$ (LMO) at 4.0V and $LiFePO_4$ (LFP) at 3.4V, the higher discharge voltages increase the energy density of the cell when used with the same anode. A common question often raised is whether electrolytes exist that would enable the operation of the high voltage cathodes. This uncertainty likely results from the generally accepted belief that the state-of-the-art Li-ion electrolytes can only be operated at voltages less than ~4.5V with respect to Li/Li^+ [3]. Therefore, the reported high voltage electrolyte developments have taken two main approaches. The first approach is the use of fluorinated versions of the carbonate solvents, taking advantage of their higher oxidation stability compared to the non-fluorinated counterparts [4–8]. Raising the oxidation potential with fluorine addition usually also raises the reduction potential of these solvents. When using a graphite anode, forming and maintaining the solid electrolyte interface (SEI) is critical to battery performance. For this reason, changing the bulk properties of the electrolyte may not lead to a suitable solution. The other approach is the use of additives that oxidize at potentials lower than the oxidation potential of the active cathode material (and lower than that of the bulk electrolyte) and thus form a protective layer on the cathode [9,10]. Lucht et al. have shown that the use of additives improve the electrolyte performance in LNMO/Graphite full cells, especially at high temperature [11–13].

Previous molecular modeling studies indicated that oxidation of electrolyte components is often complicated by decomposition or deprotonation reactions [14,15]. For example, quantum chemistry (QC) calculations predicted that the oxidation potential of the commonly used solvent ethylene carbonate (EC) in an isolated state but surrounded by implicit solvent with a dielectric constant ε=20, which is similar to the dielectric constant of an EC:EMC mixed electrolyte, is ~7V vs. Li/Li^+. When similar calculations are performed with EC in the presence of an anion such as BF_4^- the oxidation stability lowers to 6.2-6.3V when the same M05-2X density functional theory (DFT) was used. Even lower oxidation potentials of 6.1V and 5.2V were predicted for EC–BF_4 and EC_2, respectively, when the more accurate G4MP2 theory was used [16]. The oxidation potential of ethyl methyl carbonate (EMC) in the presence of PF_6^-, (EMC/PF_6^-), is predicted to be 5.9V, which is also more than one volt lower than the oxidation potential of an isolated EMC molecule [14]. The oxidation stability of the PF_6^- anion was predicted as 8.6V in an implicit solvent with a dielectric constant ε=20 [14,17], which is significantly higher than oxidation stability of EC_2, EC–PF_6 indicating that the PF_6^- anion is the most oxidatively stable component of EC:DMC/$LiPF_6$ electrolytes. EC and DMC were also found in QC calculations to undergo deprotonation on the surface of charged LNMO spinel [16]. While calculations of oxidation reactions of single molecules do not represent real world situations in which the molecules are in a complex mixture of salts and solvents polarized at the cathode interface, examination of the representative electrolyte clusters provided significant insight into oxidation stability and initial decomposition reactions. In short, these previous computational results suggest that the electrolytes made of 1M $LiPF_6$ salt in EC:EMC solvent mixtures would undergo oxidation coupled with H-transfer around 5V vs. Li/Li^+, which is close to the operational voltage of high voltage electrolytes.

The computational results could be validated by experimental results using cyclic voltammetry (CV) on an inert electrode such as glassy carbon (GC) [18], in which a voltage sweep of the GC electrode in either the oxidative or reductive direction is applied, and the current response is observed until a large current increase occurs, indicating the decomposition voltage of the electrolyte. The CV results suggest the oxidative stability of electrolytes composed of $LiPF_6$ in EC:EMC mixtures are >5V without considering the reactivity with the cathode, with the results being dependent on the scan rate.

When the working electrode is made of active material, the current observed in CV measurements is often complicated by redox or irreversible reaction currents coming from Faradaic reactions by the active material in addition to the oxidative or reductive decomposition current of the electrolyte itself. Therefore, it would be useful to examine the oxidative and reductive responses of the electrolyte on an inert electrode before examination on an active electrode.

One important point to consider for the development of electrolytes for high voltage Li-ion cells is the fact that the full cell is first formed by charging the cell to an upper voltage limit during which time the anode quickly drops in voltage to the lithium intercalation potential of graphite while the cathode remains at relatively low voltage (Figure 1). The electrolyte undergoes a reduction reaction at the graphite anode first before it is oxidized at the cathode while the full-cell voltage is increased gradually until it reaches the high voltage cutoff. The reduction reaction of the electrolyte decides the ensuing reactions of the cell. While it is important to look for a more oxidatively stable electrolyte, we believe that it is also critical to understand the reduction reactions and how they affect the subsequent oxidation reactions of the electrolyte components both experimentally and in QC calculations.

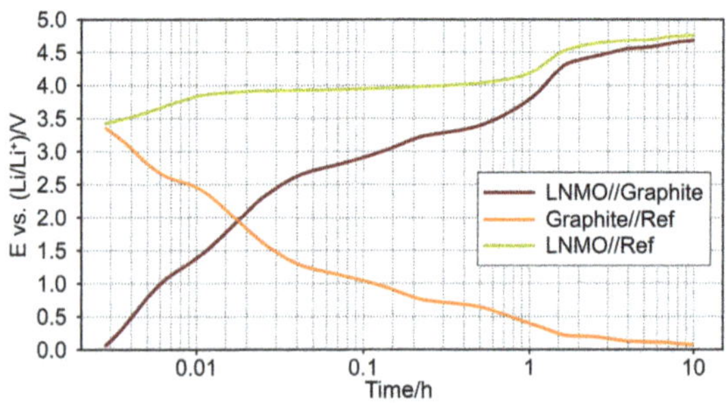

Fig. 1 Voltage vs Time plot for 3-electrode LNMO/Li/Gr (cathode/reference/anode) pouch cell showing observed voltages at cathode and anode relative to full cell voltage (x-axis on log scale for clarity).

This paper will examine the electrochemical redox reactions of the electrolyte components including salts, solvents, and additives (Figure 2) on an inert electrode (glassy carbon) and how they compare with the computational predictions. We will then explain how additives affect reductive reactions or the oxidative reactions of electrolytes and ultimately the cycling behavior of high voltage Li-ion cells with high voltage cathodes.

2. Experimental materials

2.1 Materials

Electrolytes were prepared and stored in an argon filled dry box. $LiPF_6$, EMC, EC, FEC, and DMC were purchased from BASF and used as received (H_2O < 20ppm). LiTFSI was purchased from 3M and dried under vacuum at 110°C overnight. Lithium 4,5-dicyano-2-trifluoromethyl-imidazolide (LiTDI) was synthesized according to literature procedures [19,20]. LiDFOB was synthesized according to the literature [21,22]. Tris(hexafluoroisopropyl) phosphate (HFiPP) was obtained from Argonne National Lab (MERF). Tris(trimethylsilyl) phosphate (TMSP) and 1,3-propane sultone (PS) were purchased from TCI and used as received. Vinylene carbonate (VC) was purchased from TCI and distilled under partial vacuum and inert atmosphere (N_2). $LiBF_4$ (99.99%) was purchased from Sigma-Aldrich and used as received.

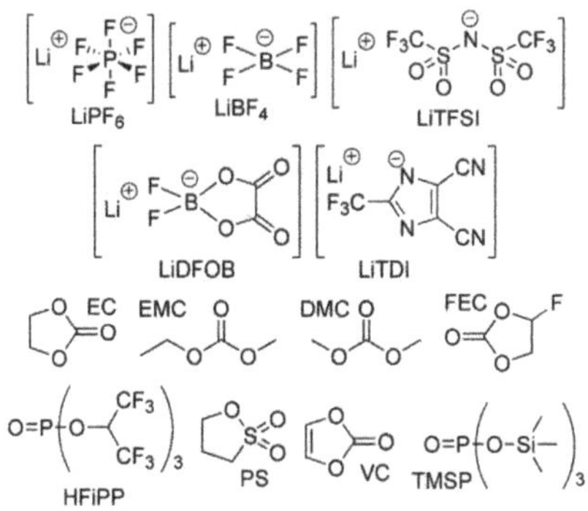

Fig. 2 Electrolyte components used in this study.

2.2 Cyclic voltammetry

Cyclic voltammetry measurements were performed using a ParStat 4000 with a three electrode setup in an argon filled dry box. Glassy carbon electrodes (BASi, 3mm diameter) were used as working electrodes. Lithium metal foil was used as the counter and reference electrodes. The electrodes were allowed to stabilize in the electrolyte solutions before the scan was started. For reduction experiments, the potential was swept from open circuit voltage (OCV) to 0V, then between 0V and 3.5V for three cycles. Oxidation experiments were cycled from OCV to 7V, then back to 3V. Experiments were performed at a scan rate of 10mV s^{-1} unless otherwise noted. Electrolytes were generally prepared using the baseline (1.0M $LiPF_6$ in EC:EMC, 3:7 wt%) with the addition of 5 wt% of the additive being investigated (this concentration was chosen to increase the current response to determine redox potentials of the additive). The reduction and oxidation stability of the electrolytes was determined by the method used by Egashira et al. where the current peak was extrapolated to the baseline and that intersection was the value [18].

2.3 Coin cell cycling

The LNMO cathodes contained 84 wt% $LiNi_{0.5}Mn_{1.5}O_4$ powder (NEI SP-10), 8 wt% Super P-Li (Timcal) and 8 wt% PVDF (Solvey 5130) on Al foil (20 micron foil with 72 micron coating) at 14 mg cm^{-2} (coated by ANL). The graphite (Gr) anodes consisted of 89.8 wt% graphite (Conoco Phillips CGP-A12), 4 wt% Super P-Li (Timcal) 6 wt% PVDF (Kureha 9300) and 0.2 wt% oxalic acid on Cu foil (10 micron foil with 40 micron coating) at 5 mg cm^{-2} (coated by Saft). The electrodes were provided by the CAMP Facility at Argonne National Laboratory. Coin cells were constructed using Hohsen 2032-size components with aluminum cladding on nickel-plating on the cathode side. Electrodes were cut into 9/16" diameter and used with 5/8" diameter microporous polyethylene separators (Celgard 2400). Two formation cycles were performed at C/10 between 3.5 and 4.9V. Charge-discharge cycling and dQ/dE data were collected using a Maccor Series 4000 battery cycler. Electrochemical impedance spectroscopy (EIS) data was acquired using a Solartron Analytical Modulab potentiostat equipped with a frequency response analyzer (FRA) at 4.6V after equilibrating the cell (at 25°C) until the internal current dropped below C/100.

2.4 Quantum chemistry studies of lithium solvation, electrolyte reduction and oxidation

The Li^+–solvent binding energy was calculated relative to the energy of isolated components, where each species was solvated by the SMD solvation model developed by the Truhlar group with acetone (ε=20.4) parameters representing a typical mixed solvent electrolyte [23]. The Li^+–EC binding energy in implicit solvent was subtracted from all investigated complexes yielding Li^+–solvent binding energies relative to the Li^+–EC as shown in Table 1. The cluster – continuum calculations were also performed for the four solvent coordinated lithium solvates $(EC)_n(FEC)_m(DMC)_kLi^+$ with n+m+k=4, in which the first Li^+ solvation shell was represented explicitly while the second solvation shell and beyond was represented implicitly using SMD(ε=20) model. The implicit model was shown to adequately reproduce solvation by highly dipolar solvents such as EC but had higher errors representing solvents with a small dipole such as DMC [24]. Gaussian g09 rev. b and rev. c software was used in all calculations. The computational framework allowing connection with databases, automatic job restart and error correction was used [25]. DFT calculations were performed using PBE functional with and without Grimmer dispersion correction with the 6-31+G(d,p) basis set as it was shown to be a good compromise of computational efficiency and accuracy for predicting Li^+–solvent binding energies [26]. M05-2X function was used in some studies as it was shown to predict the hole delocalization in oxidized EC better than B3LYP [24]. G4MP2 calculations were performed on smaller representative systems such as Li^+–solvent, solvent dimer and salt clusters to examine binding energies, reduction and oxidation of small complexes.

Table 1 Binding energies (ΔE in kJ mol^{-1}) and free energies at 298.15 K (ΔG in kJ mol^{-1}) relative to Li^+(EC) from QC calculations with all complexes immersed in implicit solvent modelled using SDM(ε=20) model.

	M05-2X/ 6-31+G(d,p)		PBE/ 6-31+G(d,p)		PBE+D3BJ/ 6-31+G(d,p)		G4MP2	
solvent/additive	ΔE	ΔG	ΔE	ΔG	ΔE	ΔG	ΔE	ΔG
EC	0.0	0.0	0.0	0.0	0.0	0.0	0.0	0.0
DMC	2.7	3.6	1.6	3.1	-0.2	2.7	2.0	4.1
VC	8.9	5.7	9.1	4.0	9.5	3.7	6.9	7.2
FEC (Li...O=)	10.2	9.8	9.4	8.0	9.5	7.2	9.7	11.3
FEC (Li...F)	41.8	39.7	37.2	34.5	36.6	33.1	37.7	41.1
PS	7.5	7.1	11.4	10.9	10.2	8.5	10.3	15.7
TMSP	-27.8	-16.0	-29.4	-22.9			-27.8	-16.0
HFiPP	17.6	20.7	23.5					

The oxidation (eq. 1) and reduction (eq. 2) potentials for the representative model systems were estimated from the absolute oxidation and reduction potentials of a complex M relative to an electron at rest in vacuum with a subsequent subtraction of 1.4V in order to convert them to the commonly used Li/Li^+ potential scale and compared with experimental data [14],

$$E^0_{oxidation}(M) = [\Delta G_e + \Delta G^0_S(M^+) - \Delta G^0_S(M)]/F - 1.4 \quad (1)$$

$$E^0_{reduction}(M) = -[\Delta G_e + \Delta G^0_S(M^-) - \Delta G^0_S(M)]/F - 1.4 \quad (2)$$

where ΔG_e is the ionization free energy or electron affinity in gas-phase at 298.15 K; $\Delta G_S(M^+)$, $\Delta G_S(M^-)$ and $\Delta G_S(M)$ are the free energies of solvation of the oxidized, reduced and initial complexes, respectively, and F is the Faraday constant. If the cluster M geometry is not allowed to relax after electron transfer, eq. 1 – eq. 2 yields the vertical reduction and oxidation potentials, while potentials that account for the geometry change as a result of electron transfer will be called adiabatic potentials or simply oxidation and reduction potentials.

Subtraction of 1.4V accounts for the conversion from the absolute electrochemical scale to the commonly used Li/Li^+ potential scale in order to compare predicted values with experimental data using the same reference electrode [15]. Using the Li/Li^+ standard electrode potential of -3.05V versus SHE for aqueous solutions, the conversion factor is established by Eq. 3.

$$E° (vs.\ Li/Li^+) = E°_{abs} - 4.42 + 3.05V = E^0_{abs} - 1.37V \approx E^0_{abs} - 1.4V \quad (3)$$

where $E°(vs.\ Li/Li^+)$ is the potential vs. Li/Li^+. Solvent variation is expected to change this factor of 0.1-0.3V as previously discussed elsewhere [15].

2.5 Quantum chemistry studies of the solvent – cathode surface interactions

Periodic planewave DFT+U calculations for the TMSP/$Ni_{0.5}Mn_{1.5}O_4$ additive/cathode interface system were performed using the Vienna ab initio Simulation Package (VASP) with the spin-polarized Perdew-Burke-Ernzerhof exchange-correlation functional and the projector-augmented wave (PAW) scheme to treat core electrons. The set of VASP supplied standard pseudopotentials (PP) was used for carbon, oxygen, silicon, phosphorous, and hydrogen atoms. The PAW PPs employed for the transition metal atoms Ni and Mn (denoted as Ni_pv and Mn_pv, respectively, in the VASP PP repository) treat p semi-core states as valence states. A large energy cutoff of 520 eV together with a 4×4×4 Monkhorst-Pack k-point mesh for bulk structures and a 2×2×1 k-point sampling for the electrolyte/cathode supercell models were used. Dipole and quadrupole corrections to the energy were included using a generalized Makov–Payne method as implemented in VASP. For the +U augmented treatment of Ni and Mn 3d orbitals, we chose a U_{eff} value of 5.96 eV for Ni and 4.5 eV for Mn.

In DFT calculations we focused on the surface reactivity of the fully delithiated $Ni_{0.5}Mn_{1.5}O_4$ [111] Fd3m spinel structured cathode interacting with TMSP molecules. For the Fd3m phase, we take the uniformly disordered transition metal cation arrangement as a representative Ni/Mn configuration of a random distribution in the disordered Fd3m

LNMO system. Among the many possible surface terminations, the Li/Ni/O termination of the LNMO [111] surface facet as used, which in previous work was shown to exhibit the lowest surface energy. The [111] surface for the fully delithiated ($Ni_{0.5}Mn_{1.5}O_4$) structure was generated by removing the Li atoms from the corresponding fully lithiated ($LiNi_{0.5}Mn_{1.5}O_4$) slab structure followed by geometry optimization. Details of the model system setup, including magnetic ordering considerations, are given elsewhere[16]. For the TMSP/NMO interface, the initial starting configuration consists of the electrolyte TMSP molecule adsorbed on the [111] surface facet of the fully delithiated NMO Fd3m-phase spinel. The barrier for the H-transfer reaction from TMSP to NMO was estimated by scanning the H(TMSP)–O(NMO) distance.

3. Results and discussion

3.1 The lithium solvation shell composition and electrolyte electrochemical stability from quantum chemistry calculations

3.1.1 Li^+ solvation shell structure

The reduction chemistry is intimately linked with the Li^+ solvation shell composition [27–29], therefore the solvent (or additive) binding energy to Li^+ was examined from QC calculations in order to estimate the probability of solvent molecules contributing to the Li^+ solvation shell in the multicomponent electrolyte mixtures. Due to the large discrepancy between the Li^+ solvation shell composition in the gas-phase and liquid for multisolvent mixtures such as EC:DMC [26], the Li^+–solvent binding energies in the gas-phase are inadequate for prediction of the composition of the Li^+ solvation shell in liquid electrolytes. For example, in gas-phase the binding energy was higher for Li^+–EC than for Li^+– DMC(cis-cis) by 26 kJ mol^{-1} suggesting that Li^+ is overwhelmingly solvated by EC and DMC contributes less than a fraction of a per cent to the solvation shell. The populations of solvents in the Li^+ first solvation shell relative to EC were estimated at room temperature using Boltzmann factors shown in eq. 4

$$p_j = p_{EC} \exp(-\Delta E / k_B T) \tag{4}$$

where p_j and p_{EC} are populations of species j and EC, ΔE is the relative energy of Li^+–solvent vs. Li–EC, k_B is Boltzmann constant and T is temperature. The hybrid cluster-continuum calculations, where the Li^+ first solvation shell is treated explicitly while the second solvation shell and beyond are treated implicitly, however, predicted the Li^+ solvation shell in a reasonable agreement with predictions from Born–Oppenheimer molecular dynamics (BOMD) simulations and experiments for EC:DMC–$LiPF_6$

electrolyte [26]. In this work, we follow this hybrid scheme and estimate the relative contributions of solvents to the Li^+ solvation shell from accurate, but computationally expensive for large systems, G4MP2 calculations where possible as shown in Table 1. For large molecules such as HFIPP, only more computationally trackable DFT calculations have been performed.

Examination of the relative free energies indicates that G4MP2 and DFT calculations using PBE, PBE+D3BJ (dispersion corrected) and M05-2X functionals predicted a following order of the solvent binding strength to Li^+, namely: TMSP > EC > DMC > VC > FEC (with Li bound to O=C group) > PS > HFiPP > FEC (with Li^+ bound to F of FEC) with an exception that for M05-2X calculations the order of PS and FEC (with Li bound to O=C group) was reversed. PBE functional predicted the Li^+–solvent relative binding energies in the best agreement with G4MP2 calculations with average errors of 0.3 kJ mol^{-1}. M05-2X yielded a higher average error of 0.9 kJ mol^{-1}, thus while the M05-2X functional systematically overestimates the absolute Li^+–solvent binding energies as was shown elsewhere [26,30], it reasonably predicts the relative binding energies as shown in Table l, albeit not as good as PBE functional. Table 1 also indicates that vibrational-rotational contribution to free energy leads to a few kJ mol^{-1} differences between the energy and free energy with the largest difference > 10kJ mol^{-1} observed for Li^+–TMSP complex. Further examination of Table 1 indicates that TMSP will dominate the solvation shell and exclude all other solvent molecules, while HFiPP contribution to be negligibly small. The contributions of PS, VC and FEC are not negligible but albeit much smaller than EC and DMC. The primary binding of Li^+ is to the carbonate oxygen of FEC and not the fluorine of FEC indicating that at low salt concentration the defluorination reaction is expected to be slow [31]. Malliakas et al. also reported from ab initio molecular dynamics (AIMD) that a $Li^+(FEC)_6$ cluster in the gas phase, with Li^+ ion coordinated to F atoms, is metastable; only Li–OCO bonding to four FEC molecules was observed after equilibration [32]. Condensed phase simulations reported by Malliakas et al. indicated the fluorine atoms in FEC molecules are preferentially located 6 Å away from Li^+ ion, indicating that FEC molecules form a star-like tetrahedral cluster with the carbonyl oxygens pointing towards the center.

We examined a relation between the Li^+–solvent binding energy and solvent dipole moments and found no strong correlation as shown in Figure S1. This indicates that, unfortunately, solvent dipole moment is not a reliable predictor of the composition of the Li^+ first solvation shell and the strength of the Li^+–solvent interactions.

Application of Boltzmann factor (eq. 4) to the relative binding energies and free energies for Li^+(EC) and Li^+(DMCcc) at the G4MP2 level from Table 1 yields the DMCcc population of 0.2 and 0.45 of the EC populations at room temperature, respectively as shown in Table S2 in SI. This estimate is in a reasonable agreement with the composition of the Li^+ first solvation shell predicted in all atom BOMD simulations and re-analyzed Raman experiments for EC:DMC(1:1)/$LiPF_6$ electrolyte [26]. We stress that despite such fortuitously accurate predictions of the EC:DMC(cc) composition of the Li^+ first solvation shell from very simplistic Li^+(solvent) calculations, solvent electrostatic and steric interactions in the first solvation shell are neglected in these calculations while they are expected to play an important role in determining relative populations of solvates. For example, the binding energy of Li^+(DMCcis-trans) conformer relative to the isolated Li^+ and DMC(cc) is 1.1 kJ mol^{-1}. It is similar to the binding energy of 1.0 kJ mol^{-1} for the Li^+(DMCcc) complex (see Table 2). When free energy is considered the Li^+(DMCct) is even slightly more stable than Li^+(DMCcc). In contrast to these simple Li^+(solvent) calculations, the $Li^+(solvent)_4$ calculations indicated that the $Li(DMCcc)(EC)_3$ and $Li(DMCcc)_2(EC)_2$ are more stable than $Li(DMCct)(EC)_3$ and $Li(DMCcc)(DMCct)(EC)_2$, respectively [26], indicating a significant preference of Li^+ for DMCcc vs. DMCct in condensed phase or when a full Li^+ solvation shell is considered. Interestingly, the first solvation shell of the contact ion pair $LiPF_6$ had slightly more DMC vs. EC compared to the solvent separated Li^+ cation [26].

In order to investigate the influence of electrostatic and steric interactions between EC, FEC and DMC(cc) in the Li^+ first solvation we compared the relative binding energies of the Li^+ cation with one solvent={EC,DMC,FEC} with the relative binding energies for the $(EC)_n(FEC)_m(DMC)_kLi^+$ solvates with n+m+k=4 as shown in Table 2. A number of possible orientations of FEC in these complexes was considered as shown in SI. Examination of Table 2 indicates that replacing EC in $Li^+(EC)_4$ with FEC costs at least 3.3 kJ mol^{-1} in energy and 8.2 kJ mol^{-1} in free energy, which is less than 9.5 and 11.4 kJ mol^{-1} the difference between Li^+(EC) and Li^+(FEC). Similarly, the energy difference of 8.2 kJ mol^{-1} between $Li^+(FEC)_2(EC)$ and $Li^+(EC)_4$ is actually smaller than the difference between Li^+(EC) and Li^+(FEC) due to solvent interactions in the Li^+ first solvation shell. Further examination of the solvates that contain EC, DMC and FEC indicates that the relative energy of $Li^+(EC)(DMCcc)_2(FEC)$ solvates is actually lower than that for $Li^+(EC)_4$ and the $Li^+(EC)(DMCcc)_2(FEC)$ solvates are the most energetically stable among solvates considered in Table 2. Free energy of $Li^+(EC)(DMCcc)_2(FEC)$ is higher than free energy of $Li^+(EC)_3(FEC)$ and $Li^+(EC)_2(DMCcc)(FEC)$ solvates, which is of the order to 7 – 9.6 kJ mol^{-1}. These solvates are expected to have populations of 6 and 2 % at room temperature as estimated from Boltzmann factors, which is only a slightly lower than the relative population of FEC vs. EC estimated from free energies of the simplified systems: Li^+(EC) and Li^+(FEC), respectively. These calculations indicate the Li^+ solvation preference EC > DMC > FEC.

Table 2 Binding energies (ΔE in kJ mol⁻¹) and free energies at 298.15 K (ΔG in kJ mol⁻¹) relative to $Li^+(EC)$ and $Li^+(EC)_4$ from PBE-GD3/6-31+G(d,p) with SDM(ε=20) model.

Complex	ΔE/kJ mol⁻¹	ΔG/kJ mol⁻¹
$Li^+(EC)$	0.0	0.0
$Li^+(DMCcc)$	1.0	3.2
$Li^+(DMCct)$	1.1	2.1
$Li^+(FEC)$	9.5	11.4
$Li^+(EC)_4$	0.0	0.0
$Li^+(EC)_3(FEC)$ (a)	6.1	7.0
$Li^+(EC)_3(FEC)$ (b)	3.3	8.2
$Li^+(EC)_2(FEC)_2$ (a)	10.1	13.7
$Li^+(EC)_2(FEC)_2$ (b)	8.2	19.0
$Li^+(EC)(FEC)_3$ (a)	15.0	22.1
$Li^+(EC)(FEC)_3$ (b)	16.3	23.5
$Li^+(EC)_2(DMCcc)(FEC)$ (a)	5.8	9.6
$Li^+(EC)_2(DMCcc)(FEC)$ (b)	6.2	13.5
$Li^+(EC)_2(DMCcc)(FEC)$ (c)	5.6	8.7
$Li^+(EC)(DMCcc)_2(FEC)$ (a)	-1.8	14.6
$Li^+(EC)(DMCcc)_2(FEC)$ (b)	-0.1	10.7
$Li^+(EC)(DMCcc)_2(FEC)$ (c)	-1.5	11.5
$Li^+(EC)(DMCcc)_2(FEC)$ (d)	-1.1	18.9

3.1.2 Reduction stability of electrolyte

As the Li^+ cation polarizes solvent and anions and increases their reduction potential, we have chosen the Li^+–solvent and Li^+–anion complexes instead of the isolated solvent and anion as model systems in order to estimate their reduction potential. In some cases at the anode interfaces anions are expected to bound to more than one Li^+ in order to account for the excess of Li^+ observed at the negatively charged electrode in MD simulations of EC:DCM/$LiPF_6$ electrolytes [33]. Predicted reduction potentials are shown in Table 3 along with the previously reported values. For a number of compounds such as EC–Li, DMC–Li, FEC–Li and $(Li^+)_2TFSI^-$ have multiple Li^+ binding locations and, in the case of DMC multiple conformations (cis-cis and cis-trans), yielded different reduction potentials as reflected by the range of values shown in Table 3. EC, VC and FEC reduction potentials are consistent with the previously reported values by other groups from DFT or MP2 calculations [34–36]. Note that increasing the number of EC in the Li^+ solvation shell did not significantly influence EC reduction potential [29]. The reduction potential from Table 3 needs to be combined with composition of the Li^+ solvation shell in order to get an adequate picture of the preferential solvent reduction. For example, higher populations of EC than DMC(cis-cis) in the first Li^+ solvation shell and higher reduction potential of EC–Li^+ compared to DMC(cis-cis)–Li^+ indicate preferential reduction of EC vs. DMC(cis-cis) as previously discussed [16,26,37,38]. FEC is expected to undergo preferential reduction in EC:DMC:FEC mixed electrolyte due to its higher reduction potential and a non-negligible contribution to the solvation shell. In addition to the robust FEC reduction around 0.9V vs. Li/Li^+, FEC was also predicted to undergo reduction at ~2V if LiF is simultaneously formed. This process, however, is expected to be very sluggish due to the improbability of Li binding to the fluorine of FEC, at least in dilute electrolytes and an additional barrier for the defluorination reaction. VC has a slightly higher reduction potential than the EC and DMC main solvents and is expected to contribute less than 10% to the population of the first solvation shell. Thus, its previously discussed ability to react with EC radicals might be its primary effect on SEI structure [39]. Due to the dominance of TMSP in the Li^+ solvation shell and its low reduction potential, the addition of TMSP is expected to reduce reduction potential of the EC:DMC-based electrolytes, modify SEI formation mechanism and SEI composition.

A number of salts such as LiDFOB, LiTDI and LiTFSI were also predicted to have higher reduction potentials than baseline EC:DMC solvents, especially when LiTDI and LiDFOB exist as aggregates. The reduction of $LiBF_4$ salt was investigated using a dimer $(LiBF_4)_2$ and $(BF_4^-)(Li^+)_3(BF_4^-)$ aggregate to reflect that it is aggregating in mixed EC:DMC solvents and an excess of Li^+ is likely to be found on the negative electrode. Reduction reactions with and without LiF formation was examined, as shown in the SI, with reduction potentials given in Table 3. The $(BF_4^-)(Li^+)_3(BF_4^-)$ aggregate had a slightly higher reduction potential than $(LiBF_4)_2$ due to stabilization of the transferred electron by the extra Li^+ cation but both reduction potentials are quite low indicating that the EC and DMC solvents will preferentially reduce before $LiBF_4$ salt. A different picture was observed for LiTFSI and $LiPF_6$ that showed quite high reduction potentials above 1.5V upon anion defluorination and LiF formation. Such reactions, however, are expected to have high barriers and might be challenging to observe electrochemically due to low currents. Particularly, reduction of $(PF_6^-)(Li^+)_3(PF_6^-)$ with LiF formation was predicted from QC to occur above 2V vs. Li/Li^+ supporting experimental finding of LiF formation due to $LiPF_6$ reduction at such high potentials [40]. These high voltage reduction peaks (> 1.5V) in CV were also attributed this to $PF_5 + H_2O$ reactions [41]. However, reduction of salt aggregates at high potentials appears to be a general phenomenon as highly concentrated LiFSI in DME was also shown experimentally and by QC to undergo reduction with LiF formation above 2.3V vs. Li/Li^+ [42]. It was also observed in highly concentrated LiTFSI and $LiCF_3SO_3$ electrolytes [43,44].

Table 3 Calculated reduction potentials vs. Li/Li^+ from QC with SMD(ε=20).

Complex	G4MP2	M05-2X/6-31+G**	Ref.
Li^+(EC)	0.48-0.61	0.45-0.61	[16]
Li^+(DMC)(cis-cis)	0.22-0.4	0.33	[16,31]
Li^+(DMC)(cis-trans)	0.6		[16]
Li^+(FEC)	0.9	0.92	[16]
Li^+(FEC) (Li–F formed)	2.25	1.7	[16,31]
Li^+(VC)	0.80	0.23	
Li^+(PS)	0.46	0.04	
Li^+(TMSP)	0.64	0	
LiDFOB	1.57	1.71	
$(LiDFOB)_2$	2.12	2.18	
LiTDI	0.75	0.72	
Li_2TDI	1.10	1.11	
$TFSI^-$	1.4		[43]
$(Li^+)_2TFSI^-$ (LiF formed)	2.1-2.9		[43]
$(LiPF_6)_2$ (LiF formed)	1.61	1.7	
$(PF_6^-)(Li^+)_3(PF_6^-)$	2.1	2.2	
$(LiBF_4)_2$	0.0	-0.4	
$(BF_4^-)(Li^+)_3(BF_4^-)$	0.2-0.3	-0.1	

Examination of the subsequent steps of the TMSP reduction at the M05-2X/6-31+G(d,p) level revealed that detachment of the –O–Si(CH_3)$_3$ radical from the reduced TMSP is isoenergetic and after its decomposition it is expected to be reduced again at 3.5V to form an anion. The anion is expected to eventually migrate to the cathode surface where it will get oxidized forming a reactive radical that is likely to participate in formation of the passivation layer at the cathode.

3.1.3 Oxidation stability of electrolyte

Prediction of the electrolyte oxidation reactions is complicated due to possible intermolecular reactions that might proceed with a negligible, or in some cases substantial, barrier. Specifically, the intra- and intermolecular H-transfer reactions were found to occur during electrolyte oxidation for a wide variety of compounds ranging from carbonates, to phosphates, to sulfones [14,45–49]. Such H-transfer reactions significantly lowered the oxidation potential of electrolyte components [15,16]. In order to adequately capture oxidation behavior of electrolyte components, we examined not only the oxidation stability of isolated solvents and anions surrounded by dielectric solvent but also studied possible H-transfer reactions in the solvent–anion complexes and solvent dimers shown in Figure 3. The smallest cluster size that would allow intermolecular H-transfer reaction to occur was chosen with an exception of a large TMSP molecule. For TMSP, the intramolecular H-transfer reaction upon oxidation was examined.

H-transfer reactions coupled with solvent oxidation were found to be exergonic for all solvents considered here (EC_2, PS_2, TMSP) with the exception of VC and the majority of solvent–anion complexes such as EC(BF_4^-), EC($TFSI^-$), EC ($DFOB^-$) and, unexpectedly, for EC(TDI^-). In order to compare all studied oxidation reactions, the oxidation potential for each model complex was plotted as a function of the difference between vertical and adiabatic oxidation energies that reflects reorganization energy (λ) as shown in Figure 4. The difference between vertical and adiabatic oxidation energies comes as result of geometry and energy change after the electron removal from the cluster with a subsequent geometry optimization. While it is an approximation, higher reorganization energy is expected to lead to slower reaction kinetics such that Figure 4 could be considered as a first approximation of the log of oxidation reaction rate vs. oxidation potentials in order to compare these predictions with CV measurements under the assumption that there are no transport limitations.

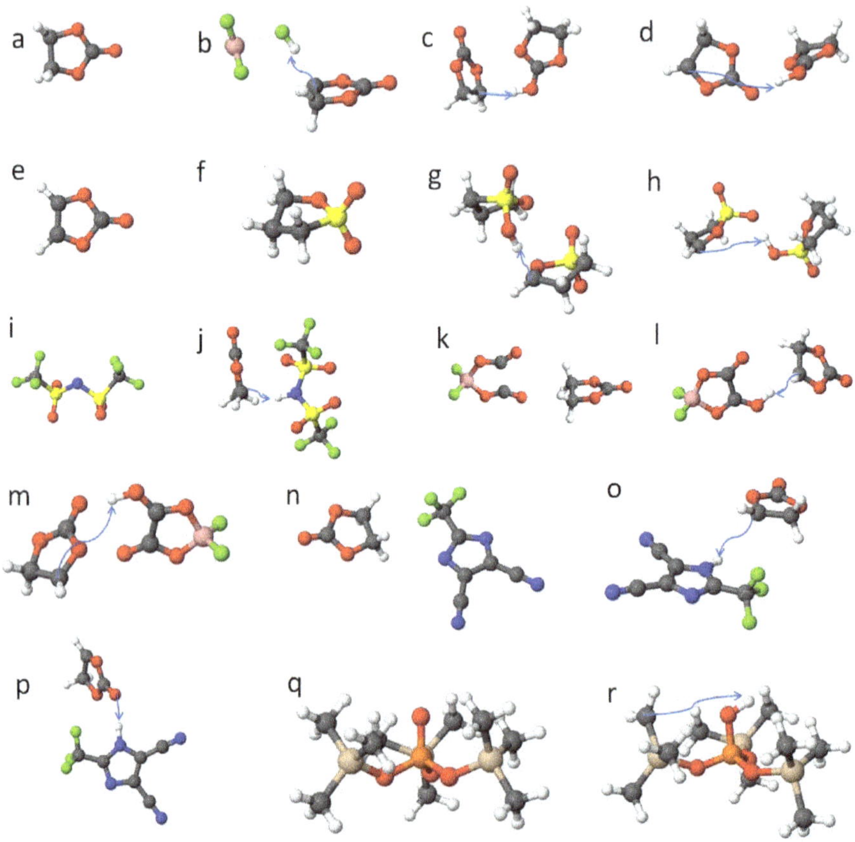

Fig. 3 Optimized geometries for the oxidized clusters from M05-2X/6-31+G(d,p) SMD(ε=20) calculations.

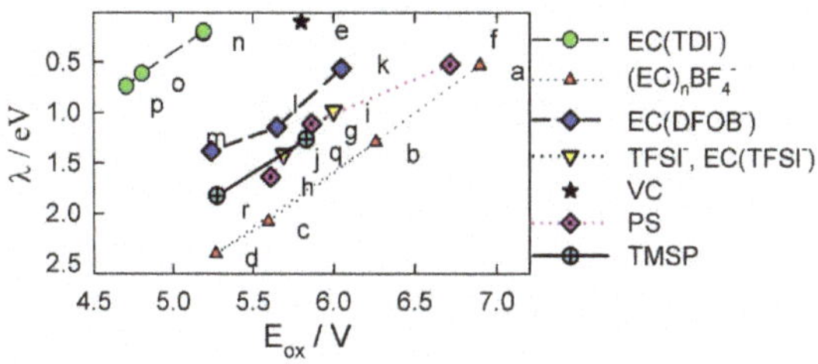

Fig. 4 The difference between vertical and adiabatic oxidation energies (λ) vs. oxidation potential calculated from M05-2X/6-31+G(d,p) SMD(ε=20) calculations for clusters shown in Figure 3. The EC(BF_4^-) and EC($TFSI^-$) data were taken from previous work [14,15].

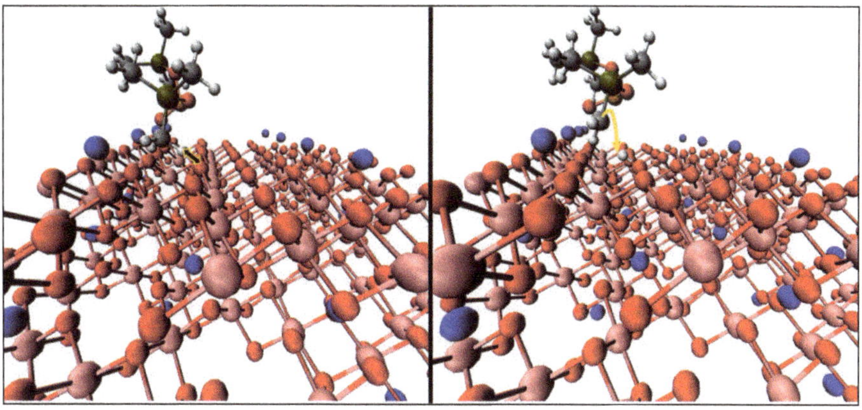

Fig. 5 Interface model for the TMSP/$Ni_{0.5}Mn_{1.5}O_4$ system. Left panel: Initial starting configuration for the potential energy surface scan. The double-ended arrow shows the hydrogen-to-oxygen distance used in the constrained optimizations. Right panel: Optimized structure for the deprotonated system where the yellow arrow displays the proton abstraction reaction.

Examination of Figure 4 indicates that the baseline EC–$LiBF_4$ electrolyte is the most oxidatively stable among the examined model clusters at the corresponding reorganization energies reflecting reaction rates. Stability of the other components is as follows: $EC_n(BF_4^-)$ > PS ≈ EC($TFSI^-$) ≈ TMSP > EC($DFOB^-$) > VC > EC(TDI^-). One should keep in mind that DFT results using M05-2X functional likely overestimate oxidation stability of the TFSI-containing clusters by as much as 0.6V and DFOB-containing clusters by 0.2V as suggested by the comparison with G4MP2 results for isolated anions and EC(TFSI$^-$) complex [14]. Oxidation potential of TMS with the H-transferred intramolecularly (see Figure 3r) also decreased from 5.3V to 4.9V when more accurate G4MP2 calculations were used instead of M05-2X/6-31+G(d,p) calculations indicating that it is expected to oxidize and undergo deprotonation reaction at potentials similar to those of the $Ni_{0.5}Mn_{1.5}O_4$ spinel cathode.

The particular phase, facet, and termination layer of the $Ni_{0.5}Mn_{1.5}O_4$ [111] (NMO) surface, consisting of an oxygen-rich surface layer, exposes a nucleophilic-dominant surface reactivity which is expected to dictate the extent and type of electrolyte/cathode interfacial decomposition reactions. For carbonate based electrolytes interacting with the $LiNi_{0.5}Mn_{1.5}O_4$ [111] (LNMO) cathode, a recent study showed a possible electrolyte-decomposition mediated mechanism for the dissolution of transition metal atoms, which led to the severe capacity fade observed for the full-cell graphite/LNMO system [50]. The addition of TMSP to a carbonate based electrolyte solvent was shown to mitigate the capacity fade exhibited by a similar high voltage material, the $LiNi_{0.5}Co_{0.2}Mn_{0.3}O_2$ cathode, by suppressing transition metal dissolution from the cathode material [49]. In order to investigate the improved electrochemical performance obtained with solvents incorporating TMSP as an additive, we examine the proton abstraction reaction between the TMSP molecule and the NMO [111] surface and compare it to the interfacial reactions between EC/DMC and NMO. Although we focus on only one type of cathode-electrolyte reaction mechanism, the comparison provides insight into the improved capacity performance gain in using TMSP as an additive. To determine the barrier associated with the proton transfer reaction for the TMSP/NMO [111] interface, we performed a potential energy surface scan with one hydrogen atom from TMSP coordinated to the NMO [111] surface at constrained distances of 1.2 Å, 1.3 Å, 1.4 Å, and 1.5 Å (Figure 5). In addition, the structures consisting of a 1.2 Å and a 1.5 Å TMSP-to-surface distance were allowed to relax without the frozen-hydrogen constraint in order to get the product (final proton abstracted structure) and reactant energies.

Unlike the EC/NMO and DMC/NMO interface systems of our previous study, which showed the decomposition state to be lower in energy by 1.36 eV and 1.82 eV, respectively, than the physisorbed state, we find that the deprotonation reaction for TMSP on the NMO [111] surface is not energetically favorable with a large barrier of 1.74 eV [16]. A different adsorption geometry was also investigated as shown in Figure S8, which gave a reaction barrier of 1.05 eV and also yielded an endothermic H-transfer. Based upon previous calculations by Kumar and co-workers for EC adsorption on the low voltage $LiMn_2O_4$ spinel which showed that the reaction barrier for EC decomposition decreases with de-lithiation, we expect the barrier to be larger for the fully lithiated spinel making TMSP deprotonation even less likely [51]. Since the formation of a stable, compact, and thin surface layer is believed to be one of the main reasons for the enhanced cycling performance of the TMSP containing solvent, our calculations suggest that the formation of a stable passivation layer on the LNMO cathode surface by TMSP-containing solvents upon cycling is, in part, a result of the significantly lower TMSP/NMO surface reactivity, where reducing the extent of electrolyte decomposition on the cathode surface at high voltage would suppress transition metal dissolution from the cathode material. Considering that TMSP oxidizes before the decomposition of carbonates and does not react with the NMO surface, it is likely that intra-molecular and inter-molecular oxidation induced decomposition reactions between TMSP molecules will generate radical species as decomposition products which may readily undergo polymerization reactions with nearby TMSP radical species to form a stable passivation layer on the cathode surface.

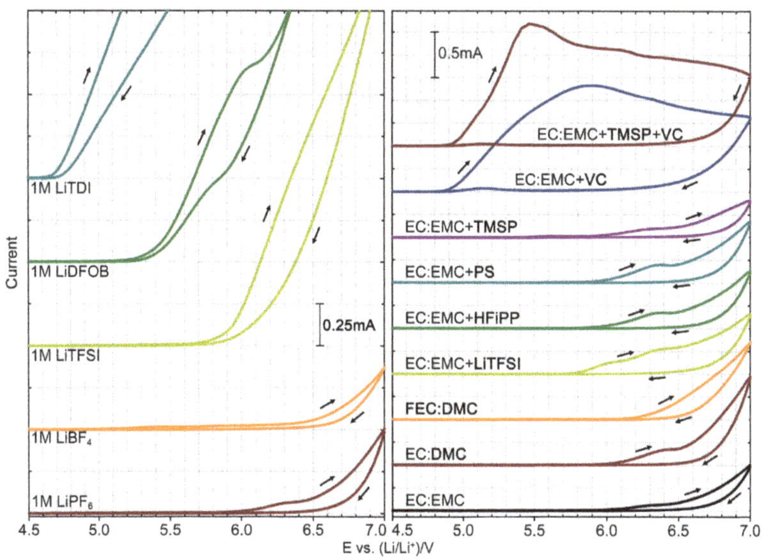

Fig. 6 Oxidation CV curves of various 1M salt in EC:EMC (3:7 wt%) solvent mixtures, where the salts are $LiPF_6$, $LiBF_4$, LiTFSI, LiDFOB and LiTDI, respectively (Left). Oxidation CV curves of electrolytes of 1M $LiPF_6$ in EC:EMC (3:7 wt%), EC:DMC (3:7 wt%), FEC:DMC (1:4 wt%) and the electrolytes of 1M $LiPF_6$ in EC:EMC (3:7 wt%) with additives LiTFSI, HFiPP, PS, TMSP, VC, and TMSP+VC (Right).

3.2 Oxidative stability of electrolytes with different salts on glassy carbon electrodes

The CV curves of electrolytes with several different lithium salts in a solvent mixture of EC:EMC (3:7 wt%) on a GC electrode are shown in Figure 6. The decomposition potentials observed in the CV curves are listed in Table 4 [14]. The experimental decomposition potentials of LiTFSI, LiDFOB and LiTDI in EC:EMC (3:7 wt%) are 5.8, 5.3 and 4.7V, respectively. These values are similar and consistent with the computational results of the anions. The experimental decomposition potential of 1M $LiPF_6$ and 1M $LiBF_4$ electrolytes, both in EC:EMC (3:7 wt%), are 6.0V and 5.5V, respectively, (experimental decomposition potential determined by onset of oxidation peak). We can reasonably assign the value of 6.0V as the oxidation potential of the EC:EMC (3:7 wt%) solvent mixture in the presence of the PF_6^- anion (Figure 6). The calculated oxidative stability of isolated PF_6^- and BF_4^- are 8.7 and 8.4V [14], respectively, which are much higher than the experimental decomposition potentials that are observed in the CV. It has been previously demonstrated, however, that the oxidation potential of EC is highly dependent on the presence of PF_6^-. The oxidation of the 1M $LiPF_6$ EC:EMC electrolyte at 6.0V suggests that the onset current in CV for the $LiPF_6$ and $LiBF_4$ electrolytes is due to the decomposition of the solvent rather than that of the anions.

3.3 Oxidative stability of electrolytes with different solvents and additives on GC

With the oxidative decomposition potential of 6.0V assigned to the electrolyte mixture of EC:EMC (in the presence of PF_6^-), we can investigate the impact of different solvents and additives on the baseline electrolytes. In order to differentiate the impact of additives compared to the baseline electrolyte, 5wt% of additives were used. The CV curves of these electrolytes are shown in Figure 6 and the values of their oxidation stability are listed in Table 4. The electrolyte with DMC, instead of EMC, has insignificant impact on the oxidation stability relative to the baseline electrolyte (DMC and EMC have similar redox properties). When EC is replaced with FEC, however, the oxidation potential of the electrolyte is increased to 6.2V, which is consistent with computational results that show that FEC has a higher oxidation potential than EC [14]. Incorporation of LiTFSI, PS, HFiPP or TMSP lowers the oxidation stability of the electrolyte slightly while VC containing electrolytes have an oxidation potential of 4.9V. There is in general a good agreement between the order of predicted oxidative stability of electrolytes from DFT calculations shown in Figure 4 and CV measurements shown in Figure 6. One should also note that DFT calculations reflect thermodynamic reduction and oxidation potentials, while experimental values are dependent on the scan rate [52].

Table 4 Experimental reduction and oxidation stability of different salts, solvents, and additives on GC in various electrolytes (p = passivated).

Salt, solvent, additive, or electrolyte	Red. E/V	Ox. E/V
1M $LiPF_6$ in EC:EMC (3:7)	1.7, 0.9 (p)	6.0
1M $LiBF_4$ in EC:EMC (3:7)	0.8	5.5
1M LiTFSI in EC:EMC (3:7)	0.8	5.8
1M LiDFOB in EC:EMC (3:7)	1.8	5.3
1M LiTDI in EC:EMC (3:7)	1.6	4.7
1M $LiPF_6$ in EC:DMC (3:7)	1.7, 0.9 (p)	6.0
1M $LiPF_6$ in FEC:EMC (1:4)	1.4 (p)	6.2
1M $LiPF_6$ in EC:EMC (3:7) + 5% LiTFSI	1.8, 1.1 (p)	5.8
1M $LiPF_6$ in EC:EMC (3:7) + 5% HFiPP	1.8, 0.9 (p)	6.0
1M $LiPF_6$ in EC:EMC (3:7) + 5% PS	1.8, 0.5	6.0
1M $LiPF_6$ in EC:EMC (3:7) + 5% VC	0.5	4.9
1M $LiPF_6$ in EC:EMC (3:7) + 5% TMSP	0.7	5.9
1M $LiPF_6$ in EC:EMC (3:7) + 5% (TMSP+VC)	0.7	4.9

3.4 Reduction of electrolytes with different salts in EC:EMC (3:7 wt%) on GC

The reduction of 1M solutions of various salts including $LiPF_6$, $LiBF_4$, LiTFSI, LiDFOB, and LiTDI in EC:EMC (3:7 wt%) solvent mixtures on a GC electrode are shown as CV curves in Figure 7. The $LiPF_6$-based electrolyte displays a reduction peak at ~1.7V that is consistent with our calculations for $(LiPF_6)_2$ (see Table 3). Leung et al. also observed that LiF formed as a result of $LiPF_6$ reduction [35]. The LiDFOB-based electrolyte also displays a reduction peak at 1.8V that is attributed to the $DFOB^-$ anion reduction, this value is also consistent with DFT-calculated values (Table 3). The BF_4^- and $TFSI^-$ do not display anion reduction peaks. The absence of the BF_4^- reduction peak is in good agreement with DFT predictions (see Table 3) as it also suggests that participation of $LiPF_6$ in the electrolyte reduction unlike $LiBF_4$ likely explains the difference in electrolyte passivation layer formation on GC. TDI^- shows a very small reduction peak at 1.6V. DFT calculations predict Li_2TDI^- reduction at slightly lower potential of 1.1V.

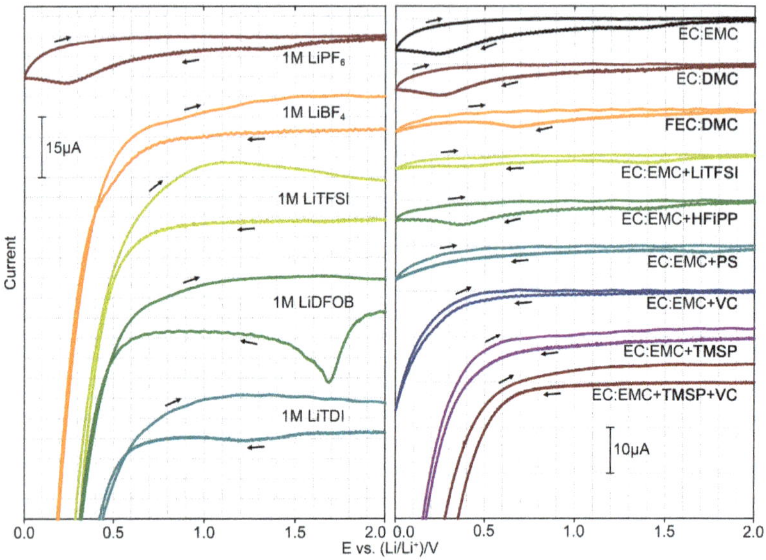

Fig. 7 Reduction CV curves of various 1M salt in EC:EMC (3:7 wt%) solvent mixtures, where the salts are $LiPF_6$, $LiBF_4$, LiTFSI, LiDFOB and LiTDI, respectively (Left). Reduction CV curves of electrolytes of 1M $LiPF_6$ in EC:EMC (3:7 wt%), EC:DMC (3:7 wt%), FEC:DMC (1:4 wt%) and the electrolytes of 1M $LiPF_6$ in EC:EMC (3:7 wt%) with additives LiTFSI, HFiPP, PS, TMSP, VC, and TMSP+VC (Right).

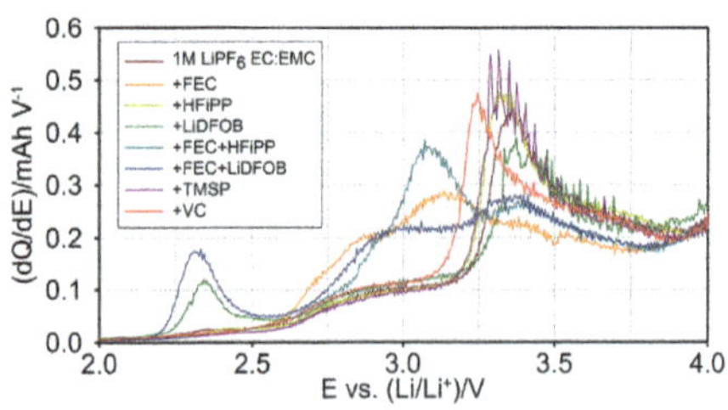

Fig. 8 dQ/dE vs. Voltage data for initial cycle of LNMO/Gr coin cells with various electrolytes showing reduction of electrolyte components occurring before oxidation.

The most notable feature is that all electrolytes except for the $LiPF_6$-based electrolyte show a large reduction current with an onset potential at about 0.7V that corresponds to the predicted reduction potential of EC of ~0.6V (see Table 3). The electrolyte containing $LiPF_6$ passivated the inert GC electrode. In fact, on subsequent scans there was even less current observed. Salts such as $LiBF_4$, LiTFSI, LiTDI and LiDFOB would work in practical Li-ion cells but when used with carbonate solvent systems the electrolytes seldom yield a quality SEI passivation layer for good performance. The reduction of EC is often credited for the SEI formation, but it is essential to have PF_6^- present so that the solvent is not perpetually reduced (as is the case with the other salts). The enhanced reduction stability of the EC:EMC–$LiPF_6$ mixtures enables us to examine additives/cosolvents that can be preferentially reduced (sacrificially) before the baseline electrolyte is reduced.

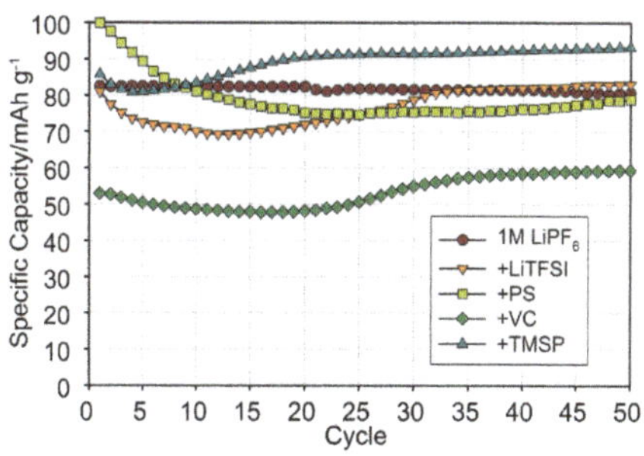

Fig. 9 Cycling data for LNMO/Graphite cells with various additives (1C rate at 30°C).

3.5 Reduction of electrolytes with different solvents and additives on GC

The reduction reactions of the electrolytes containing $LiPF_6$ in FEC:DMC, EC:DMC, and EC:EMC with different additives are shown as CV curves in Figure 7. EC:EMC and EC:DMC display very similar reduction potentials, which is not altered by the addition of HFiPP. This is consistent with HFiPP having minor contribution to the Li^+ solvation shell (see Table 1). FEC:DMC mixtures display a lower reduction stability (1.4V vs. 0.9V), due to the reduction of FEC. QC calculations predicted a higher reduction potential for FEC compared to EC. EC:EMC with LiTFSI shows very little current, suggesting the decomposition of EC, EMC, and/or PF_6^- is reduced by the addition of the $TFSI^-$ anion. PS, VC, and TMSP each display reduction curves that are not consistent with immediate passivation of the GC electrode as is the case with the baseline electrolyte. From this data, it is apparent that electrolytes containing PS, VC and TMSP each undergo reduction reactions with onset potentials near that of EC:EMC in the presence of PF_6^-. Lower PS binding energy to Li^+ and lower reduction potential compared to EC from QC calculations suggest that the onset of the reduction peak is due to EC, followed to a smaller extent at lower potentials by DMC and at even lower potentials by PS.

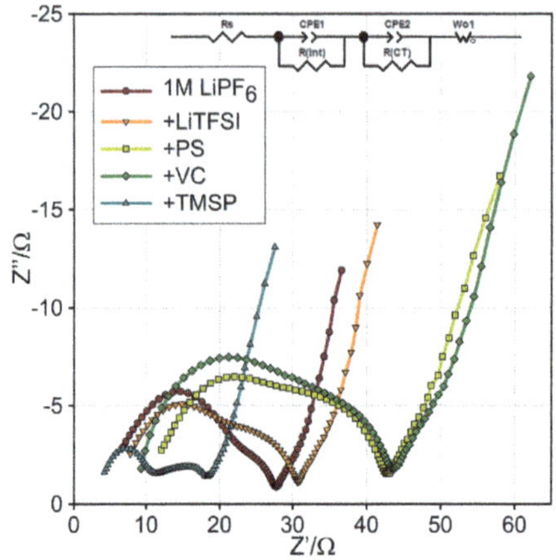

Fig. 10 EIS data acquired at 4.6V for cycled cells with various electrolytes (note x- and y-axes on different scales). First semi-circle is from interfacial resistance and second semi-circle is due to charge transfer resistance. Electrical circuit used for modeling data below plot.

Interestingly, the TMSP mixture yields a large reduction current at 0.7V, even in the presence of EC and PF_6^-. The amplitude of the reduction current decreases with increasing CV scans, indicating that TMSP gradually reacts with EC and PF_6^- to form a passivation layer on the GC electrodes after several cycles. DFT calculations predicted that addition of TMSP significantly disrupt the Li^+ solvation shell by partially replacing EC and DMC from it and, thus, changing the SEI formation mechanism. Reduction potential of Li^+(TMSP) was also predicted to be similar to than for Li^+(EC), see Table 3.

3.6 Redox reactions in a full cell

A freshly assembled full cell must first undergo SEI formation. This is done by charging at a slow rate (C/10 or slower) to form a more uniform and robust SEI layer. Figure 1 shows the voltage vs. time plot, which includes full cell voltage, cathode vs. Li/Li^+ voltage, and anode vs. Li/Li^+ voltage of the first charging curves of a LNMO/Li/Gr full (3-electrode) cell at C/10 rate with a 1M $LiPF_6$-EC:EMC (3:7) electrolyte. When the full cell is at 3.5V, the cathode is at 4.1V, where the cathode has not begun delithiation, and the anode is at 0.6V, where reduction reactions have already proceeded (according to the data from Tables 3 and 4).

Figure 8 shows the initial charge cycle dQ/dE vs. E plot for LNMO/Gr cells containing the baseline electrolyte with different additives including VC, TMSP, FEC, HFiPP, LiDFOB, FEC:HFiPP (1:1), and FEC:LiDFOB (1:1). All of the peaks observed below a cell voltage of 3.5V in Figure 8 indicate reduction of the electrolyte, not oxidation, according to the potentials in Figure 7. This observation is also supported by both Gr/Li and LNMO/Li half-cell data that showed reduction peaks on graphite before lithiation (0.8V) but no oxidation peaks on LNMO before delithiation (4.4V). Once delithiation begins, the oxidation reactions of the electrolyte are masked by the intensity of the delithiation current from the cathode. The baseline electrolyte has a peak at 3.3V due to the SEI formation plateau on graphite electrodes (observed when the graphite electrode has a potential of 0.8V). FEC-containing electrolytes have a peak at 3.1V (corresponding to 0.9V on graphite) and LiDFOB-containing electrolytes have a peak at 2.3V (corresponding to 1.7V on graphite) meaning that the electrolytes are reduced at higher voltages than the baseline electrolyte. These values agree with both the computational and experimental results from Tables 3 and 4.

It is clear from this data that the reduction stability of additives and co-solvents are of primary importance because the electrolyte is subjected to reducing potentials on the anode during charging before the cell reaches oxidative potentials. It is reasonable to assume that diffusion of reacted species can occur during charging at slow rates and thus, reduction products may migrate from the anode to the cathode during charging and be present as the passivation layer is formed on the cathode.

3.7 Impact of TMSP on cycling of LNMO/graphite cells

Full LNMO/Graphite cells were cycled at a 1C rate at 30°C for 50 cycles with various electrolytes (Figure 9). The cells with TMSP as an additive had the highest discharge capacity after cycling. Electrochemical impedance spectroscopy measurements were performed on the cells to determine the effect the additives had on the overall cell resistance (Figure 10). Higher impedance is indicative of thicker, more resistive SEI layers on the anode and cathode surfaces, while lower impedance suggests thinner or more ionically conductive SEI layers. The cells containing VC had the highest interfacial resistance (35.4 Ω) and lowest discharge capacity. This is in agreement with the CV data that shows not only large current due to reduction reactions, but also large current during oxidation scans (see Figures 6 and 7). The electrolyte containing PS had slightly lower interfacial resistance (32.7 Ω) but was more stable towards redox reactions compared to VC. The TMSP electrolyte had the lowest interfacial resistance (16.3 Ω) compared to the baseline electrolyte (23.6 Ω). According to the CV data, however, there was more current during reduction scans than the baseline electrolyte. Since the CV scans were performed using 5 wt% additive, and the ratio of surface area to electrolyte for the GC electrode was significantly lower than the ratio of surface area to electrolyte of the LNMO electrode in the LNMO/Graphite coin cells, it is possible that TMSP reduces to form a thin SEI on graphite and would eventually be consumed in a battery configuration as opposed to a CV setup. Thus, it is believed the TMSP sacrificially reduces to form a low impedance SEI layer in the initial formation cycles that enables high efficiency cycling due to the stability of the remaining 1M $LiPF_6$ EC:EMC electrolyte as mentioned in other work [53–55], which is consistent with the QC results shown above. After TMSP reduction and P–O bond breaking a second reduction of radicals yields a species that would diffuse to the cathode, oxidize and initiate formation of the passivation layer. Additionally, deprotonation of TMSP upon oxidation with H-transfer to another TMSP instead of cathode oxidations indicates a path to condensation polymerization at the cathode without making the surface more acidic due to high barrier of H-transfer to TMSP to cathode oxygens.

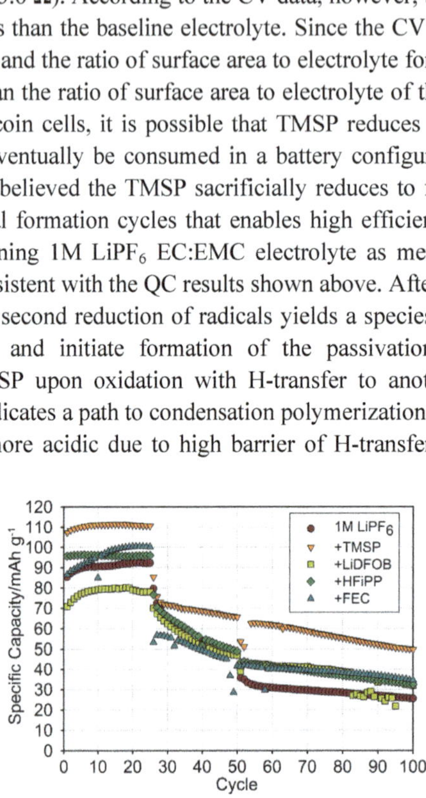

Fig. 11 Cycling data for LNMO/Graphite cells at C/3 rate for 25 cycles at 25°C and 75 cycles at 55°C.

The effects of elevated temperature (55°C) were also investigated in full cells. The cells were initially cycled at a C/3 rate for 25 cycles at 25°C before cycling for 75 cycles at 55°C. All cells experienced a drop in discharge capacity upon switching from room temperature to elevated temperature. While the baseline and other additives had a higher capacity fade, the TMSP electrolyte was more stable (Figure 11). The differential capacity (dQ/dE vs. E) was investigated for Cycle 25 (last cycle at 25°C), Cycle 26 (first cycle at 55°C) and Cycle 50 (25th cycle at 55°C) and plotted in Figure 12. For the baseline electrolyte, Cycle 25 shows defined peaks for the charge and discharge scans.

Cycle 26 shows a shift to a lower voltage during charging and higher voltage during discharging, most likely due to kinetic effects of raising the temperature. Cycle 50 has peaks shifted to a higher voltage during charging and lower voltage during discharging, due to impedance buildup. The peaks are also broader and less defined. The TMSP cells have similar features for Cycles 25 and 26 but the Cycle 50 data has retained more of the features than the baseline cells at Cycle 50. The peaks have only slightly shifted higher for charging and slightly lower for discharging while maintaining the overall characteristics. The TMSP additive is able to stabilize the electrodes and electrolyte at 55°C cycling enabling improved capacity retention.

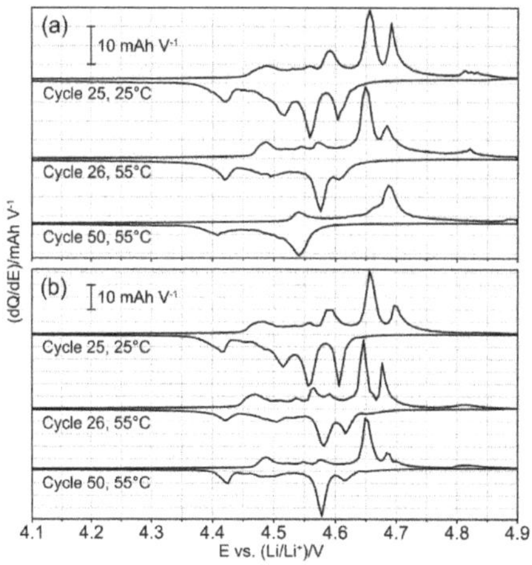

Fig. 12 dQ/dE data for baseline electrolyte (a) and TMSP additive (b) after 25 cycles at 25°C (top traces), 1 cycle at 55°C (middle traces) and 25 cycles at 55°C (bottom traces).

4. Conclusions

Pursuing more oxidatively stable electrolyte components for high voltage lithium ion batteries is not the only path forward. It is important to also understand electrolyte reduction for development of more stable high voltage electrolytes that could also benefit from kinetic stabilization due to passivation layer formation on the cathode. A series of electrolytes that incorporated different lithium salts, solvents and additives was analyzed using CV on GC and coin cell cycling at room and elevated temperature. EIS measurements and dQ/dE analysis were also performed on the coin cells. QC calculations were utilized to estimate the composition of the Li^+ solvation shell, indicating that TMSP will displace EC from the Li^+ solvation shell while other solvents will contribute in the following order EC > DMC > VC > FEC > PS >> HFiPP. Reduction potentials were also predicted for the Li^+–solvent and Li^+–anion complexes surrounded by implicit solvent. While improbable in dilute solutions, the solvent and anion defluorination reactions resulting in LiF formation were predicted to significantly increase reduction potentials for solvents such as FEC and salts such as $LiPF_6$ and LiTFSI but not $LiBF_4$. The $LiPF_6$ salt participation in SEI formation was attributed to the differences of GC passivation by EC:DMC–$LiPF_6$ and EC:DMC–$LiBF_4$ electrolytes. In order to adequately predict electrolyte oxidation stability, a variety of complexes were considered with a particular attention the influence of H-transfer during oxidation. When plotted as a function of the difference between the vertical and adiabatic oxidation energies, calculated oxidation potentials were found to largely follow the results of CV measurements indicating that reorganization energy during redox reactions is an important parameter for understanding electrolyte redox stability.

The electrolyte containing 1M $LiPF_6$ in EC:EMC was able to passivate the GC electrode during reduction scans. Addition of VC to that electrolyte inhibits the ability for the electrolyte to passivate the GC electrode upon scanning down to 0V. During high voltage scans with the VC electrolyte, oxidation occurs well before that of the baseline electrolyte, precluding its use in high voltage systems, such as LNMO. The electrolyte containing TMSP also inhibits the passivation of GC electrodes during reduction scans, however, the oxidative stability is closer to that of the baseline electrolyte. Surface calculations on delithiated LNMO indicate that TMSP is not likely to transfer a proton to the surface, instead it will transfer a proton to another TMSP or intramolecularly upon oxidation resulting in the formation of a passivation layer without making the cathode surface more acidic. The electrolyte containing 1% TMSP showed lower impedance after cycling and was able to maintain a higher capacity than the other electrolytes tested. Analysis of the dQ/dE vs. E plots show that the traces of the cells with TMSP containing electrolyte maintain the RedOx characteristics (peak shape, position, and area) better than the baseline electrolyte.

Acknowledgements

SAD and MO were funded by the Army Research Laboratory postdoctoral program administered through Oak Ridge Associated Universities. Modeling work was supported by ARL Enterprise for Multiscale Research.

The work has been previously published by Elsevier:

Electrochimica Acta Volume 209, 10 August 2016, Pages 498-510 Importance of Reduction and Oxidation Stability of High Voltage Electrolytes and Additives Samuel A.

Delp, Oleg Borodin, Marco Olguin, Claire G. Eisner, Joshua L. Allen, T. Richard Jow https://doi.org/10.1016/j.electacta.2016.05.100

References

[1] A. Manthiram, Materials Challenges and Opportunities of Lithium Ion Batteries, J. Phys. Chem. Lett. 2 (2011) 176–184. https://doi.org/10.1021/jz1015422.

[2] J.L. Allen, T.R. Jow, J. Wolfenstine, Improved cycle life of Fe-substituted $LiCoPO_4$, J. Power Sources. 196 (2011) 8656–8661. https://doi.org/10.1016/j.jpowsour.2011.06.057.

[3] D. Aurbach, B. Markovsky, G. Salitra, E. Markevich, Y. Talyossef, M. Koltypin, et al., Review on electrode–electrolyte solution interactions, related to cathode materials for Li-ion batteries, J. Power Sources. 165 (2007) 491–499. https://doi.org/10.1016/j.jpowsour.2006.10.025.

[4] M. Ue, Y. Sasaki, Y. Tanaka, M. Morita, Ch 2. Nonaqueous Electrolytes with Advances in Solvents, in: T.R. Jow, K. Xu, O. Borodin, M. Ue (Eds.), Electrolytes Lithium Lithium-Ion Batter., Springer, New York, 2014: pp. 93–165.

[5] Z. Zhang, L. Hu, H. Wu, W. Weng, M. Koh, P.C. Redfern, et al., Fluorinated electrolytes for 5 V lithium-ion battery chemistry, Energy Environ. Sci. 6 (2013) 1806. https://doi.org/10.1039/c3ee24414h.

[6] L. Hu, Z. Zhang, K. Amine, Fluorinated electrolytes for Li-ion battery: An FEC-based electrolyte for high voltage $LiNi_{0.5}Mn_{1.5}O_4$/graphite couple, Electrochem. Commun. 35 (2013) 76–79. https://doi.org/10.1016/j.elecom.2013.08.009.

[7] L. Hu, Z. Xue, K. Amine, Z. Zhang, Fluorinated Electrolytes for 5-V Li-Ion Chemistry: Synthesis and Evaluation of an Additive for High-Voltage $LiNi_{0.5}Mn_{1.5}O_4$/Graphite Cell, J. Electrochem. Soc. 161 (2014) A1777–A1781. https://doi.org/10.1149/2.0141412jes.

[8] E. Markevich, G. Salitra, K. Fridman, R. Sharabi, G. Gershinsky, A. Garsuch, et al., Fluoroethylene Carbonate as an Important Component in Electrolyte Solutions for High-Voltage Lithium Batteries: Role of Surface Chemistry on the Cathode, Langmuir. 30 (2014) 7414–7424. https://doi.org/10.1021/la501368y.

[9] K. Chiba, T. Ueda, Y. Yamaguchi, Y. Oki, F. Shimodate, K. Naoi, Electrolyte Systems for High Withstand Voltage and Durability I. Linear Sulfones for Electric Double-Layer Capacitors, J. Electrochem. Soc. 158 (2011) A872–A882. https://doi.org/10.1149/1.3593001.

[10] K. Abe, T. Takaya, H. Yoshitake, Y. Ushigoe, M. Yoshio, H. Wang, Functional Electrolyte: Additives for Improving the Cyclability of Cathode Materials, Electrochem. Solid-State Lett. 7 (2004) A462–A465. https://doi.org/10.1149/1.1809557.

[11] M. Xu, N. Tsiouvaras, A. Garsuch, H.A. Gasteiger, B.L. Lucht, Generation of Cathode Passivation films via Oxidation of Lithium Bis(oxalato) Borate on High Voltage Spinel ($LiNi_{0.5}Mn_{1.5}O_4$), J. Phys. Chem. C. (2014). https://doi.org/10.1021/jp501970j.

[12] M. Xu, L. Zhou, Y. Dong, U. Tottempudi, J. Demeaux, A. Garsuch, et al., Improved Performance of High Voltage Graphite/$LiNi_{0.5}Mn_{1.5}O_4$ Batteries with Added Lithium Tetramethyl Borate, ECS Electrochem. Lett. 4 (2015) A83–A86. https://doi.org/10.1149/2.0021508eel.

[13] M. Xu, L. Zhou, Y. Dong, Y. Chen, J. Demeaux, A.D. MacIntosh, et al., Development of novel lithium borate additives for designed surface modification of high voltage $LiNi_{0.5}Mn_{1.5}O_4$ cathodes, Energy Environ. Sci. (2016). https://doi.org/10.1039/C5EE03360H.

[14] O. Borodin, W. Behl, T.R. Jow, Oxidative Stability and Initial Decomposition Reactions of Carbonate, Sulfone, and Alkyl Phosphate-Based Electrolytes, J. Phys. Chem. C. 117 (2013) 8661–8682. https://doi.org/10.1021/jp400527c.

[15] O. Borodin, Ch 8. Molecular Modeling of Electrolytes, in: T.R. Jow, K. Xu, O. Borodin, M. Ue (Eds.), Electrolytes Lithium Lithium-Ion Batter., Springer, New York, 2014: pp. 371–401.

[16] O. Borodin, M. Olguin, C.E. Spear, K.W. Leiter, J. Knap, Towards high throughput screening of electrochemical stability of battery electrolytes, Nanotechnology. 26 (2015) 354003. https://doi.org/10.1088/0957-4484/26/35/354003.

[17] E. Jónsson, P. Johansson, Electrochemical oxidation stability of anions for modern battery electrolytes: a CBS and DFT study, Phys. Chem. Chem. Phys. 17 (2015) 3697–3703. https://doi.org/10.1039/C4CP04592K.

[18] M. Egashira, H. Takahashi, S. Okada, J. Yamaki, Measurement of the electrochemical oxidation of organic electrolytes used in lithium batteries by microelectrode, J. Power Sources. 92 (2001) 267–271. https://doi.org/10.1016/S0378-7753(00)00553-X.

[19] L. Niedzicki, G.Z. Żukowska, M. Bukowska, P. Szczeciński, S. Grugeon, S. Laruelle, et al., New type of imidazole based salts designed specifically for lithium ion batteries, Electrochimica Acta. 55 (2010) 1450–1454. https://doi.org/10.1016/j.electacta.2009.05.008.

[20] D. McOwen, S. Delp, E. Paillard, C. Herriot, S.-D. Han, P.D. Boyle, et al., Anion Coordination Interactions in Solvates with the Lithium Salts LiDCTA and LiTDI, J. Phys. Chem. C. (2014). https://doi.org/10.1021/jp412601x.

[21] S. Shui Zhang, An unique lithium salt for the improved electrolyte of Li-ion battery, Electrochem. Commun. 8 (2006) 1423–1428. https://doi.org/10.1016/j.elecom.2006.06.016.

[22] J.L. Allen, S.-D. Han, P.D. Boyle, W.A. Henderson, Crystal structure and physical properties of lithium difluoro(oxalato)borate (LiDFOB or $LiBF_2Ox$), J. Power Sources. 196 (2011) 9737–9742. https://doi.org/10.1016/j.jpowsour.2011.07.065.

[23] A.V. Marenich, C.J. Cramer, D.G. Truhlar, Universal Solvation Model Based on Solute Electron Density and on a Continuum Model of the Solvent Defined by the Bulk Dielectric Constant and Atomic Surface Tensions, J. Phys. Chem. B. 113 (2009) 6378–6396. https://doi.org/10.1021/jp810292n.

[24] T.A. Barnes, J.W. Kaminski, O. Borodin, T.F. Miller, Ab Initio Characterization of the Electrochemical Stability and Solvation Properties of Condensed-Phase Ethylene Carbonate and Dimethyl Carbonate Mixtures, J. Phys. Chem. C. 119 (2015) 3865–3880. https://doi.org/10.1021/jp510882g.

[25] J. Knap, C.E. Spear, O. Borodin, K.W. Leiter, Advancing a distributed multi-scale computing framework for large-scale high-throughput discovery in materials science, Nanotechnology. 26 (2015) 434004. https://doi.org/10.1088/0957-4484/26/43/434004.

[26] O. Borodin, M. Olguin, P. Ganesh, P.R.C. Kent, J.L. Allen, W.A. Henderson, Competitive lithium solvation of linear and cyclic carbonates from quantum chemistry, Phys. Chem. Chem. Phys. 18 (2015) 164–175. https://doi.org/10.1039/C5CP05121E.

[27] K. Xu, "Charge-Transfer" Process at Graphite/Electrolyte Interface and the Solvation Sheath Structure of Li^+ in Nonaqueous Electrolytes, J. Electrochem. Soc. 154 (2007) A162. https://doi.org/10.1149/1.2409866.

[28] K. Xu, Y. Lam, S.S. Zhang, T.R. Jow, T.B. Curtis, Solvation Sheath of Li^+ in Nonaqueous Electrolytes and Its Implication of Graphite/Electrolyte Interface

Chemistry, J. Phys. Chem. C. 111 (2007) 7411–7421. https://doi.org/10.1021/jp068691u.

[29] A. von Wald Cresce, O. Borodin, K. Xu, Correlating Li^+ Solvation Sheath Structure with Interphasial Chemistry on Graphite, J. Phys. Chem. C. 116 (2012) 26111–26117. https://doi.org/10.1021/jp303610t.

[30] O. Borodin, G.D. Smith, Quantum Chemistry and Molecular Dynamics Simulation Study of Dimethyl Carbonate: Ethylene Carbonate Electrolytes Doped with $LiPF_6$, J. Phys. Chem. B. 113 (2009) 1763–1776. https://doi.org/10.1021/jp809614h.

[31] O. Borodin, M. Olguin, C. Spear, K. Leiter, J. Knap, G. Yushin, et al., (Invited) Challenges with Quantum Chemistry-Based Screening of Electrochemical Stability of Lithium Battery Electrolytes, ECS Trans. 69 (2015) 113–123. https://doi.org/10.1149/06901.0113ecst.

[32] C.D. Malliakas, K. Leung, K.Z. Pupek, I.A. Shkrob, D.P. Abraham, Spontaneous aggregation of lithium ion coordination polymers in fluorinated electrolytes for high-voltage batteries, Phys. Chem. Chem. Phys. 18 (2016) 10846–10849. https://doi.org/10.1039/C6CP01157H.

[33] J. Vatamanu, O. Borodin, G.D. Smith, Molecular Dynamics Simulation Studies of the Structure of a Mixed Carbonate/$LiPF_6$ Electrolyte near Graphite Surface as a Function of Electrode Potential, J. Phys. Chem. C. 116 (2012) 1114–1121. https://doi.org/10.1021/jp2101539.

[34] Y.-K. Han, S.U. Lee, Performance of density functionals for calculation of reductive ring-opening reaction energies of Li^+-EC and Li^+-VC, Theor. Chem. Acc. 112 (2004) 106–112. https://doi.org/10.1007/s00214-004-0572-5.

[35] K. Leung, Predicting the voltage dependence of interfacial electrochemical processes at lithium-intercalated graphite edge planes, Phys. Chem. Chem. Phys. 17 (2014) 1637–1643. https://doi.org/10.1039/C4CP04494K.

[36] K. Leung, S.B. Rempe, M.E. Foster, Y. Ma, J.M.M. del la Hoz, N. Sai, et al., Modeling Electrochemical Decomposition of Fluoroethylene Carbonate on Silicon Anode Surfaces in Lithium Ion Batteries, J. Electrochem. Soc. 161 (2014) A213–A221. https://doi.org/10.1149/2.092401jes.

[37] M. Xu, W. Li, B.L. Lucht, Effect of propane sultone on elevated temperature performance of anode and cathode materials in lithium-ion batteries, J. Power Sources. 193 (2009) 804–809. https://doi.org/10.1016/j.jpowsour.2009.03.067.

[38] A. von Cresce, K. Xu, Preferential Solvation of Li^+ Directs Formation of Interphase on Graphitic Anode, Electrochem. Solid-State Lett. 14 (2011) A154–A156. https://doi.org/10.1149/1.3615828.

[39] K. Ushirogata, K. Sodeyama, Y. Okuno, Y. Tateyama, Additive Effect on Reductive Decomposition and Binding of Carbonate-Based Solvent toward Solid Electrolyte Interphase Formation in Lithium-Ion Battery, J. Am. Chem. Soc. 135 (2013) 11967–11974. https://doi.org/10.1021/ja405079s.

[40] T. Kawaguchi, K. Shimada, T. Ichitsubo, S. Yagi, E. Matsubara, Surface-layer formation by reductive decomposition of $LiPF_6$ at relatively high potentials on negative electrodes in lithium ion batteries and its suppression, J. Power Sources. 271 (2014) 431–436. https://doi.org/10.1016/j.jpowsour.2014.08.010.

[41] S. Laruelle, S. Pilard, P. Guenot, S. Grugeon, J.-M. Tarascon, Identification of Li-Based Electrolyte Degradation Products Through DEI and ESI High-Resolution Mass Spectrometry, J. Electrochem. Soc. 151 (2004) A1202–A1209. https://doi.org/10.1149/1.1760992.

[42] H. Kim, F. Wu, J.T. Lee, N. Nitta, H.-T. Lin, M. Oschatz, et al., In Situ Formation of Protective Coatings on Sulfur Cathodes in Lithium Batteries with LiFSI-Based Organic Electrolytes, Adv. Energy Mater. 5 (2015). https://doi.org/10.1002/aenm.201401792.

[43] L. Suo, O. Borodin, T. Gao, M. Olguin, J. Ho, X. Fan, et al., "Water-in-salt" electrolyte enables high-voltage aqueous lithium-ion chemistries, Science. 350 (2015) 938–943. https://doi.org/10.1126/science.aab1595.

[44] L. Suo, O. Borodin, W. Sun, X. Fan, C. Yang, F. Wang, et al., Advanced High Voltage Aqueous Li-ion Battery Enabled by "Water-in-Bisalt" Electrolyte, Angew. Chem. Int. Ed. (2016) in press.

[45] L. Xing, O. Borodin, G.D. Smith, W. Li, Density Functional Theory Study of the Role of Anions on the Oxidative Decomposition Reaction of Propylene Carbonate, J. Phys. Chem. A. 115 (2011) 13896–13905. https://doi.org/10.1021/jp206153n.

[46] O. Borodin, T.R. Jow, Quantum Chemistry Studies of the Oxidative Stability of Carbonate, Sulfone and Sulfonate-Based Electrolytes Doped with BF_4^-, PF_6^- Anions, ECS Trans. 33 (2011) 77–84. https://doi.org/10.1149/1.3563092.

[47] L. Xing, O. Borodin, Oxidation induced decomposition of ethylene carbonate from DFT calculations – importance of explicitly treating surrounding solvent, Phys.

Chem. Chem. Phys. 14 (2012) 12838–12843. https://doi.org/10.1039/C2CP41103B.

[48] O. Borodin, T.R. Jow, Quantum Chemistry Study of the Oxidation-Induced Decomposition of Tetramethylene Sulfone (TMS) Dimer and TMS/BF_4^-, ECS Trans. 50 (2013) 391–398. https://doi.org/10.1149/05026.0391ecst.

[49] Y. Wang, L. Xing, O. Borodin, W. Huang, M. Xu, X. Li, et al., Quantum chemistry study of the oxidation-induced stability and decomposition of propylene carbonate-containing complexes, Phys. Chem. Chem. Phys. 16 (2014) 6560–6567. https://doi.org/10.1039/C3CP54728K.

[50] A. Jarry, S. Gottis, Y.-S. Yu, J. Roque-Rosell, C. Kim, J. Cabana, et al., The Formation Mechanism of Fluorescent Metal Complexes at the $Li_xNi_{0.5}Mn_{1.5}O_4-\delta$/Carbonate Ester Electrolyte Interface, J. Am. Chem. Soc. 137 (2015) 3533–3539. https://doi.org/10.1021/ja5116698.

[51] N. Kumar, K. Leung, D.J. Siegel, Crystal Surface and State of Charge Dependencies of Electrolyte Decomposition on $LiMn_2O_4$ Cathode, J. Electrochem. Soc. 161 (2014) E3059–E3065. https://doi.org/10.1149/2.009408jes.

[52] M. Ue, A. Murakami, S. Nakamura, Anodic Stability of Several Anions Examined by Ab Initio Molecular Orbital and Density Functional Theories, J. Electrochem. Soc. 149 (2002) A1572–A1577. https://doi.org/10.1149/1.1517579.

[53] H. Rong, M. Xu, L. Xing, W. Li, Enhanced cyclability of $LiNi_{0.5}Mn_{1.5}O_4$ cathode in carbonate based electrolyte with incorporation of tris(trimethylsilyl)phosphate (TMSP), J. Power Sources. 261 (2014) 148–155. https://doi.org/10.1016/j.jpowsour.2014.03.032.

[54] G. Yan, X. Li, Z. Wang, H. Guo, C. Wang, Tris(trimethylsilyl)phosphate: A film-forming additive for high voltage cathode material in lithium-ion batteries, J. Power Sources. 248 (2014) 1306–1311. https://doi.org/10.1016/j.jpowsour.2013.10.037.

[55] J. Zhang, J. Wang, J. Yang, Y. NuLi, Artificial Interface Deriving from Sacrificial Tris(trimethylsilyl)phosphate Additive for Lithium Rich Cathode Materials, Electrochimica Acta. 117 (2014) 99–104. https://doi.org/10.1016/j.electacta.2013.11.024.

Chapter 3

Factors Limiting Li^+ Charge Transfer Kinetics in Li-ion Batteries

T. Richard Jow[1,*], Samuel A. Delp[1], Jan L. Allen[1], John-Paul Jones[2], Marshall C. Smart[2]
[1]U.S. Army Research Laboratory, Adelphi, MD 20783
[2]Jet Propulsion Laboratory, California Institute of Technology, Pasadena, CA 91109
t.r.jow.civ@mail.mil

Abstract

Understanding the factors limiting Li^+ charge transfer kinetics in Li-ion batteries is essential in improving the rate performance, especially at lower temperatures. The Li^+ charge transfer process involved in the lithium intercalation of graphite anode includes the step of de-solvation of the solvated Li^+ in the liquid electrolyte and the step of transport of Li^+ in the preformed solid electrolyte interphase (SEI) on electrodes until the Li^+ accepts an electron at the electrode and becomes a Li in the electrode. Whether the de-solvation process or the Li^+ transport through the SEI is a limiting step depends on the nature of the interphases at the electrode and electrolyte interfaces. Several examples involving the electrode materials such as graphite, lithium titanate (LTO), lithium iron phosphate (LFP), lithium nickel cobalt aluminum oxide (NCA) and solid Li^+ conductor such as lithium lanthanum titanate or Li-Al-Ti-phosphate are reviewed and discussed to clarify the conditions at which either the de-solvation or the transport of Li^+ in SEI is dominating and how the electrolyte components affect the activation energy of Li^+ charge transfer kinetics. How the electrolyte additives impact the Li^+ charge transfer kinetics at both the anode and the cathode has been examined at the same time in 3-electrode full cells. The resulting impact on Li^+ charge transfer resistance, R_{ct}, and activation energy, E_a, at both electrodes are reported and discussed.

Keywords

Lithium-Ion Charge Transfer Kinetics, Limiting Factor, De-Solvation, SEI, Lithium-Ion Batteries

The work has been previously published by ECS: T. Richard Jow, Samuel A. Delp, Jan L. Allen, John-Paul Jones, and Marshall C. Smart, Factors Limiting Li+ Charge Transfer Kinetics in Li-Ion Batteries, Journal of The Electrochemical Society, 165 (2) A361-A367 (2018). https://doi.org/10.1149/2.1221802jes

Contents

1. Introduction

To improve the power performance of Li-ion batteries, it is important to understand the factors that limit the Li^+ charge transfer kinetics. Li-ion batteries comprised of a graphite anode and a lithium cobalt oxide cathode in an electrolyte consisting of 1 M $LiPF_6$ in ethylene carbonate (EC)-dimethyl carbonate (DMC)-diethyl carbonate (DEC) carbonate solvent mixture could not deliver their room temperature capacity at a rate of C/2 at -30 and -40 °C [1]. When DEC was replaced by a linear ester solvent, such as ethyl acetate (EA) or methyl butyrate (MB), the Li-ion batteries at -30 and -40 °C could deliver over 80% of their room temperature capacity at the same rate [2]. When the $LiPF_6$ salt is replaced by lithium bis(oxalato)borate (LiBOB) in EC-ethyl methyl carbonate (EMC) (1:1 wt. ratio) carbonate solvent mixture, the impedance of the graphite-electrolyte interface measured using graphite/Li half cells in the electrolyte with LiBOB is three times that in the electrolyte with $LiPF_6$ [3]. These examples illustrate that the electrolyte components play crucial roles in affecting Li^+ charge transfer kinetics in Li-ion batteries.

The Li^+ charge transfer process starts from the solvated Li^+ in the electrolyte to the reception of an electron (e^-) from the electrode and becomes Li (i.e., Li_xC in graphitic anodes). This involves the de-solvation step of Li^+ before entering into a layer of solid electrolyte interphases, or SEI, that is often referred to that on the anode such as carbonaceous materials, and the diffusion step of Li^+ through the SEI, which is pre-formed on the electrode at the electrode and electrolyte interfaces, before receiving an e^-

from the electrode at the electrode and SEI interface. Abe et al. [4-6] believe that the Li^+ de-solvation process is the rate limiting step, which is supported by the results from the investigation of the electrolytes of different solvent systems using $LiClO_4$ salt at highly oriented pyrolytic graphite (HOPG), $Li_4Ti_5O_{12}$ and Li^+ solid conductors interfaces. The question is whether the de-solvation as a limiting step can be extended to the cathode-electrolyte interface. Jow et al. [7] examined the Li^+ charge transfer kinetics at the graphite anode-electrolyte interface and $LiFePO_4$ (LFP) cathode-electrolyte interface at the same time in a full cell, Gr/LFP, using Li as a reference electrode in 1 M $LiPF_6$ in EC-DMC-MB with vinylene carbonate (VC) as an additive. It is found that the activation energy (E_a) at the graphite-electrolyte is about 67 kJ mol^{-1}, which is much higher than 33 kJ mol^{-1} found at the LFP-electrolyte interface. It is concluded that the electrodes and their associated electrode-electrolyte interfacial layer as SEI is controlling the Li^+ charge transfer kinetics.

Smart et al. [8] studied the effect of additives on the performance of $MCMB/LiNi_xCo_{1-x}O_2$ Li-ion cells. It was determined that the discharge capacity at -40 °C can be improved by the addition of additives such as VC and LiBOB. As additives change the characteristics of the SEI (but presumably do not significantly impact the bulk electrolyte properties), it is concluded that the SEI plays a crucial role in the Li-ion cell performance. The impact of additives such as VC, LiBOB, and lithium bis(fluorosulfonyl)imide (LiFSI), etc. on the impedance of the anode-electrolyte and the cathode-electrolyte interfaces in $Gr/LiNiCoAlO_2$ cylindrical cells in 1.0 M $LiPF_6$ in EC-EMC-methyl propionate (MP) (20:20:60 vol %) at low temperatures was studied by Jones et al. [9]. It was found that the additives impact the impedance at the anode differently from that at the cathode. Different electrolyte components, including additives, result in different reduction and oxidation reactions during cell formation and cycling at the anode and cathode [10] and therefore, different SEIs.

Both the de-solvation step and the transport of Li^+ in SEI are important steps in controlling the rate. The question of which is the dominant one is an interesting one to answer. This paper will first review how we define the charge transfer process and how the charge transfer resistance is measured. This paper intends to provide a review of the Li^+ charge transfer process in Li-ion batteries, when the de-solvation step is dominant and when the diffusion of Li^+ in SEI is dominant, followed by a discussion on possible reasons behind them and a conclusion.

2. Li^+ charge transfer process

The Li^+ charge transfer process is defined as the process of turning a solvated Li^+ in the electrolyte into Li in the electrode by accepting an electron from the electrode. At the graphite anode, during charge, as shown in Fig. 1(a), the Li^+ charge transfer process is made of the following three steps ignoring the transport of the solvated Li^+ in the electrolyte before reaching at the SEI and electrolyte interface. This is reasonable as the electrolytic conduction of 1 M $LiPF_6$ in EC:EMC (30:70 vol.%) varied from 1.1 mS cm^{-1} at -40 °C to 9.6 mS cm^{-1} at 25 °C with an activation energy of Li^+ conduction of only 22 kJ mol^{-1} or less [11,12].

1. $Li^+ \cdot (solv.)_n \rightarrow Li^+$, (De-solvation of Li^+ at the SEI/electrolyte interface)
2. $Li^+ \rightarrow Li^+$, (Li^+ transport through anode SEI towards anode)
3. $x\,Li^+ + x\,e^- + C_6 \rightarrow Li_xC_6$. ($Li^+$ charge transfer: Li^+ reaching the anode and accepting an electron at the anode-SEI interface or at the anode-electronic conductor/SEI interface. The electron is supplied through the external circuit during charge.)

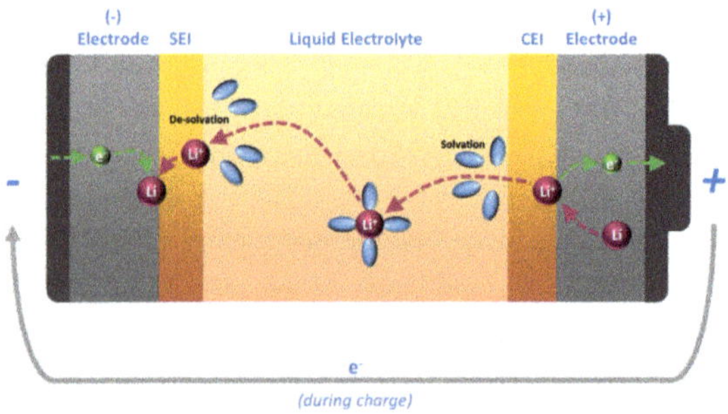

Figure 1(a) A schematic view of the Li+ charge transfer process during charge.

The kinetics of the Li^+ transport in the SEI is determined by the nature of SEI, which is made of mixtures of inorganic and organic Li^+ conductors [13].

At the cathode, there is a Li de-intercalation process resulting in a solvated Li^+ in the electrolyte, which is described in the following steps.

4. $LiCoO_2 \rightarrow x\,Li^+ + x\,e^- + Li_{1-x}CoO_2$, (Li de-intercalation, dissolution, from $LiCoO_2$ to become Li^+ and e^-)
5. $Li^+ \rightarrow Li^+$, (Li^+ transport through cathode/electrolyte interphases, CEI)
6. $Li^+ \rightarrow Li^+ \cdot (solv.)_n$ (Li^+ solvation in electrolyte)

During discharge as shown in Fig. 1(b), the Li^+ charge transfer process that involves the Li^+ de-solvation step from the solvated Li^+ in the electrolyte is occurring at the cathode and the Li de-intercalation step, $Li_xC_6 \rightarrow x\ Li^+ + x\ e^- + C_6$, is occurring at the anode.

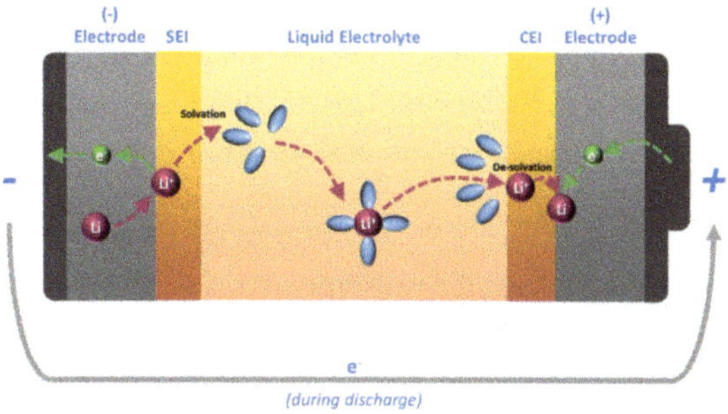

Figure 1(b) A schematic view of the Li+ charge transfer process during discharge.

Step 3, in which the Li^+ accepts an electron and becomes Li intercalated graphite, is considered a fast step. Step 4, the de-intercalation kinetics at the electrode and electrolyte interface, is also considered fast. The subsequent diffusion of Li in the electrode bulk active material is often the slower and rate limiting step in the battery charge discharge processes at normal ambient temperatures. Although the diffusion of Li within the active cathode or anode material is not our main focus, it is worth of our understanding as it links to the Li^+ charge transfer kinetics at the electrode and electrolyte interface closely. The relative facility of each of these steps changes as a function of temperature, so the rate limiting step may change at different temperatures. For example, studies by Wang et al. [14], involving the low temperature characterization of graphite and MCMB coke anodes, have shown that the stage transformation resistance, or diffusion in the bulk electrode, is rate determining at 25 °C, whereas the SEI resistance is rate limiting at −30 °C. As described, the researchers reported over a 27-fold increase in the SEI resistance from 25 °C to −30 °C, which was the primary cause for the high polarization observed at low temperature. Smart et al. [15] have previously reported that cells consisting of MCMB anodes and $LiNiCoO_2$ cathodes with various electrolytes display poorer cathode kinetics compared to the anode. However, at temperatures below −30 °C the lithium intercalation kinetics at the MCMB appear to become performance limiting, illustrating that the kinetics change at different temperatures for the different electrodes. Others have also reported MCMB anodes to be performance limiting at temperatures < -20 °C compared to the $LiCoO_2$ cathodes [16]. These studies illustrate that broad conclusions of which electrode is rate determining over a wide range of temperatures may not be valid. Nonetheless, the diffusion of Li in the electrode changes with temperature and can be dominating in certain systems depending on temperature.

Therefore, the investigation on whether the de-solvation step, step 1, and the diffusion of Li^+ across the SEI and CEI, steps 2 and 5, respectively, occurring at the electrode/electrolyte interfaces are possible rate limiting steps will be our main focus. How do we characterize the charge transfer kinetics? How do we determine which step is rate limiting?

3. Electrochemical charge transfer resistance

The resistance resulting from the charge transfer process is denoted as R_{ct}, which results from contributions from both the de-solvation step and the diffusion in the SEI step. The R_{ct} follows the relationship based on the thermally activated process [17],

$$\frac{1}{R_{ct}} = A_o\, e^{-E_a/RT}, \qquad [1]$$

where A_o, E_a, R and T are a frequency factor, the activation energy, the gas constant and the temperature in Kelvin, respectively. The E_a of the Li^+ charge transfer process, or the lithium-ion kinetics, represents the barrier that the Li^+ needs to overcome to cross the interface between the electrolyte and the electrode. The value of E_a can be obtained from the slope of a $\log(1/R_{ct})$ versus the inverse of temperature (1/T) plot.

Electrochemical impedance spectroscopy (EIS) measurements [18] can be used to characterize the resistance of different processes occurring in an electrochemical cell through the fitting of the impedance spectra, Nyquist plots, to an equivalent circuit.

Which step is rate limiting appears to depend on the cell components that are used in the formation of interfaces or interphases, including the electrode and electrolyte type selections. The cell interfaces that can potentially lead to the de-solvation process being a rate limiting step is to be considered first, and then the conditions under which the Li transport in SEI as a rate limiting step will be discussed.

4. De-solvation of solvated Li^+ as a rate limiting step

Abe et al. [4] reported the study of the activation energy of the Li^+ charge transfer process across the HOPG electrode and electrolyte interface. When the HOPG electrode is held at 0.4 V or lower vs. Li/Li^+ in 1 M $LiClO_4$ in EC:DEC electrolyte, there is a SEI formed on HOPG and the solvated Li^+ needs to de-solvate to diffuse through SEI and reach the electrode before accepting an electron. When the HOPG is held at 1.1 V in 1 M

$LiCF_3SO_3$ in 1,2-dimethoxyethane (DME) electrolyte, there is no SEI formed on HOPG, and the solvated Li^+ can intercalate into graphite directly. The E_a for this process, which involves no de-solvation, is only 25 kJ mol^{-1}. This is much lower than an E_a of 53 to 59 kJ mol^{-1} for the prior process, which involves de-solvation because of the existence of the SEI. It is therefore concluded that the rate limiting step is de-solvation. Further studies support this conclusion. Two of them are briefly reviewed below.

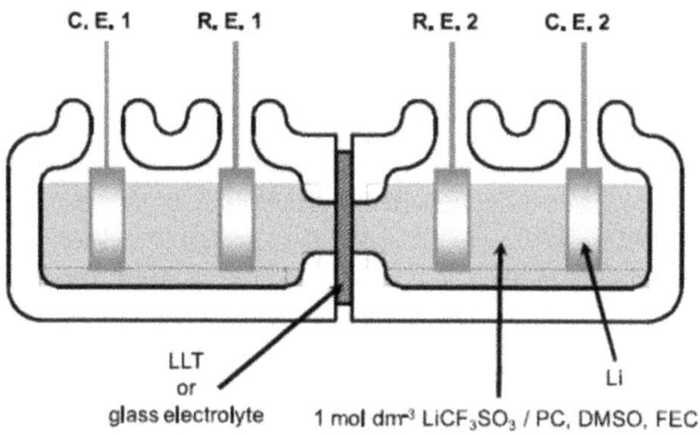

Figure 2. Experimental set up for the measurement of the R_{ct} between the solid-state ionic conductor such as LLT or glass electrolyte and liquid electrolyte such as 1 mol dm^{-3} of $LiCF_3SO_3$ in PC, DMSO or FEC (reproduced with permission by Electrochemical Society, Inc. from [6]).

Li^+ charge transfer at the interface between solid-state ceramic electrolyte and liquid electrolyte

The study of the solid-state (SS) electrolyte and liquid electrolyte interface was first reported by Abe et al. [6]. This study uses a four-Li electrode cell as shown in Fig. 2 [6], in which the SS electrolyte disk is in the center and in contact with the liquid electrolyte only. Two Li electrodes on each side of the SS electrolyte with one being the reference electrode and other one being the counter electrode are designed. This design allows the measurement of Li^+ de-solvation and solvation into and out of the SS electrolyte without encountering other complications. The SS electrolytes used in this study are crystalline lithium lanthanum titanate, $Li_{0.35}La_{0.55}TiO_3$ (LLT), [19,20] and Li-Al-Ti-phosphate ($Li_{1+x}Al_xTi_{2-x}(PO_4)_3$) based glass electrolyte (Ohara glass) [21]. The activation energy of Li^+ charge transfer at the interfaces of the Ohara glass and the liquid electrolytes including 1 M $LiCF_3SO_3$ in dimethyl sulfoxide (DMSO), propylene carbonate (PC), and mono-fluoroethylene carbonate (FEC) are 69.9, 57.3 and 31.5 kJ mol^{-1}, respectively, based on the charge transfer resistance obtained from the EIS measurements, which are consistent with the DFT calculated reaction enthalpy, ΔH, values of -235.5, -217.1, and -190.3 kJ mol^{-1}, respectively. The ΔH is defined as H (Li^+ - solvent) – [H(solvent) + H (Li^+)]. The de-solvation step as a rate limiting step is clearly demonstrated in this interface. These results suggest that the higher reaction enthalpy energy of solvent or the higher solvation ability of the solvent with Li^+ leads to higher activation energy for de-solvation.

Li^+ charge transfer at the interface between $Li_4Ti_5O_{12}$ (LTO) and liquid electrolyte

The study of the Li^+ charge transfer at the lithium titanate, $Li_4Ti_5O_{12}$ (LTO), and liquid electrolyte interface was recently reported by Ishihara et al. [5]. The liquid electrolytes studied include 1 M $LiClO_4$ in PC, EC:DMC (1:1) and DMC, respectively. The charge transfer resistance was measured at 1.56 V vs. Li/Li^+, at which potential LTO has the highest redox capacity and the lowest charge transfer resistance. The E_as for the Li^+ charge transfer at the interfaces between LTO and the above electrolytes are 64, 54 and 37 kJ mol^{-1}, respectively. Similarly, the E_a values are consistent with the solvent reaction enthalpies of PC, EC and DMC with Li^+, which are -217.1, -210, and -190 kJ mol^{-1}, respectively.

In the above two examples, there are not traditional SEI layers present at the interfaces between the ceramic Li^+ conductor and the liquid electrolyte and between LTO and the liquid electrolytes. Yet, in both cases the activation energy correlates well with the solvation ability of the solvent to Li^+.

5. Li^+ transport in the sei as a rate limiting step

In Li-ion batteries, the SEI formed between the graphite anode and the electrolyte allows the repeated charge/discharge operation of the batteries achieving long cycle life with little loss of capacity. The charge and discharge rates achievable are increasingly limited with decreasing temperature. A study of low temperature impedance of Li-ion batteries indicates that the impedance of the cell is dominated by the charge transfer resistance, R_{ct} [22]. The R_{ct} includes mainly the resistance for the de-solvation step and the diffusion of Li^+ in SEI. The de-solvation (and solvation) step should occur at both the anode and the cathode before a Li^+ can intercalate into the electrodes. If the de-solvation process was the rate limiting step, it should be reasonable to expect the E_a at both the anode and cathode should be similar. If they are different, the roles of the SEI and CEI need to be considered. Three studies that are used as examples to clarify which step is the rate limiting step in Li-ion batteries are presented below.

Differentiating Li^+ charge transfer at graphite/electrolyte and $LiFePO_4$/electrolyte interfaces in a full cell

Jow et al. [7] studied the Li^+ charge transfer kinetics at both the graphite anode/electrolyte interface and the cathode/electrolyte interface at the same time using a full cell, graphite (Gr) as an anode and $LiFePO_4$ (LFP) as a cathode, with Li as a reference electrode in the cell containing an electrolyte consisting of 1 M $LiPF_6$ in EC-DMC-MB with VC as an additive. It was observed that the activation energy (E_a) at the graphite-electrolyte is about 67 kJ mol^{-1}, which is much higher than the 33 kJ mol^{-1} found at the LFP-electrolyte interface.

The large difference in E_a cannot be explained by the de-solvation process, since the electrolyte is the same in the cell. The difference could be attributed to the fact that the interphase formed on the graphite anode (SEI), resulting from the reduction of the electrolyte, is different from the interphase formed on the LFP electrode, or the cathode-electrolyte interphases (CEI), generally resulting from being in contact with organic-based electrolyte (i.e., such as decomposition products being formed at high potential or involving species formed at the anode) [23,24,25]. In the case of the LFP electrode, the surface film is expected to be much less significant compared to the anode, primarily due to the low operating potential which limits the onset of electrolyte oxidation [26]. It is possible that the different compositions of SEI and CEI on the electrodes can lead to different energies for the de-solvation process. It is reasonable to assume that, even if the solvation energy is affected by the different SEI and CEI interphases, the cause for large E_a difference should still be attributed to the different SEI and CEI associated with their respective electrodes. Therefore, it is concluded that the electrodes and their associated electrode-electrolyte interfacial layers, SEI or CEI, is controlling the Li^+ charge transfer kinetics.

Impact of additives on Li^+ charge transfer kinetics on both anode and cathode

Jones et al. [9] investigated the impact of electrolyte compositions on the charge discharge behavior of Li-ion cells consisting of $LiNi_{0.80}Co_{0.15}Al_{0.05}O_2$ (NCA) cathodes and graphite anodes with Tonen separator (polyethylene) in a baseline electrolyte of 1.0 M $LiPF_6$ in EC:EMC:MP (20:20:60 vol. ratio), where MP is methyl propionate. The test cells are in a spiral wound configuration with a capacity of about 105 mAh incorporating a Li reference electrode allowing the monitoring the charge transfer resistance at the anode and the cathode at the same time. The schematics of the cell construction are shown in Fig. 3. By introducing different additives, it is reasonable to expect that the nature of the SEI and CEI formed during formation resulting from the reduction on the anode and oxidation on the cathode with the electrolyte would vary from one additive to the other. However, the bulk electrolyte in the cells should remain relatively constant, since the additives are introduced in low concentration. The additives with their concentrations selected for this study include lithium bis(oxalato)borate (LiBOB) in 0.1 M concentration, vinylene carbonate (VC) in 2 wt %, 1,3-propanesultone (PS) in 2%, lithium difluoro(oxalato)borate (LiDFOB) in 0.1 M, and lithium bis(fluorosulfonyl)imide (LiFSI) in 0.1 M. The structures of these additives are shown in Fig. 4.

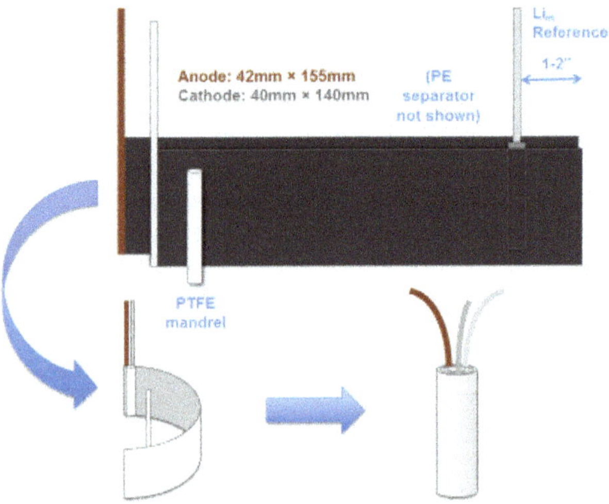

Figure 3. Schematics of a spirally wound full cell with a Li metal reference electrode placed between the anode and the cathode for monitoring the charge transfer resistance at both electrodes at the same time.

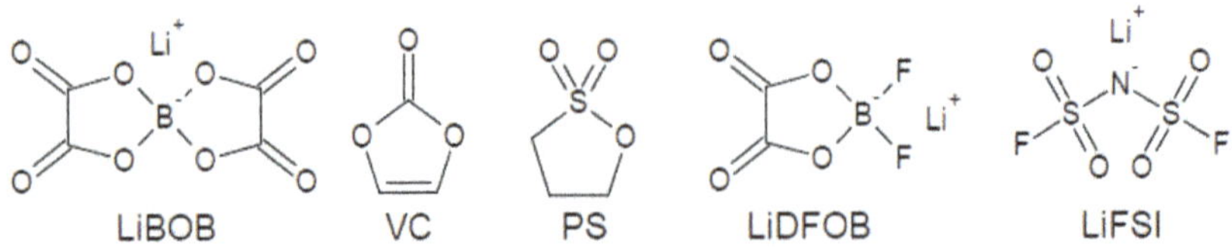

Figure 4. Additives studied in this example.

The impedance of the cells at the temperatures of 23, −20, −30, −40 °C are measured at the cell voltage set at 4.07 V, at which the potential of the graphite anode is of about 0.090 to 0.105 V vs. Li/Li^+ and the potential of the NCA cathode is about 4.179 V vs. Li/Li^+. The impedance, including that at the anode and that at the cathode, for the baseline electrolyte with and without these additives have been measured. The impedance of both the graphite anode and the NCA cathode at 23 °C are shown as Nyquist plots in Figures 5(a) and 5(b), respectively. The R_{ct}s at this temperature judging from the diameter of the semicircle are quite small at both the anode and cathode, with the values at the anode being about three times larger than that at the cathode. The straight lines after the semicircle shown Fig. 5 suggest that a diffusion limited process most likely occurring in the electrodes is limiting the rate performance of the cell.

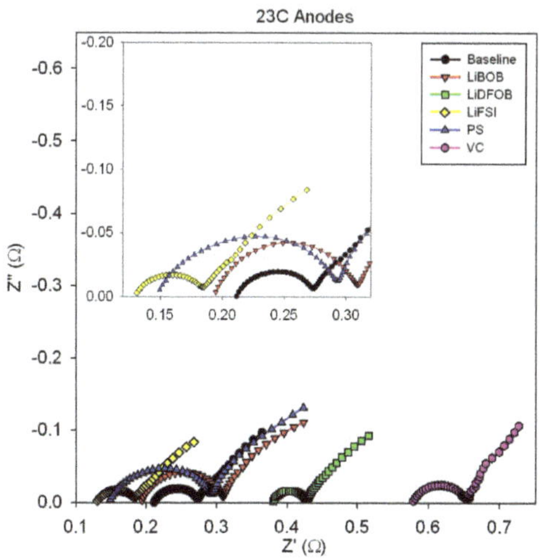

Figure 5(a) Nyquist plots of the impedance of the graphite anode in Gr/NCA cell in the electrolytes with and without various additives at 23 °C.

The impedance of both the graphite anode and the NCA cathode at −30 °C are shown as Nyquist plots in Figures 6(a) and 6(b), respectively. At this temperature, the cell resistance is dominated by the resistance at the graphite anode. Fig. 6(a) shows that the electrolytes with LiDFOB and LiFSI produce the SEI layers with resistance similar to the baseline electrolyte, while the other additives, VC, LiBOB, and PS, produce more resistive SEIs on the graphite anode. Fig. 6(b) shows that the electrolytes with VC,

LiBOB, LiDFOB, and PS additives produce less resistive CEIs than the baseline electrolyte on the NCA cathode.

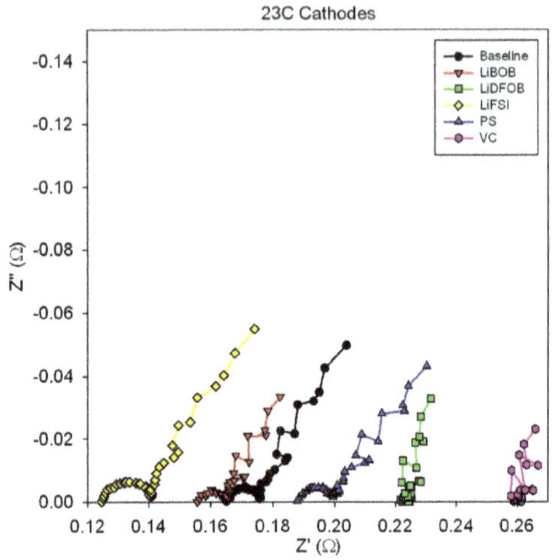

Figure 5(b) Nyquist plots of the impedance of the NCA cathode in Gr/NCA cell in the electrolytes with and without various additives at 23 °C.

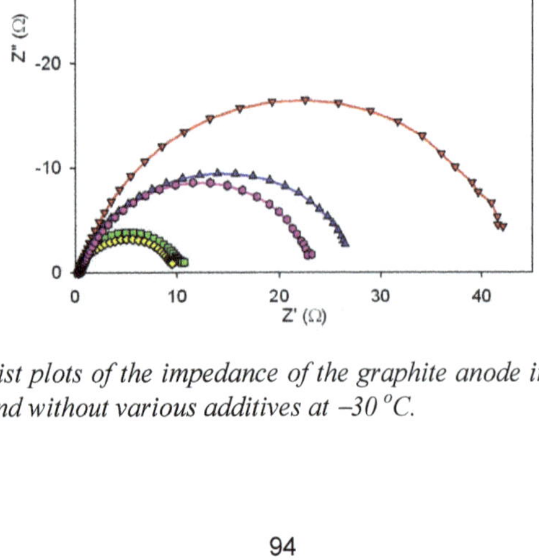

Figure 6(a) Nyquist plots of the impedance of the graphite anode in Gr/NCA cell in the electrolytes with and without various additives at –30 °C.

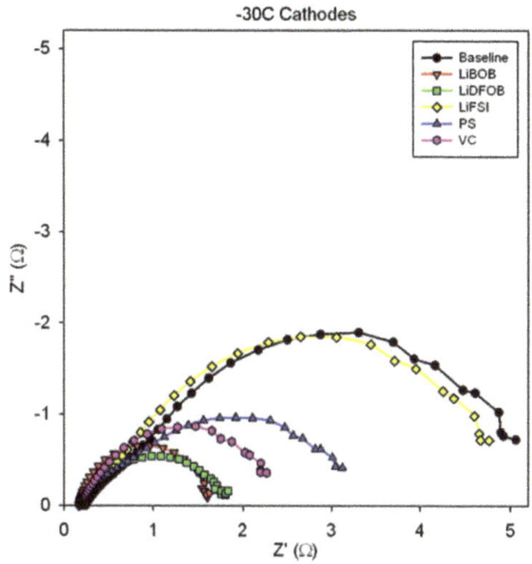

Figure 6(b) Nyquist plots of the impedance of the NCA cathode in Gr/NCA cell in the electrolytes with and without various additives at –30 °C.

The equivalent circuit as shown in Fig. 7 is used to analyze the charge transfer resistance, R_{ct}, which is equivalent to R2 as shown in the figure. The activation energy, E_a, of the Li^+ charge transfer process at the anode and at the cathode, can be determined from the slopes of the log ($1/R_{ct}$) vs. 1000/T plots as shown in Fig. 8(a) and 8(b). These plots all display a good linear relationship. The room temperature R_{ct} values at the cathode in the electrolytes with VC and LiDFOB are not shown, as the numbers are too small to be accurate. However, the estimated values would fall within the linear relationship with the values at lower temperatures. These E_a values at both graphite anode and NCA cathode are summarized in Table I.

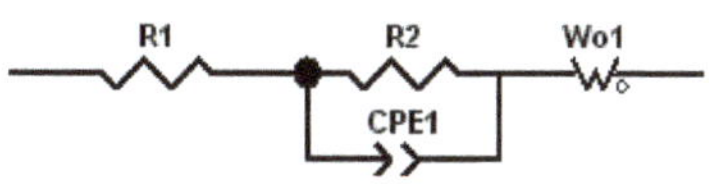

Figure 7. Equivalent circuit used to match the impedance spectra.

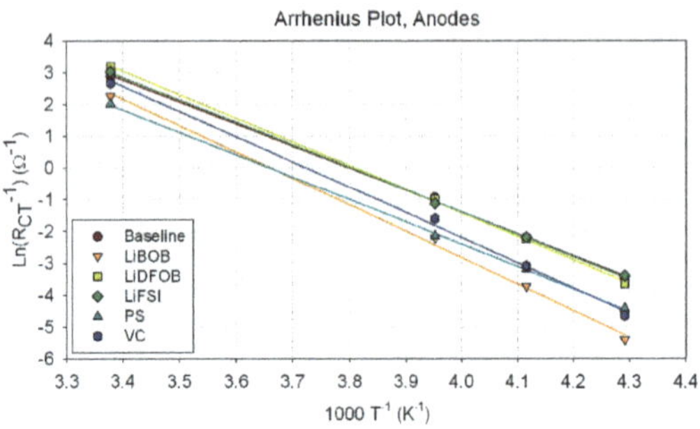

Figure 8(a) Log (1/R_{ct}) vs. 1000/T °K plots for the graphite anode for the Gr/NCA cells containing baseline electrolytes with and without additives.

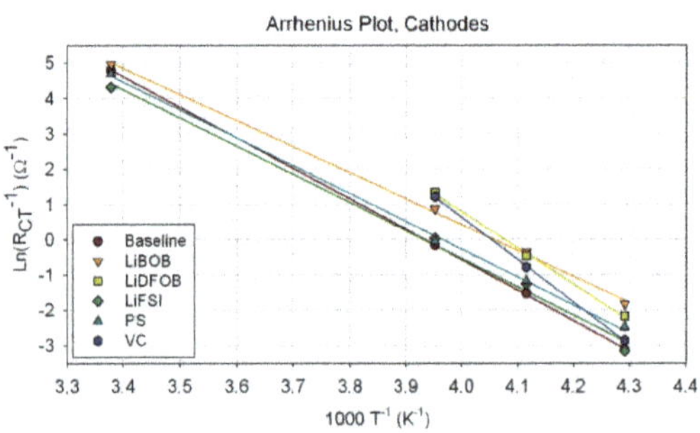

Figure 8(b) Log (1/R_{ct}) vs. 1000/T °K plots for the NCA cathode for the cells containing baseline electrolytes with and without additives.

Table 1 Activation energy of Li^+ charge transfer at the graphite anode/electrolyte interface and the NCA cathode/electrolyte interface

Electrolyte	E_a, kJ mol^{-1}	
	At Graphite Anode	At NCA Cathode
Baseline	57.7	71.8
+ 2% VC	66.1	100.0
+ 0.1 M LiBOB	69.1	61.2
+ 0.1 M LiDFOB	61.9	85.9
+ 2% PS	58.7	65.7
+ 0.1 M LiFSI	58.7	66.4

Baseline electrolyte: 1 M $LiPF_6$ in EC:EMC:MP (20:20:60 v/o)

The effects of LiFSI and LiBOB additives on R_{ct} and E_a are especially noteworthy. The addition of LiFSI in the baseline electrolyte results in lower R_{ct} at the graphite anode than that in the baseline electrolyte and an E_a similar to that of the baseline electrolyte. It also results in lower R_{ct} at the cathode than that in electrolyte without FSI and, at the same time, slightly lower E_a for Li^+ charge transfer at the cathode. The addition of LiBOB results in a lower R_{ct} and E_a on the NCA cathode than that in the baseline electrolyte. However, the presence of LiBOB results in higher R_{ct} and E_a on the graphite anode compared to the baseline electrolyte. Different additives lead to different reduction and oxidation reactions at the anode and the cathode and, therefore, different SEI and CEI, respectively [10]. The bulk electrolyte compositions remain about the same. Therefore, we conclude that the SEI is the rate limiting step in the Li-ion batteries.

Li^+ charge transfer between HOPG and 1 M $LiClO_4$ in DMC electrolyte

Yamada et al. [27] studied the Li^+ charge transfer kinetics at the interface between graphite electrodes and liquid electrolytes at 0.2 V vs. Li/Li^+. One interesting part of this study is the investigation of the Li^+ charge transfer between HOPG and 1 M $LiClO_4$ in DMC electrolyte across interfaces or SEIs that are pre-formed in different electrolytes. The E_as for two such preformed SEI are different and are summarized in the Table 2 below.

Table 2 E_a for Li^+ charge transfer at the interface between HOPG, which has different preformed SEI, and the electrolyte of 1 M $LiClO_4$ in DMC [27].

Electrode / SEI / Electrolyte Systems			E_a, kJ mol^{-1}
Electrode	SEI preformed in	Electrolyte	
HOPG at 0.2 V vs. Li/Li^+	1 M $LiClO_4$ in EC:DMC (1:1 v/o)	1 M $LiClO_4$ in DMC	40
	1 M $LiClO_4$ in EC:DMC (1:1 v/o) + 3 w/o PFS*		52

* PFS: Pentafluorostyrene, 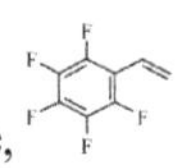

The SEI on HOPG varies with the electrolyte compositions including solvents, salts, and additives that undergo reduction reactions for forming the SEI. The cyclic voltammogram (CV) of HOPG in 1 M $LiClO_4$ in EC:DMC (1:1) + 3 w/o PFS indicates reduction reactions starting from about 2.0 V to 1.3 V vs. Li/Li^+ while the CV of HOPG in 1 M $LiClO_4$ in EC:DMC (1:1) shows no reduction reactions until the voltage decreases to less than 0.7 V vs. Li/Li^+ [27]. This suggests that a more robust SEI is formed from the reduction of PFS, which could contain Li fluorinated compounds, than that formed in the electrolyte without PFS. With the same electrolyte but different SEIs on HOPG, this demonstrates that SEI dominate the Li^+ charge transfer kinetics.

6. Discussion

At the interfaces between SS ceramic electrolyte, LLT or Ohara glass, and liquid electrolyte and between the LTO and liquid electrolyte interface, the E_a for Li^+ charge transfer correlates well with the ability of these solvents binding the Li^+ [6,5]. At the SS electrolyte and the liquid electrolyte interface, we can consider that the SS electrolyte is equivalent to a SEI as in Li-ion batteries. In this case, LLT or Ohara glass is not a rate limiting factor. Considering both LLT and Ohara glass are fast Li^+ conductors, with high Li^+ conductivity of around $1.0x10^{-3}$ S cm^{-1} and an activation energy of around 33 kJ mol^{-1} [19-21], it is reasonable that the rate limiting step is dominated by de-solvation in these systems. At the interface between the LTO and liquid electrolyte at 1.5 V vs. Li/Li^+, there is no electrolyte reduction reaction and therefore no SEI is formed on LTO at 1.5 V vs. Li/Li^+ in $LiClO_4$ in EC:EMC electrolyte [5]. Furthermore, LTO is a good Li^+ conductor near 1.5 V vs. Li/Li^+, where LTO shows the highest Li^+ redox reactions. Li^+ conduction in LTO has been studied and shown an E_a of around 27 kJ mol^{-1} using electrochemical methods [28] or varying between 0.30-0.48 eV for the spinel phase $Li_4Ti_5O_{12}$ and between 0.20-0.51 eV in the lithiated rock salt phase $Li_7Ti_5O_{12}$ using first principle calculations [29]. With no SEI and fast Li^+ conduction in LTO, the Li^+ de-solvation process is a rate limiting step at the LTO and electrolyte interface.

The studies of Li^+ charge transfer at Gr/electrolyte interface and at the LFP/electrolyte interface in a Gr / electrolyte / LFP full cell in the same electrolyte shows a large difference in E_a between the two interfaces [7]. This large difference is attributed to the difference in the nature of the SEI and CEI on the graphite anode and LFP cathode, respectively. It is worth noting that the E_a for Li^+ charge transfer across the LFP/electrolyte interface is 33 kJ mol^{-1}, which is very similar to that for the electrochemical $FePO_4$ to $LiFePO_4$ phase transformation, which is found to be about 30 kJ mol^{-1} [30-32]. This number is derived from using the Avrami-Johnson-Mehl-Eroofev equation [30-32]. A recent study by Johannes et al. [33] on hole polaron formation and migration in olivine phosphate materials using DFT method is also worth noting as a reference. It is found that the polaron migration barrier in LFP, which is closely linked to the Li^+ conduction kinetics, is 0.20 eV or 19.3 kJ mol^{-1}. This number is lower than the activation energy for the electrochemical $FePO_4$ to $LiFePO_4$ transformation. It could suggest that there is a CEI on LFP at the voltage less than 3.6 V, even if it is insignificant, that can affect the charge transfer kinetics.

The additives in the electrolyte impact not only anode, but also cathode, Li^+ charge transfer kinetics in Li-ion cells, such as Gr / electrolyte / NCA cell studies [9], where the reduction and oxidation reactions of the electrolytes varied with the selection of additives. This agrees with the Li^+ charge transfer kinetics studies of HOPG in 1 M $LiClO_4$ in DMC electrolyte, in which HOPG has a different preformed SEI. In summary, all components in the electrolyte such as salt, e.g. $LiClO_4$ vs. $LiPF_6$, solvent, e.g. carbonate vs. ester, additive, e.g. LiFSI vs. VC, can change the nature of SEI and CEI, and therefore influence the Li^+ charge transfer kinetics [10,34]. Table 3 summarizes the impact of salt, solvent and additive on the activation energy of Li^+ charge transfer at the graphite anode using limited examples for illustration only.

Table 3 Activation energy of Li^+ charge transfer at the graphite electrode with electrolytes with different salts, solvents and additives.

Electrode @ Potential vs. Li/Li^+	Electrolyte	E_a, kJ mol^{-1}
HOPG @ 0.2 V	1 M $LiClO_4$ in EC:DMC (1:1 v/o)	55[a]
Graphite @ 0.1 V	1 M $LiPF_6$ in EC:EMC (30:70 w/o)	68[b]
Graphite @ 0.1 V	1 M $LiPF_6$ in EC:EMC:MP	57
Graphite @ 0.1 V	1 M $LiPF_6$ in EC:EMC:MP[c] + VC (2 wt.%)	66
Graphite @ 0.1 V	1 M $LiPF_6$ in EC:EMC:MP[c] + LiFSI (0.1 M)	58

Note: a – Ref. 27, b – Ref. 35, c – EC:EMC:MP (20:20:60 v/o)

Recent computational studies of the de-solvation kinetics and the transport properties of the SEI are worth mentioning as references for the above experimental studies [36,37]. Molecular Dynamics simulations were performed to evaluate the de-solvation energy of an electrolyte consisting of 1 M $LiPF_6$ / EC:DMC (3:7 mole ratio) in contact with a SEI made of Li_2EDC, $(CH_2OCO_2^-/Li^+)_2$, or Li_2BDC, key SEI components [36]. The many-body polarizable APPLE&P force field was used in the simulations. E_a, the activation energies, for the Li^+ solvation–de-solvation reactions are 0.42–0.46 eV (40-44 kJ mol^{-1}) for the SEI–electrolyte system. E_a for the Li^+ diffusion in bulk Li_2EDC is 0.64 eV (62 kJ mol^{-1}) [37]. The activation energy for the lithium diffusion in the other common SEI compound Li_2CO_3 for the combined process of the formation the Li^+ defect and Li^+ hopping was estimated to be around 0.7 eV (67.5 kJ mol^{-1}) at low voltages close to lithium metal potential, indicating that many common SEI components have similar activation energy [38]. However, it should be noted that the SEI in Li-ion batteries is more complicated than that assumed in the simulation. The computational results in general agree with what have been observed.

7. Conclusion

As mentioned previously in the section of Li^+ CHARGE TRANSFER PROCESS [14,15,16] and a recent study on $LiNi_{1/3}Co_{1/3}Mn_{1/3}O_2$ by Kasnatscheew et al. [39], the lithium migration through the solid bulk active electrode materials may be rate limiting in some systems under certain conditions, especially at warmer temperatures. In the context of the above results and discussion, Li^+ charge transfer kinetics is mostly dominated by the SEI in Li-ion batteries. In the case when the SEI is very conductive and of low activation energy for Li^+ conduction, such as LLT and Ohara glass, the Li^+ charge transfer kinetics is dominated by the de-solvation process. In the case when there is no SEI or CEI, or insignificant SEI/CEI, such as with the LTO/electrolyte interface or LFP/electrolyte interface, the Li^+ charge transfer kinetics is dominated by de-solvation or by the Li transport in the electrode.

When the additive is introduced in the electrolyte, the redox chemistries on the electrodes are modified resulting in different SEI and CEI from the original. Using full cells, such as Gr / 1 M $LiPF_6$ in EC:EMC:MP / NCA with a Li reference electrode, allows one to examine the full impact of additives on the Li^+ charge transfer kinetics at both the anode and the cathode at the same time. In this work, we found that the LiBOB additive leads to a reduced R_{ct} and E_a at the NCA cathode, but results in much increased R_{ct} and E_a at the graphite anode. The LiFSI additive amongst those studied in this work provides the best overall benefits in terms of Li^+ charge transfer kinetic improvement.

Acknowledgement

Some work described here was carried out at the Jet Propulsion Laboratory, California Institute of Technology, under contract with the National Aeronautics and Space Administration (NASA) and supported by the NASA Planetary Sciences Division (PSD). The authors also wish to thank Oleg Borodin of the U.S. Army Research Laboratory for helpful discussion.

The work has been previously published by ECS:

T. Richard Jow, Samuel A. Delp, Jan L. Allen, John-Paul Jones, and Marshall C. Smart, Factors Limiting Li+ Charge Transfer Kinetics in Li-Ion Batteries, Journal of The Electrochemical Society, 165 (2) A361-A367 (2018). https://doi.org/10.1149/2.1221802jes

Referencs

[1] M. C. Smart, B. V. Ratnakumar, and S. Surampudi, J. Electrochem. Soc., 146 (2), 486 (1999). https://doi.org/10.1149/1.1391633

[2] S. Herreyre, O. Huchet, S. Barusseau, F. Peron, J. M. Bodet, Ph. Biensan, J. Power Sources, 97-98, 576 (2001). https://doi.org/10.1016/S0378-7753(01)00670-X

[3] K. Xu, J. Electrochem. Soc., 155 (10), A733 (2008). https://doi.org/10.1149/1.2961055

[4] T. Abe, H. Fukuda, Y. Iriyama, Z. Ogumi, J. Electrochem. Soc., 151 (8), A1120 (2004). https://doi.org/10.1149/1.1763141

[5] Y. Ishihara, K. Miyazaki, T. Fukutsuka, T. Abe, ECS Electrochemistry Lett., 3 (8), A83 (2014). https://doi.org/10.1149/2.0011408eel

[6] T. Abe, F. Sagane, M. Ohtsuka, Y. Iriyama, and Z. Ogumi, J. Electrochem. Soc., 152 (11), A2151 (2005). https://doi.org/10.1149/1.2042907

[7] T. R. Jow, M. B. Marx, J. L. Allen, J. Electrochem. Soc., 159 (5), A604 (2012). https://doi.org/10.1149/2.079205jes

[8] M. C. Smart, B. L. Lucht, S. Dalavi, F. C. Krause, and B. V. Ratnakumar, J. Electrochem. Soc., 159 (6), A739 (2012). https://doi.org/10.1149/2.058206jes

[9] J.-P. Jones, M. C. Smart, F. C. Krause, B. V. Ratnakumar, and E. J. Brandon, ECS Trans., 75, 1 (2017). https://doi.org/10.1149/07521.0001ecst

[10] S. A. Delp, O. Borodin, M. Olguin, C. G. Eisner, J. L. Allen, T. R. Jow, Electrochimica Acta, 209, 498 (2016). https://doi.org/10.1016/j.electacta.2016.05.100

[11] M. Ue, Y. Sasaki, Y. Tanaka, M. Morita, in Electrolytes for Lithium and Lithium-Ion Batteries, T. R. Jow, K. Xu, O. Borodin, M. Ue, Editors, p. 93, Springer Series: Modern Aspects of Electrochemistry, Vol. 58 (2014).

[12] M. S. Ding, J. Chem. Eng. Data, 48, 519 (2003). https://doi.org/10.1021/je020219o

[13] E. Peled, D. Golodnitsky, G. Ardel, J. Electrochem. Soc., 144, L208 (1997). https://doi.org/10.1149/1.1837858

[14] C. Wang, A. J. Appleby, F. E. Little, J. Electrochem. Soc., 149(6), A754 (2002). https://doi.org/10.1149/1.1474427

[15] M. C. Smart, B. V. Ratnakumar, K. B. Chin, and L. D. Whitcanack, J. Electrochem. Soc., 157 (12), A1361 (2010). https://doi.org/10.1149/1.3501236

[16] H. -P. Lin, D. Chua, M. Salomon, H. –C. Shiao, M. Hendrickson, E. Plichta, and S. Slane, Electrochem. Solid State Lett., 4 (6), A71 (2001). https://doi.org/10.1149/1.1368736

[17] Z. Ogumi, Electrochemistry, 78 (5), 319 (2010). https://doi.org/10.5796/electrochemistry.78.319

[18] M. E. Orazem and B. Tribollet, Electrochemical Impedance Spectroscopy, John Wiley & Sons, Inc., Hoboken, New Jersey (2008). https://doi.org/10.1002/9780470381588

[19] Y. Inaguma, L. Chen, M. Itoh, T. Nakamura, Solid State Ionics, 70/71, 196 (1994). https://doi.org/10.1016/0167-2738(94)90309-3

[20] Y. Inaguma, L. Chen, M. Itoh, T. Nakamura, T. Uchida, H. Ikuta, M. Wakihara, Solid State Communications, 86 (10), 689 (1993). https://doi.org/10.1016/0038-1098(93)90841-A

[21] J. Fu, Solid State Ionics, 96, 195 (1997). https://doi.org/10.1016/S0167-2738(97)00018-0

[22] S. S. Zhang, K. Xu, T. R. Jow, Electrochimica Acta, 49, 1057 (2004). https://doi.org/10.1016/j.electacta.2003.10.016

[23] K. Xu, Chem. Rev., 104, 4303−4417 (2004). https://doi.org/10.1021/cr030203g

[24] D. Aurbach, M. D. Levi, E. Levi, H. Teller, B. Markovsky, G. Salitra, U. Heider, L. Heider, J. Electrochem. Soc., 145, 3024 (1998). https://doi.org/10.1149/1.1838758

[25] J. Kasnatscheew, M. Evertz, B. Streipert, R. Wagner, S. Nowak, I. C. Laskovic, M. Winter, J. Phys. Chem. C, 121, 1521 (2017). https://doi.org/10.1021/acs.jpcc.6b11746

[26] J. Kasnatscheew, B. Streipert, S. Röser, R. Wagner, I. C. Laskovic and M. Winter, Phys. Chem. Chem. Phys., 19, 16078 (2017). https://doi.org/10.1039/C7CP03072J

[27] Y. Yamada, Y. Iriyama, T. Abe, and Z. Ogumi, Langmuir, 25, 12766 (2009). https://doi.org/10.1021/la901829v

[28] D. Wu, Ionics, 18 (6), 559 (2012). https://doi.org/10.1007/s11581-011-0656-x

[29] B. Ziebarth, M. Klinsman, T. Ecki, C. Elsasser, Physical Review B, 89 (17):174301 (2014). https://doi.org/10.1103/PhysRevB.89.174301

[30] J. L. Allen, T. R. Jow, J. Wolfenstine, Chem. Mater., 19 (8), 2108 (2007). https://doi.org/10.1021/cm062963o

[31] J. L. Allen, T. R. Jow, J. Wolfenstine, J. Solid State Electrochem., 12 (7-8), 1031 (2008). https://doi.org/10.1007/s10008-007-0459-1

[32] J. L. Allen, T. R. Jow, J. Wolfenstine, Chem. Mater., 24 (7), 1400 (2012). https://doi.org/10.1021/cm300743n

[33] M. D. Johannes, K. Hoang, J. L. Allen, and K. Gaskell, Phys. Rev. B, 85, 115106 (2012). https://doi.org/10.1103/PhysRevB.85.115106

[34] M. Gauthier, T. J. Carney, A. Grimaud, L. Giordano, N. Pour, H.-H. Chang, D. P. Fenning, S. F. Lux, O. Paschos, C. Bauer, F. Maglia, S. Lupart, P. Lamp, and S.-H. Yang, J. Phys. Chem. Lett. 6, 4653 (2015). https://doi.org/10.1021/acs.jpclett.5b01727

[35] K. Xu, Y. Lam, S. S. Zhang, T. R. Jow, T. B. Curtis, J. Phys. Chem. C, 111, 7411 (2007). https://doi.org/10.1021/jp068691u

[36] O. Borodin, D. Bedrov, J. Phys. Chem. C, 118, 18362 (2014). https://doi.org/10.1021/jp504598n

[37] O. Borodin, G. V. Zhuang, P. N. Ross, K. Xu, J. Phys. Chem. C, 117, 7433 (2013). https://doi.org/10.1021/jp4000494

[38] O. Borodin, in Electrolytes for Lithium and Lithium-Ion Batteries, T. R. Jow, K. Xu, O. Borodin, M. Ue, Editors, p. 371, Springer Series: Modern Aspects of Electrochemistry, Vol. 58 (2014).

[39] J. Kasnatscheew, U. Rodehorst, B. Streipert, S. Wiemers-Meyer, R. Jakelski, R. Wagner, I. C. Laskovic and M. Winter, J. Electrochem. Soc., 163, A2943 (2016). https://doi.org/10.1149/2.0461614jes

Chapter 4

In-situ and Quantitative Characterization of the Solid Electrolyte Interphase

Arthur v. Cresce[1], Selena M. Russell[1], David R. Baker[1], Karen J. Gaskell[2], Kang Xu[1]*

[1] Electrochemistry Branch, Power and Energy Division, Sensor and Electron Devices Directorate, U. S. Army Research Laboratory, Adelphi, MD 20783, USA

[2] Department of Chemistry and Biochemistry, University of Maryland, College Park, MD 20742, USA

* conrad.k.xu.civ@mail.mil

Abstract

Despite its importance in dictating electrochemical reversibility and cell chemistry kinetics, the solid electrolyte interphase (SEI) on graphitic anodes remains the least understood component in Li ion batteries due to its trace presence, delicate chemical nature, heterogeneity in morphology, elusive formation mechanism, and lack of reliable in-situ and quantitative tools to characterize it. This work summarizes our systematic approach to understand the SEI during its live-formation, in which in-situ electrochemical atomic force microscopy was used as both topographic imaging tool and nano-probe to provide quantitative information about the structure, hierarchy, and thickness of interphases as function of electrolyte compositions. Complemented by an ex-situ chemical analysis, a comprehensive and dynamic picture of interphase formation during the first lithiation cycle of the graphitic anode is described. This combined approach provides an in-situ and quantitative tool to conduct "quality control" over interphases formed.

Keywords

Interphase, SEI, In-situ Characterization, HOPG, Li Ion Battery, Electrolyte, AFM

The work has been previously published by ACS Publications: In Situ and Quantitative Characterization of Solid Electrolyte Interphases, Arthur v. Cresce, Selena M. Russell, David R. Baker, Karen J. Gaskell, and Kang Xu Nano Letters 2014 14 (3), 1405-1412. https://doi.org/10.1021/nl404471v

Contents

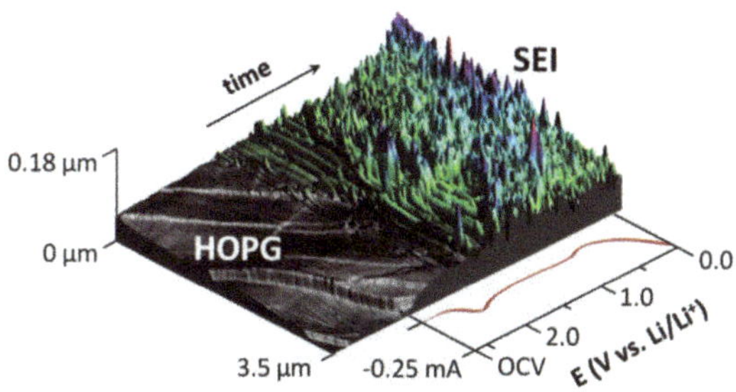

1. Introduction

In state-of-the-art lithium ion batteries, the graphitic anode typically operates at potentials (-3.03 V vs. SHE) where nearly no organic electrolyte component could remain thermodynamically stable against electrochemical reduction.[1] To enable reversible Li^+ intercalation/deintercalation with graphitic anodes, electrolyte solvents were chosen to maximize lithium salt solubility and Li^+ conductivity, but more importantly for their ability to stabilize the electrolyte-graphite interface with the formation of a new material subcomponent.[2] This new phase, consisting of sacrificial reduction products from the electrolyte solution,[3] prevents continuous electrolyte decomposition and is known as the solid electrolyte interphase (SEI) due to its electrolyte-like nature.[3] The physical and chemical properties of the SEI not only ensure the reversibility of Li^+ intercalation at the graphitic anode, but also strongly influence the kinetics of Li^+ transport across the electrolyte-electrode interface.[4] The former manifests as cycle/calendar life stability of Li ion batteries as well as resilience against chemical and thermal degradation; the latter dictates rate capability and power density.

Given the importance of the SEI, numerous efforts were dedicated to understanding its chemistry, physical properties, and formation process. Even today, many aspects of the interphase are still under debate, including the electrochemical potential at which it forms, whether it consists of a singular or multiple components from different mechanisms, or if these components are arranged in a homogenous manner or distributed hierarchically in a multi-layer structure.[5-7] Most such studies relied on ex-situ instrumentation, such as X-ray photoelectron spectroscopy (XPS), Fourier-transform infrared or Raman spectroscopy (FTIR, FT-Raman), and scanning or transmission electron microscopy (SEM, TEM), which provide rich information about interphasial chemistry and morphology, but inevitably introduce significant disturbances to the delicate SEI due to exposure to environment or high energy beams, as well as mechanical and chemical deteriorations through sample preparation and handling. On the other hand, in-situ investigations of SEI formation have been restricted by instrumentation, which are either too phenomenological in nature to reveal key characteristics, as in the case of electrochemical impedance spectroscopy, or are subject to excessive constraints in simulating an actual electrochemical cell, as can occur with ellipsometry and neutron reflectometry.

2. SEI formation and structure

In-situ analysis by atomic force microscopy (AFM) with electrochemical control on the target surface allows direct probing of SEI formation and structure while maintaining conditions that closely simulate real-life Li ion devices with minimal destructive impact. The quantitative nature of the force microscope, which is rare for most in-situ tools, further provides the capability to directly monitor the changes in mechanical strengths of the interphasial layers as functions of the electrochemical potentials applied or the electrolyte compositions/additives used, making it not only an in-situ imaging tool but also potentially a standard diagnostic methodology to evaluate the quality of SEI thus formed. This paper represents the first part of a systematic approach to use such a tool to understand the live-formation process of the SEI on a graphitic surface. In addition, ex-situ surface analysis by X-ray photoelectron spectroscopy (XPS) was used as complementary means to provide chemical insights.

Details of both the AFM and XPS experiments are summarized in the Supporting Information (SI) section. Highly oriented pyrolytic graphite (HOPG), used as a surrogate anode material in the work, served as a compromise between satisfying instrumentation requirement of a "flat" surface and closely simulating real graphitic anode surface used in Li ion batteries, thanks to the smooth and chemically homogeneous surface of HOPG that displays both graphite basal planes and edge sites. Instead of a typical carbonate electrolyte, pure ethylene carbonate (EC) was used as the single solvent in order to avoid the disturbance of AFM tip stability caused by solvent evaporation during the experiment. The absence of linear carbonates is not expected to significantly alter interphasial chemistry, as EC has been shown to be the main contributor to SEI formation.[5, 8] Lithium bis(trifluoromethanesulfonyl) imide (LiTFSI) was used in place of the industry-standard lithium hexafluorophosphate ($LiPF_6$) to circumvent the degradation of the AFM sample stage by HF, with the understanding that the $TFSI^-$ anion tends to participate in the SEI formation chemistry more than PF_6^- does, leaving its chemical signatures of sulfur and fluorine containing species in the interphase. LiTFSI was added at a high concentration (1.32M to 1.5M) to maintain the electrolyte in a liquid state at room temperature.

Figure 1 presents a typical live-formation of interphase on HOPG as its potential was scanned from open circuit voltage (OCV, ~ 2.5 V vs. Li) to lithiation potential (~0.0 V vs. Li) in control electrolyte (1.5 M LiTFSI/EC). During this first cathodic scan, the HOPG surface was continuously imaged with low contact force (F_c) to minimize disturbance to the growing SEI. The differentiated topographic image highlights variations in height with strong contrast. From OCV (V_1) to 2.1 V (V_2), the HOPG surface appears to be interphase-free, with both basal planes and edge sites clearly exposed. The surface morphology changes significantly negative of ~1.5 V, where pronounced accumulation of interphasial species occurs along edge sites, as highlighted by the yellow color spots that visibly outline the line of edge sites. This process likely corresponds to the initial intercalation of solvated Li^+ into the top graphene layers of HOPG and the subsequent reduction of EC molecules that coordinate with the intercalated Li^+,[9, 10] Around 1.0 V (V_3), surface species engulf the basal planes, and the extensive presence of surface species below 0.6 V obscures the view of both basal planes and edge sites.

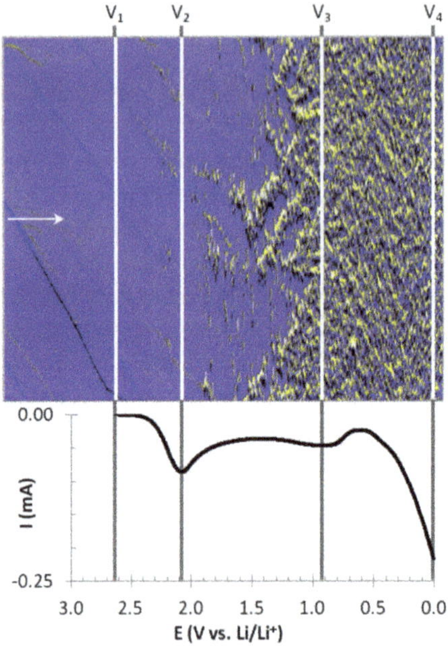

Fig. 1 Differentiated topographic AFM image of SEI live-formation on the HOPG surface (top) in 1.5 M EC/LiTFSI (control electrolyte) as a function of electrochemical potential (bottom) from OCV at V_1 to 0 V at V_4. Vertical lines correspond to key features described in the text. The arrow indicates the slow scan direction (rotated 90° clockwise from the vertical convention used in AFM images elsewhere in this paper). Image parameters: 3.5 × 4.5 μm, $F_{c,\ HOPG}$ = 1.2 nN, 1024 × 1024 pixels. The topographic image was differentiated with height variations highlighted in strong contrast, in which the leading edges and rough areas appear more yellow, while descending edges and smooth areas appear bluer.

The surface roughness increases by a factor of 5 ± 4 (N = 16) relative to the pristine HOPG surface between OCV and 0 V (V_4). In the accompanying voltammetry of Figure 1, the accumulation of these species correspond to two peaks at approximately 2.1 V and 0.95 V, respectively, labeled as V_2 and V_3 in Figure 1 respectively. While the V_2 process likely arises from reduction of dissolved O_2 contamination in the electrolyte,[11-16] V_3 correlates to typical reduction of carbonate species such as EC.[2] The implication of these observations is that the visible surface deposits, most likely the SEI precursors, appear on graphitic surfaces well before Li^+ intercalation occurs. Between 0.90 V (V_3) and 0.60 V, the decrease in cathodic current indicates that the surface might be electronically passivated, and the substantial increase of cathodic current negative of 0.60 V marks the start of HOPG lithiation, likely accompanied with further reduction of EC molecules or precursors of the SEI ingredients formed above 0.95 V. None of the peaks positive of 0.5 V were observed in subsequent voltammetry cycles of the HOPG anode (Figure S1), which is expected after the HOPG surface had been covered with a functional SEI.

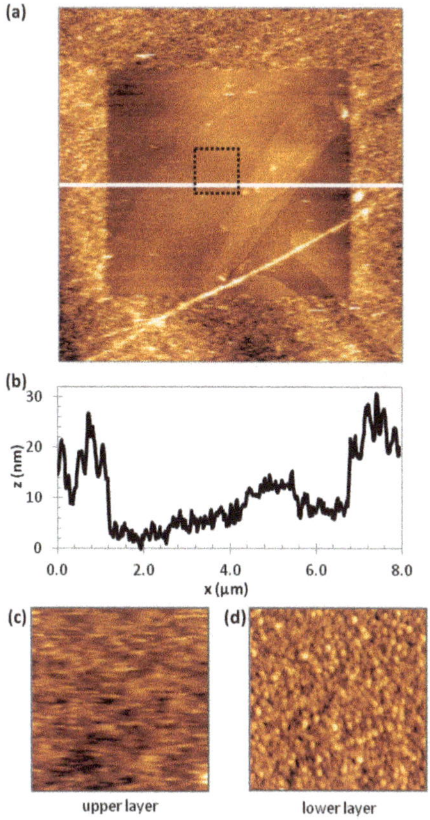

Fig. 2 (a) Topographic AFM image at 0 V in the control electrolyte, after repeated scanning of the inner, ca. 5 × 5 μm, region. (b) Line profile corresponding to the white line in (a) illustrating the height variation of the SEI. (c, d) Magnifications (black dotted square in a) showing the features of the upper (c) and lower (d) SEI layers. Image parameters: (a) 8.0 × 8.0 μm, (c-d) 1.0 × 1.0 μm, all F_c ~ 14 - 18 nN.

The interphase formed on HOPG appeared heterogeneously structured, with little correlation to the underlying HOPG features, as illustrated in Figures 1 and 2. Streaking during imaging indicated that the probe interacted with the surface species; hence even with minimal contact force the probe disturbed the uppermost SEI. The observation that the top-surface of the SEI was soft and interacted with the AFM probe agreed with previously published AFM analysis of the SEI. A heterogeneous multi-layer SEI structure has been proposed by previous studies, including the preliminary hypothesis made by Peled *et al.*[17] and observation by Lu *et al.* based on ex-situ TOF-SIMS analyses of SEI formed on a Cu substrate.[7] Figure 2c shows a magnification of the upper layer with streaking. Figure 2a presents the results of deliberate removal of the upper layer, in which an area (ca. 5 × 5 μm) was repeatedly scanned until no further morphology changes were observed.

The poorly adhered nature of this upper layer and its "sticky" interactions with the probe suggest that this part of SEI cannot be imaged in an unperturbed state with contact-mode AFM. The region surrounding the repeatedly scanned area in Figure 2a shows the nearly pristine interphase that was imaged one time, while Figures 2c-d compare the upper and lower layer morphologies. The lower layer consists of nanometer-sized, nodular features that decorate both basal planes and edges sites, and were termed nanotuffets hereafter. Thus, the SEI has an approximately bilayer structure; an upper, pliable, and poorly adhered phase, which is likely permeable to electrolyte, and a lower, hard, well-adhered phase near the HOPG surface.

The line profile in Figure 2a-b illustrates the bilayer structure of the SEI, and appears to measure an upper layer thickness of 3 to 20 nm. Reports from previous AFM experiments by Domi et al. also demonstrated the removal of soft material exposing a rough surface to evaluate SEI thickness.[18] Such line profiles from topographic AFM images as shown in Figure 2b constituted the basis for the thickness measurement of the interphase, which generated values ranging between 1 and 60 nm.[18-23] However, there are several uncertainties associated with this method as line profiles provide neither the absolute height nor the absolute depth of the SEI because the probe constantly interacts with the viscoelastic upper layer and cannot penetrate the hard lower layer.[16] In addition, the line profile in Figure 2b illustrates the varied height and roughness of both layers, increasing the uncertainty of thickness measurements derived from this method. On the other hand, SEI thickness values as estimated from ex-situ approaches were associated with their own uncertainties as well, which ranged from 10 nm for TEM,[5] 2 - 90 nm for XPS,[24-26] and up to 230 nm for secondary-ion mass spectrometry (SIMS) approaches derived from assumed sputtering rates.[26] Intrinsic inaccuracy occurred in all these methodologies due to the heterogeneous nature of SEI,[27] and it is very likely that the ex-situ nature and the corresponding process of sample preparation excluded some, or all, of the upper layer from the thickness measurement, either through collapse or loss of the delicate upper layer interphasial material through solvent removal or mechanical separation.

To address the representative issue of localized measurement by line profile, and as the first step in attempting to differentiate the contribution to SEI thickness from upper/soft and lower/hard layers, a new statistical-based approach was developed utilizing force-displacement spectroscopy (F-d). Figure 3 provides as an example the experimental F-d approaching curve recorded on an SEI in control electrolyte, in which repulsive interactions between the probe and the sample surface caused the cantilever to deflect near 0.25 μm. As the force increased (moving right to left in Figure 3) the probe pushed against the pliable upper layer, eventually penetrating it to reach the hard lower layer, indicated by the linear deflection below 0.12 μm. As noted earlier in the discussion of Figure 2, the probes used in this study cannot displace the lower layer SEI of nanotuffets, thus these particular F-d measurements only estimated the thickness of the upper layer section of SEI, instead of the thickness of the whole interphase. The upper layer thickness was found to range from 10 to 480 nm (102 ± 119 nm, N = 15), which was up to an order of magnitude greater than previously published values based on the line profile and spectroscopic methods discussed in the previous paragraph; nevertheless, we believe that the wide variance of these values truthfully reflect the heterogeneous nature of the interphase and its uneven distribution across the graphitic surface, thus is more reliable and representative than local or ex-situ measurements as provided by simple AFM line scans or XPS, TEM and SIMS.

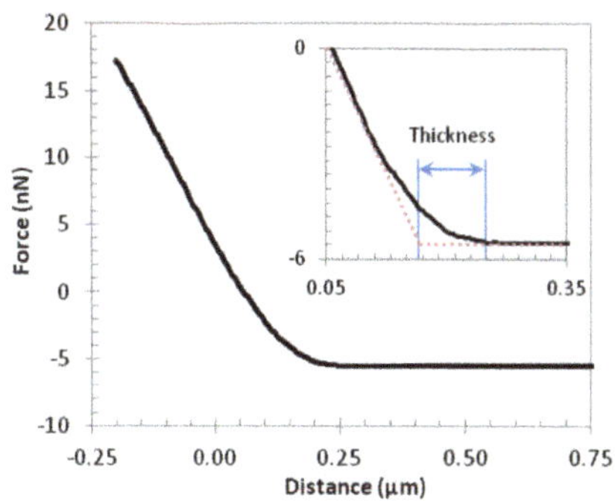

Fig. 3 Force-distance approaching curve of AFM tip toward an SEI formed in the control system. As comparison dotted lines show the tip behavior on an SEI-free HOPG in electrolyte at OCV, before SEI formation. While the curve only represents a general response when interphase is present, there is certain variability in each individual curve as the location varies. X and Y axes of the inset are μm and nN.

It is recognized that certain sacrificial electrolyte additives, even at trace concentrations, might significantly alter the interphasial chemistry on graphitic surfaces. Three compounds were known to reduce electrochemically before EC does, thus forming an SEI of particular chemical signature that improve battery performance by increasing first-cycle efficiency or reducing capacity fading during the subsequent cycles. Vinylene carbonate (VC)[28] and fluoroethylene carbonate (FEC)[29, 30] are such additives well described in the literature, while tris(hexafluoro-isopropyl)phosphate (HFiPP), developed originally to stabilize electrolytes at high voltage cathode surfaces ($LiNi_{0.5}Mn_{1.5}O_4$ or $LiCoPO_4$), has also been shown to form a protective SEI on graphite, even under severe exfoliating conditions using a propylene carbonate (PC)- based electrolyte.[31] Table S1 in the Supplementary section contains the structural and compositional details of the additives.

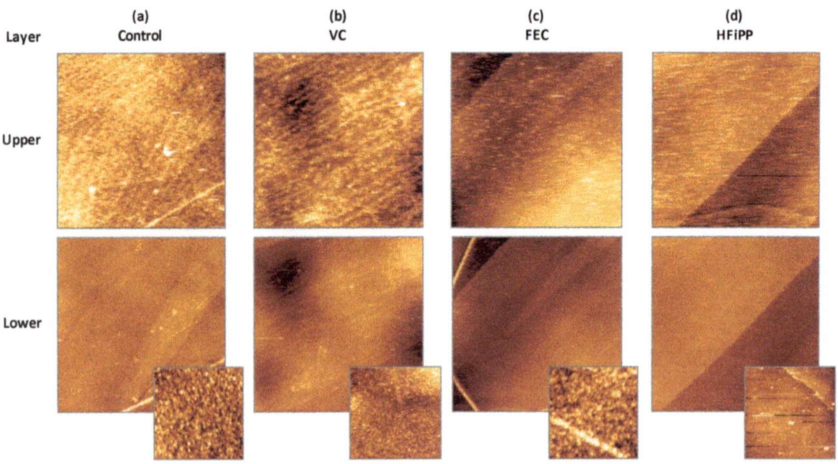

Fig. 4 Comparison of topographic AFM images of the upper and lower layers of the interphase formed on HOPG in electrolytes with or without additives at 0 V. Top: first image of a given region before removing the upper layer, and bottom: image of the same region after multiple scans until the lower layer was exposed and no further topography changes were observed. Insets: Magnifications of the lower layers. Image parameters: 5.0 × 5.0 μm, insets 1.0 × 1.0 μm, all F_c ~ 14 to 18 nN, with excursions up to 30 nN.

Column (a) in Figure 4 shows the first and the last scan of a 5 × 5 μm region of SEI formed in the control electrolyte, without additives, which is the same region as in Figure

2. While the upper layer initially appeared rough with several larger/taller formations (top), the lower layer looks smoother but still rougher than pristine HOPG. The additive effect cannot be easily differentiated from the control electrolyte based on topography alone as shown by Figure 4, in which each surface displays a diverse and heterogeneous soft upper layer and a hard lower layer comprised of nanotuffets; however, substantial additive effects can be revealed by F-d measurements in the thicknesses of the upper layer interphase. All additives tested led to upper layer SEIs that are an order of magnitude thinner than that of the control, implying that the presence of these additives participates in the formation of SEI and significantly interferes with the EC-reduction chemistry. The thickness values from all experiments are plotted in Figure 5, which showed the upper layer thickness of (102 ± 119, N = 15) nm for control electrolyte, and an average thickness between 9 and 17 nm for electrolytes containing additives, with narrower standard deviations and smaller maximum thicknesses. In particular, FEC generated a conspicuously thinner upper layer, ranging from 2 to 24 nm (9 ± 4 nm, N = 22), while VC and HFiPP generated comparable upper layer SEI heights of 17 ± 8 nm (N = 10) and 16 ± 7 nm (N = 7), respectively. Table S2 details the information presented in Figure 5.

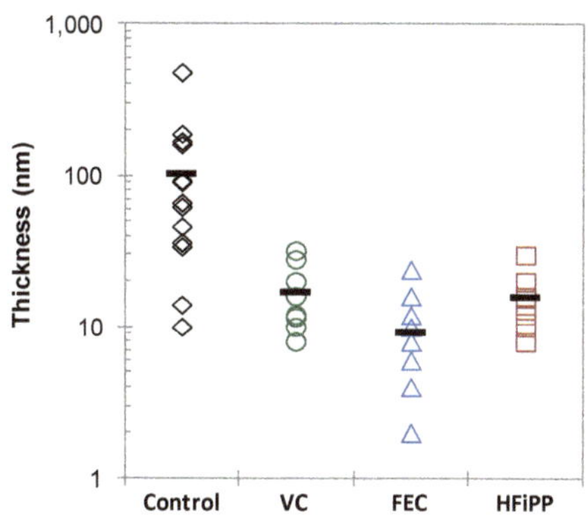

Fig. 5 SEI upper layer thickness derived from force-distance curves on HOPG with different additives at 0 V.

Further analysis of images like those presented in Figure 4 reveals a slight morphological difference between the electrolyte systems in terms of the root mean square surface roughness (R_{rms}) and applied potential. Compared to the typical HOPG surface, R_{rms} increases by a factor of 8 ± 6 in the control system, but only by 4 ± 3 with additives between OCV and 0.8 V. With or without additives, with the exception of FEC, and between 0.8 and 0 V, the upper layer R_{rms} does not significantly change and the lower layer R_{rms} increases by a factor of 3 ± 1. The change is much larger with FEC; R_{rms} increases by a factor of 8 ± 8 in the same voltage range. The variability in the data originates from the highly variable nature of the SEI. Detailed roughness data are presented in the SI. Among these additives FEC is the outlier, as it was observed that the electrolyte with FEC additive not only generated a thinner upper layer interphase, its lower layer was also rougher than those generated in the other electrolyte systems.

Because the lower SEI layer was impenetrable by the AFM probe, the quantitative information about the thickness or mechanical properties of the hard lower SEI layer cannot be obtained from the above force-distance based technique. However, even without explicit thickness data, insight on this key SEI substructure can still be extracted by comparing the lower layers generated in control and electrolytes containing additives. Compared to electrolytes with additives present, the SEI originating from the control electrolyte grows much thicker. Since the presence of additive in electrolyte systems led to much thinner upper layers, it is reasonable to assume that the hard lower layer must have been formed at a slower rate during the reductive decomposition of EC in the control electrolyte. These additives, FEC in particular, may very likely accelerated the formation of an insulating lower layer before a thick (> 30 nm) upper layer could accumulate through reduction by electrons from the anode. The SEI-limiting effect of additives correlates well with the increased first cycle Columbic efficiency frequently reported with the use of additive in Li ion battery electrolytes. This inference suggests that the lower layer of the visualized interphase is responsible for the electrically insulating nature of the SEI and has a direct effect on the overall thickness of the SEI. The chemistry, morphology, thickness, and other mechanical properties of the lower layer should be the focus of study in the future efforts.

Since AFM cannot provide chemical information, XPS was employed as complementary tool to gain chemical insight into the observed SEI structures. SEI samples of each electrolyte origin were grown on the same HOPG surfaces as the AFM portion of the study. Progressive rinsing with dimethyl carbonate (DMC) was used to remove the upper layer SEI material, allowing an XPS comparison of unrinsed SEI with an SEI that had been stripped of its softer upper layer. DMC has been known to effectively remove residual electrolyte and polymer-like interphasial species from electrodes, but inorganic species, such as Li_2CO_3 or LiF, are negligibly soluble in it and should remain on the rinsed electrode surface. The solubility of semi-carbonate species, in this case dilithium ethylene dicarbonate (LEDC) as the reduction product of EC via a single-electron pathway,[32, 33] lies somewhere between polymers and inorganic Li_2CO_3 or LiF.[2, 34-36] In this way, unrinsed XPS samples corresponded to an SEI with an intact upper layer, while the rinsed XPS samples represented SEIs where the upper layer had been partially or completely removed.

Figure 6 shows the C_{1s} spectra of HOPG surfaces cycled in various electrolytes and in different rinsed states. For the unrinsed interphase formed in control electrolyte, 4 peaks were found in the C_{1s} spectrum (Figure 6a). The peak at ~293 eV was generated by CF_3 groups from LiTFSI from residual electrolyte on the surface, a characteristic pair of peaks at ~290.2 and ~286.8 eV arose from carbonyl (CO_3) and ether (CO) species, respectively. The roughly 1:1 abundance ratio of CO_3:CO listed in Table S4 agrees well with the relative abundance of carbonyl and ethereal carbons in LEDC.

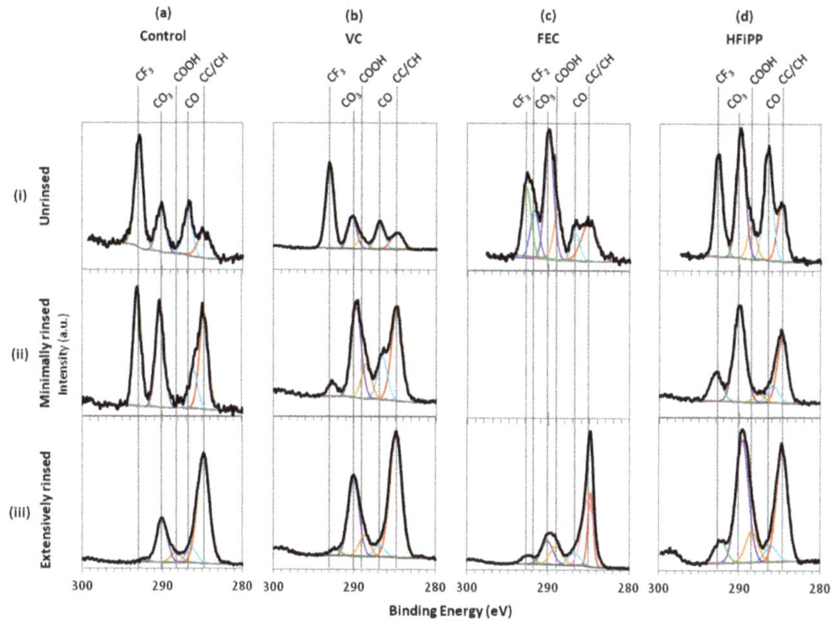

Fig. 6 C_{1s} spectra of different HOPG anodes after SEI formation from different electrolyte additives without rinsing (top), after minimal rinsing (middle), and extensive rinsing (bottom) with DMC.

After partial rinsing of samples identically cycled for the purposes of comparison, the same 4 peaks in the C_{1s} spectra were observed, although with different relative intensities. The change in observed intensities is consistent with the thinning of upper interphase layer, along with the increased intensity of carbon (~284.8 eV), and inorganic species. The CO_3:CO abundance ratio of the control SEI rose to 2.9:1. The increase in carbonate was assigned to the appearance of Li_2CO_3 close to the anode surface with the possibility of residual semi-carbonate LEDC. Unfortunately XPS was unable to differentiate the carbonate contributions from LEDC or Li_2CO_3. In spite of this spectroscopic limitation, it was concluded with confidence that as the CO3:CO ratio increased above 1:1, the contribution from organic species as indicated by CO must have decreased. The increasing CO_3:CO ratio was accompanied by a decreased presence of LiTSFI, also indicated by a CF_3 species at 293 eV (Figure 6a,ii). Extensive rinsing resulted in an lower layer as shown in Figure 2d and the XPS shown in Figure 6a-iii. The absence of CF_3 (LiTSFI) and decimated presence of CO (286.8 eV) suggested that lower layer mainly consisted of inorganic species such as Li_2CO_3.

Upon examining both Figure 6 and Table S4, it can be seen that the abundance ratios between CO_3 and CO for the unrinsed surfaces generated in the presence of VC and HFIPP additives, respectively, were also near 1:1, indicating an upper layer consisting mainly of LEDC as well. However, the surface generated in the presence of FEC significantly differed from the others with a much larger CO_3:CO ratio, along with additional peaks arising from carboxylate and CF_2 species. Partial rinsing removed significant amounts of the upper layer from all surfaces, as evidenced by the significant reduction in CF_3 peak (293 eV) and the corresponding increase of the carbon signal at 284.8 eV relative to the CO_3 and CO species (Figure 6ii).

After thorough rinsing, all surfaces generated in the presence of additives were deprived of their organic upper layer, similar to the surface generated in control electrolyte, with comparable abundance of CO_3 species at 290 eV and carbon at 284 eV as shown in their C_{1s} spectra (cf. Figure 6iii, a and c-d). Again, the interphase generated in the presence FEC differed in that the abundance of CO_3 species was relatively low. The appearance of fluorinated species in the lower layer along with the closer ratio between CO and CO_3 species suggests that FEC is more involved in forming the lower layer interphase than are other additives. This conspicuous behavior of FEC is consistent with topography and F-d measurement results shown in Figure 5.

In addition to the C_{1s} spectra, shells of other nuclei provide supplemental information about the chemical nature of both the upper and lower layer of the interphase. F_{1s} spectra (Figure S3) revealed that CF_3 species only exist at the surface and vanish upon rinsing, indicating that the $TFSI^-$ anion may remain intact in the upper layer, probably existing as residual salt. DMC rinses of identical samples exposed increasing abundance of F species, demonstrating that LiF could co-exist with Li_2CO_3 as a lower layer component. In the lower layer of the SEI, exposed by thorough rinsing, both sulfonyl and sulfide (S^{2-}) species were detected in S_{2p} spectra (Figure S4), indicating that $TFSI^-$ clearly underwent reductive decomposition. O_{1s} spectra (Figure S5), on the other hand, confirmed the major presence of Li_2CO_3 in lower layer with a peak at ~532.0 eV. The presence of Li_2O (528.5 eV) species was only identified at the upper layer surface of the control sample, but progressive rinsing with DMC revealed increasing amount of Li_2O in the lower layer, further supporting the inorganic nature of the SEI section closer to the electrode. Based on the intensity of the CO_3, S, O and F species detected in corresponding spectra, the relative concentrations of each inorganic species in the lower layer was calculated and are summarized in Table S5.

With the combination of in-situ AFM and ex-situ XPS, a comprehensive and dynamic picture of the SEI formation on graphitic surfaces has been depicted. While interphasial species start to accumulate preferentially along graphite edges at ~ 1.5 V, the eventual SEI actually consists of two parts, i.e. a soft/polymeric upper layer and a hard/salt-like lower layer, covering both edges and basal planes. A new AFM force-displacement method was established to accurately estimate the upper layer thickness, which revealed that this upper layer could be up to 480 nm thick, and would be significantly reduced by the use of electrolyte additives. Complementary chemical analysis of the two layers confirmed the organic nature of the upper layer SEI and corroborated that the lower, hard SEI layer was indeed populated by inorganic Li-containing species. This combined diagnostic approach could potentially serve as a universal quantitative tool to evaluate the quality of the interphases formed on battery electrode surfaces. Further efforts should expand the use of the in-situ AFM technique to the study of the lower interphase layer, and the effect of other electrolyte/salt systems and additives on this key section of the SEI, so that a more detailed picture of how electrodes in Li ion batteries interface with the electrolytes could be provided.

3. Associated content

Supporting Information Available: Electrolyte composition and structure of additives used; instrumentation and methods; cyclic voltammetry, thickness and roughness data, and XPS F_{1s}, S_{2p} and O_{1s} spectra. This material is available free of charge via the Internet at http://pubs.acs.org

3.1 Supporting information: in-situ and quantitative characterization of solid

3.1.1 Environment

All AFM and electrochemical experiments were performed in a dry room environment with a measured dew point of -80 °C or lower (corresponding to a moisture level < 5 ppm) and at room temperature (20 ± 1) °C.

3.1.2 Electrolytes

The electrolytes used in this study are listed in Table S1. Ethylene carbonate (EC; Ferro, Mayfield Heights, OH) and lithium bis(trifluoromethane sulfonyl)imide (LiTFSI) salt (HQ115; 3M, St. Paul, MN) were stored in an argon-filled glove box and used as received. Vinylene carbonate (VC; TCI America, Portland, OR) was made inhibitor-free by distillation in the dry room and was stored in a -20 °C freezer to prevent polymerization. Fluoroethylene carbonate (FEC; Solvay Fluor, Belgium) was dried, distilled, and stored in the dry room. Tris(hexafluoroisopropyl)phosphate (HFiPP) was prepared according to literature procedures[31], purified twice by sublimation, and stored in the dry room.

Table S1 Electrolytes

Electrolyte or additive	structure	Composition AFM and XPS experiments
Control	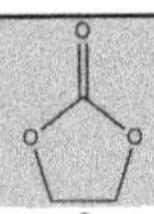	EC + 1.32 to 1.5 M LiTFSI
VC	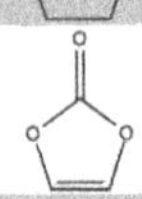	EC + 1.32M LiTFSI + 160mM VC (1 vol%)
FEC	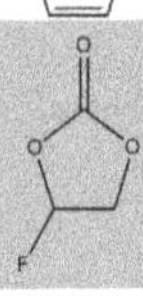	EC + 1.32M LiTFSI + 700mM FEC (5 vol%)
HFiPP	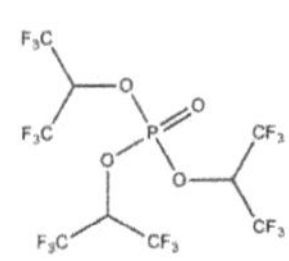	EC + 1.32M LiTFSI + 5mM HFiPP(0.3 vol%)

3.1.3 Cyclic voltammetry

Figure S1 shows an example CV of the control electrolyte on HOPG in the AFM liquid cell. The features in the first cathodic cycle near 2 and 1.5 V disappear in subsequent cycling.

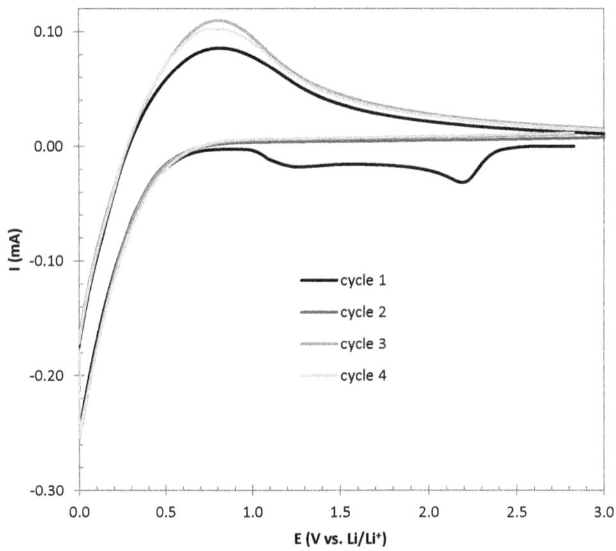

Fig. S1 Cyclic voltammogram of the HOPG basal plane in the control electrolyte (1.32 M LiTFSI dissolved in EC) at room temperature with a sweep rate of 5 mV/s.

3.1.4 Atomic force microscopy

AFM was performed with a liquid immersion sample stage equipped with electrochemical connectors for working, counter, and reference electrodes (5500 AFM/SPM; Agilent Technologies, Chandler, AZ). AFM topography and force-distance (F-d) curves were collected in contact mode using pyramidal Si_3N_4 (DNP-10; Bruker, Camarillo, CA) probes with a nominal tip radius of 20 nm on a triangular cantilever with a nominal spring constant of 0.12 N/m. The spring constant was calibrated via the thermal method (Thermal-K; Agilent Technologies, Chandler, AZ) for each probe in this study. Contact force (F_c) ranged from approximately 14 to 18 nN, with excursions up to 30 nN. Typical 5×5 μm^2 images were obtained at a scan rate of 1.0 line/s and a resolution of 512 points/line. The AFM fast scan direction was horizontal in the frame unless otherwise specified. Basal plane HOPG wafers (SPI-2 grade, SPI Supplies, West Chester, PA) were mechanically exfoliated with tape at least 15 times prior to each experiment to obtain a clean graphite surface. Approximately 188 mm^2 of the HOPG surface was exposed to the electrolyte and served as the working electrode in the AFM in-situ electrochemical cell. An approximately 12 mm × 4 mm strip of lithium foil (FMC) was used as the counter/reference electrode and was partially submerged in the electrolyte. Electrochemical measurements and control were managed by a Solartron 1278 potentiostat/galvanostat with a CorrWare control interface (Solartron Analytical, Hampshire, UK). Potentiostatic ramp rates were 5mV/s.

Force-distance (F-d) measurements were conducted with a vertical probe speed of 0.3 µm/s, a resolution of 1000 points, and 16 points per analyzed region. F-d curves were always collected in areas where no topographical scanning had occurred, ensuring that F-d data reflected the thickness of the SEI absent the removal of SEI material by the probe during imaging. F-d data analysis was performed using an in-house MATLAB software (MathWorks, Natick, MA) routine to estimate the position of the hard lower SEI layer relative to the initial probe-sample interaction on approach to determine the SEI thickness. For each F-d curve, the routine linearly fit the free approach and contact regions, the intersection of these two lines gave the estimated location of the hard subsurface SEI layer. A schematic of this process is presented in Figure 3 and the results are summarized in Table S2.

Table S2 Upper SEI layer thickness measured after lithiation, at 0 V.

(nm)	Control	VC	FEC	HFiPP
average	102	17	9	16
std. dev.	119	8	4	7
max.	480	32	24	30
min.	10	8	2	8
N	15	10	22	7

The root mean square surface roughness (R_{rms}) was calculated in WSxM software (Nanotec Electronica, Madrid, Spain) for a variety of 1.0×1.0 μm^2 regions imaged during AFM experiments. Below 1 V, the surface roughness increases by a factor of 5 ± 4 (upper layer, N=16) relative to the pristine HOPG surface. Between 0.8 and 0 V, the upper layer R_{rms} does not significantly change, with the exception of FEC. In the same voltage range the lower layer R_{rms} increases by a factor of 3 ± 1 for Control, VC, and HFiPP. The change is much larger with FEC, with the R_{rms} value increasing by a factor of 8 ± 8. FEC is the outlier; the lower layer R_{rms} increases significantly between 0.8 and 0 V, unlike the other electrolyte systems.

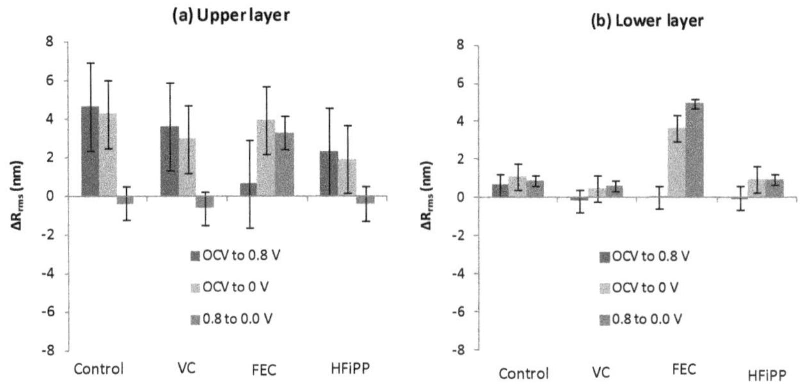

Fig. S2 Change in the root mean square surface roughness (R_{rms}) of 1.0×1.0 μm^2 regions recorded with AFM before, during, and after the 1st cathodic cycle.

Table S3 SEI R_{rms} of 1.0×1.0 μm^2 regions during and after the 1st cathodic cycle corresponding to data in Figure S2.

Layer	V vs. Li/Li$^+$	ΔR_{rms} (nm)					
		Control	VC	FEC	HFiPP	All	All except FEC
Upper	OCV to 0.8 V	5 ± 4	4 ± 1	0.6 ± 1.3	2 ± 2	3 ± 2	3 ± 2
	OCV to 0 V	4 ± 2	3.0 ± 0.9	4 ± 5	2 ± 2	3 ± 2	3 ± 2
	0.8 to 0.0 V	-0.4 ± 1.2	-0.6 ± 0.9	3 ± 4	-0.4 ± 0.9	0.3 ± 2.2	-0.4 ± 0.9
	N	3	4	4	6	16	13
Lower	OCV to 0.8 V	0.7 ± 1	-0.2 ± 0.4	0.02 ± 0.13	-0.006 ± 0.183	0.1 ± 0.5	0.1 ± 0.6
	OCV to 0 V	1 ± 1	0.5 ± 0.6	4 ± 4	1.0 ± 0.4	2 ± 2	0.9 ± 0.7
	0.8 to 0.0 V	0.90 ± 0.07	0.6 ± 0.2	5 ± 5	1.0 ± 0.4	2 ± 3	0.8 ± 0.3
	N	2, 3, 2	3	2, 3, 2	2, 4, 2	9, 13, 9	7, 10, 7
Upper to lower	0.8 V	-6 ± 1	-4 ± 2	-1.3 ± 0.5	-2 ± 1	-3 ± 2	-4 ± 2
	0 V	-3 ± 1	-3 ± 2	-0.3 ± 0.3	-0.5 ± 1.1	-2 ± 2	-2 ± 2
	N	2, 3	3	2, 3	2, 4	9, 13	13, 7

3.1.5 X-ray photoelectron spectroscopy

HOPG anode samples for XPS analysis were prepared in the AFM in-situ electrochemical cell with the appropriate electrolyte. After the first cathodic cycle, the samples were removed from the cell. Some were rinsed with DMC as needed. All were dried overnight in the dry room. Samples were transported in double heat-sealed-pouches. In an Ar glove box, the samples were removed from the pouches, mounted, and placed in sealed bags. Samples were introduced to the XPS chamber under inert conditions in an argon-filled glove bag.

X-ray photoelectron spectra and analysis was collected while operating in hybrid mode (combination of magnetic and electrostatic lenses) with either monochromatic Al Kα (1486.7 eV; not rinsed and partially rinsed samples) and non-monochromated Mg Kα (1253.6 eV; rinsed 2 samples) radiation (Axis 165 X-Ray photoelectron spectrometer, Kratos Analytical, Manchester, UK). The working pressure of the instrument was 5×10^{-8} Torr (6.6×10^{-6} Pa) or better throughout data collection. The analysis area as defined by the analyzer was approximately 700 μm by 300 μm. Charge neutralization was required to minimize surface charge buildup. All samples were calibrated to the graphitic peak at 284.6 eV. Peak fitting was done using CasaXPS software, after removal of a Shirley background and using peaks with a 70% Gaussian/ 30% Lorentzian product function. For the C 1s peak fit all peaks were fixed to have equal FWHM (full width at half maximum) with the exception of the graphitic/hydrocarbon peak which was not constrained.

For the depth profile expereiments, spectra were recorded of the sample surface before sputtering and between sputtering cycles. During Ar^+ sputtering, sample current density and Ar pressure were constant. A 1 mm ion beam spot was quickly rastered over a 5×5 mm^2 area.

Sputtering experiments were also conducted to probe how upper- transforms into lower layer. It was revealed that there is no clear demarcation between the two phases. Instead, a gradual transition of chemical compositions occurs, indicating that the two layers mingled in a middle section, before the soft layer is completely replaced by nanotuffets. As mentioned earlier, sputtering of heterogeneous materials gives ambiguous depth estimations, so no attempt has been made to quantify layer thickness based on sputtering rate. In addition, the thickness of the salt-like lower layer cannot be estimated from the attenuation of the graphite signal as the films are not smooth or continuous, but rather nanotuffeted with varying roughness and an unclear relationship to the substrate (the heterogeneities would attenuate the signal differently).

Table S4 Relative atomic composition from XPS.

		%					ratio	
		C	O	Li	F	S	CO_3 / CO	CF / F^-
Control	not rinsed	20.7	30.7	11.0	26.8	10.7	1.0	41.0
	rinsed 1	22.9	30.4	18.2	21.8	6.7	2.7	8.8
	rinsed 2	31.9	30.9	29.1	5.2	3.0	2.9	0.74
VC	not rinsed	20.8	28.6	11.7	29.5	9.3	1.2	7.9
	rinsed 1	28.6	35.4	23.9	9.3	2.8	2.1	1.2
	rinsed 2	31.8	35.2	25.3	5.1	2.5	6.8	1.3
FEC	not rinsed	20.4	36.4	16.9	19.4	3.6	3.1	34.5
	rinsed 1	...	...	...	...	...	...	...
	rinsed 2	35.7	30.3	20.2	10.6	3.2	1.8	1.8
HFiPP	not rinsed	36.0	36.0	15.8	17.3	3.3	1.1	7.2
	rinsed 1	24.2	35.5	24.3	12.5	3.6	5.6	6.0
	rinsed 2	24.1	37.8	26.5	8.6	3.0	7.6	0.42

Table S5 Relative composition of the inorganic species in the lower SEI layer from XPS.

	%				ratio
	Li_2CO_3	LiF	Li_2S	Li_2O	Li_2CO_3 / LiF
Control	60.1	13.3	12.9	13.7	4.5
VC	82.1	9.8	8.0	0	8.4
FEC	65.2	25.2	9.7	0	2.6
HFiPP	88.0	6.2	5.8	0	14.2

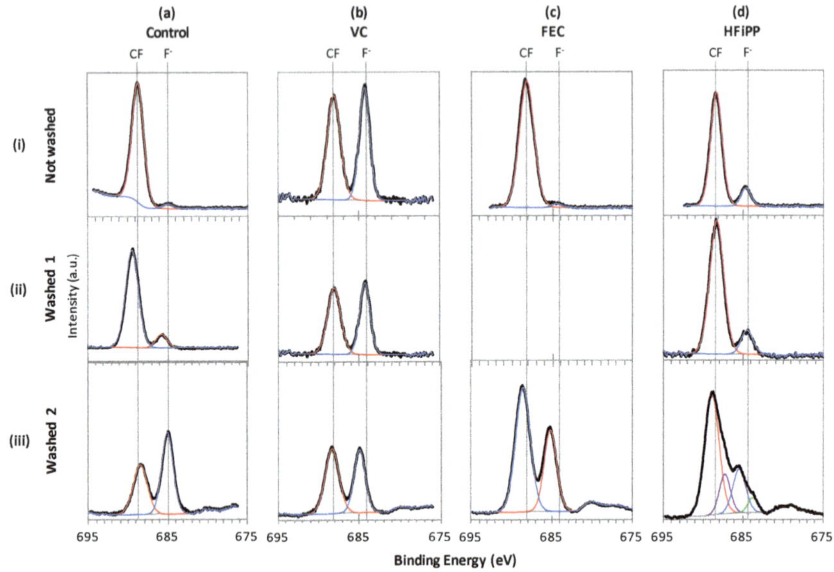

Fig S3 F_{1s} spectra of different HOPG anodes after SEI formation from different electrolyte additives without washing (i), after minimal washing (ii), and extensive washing (iii) with DMC.

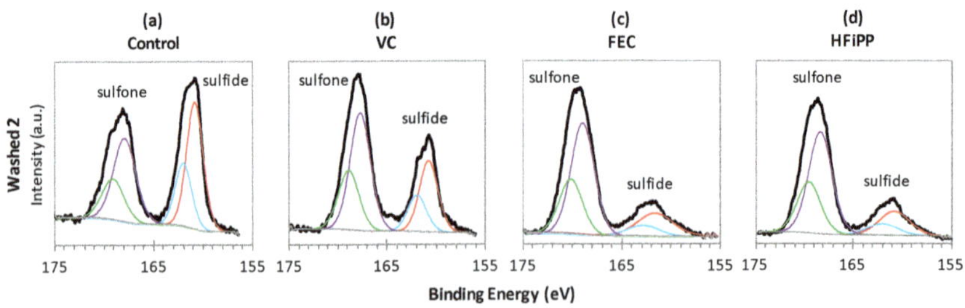

Fig. S4 S_{2p} spectra of different HOPG anodes after SEI formation from different electrolyte additives extensive washing with DMC.

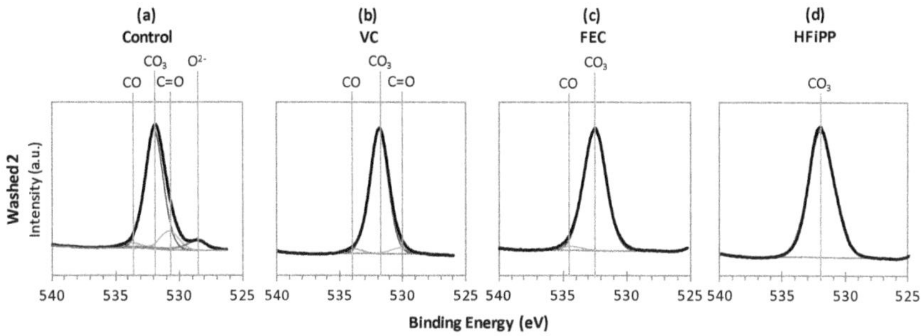

Fig. S5 O_{1s} spectra of different HOPG anodes after SEI formation from different electrolyte additives after extensive washing with DMC.

4. Author Contributions

Arthur Cresce and David Baker originated the concept of the study, designed the framework test protocol, and collected part of the data used. Selena Russell contributed significant AFM expertise in data gathering and analysis, as well as data validation and advanced test methods. Karen Gaskell performed XPS data gathering and contributed her written analysis to this document. Kang Xu provided senior leadership, expert advice, and contributed significantly to the writing of this paper.

Notes: The authors declare no competing financial interest.

Acknowledgements

The authors thank Dr. Kyle Grew (ARL) for assistance in composing a MATLAB routine to analyze AFM F-d data. Partial financial support from Department of Energy Applied Battery Research (DOE-ABR) Program is appreciated. This research was supported in part by an appointment to the U.S. Army Research Laboratory Postdoctoral Fellowship Program administered by the Oak Ridge Associated Universities through a cooperative agreement with the U.S. Army Research Laboratory.

The work has been previously published by ACS Publications:

In Situ and Quantitative Characterization of Solid Electrolyte Interphases, Arthur v. Cresce, Selena M. Russell, David R. Baker, Karen J. Gaskell, and Kang Xu Nano Letters 2014 14 (3), 1405-1412. https://doi.org/10.1021/nl404471v

References

[1] Peled, E., The Electrochemical Behavior of Alkali and Alkaline Earth Metals in Nonaqueous Battery Systems—The Solid Electrolyte Interphase Model. *Journal of The Electrochemical Society* 1979, 126, 2047-2051. https://doi.org/10.1149/1.2128859

[2] Xu, K., Nonaqueous Liquid Electrolytes for Lithium-Based Rechargeable Batteries. *Chemical Reviews* 2004, 104, 4303-4418. https://doi.org/10.1021/cr030203g

[3] Winter, M., The Solid Electrolyte Interphase – The Most Important and the Least Understood Solid Electrolyte in Rechargeable Li Batteries *Zeitschrift fur Physikalische Chemie* 2009, 223.

[4] Peled, E., Lithium stability and film formation in organic and inorganic electrolyte for lithium battery systems. *Lithium Batteries, J.P. Gabano, Ed.* 1983, 43-72.

[5] Nie, M.; Abraham, D. P.; Seo, D. M.; Chen, Y.; Bose, A.; Lucht, B. L., Role of Solution Structure in Solid Electrolyte Interphase (SEI) Formation on Graphite with LiPF6 in Propylene Carbonate (PC). *The Journal of Physical Chemistry C* 2013. https://doi.org/10.1021/jp409765w

[6] Shkrob, I. A.; Zhu, Y.; Marin, T. W.; Abraham, D., Reduction of Carbonate Electrolytes and the Formation of Solid-Electrolyte Interface (SEI) in Lithium-Ion Batteries. 1. Spectroscopic Observations of Radical Intermediates Generated in One-Electron Reduction of Carbonates. *The Journal of Physical Chemistry C* 2013. https://doi.org/10.1021/jp406274e

[7] Lu, P.; Harris, S. J., Lithium transport within the solid electrolyte interphase. *Electrochemistry Communications* 2011, 13, 1035-1037. https://doi.org/10.1016/j.elecom.2011.06.026

[8] Xu, K.; Lam, Y.; Zhang, S. S.; Jow, T. R.; Curtis, T. B., Solvation Sheath of Li+ in Nonaqueous Electrolytes and Its Implication of Graphite/Electrolyte Interface Chemistry. *The Journal of Physical Chemistry C* 2007, 111, 7411-7421. https://doi.org/10.1021/jp068691u

[9] Besenhard, J. O.; Winter, M.; Yang, J.; Biberacher, W., Filming mechanism of lithium-carbon anodes in organic and inorganic electrolytes. *Journal of Power Sources* 1995, 54, 228-231. https://doi.org/10.1016/0378-7753(94)02073-C

[10] Wagner, M. R.; Albering, J. H.; Moeller, K. C.; Besenhard, J. O.; Winter, M., XRD evidence for the electrochemical formation of Li+(PC)yCn- in PC-based

electrolytes. *Electrochemistry Communications* 2005, 7, 947-952. https://doi.org/10.1016/j.elecom.2005.06.009

[11] Aurbach, D.; Daroux, M.; Faguy, P.; Yeager, E., The electrochemistry of noble metal electrodes in aprotic organic solvents containing lithium salts. *Journal of Electroanalytical Chemistry and Interfacial Electrochemistry* 1991, 297, 225-244. https://doi.org/10.1016/0022-0728(91)85370-5

[12] Tang, M.; Miyazaki, K.; Abe, T.; Newman, J., Effect of Graphite Orientation and Lithium Salt on Electronic Passivation of Highly Oriented Pyrolytic Graphite. *Journal of The Electrochemical Society* 2012, 159, A634-A641. https://doi.org/10.1149/2.073205jes

[13] Chu, A. C.; Josefowicz, J. Y.; Farrington, G. C., Electrochemistry of Highly Ordered Pyrolytic Graphite Surface Film Formation Observed by Atomic Force Microscopy. *Journal of The Electrochemical Society* 1997, 144, 4161-4169. https://doi.org/10.1149/1.1838160

[14] Jeong, S.-K.; Inaba, M.; Abe, T.; Ogumi, Z., Surface Film Formation on Graphite Negative Electrode in Lithium-Ion Batteries: AFM Study in an Ethylene Carbonate-Based Solution. *Journal of The Electrochemical Society* 2001, 148, A989-A993. https://doi.org/10.1149/1.1387981

[15] El Abedin, S. Z.; Garsuch, A.; Endres, F., Aluminium Nanowire Electrodes for Lithium-Ion Batteries. *Australian Journal of Chemistry* 2012, 65, 1529-1533. https://doi.org/10.1071/CH12330

[16] Hirasawa, K. A.; Sato, T.; Asahina, H.; Yamaguchi, S.; Mori, S., In Situ Electrochemical Atomic Force Microscope Study on Graphite Electrodes. *Journal of The Electrochemical Society* 1997, 144, L81-L84. https://doi.org/10.1149/1.1837560

[17] Peled, E.; Golodnitsky, D.; Ardel, G., Advanced Model for Solid Electrolyte Interphase Electrodes in Liquid and Polymer Electrolytes. *Journal of The Electrochemical Society* 1997, 144, L208-L210. https://doi.org/10.1149/1.1837858

[18] Domi, Y.; Ochida, M.; Tsubouchi, S.; Nakagawa, H.; Yamanaka, T.; Doi, T.; Abe, T.; Ogumi, Z., In Situ AFM Study of Surface Film Formation on the Edge Plane of HOPG for Lithium-Ion Batteries. *The Journal of Physical Chemistry C* 2011, 115, 25484-25489. https://doi.org/10.1021/jp2064672

[19] Matsuoka, O.; Hiwara, A.; Omi, T.; Toriida, M.; Hayashi, T.; Tanaka, C.; Saito, Y.; Ishida, T.; Tan, H.; Ono, S. S.; Yamamoto, S., Ultra-thin passivating film

induced by vinylene carbonate on highly oriented pyrolytic graphite negative electrode in lithium-ion cell. *Journal of Power Sources* 2002, 108, 128-138. https://doi.org/10.1016/S0378-7753(02)00012-5

[20] Jeong, S.-K.; Inaba, M.; Mogi, R.; Iriyama, Y.; Abe, T.; Ogumi, Z., Surface Film Formation on a Graphite Negative Electrode in Lithium-Ion Batteries: Atomic Force Microscopy Study on the Effects of Film-Forming Additives in Propylene Carbonate Solutions. *Langmuir* 2001, 17, 8281-8286. https://doi.org/10.1021/la015553h

[21] Jeong, S.-K.; Inaba, M.; Iriyama, Y.; Abe, T.; Ogumi, Z., Surface film formation on a graphite negative electrode in lithium-ion batteries: AFM study on the effects of co-solvents in ethylene carbonate-based solutions. *Electrochimica Acta* 2002, 47, 1975-1982. https://doi.org/10.1016/S0013-4686(02)00099-3

[22] Domi, Y.; Ochida, M.; Tsubouchi, S.; Nakagawa, H.; Yamanaka, T.; Doi, T.; Abe, T.; Ogumia, Z., Electrochemical AFM Observation of the HOPG Edge Plane in Ethylene Carbonate-Based Electrolytes Containing Film-Forming Additives. *Journal of The Electrochemical Society* 2012, 159, A1292-A1297. https://doi.org/10.1149/2.059208jes

[23] Tsubouchi, S.; Domi, Y.; Doi, T.; Ochida, M.; Nakagawa, H.; Yamanaka, T.; Abe, T.; Ogumi, Z., Spectroscopic Analysis of Surface Layers in Close Contact with Edge Plane Graphite Negative-Electrodes. *Journal of The Electrochemical Society* 2013, 160, A575-A580. https://doi.org/10.1149/2.025304jes

[24] Niehoff, P.; Passerini, S.; Winter, M., Interface Investigations of a Commercial Lithium Ion Battery Graphite Anode Material by Sputter Depth Profile X-ray Photoelectron Spectroscopy. *Langmuir* 2013, 29, 5806-5816. https://doi.org/10.1021/la400764r

[25] Ota, H.; Sakata, Y.; Inoue, A.; Yamaguchi, S., Analysis of Vinylene Carbonate Derived SEI Layers on Graphite Anode. *Journal of The Electrochemical Society* 2004, 151, A1659-A1669. https://doi.org/10.1149/1.1785795

[26] Lee, J. T.; Nitta, N.; Benson, J.; Magasinski, A.; Fuller, T. F.; Yushin, G., Comparative study of the solid electrolyte interphase on graphite in full Li-ion battery cells using X-ray photoelectron spectroscopy, secondary ion mass spectrometry, and electron microscopy. *Carbon* 2013, 52, 388-397. https://doi.org/10.1016/j.carbon.2012.09.049

[27] Wagner, M. S.; Graham, D. J.; Ratner, B. D.; Castner, D. G., Maximizing information obtained from secondary ion mass spectra of organic thin films using

multivariate analysis. *Surface Science* 2004, 570, 78-97. https://doi.org/10.1016/j.susc.2004.06.184

[28] Aurbach, D.; Gamolsky, K.; Markovsky, B.; Gofer, Y.; Schmidt, M.; Heider, U., On the use of vinylene carbonate (VC) as an additive to electrolyte solutions for Li-ion batteries. *Electrochimica Acta* 2002, 47, 1423-1439. https://doi.org/10.1016/S0013-4686(01)00858-1

[29] McMillan, R.; Slegr, H.; Shu, Z. X.; Wang, W., Fluoroethylene carbonate electrolyte and its use in lithium ion batteries with graphite anodes. *Journal of Power Sources* 1999, 81–82, 20-26. https://doi.org/10.1016/S0378-7753(98)00201-8

[30] Zhang, S. S., A review on electrolyte additives for lithium-ion batteries. *Journal of Power Sources* 2006, 162, 1379-1394. https://doi.org/10.1016/j.jpowsour.2006.07.074

[31] von Cresce, A.; Xu, K., Electrolyte Additive in Support of 5 V Li Ion Chemistry. *Journal of The Electrochemical Society* 2011, 158, A337. https://doi.org/10.1149/1.3532047

[32] Aurbach, D.; Daroux, M. L.; Faguy, P. W.; Yeager, E., Identification of Surface Films Formed on Lithium in Propylene Carbonate Solutions. *Journal of The Electrochemical Society* 1987, 134, 1611-1620. https://doi.org/10.1149/1.2100722

[33] Aurbach, D.; Ein-Ely, Y.; Zaban, A., The Surface Chemistry of Lithium Electrodes in Alkyl Carbonate Solutions. *Journal of The Electrochemical Society* 1994, 141, L1-L3. https://doi.org/10.1149/1.2054718

[34] Tasaki, K.; Goldberg, A.; Lian, J.-J.; Walker, M.; Timmons, A.; Harris, S. J., Solubility of Lithium Salts Formed on the Lithium-Ion Battery Negative Electrode Surface in Organic Solvents. *Journal of The Electrochemical Society* 2009, 156, A1019-A1027. https://doi.org/10.1149/1.3239850

[35] Jones, J.; Anouti, M.; Caillon-Caravanier, M.; Willmann, P.; Sizaret, P.-Y.; Lemordant, D., Solubilization of SEI lithium salts in alkylcarbonate solvents. *Fluid Phase Equilibria* 2011, 305, 121-126. https://doi.org/10.1016/j.fluid.2011.03.007

[36] Xu, K., Whether EC and PC Differ in Interphasial Chemistry on Graphitic Anode and How. *Journal of The Electrochemical Society* 2009, 156, A751-A755. https://doi.org/10.1149/1.3166182

Chapter 5

$LiCoPO_4$ 5 V Li-ion Cathode

Jan L. Allen*, Joshua L. Allen, Samuel A. Delp III, Jeff Wolfenstine, T. Richard Jow
U.S. Army Research Laboratory
*jan.l.allen8.civ@mail.mil

Abstract

The use of $LiCoPO_4$ as a Li-ion cathode material can enable high energy 5 V batteries. However unmodified $LiCoPO_4$ shows a severe loss of discharge capacity upon multiple charge – discharge cycles owing to structure deterioration and electrolyte decomposition. The intense development of high voltage electrolytes and strategies that improve the cycle life, discharge capacity and rate capability have increased the interest in this material. Recent advances in chemical substitution of $LiCoPO_4$ now show that capacity fade can be significantly reduced enabling the development of the next generation of Li-ion batteries based on 5 V cathodes.

Keywords

Li-ion, Cathode, Energy Storage, Battery, LiCoPO4, 5 V

Contents

The work has been previously published by Elsevier: Jan L. Allen, Joshua L. Allen, Travis Thompson, Samuel A. Delp, Jeff Wolfenstine, T. Richard Jow, "Cr and Si Substituted-$LiCo_{0.9}Fe_{0.1}PO_4$: Structure, full and half Li-ion cell performance," 2016 J. Power Sources 327 229, https://doi.org/10.1016/j.jpowsour.2016.07.055

1. Introduction

5 V Li-ion cathodes offer a new pathway along the continuing development of Li based high energy density electrical energy storage. An electrochemical cell stores electrical energy in the chemical bonds of the electrode materials. In a Li-ion battery, the difference in the chemical energy or stability of the Li ion in the cathode relative to the anode creates a voltage and the electrodes are redox active – one electrode acts as reductant the other as oxidant. Li is transferred by a liquid or solid electrolyte between the electrodes as the redox reaction occurs and the electrons are transferred outside the cell to do work. The process needs to be fully reversible so that the stored energy can be released and stored again many times. This limits the electrode structures to materials which can undergo lithium extraction and insertion without a major structural change. Additionally, the electrode must possess adequate electronic conductivity and be redox active.

The pioneering work of Whittingham used layered TiS_2 a layered structure with good electronic conductivity which can readily insert Li through Ti reduction from Ti(IV) to Ti(III). A TiS_2/Li cell was commercialized wherein the Li inserted into TiS_2 during the discharge and re-plated back onto Li during the charge. Unfortunately, the re-plating process did not form a smooth surface but grew in needle shaped, dendritic shapes which eventually penetrated the separator, shorting the cell and sometimes leading to explosive consequences. This eventually led researchers to replace the Li anode with graphite which readily undergoes reversible Li intercalation however at a loss in discharge capacity relative to metallic Li. This change, enabled by an electrolyte reformulation was spectacularly successful and the graphite anode, which can readily insert one Li ion per 6 carbons, is used in almost all current commercial Li-ion cells.

The difference in the electrochemical potential of Li in the two electrodes is the open circuit voltage and the amount of energy which can be stored in a cell is a product of the reversible range of Li in the electrode structure, the discharge capacity measured in mAh g^{-1}, and the discharge voltage, which is lower than the open circuit voltage owing to internal battery resistance. Thus, to increase the energy storage density it is necessary to either increase the amount of Li that can be reversibly exchanged and / or increase the voltage difference between the electrodes. Using TiS_2 gives a cell voltage versus Li of about 2.2 V which against graphite would be about 2 V and using it versus graphite would require difficult pre-lithiation of the graphite.

The Goodenough research group discovered that $LiCoO_2$ can undergo a reversible Li removal allowing a cell voltage of about 3.7 V versus graphite. The cell is assembled in a discharged state and is charged prior to use and since the cathode contains Li there is no need for pre-lithiation of the graphite anode. The much higher voltage leads to a much higher energy density. The prototypical $LiCoO_2$ is a layered structure, with LiO_6 octahedral layers alternating with CoO_6 octahedral layers, thereby allowing Li to be transported out during the charge up until about $Li_{0.45}CoO_2$. A few other structures have been commercialized including the spinel, $LiMn_2O_4$ and the olivine, $LiFePO_4$ which likewise are able to reversibly lose and gain Li ions during charge and discharge, respectively. The Li-ion battery has led to a revolution in portable electronic devices displacing earlier battery technologies. However, the rising commercial and military demand for dense electrochemical energy storage for longer and longer run times has led to the search for new cathodes with a higher voltage and /or capacity and therefore higher energy storage capability. Herein, we will focus on the development of a higher voltage analog of the commercialized $LiFePO_4$ cathode as an answer to this demand.

$LiFePO_4$ (LFP) with the phospho-olivine structure is a well-known 3.5 V, commercialized positive electrode for Li-ion batteries. Complete substitution of Fe by Mn, Co or Ni changes the voltage of discharge to 4.1, 4.8 and 5.1 V, respectively [1-3] thus increasing the maximum theoretical energy storage from 585 Wh kg^{-1} for Fe to 685, 802 and 852 Wh kg^{-1}, respectively. $LiMnPO_4$ therefore seems to be a next logical step from LFP to boost the energy density of resultant Li-ion batteries since it falls within the region where the electrolyte is stable. However, initial work on $LiMnPO_4$ phospho-olivines [1] made it clear that the rate capability was extremely poor. It was only through elemental substitution of part of the Mn by Fe, $LiFe_{1-x}Mn_xPO_4$, that Pahdi et al [1] was able to obtain a reasonable discharge capacity. The poor rate capability can be explained by the electronic structure of $LiMnPO_4$ relative to $LiFePO_4$. Since the electronic conduction in phospho-olivines is polaronic the conduction is through hopping of a polaron which is a positive charge localized on the transition metal [4], a local M^{3+}, from site to site through the crystal. In the case of Mn^{3+}, owing to the fact that Mn^{3+} is a Jahn-Teller ion a distortion of the Mn-O bonding occurs in octahedral coordination. During electronic conduction the entire distortion must jump from site to site and thus the electron transport is slow. The substitution of Fe for Mn creates the possibility for Fe^{3+} charge carriers that much more readily move throughout the structure. The 5.1 V $LiNiPO_4$ cathode might also be if interest to use owing to its very high theoretical energy density but it is unfavorable to use as a Li-ion cathode because its electronic structure's valence band maximum (VBM) is heavily hybridized with oxygen [4]. Thus, instead of oxidizing Ni^{2+} to Ni^{3+} during charging in a Li-ion cell, a peroxide is formed which is unstable and therefore likely to be evolved as molecular oxygen. This leave $LiCoPO_4$ which has mainly transition metal character at its VBM though less so than Fe and it therefore readily forms polarons. It is also not susceptible to the Jahn-Teller effect. The 4.8V discharge potential requires a charge to 5 V which is problematic for the electrolyte. However, $LiCoPO_4$ in Li-ion cells does not cycle well [5-7]. The charged species, $CoPO_4$ is thermally unstable with respect to $Co_2P_2O_7$ [8].

$2\ CoPO_4 \rightarrow Co_2P_2O_7 + \frac{1}{2}O_2.$

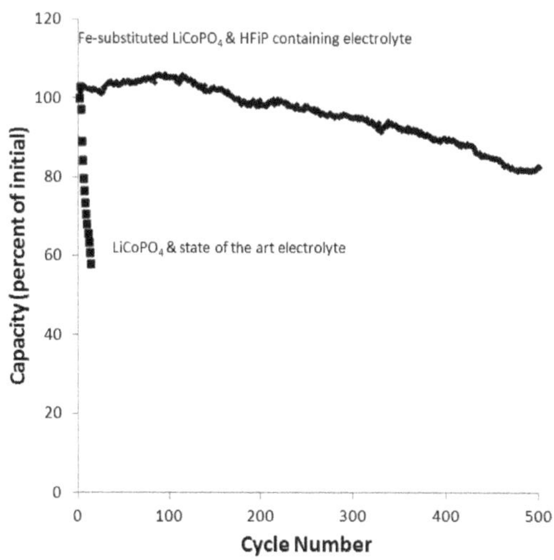

Fig. 1 Comparison of cycle life of unsubstituted $LiCoPO_4$ with standard electrolyte versus Fe-substituted $LiCoPO_4$ with an electrolyte additive.

The elemental substitution of Fe^{2+}/Fe^{3+} for Co^{2+}, $Li_{1-\delta}Co_{1-x}Fe_xPO_4$ reduces this instability and greatly improves the cycle life [9]. This may be a result of the higher transition metal character at the VBM for $LiFePO_4$ relative to $LiCoPO_4$ [4] which reduces the likelihood of oxygen evolution to form $Co_2P_2O_7$. The use of an electrolyte additive [10] further improves the cycle life. Fe substituted $LiCoPO_4$ with an electrolyte additive is compared to $LiCoPO_4$ with standard electrolyte in Figure 1. Subsequent characterization of the transport properties of $LiCo_{1-x}Fe_xPO_4$ shows improved electronic and ionic conductivity [11]. The electronic conductivity increases because Fe^{3+} type polarons have higher mobility than Co^{3+} type polarons [4]. The ionic conductivity increases owing to the creation of Li^+ vacancies as charge compensation for the Fe^{3+} substitution for Co^{2+}. However, the capacity of $Li_{1-\delta}Co_{1-x}Fe_xPO_4$ is much less than the theoretical and elemental substitutions were further explored in order to try to increase the discharge capacity. The fact that Fe^{3+} was found to substitute for Co^{2+} in $Li_{1-\delta}Co_{1-x}Fe_xPO_4$ has led to an exploration of aliovalent cations which might substitute for Co^{2+} and improve the discharge capacity. Furthermore, elemental substitution can be used to reduce the reactivity of the $LiCoPO_4$ based with the electrolyte in order to improve the coulombic efficiency and thereby increase cycle life. The improved discharge capacity and cycle life of this multi-elemental substitution of $LiCoPO_4$ are discussed here.

2. Experimental

Samples of $Li_{1+y/2}Co_{1-x-y}Fe_xM_y(PO4)_{1+y/2}$ with M = Cr, Ti, Al and / or Ga, $0 < y \leq 0.10$ and $0 \leq x \leq 0.2$ were prepared via a solid state route. Stoichiometric amounts of $Co(OH)_2$, LiH_2PO_4, Cr_2O_3, TiO_2, $Al(OH)_3$, Ga_2O_3, $FeC_2O_4.2H_2O$ and/or acetylene black (5 wt.% of final product) were ball-milled for 90 minutes then the mixture was heated at a heating rate of 10 °C min-1 to 700 °C under flowing N_2 and then the reactant mixture was held at this temperature for 12 h Samples of $Li_{1.025}Co_{0.99}Si_{0.01}Fe_{0.10}Cr_{0.05}(PO_4)_{1.025}$ (Fe, Cr- substituted $LiCoPO_4$) were prepared via the same method using $Si(OOCCH_3)_4$ as the Si source. The resulting crystalline phase(s) were identified using X-ray powder diffraction. X-ray data were collected using a Rigaku Ultima III diffractometer. For electrochemical testing, a composite electrode was fabricated by a slurry coating method. Using N-methylpyrrolidone (NMP) as solvent, a slurry was used to coat an Al foil substrate to produce a composite electrode of ca. 80 wt.% active, 10 wt.% polyvinylidene fluoride (PVDF) and 8 wt.% super-P carbon and 2 wt.% conductive carbon nanotube composite (CheapTubes.com). The electrode film was cut into small discs with an area of 0.97 cm^2, dried in drying oven at 80 °C in air before use. In a dry room (Dew point < -80 °C), Li/active coin cells (Hohsen Al-clad CR2032) were assembled using 3 layers of Celgard® 2400 as the separator and a 1.0 molal $LiPF_6$ solution in a 3:7 (wt.%) mixture of ethylene carbonate (EC) and ethyl methyl carbonate (EMC) electrolyte with 1 wt.% HFiPP (tris-hexafluoroisopropyl phosphate) or 2% TTSP (tris-trimethylsilyl phosphite). 100-150 μL of electrolyte was used per coin cell. Electrochemical testing was performed using a Maccor Series 4000 tester. More detailed structural information was obtained from a combined neutron and synchrotron X-ray Rietveld refinement of diffraction data. Time of flight neutron powder diffraction data were collected at the POWGEN beamline at the Spallation Neutron Source (SNS), Oak Ridge National Laboratory (ORNL) with center wavelengths of 1.066 Å at 298K to obtain diffraction patterns that spans from d-spacing of 0.35 to 4.6 Å. High resolution synchrotron powder diffraction data were collected using beamline 11-BM at the Advanced Photon Source (APS), Argonne National Laboratory (ANL) using wavelength of 0.414 Å at 300K to obtain diffraction patterns that spans from d-spacing of 0.7 Å to 9 Å. Rietveld structural refinements of the synchrotron X-ray and neutron diffraction data were performed using GSAS with EXPGUI [12,13].

3. Results and Discussion

3.1 Substitutional Improvements to $LiCoPO_4$

Singly substituted $LiCoPO_4$ in the absence of Fe did not show long cycle life therefore we focused on substituting elements in addition to Fe for Co including Cr, Ti, Al and Ga in order to find compositional modifications that can increase the discharge capacity of Fe-substituted $LiCoPO_4$ while maintaining long cycle life. We chose a nominal stoichiometry, $Li_{1+y/2}\,Co_{1-x-y}Fe_xM_y(PO_4)_{1+y/2}$, (M = Cr, Ti, Al and / or Ga) that favors substitution on the Co site and cation vacancies since substitution on the Li site would block Li-ion conduction. Ti, Fe and Cr, Fe double modifications were found to be most favorable with respect to discharge capacity and cycle life. Furthermore, a composition level of 0.10 was found to be optimal for Fe and 0.05 for Ti or Cr based on longer cycling data not shown here. All samples show superior cycle life to $LiCoPO_4$ and increased discharge capacity relative to singly Fe-substituted $LiCoPO_4$ which corresponds to nominal stoichiometry of $Li_{1-\delta}Co_{0.8}Fe_{0.2}PO_4$. The electrochemical performance and nature of this Fe-only substituted material as well as its electronic and ionic conductivity is described in detail by Allen et al. [9,11].The most favorable properties in terms of high discharge capacity and long cycle life were observed for the composition $Li_{1.025}Co_{0.85}Fe_{0.10}Cr_{0.05}(PO_4)_{1.025}$. Its cycle life and discharge capacity compared to $Li_{1-\delta}Co_{0.8}Fe_{0.2}PO_4$ are shown in Fig. 2.

It is clear from Fig. 2 that the double substitution of Fe and Cr for Co leads to increased discharge capacity and better cycle life. However, even with the use of electrolyte additives the coulombic efficiency is too low at around 97.5% This inefficiency meaning that the charge capacity is greater than the discharge capacity is generally indicative of the electrolyte being decomposed on each cycle. Elemental substitution of Si for Co was then tried in order to reduce the reactivity of the electrode with the electrolyte and improve the coulombic efficiency owing to the increased bond strength of the Si-O bond relative to the Co-O bond which might therefore reduce its reactivity with the electrolyte. The improvement was substantial, increasing the coulombic efficiency to greater than

99% with this substitution of Si for 1% of Co. Using additional Si did not further improve the efficiency. Fe substitution improves cycle life. Cr substitution further improves cycle life and boosts the discharge capacity. Si substitution reduces the destructive reaction of the electrolyte with the high voltage cathode materials. In combination, the multi-element substituted $LiCoPO_4$ is able to achieve an energy density much greater than $LiFePO_4$ with good cycle life.

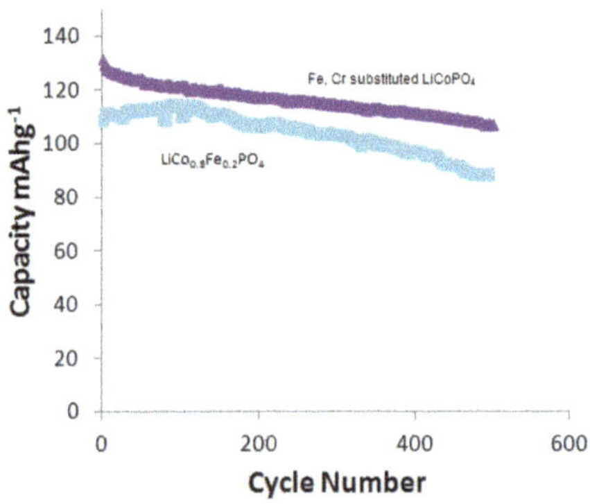

Fig. 2 Comparison of cycle life and discharge capacity of Fe only substituted $LiCoPO_4$ to Fe, Cr - substituted $LiCoPO_4$.

3.2 Structure of Cr and Si Substituted $LiCo_{0.9}Fe_{0.1}PO_4$

The initial phase purity was evaluated with a Rigaku Ultima III powder X-ray diffractometer and within the detection limits the material appeared to be single phase. For structure determination synchrotron X-ray and neutron data were collected and from these high resolution data sets the structure and phase composition were modelled through Rietveld refinement. Since 5 atoms (Li, Co, Cr, Fe, and Si) and vacancies can be distributed over two octahedral sites and the tetrahedral P site in $LiCoPO_4$, several structural models are possible. The substituting atoms (Fe, Cr, Si) and vacancies were introduced one at a time. The goodness of fit (χ^2) and site occupancies were used to rank the structure solutions. If a model resulted in a solution with a low χ^2 (indicating a good fit of the calculated pattern to the experimental data), but also resulted in unrealistic site occupancies, then the model was discarded. The 300K neutron data set, its refinement, Bragg reflections and difference pattern are shown in Fig. 3.

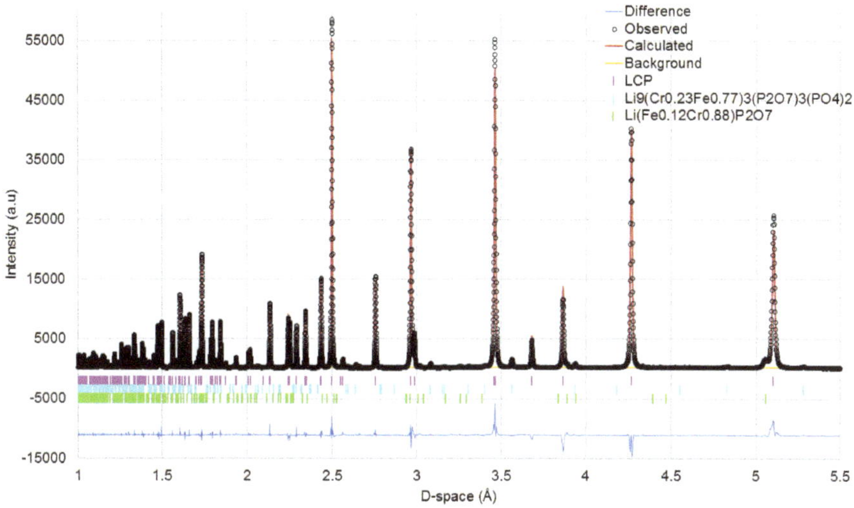

Fig. 3 Example of neutron diffraction pattern of Cr, Si-$LiCo_{0.9}Fe_{0.1}PO_4$ *recorded at 300K.*

Table 1 Final structural model of Cr,Si-$LiCo_{0.9}Fe_{0.1}PO_4$ from Rietveld refinement of neutron and synchrotron diffraction data.

Atom	Wyckoff	x/a	y/b	z/c	Ui/Ue*100	Occ.
Li	4*a*	0.00000	0.00000	0.00000	0.7009(434)	0.930(5)
Fe	4a	0.00000	0.00000	0.00000	0.7009(434)	0.001(0)
Cr	4a	0.00000	0.00000	0.00000	0.7009(434)	0.005(1)
Co	4*c*	0.27914(23)	0.2500	0.97817(5)	0.2016(434)	0.822(0)
Fe	4c	0.27914(23)	0.2500	0.97817(5)	0.2016(31)	0.107(0)
Cr	4c	0.27914(23)	0.2500	0.97817(5)	0.2016(31)	0.056(0)
Li	4c	0.27914(23)	0.2500	0.97817(5)	0.2016(31)	0.015(4)
Si	4c	0.27914(23)	0.2500	0.97817(5)	0.2016(31)	0.012(1)
O1	4c	0.0992(8)	0.2500	0.7431(17)	0.3742(22)	1
O2	4c	0.4538(8)	0.2500	0.2136(17)	0.3742(22)	1
O3	8*d*	0.1630(6)	0.0497(8)	0.2810(11)	0.3742(22)	1
P	4c	0.09437(4)	0.2500	0.4197(9)	0.252(6)	1

(S.G. *Pnma*); a = 10.2032(18), b = 5.9304(10), c = 4.6997(9), R_{wp} = 6.68, χ^2 = 5.133
calculated density = 3.725 g mL^{-1}

The data are shown as open black circles, the calculated pattern is the continuous red line. The Bragg reflections of the majority, 95.9 wt. %, LCP olivine phase, are indicated with purple bars. Two minor phases were evidenced in the high resolution data and included in the Rietveld refinement, ~3.0 wt.% $Li_9(Cr_{0.23}Fe_{0.77})_3(P_2O_7)_3(PO_4)_2$, Bragg reflections indicated in light blue and ~1.1 wt.% $Li(Fe_{0.12}Cr_{0.88})P_2O_7$, Bragg reflections indicated in green. The goodness of the calculated fit is indicated by a final R_{wp} = 6.68, the $\chi2$ = 5.133 and from observation of Fig. 1. The structural refinement led to the model with atomic positions listed in Table 1 with stoichiometry of $(Li_{0.930}Cr_{0.005}Fe_{0.001})_{M1}(Co_{0.822}Fe_{0.107}Cr_{0.056}Si_{0.012}Li_{0.015})_{M2}PO_4$.

From the refinement a couple of points can be made. The pattern confirms that a phospho-olivine structure was prepared in high purity (~96%). That is, the predominant phase is isostructural with triphylite ($LiFePO_4$, space group *Pnma* [14]) having a hexagonal close-packed oxygen array with Li occupying octahedral sites, denoted as M1 sites, forming edge-sharing chains and Co and its substituents (Fe, Cr, Si) occupying corner-sharing octahedral sites, denoted as M2 sites and P occupying tetrahedral sites. The occupancy of the Li site (M1) shows the presence of vacancies, 6.4 atom %, and very low, 0.6 atom % non-Li occupancy, e.g., low anti-site disorder. The presence of vacancies should lead to an improvement in Li-ion conduction through an increase in the Li-ion mobility since ionic mobility is, in part, dependent on having empty sites available. Furthermore, the very low anti-site disorder on the M1 site, suggests high Li-ion conduction along the 1-dimensional Li channels of the $LiCoPO_4$ structure. Anti-site disorder is a general concern for this structure since the presence of Fe, Co or other non-Li ions on the M1 site will impede the Li conduction along an entire chain of edge-sharing octahedral and as a consequence would reduce the discharge capacity and rate. A small amount of Li is found on the M2 site. Similarly, the presence of Li on the M2 site of the isostructural Cu substituted $LiFePO_4$ has also been reported [15]. The substituting elements Cr, Fe, Si are almost exclusively located on the M2, Co site and not the M1, Li site. The substitution of Si in an octahedral site versus the tetrahedral phosphorus site is somewhat surprising from a crystal chemistry viewpoint but octahedral coordination of Si is not unprecedented and the cation ratio was controlled during synthesis in an attempt to achieve Si substitution on the octahedral Co (M2) site. Amin et al. [16] reported 1 atom % Si substitution into $LiFePO_4$ single crystals and proposed that the donor effect observed upon substitution was consistent with Si substituting for Fe to create a Li deficiency. No synchrotron data was obtained by Amin et al. [16], but the increase in unit cell is characteristic of Li vacancies on the M1 site of $LiFePO_4$ as charge compensation and is therefore also consistent with Si substitution for Fe [16]. Additionally, Jin et al. reported silicon substitution on the octahedral Co site in $LiCoO_2$ [17]. The refined stoichiometry allows us to determine the defect charge compensation mechanism that results from the aliovalent substitutions using Kröger-Vink notation [18]: 0.06 $[Cr_{Co}^{\bullet}]$ + 0.01 $[Si_{Co}^{\bullet\bullet}]$ = 0.07 $[V_{Li}']$ + 0.01 $[Li_{Co}']$. Table 2 shows the lattice constants along with other compounds prepared for comparison. From Table 2, several points can be made. First, the lattice constants show a decrease in the unit cell volume of $LiCo_{0.829}Fe_{0.0976}Cr_{0.0488}PO_4$ relative to $LiFe_{0.10}Co_{0.90}PO_4$ consistent with the substitution of the smaller Cr^{3+} (effective octahedral ionic radius of 0.615Å) [19] for Co^{2+} (0.745 Å) [19]. Second, the lattice constant of $LiCo_{0.820}Fe_{0.0976}Cr_{0.0488}Si_{0.00976}PO_4$ increases relative to the Si free $LiCo_{0.829}Fe_{0.0976}Cr_{0.0488}PO_4$. This increase in unit cell is consistent with the results of Amin et al. [16] for Si substituted $LiFePO_4$ and most likely is a result of the creation of Li vacancies on the Li (M1) site in order to charge compensate the Si^{4+} on the Co^{2+} (M2) site. Li vacancies on the M1 site of phospho-olivines caused by aliovalent doping are known to increase the lattice volume [20].

Table 2 Lattice constants for substituted $LiCoPO_4$ compounds from Rietveld refinement of data obtained on Rigaku Ultima III XRD using parallel beam configuration.

Starting composition	***a*, Å**	***b*, Å**	***c*, Å**	**Vol., Å^3**
LiCoPO4	10.1950	5.9179	4.6972	283.4
$LiCo_{0.9}Fe_{0.1}PO_4$	10.2175	5.9335	4.7025	285.1
$LiCo_{0.829}Fe_{0.0976}Cr_{0.0488}PO_4$	10.1703	5.9204	4.6991	282.9
$LiCo_{0.820}Fe_{0.0976}Cr_{0.0488}Si_{0.00976}PO_4$	10.2002	5.9291	4.6991	284.2

3.3 Discharge Capacity and Cycle Life of Cr and Si Substituted $LiCo_{0.9}Fe_{0.1}PO_4$

Fig. 4 shows the cycling performance of Li / Cr,Si-$LiCo_{0.9}Fe_{0.1}PO_4$ coin cells (Fig. 2a) and graphite / Cr,Si-$LiCo_{0.9}Fe_{0.1}PO_4$ coin cells (Fig. 2b). Every fifth cycle is shown for clarity but the full cycling data is reported in the supporting information. From Fig. 4 several points can be made: 1) the discharge capacity of Cr,Si-$LiCo_{0.9}Fe_{0.1}PO_4$ is demonstrated to be ~140 mAh g^{-1} at the C/3 rate, 2) the overlap of the charge and discharge capacity in both half and full cells demonstrates the high coulombic efficiency, calculated to be greater than 99.5%, 3) no fade is evident in the half cell, 4) the full cell shows some fade, which we attribute to the loss of cyclable Li owing to the Li-consuming growth of the graphite solid electrolyte interphase (SEI) with each cycle [21] and increasing cell polarization. The performance can be compared to the discharge capacity fade of $LiCoPO_4$ half cells reported in the literature. For example, Tadanga et al. [22] reported a 10th cycle discharge capacity of ~52% of the initial capacity, Bramnik et al. [23] showed a 10th cycle discharge of ~59% of the initial capacity and Wolfenstine et al. [24] a ~53% capacity retention at the 10th cycle. Furthermore, it is also an improvement in capacity retention over our previously reported cycling results for Fe for Co substituted $LiCoPO_4$ [11] where we observed 97% capacity retention at the 250th half cell cycle. We observe 100% capacity retention for Cr,Si-$LiCo_{0.9}Fe_{0.1}PO_4$ at the 250th cycle in half cells and greater than 80% capacity retention at the 250th cycle of the full cell. The half cell energy density at the initial cycle and 250th cycle changes only from 665 to 647 Wh kg^{-1} indicating that in addition to capacity stability the voltage remains stable in the half cell.

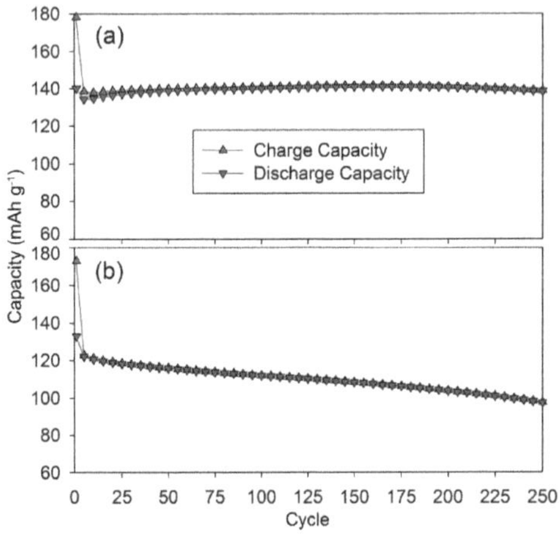

Fig. 4 Representative plot of charge and discharge capacity of Cr,Si-$LiCo_{0.9}Fe_{0.1}PO_4$ *(a) half cells and (b) full cells as a function of cycle number. Only every 5th cycle is shown for clarity.*

Fig. 5 shows the charge and discharge curves obtained at the 2nd (1st cycle after formation, blue solid lines) and the 250th cycle (red dashed lines) for Li / Cr,Si-$LiCo_{0.9}Fe_{0.1}PO_4$ half cells (Fig. 5a) and graphite / Cr,Si-$LiCo_{0.9}Fe_{0.1}PO_4$ full cells (Fig. 5c), respectively, at a 1C rate. The corresponding differential capacity as a function of voltage (dQ / dV) of the half and full cell is shown in Fig. 5b and 5d, respectively. More cycles are omitted from Fig. 5 for clarity. From Fig. 5, the following points can be made. For the half cell (Fig. 5a), there is an increase in the charge voltage and decrease in the discharge voltage over the 250 cycles with slight increase in discharge capacity. For the full cell (Fig. 5c), there is a similar change in voltage with a larger decrease in discharge capacity over the 250 cycles. The increase in voltage polarization with cycling in the half and full cells is attributable to impedance buildup at the Li and graphite electrodes, respectively [25]. The full cell's (Fig. 5d) gradual loss in capacity can be attributed to the formation and reformation of the SEI layer on the graphite anode which irreversibly consumes Li [21].

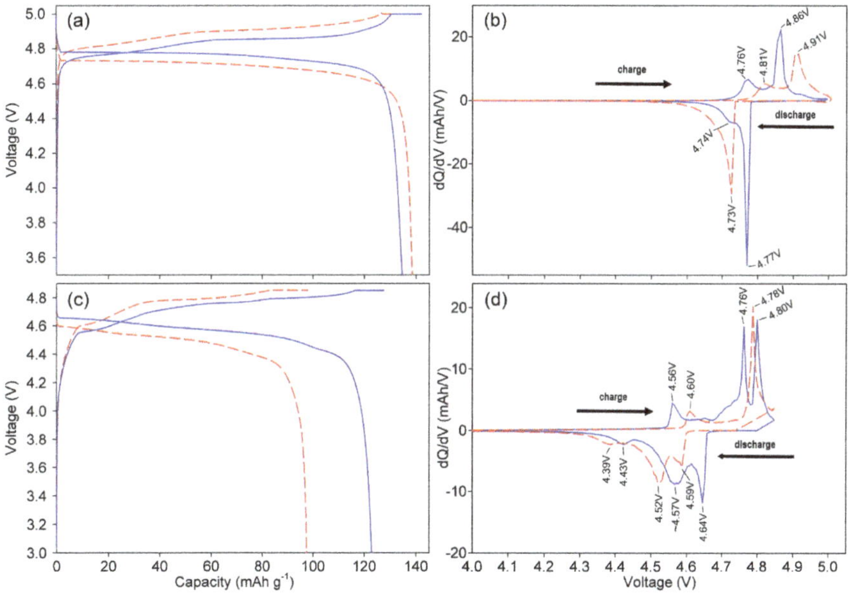

Fig. 5 Representative charge and discharge curves of Cr,Si-$LiCo_{0.9}Fe_{0.1}PO_4$ *(a) half cells and (c) full cells and corresponding differential capacity (dQ/dV) plots, (b) and (d), respectively at 1C rate. Solid blue and dashed red lines indicate the 2nd and 250th cycle, respectively.*

The differential capacity plots dQ / dV for the 2nd and 250th cycle are plotted in Fig. 5b and Fig. 5d for the half and full cells, respectively. Solid blue and dashed red lines indicate the 2nd and 250th cycle, respectively. In the half cell, for the 2nd cycle, two charge peaks are observed at 4.76 V and 4.86V and two discharge peaks are observed at 4.77 V and 4.74V. The two charge and two discharge peaks are a known feature of $LiCoPO_4$-based cells and evidence the reversible phase transition from $LiCoPO_4$ to $Li_{2/3}CoPO_4$ and subsequent reversible phase transition to $CoPO_4$ [26,27]. At the 250th cycle of the half cell as impedance builds up the charge peaks' voltages increase to 4.81 V and 4.91 V and only 1 discharge peak at 4.73V is observed. The 2nd discharge peak of the 250th cycle is observed as a shoulder on the 4.73V peak without a clear maximum. In the 2nd cycle of the full cell (Fig. 5d), three charge (4.56 V, 4.76 V and 4.80V) and three discharge (4.64V, 4.57V and 4.43V) peaks are observed. The additional peaks compared to the half cell result from the differing voltage plateaus of staged phases of graphite as a function of Li content [27]. Over 250 cycles the two graphite plateaus become less distinguishable and only two charge peaks (4.60V and 4.78V) and are observed, however, three discharge peaks (4.59V, 4.52V and 4.39V) are still observed.

3.4 Specific Energy of Full 5 V Li-ion Cells

The specific energy of the graphite / Cr,Si-$LiCo_{0.9}Fe_{0.1}PO_4$ Li-ion cell (Fig. 5b) during the initial 4-5 discharges is approximately 550 Wh kg^{-1} based on the cathode active mass at 1C rate. After 250 cycles the energy supplied is 510 Wh kg^{-1} so there is an approximately 0.16 Wh kg^{-1} loss in energy stored per cycle. The performance can be compared to the specific energy of $LiCoPO_4$ based cells reported in the literature. By comparison, Di Lecce et al. [28] recently reported an initial cycle energy density of 410 Wh kg^{-1} for a Sn-C / $LiFe_{0.1}Co_{0.9}PO_4$ full cell discharged at a 0.05 C rate. However, by the 7th cycle the energy density drops to about 315 Wh kg^{-1} as estimated from Fig. 3c of Di Lecce et al. [28], a drop of approximately 13.6 Wh kg^{-1} per cycle. Similarly, from Du et al. [29] we estimate an initial cycle energy density of 434 Wh kg^{-1} for a $Li_4Ti_5O_{12}$ / $LiCoPO_4$ full cell discharged at a 0.1 C rate based on the reported discharge capacity and voltage [29] but by the 10th cycle the energy density drops to ~ 407 Wh kg^{-1}, a drop of energy density of 2.7 Wh kg^{-1} per cycle. Furthermore, Ni et al. [30] report full cell energy density for a $Li_4Ti_5O_{12}$ / $LiCoPO_4$ full cell of 378 Wh kg^{-1} at 0.1C which drops to 287 Wh kg^{-1} by the 20th cycle. Furthermore, on a weight basis, the energy stored is greater than the most commonly used Li-ion cell: graphite / $LiCoO_2$: 140 mAh g^{-1} x 3.8 V = 532 Wh kg^{-1}.

3.5 Rate Capability of Cr and Si Substituted $LiCo_{0.9}Fe_{0.1}PO_4$

Fig. 6 shows the rate capability of the Li / Cr,Si-$LiCo_{0.9}Fe_{0.1}PO_4$ half cells. The upper chart shows the charge capacity (red upward triangles) and discharge capacity (blue downwards triangles) at C/10, C/5, C/2, 1C, 3C, 5C and 10C discharge rates.

Charging was at a C/10 rate to 5 V for cycle 1. Subsequent cycles used a C/2 charge rate to 5 V and the cells were held at this voltage until the current was less than C/10. The discharge capacity remains approximately the same from C/10 to 1C with only a slight reduction of about 5%. At higher rates, there is a larger decrease in the capacities but even at 3C, a capacity of ~120

mAh g^{-1} is observed. The lower portion of Fig. 6 shows the same data in a semi-logarithmic plot with rate plotted on the x-axis on a logarithmic scale and the capacity plotted on the y-axis on a linear scale. As a comparison, Fe only substituted $LiCoPO_4$ shows a capacity of ~100 mAh g^{-1} at

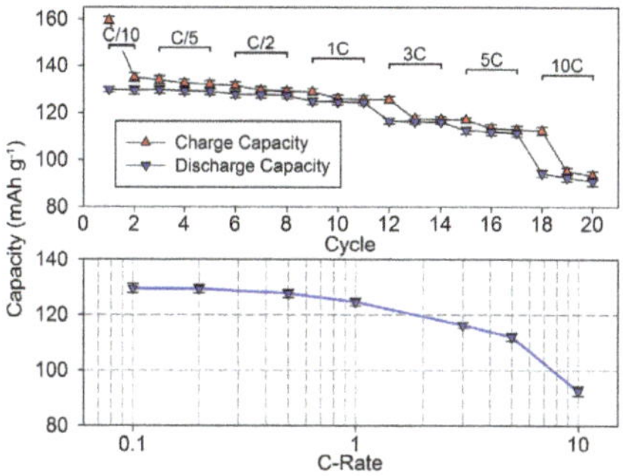

Fig. 6 Rate capability of Li / Cr,Si-$LiCo_{0.9}Fe_{0.1}PO_4$ *cells shown as discharge and charge capacity as a function of the rate of discharge.*

1C rate versus 125 mAh g^{-1} for Cr,Si-$LiCo_{0.9}Fe_{0.1}PO_4$ at the same rate. From observation of Fig. 6, it is evident that the material is capable for use in high rate applications. The high rate capability compared to unsubstituted $LiCoPO_4$ is likely a result of improved Li-ion mobility resulting from vacancies on the Li, M1 site [V_{Li}'] that form as charge compensating mechanism from the introduction of Cr, [$Cr_{Co}^{\bullet}$] and Si, [$Si_{Co}^{\bullet\bullet}$] into $LiCoPO_4$ on the Co, M2 site and from the nearly complete absence of anti-site disorder on the Li site. A similar enhancement in rate capability of Cr substituted $LiFePO_4$ was reported by Park et al. [31] which the authors attributed to increased lithium diffusivity. The rate capability is comparable to similar, laboratory grade $LiFePO_4$ based electrodes reported by Striebel et al. [32].

4. Conclusion

Graphite / Cr,Si-$LiCo_{0.9}Fe_{0.1}PO_4$ cells show a specific energy density of 550 Wh kg^{-1} based on the cathode active mass, a specific energy density is higher than Graphite / $LiCoO_2$. The high energy storage performance and rate capability is believed to result from the defect olivine structure of Cr,Si-$LiCo_{0.9}Fe_{0.1}PO_4$ that suggests higher Li-ion conductivity relative to $LiCo_{0.9}Fe_{0.1}PO_4$ and unsubstituted $LiCoPO_4$ as modelled from neutron and synchrotron diffraction data. In half cells, the well-known capacity fade of $LiCoPO_4$ based cathodes has been almost completely eliminated through 250 cycles. In full cells, the energy density decreases about 0.16 Wh kg^{-1}. Furthermore, the rate capability is on par with similar lab grade $LiFePO_4$ electrodes. It is evident that substituted $LiCoPO_4$ represents a potential jump in energy density compared to currently commercialized chemistries and represents a large step towards a new generation of higher voltage Li-ion.

Acknowledgements

The authors acknowledge the Army Research Laboratory for funding. Saft America provided graphite anodes. SAD was funded by the Army Research Laboratory postdoctoral program administered through Oak Ridge Associated Universities.

The work has been published previously by Elsevier:

Jan L. Allen, Joshua L. Allen, Travis Thompson, Samuel A. Delp, Jeff Wolfenstine, T. Richard Jow, "Cr and Si Substituted-$LiCo_{0.9}Fe_{0.1}PO_4$: Structure, full and half Li-ion cell performance," 2016 J. Power Sources 327 229, https://doi.org/10.1016/j.jpowsour.2016.07.055

References

[1] A.K. Padhi, K.S. Nanjundaswamy, J.B. Goodenough, Phospho-olivines as positive electrode materials for rechargeable lithium batteries, J. Electrochem. Soc. 144 (1997) 1188-1194. https://doi.org/10.1149/1.1837571

[2] K. Amine, H. Yasuda, M. Yamachi, Olivine $LiCoPO_4$ as 4.8 V electrode material for lithium batteries, Electrochem. Solid State Lett. 3 (2000) 178-179. https://doi.org/10.1149/1.1390994

[3] J.L. Wolfenstine, J.L. Allen, Ni^{3+}/Ni^{2+} redox potential in $LiNiPO_4$, J. Power Sources 142 (2005) 389-390. https://doi.org/10.1016/j.jpowsour.2004.11.024

[4] M.D. Johannes, K. Hoang, J.L. Allen, K. Gaskell, Hole polaron formation in olivine phosphate materials, Phys. Rev. B 85 (2012) 115106-11512. https://doi.org/10.1103/PhysRevB.85.115106

[5] K. Tadanaga, F. Mizuno, A. Hayashi, T. Minami, M. Tatsumisago, Preparation of $LiCoPO_4$ for lithium battery electrodes through solution process, Electrochemistry 71 (2003) 1192-1195.

[6] N.N. Bramnik, K.G.Bramnik, T. Buhrmester, C. Baehtz, H. Ehrenberg and H. Fuess, Electrochemical and structural study of $LiCoPO_4$-based electrodes, J. Solid State Electrochem. 8 (2004) 558-564. https://doi.org/10.1007/s10008-004-0497-x

[7] J. Wolfenstine, U. Lee, B. Poese, J.L. Allen, Effect of oxygen partial pressure on the discharge capacity of $LiCoPO_4$, J. Power Sources 144 (2005) 226-230. https://doi.org/10.1016/j.jpowsour.2004.12.013

[8] N.N. Bramnik, K. Nikolowski, D.M. Trots, H. Ehrenberg, Thermal stability of $LiCoPO_4$ cathodes, Electrochem. Solid-State Lett, 11 (2008) A89-A93. https://doi.org/10.1149/1.2894902

[9] J.L. Allen, T.R. Jow, J. Wolfenstine, Improved Cycle life of Fe-substituted $LiCoPO_4$ cathodes, J. Power Sources 196 (2011) 8656-8661. https://doi.org/10.1016/j.jpowsour.2011.06.057

[10] A.V. Cresce, K.Xu, Electrolytes in support of 5 V Li ion chemistry, J. Electrochem. Soc. 158 (2011) A337-A342. https://doi.org/10.1149/1.3532047

[11] J.L. Allen, T. Thompson, J. Sakamoto, C.R. Becker, T.R. Jow, J. Wolfenstine, Transport Properties of $LiCoPO_4$ and Fe-substituted $LiCoPO_4$, J. Power Sources 254 (2014) 204-208. https://doi.org/10.1016/j.jpowsour.2013.12.111

[12] A. C. Larson, R.B. von Dreele, General Structure Analysis System (GSAS), Report LAUR86-748, Los Alamos National Laboratory, Los Alamos, NM USA 2000.

[13] B.H. Toby, EXPGUI, a graphical user interface for GSAS, J. Appl. Crystallogr. 34 (2001) 210. https://doi.org/10.1107/S0021889801002242

[14] G. Rousse, J. Rodriguez-Carvajal, S. Patoux, C. Masquelier, Magnetic structures of the triphylite $LiFePO_4$ and of its delithiated form $FePO_4$, Chem. Mater. 15 (2003) 4082-4090. https://doi.org/10.1021/cm0300462

[15] S. Upreti, N.A. Chernova, J. Xiao, J.K. Miller, O.V. Yakubovich, J. Cabana, C.P. Grey, V.L. Chevrier, G. Ceder, J.L. Musfeldt, M.S. Whittingham, Crystal Structure, Physical Properties, and Electrochemistry of Copper Substituted $LiFePO_4$ Single Crystals, Chem. Mater. 24 (2012) 166-173. https://doi.org/10.1021/cm2026619

[16] R. Amin, C. Lin, J. Peng, K. Weichert, T. Acartürk, U. Starke, J. Maier, Silicon-Doped $LiFePO_4$ Single Crystals: Growth, Conductivity Behavior, and Diffusivity Adv. Funct. Mater. 19 (2009) 1697-1704. https://doi.org/10.1002/adfm.200801604

[17] Y. Jin, P. Lin, C.H. Chen, An investigation of silicon-doped $LiCoO_2$ as cathode in lithium-ion secondary batteries, Sol. State Ionics 177(2006) 317-322. https://doi.org/10.1016/j.ssi.2005.11.009

[18] F.A. Kröger, H.J. Vink, Relations between the concentrations of imperfections in crystalline solids, Sol. State Phys. 3 (1956) 307-435. https://doi.org/10.1016/S0081-1947(08)60135-6

[19] R.D. Shannon, Revised effective ionic-radii and systematic studies of interatomic distances in halides and chalcogenides, Acta. Crystallogr. A 32 (1976) 751-767. https://doi.org/10.1107/S0567739476001551

[20] N. Meethong, Y.-H. Kao, S.A. Speakman, Y.-M. Chiang, Aliovalent Substitutions in Olivine Lithium Iron Phosphate and Impact on Structure and Properties, Adv. Funct. Mater. 19 (2009) 1060-1070. https://doi.org/10.1002/adfm.200801617

[21] Y.C. Zhang, C.Y. Wang, X.D. Tang, Cycling degradation of an automotive $LiFePO_4$ lithium-ion battery J. Power Sources 196 (2011) 1513-1520. https://doi.org/10.1016/j.jpowsour.2010.08.070

[22] K. Tadanaga, F. Mizuno, A. Hayashi, T. Minami, M. Tatsumisago, Preparation of $LiCoPO_4$ for lithium battery cathodes through solution process, Electrochemistry 71(2003) 1192-1195.

[23] N.N. Bramnik, K.G. Bramnik, T. Buhrmester, C. Baehtz, H. Ehrenberg, H. Fuess, Electrochemical and structural study of $LiCoPO_4$-based electrodes, J. Solid State Electrochem. 8 (2004) 558-564. https://doi.org/10.1007/s10008-004-0497-x

[24] J. Wolfenstine, U. Lee, B. Poese, J.L. Allen, Effect of oxygen partial pressure on the discharge capacity of LiCoPO4, J. Power Sources 144 (2005) 226-230. https://doi.org/10.1016/j.jpowsour.2004.12.013

[25] W. Xu, J. Wang, F. Ding, X. Chen, E. Nasybulin, Y. Zhang, J.-G. Zhang, Lithium metal anodes for rechargeable batteries, Energy & Environmental Science 7 (2014) 513-537. https://doi.org/10.1039/C3EE40795K

[26] F.C. Strobridge, R.J. Clément, M. Leskes, D.S. Middlemiss, O.J. Borkiewicz, K.M. Wiaderek, K.W. Chapman, P.J. Chupas, C.P. Grey, Identifying the Structure of the Intermediate, $Li_{2/3}CoPO_4$, Formed during Electrochemical Cycling of $LiCoPO_4$, Chem. Mater. 26 (2014) 6193-6205. https://doi.org/10.1021/cm502680w

[27] J.R. Dahn, Phase-diagram of Li_xC_6, Phys. Rev. B 44 (1991) 9170-9177. https://doi.org/10.1103/PhysRevB.44.9170

[28] D. Di Lecce, S. Brutti, S. Panero, J. Hassoun, A new Sn-C/$LiFe_{0.1}Co_{0.9}PO_4$ full lithium-ion cell with ionic liquid-based electrolyte, Materials Lett. 139 (2015) 329-332. https://doi.org/10.1016/j.matlet.2014.10.089

[29] C.Q. Du, Z.Y. Tang, J.W. Wu, H.Q. Tang, X.H. Zhang, A three volt lithium ion battery with $LiCoPO_4$ and zero-strain $Li_4Ti_5O_{12}$ as insertion material, Electrochim. Acta 125 (2014) 58-64. https://doi.org/10.1016/j.electacta.2014.01.093

[30] J. Ni, W. Liu, J. Liu, L. Gao, J. Chen, Investigation on a 3.2 V $LiCoPO_4/Li_4Ti_5O_{12}$ full battery, Electrochem. Commun. 35 (2013) 1-4. https://doi.org/10.1016/j.elecom.2013.07.030

[31] C.-K. Park, S.-Bin Park, H.-C. Shin, W.-I. Cho, J. Ho., Li Ion Diffusivity and Rate Performance of the $LiFePO_4$ Modified by Cr Doping, Bull. Korean Chem. Soc. 32 (2011) 191-195. https://doi.org/10.5012/bkcs.2011.32.1.191

[32] K. Striebel, J. Shim, V. Srinivasan, J. Newman, Comparison of $LiFePO_4$ from different sources, J. Electrochem. Soc. 152 (2005) A664-A670. https://doi.org/10.1149/1.1862477

Chapter 6

"Water-in-Salt" Electrolyte Enables High Voltage Aqueous Li-ion Chemistries

Liumin Suo,[1] Oleg Borodin,[2] Tao Gao,[1] Marco Olguin,[2] Janet Ho,[2] Xiulin Fan,[1] Chao Luo,[1] Chunsheng Wang,[1,*] Kang Xu [2,*]

[1]Department of Chemical and Biomolecular Engineering, University of Maryland College Park, Maryland, 20740, USA

[2] Electrochemistry Branch, Sensor and Electron Devices Directorate, Power and Energy Division U.S. Army Research Laboratory Adelphi, Maryland, 20783, USA

*Correspondence to: E-mail: cswang@umd.edu, conrad.k.xu.civ@mail.mil

Abstract

Li-ion batteries still cause safety, environmental and cost concerns, which mostly arise from the non-aqueous electrolytes. The use of aqueous alternatives is limited by their narrow electrochemical stability window (1.23 V), which sets an intrinsic limit on the practical voltage and energy output. We report a highly-concentrated aqueous electrolyte, whose window was expanded to ~3.0 V with the formation of an electrode/electrolyte interphase. A full Li-ion battery of 2.3 V using such aqueous electrolyte was demonstrated to cycle up to 1000 times, with nearly 100% Coulombic efficiency at both low (0.15 C) and high (4.5 C) rates.

Keywords

Li-ion Batteries, Aqueous, Electrolytes, High Voltage

Contents

The work has been previously published by Science: Liumin Suo, Oleg Borodin, Tao Gao, Marco Olguin, Janet Ho, Xiulin Fan, Chao Luo, Chunsheng Wang, Kang Xu, Water-in-salt" electrolyte enables high-voltage aqueous lithium-ion chemistries, Science, 20 November 2015 • Vol 350 Issue 6263. https://doi.org/10.1126/science.aab1595

1. Introduction

Li-ion batteries power much of our digital and mobile lifestyle (*1, 2*). However, their adoption in more strategically important applications such as vehicle-electrification and grid-storage has been slower, mainly due to the concerns raised over their safety, cost, and environmental impact (*3*). Most of these concerns come from the non-aqueous electrolytes needed to withstand the high voltages (> 3.0 V) of the chemistries (*4*), because the ester-based solvents are highly flammable and reactive with the charged electrodes (*5*), and the lithium salt ($LiPF_6$) is thermally unstable and extremely toxic (*6*). Significant cost is incurred, not only directly by these electrolyte components, but also to larger degree by the stringent moisture-free process and safety management required for the dangerous combination of flammable electrolytes and energy-intense electrodes (*4, 7, 8*).

Aqueous electrolytes could resolve these concerns (*9-11*), however their electrochemical stability window (1.23 V) is too narrow to support most of the electrochemical couples used in Li-ion batteries. In particular, hydrogen evolution at the anode presents the gravest challenge as it occurs at a potential (2.21~3.04 V vs. Li depending on pH value) far above where most Li-ion battery anode materials operate. Even in trace presence, hydrogen severely deteriorates the electrode structure during cycling. A common practice to suppress hydrogen evolution in aqueous electrochemistry is to adjust the alkalinity (*12, 13*), so that the water reduction potential shifts downward to allow the use of anode materials otherwise prohibited under neutral or acidic conditions. However, as the overall electrochemical stability window of aqueous electrolytes remains constant, anodic stability against oxygen evolution suffers a corresponding compromise as illustrated on a Pourbaix diagram *(9)*. A maximum voltage of 1.5 V was achieved in aqueous Li-ion batteries, where the residual currents for H_2 or O_2 evolution still brought high self-discharge rates despite the moderate cell voltage (*9-11*).

In contrast to non-aqueous electrolyte systems where cathode and anode materials often operate far beyond thermodynamic stability limits of electrolyte components (*4, 14*), kinetic protection from a solid electrolyte interphase (SEI) in aqueous media has never been considered possible. Such interphases, situating between electrode surfaces and electrolyte, are formed by sacrificial electrolyte decomposition during the initial charging, and constitute a barrier allowing ionic conduction but forbidding electronic conduction. Their presence significantly expands the usable electrochemical stability window of electrolytes. The absence of a protective interphase in conventional aqueous electrolytes is because none of the decomposition products from water (H_2, O_2 or OH^-) can deposit in dense solid state. In the absence of interphases, aqueous Li-ion batteries were typically limited to low voltage (< 1.5 V) and low energy density (< 70 Wh/Kg), often with rapid capacity fading and low Coulombic efficiency. The latter became especially pronounced at low C rates.

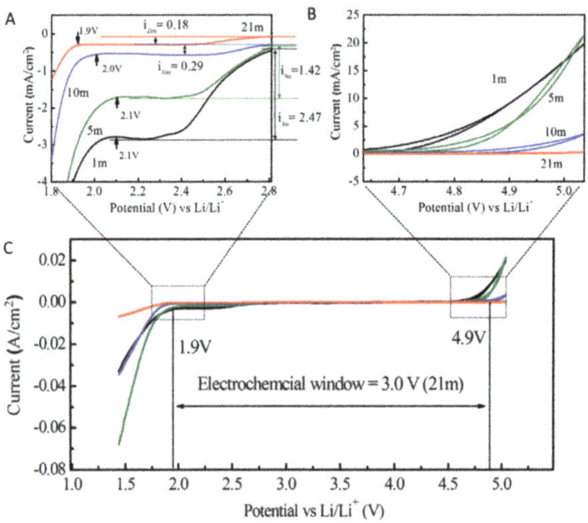

Fig. 1. The electrochemical stability window of LiTFSI-H_2O electrolytes on non-active electrodes: measured at different concentrations (molarity) with cyclic voltammetry (CV) on stainless steel working electrodes between -1.8 V and 1.8 V vs. Ag/AgCl at 10 mV/s), wherein the potential has been converted to Li reference for convenience. The regions outlined near anodic and cathodic extremes in (C) are zoomed-in in (A) and (B).

We report the formation of such interphases in aqueous electrolyte by manipulating the source of electrolyte decomposition during the initial charging processes. A "Water-in-Salt" electrolyte is obtained by dissolving lithium bis(trifluoromethane sulfonyl)imide (LiTFSI) at extremely high concentrations (> 20 m) in water. This leads to an anion-containing Li^+-solvation sheath, which results in the formation of a dense interphase on the anode surface mainly arising from anion reduction. Combined with the significantly reduced electrochemical activity of water at such a high concentration, this highly concentrated "Water-in-Salt" electrolyte provides an expanded electrochemical stability window of ~3.0 V. A full aqueous Li-ion battery constructed with a model electrochemical couple ($LiMn_2O_4$ and Mo_6S_8) demonstrated an open circuit voltage (OCV) of 2.3 V, and was cycled at nearly 100% Coulombic efficiency for up to 1000 cycles at both low (0.15 C) and high (4.5 C) rates.

2. Results

2.1 "Water-in-Salt" Electrolytes

Lithium bis(trifluoromethane sulfonyl)imide (LiTFSI) was chosen as the salt because of its high solubility in water (> 20 m at 25 °C), and high stability against hydrolysis (*15*). When the LiTFSI concentration is above 5 m, the "Water-in-Salt" definition applies as the salt outnumbers the solvent in this binary system by both weight and volume (Fig. S1). In these solutions, the average number of water molecules available to solvate each ion is far below the "solvation numbers" well-established in conventional electrolytes (~1.0 m). Instead, interionic attractions become more pronounced as compared with solvent-ion interactions, with unusual physicochemical properties arising (*16-20*). More importantly, the interphasial chemistry on electrode surfaces could be altered as the direct consequence of the different cation solvation sheath structure (*14, 21-24*). This "cation solvation-interphase chemistry" correlation has been leveraged to manipulate the electrochemical behavior of non-aqueous electrolytes on both graphitic anode and metal oxide cathodes (*17, 23, 25, 26*), but its effect on aqueous electrolytes remained unexplored. Fig. S2~S4 summarize the physical and transport properties of LiTFSI in water at varying concentrations. According to the thermal stability of the highly-concentrated solutions, solutions at both 20 and 21 m are "true" liquids at room temperature (25 °C) that can be supercooled down to -90 °C with negligible crystallinity (*27*). The room temperature conductivity of 21 m solution remains ~10 mS/cm, comparable to non-aqueous electrolytes (9.0 mS/cm) used in commercial Li-ion batteries (Fig. S3 and S4).

The electrochemical stability window for these aqueous electrolytes was evaluated with cyclic voltammetry (CV) on stainless steel electrodes, whose first cathodic and anodic scans are shown in Fig. 1A, B and C. The overall stability window expands as the LiTFSI concentration increases, with both oxygen and hydrogen evolution potentials pushed well beyond the thermodynamic stability limits of water. Upon closer examination of Fig. 1A, the onset of reduction at all concentrations occurs at ~2.80 V, which rapidly accelerates at 2.63 V and then reaches a plateau. Quantum chemistry calculations predict the onset of LiTFSI-reduction at such high concentrations around 2.7~2.9 V, which is slightly higher than hydrogen evolution process (2.63V) corresponding to the exponential current increase. Thus, prior to hydrogen evolution, the reduction of TFSI seems to result in a passivation process that intensifies with salt concentration, and reduces the plateau current by more than an order of magnitude, from 2.47 mA/cm^2 (1 m) to 0.18 mA/cm^2 (21 m). This passivation eventually suppresses hydrogen evolution, pushing its onset from 2.63 V to 1.90 V. On the cathode side (Fig. 1B), the oxygen evolution also seems to be suppressed with increasing salt concentration, but without a clear passivation process, probably due to both the reduced water activity when coordinated to Li^+ at high concentration and an inner-Helmhotz layer increasingly populated by TFSI anions (*28*). Overall, a stability window of ~3.0 V is achieved due to the extra-high concentration of LiTFSI (Fig. 1C), with both cathodic (~ 1.9 V vs. Li) and anodic (~ 4.9 V vs. Li) limits beyond those defined by the Pourbaix-diagram of water at pH=7 (cathodic ~ 2.63 V vs. Li, and anodic ~ 3.86 V vs. Li).

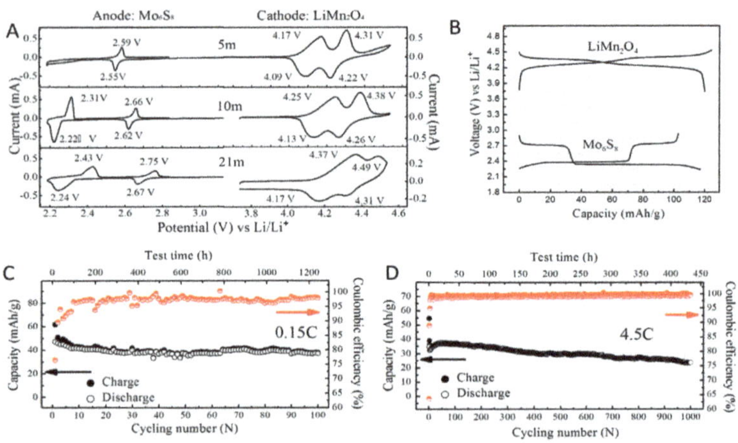

Fig. 2. Electrochemical performance on active electrodes: (A) The electrochemical stability window of electrolytes at various LiTFSI concentrations as measured with CV on active ($LiMn_2O_4$ and Mo_6S_8) electrode surfaces at 0.1 mV/S; and (B) The typical voltage profile of $LiMn_2O_4$ and Mo_6S_8 electrodes in 21m LiTFSI solution at constant current (0.2 C). Data in both (A) and (B) were collected in a three-electrode device including Mo_6S_8 and $LiMn_2O_4$ as working electrodes and Ag/AgCl as reference electrode,

respectively; (C) and (D) The cycling stability and Coulombic efficiencies of full aqueous Li-ion cells employing Mo_6S_8 and $LiMn_2O_4$ as anode and cathode materials in 21m LiTFSI solution, at low (0.15 C) and high (4.5 C) rates, respectively.

Model cathode and anode materials $LiMn_2O_4$ and Mo_6S_8 (Fig. S5) were chosen as active working electrodes in CV experiments to verify the expanded stability window. Fig. 2A demonstrates two redox couples of $LiMn_2O_4$, which gradually shift from 4.09/4.17 V and 4.22/4.31 V in 5 m solution to 4.17/4.37 V and 4.31/4.49 V in 21 m solution, while the reaction kinetics become slower at high salt concentrations. For Mo_6S_8, a single lithiation/delithiation process was observed at ~2.5 V in dilute solutions, and only when salt concentrations reached 10 m or above did the second lithiation/delithiation at ~2.3 V appear. The modulation of redox processes toward positive potentials, observed in all cases in Fig. 2A, is attributed to the Li^+-activity change in the solution according to the Nernst equation (Supporting Information, Eq. 1~8). It is this modulation that moves the second redox process of Mo_6S_8 into the expanded electrochemical stability window of the "Water-in-Salt" electrolyte, allowing the full and reversible utilization of its Li storage sites otherwise inaccessible in diluted aqueous solutions (Fig. S6). Hence, a 2.3 V aqueous $LiMn_2O_4$ /Mo_6S_8 full Li-ion cell is enabled by the "Water-in-Salt" electrolyte (Fig. 2B). In control electrolytes using $LiNO_3$ and Li_2SO_4 (Fig. S7 and S8), similar positive modulations of Mo_6S_8 were observed; however, their limited solubility prohibited the appearance of the second lithiation/delithiation process. Even at the highest possible $LiNO_3$ concentration (12.5 m), hydrogen evolution still overwhelms the second lithiation of Mo_6S_8 at ~2.5 V vs. Li (Fig. S7), without any sign of passivation. Thus, the high concentration of LiTFSI in water not only reduces water activity and modulates redox potentials, but more importantly, suppresses the hydrogen evolution through formation of an interphase.

2.2 A High Voltage Aqueous Li-ion Battery

A full Li-ion cell using $LiMn_2O_4$ cathode and Mo_6S_8 anode was assembled to evaluate the viability of "Water-in-Salt" electrolyte at 21 m (Fig 2B, C, D and Fig. S9~S12). The cathode/anode mass ratio was set to 2:1 in order to compensate for the irreversible capacity at the Mo_6S_8 anode due to formation of SEI during the initial cycles. As Fig. S9 shows, at 0.15 C the cell displays two voltage plateaus at 1.52 V and 2.10 V, respectively, delivering a discharge capacity of 47 mAh/g (of total electrode mass). A conservative estimate made on the basis of average voltage and capacity places the energy density in the vicinity of 84 Wh/Kg (of total electrode mass). To further explore the energy density achievable in the ideal scenario where Li^+-consumption is minimized during the initial interphase formation, a Mo_6S_8 anode recovered from a cycled full cell is matched with a fresh $LiMn_2O_4$ cathode at 1:1 ratio. Such a full cell delivers an energy density of 100 Wh/Kg (of total electrode mass) (Fig. S11). A full aqueous Li-ion cell using Mo_6S_8 and $LiNi_{0.8}Co_{0.15}Al_{0.05}O_2$ cathode also cycles reversibly but yields a lower capacity (Fig. S13).

Fig. 2C and D display the cycling stability and Coulombic efficiency of $LiMn_2O_4/Mo_6S_8$ full cells at both high (4.5 C) and low (0.15 C) rates. Excellent stability with high capacity retention (68% with 1000 cycles at 4.5C and 78% with 100 cycles at 0.15C) and near 100% Coulombic efficiency are observed at both rates (Fig. S14). As taught by Dahn et al, the most rigorous proof of stability does not come from the number of cycles, but from the time spent by a system at a fully charged state (*29*) as well as from high Coulombic efficiency at low C rates. In previous studies, the cycling stability of aqueous Li-ion cells was often tactically evaluated at high instead of low rates, so that the effect of residual hydrogen/oxygen evolution on cycling stability would be less apparent. To reveal how much impact the trace parasitic reactions in "Water-in-Salt" electrolyte exert on full Li-ion cell performance, the open-circuit voltage decay of fully charged cells was monitored upon storage, which was followed by an immediate measurement of recoverable capacity. The results are shown in Fig. S15 and S16, confirming the negligible effect of either hydrogen or oxygen evolution. Only in longer term cycling tests would the effect of parasitic reactions become more apparent, as evidenced by the slow but steady capacity fading shown in Fig. 2C and D, suggesting that the interphasial chemistry needs to be tailored for more effective protection. Elevated temperature (45 °C) did not induce accelerated capacity fading, indicating that the SEI formed should be stable against dissolution (Fig. S12).

2.3 Interphasial Chemistry and its Significance

High salt concentration deviates solution dynamics from an ideal state (*30*). To quantify such deviation, we calculated the relative Li^+ activity coefficients in electrolytes based on the peak potential shifts as measured from CV (Fig. 2A). The details are given in supplementary section (Table S1 and S2), while Fig. S17 summarizes the dependence of these coefficients on salt concentration, which experiences a steady rise from 5 m to 21 m. According to Bockris et al., this increase of activity coefficients at high concentration reflects the deprived population of water as a free solvent and the intensified cation-anion interaction (*30*). The interplay between Li^+, $TFSI^-$ and water was also investigated with spectroscopies. Major Raman vibrational signals are summarized in Fig. S18 and Table S3, while Fig. 3A compares the shift of the most intense peak at 744 cm^{-1} as the salt concentration increases, with crystalline LiTFSI as the reference (top trace). Evidently this vibration mode is rather sensitive to the change of the anion environment, drifting from 744.6 cm^{-1} at 5.0 m to 748.5 cm^{-1} at 21 m. The latter is essentially identical to that in a crystalline lattice. Based on previous assignment of 748 cm^{-1} peak in non-aqueous electrolytes (*31*), we attribute this band to the formation of a percolated $TFSI^-$-anionic network in the highly concentrated solutions. In the ^{17}O-NMR spectra, the two sets of oxygen nuclei were assigned to water (~ 0 ppm) and $TFSI^-$ (~154 ppm), respectively, and Fig. 3B displays the change in ^{17}O-chemical shifts for both nuclei with LiTFSI concentration. The water ^{17}O-signal is rather sensitive to the presence of salt, because Li^+ directly interacts with the lone pair electrons of water oxygen. Its coordination with Li^+ results in the deshielding as indicated by the downshift displacement in its chemical shift, while the increasing salt concentration above 10 m intensifies this effect. On the other hand, the ^{17}O-signal at 154 ppm seems to be rather insensitive to salt presence although its successive downshift is still visible. This can be attributed to the fact that oxygen on $TFSI^-$ does not directly coordinate with Li^+. Interestingly, both oxygen nuclei display obvious peak broadening at salt concentrations > 10 m, as if the anion environment gradually transitions into a semi-solid state similar to a LiTFSI crystal. Such an intimate interaction between cation and anion as revealed by Raman and NMR bears close resemblance to an ionic liquid, where on average each Li^+ is surrounded by at least one anion. Naturally, when Li^+ in this coordination environment is brought into the vicinity of the anode surface, $TFSI^-$ would be reasonably expected to display its own electrochemical features.

The relationship between solution structure and electrochemical properties is further revealed from molecular dynamics (MD) simulations. For dilute solutions $\leq$5 m, Li^+ remains well hydrated in its primary solvation sheath with sufficient free water available (Fig. 3C). In such electrolytes, attempts to lithiate an anode, whose lithiation potential is below that of water reduction, would lead to preferential reduction of water and the sustained hydrogen-evolution, which prevents not only any Li^+-intercalation but also any possible reduction of $TFSI^-$. However, with LiTFSI concentrations increased > 20 m, MD simulation predicts that on average 2 TFSI anion would be observed in each Li^+ primary solvation sheath (Fig. 3D), and such high probability of $TFSI^-$ leads to an interphasial chemistry dominated by the reduction of $TFSI^-$. The reduction potential of $TFSI^-$is also altered by its intimate interaction with Li^+. According to DFT calculations (Fig. 4A and S19), aggregates like $Li_2(TFSI)(H_2O)_x$ become reductively unstable below 2.9 V vs. Li, which is significantly higher than the reduction potential for the isolated TFSI-anion at 1.4 V (Fig. 4A) and hydrogen evolution at 2.63 V. Further corroborating the preferential reduction of $TFSI^-$ over water, a density-of-states analysis obtained from DFT calculations indicates that both the TFSI conduction band minimum and the water valence band maximum shift to lower potentials with increasing salt concentration. At >

20 m, this trend leads to a premature TFSI$^-$-reduction and delayed water oxidation (Fig. 4B and supporting information). We therefore conclude that this reduction process generates sufficient LiF from TFSI$^-$ to form an anode/electrolyte interphase, which kinetically prevents sustained reduction of both water and TFSI$^-$ in a similar manner as an SEI in non-aqueous electrolytes does. The earlier observation that TFSI$^-$ was electrochemically reduced to LiF on an anode surface in non-aqueous electrolyte *(32, 33)* in conjunction with limited solubility of LiF in water make this speculation reasonable. In retrospect, an ideal lithium salt that could enable the formation of an aqueous SEI should be not only highly soluble and chemically stable in water, but also be susceptible to electrochemical reduction at a desired potential, i.e., a potential higher than that of H_2-evolution, producing a solid product insoluble in aqueous media. Lithium salts meeting these requirements include those based on fluoroalkyl sulfonylimide anions, to which LiTFSI belongs, as well as fluoroalkyl sulfonate anions. Compared with the higher cost of imides, sulfonates might make more appropriate candidates, whose effect on aqueous electrolytes has been found to be nearly identical to that of LiTFSI (Fig. S20).

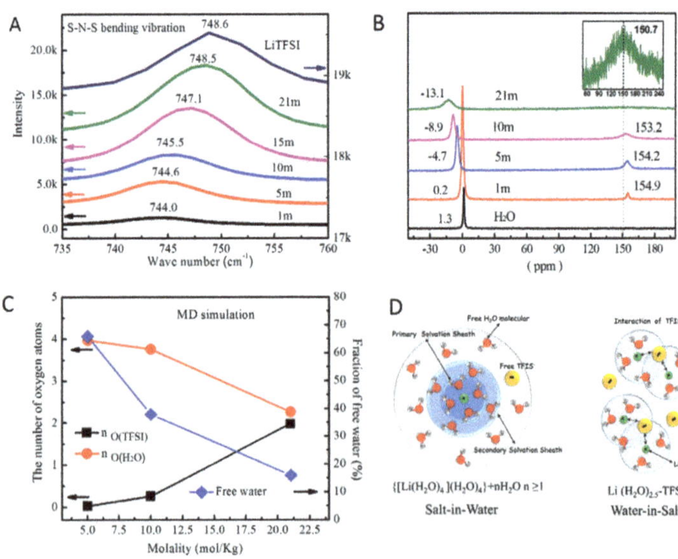

Fig. 3. The effect of LiTFSI concentration on ion-solvent and ion-ion interactions: (A) The progression of Raman vibration at ~744 cm^{-1} with salt concentration; (B) The

change of chemical shifts for [17]O-nuclei in both solvent (water) and anion (TFSI), wherein the inset shows the TFSI⁻ peak at 21m; (C) The number of water and TFSI⁻ oxygen atoms within the Li^+-primary solvation sheath of 0.27 nm and the fraction of "free" water not bound to any Li^+, all obtained from MD simulations; and (D) The schematic illustration of the evolution of Li^+-primary solvation sheath in diluted and "Water-in-Salt" solutions.

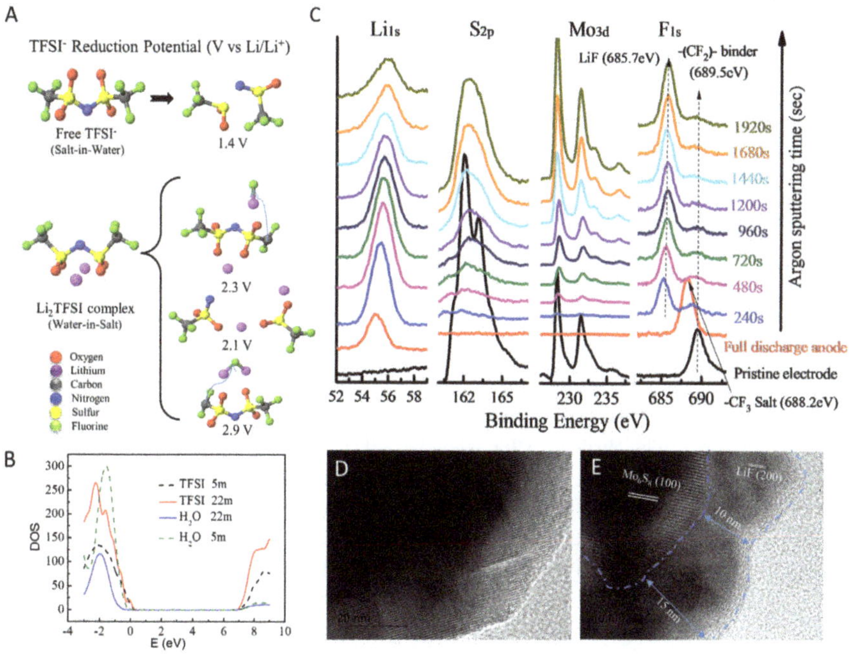

Fig. 4. SEI formation in aqueous electrolyte: (A) Predicted reduction potentials from G4MP2 quantum chemistry calculations; (B) Projected density of states (DOS) for H_2O-LiTFSI electrolyte from HSE06 DFT calculations; (C) X-ray photoelectron spectroscopy (XPS) spectrum of pristine (bottom) and cycled Mo_6S_8 at full lithiation state after various durations of Ar^+-sputtering; and TEM images of both pristine (D) and cycled Mo_6S_8 (E).

To demonstrate the existence of an interphase in our aqueous electrolytes, X-ray photoelectron spectroscopy (XPS) was conducted on an anode recovered after a complete lithiation cycle. Fig. 4C shows the valence electrons binding energies for a few detected elements. Included as comparisons are the spectra of the pristine Mo_6S_8, whose Mo 3d and S 2p signals serve as references, and F 1s detected therein at 689.5 eV came from poly(tetrafluoroethylene) (PTFE) used as binder in the composite electrode. The most conspicuous change after being charged to 2.3 V (corresponding to full lithiation of Mo_6S_8) is the disappearance of both S 2p and Mo 3d signals, and the appearance of Li 1s and F 1s signals, the latter of which is identified to be CF_3-species. Hence, lithiation left the anode surface completely covered with species originated from $TFSI^-$, while the CF_3-species could arise from either excess LiTFSI remaining on the anode surface, or its incomplete reduction products. Etching this surface by Ar^+ reveals an underlying interphase consisting overwhelmingly of LiF, which not only precedes the re-appearance of the original S 2p and Mo 3d signals, but more interestingly, remains even after prolonged (1920 s) sputtering. This resistance against sputtering is undoubtedly the consequence of a very dense surface interphase. Based on earlier reports that $TFSI^-$ is electrochemically reduced to LiF (*32, 33*), we believe that in the current "Water-in-Salt" electrolytes, this LiF-rich interphase serves as an electron barrier preventing the reduction of water while allowing Li^+-migration. Evidence also comes from transmission electron microscopy (TEM) pictures of the cycled Mo_6S_8, whose surface, when compared with pristine state, is found to be covered with a crystalline phase of 10~15 nm thickness (Fig. 4D and E). Interplanar spacing of this crystalline phase identifies itself as imperfect crystalline LiF. Further confirmation comes from TEM energy-dispersive X-ray spectroscopy (EDX) analysis (Fig. S21), whose line-scan reveals that F evenly distributes on the Mo_6S_8 particle surface. This aqueous LiF-based SEI, originated from $TFSI^-$-reduction and crystalline in nature, seems to differ chemically from the composite interphases well characterized in non-aqueous electrolytes *(4, 14)*, where solvent-reduction products contribute overwhelmingly as the chemical building blocks. On the cathode surface, however, no such LiF-based SEI was found even after extended cycling (Fig. S22 and S23).

Summarizing all the spectra information, an overall picture of Li^+-solvation sheath and its role in interphasial chemistry emerges. In dilute solutions (< 5 m) (*34, 35*), the abundance of water can afford a conventional solvation sheath structure, i.e., four water molecules per Li^+ in the primary and more in the loosely-bound secondary sheath (*36*), where Li^+ and $TFSI^-$ are well-separated by water. When salt concentration increases to a certain level (> 10 m), the insufficient water population leads to a drastic change in Li^+ solvation sheath structure. In particular, at 21 m, there are only 2.6 water molecules per Li^+, which can no longer effectively neutralize the electrostatic field created by the formal charge on Li^+; consequently, $TFSI^-$ enters the Li^+ solvation sheath, leading to intimate Li^+-$TFSI^-$ interactions as observed in Raman and ^{17}O-NMR. A direct result of this anion-containing solvation sheath is an interphasial chemistry on the anode consisting of mainly LiF from $TFSI^-$-reduction, which is enabled not only by the high probability of $TFSI^-$ in the Li^+-

solvation sheath, but also by its now higher-than-water reduction potential. Simultaneously, the scarcity of free water molecules due to high salt concentration contributes to the its inactivity that benefits both cathodic and anodic stability limits. These factors work in synergy to realize an expanded electrochemical stability window of 3.0 V (Fig. 1C).

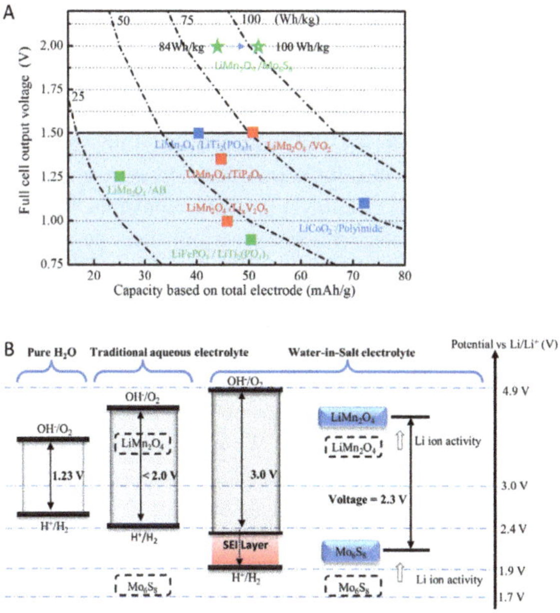

Fig. 5. Performance of aqueous Li-ion chemistries benefitted from expanded electrochemical stability window of aqueous electrolytes. (A) Performances for aqueous Li-ion batteries based on various electrochemical couples; and (B) Schematic illustration of expanded electrochemical stability window for "Water-in-Salt" electrolytes together with the modulated redox couples of both $LiMn_2O_4$ cathode and Mo_6S_8 anode caused by high salt concentration.

Battery performance can be quantified with four parameters, i.e., cell voltage, capacity, Coulombic efficiency and cycling stability. While the first two determine the energy density, the latter two dictate the life and energy efficiency. For comparison, we plotted our "Water-in-Salt" battery in Fig. 5A against other aqueous systems previously investigated using these parameters, in which the cycling stability is color-coded with red, blue and green representing <100 cycles, 100~200 cycles, and >1000 cycles, respectively. More than 1000 cycles were reported for electrochemical couples of $LiMn_2O_4$/acetylene black (*37*) and $LiFePO_4/LiTi_2(PO_4)_3$ (*13*). Their excellent stability, however, was achieved at the expense of voltage (< 1.25 V) and energy density (< 50 Wh/kg). On the other hand, efforts to increase cell voltage to ~1.50 V were accompanied by an appreciable compromise in cycling stability (*10, 38-41*). In all cases, energy densities were below 75 Wh/Kg.The formation of an anode/electrolyte interphase in "Water-in-Salt" electrolyte enables us to decouple voltage from cycling stability and achieve high in both. It should be emphasized that the electrochemical couple used in the current work ($LiMn_2O_4$ and Mo_6S_8) did not actually fully utilize the expanded electrochemical stability window, thus further advances in exploring more powerful and energetic battery chemistry for this electrolyte system are possible. From our perspective, perhaps more important than the electrochemical performances described in Fig. 2C and D is the fact that an interphase could form and function in aqueous electrolytes, which opens avenues to aqueous electrochemistry (Fig. 5B).

Acknowledgments

The PIs (CW and KX) gratefully acknowledge both funding and inspiration from Dr. Ping Liu of DOE ARPA-E (DEAR0000389). We also acknowledge the support of the Maryland Nano Center and its NispLab. The NispLab is supported in part by the NSF as a MRSEC Shared Experimental Facility. We acknowledge Dr. Karen Gaskell, Dr. Yiu-Fai Lam, Dr. Sz-Chian Liou, Mr.Ankit Goyal and Prof. Srinivasa R. Raghavan for their technical support and helpful discussion. Modeling efforts were supported by ARL Enterprise for Multiscale Research of Materials, and MO was supported by an Oak Ridge Associated Universities (ORAU) Postdoctoral Fellowship.

The work has been publihsed previously by Science:

Liumin Suo, Oleg Borodin, Tao Gao, Marco Olguin, Janet Ho, Xiulin Fan, Chao Luo, Chunsheng Wang, Kang Xu, Water-in-salt" electrolyte enables high-voltage aqueous lithium-ion chemistries, Science, 20 November 2015 • Vol 350 Issue 6263. https://doi.org/10.1126/science.aab1595

Supplemental Materials

Supplement figures and tables can be accessed at the following website for free:

www.sciencemag.org/content/350/6263/938/suppl/DC1

References

[1] M. Armand, J. M. Tarascon, Building better batteries. *Nature* 451, 652-657 (2008). https://doi.org/10.1038/451652a

[2] M. S. Whittingham, Lithium batteries and cathode materials. *Chem. Rev.* 104, 4271-4301 (2004). https://doi.org/10.1021/cr020731c

[3] B. Dunn, H. Kamath, J. M. Tarascon, Electrical Energy Storage for the Grid: A Battery of Choices. *Science* 334, 928-935 (2011). https://doi.org/10.1126/science.1212741

[4] K. Xu, Nonaqueous liquid electrolytes for lithium-based rechargeable batteries. *Chem.Rev.* 104, 4303-4417 (2004). https://doi.org/10.1021/cr030203g

[5] L. Hu, K. Xu, Nonflammable electrolyte enhances battery safety. *Proc. Natl. Acad. Sci.* 111, 3205-3206 (2014). https://doi.org/10.1073/pnas.1401033111

[6] A. Hammami, N. Raymond, M. Armand, Lithium-ion batteries: Runaway risk of forming toxic compounds. *Nature* 424, 635-636 (2003). https://doi.org/10.1038/424635b

[7] J. B. Goodenough, Y. Kim, Challenges for Rechargeable Li Batteries. *Chem.Mater.* 22, 587-603 (2010). https://doi.org/10.1021/cm901452z

[8] J. M. Tarascon, M. Armand, Issues and challenges facing rechargeable lithium batteries. *Nature* 414, 359-367 (2001). https://doi.org/10.1038/35104644

[9] H. Kim *et al.*, Aqueous Rechargeable Li and Na Ion Batteries. *Chem. Rev.* 114, 11788-11827 (2014). https://doi.org/10.1021/cr500232y

[10] W. Li, J. R. Dahn, D. S. Wainwright, Rechargable lithium batteries with aqueous electrolytes. *Science* 264, 1115-1118 (1994). https://doi.org/10.1126/science.264.5162.1115

[11] Y. G. Wang, J. Yi, Y. Y. Xia, Recent progress in aqueous lithium-ion batteries. *Adv. Energy Mater.* 2, 830-840 (2012). https://doi.org/10.1002/aenm.201200065

[12] W. Li, W. R. McKinnon, J. R. Dahn, Lithium intercalation from aqueous solutions. *J. Electrochem. Soc.* 141, 2310-2316 (1994). https://doi.org/10.1149/1.2055118

[13] J-Y. Luo, W-J. Cui, P. He, Y-Y. Xia, Raising the cycling stability of aqueous lithium-ion batteries by eliminating oxygen in the electrolyte. *Nat. Chem.* 2, 760-765 (2010). https://doi.org/10.1038/nchem.763

[14] K. Xu, Electrolytes and interphases in Li-Ion Batteries and Beyond. *Chem. Rev.* 114, 11503-11618 (2014). https://doi.org/10.1021/cr500003w

[15] S. F. Lux *et al.*, LiTFSI stability in water and its possible use in aqueous lithium-ion batteries: pH dependency, electrochemical window and temperature stability. *J. Electrochem. Soc.* 160, A1694-A1700 (2013). https://doi.org/10.1149/2.039310jes

[16] C. A. Angell, C. Liu, E. Sanchez, Rubbery solid electrolytes with dominant cationic transport and high ambient conductivity. *Nature* 362, 137-139 (1993). https://doi.org/10.1038/362137a0

[17] K. Yoshida, M. Tsuchiya, N. Tachikawa, K. Dokko, M. Watanabe, Change from glyme solutions to quasi-ionic liquids for binary mixtures consisting of lithium bis(trifluoromethanesulfonyl)amide and glymes. *J. Phys. Chem. C.* 115, 18384-18394 (2011). https://doi.org/10.1021/jp206881t

[18] L. Suo, Y-S. Hu, H. Li, M. Armand, L. Chen, A new class of Solvent-in-Salt electrolyte for high-energy rechargeable metallic lithium batteries. *Nat. Commun.* 4, 1481 (2013). https://doi.org/10.1038/ncomms2513

[19] K. Ueno, K. Yoshida, M. Tsuchiya, N. Tachikawa, K. Dokko, M. Watanabe, Glyme-Lithium Salt Equimolar Molten Mixtures: Concentrated Solutions or Solvate Ionic Liquids?. *J. Phys. Chem. B.* 116, 11323–11331 (2012). https://doi.org/10.1021/jp307378j

[20] J-W. Park *et al.*, Solvent Effect of Room-Temperature Ionic Liquids on Electrochemical Reactions in Lithium-Sulfur Batteries. *J. Phys, Chem. C.* 117, 4431-4440 (2013). https://doi.org/10.1021/jp400153m

[21] K. Xu, Y. Lam, S. S. Zhang, T. R. Jow, T. B. Curtis, Solvation sheath of Li^+ in nonaqueous electrolytes and its implication of graphite/electrolyte interface chemistry. *J. Phys. Chem. C.* 111, 7411-7421 (2007). https://doi.org/10.1021/jp068691u

[22] K. Xu, A. von Cresce, Interfacing electrolytes with electrodes in Li ion batteries. *J. Mater. Chem.* 21, 9849-9864 (2011). https://doi.org/10.1039/c0jm04309e

[23] M. Nie *et al.*, Role of solution structure in solid electrolyte interphase formation on graphite with $LiPF_6$ in propylene carbonate. *J. Phys. Chem. C.* 117, 25381-25389 (2013). https://doi.org/10.1021/jp409765w

[24] S-K. Jeong *et al.*, A simple method of electrochemical lithium intercalation within graphite from a propylene carbonate-based solution. *Electrochem. Commun.* 31, 24-27 (2013). https://doi.org/10.1016/j.elecom.2013.02.019

[25] Y. Yamada *et al.*, Unusual stability of acetonitrile-based superconcentrated electrolytes for fast-charging lithium-Ion batteries. *J. Am. Chem. Soc.* 136, 5039-5046 (2014). https://doi.org/10.1021/ja412807w

[26] K. Yoshida *et al.*, Oxidative-stability enhancement and charge transport mechanism in glyme–Lithium salt equimolar complexes. *J. Am. Chem. Soc.* 133, 13121-13129 (2011). https://doi.org/10.1021/ja203983r

[27] G. Perron, D. Brouillette, J. E. Desnoyers, Comparison of the thermodynamic and transport properties of lithium bis(trifluoromethylsulfonyl)imide (LiTFSI) with $LiClO_4$ and Bu_4NBr in water at 25 degrees C. *Can. J. Chem.* 75, 1608-1614 (1997). https://doi.org/10.1139/v97-191

[28] D. W. McOwen *et al.*, Concentrated electrolytes: decrypting electrolyte properties and reassessing Al corrosion mechanisms. *Energ. Environ. Sci.* 7, 416-426 (2014). https://doi.org/10.1039/C3EE42351D

[29] J. C. Burns *et al.*, Evaluation of effects of additives in wound Li-Ion cells through high precision coulometry. *J. Electrochem. Soc.* 158, A255 (2011). https://doi.org/10.1149/1.3531997

[30] J. O. M. Bockris, A. K. N. Reddy, Mosern Electrochemistry, 2nd Ed, Kluwer Academic/Plenum Publishers. (2000).

[31] D. M. Seo, O. Borodin, S.-D. Han, P. D. Boyle, W. A. Henderson, Electrolyte solvation and ionic association II. acetonitrile-lithium salt mixtures: highly dissociated salts. *J. Electrochem. Soc.* 159, A1489-A1500 (2012). https://doi.org/10.1149/2.035209jes

[32] E. Krämer *et al.*, Mechanism of anodic dissolution of the aluminum current collector in 1 M LiTFSI EC:DEC 3:7 in rechargeable lithium batteries. *J. Electrochem. Soc.* 160, A356-A360 (2013). https://doi.org/10.1149/2.081302jes

[33] A. v. Cresce, S. M. Russell, D. R. Baker, K. J. Gaskell, K. Xu, In situ and quantitative characterization of solid electrolyte interphases. *Nano Letters* 14, 1405-1412 (2014). https://doi.org/10.1021/nl404471v

[34] Y. Marcus, Effect of ions on the structure of water: structure making and breaking. *Chem. Rev.* 109, 1346-1370 (2009). https://doi.org/10.1021/cr8003828

[35] H. Ohtaki, T. Radnai, Structure and dynamics of hydrated ions. *Chem. Rev.* 93, 1157-1204 (1993). https://doi.org/10.1021/cr00019a014

[36] X. Bogle, R. Vazquez, S. Greenbaum, A. V. Cresce, K. Xu, Understanding Li^+-Solvent Interaction in Nonaqueous Carbonate Electrolytes with O-17 NMR. *J. Phys.Chem. Lett.* 4, 1664-1668 (2013). https://doi.org/10.1021/jz400661k

[37] Y-G. Wang, Y-Y. Xia, Hybrid Aqueous energy storage cells using activated carbon and lithium-intercalated compounds: I. the $C/LiMn_2O_4$ System. *J. Electrochem. Soc.* 153, A450-A454 (2006). https://doi.org/10.1149/1.2140678

[38] J. Y. Luo, Y. Y. Xia, Aqueous lithium-ion battery $LiTi_2(PO_4)_3/LiMn_2O_4$ with high power and energy densities as well as superior cycling stability. *Adv. Funct. Mater.* 17, 3877-3884 (2007). https://doi.org/10.1002/adfm.200700638

[39] H. Wang, Y. Zeng, K. Huang, S. Liu, L. Chen, Improvement of cycle performance of lithium ion cell $LiMn_2O_4/Li_xV_2O_5$ with aqueous solution electrolyte by polypyrrole coating on anode. *Electrochim. Acta.* 52, 5102-5107 (2007). https://doi.org/10.1016/j.electacta.2007.02.004

[40] H. Wang, K. Huang, Y. Zeng, S. Yang, L. Chen, Electrochemical properties of TiP_2O_7 and $LiTi_2(PO_4)_3$ as anode material for lithium ion battery with aqueous solution electrolyte. *Electrochim. Acta.* 52, 3280-3285 (2007). https://doi.org/10.1016/j.electacta.2006.10.010

[41] H. Qin, Z. P. Song, H. Zhan, Y. H. Zhou, Aqueous rechargeable alkali-ion batteries with polyimide anode. *J. Power.Sources.* 249, 367-372 (2014). https://doi.org/10.1016/j.jpowsour.2013.10.091

[42] T. R. Ferguson, M. Z. Bazant, Nonequilibrium thermodynamics of porous electrodes. *J. Electrochem. Soc.* 159, A1967-A1985 (2012). https://doi.org/10.1149/2.048212jes

[43] O. N. Starovoytov, O. Borodin, D. Bedrov, G. D. Smith, Development of a polarizable force field for molecular dynamics simulations of poly (Ethylene Oxide) in aqueous Solution. *J. Chem. Theory Comput.* 7, 1902-1915 (2011). https://doi.org/10.1021/ct200064u

[44] O. Borodin, Polarizable force field development and molecular dynamics simulations of ionic liquids. *J. Phys. Chem. B* 113, 11463-11478 (2009). https://doi.org/10.1021/jp905220k

[45] C. M. Baker, A. D. MacKerell, Polarizability rescaling and atom-based Thole scaling in the CHARMM Drude polarizable force field for ethers. *J. Mol. Model* 16, 567-576 (2010). https://doi.org/10.1007/s00894-009-0572-4

[46] H. B. Yu *et al.*, Simulating Monovalent and Divalent Ions in Aqueous Solution Using a Drude Polarizable Force Field. *J. Chem. Theory. Comput.* 6, 774-786 (2010). https://doi.org/10.1021/ct900576a

[47] C. W. Bauschlicher, J. B. Haskins, E. W. Bucholz, J. W. Lawson, O. Borodin, Structure and Energetics of Li^+-$(BF_4^-)_n$, Li^+-$(FSI^-)_n$, and Li^+-$(TFSI^-)_n$: Ab Initio and Polarizable Force Field Approaches. *J. Phys. Chem. B.* 118, 10785-10794 (2014). https://doi.org/10.1021/jp506422p

[48] P. T. Kiss, A. Baranyai, A new polarizable force field for alkali and halide ions. *J. Chem. Phys.* 141, (2014). https://doi.org/10.1063/1.4895129

[49] B. Dunweg, K. Kremer, Molecular dynamics simulation of a polymer chain in solution. *J. Chem. Phys.* 99, 6983-6997 (1993). https://doi.org/10.1063/1.465445

[50] D. M. Seo *et al.*, Electrolyte Solvation and Ionic Association III. Acetonitrile-Lithium Salt Mixtures–Transport Properties. *J. Electrochem. Soc* 160, A1061-A1070 (2013). https://doi.org/10.1149/2.018308jes

[51] H. Kim *et al.*, In Situ Formation of Protective Coatings on Sulfur Cathodes in Lithium Batteries with LiFSI-Based Organic Electrolytes. *Adv. Energy Mater.* 5, 1401792 (2014). https://doi.org/10.1002/aenm.201401792

[52] O. Borodin, W. Behl, T. R. Jow, Oxidative Stability and Initial Decomposition Reactions of Carbonate, Sulfone, and Alkyl Phosphate-Based Electrolytes. *J. Phys. Chem. C* 117, 8661-8682 (2013). https://doi.org/10.1021/jp400527c

[53] J. Vatamanu, O. Borodin, G. D. Smith, Molecular Dynamics Simulation Studies of the Structure of a Mixed Carbonate/$LiPF_6$ Electrolyte near Graphite Surface as a Function of Electrode Potential. *J. Phys. Chem. C* 116, 1114-1121 (2012). https://doi.org/10.1021/jp2101539

[54] G. Lippert, J. Hutter, M. Parrinello, A hybrid Gaussian and plane wave density functional scheme. *Mol. Phys.* 92, 477-487 (1997). https://doi.org/10.1080/002689797170220

[55] J. VandeVondele *et al.*, Quickstep: Fast and accurate density functional calculations using a mixed Gaussian and plane waves approach. *Comput. Phys. Commun.* 167, 103-128 (2005). https://doi.org/10.1016/j.cpc.2004.12.014

[56] C. Hartwigsen, S. Goedecker, J. Hutter, Relativistic separable dual-space Gaussian pseudopotentials from H to Rn. *Phys. Rev. B.* 58, 3641-3662 (1998). https://doi.org/10.1103/PhysRevB.58.3641

[57] J. Heyd, G. E. Scuseria, M. Ernzerhof, Hybrid functionals based on a screened Coulomb potential. *J. Chem. Phys.* 118, 8207-8215 (2003). https://doi.org/10.1063/1.1564060

[58] A. V. Krukau, O. A. Vydrov, A. F. Izmaylov, G. E. Scuseria, Influence of the exchange screening parameter on the performance of screened hybrid functionals. *J. Chem. Phys.* 125, (2006). https://doi.org/10.1063/1.2404663

[59] M. Guidon, J. Hutter, J. VandeVondele, Robust periodic Hartree-Fock exchange for large-Scale simulations using Gaussian basis sets. *J. Chem. Theory Comput.* 5, 3010-3021 (2009). https://doi.org/10.1021/ct900494g

[60] M. Guidon, J. Hutter, J. VandeVondele, Auxiliary density matrix methods for Hartree-Fock exchange calculations. *J. Chem. Theory Comput.* 6, 2348-2364 (2010). https://doi.org/10.1021/ct1002225

[61] J. Paier *et al.*, Accurate Hartree-Fock energy of extended systems using large Gaussian basis sets. *Phys. Rev. B.* 80, (2009). https://doi.org/10.1103/PhysRevB.80.174114

Chapter 7

Pyrite FeS_2 as an Efficient Adsorbent of Lithium Polysulfide for Improved Lithium–Sulphur Batteries

Sheng S. Zhang and Dat T. Tran

Energy and Power Division, Sensors and Electron Devices Directorate, U.S. Army Research Laboratory, Adelphi, MD 20783-1138, USA

shengshui.zhang.civ@mail.mil or shengshui@gmail.com

Abstract

Pyrite FeS_2 chemically combines lithium polysulphide (Li_2S_n) to form active Li_2FeS_{2+n} complexes that substantially reduce the out-diffusion of dissolved Li_2S_n from sulphure cathode and consequently improve the cycling performance of lithium-sulphur batteries. The same functions are also applicible to other tranisition metal sulfides, which opens a new direction for polysulfide sequestration in lithium-sulphur batteries.

Keywords

Li-S Battery, FeS_2, Polysulphide, Chemical Adsorption, Li Anode

Contents

1. Introduction

Lithium-sulphur (Li-S) batteries have recently been under intensive investigation due to their potential to provide higher energy density and lower cost than state-of-the-art Li-ion batteries. In Li-S batteries, the dissolution of long-chain lithium polysulphide (PS, Li_2S_n with n>2), a series of intermediates of sulphur reduction, in organic electrolytes has been

This work has been previously published by the Royal society of chemistry: Sheng S. Zhang, Dat T. Tran, "Pyrite FeS2 as an efficient adsorbent of lithium polysulphide for improved lithium–sulphur batteries", J. Mater. Chem. A, 2016, 4, 4371-4374. https://doi.org/10.1039/C6TA01214K

recognized to be unavoidable and essential for enabling the fast reaction kinetics and high utilization of electrically insulating elemental sulphur.[1] Such dissolution was also identified to be the source for a number of known problems with the Li-S batteries, including fast capacity fading, low energy efficiency, high self-discharge rate and short cycle life. Previous effort on the subject has been focused on the sequestration of dissolved Li_2S_n from diffusing out of the cathode on the basis of a concept to form sulphur composites with a porous carbon material or a porous metal oxide.[2-5] These materials sequestrate PS mainly through weak van der Waals interaction (i.e., the physical absorption) and features temporary sequestration. Upon long-term operation, the Li-S cells still suffer significant capacity fading. More recently, the attention has moved to the chemical adsorption that traps PS species via strong chemical binding and features long-term PS sequestration.[6-8] Typical materials studied for the chemical adsorption are single- or co-heteroatom doped carbon materials,[6] such as those doped or co-doped by the electron-donated oxygen,[9, 10] sulphur,[11, 12] nitrogen,[13-15] as well as the electron-accepted boron.[16, 17] These materials sequestrate PS either through the interaction of heteroatom (HA) with positive Li^+ cation in Li_2S_n molecules to form HA···Li_2S_n binding for the electron-donated HA or through the interaction with negative PS anion to form Li_2S_n···HA binding for the electron-accepted HA.[16, 17] Beside the above, transition metal oxides[18] such as TiO_2,[19] Ti_4O_7,[20] MnO_2,[21] SnO_2,[22] and $NiFe_2O_4$;[23] transition metal sulphides[24] such as CuS,[25] Co_9S_8,[26] CoS_2,[27] and TiS_2;[24, 28, 29] MXene;[30] or even metal particles[31-33] that are able to *in-situ* form metal sulphide with sulphur or Li_2S_n also show the ability to chemically adsorb Li_2S_n species, whose PS sequestration is generally based on interactions between the hard Lewis acid (M^{n+} cation) and the soft Lewis base (PS anion). Moreover, such interactions were also shown to promote preferential deposition of the ultimate Li_2S product on the host surface, which further improves the Li-S cell's performance.[20, 22] In addition to the Lewis acid-base interaction, some transition metal oxides with their reduction potential slightly higher than the oxidation potential of Li_2S_n, such as MnO_2[34] and VO_2,[35] have been reported to sequestrate Li_2S_n by oxidizing PS anions to form active surface-bound polythionate species. However, all the above host materials are not inexpensive, which require either a nanoscale or a highly porous structure for efficient PS sequestration.

Aiming to sequestrate PS effectively, we notice that many earth alkali metal and transition metal cations can combine with PS anions to form stable and stoichiometric MS_n complexes,[36-38] and that the resulting complexes can reversibly decompose to MS and sulphur.[39, 40] Based on this knowledge, in the present work we first explored a new approach for PS sequestration according to the formation of MS_n complexes by starting with an inexpensive and earth-abundant pyrite FeS_2. We believe that the conclusions obtained from FeS_2 are also applicable to other earth alkali metal and transition metal sulphides.

2. Results and Discussion

In order to remain the low cost advantage, commercial elemental sulphur and pyrite FeS_2 were used without any treatments. The scanning electron microscopy (SEM) image (Fig. 1a) shows that pristine FeS_2 particles are ranged in size from several to tens micrometers and distributed in very irregular shapes. A Li_2S_9-triglyme solution was prepared and used to examine the chemical adsorption of Li_2S_n to FeS_2 host. A 0.5 mol kg^{-1} Li_2S_9 solution was in red-dark colour and turned to slight green colour when diluted to 0.005 mol kg^{-1} because some of the S_9^{2-} anions are dissociated into short polysulphide radical anions ($S_n^{\cdot-}$) in a diluted solution as suggested by, for example, eqn. 1.[41]

$$S_9^{2-} \leftrightarrow S_5^{\cdot-} + S_4^{\cdot-} \quad (1)$$

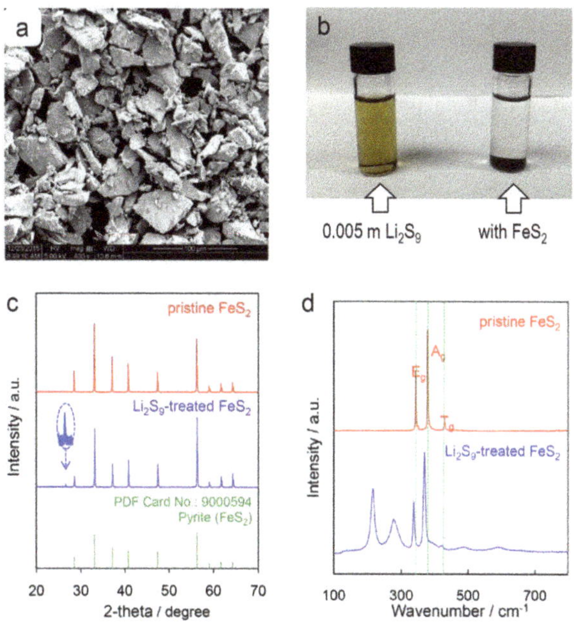

Fig. 1 (a) SEM image of FeS_2, (b) visual photo of a 0.005 mol kg^{-1} Li_2S_9 solution, (c) XRD patterns of FeS_2 before and after Li_2S_9-treatment, and (d) Raman spectra of FeS_2 before and after Li_2S_9-treatment.

As indicated by Fig. 1b, the 0.005 mol kg^{-1} Li_2S_9 solution was completely decoloured in minutes after adding small amount of FeS_2 powder, indicating strong PS adsorption. In order to understand the mechanism of PS adsorption, the FeS_2 powder was treated with a 0.5 mol kg^{-1} Li_2S_9-triglyme solution at 80 °C for overnight, then separated and washed using a 1:1 (wt.) mixed solvent of dimethyl ether (DME) and 1,3-dioxolane (DOL). The SEM images indicate that there were no visible changes in the morphology of FeS_2 particles before and after Li_2S_9 treatment. An energy-dispersive X-ray spectroscopy (EDX) analysis on the Li_2S_9-treated FeS_2 indicates that the S/Fe weight ratio was increased to 1.003 from 0.9649 (calc. 1.1484) of the pristine FeS_2. Accordingly, in the X-ray diffraction (XRD) pattern an extra peak at 27° as indicated by a circle inset in Fig. 1c shows up while all others indexed to pyrite FeS_2 did not change compared with the pristine FeS_2, suggesting that a new phase associated with the Li_2S_9 had been formed on the FeS_2 surface. Raman spectroscopy (Fig. 1d) further reveals that new broad peaks at 215, 275, 477, and 580 cm^{-1}, which were reported to be the characteristic absorbance of S−S bonds in polysulphide species,[42, 43] are present in the Li_2S_9-treated FeS_2 in addition to slight blueshift on two strong absorbances at 363 and 375 cm^{-1} and a weak absorbance at 433 cm^{-1} for the S_2 vibration (E_g), S−S in-phase stretch (A_g), and coupled vibration and stretch (T_g) modes of pyrite FeS_2.[44, 45] The blueshift of Raman absorbance provides excellent evidence for the formation of FeS_2-Li_2S_n complexes. Additionally, the Li_2FeS_{2+n} complexes are thermodynamically supported by the Fe-S binary phase diagram, which shows that the FeS_2 composition is under an arch-shaped liquidus line, and that the melt with a composition of FeS_n separates into two liquids of FeS_x and FeS_y when cooling down. Based on the above results and the known fact that the stable and stoichiometric MS_n complexes can be readily formed,[36-38] it may be concluded that the FeS_2 adsorbs PS species through the FeS_2-Li_2S_n (or written as Li_2FeS_{2+n}) complexes formed by the S−S covalent bonds between the FeS_2 and Li_2S_n, as proposed by eqn. 2.

$$FeS_2 + Li_2S_n \leftrightarrow Li_2FeS_{2+n} \qquad (2)$$

The reaction mechanism of eqn. 2 is rather complicated, which may start with the cleavage of S−S bond in Li_2S_n into short polysulphide radical anions (as suggested by eqn. 1) that subsequently attack the S−S bond in FeS_2 to combine into Li_2FeS_{2+n} complex. As a support for this hypothesis, the polysulphide radical anions ($S_n^{\cdot-}$, especially with n=3 and 4) have been frequently observed not only *ex-situ* from the electrolyte solutions[38, 41, 46] but also *in-situ* from the Li-S cells.[42, 43, 47]

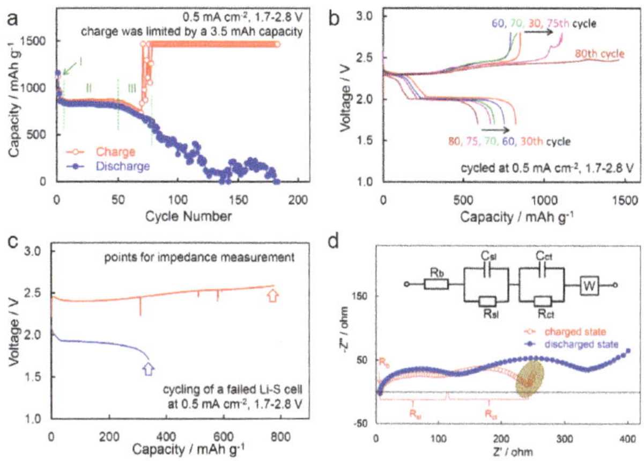

Fig. 2 (a) Charge and discharge capacities of a Li-S cell, (b) selected voltage profiles of the Li/S cell, (c) voltage curve of a failed Li-S cell indicating the points where ac-impedance was measured, and (d) impedance spectra of a failed Li-S cell at charged and discharged states.

Using a sulphur cathode with a 77% of sulphur content in a ~2 mg cm^{-2} of sulphur loading and an electrolyte of 0.25 mol kg^{-1} LiTFSI-0.25 mol kg^{-1} $LiNO_3$ dissolved in a 1:1 (wt.) DME-DOL mixed solvent, a Li-S cell was assembled and galvanostatically cycled to understand the capacity fading mechanism. As shown in Fig. 2a, the cycle life of a Li-S cell can be briefly divided into three regions, i.e., the fast capacity fading Region I, slow capacity fading Region II, and accelerated capacity fading Region III. Region I is attributed to the redistribution of sulphur active material from the cathode to the electrolyte throughout the entire cell as a result of the PS dissolution; Region II is mainly due to the progressive loss of sulphur active material and the resultant growth of resistive surface layer on the Li anode as a result of the chemical or electrochemical reduction of dissolved Li_2S_n to insoluble Li_2S_2 or Li_2S on the Li surface; Region III, which features an increasing decrease followed by a dramatic decrease in the cell's coulombic efficiency with the cycle number, is ascribed to the localized electric circuit shortening caused by the pulverized Li particles. The last judgement is well evidenced by observing the charging voltage profile. Fig. 2b compares the voltage profiles of a Li-S cell cycling in Region II and Region III. It can be seen that in Region III, with an increase in the cycle number the charging capacity unexpected increases in contrast to a decrease in the discharge capacity. Moreover, the increase of charging voltage experiences irregular fluctuations (see those at 75th and 80th charging), being indicative of the localized electric circuit shortening, other than the PS redox shuttle that has been frequently observed in the conventional Li-S cells and features to reach a constant charging voltage. In order to further confirm the localized electric circuit shortening, the impedance spectra of a failed Li-S cell (i.e., in Region III) at the charged state and discharged state, as indicated by two arrows in Fig. 2c, are measured and compared. It is distinctly observed in Fig. 2d that the charged state not only has significantly lower R_{ct} but also has much large slope for the Warburg impedance (W) compared with the discharged state, being excellent evidence for the localized electric circuit shortening caused by the pulverized Li particles that penetrate across the separator to electrically connect the sulphur cathode and Li anode. The result that the failed cell has much lower R_{ct} and larger Warburg impedance slope in the charged state than in the discharged state is highly reproducible although their actual values may change from cell to cell.

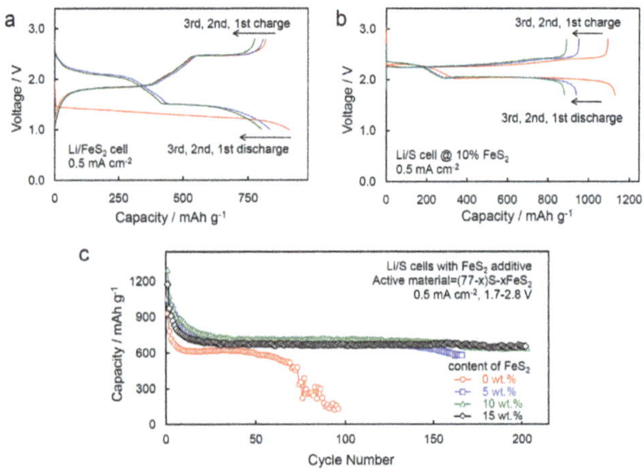

Fig. 3 (a) Voltage profile of the first three cycles of a Li-FeS_2 cell, (b) voltage profile of the first three cycles of a Li-S cell with 10% FeS_2 additive, and (c) cycling performance of Li-S cells with different content of FeS_2 additive.

In order to employ FeS_2 as the cathode additive for Li-S cells, the electrochemical property of FeS_2 is examined in Fig. 3a. It is unique that the first discharge does not initiate until the cell's voltage reaches ~1.5 V, which is due to the large grain boundary resistance of the solid FeS_2 particles[48] and the similar phenomena have been frequently observed from other solid electrode materials.[49, 50] Upon the first discharge during which

FeS_2 is ultimately reduced to metallic Fe and Li_2S, the Li-FeS_2 cell converts to a hybrid of two conversion-type cells, namely a Li-S cell and a Li-FeS cell, as suggested by two characteristic discharge voltage plateaus at ~2 V and ~1.5 V for the Li-Li_2S_n cell and Li-FeS cell, respectively.[48, 51] The above results reveal that when used as the cathode additive, the FeS_2 is able to retain its original structure as long as the cell is not discharged to or below 1.5 V. Fig. 3b shows that voltage profiles of the first three cycles for a Li-S cell with 10% FeS_2 as the cathode additive (i.e., 10% S out of the 77% S cathode is replaced with FeS_2). When cutting the discharge off at 1.7 V, the Li-S cell delivers a capacity of 1129 mAh g^{-1} in the first discharge, and declines to 940 and 881 mAh g^{-1} in the second and third discharges, respectively, exhibiting the normal behaviours of a Li-S cell without any contributions of FeS_2 to the discharge capacity. However, the addition of FeS_2 remarkably enhances the capacity retention of Li-S cells. As shown in Fig. 3c, the control cell stabilized in a range of 620~630 mAh g^{-1} capacities and was only able to cycle for 50 cycles before the capacity started to decline dramatically. With the addition of FeS_2 into the sulphur cathode, however, the capacity slightly increased to 680~720 mAh g^{-1} while the capacity retention was improved remarkably. It is also observed in Fig. 3c that with an increase in the content of FeS_2 from 5% to 15%, the improvement on the capacity was very limited, however, the capacity retention was enhanced dramatically. For example, the Li-S cell with 5% FeS_2 ran only for ~150 cycles, while those with 10% and 15% FeS_2 for at least 200 cycles without visible fading in the capacity. From a comprehensive view of the discharge capacity and capacity retention, 10% FeS_2 seems to be the optimum, which was able to stabilize the capacity of a Li-S cell at ~700 mAh g^{-1} for at least 200 cycles. It is proven that 5% FeS_2 is insufficient to enable long-term cycling whereas 15% FeS_2 leads to relatively low overall capacities due to a decrease in the content of sulphur (since total content of sulphur and FeS_2 was fixed at 77%). Since the S to FeS_2 ratio is the only variation in the composition of the sulphur cathodes used in Fig. 3c, the observed improvement on the capacity retention of Li-S cells can be reasonably to the efficiently chemical adsorption of the dissolved Li_2S_n to the FeS_2 host, which consequently reduces the loss of sulphur active material in relation to the dissolution of long-chain Li_2S_n.

3. Conclusions

In summary, we have shown that pyrite FeS_2 has exceptional ability in chemically adsorbing Li_2S_n, and that it can be used as an efficient polysulphide adsorbent to sequestrate the dissolved Li_2S_n from diffusing out of the cathode in Li-S batteries. It is shown that the Li_2S_n sequestration involves the formation of active Li_2FeS_{2+n} complexes via the interactions between the polysulphide radical anions produced during sulphur redox and the sulphur sites in FeS_2 host. The Li_2S_n sequestration reduces the loss of sulphur active material, which consequently stabilizes the capacity retention of Li-S cells. The similar Li_2S_n sequestration is also applicable to other transition metal sulphides. The present strategy may open a new direction for the Li_2S_n sequestration in Li-S batteries.

References

[1] S. S. Zhang, *J. Power Sources*, 2013, **231**, 153. https://doi.org/10.1016/j.jpowsour.2012.12.102

[2] Y. Yang, G. Zheng and Y. Cui, *Chem. Soc. Rev.*, 2013, **42**, 3018. https://doi.org/10.1039/c2cs35256g

[3] A. Manthiram, Y. Fu, S. H. Chung, C. Zu and Y. S. Su, *Chem. Rev.*, 2014, **114**, 11751. https://doi.org/10.1021/cr500062v

[4] J. G. Wang, K. Xie and B. Wei, *Nano Energy*, 2015, **15**, 413. https://doi.org/10.1016/j.nanoen.2015.05.006

[5] Z. Li, Y. Huang, L. Yuan, Z. Hao and Y. Huang, *Carbon*, 2015, **92**, 41. https://doi.org/10.1016/j.carbon.2015.03.008

[6] S. S. Zhang, *Inorg. Chem. Front.*, 2015, **2**, 1059. https://doi.org/10.1039/C5QI00153F

[7] L. Ma, K. E. Hendrickson, S. Wei and L. A. Archer, *Nano Today*, 2015, **10**, 315. https://doi.org/10.1016/j.nantod.2015.04.011

[8] H. J. Peng and Q. Zhang, *Angew. Chem., Int. Ed.*, 2015, **54**, 11018. https://doi.org/10.1002/anie.201505444

[9] L. Ji, M. Rao, H. Zheng, L. Zhang, Y. Li, W. Duan, J. Guo, E. J. Cairns and Y. Zhang, *J. Amer. Chem. Soc.*, 2011, **133**, 18522. https://doi.org/10.1021/ja206955k

[10] B. Wang, S. M. Alhassan and S. T. Pantelides, *Phys. Rev. Appl.*, 2014, **2**, 034004. https://doi.org/10.1103/PhysRevApplied.2.034004

[11] Q. Pang, J. Tang, H. Huang, X. Liang, C. Hart, K. C. Tam and L. F. Nazar, *Adv. Mater.*, 2015, **27**, 6021. https://doi.org/10.1002/adma.201502467

[12] G. Zhou, E. Paek, G. S. Hwang and A. Manthiram, *Nat. Commun.*, 2015, **6**, 7760. https://doi.org/10.1038/ncomms8760

[13] H. J. Peng, T. Z. Hou, Q. Zhang, J. Q. Huang, X. B. Cheng, M. Q. Guo, Z. Yuan, L. Y. He and F. Wei, *Adv. Mater. Inter.*, 2014, **1**, 1400227. https://doi.org/10.1002/admi.201400227

[14] Y. Qiu, W. Li, W. Zhao, G. Li, Y. Hou, M. Liu, L. Zhou, F. Ye, H. Li, Z. Wei, S. Yang, W. Duan, Y. Ye, J. Guo and Y. Zhang, *Nano Lett.*, 2014, **14**, 4821. https://doi.org/10.1021/nl5020475

[15] A. Schneider, C. Weidmann, C. Suchomski, H. Sommer, J. Janek and T. Brezesinski, *Chem. Mater.*, 2015, **27**, 1674. https://doi.org/10.1021/cm504460p

[16] C. P. Yang, Y. X. Yin, H. Ye, K. C. Jiang, J. Zhang and Y. G. Guo, *ACS Appl. Mater. Interfaces*, 2014, **6**, 8789. https://doi.org/10.1021/am501627f

[17] Y. Xie, Z. Meng, T. Cai and W. Q. Han, *ACS Appl. Mater. Interfaces*, 2015, **7**, 25202. https://doi.org/10.1021/acsami.5b08129

[18] Q. Zhang, Y. Wang, Z. W. Seh, Z. Fu, R. Zhang and Y. Cui, *Nano Lett.*, 2015, **15**, 3780. https://doi.org/10.1021/acs.nanolett.5b00367

[19] S. Evers, T. Yim and L. F. Nazar, *J. Phys. Chem. C*, 2012, **116**, 19653. https://doi.org/10.1021/jp304380j

[20] Q. Pang, D. Kundu, M. Cuisinier and L. F. Nazar, *Nat. Commun.*, 2014, **5**, 4759. https://doi.org/10.1038/ncomms5759

[21] Z. Li, J. Zhang and X. W. Lou, *Angew. Chem., Int. Ed.*, 2015, **54**, 12886. https://doi.org/10.1002/anie.201506972

[22] H. Yao, G. Zheng, P.-C. Hsu, D. Kong, J. J. Cha, W. Li, Z. W. Seh, M. T. McDowell, K. Yan, Z. Liang, V. K. Narasimhan and Y. Cui, *Nat. Commun.*, 2014, **5**, 3943.

[23] Q. Fan, W. Liu, Z. Weng, Y. Sun and H. Wang, *J. Am. Chem. Soc.*, 2015, **137**, 12946. https://doi.org/10.1021/jacs.5b07071

[24] Z. W. Seh, J. H. Yu, W. Li, P. C. Hsu, H. Wang, Y. Sun, H. Yao, Q. Zhang and Y. Cui, *Nat. Commun.*, 2014, **5**, 5017. https://doi.org/10.1038/ncomms6017

[25] K. Sun, D. Su, Q. Zhang, D. C. Bock, A. C. Marschilok, K. J. Takeuchi, E. S. Takeuchi and H. Gan, *J. Electrochem. Soc.*, 2015, **162**, A2834. https://doi.org/10.1149/2.1021514jes

[26] Q. Pang, D. Kundu and L. F. Nazar, *Materials Horizons*, 2016.

[27] Z. Yuan, H. J. Peng, T. Z. Hou, J. Q. Huang, C. M. Chen, D. W. Wang, X. B. Cheng, F. Wei and Q. Zhang, *Nano Letters*, 2016, **16**, 519. https://doi.org/10.1021/acs.nanolett.5b04166

[28] A. Garsuch, S. Herzog, L. Montag, A. Krebs and K. Leitner, *ECS Electrochem. Lett.*, 2012, **1**, A24.

[29] L. Ma, S. Wei, H. L. Zhuang, K. E. Hendrickson, R. G. Hennig and L. A. Archer, *J. Mater. Chem. A*, 2015, **3**, 19857. https://doi.org/10.1039/C5TA06348E

[30] X. Liang, A. Garsuch and L. F. Nazar, *Angew. Chem., Int. Ed.*, 2015, **54**, 3907. https://doi.org/10.1002/anie.201410174

[31] S. Zheng, P. Han, Z. Han, P. Li, H. Zhang and J. Yang, *Adv. Energy Mater.*, 2014, **4**, 1400226. https://doi.org/10.1002/aenm.201400226

[32] G. Babu, K. Ababtain, K. Y. S. Ng and L. M. R. Arava, *Sci. Rep.*, 2015, **5**, 8763. https://doi.org/10.1038/srep08763

[33] H. Al Salem, G. Babu, C. V. Rao and L. M. R. Arava, *J. Am. Chem. Soc.*, 2015, **137**, 11542. https://doi.org/10.1021/jacs.5b04472

[34] X. Liang, C. Hart, Q. Pang, L. F. Nazar, A. Garsuch and T. Weiss, *Nat Commun*, 2015, **6**, 5682. https://doi.org/10.1038/ncomms6682

[35] X. Liang, C. Y. Kwok, F. Lodi-Marzano, Q. Pang, M. Cuisinier, H. Huang, C. J. Hart, D. Houtarde, K. Kaup, H. Sommer, T. Brezesinski, J. Janek and L. F. Nazar, *Adv. Energy Mater.*, 2015, **5**, DOI: 10.1002/aenm.201501636. https://doi.org/10.1002/aenm.201501636

[36] S. Dev, E. Ramli, T. B. Rauchfuss and S. R. Wilson, *Inorg. Chem.*, 1991, **30**, 2514. https://doi.org/10.1021/ic00011a011

[37] A. K. Verma, T. B. Rauchfuss and S. R. Wilson, *Inorg. Chem.*, 1995, **34**, 3072. https://doi.org/10.1021/ic00115a038

[38] S. J. Chadwell, D. Rickard and G. W. Luther, III, *Electroanal.*, 2001, **13**, 21. https://doi.org/10.1002/1521-4109(200101)13:1<21::AID-ELAN21>3.0.CO;2-P

[39] W. L. Bowden, L. H. Barnette and D. L. DeMuth, *J. Electrochem. Soc.*, 1988, **135**, 1. https://doi.org/10.1149/1.2095554

[40] D. Rickard and G. W. Luther, *Chem. Rev.*, 2007, **107**, 514. https://doi.org/10.1021/cr0503658

[41] R. Steudel, *Top. Curr. Chem.*, 2003, **231**, 127. https://doi.org/10.1007/b13183

[42] M. Hagen, P. Schiffels, M. Hammer, S. Doerfler, J. Tuebke, M. J. Hoffmann, H. Althues and S. Kaskel, *J. Electrochem. Soc.*, 2013, **160**, A1205. https://doi.org/10.1149/2.045308jes

[43] H. L. Wu, L. A. Huff and A. A. Gewirth, *ACS Appl. Mater. Interfaces*, 2015, **7**, 1709. https://doi.org/10.1021/am5072942

[44] H. Vogt, T. Chattopadhyay and H. J. Stolz, *J. Phys. Chem. Solids*, 1983, **44**, 869. https://doi.org/10.1016/0022-3697(83)90124-5

[45] A. K. Kleppe and A. P. Jephcoat, *Mineral. Mag.*, 2004, **68**, 433. https://doi.org/10.1180/0026461046830196

[46] F. Gaillard, E. Levillain and J. P. Lelieur, *J. Electroanal. Chem.*, 1997, **432**, 129. https://doi.org/10.1016/S0022-0728(97)00192-7

[47] M. Cuisinier, C. Hart, M. Balasubramanian, A. Garsuch and L. F. Nazar, *Adv. Energy Mater.*, 2015, **5**, 1401801. https://doi.org/10.1002/aenm.201401801

[48] S. S. Zhang and D. T. Tran, *Electrochim. Acta*, 2015, **176**, 784. https://doi.org/10.1016/j.electacta.2015.07.096

[49] J. Cabana, L. Monconduit, D. Larcher and M. R. Palacin, *Adv. Mater.*, 2010, **22**, E170. https://doi.org/10.1002/adma.201000717

[50] T. Stephenson, Z. Li, B. Olsen and D. Mitlin, *Energy Environ. Sci.*, 2014, **7**, 209-231. https://doi.org/10.1039/C3EE42591F

[51] S. S. Zhang, *J. Mater. Chem. A*, 2015, **3**, 7689. https://doi.org/10.1039/C5TA00623F

Chapter 8

Electrochemical Stability of $Li_{6.5}La_3Zr_2M_{0.5}O_{12}$ (M = Nb or Ta) Against Metallic Lithium

Yunsung Kim[1], Aeri Yoo[2], Robert Schmidt[1], Asma Sharafi[1], Heechul Lee[2], Jeff Wolfenstine[3], and Jeff Sakamoto[1*]

[1]Department of Mechanical Engineering, University of Michigan, Ann Arbor, MI 48109-2125, USA

[2]Department of Advanced Materials Engineering, Korea Polytechnic University, Siheung, Gyeonggi, 15073 Republic of Korea

[3]Army Research Laboratory, RDRL-SED-C, 2800 Powder Mill Road, Adelphi, MD 20783, USA

Abstract

The electrochemical stability of $Li_{6.5}La_3Zr_{1.5}Nb_{0.5}O_{12}$ (LLZNO) and $Li_{6.5}La_3Zr_{1.5}Ta_{0.5}O_{12}$ (LLZTO) against metallic Li was studied using direct current (DC) and electrochemical impedance spectroscopy (EIS). Dense polycrystalline LLZNO (ρ=97 %) and LLZTO (ρ=92 %) were made using sol-gel synthesis and rapid induction hot-pressing at 1100 °C and 15.8 MPa. During DC cycling tests at room temperature (±0.01 mA/cm^2 for 36 cycles), LLZNO exhibited an increase in Li-LLZNO interface resistance and eventually short-circuiting while the LLZTO was stable. After DC cycling, LLZNO appeared severely discolored while the LLZTO did not change in appearance. We believe the increase in Li-LLZNO interfacial resistance and discoloration are due to reduction of Nb^{5+} to Nb^{4+}. The negligible change in interfacial resistance and no color change in LLZTO suggest that Ta^{5+} may be more stable against reduction than Nb^{5+} in cubic garnet versus Li during cycling.

Keywords

Li Anode, DC Cycling, Hot-Pressing, Reduction, Interface

Contents

The work has been previously published by Frontiers: Yunsung Kim, Aeri Yoo, Robert Schmidt, Asma Sharafi, Heechul Lee, Jeff Wolfenstine and Jeff Sakamoto, Electrochemical Stability of $Li_{6.5}La_3Zr_{1.5}M_{0.5}O_{12}$ (M = Nb or Ta) against Metallic Lithium, Frontiers in Energy Research May 2016 Vol 4 pg 1-7, https://doi.org/10.3389/fenrg.2016.00020

1. Introduction

Improving the performance and safety of batteries may be achieved through the development and integration of solid-state ceramic electrolytes into solid-state batteries (Salam et al., 1999; Dudney et al., 2015). While numerous solid-electrolytes exhibit high conductivity, few examples of viable bulk-scale solid-state batteries have been reported (Knauth et al., 2009; Dudney et al., 2015). One of the challenges in developing solid-state batteries stems from the lack of understanding of solid electrode-solid electrolyte interface stability, specifically of the Li metal-solid electrolyte interface. Owing to the electropositive nature of Li, there are few examples of bulk oxide solid-electrolytes that are stable at 0 V vs Li/Li^+. The Perovskite-type Li-ion conductor, lithium lanthanum titanate (LLTO), exhibits one of the highest bulk ionic conductivities (~1 mS/cm) at room temperature (RT)(Inaguma et al., 1993). Similarly, Aono et al., 1990 reported NASICON (Na superionic conductor) type $Li_{1.3}M_{0.3}Ti_{1.7}(PO_4)_3$ (M=Al or Sc, LATP) also exhibits high bulk ionic conductivity (0.7 mS/cm) at RT. However, LLTO and LATP contain Ti^{4+}, which spontaneously reduces to Ti^{3+} upon contact with metallic Li. In addition, and despite its known stable cycling behavior, Schwöbel et al., 2015 found LIPON decomposed into Li_3PO_4, Li_3P, Li_3N, and Li_2O when paired with a metallic Li anode. Thus, cells using LIPON to enable Li anodes likely form a kinetically limited passivation layer. Recent reports of garnet type electrolyte suggest that the formulation consisting of lithia, lanthania, and zirconia is stable against metallic Li (Murugan et al., 2007).

However, it is known that stoichiometric $Li_7La_3Zr_2O_{12}$ (LLZO) results in the tetragonal polymorph with Li-ion conductivities in the 10^{-5} S/cm range at 25 °C (Awaka et al., 2009; Wolfenstine et al., 2012; Thompson et al., 2014). We and others have demonstrated that ~0.4-0.5 moles of Li vacancies are required to stabilize the higher conductivity (~10^{-4} to 10^{-3} S/cm at 25 °C) cubic garnet type polymorph (Geiger et al., 2011; Rangasamy et al., 2012; Thompson et al., 2014; Thompson et al., 2015). For example, when approximately >0.2 moles of Al^{3+} or Ga^{3+} substitute for Li, >0.4 moles of Li vacancies are created in the LLZO lattice, thus stabilizing the cubic garnet type polymorph (Geiger et al., 2011; Rangasamy et al., 2012). Similarly, when approximately 0.25-0.5 moles of Nb or Ta substitute for Zr, 0.25-0.5 moles of Li vacancies are created, respectively, thus stabilizing the cubic garnet type polymorph (Ohta et al., 2011; Adams et al., 2012; Miara et al., 2013). It has been shown that the latter doping scheme (doping on the Zr site) is the approach that results in the highest bulk ionic conductivities approaching 1 mS/cm at 25 °C (Ohta et al., 2011; Miara et al., 2013; Thompson et al., 2015). Overall, correlating LLZO formulations with conductivity is reasonably well understood, but understanding the effect of cubic garnet type stabilizing dopants on stability against Li metal is not. Thus, the purpose of this work was to study the electrochemical stability against Li metal for the highest known conductivity garnet type formulations; $Li_{6.5}La_3Zr_{1.5}Nb_{0.5}O_{12}$ (LLZNO) and $Li_{6.5}La_3Zr_{1.5}Ta_{0.5}O_{12}$ (LLZTO).

In this study, cubic garnet type LLZNO and LLZTO powders were synthesized using a sol-gel method taking care to prevent Al contamination. The powders were densified using rapid induction hot-pressing (RIHP) to produce $\geq$92 % relative density. To achieve high phase purity (e.g. limit the commonly observed pyrochlore $La_2Zr_2O_7$ phase) excess Li_2CO_3 was added after calcination and before hot-pressing. The resulting pellets were characterized in Li-garnet-Li cells for electrochemical stability. Electrochemical impedance spectroscopy (EIS) and direct current (DC) electrochemical techniques were used to characterize interfacial stability. It will be shown that of these two high conductivity garnet type formulations, the Ta dopant is more electrochemically stable against Li metal than Nb.

2. Materials and methods

2.1 Powder preparation

Cubic Al-free LLZTO and LLZNO with the nominal composition $Li_{6.5}La_3Zr_{1.5}Ta_{0.5}O_{12}$ and $Li_{6.5}La_3Zr_{1.5}Nb_{0.5}O_{12}$ were prepared using a sol-gel synthesis method, respectively (Sakamoto et al., 2013). $LiNO_3 \cdot x6H_2O$ (x = 0.5, 99.999 %, Alfa Aesar), $La(NO_3)_3 \cdot 6H_2O$ (99.9 % Sigma Aldrich), $Zr(OH_7C_3)$ (70 wt.% in 1-propanol, Sigma Aldrich), and

$Nb(OCH_2CH_3)_5$ (99.95 %, Sigma Aldrich) and/or $Ta(CH_3C_2H_2O)_5$ (99.98 %, Sigma Aldrich) were used as the sol-gel precursors. 1-propanol (anhydrous 99.7 %, Sigma Aldrich) and $CH_3OCH_2CH_2OH$ (2-MOE, anhydrous 99.8 %, Sigma Aldrich) were used as a solvent and acetic acid (AC, 1.0M CH_3COOH, Fluka) was used as a chelating agent. The sol-gel process flow chart is shown in Figure 1. The Li and La precursors were dissolved in 1-propanol and Zr and Nb (or Ta) precursors were dissolved in 2-MOE. After the precursors were dissolved, the solutions were combined and stirred until gelation occurred (approximately 45 min) followed by 12 h of ageing to assure the reaction was complete. The gels were dried at room temperature for 9 h followed by drying at 450 °C for 4 h under air to eliminate organics. The dried powder was cold-pressed into pellets at 10 MPa in a 2 cm diameter stainless steel die. The cold-pressed pellets were then calcined at 900 °C for 4 h in an MgO crucible under air. The heating rate was 5 °C/min. After calcination, the pellets were manually crushed with an agate mortar and pestle followed by ball-milling for 15 min at 350 rpm using a planetary mill (PM 100; Retsch, Haan, Germany); 80 mL agate vial and 6 agate balls of 10 mm diameter. 6 and 11 wt.% excess Li_2CO_3 (99.9% Alfa Aesar) were added to LLZNO and LLZTO, respectively, to compensate for Li loss during calcination and densification.

2.2 Densification

Al-free LLZNO and LLZTO powders were hot-pressed at 1100 °C using a rapid induction hot-press. A 1.27 cm bore graphite die was used as the susceptor in flowing argon (Rangasamy et al., 2012; David et al., 2015). Since LLZNO and LLZTO powders included excess Li_2CO_3, which melts at ~725 °C, a 2-step heating profile was used. First, the powder was heated at 1100 °C for 15 min without pressure to prevent expulsion of molten Li_2CO_3, followed by the application of 15.8 MPa pressure for 45 min. The cooling rate was 6 °C/min. After hot-pressing, each pellet was mounted in Crystalbond® wax and cut into two discs using a diamond saw. To ensure parallel surfaces, the discs were ground with sand paper (400 grit Black ice dry/wet sand paper, Norton Corporation, USA) using a lapping fixture (Model 900; Southbay Technologies, San Clemente, CA)(Kim et al., 2016). The discs were stored in an argon-filled glove box (<1 ppm O_2, <1 ppm H_2O) to minimize surface contamination (Jin et al., 2013; Larraz et al., 2015).

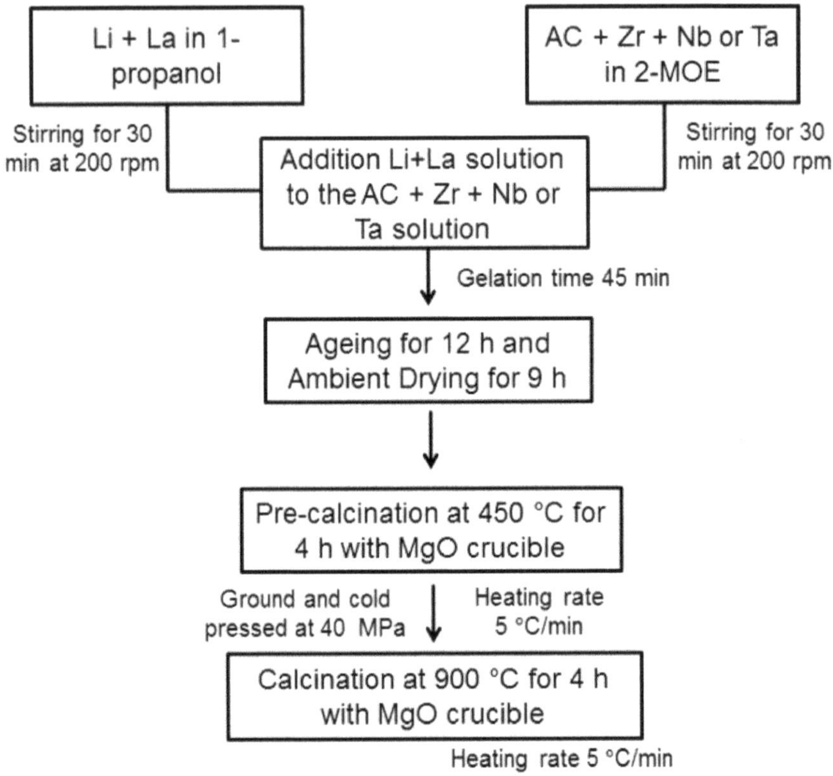

Fig. 1 Process flow chart of sol-gel synthesis for $Li_{6.2}La_3Zr_{1.5}Ta_{0.5}O_{12}$ *and* $Li_{6.2}La_3Zr_{1.5}Nb_{0.5}O_{12}$ *powder.*

2.3 Characterization

The relative densities were defined by dividing the geometric by the theoretical density. The theoretical density was determined by dividing the mass of the atoms in a unit cell by the volume of unit cell determined using X-ray diffraction refinement (XRD, Rigaku Miniflex 600 system, 40 kV and 25 mA).

The phase purity was determined using XRD and Raman spectroscopy (inVia confocal Raman microscope, UK) on of LLZNO and LLZTO before and after cycling test. Raman spectroscopy was performed using a 532 nm green laser and an 1800 line per mm holographic grating.

To characterize the stability of the hot-pressed LLZTO and LLZNO samples against Li metal, Li–LLZNO (or LLZTO)–Li symmetric cells were assembled in an argon-filled glove box. To remove the Li_2CO_3 and LiOH surface contamination layers (Jin et al., 2013; Larraz et al., 2015), the hot-pressed LLZNO and LLZTO discs were dry polished with sand paper on a glass plate in the argon-filled glove box. Li foil (Alfa Aesar) was scraped using a spatula to remove the oxide surface layer. A 465 N uniaxial force was applied to the cells and measured using a compression load cell (Omega, LC304-1k, New England, USA).

DC cycling was conducted using a potentiostat (VMP300, Bio-Logic, Knoxville) at 23.5± 1 °C in the argon-filled glove box. Prior to cycling tests, pre-conditioning cycles were conducted to lower the cell impedance between the LLZNO (or LLZTO) and the Li electrodes. The pre-conditioning cycles were carried out using ±0.01 mA/cm^2 at 70 °C for 10 cycles (each cycle for 2 h). After pre-conditioning, DC cycling tests were performed at 23.5±1 °C for 36 cycles using ±0.01 mA/cm^2. The cell impedance was characterized using EIS between 1 Hz to 7 MHz using a 100 mV perturbation amplitude.

The microstructure of the hot-pressed LLZNO and LLZTO pellets was examined using an optical microscope (Meiji EMZ-13TR, Japan) in the argon-filled glove box.

3. Results and discussion

3.1 Materials characterization

The XRD patterns after hot-pressing for LLZNO and LLZTO with/without 6 and 11 wt.% Li_2CO_3 and the reference pattern for cubic garnet LLZO are shown in Figure 2. All the hot-pressed samples consisted of the cubic garnet phase, however some secondary phases were observed in the hot-pressed LLZNO and LLZTO without excess Li_2CO_3. The predominant secondary phase was pyrochlore ($La_2Zr_2O_7$) and was present at ~6.2 and 17.7 wt.% estimated from XRD in LLZNO and LLZTO, respectively. Based on the wt.% fraction of pyrochlore present, the excess Li_2CO_3 required to compensate for Li loss was empirically determined to be 6 and 11 wt.% for LLZNO and LLZTO, respectively. Consequently, the hot-pressed cubic LLZNO and LLZTO with no observable secondary phases were obtained except for a small amount of an unknown phase that was present in the hot-pressed LLZNO with 6 wt.% excess Li_2CO_3 (Figure 2). The relative densities of the hot-pressed samples were 97 % for LLZNO and 92 % for LLZTO, respectively.

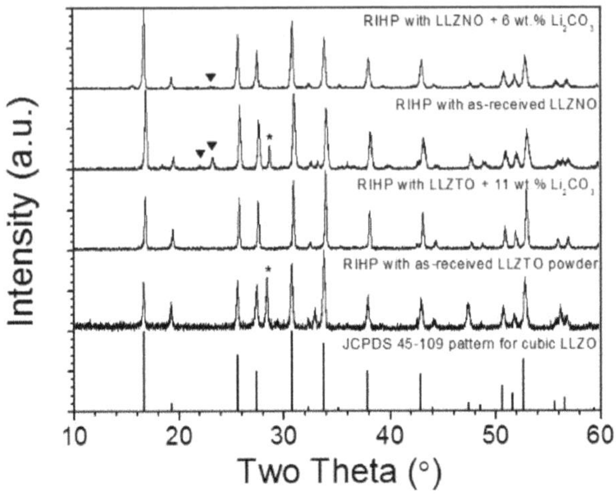

*Fig. 2 X-ray diffraction patterns of hot-pressed $Li_{6.5}La_3Zr_{1.5}Ta_{0.5}O_{12}$ with 0 and 11 wt.% excess Li_2CO_3, and $Li_{6.5}La_3Zr_{1.5}Nb_{0.5}O_{12}$ with 0 and 6 wt.% excess Li_2CO_3, respectively. * Pyrochlore ($La_2Zr_2O_7$) and ▾ (unknown phase)*

3.2 Initial and after pre-conditioning electrochemical impedance

Dramatically different cell impedance was observed when comparing LLZNO and LLZTO (Figure 3). The impedance spectra are composed of a small semicircle in the high frequency range from 120 kHz to 7 MHz, which is attributed to the total resistance (R_{total}) of the bulk and grain boundaries for the hot-pressed LLZNO (or LLZTO). The relatively large semicircle in the low frequency range between 1 Hz and 120 kHz is attributed to the interfacial resistance (R_{int}) between the hot-pressed LLZNO (or LLZTO) and the Li electrode (Figure 3). To determine each impedance and capacitance component, equivalent circuit modeling was conducted. The refined capacitance values for the total resistance of LLZNO (or LLZTO) fall within the range between 10^{-12}-10^{-8} F/cm^2. In addition, the refined capacitance values for the interfacial resistance between LLZNO (or LLZTO) and the Li electrodes fall within the range of 10^{-7}-10^{-5} F/cm^2. These values are in agreement with the expected capacitance values of the respective charge transport phenomena (Irvine et al., 1990; Thompson et al., 2014; Sharafi et al., 2016). In addition, the total ionic conductivities were 0.48 and 0.66 mS/cm for the hot-pressed

LLZTO and LLZNO calculated from the resistance values (R_{LLZTO}=156.1 Ω·cm^2 and R_{LLZNO}=280.9 Ω·cm^2), respectively. These values are in agreement with the reported values in the literature (Ohta et al., 2011; Liu et al., 2014). From Figure 3, it is observed that interfacial resistances ($R_{int,LLZTO}$=373.6 Ω·cm^2 and $R_{int,LLZNO}$=22,924 Ω·cm^2) are relatively large compared to the total cell resistances. The large difference in Li electrode interfacial resistance between the hot-pressed Li-LLZTO and Li-LLZNO could be affected by the difference in relative density (ρ_{LLZTO}=92 % and ρ_{LLZNO}=97 %) of the hot-pressed samples, which would enhance wettability and increase surface area. However, it cannot fully explain the ~61 times higher interfacial resistance of Li-LLZNO compared to that of Li-LLZTO. The cause of this difference will be discussed later. It is known that the Li-garnet interfacial resistance can be reduced by simultaneously heating and cycling (Sharafi et al., 2016). It was suggested that the heat and cycling increased wetting and physical contact between LLZTO (or LLZNO) and the Li electrodes. To emulate previous work, pre-conditioning was performed at low current density (±0.01 mA/cm^2) for 10 cycles at 70 °C. The EIS results after pre-conditioning are shown in Figure 4A and 4B. A comparison of Figures 3 and 4 reveals that pre-conditioning reduced the interfacial resistances by ~39 % for the hot-pressed LLZTO (Figure 4A) and by ~7 % for the hot-pressed LLZNO (Figure 4B).

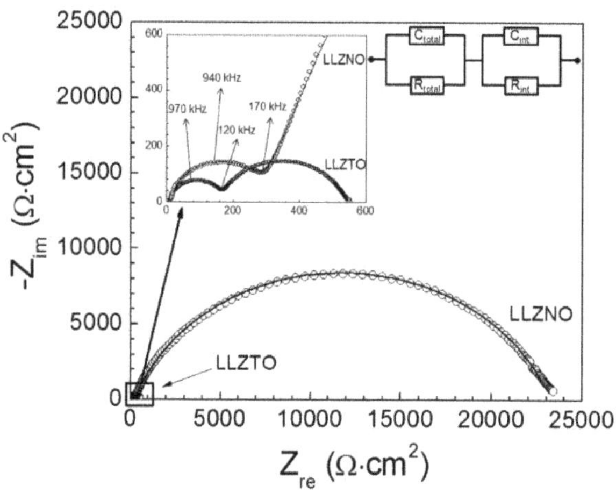

Fig. 3 The impedance spectrum with fitted line for as-fabricated Li-LLZNO (or LLZTO)-Li symmetric cells at 23.5±1 °C. The solid line corresponds to the fitted the result of the equivalent circuit. (○: $Li_{6.5}La_3Zr_{1.5}Nb_{0.5}O_{12}$, □: $Li_{6.5}La_3Zr_{1.5}Ta_{0.5}O_{12}$)

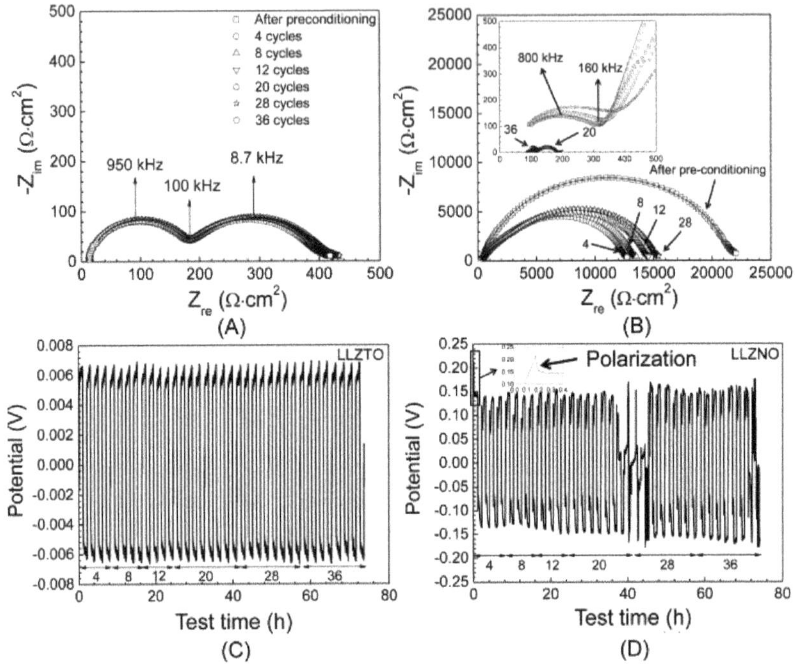

Fig. 4 The impedance spectra with fitted line of (A) the Li- $Li_{6.5}La_3Zr_{1.5}Ta_{0.5}O_{12}$-Li cell and (B) the Li- $Li_{6.5}La_3Zr_{1.5}Nb_{0.5}O_{12}$-Li cell as a function of the DC cycling (0, 4, 8, 12, 28, and 36 cycles) at 23.5±1 °C, respectively. Since the complexity of the modeling with the $Li_{6.5}La_3Zr_{1.5}Nb_{0.5}O_{12}$ after 4 DC cycles, the modeling with LLZNO was not performed during the cycling tests. The results of the DC cycling for Li- $Li_{6.5}La_3Zr_{1.5}Ta_{0.5}O_{12}$-Li (C) and for Li- $Li_{6.5}La_3Zr_{1.5}Nb_{0.5}O_{12}$-Li (D) with the 0.01 mA/cm^2 at 23.5±1 °C were plotted, respectively. The number of DC cycles was noted below the DC cycling plot.

3.3 Cycling and electrochemical impedance

The results of DC cycling and EIS of the LLZNO and LLZTO cells are shown in Figure 4. The EIS spectra were measured after pre-conditioning (0 cycle), 4, 8, 12, 20, 28 and 36 cycles. From Figure 4A, it can be observed that the interfacial resistance of the LLZTO cells remained nearly constant during cycling. It can also be observed that the LLZTO sample exhibited stable DC cycling behavior up to 36 cycles (Figure 4C). This suggests that LLZTO is stable against Li during cycling. In contrast, the impedance of the LLZNO cell was dramatically reduced, compared to before pre-conditioning, after 4 DC cycles (Figure 4B). We believe that the decrease in interfacial resistance during the first four cycles was due to the plating of Li at the Li-LLZTO interface, which improved contact and reduced interfacial impedance as suggested by Gibson (1977). Gibson (1977) observed a reduction in interfacial impedance after the first few cycles of Na-beta alumina. The increase in impedance with continued cycling suggests that some reaction between LLZNO and Li electrodes had occurred. It is known that Nb^{5+} can be reduced to Nb^{4+} at relatively low potentials (Kodama et al., 2006; Wang et al., 2011). Therefore, the reduction of Nb^{5+} may adversely affect charge transport in LLZNO and/or the formation of a passivating layer. Previous work observed similar behavior in $Li_{6.25}La_3Zr_{1.25}Nb_{0.75}O_{12}$ compared to $Li_{6.25}La_3Zr_{1.25}Ta_{0.75}O_{12}$ (Nemori et al., 2015). Further proof of some interaction between LLZNO and the Li electrodes can observed in Figure 4D where short-circuiting of the Li-LLZNO-Li cell occurred during cycling. The potential of the Li-LLZNO-Li cell dramatically decreased after 18 cycles. The occurrence of short-circuiting during DC cycling was in agreement with the EIS measurements (inset in Figure 4B) after 20 cycles, where a significant reduction in LLZNO impedance was observed. We believe the short-circuiting was due to the propagation of a Li dendrite through LLZNO (Ishiguro et al., 2013). After 20 cycles, the LLZNO DC cycling behavior was erratic and likely due to Li dendrite forming and breaking-up during continued cycles as observed by Buqa et al., (2005). At cycle 36, a stable Li dendrite likely formed, thus resulting in a short-circuiting.

3.4 Characterization after cycling

The optical images of the hot-pressed LLZTO and LLZNO after 36 cycles of DC cycling are shown in Figure 5. Aside from the shiny metallic spots (Li metal) embedded in pores (~8 % porosity), there was no evidence of severe discoloration on the surface of the hot-pressed LLZTO before and after DC cycling (Figure 5A). Conversely, severe macroscopic discoloration was observed on the surface of the hot-pressed LLZNO (Figure 5B). The optical analysis is consistent with the result previously reported by Nemori et al., 2015 and observance of a significant change in the Li-LLZNO impedance spectra (Figure 4). The color change in the DC cycled hot-pressed LLZNO confirms that the LLZNO is unstable in contact with Li (Figure 5B). Similar discoloration phenomena has been observed in Li_3NbO_4 where darkening, such as the formation of yellow and black regions, resulted from the reduction of Nb^{5+} to Nb^{4+} (Zverev et al., 1972; DeLeo et al., 1988; Nyman et al., 2010). The color change of the hot-pressed LLZNO may be similar to what is observed in the reduction of Nb^{5+} in Li_3NbO_4, which resulted in the loss of Li and/or O. The reduction of Nb^{5+} can affect transport properties (Ishiguro et al.,

2013; Nemori et al., 2015), thus a similar phenomenon may occur in the LLZNO causing the interfacial resistance to increase. Conversely, the reduction of Ta^{5+} is less likely compared to Nb^{5+}, which is consistent with the observation that no apparent change in LLZTO was noted in this work (Zverev et al., 1972).

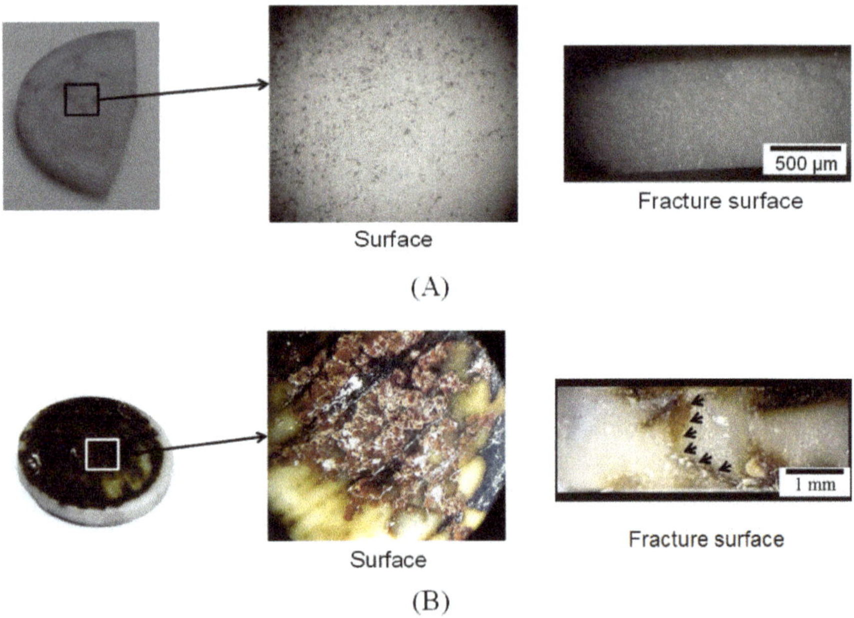

Fig. 5 The optical images of the pellet faces and fracture surfaces for the DC cycled hot-pressed (A) $Li_{6.5}La_3Zr_{1.5}Ta_{0.5}O_{12}$ and (B) $Li_{6.5}La_3Zr_{1.5}Nb_{0.5}O_{12}$ after 36 cycles at 23.5±1 °C. The arrows indicate the extended dark region across the hot-pressed $Li_{6.5}La_3Zr_{1.5}Nb_{0.5}O_{12}$.

The short-circuiting phenomenon observed during electrochemical characterization (Figure 4BD) also appears to correspond with the optical image of the fracture surface (Figure 5B). Macroscopic discoloration was observed on the face and fracture surfaces the hot-pressed LLZNO (Figure 5B). Thus, the discoloration through the entire LLZNO pellet could indicate the Li dendrite grew through the hot-pressed LLZNO during cycling test. On the contrary, no color change and/or an evidence of Li dendrite growth were observed on the fracture surface of the hot-pressed LLZTO (Figure 5A). These results are in good agreement with the stable DC cycling and EIS behavior up to 36 cycles (Figure 4AC).

XRD patterns after the 36th cycle for LLZNO and LLZTO are shown in Figure 6. No impurity peaks were detected even though the appearance of the hot-pressed LLZNO had undergone significant discoloration after cycling test.

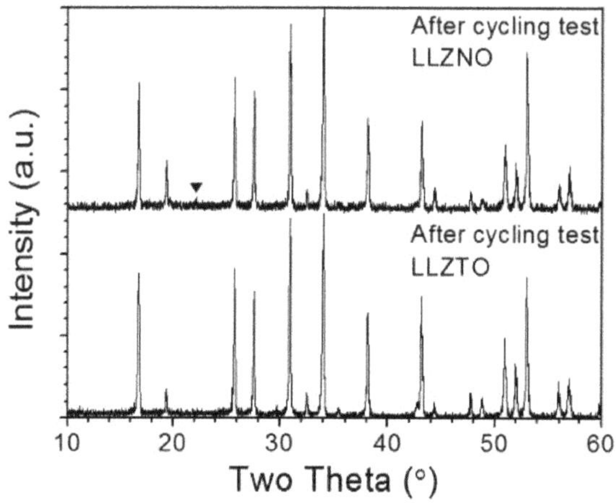

Fig. 6 X-ray diffraction patterns of hot-pressed $Li_{6.5}La_3Zr_{1.5}Ta_{0.5}O_{12}$ and $Li_{6.5}La_3Zr_{1.5}Nb_{0.5}O_{12}$ after DC cycling at 23.5±1 °C, respectively. ▼ (unknown phase)

Raman spectroscopy was also conducted to determine if the surface chemistry of LLZNO and LLZTO changed before and after DC cycling (Figure 7). The Raman spectra were in good agreement with the results of XRD, which determined that only the cubic garnet phase was present on the surface. In addition, the Raman spectra were consistent with the spectra of cubic LLZTO reported by Thompson et al., 2014. The LLZTO band at ~640 cm^{-1} and ~740 cm^{-1} are related with Zr-O bond stretching (Tietz et al., 2013 and Larraz et al., 2013) and Ta-O bond stretching (Thompson et al., 2014), respectively. Therefore, the LLZNO band at ~640 cm^{-1} and ~720 cm^{-1} of LLZNO can be assigned to the stretching of ZrO_6 and NbO_6 octahedron, respectively. In addition, no peak separation, which could have resulted from the lower symmetry found in the tetragonal phase, was observed in low frequency region (<350 cm^{-1}) after DC cycling. The results of the Raman analysis indicate that the cubic garnet phase was present despite the color change. However, it was possible that the fraction of impurity phases present were too low for detection using conventional benchtop XRD and Raman spectroscopy.

To determine if the LLZNO discoloration after DC cycling was associated with the reduction of Nb^{5+} to Nb^{4+}, the sample was heated in ambient air at 600 °C for 2 h. After heating in air, the dark discoloration was converted back to the original white color. In addition, Raman showed no change in the structure after cycling and re-oxidation (Figure 7). This suggests that the color change could be associated with a change in valence from Nb^{5+} to Nb^{4+}.

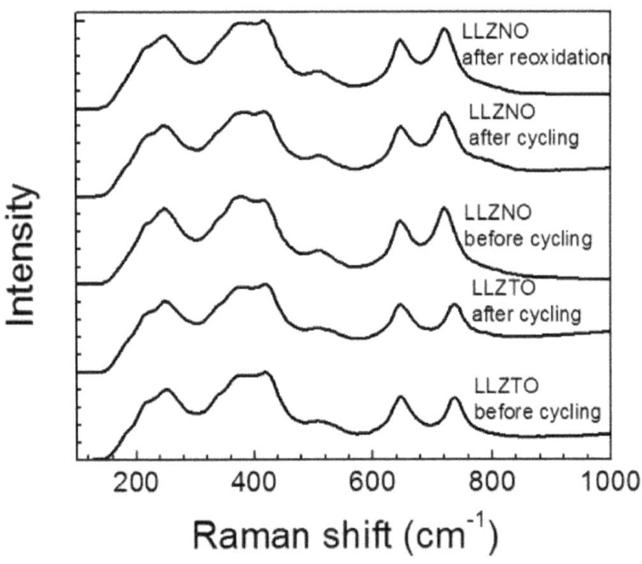

Fig. 7 Raman spectra of $Li_{6.5}La_3Zr_{1.5}Ta_{0.5}O_{12}$ and $Li_{6.5}La_3Zr_{1.5}Nb_{0.5}O_{12}$ before and after DC cycling at 23.5±1 °C; (Top spectrum) LLZNO after DC cycling and after heat treatment at 600 °C in air for 2 h.

4. Conclusions

The electrochemical stability of the LLZTO and LLZNO against Li was investigated. The LLZTO and LLZNO powders, which were synthesized using sol-gel, were hot-pressed for 1 h at 1100 °C, resulting in 92 and 97 % relative densities, respectively. The cubic garnet type phase for both LLZTO and LLZNO were obtained by adding excess Li_2CO_3 (6 wt.% for LLZNO and 11 wt.% for LLZTO).

DC and EIS were conducted on Li-LLZNO (or LLZTO)-Li cells. Several observations were made from these tests. First, pre-conditioning at 70 °C reduced cell impedance likely due to improved physical contact and wettability between the LLZNO (or LLZTO) and the Li electrode. Second, the resistance of the Li-LLZNO interface significantly decreased after 4 DC cycles. We believe that electrochemical Li deposition onto the surface resulted in enhanced physical contact between Li-LLZNO compared to the as-assembled cell impedance. Subsequently, the resistance of Li-LLZNO increased with continued DC cycling while that of Li-LLZTO interface remained constant up to 36 DC cycles at 23.5±1 °C. Third, LLZNO severely discolored while LLZTO did not change after DC cycling. We believe that the increase in interfacial resistance and discoloration in LLZNO are a result of the reduction of Nb^{5+} to Nb^{4+}. The color change of DC cycled LLZNO from dark to original white after heat treatment at 600 °C for 2 h supports the reduction of Nb^{5+} to Nb^{4+} during DC cycling. In contrast, there was no evidence of Ta^{5+} instability in LLZTO after DC cycling. Overall, the results of this study indicate while numerous supervalent dopants can stabilize the high Li-ion conducting garnet phase, not all are stable against reduction in contact with metallic Li.

Acknowledgments

YK, RS, AS, JW and JS would like to acknowledge support from the Advanced Research Projects Agency-Energy (DE-AR0000399) and from Department of Energy (DE-EE-00006821).

The work has been previously published by Frontiers:

Yunsung Kim, Aeri Yoo, Robert Schmidt, Asma Sharafi, Heechul Lee, Jeff Wolfenstine and Jeff Sakamoto, Electrochemical Stability of $Li_{6.5}La_3Zr_{1.5}M_{0.5}O_{12}$ (M = Nb or Ta) against Metallic Lithium, Frontiers in Energy Research May 2016 Vol 4 pg 1-7, https://doi.org/10.3389/fenrg.2016.00020

References

[1] Adams, S., and Rao, R.P. (2012). Ion transport and phase transition in $Li_{7-x}La_3(Zr_{2-x}M_x)O_{12}$ ($M=Ta^{5+}$, Nb^{5+}, x=0, 0.25). *J. Mater. Chem. 22*, 1426-1434. http://dx.doi.org/10.1039/C1JM14588F

[2] Aono, H., Sugimoto, E., Sadaoka, Y., Imanaka, N., and Adachi, G. (1990). Ionic conductivity of solid electrolytes based on lithium titanium phosphate. *J. Electrochem. Soc.* 137, 1023-1027. http://dx.doi.org/10.1149/1.2086597

[3] Awaka, J., Kijima, N., Hayakawa, H., and Akimoto, J. (2009). Synthesis and structure analysis of tetragonal $Li_7La_3Zr_2O_{12}$ with the garnet-related type structure. *J. Solid State Chem.* 182, 2046-2052. doi:10.1016/j.jssc.2009.05.020

[4] Buqa, H., Goers, D., Holzapfel, M., Spahr, M.E., and Novak, Petr. (2005). High rate capability of graphite negative electrodes for lithium-ion batteries. *J. Electrochem. Soc.* 152, A474-A481. http://dx.doi.org/10.1149/1.1851055

[5] David, N.I., Thompson, T., Wolfenstine, J., Allen, J.L., and Sakamoto, J. (2015). Microstructure and Li-ion conductivity of hot-pressed cubic $Li_7La_3Zr_2O_{12}$. *J. Am. Ceram. Soc.* 98, 1209-1214. http://dx.doi.org/10.1111/jace.13455

[6] DeLeo, G.G., Dobson, J.L., Masters, M.F., and Bonjack, L.H. (1988). Electronic structure of an oxygen vacancy in lithium niobate. *Phys. Rev. B* 37, 8394-8400. http://dx.doi.org/10.1103/PhysRevB.37.8394

[7] Dudney, N.J., West, W.C., and Nanda, J. (2015). Handbook of solid state batteries. Singapore: World Scientific.

[8] Geiger, C.A., Alekseev, E., Lazic, B., Fisch, M., Armbruster, T., Langner, R., Fechtelkord, M., Kim, N., Pettke, T., and Weppner, W. (2011). Crystal chemistry and stability of "$Li_7La_3Zr_2O_{12}$" garnet: A fast lithium-ion conductor. *Inorg. Chem.* 50, 1089-1097. http://dx.doi.org/10.1021/ic101914e

[9] Gibson, A. (1976). "A study of the electrode interface formed between sodium metal and beta alumina solid electrolyte," in *Proc. of the 10th Int. Power Sources Sym.* (Brigton, England).

[10] Inaguma, Y., Liquan, C., Itoh, M., Nakamura, T., Uchida, T., Ikuta, H., and Wakihara, M. (1993). High ionic conductivity in lithium lanthanum titanate. *Solid State Ionics* 86, 689-693. http://dx.doi.org/10.1016/0038-1098(93)90841-A

[11] Irvine, J.T.S., Sinclair, D.C., and West, A.R. (1990). Electroceramics: Characterization by impedance spectroscopy. *Adv. Mater.* 2, 132-138. http://dx.doi.org/10.1002/adma.19900020304

[12] Ishiguro, K., Nakata, Y., Matsui, M., Uehi, I., Takeda, Y., Yamamoto, O., and Imanishi, N. (2013). Stability of Nb-doped cubic $Li_7La_3Zr_2O_{12}$ with lithium metal. *J. Electrochem. Soc.* 160, A1690-A1693. http://dx.doi.org/10.1149/2.036310jes

[13] Jin, Y. and McGinn, P.J. (2013). Li7La3Zr2O12 electrolyte stability in air and fabrication of a Li/Li7La3Zr2O12/Cu0.1V2O5 solid-state battery. *J. Power Sources* 239, 326-331. http://dx.doi.org/10.1016/j.jpowsour.2013.03.155

[14] Knauth, P. (2009). Inorganic solid Li ion conductors: An overview. *Solid State Ionics* 180, 911-916. http://dx.doi.org/10.1016/j.ssi.2009.03.022

[15] Kodama, R., Terada, Y., Nakai, I., Komaba, I., Komaba, S., and Kumagai, N. (2006). Electrochemical and in situ XAFS-XRD investigation of Nb_2O_5 for rechargeable lithium batteries. *J. Electrochem. Soc.* 153, A583-A588. http://dx.doi.org/10.1149/1.2163788

[16] Kim, Y., Jo, H., Allen, J.L., Choe, H., Wolfenstine, J., and Sakamoto, J. (2016). The effect of relative density on the mechanical properties of hot-pressed cubic $Li_7La_3Zr_2O_{12}$. *J. Am. Ceram. Soc.* http://dx.doi.org/10.1111/jace.14084

[17] Larraz, G., Orera, A., Sanz, J., Sobrados, I., Diez-Gomez, V., and Sanjuan, M.L. (2015). NMR study of Li distribution in $Li_{7-x}H_xLa_3Zr_2O_{12}$ garnets. *J. Mater. Chem. A* 3, 5683-5691. http://dx.doi.org/10.1039/C4TA04570J

[18] Liu, K., Ma, J.T., and Wang, C.A. (2014). Excess lithium salt functions more than compensating for lithium loss when synthesizing $Li_{6.5}La_3Ta_{0.5}Zr_{1.5}O_{12}$ in alumina crucible. *J. Power Sources* 260, 109-114. doi: http://dx.doi.org/10.1016/j.jpowsour.2014.02.065

[19] Miara, L.J., Ong, S.P., Mo, Y., Richards, D.R., Park, Y., Lee, J.M., Lee, H.S., and Ceder, G. (2013). Effect of Rb and Ta doping on the ionic conductivity and stability of the garnet $Li_{7+2x-y}(La_{3-x}Rb_x)(Zr_{2-y}Ta_y)O_{12}$ ($0\leq x\leq 0.375$, $0\leq y\leq 1$) superionic conductor: A first principles investigation. *Chem. Mater.* 25, 3048-3055. http://dx.doi.org/10.1021/cm401232r

[20] Murugan, R., Thangadurai, V., and Weppner, W. (2007). Fast lithium ion conduction in garnet-type $Li_7La_3Zr_2O_{12}$. *Angew. Chem. Int. Ed.* 46, 7778-7781. http://dx.doi.org/10.1002/anie.200701144

[21] Nemori, H., Matsuda, Y., Mitsuoka, S., Matsui, M., Yamamoto, O., Takeda, Y., and Imanishi, N. (2015). Stability of garnet-type solid electrolyte $Li_xLa_3A_{2-y}B_yO_{12}$ (A=Nb or Ta, B=Sc or Zr). *Solid State Ionics* 282, 7-12. doi: http://dx.doi.org/10.1016/j.ssi.2015.09.015

[22] Nyman, M., Alam, T.M., McIntyre, S.K., Bleier, G.C., and Ingersoll, D. (2010). Alternative approach to increasing Li mobility in Li-LaNb/Ta garnet electrolytes. *Chem. Mater.* 22, 5401-5410. http://dx.doi.org/10.1021/cm101438x

[23] Ohta, S., Kobayashi, T., and Asaoka, T. (2011). High lithium ionic conductivity in the garnet-type oxide $Li_{7-x}La_3(Zr_{2-x}, Nb_x)O_{12}$ (X=0-2). *J. Power Sources* 196, 3342-3345. http://dx.doi.org/10.1016/j.jpowsour.2010.11.089

[24] Rangasamy, E., Wolfenstine, J., and Sakamoto, J. (2012). The role of Al and Li concentration on the formation of cubic garnet solid electrolyte of nominal composition $Li_7La_3Zr_2O_{12}$. *Solid State Ionics* 206, 28-32. http://dx.doi.org/10.1016/j.ssi.2011.10.022

[25] Ren, Y., Shen, Y., Lin, Y., and Nan, C.W. (2015). Direct observation of lithium dendrites inside garnet-type lithium-ion solid electrolyte. *Electrochem. Commun.* 57, 27-30. doi: http://dx.doi.org/10.1016/j.elecom.2015.05.001

[26] Sakamoto, J., Rangasamy, E., Kim, H., Kim, Y., and Wolfenstine, J. (2013). Synthesis of nano-scale fast ion conducting cubic $Li_7La_3Zr_2O_{12}$. *Nanotech.* 24, 424005 (8pp). http://dx.doi.org/10.1088/0957-4484/24/42/424005

[27] Salam, F., Brike, P., and Weppner, W. (1999). Solid-state CO_2 sensor with Li_2CO_3-MgO electrolyte and $LiMn_2O_4$ as solid reference electrode. *J. Electrochem. Solid state Lett.* 2, 201-204. http://dx.doi.org/10.1149/1.1390783

[28] Sharafi, A., Meyer, H.M., Nanda, J., Wolfenstine, J., and Sakamoto, J. (2016). Characterizing the Li-$Li_7La_3Zr_2O_{12}$ interface stability and kinetics as a function of temperature and current density. *J. Power Sources* 302, 135-139. doi: http://dx.doi.org/10.1016/j.jpowsour.2015.10.053

[29] Schwöbel, A., Hausbrabd, R., and Jaegermann, W. (2015). Interface reaction between LIPON and lithium studied by in-situ X-ray photoemission. *Solid State Ionics*. 273, 51-54. http://dx.doi.org/10.1016/j.ssi.2014.10.017

[30] Thompson T., Wolfenstine, J., Allen, J.L., Johannes, M., Huq, A., David, I.N., and Sakamoto, J. (2014). Tetragonal vs. cubic phase stability in Al-free Ta doped $Li_7La_3Zr_2O_{12}$. *J. Mater. Chem. A*. 2, 13431-13436. http://dx.doi.org/10.1039/C4TA02099E

[31] Thompson, T., Sharafi, A., Johannes, M.D., Huq, A., Allen, J.L., Wolfenstine, J., and Sakamoto, J. (2015). A tale of two sites: On defining the carrier concentration in garnet-based ionic conductors for advanced Li batteries. *Adv. Energy Mater.* 1500096, 1-9. http://dx.doi.org/10.1002/aenm.201500096

[32] Wang, X.J., Krumeich, F., Worle, M., Nesper, R., Jantsky, L., and Fiellvag, H. (2011). Niobium(V) oxynitride: Synthesis, characterization, and feasibility as anode material for rechargeable lithium-ion batteries. *Chem. Eur. J.* 18, 5970-5978. http://dx.doi.org/10.1002/chem.201102653

[33] Wolfenstine, J., Rangasamy, E., Allen, J.L., and Sakamoto, J. (2012). High conductivity of dense tetragonal $Li_7La_3Zr_2O_{12}$. *J. Power Sources* 208, 193-196. http://dx.doi.org/10.1016/j.jpowsour.2012.02.031

[34] Zverev, G.M., Levchuk, E.A., Pashkov, V.A., and Poryadin, Y.D. (1972). Laser-radiation-induced damage to the surface of lithium niobate and tantalite single crystals. *Sov. J. Quantum Electron.* 2, 167-169. http://iopscience.iop.org/0049-1748/2/2/L15

Chapter 9

In-Situ Studies on the Electrochemical Intercalation of Hexafluorophosphate Anion in Graphite with Selective Co-intercalation of Solvent

Jeffrey A. Read
U.S. Army Research Laboratory, 2800 Powder Mill Road, Adelphi, MD 20783-1138, USA
jeffrey.a.read4.civ@mail.mil

Abstract

Electrochemical cells utilizing graphite intercalation compounds at both electrodes have been proposed as an energy storage technology where the electrolyte salt is split and stored in the electrodes on charge and reformed on discharge. The anion intercalation compounds of graphite proposed as cathodes in these systems have been studied in electrolytes that are resistant to oxidation at 5V but that are incompatible with graphite anodes. Recent work has demonstrated that electrolytes based on monofluoroethylene carbonate (FEC) and ethylmethyl carbonate (EMC) have superior oxidative stability on graphite cathodes over previously studied electrolytes and form a stable solid electrolyte interphase (SEI) on graphite anodes that allow for full dual-graphite cells to be evaluated for energy storage applications. There is still a limited understanding as to structure of the anion intercalate formed in these electrolyte systems and the effect of solvent co-intercalation on cathode performance. This effort was undertaken using a number of in-situ techniques to better characterize the fully intercalated composition as well as to investigate the process of solvent co-intercalation. It was shown that a series of stages based on the $C_{24}PF_6$ composition are formed until, upon reaching full charge, the structure approaches a $C_{20}PF_6$ stage I composition with PF_6^- anion in close contact with the graphite layers and 0.7 molecules of co-intercalated solvent. For the first time, we have shown that solvent molecules move with anion during the intercalation/de-intercalation process while analysis of fully intercalated crystals demonstrated that there is an unusually strong preference for EMC over FEC to co-intercalate in this anion intercalation compound.

The work has been previously published by ACS Publications: Jeffrey A. Read, "In-Situ Studies on the Electrochemical Intercalation of Hexafluorophosphate Anion in Graphite with Selective Cointercalation of Solvent", *J. Phys. Chem. C*, 2015, 119 (16), pp 8438–8446. https://doi.org/10.1021/jp5115465

Keywords

Anion Intercalation, Graphite Staging, In-situ X-ray Diffraction, In-situ Dilatometry, In-situ Gravimetry, Dual-graphite

Contents

1. Introduction

Electrochemical cells that utilize a graphite intercalation compound (GIC) as cathode have been demonstrated by a number of researchers.[1-9] Anions that can be electrochemically intercalated into graphite include among others ClO_4^-, BF_4^-, PF_6^-, $AlCl_4^-$, $CF_3SO_3^-$, $(CF_3SO_2)_2N^-$, as well as the fluoride-anion acceptor complexes.[1-6,9] The dual-carbon cell with non-aqueous electrolyte was first proposed as a rechargeable battery in patents by McCullough et al.[10-13] In a dual-carbon cell, the lithium salt is electrolyzed on charge with the Li^+ being stored in the anode and the anion intercalating into the cathode. The salt is depleted from the electrolyte on charge and replenished upon discharge with the energy capable of being stored in practical cells depending on the concentration and molecular weight of the electrolyte salt. [14]

Anion intercalation into graphite occurs at potentials near 5V and therefore the oxidation of both electrolyte solvent and anion is of major concern.[5] The electrolyte solvents that have been investigated previously include carbonates such as propylene carbonate (PC), ethylene carbonate (EC), diethyl carbonate (DEC) and dimethyl carbonate (DMC), as well as dimethyl sulfoxide (DMSO), dimethylformamide (DMF), ethyl methyl sulfone (EMS), and the ionic liquid N-butyl-N-methylpyrrolidinium bis(trifluoromethanesulfonyl) imide ($Pyr_{14}TFSI$). These solvents show varying degrees of stability at high potential, with EMS giving the best overall results in terms of capacity and efficiency.[5,9] The oxidative stability of both the electrolyte solvent and anion have been cited as possible reasons for cycling efficiencies of less than 100% and the poor storage performance of anion intercalated graphite.[9] The electrolyte solvents PC, DMSO, DMF, EMS, and $Pyr_{14}TFSI$ are poor solvent choices for intercalation of lithium ions into graphite anodes as they co-intercalate causing exfoliation and destruction of the graphite anode lattice, although recent work suggests that SEI forming additives may be used to stabilize the graphite anode towards the ionic liquid electrolyte based on $Pyr_{14}TFSI$.[15] In order to make a dual-graphite energy storage system feasible, an electrolyte system capable of high efficiencies on both the anode and cathode needed to be found. Recent work has demonstrated the higher oxidative stability of FEC based electrolytes on both 5V spinel[16] and on graphite cathodes.[1] These results are supported by DFT calculations[17] showing that higher oxidation potential is imparted to EC by fluorine substitution. Our Recent work[1] has demonstrated that electrolytes based on monofluoroethylene carbonate (FEC) and ethylmethyl carbonate (EMC) have superior oxidative stability on graphite cathodes over previously studied electrolytes[5,9] and form a stable solid electrolyte interphase (SEI) on graphite anodes that allow for full dual-graphite cells to be evaluated for energy storage applications.

Studies on the exact nature of the PF_6^- intercalate of graphite have been undertaken in various solvent systems. These efforts have used X-Ray diffraction, dilatometry, chemical analysis and NMR to try and determine the intercalate layer spacing (d_i), molecular orientation, and chemical composition of the intercalated anion. Billaud et.al. prepared PF_6^- intercalated Highly Oriented Pyrolytic Graphite (HOPG) by chemical oxidation using NO_2PF_6 dissolved in nitromethane.[18] In-situ XRD studies, chemical analysis, and thickness measurements all indicated that a stage II intercalate, $C_{48}PF_6(CH_3NO_2)_2$, where every other graphite layer is occupied with PF_6^- anion, had formed as the end product with an intercalate layer spacing (d_i) of 7.75Å. Billaud and Chenite later prepared $C_{24}PF_6(CH_3NO_2)_2$ electrochemically[19] using $LiPF_6$ in nitromethane, although XRD of the final product was not done, the composition was inferred from the charge curve. Touzain and Bonnetain studied the electrochemical formation of the PF_6^- intercalated from PC[20] and concluded that the final product of charge at 5.2V was the Stage I compound $C_{24}PF_6\,(PC)_4$ where every layer is intercalated with anion. The Stage I product could not be isolated for analysis, but the Stage II compound, $C_{48}PF_6$ $(PC)_4$ was analyzed using ex-situ XRD and was determined to have a d_i of 14.0Å. Zhang and Lerner studied PF_6^- intercalation from $NaPF_6$ in PC and nitromethane using ex-situ XRD.[21] The stage I product $C_{24}PF_6$ was formed with the amount of solvent co-intercalation not determined. The stage I product intercalated from nitromethane was noted to be particularly stable with a d_i of 8.0 Å. Seel and Dahn studied the electrochemical intercalation of PF_6^- into graphite from $LiPF_6$/EMS electrolyte.[5] In-situ XRD was performed and the stage II compound, $C_8(PF_6)_{0.5}$ was the predicted final composition with a d_i of 4.47Å. The degree of solvent co-intercalation was not determined.

These previous studies indicate three possible models for the graphite intercalated PF_6^- based on the Daumas-Herold domain model[22] for graphite, a stage II intercalate with d_i = 4.37Å, a stage I intercalate with d_i = 7.73Å, and a stage I intercalate with d_i = 14.0Å. The true structure of the anion intercalate is important to know as this will set limits on critical performance parameters for the dual-graphite cell. The structure determines the discharge capacity and the cathode swelling behavior, while co-intercalation of solvent will affect rate capability, charge transfer resistance, and will determine electrolyte composition and volume. We characterized the PF_6^- anion intercalate in FEC based electrolyte using in-situ XRD, in-situ dilatometry, gravimetry, and analysis of fully charged HOPG crystals that serve as a model for more practical graphite materials. The use of FEC based electrolytes and the added stability they provide allow for slower scans to higher potentials. This allows us to investigate the GIC structure in more detail using in-situ techniques closer to equilibrium conditions.

2. Experimental

2.1 Materials

10 x 10 x 1 mm HOPG crystals (SPI-2 grade) from SPI Supplies were cleaved using a razor blade to a final thickness of 100-150 µm and mass of 12-18 mg. 100 mm^2 crystals were used for the XRD, dilatometry, and in-situ gravimetry, while 50 mm^2 crystals were used for ex-situ gravimetry and GC-MS analysis. Crystals were cleaned and thinned by repeatedly placing the crystal on adhesive tape, and peeling off the adhered graphite layers until the desired thickness was achieved. For the in-Situ XRD and dilatometry measurements, the initial crystal thicknesses were measured with a micrometer in 6 places. Electrolyte solvents FEC (Daikin) and EMC (BASF) were tested by Karl-Fisher and determined to be < 20 ppm H_2O. Electrolyte was prepared in a glove box (< 1 ppm H_2O) by dissolving $LiPF_6$ salt (Stella) in a 1:1 wt/wt mixture of FEC and EMC at a concentration of 1.7 molal.

2.2 In-situ XRD

Pouch cells used for in-situ XRD were constructed with a foil laminate pouch material nearly transparent to the X-ray wavelength and with the HOPG crystal in direct contact with the pouch material. Details on pouch cell construction are shown in the supporting information. Cells were mounted on an aluminum cell holder in a Rigaku Ultima III X-ray diffractometer with CuK_{α} radiation in Bragg-Brentano configuration. The aluminum cell holder was offset such that when the cell (~ 5 mm thick) was placed in the holder, the HOPG crystal was at the correct height for diffraction to take place. Once mounted in the diffractometer, the cell was connected to a Maccor cycler for testing. The XRD was set to automatically record spectra over the range from 5-95° 2θ at 2° 2θ/minute every 45 minutes for 48 hours while the cell was charged and discharged at 0.08 mA/cm^2, equivalent to a 20 hour charge / discharge rate. The in-situ XRD was carried out for 1.5 cycles after which the cell was allowed to cycle a total of 10 cycles. At the completion of cycling, a standard (NIST Silicon Powder Standard 640c) was placed on top of the pouch cell and the spectrum was measured. A correction based on this silicon standard was applied to the spectra measured during cycling for the peak position analysis. Peaks were fit to determine position using JADE software.

2.3 In-situ dilatometry

Pouch cells were mounted on the aluminum cell holder and placed in a Measureitall model LE1000-2MT12 thickness gauge with ±0.2 μm accuracy and a gauging force of 0.60 to 0.85N. An 18mm diameter glass slide was placed between the tip of the thickness gauge and the pouch cell to spread the force evenly over the $100mm^2$ crystal. The gauge was set to record data every 120 seconds while the cell was cycled at a 20 hour charge / discharge rate on the Maccor cycler. The in-situ dilatometry was carried out for 7-8 cycles after which the cell was allowed to complete a total of 10 cycles. Several cells were used for the in-situ XRD and dilatometry measurements. The first cell was initially run on the in-situ XRD followed by further cycling during in-situ dilatometry. This cell was cycled between 4.0 and 5.2V. A second cell was initially used for in-situ dilatometry followed by in-situ XRD. This cell was cycled between 2.0 and 5.2V. After the in-situ experiments were complete, the cells were charged to 5.2V and taken into the drybox. Cells were disassembled and the HOPG crystals were measured for thickness. The crystals were then weighed, wiped with an EMC damp Kimwipe, and re-weighed. Fully charged crystals were blue in color.

2.4 In-situ gravimetry

Cells were constructed to measure the change in mass as an HOPG crystal was charged and discharged. The cell was constructed as shown in Figure S-1. The HOPG crystal was held in a Pt mesh suspended from a Pt wire. The electrical connection was made to the Pt wire for charge and discharge cycling. A lithium counter electrode was placed at the bottom of the tube and electrolyte was added to the point where only a 1 mm portion of the HOPG crystal edge was in contact with the electrolyte. This allowed for measurement of the mass change while minimizing the buoyancy effects brought about by the HOPG crystal volume change during intercalation. The platinum wire was connected to the digital scale through an electrically insulated wire loop. The digital scale and cell was isolated in an enclosure to reduce the influence of air movement but the temperature of the cell was not actively controlled and significant baseline drift was experienced. The mass change was measured while the cell was cycled between 4.0 V and 5.2V at C/20, C/5 and C/40 rates on a Maccor cycler.

2.5 Ex-situ gravimetry and GC-MS

Coin cells were built with ~50 mm^2 HOPG crystals thinned to a mass of 7-15 mg. Details on cell construction and cycling is given in the supplemental information. Once cell cycling was complete, fully charged cells were taken into the drybox and disassembled to recover the intercalated HOPG crystal. The mass of the HOPG crystal was measured, wiped with an EMC dampened Kimwipe to remove any electrolyte from the surface of the crystals and measured again, then recorded a third time after several hours to determine stability of the mass. Fully charged cathodes were cut into small pieces in the drybox and placed in a 2.0 ml vial with a measured quantity of dimethyl carbonate (DMC) (~1.5g). These vials were placed on a shaker mill for 1 hour followed by 1 hour of ultrasonication to break up the crystal and extract the intercalated solvent. The GC-MS was then run on an Agilent 7890A GC with a 5975C mass spectrum detector to determine the solvent content of the crystal based on total ion current from the MS detector. Calibration curves for EMC and FEC solvents were prepared in the expected range of solvent concentration determined from the change in crystal mass.

Cells were labeled with prefixes starting with X for the XRD and dilatometry experiments, G for the in-situ gravimetry experiments, and M for the ex-situ gravimetry / GC-MS experiments.

3. Discussion

The voltage versus specific capacity curve for the initial charge of an in-situ XRD cell is shown in Figure 1. The corresponding XRD spectra are shown in Figure 2 with identifiable single phases labeled A-E′. Several small peaks at 45° and 65°, with constant position and intensity, are from the aluminum laminate pouch material used for the cell packaging. The XRD scans take place continuously while the cell is being charged and discharged and require a 45 minute period to complete, with a total of 35 scans to complete the initial charge. The scan number is labeled at the top of Figure 1 and corresponds to the point when that particular scan finished its 45 minute collection period. Scan 1 was taken at open circuit, before charge was initiated, and shows the (002), (004), and (006) peaks expected for HOPG which has been labeled as phase A. Once charge is initialed, the peaks due to HOPG begin to decline and new peaks to either side of the HOPG peaks begin to appear as shown in Figure 2. The appearance of these new peaks on either side of the (002), (004), and (006) peaks Figure 3 shows a topographical representation of the XRD peaks from 22° - 27° 2θ and the corresponding voltage-capacity curve for the first and second charge cycles as well as the first discharge cycle. of HOPG is a clear indication of graphite staging.

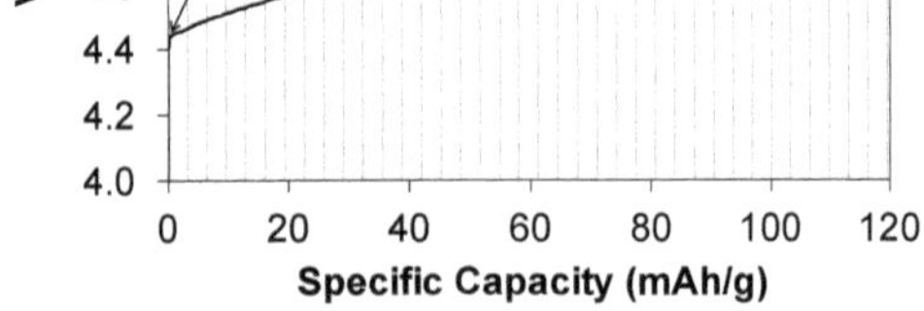

Fig. 1 1st charge curve of cell X4 during anion intercalation. Labels A-E' correspond to the XRD spectra shown in Figures 2 and 3.

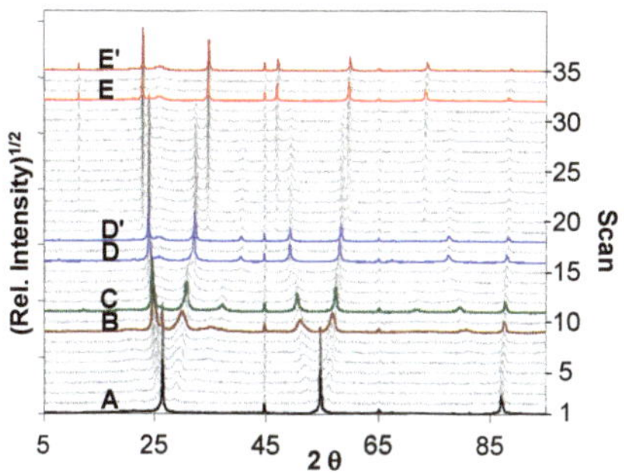

Fig. 2 In-Situ XRD Spectra for scans 1-35 from 5° - 90° 2θ on cell X4.

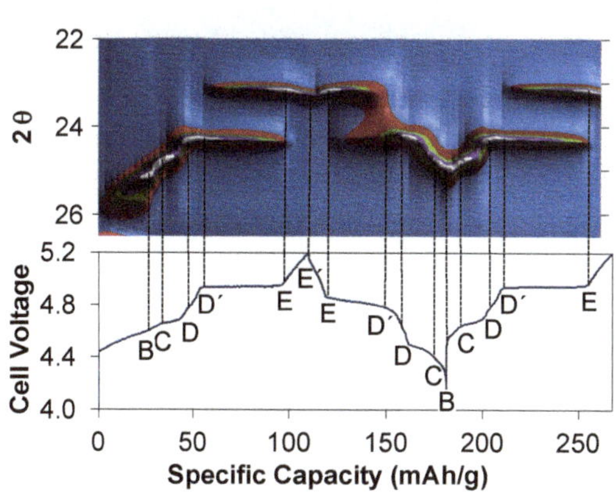

Fig. 3 In-Situ XRD Spectra for scans 1-82 from 22° - 27° 2θ in topographical format with corresponding voltage-capacity curves. (Relative Intensity)$^{1/2}$ represented by colors in the legend. □*500-400* ■*400-300* ■*300-200* ■*200-100* ■*100-0.*

The colors in the topographical graph represent different ranges of (relative intensity)$^{1/2}$ with values given in the caption. The figure shows how phases C-E′ are visible upon discharge and charge while phase B is not fully formed on discharge but is mixed with phase C at 4.0V. The noticeable difference in charge and discharge capacity on the initial cycle is explained in part by the fact that discharge to 4V does not result in the formation of pure graphite, but instead a mixture of phase C and B and accounts for a 20-25% loss in discharge capacity. Another source of irreversible loss includes the formation of an electrochemical shuttle at high potential.[1] Figure 4 shows a 3-D plot of the first charge in a format that more clearly shows the formation of each of the phases B-E′ and how the intensity of each phase grows then falls with the formation of the next phase.

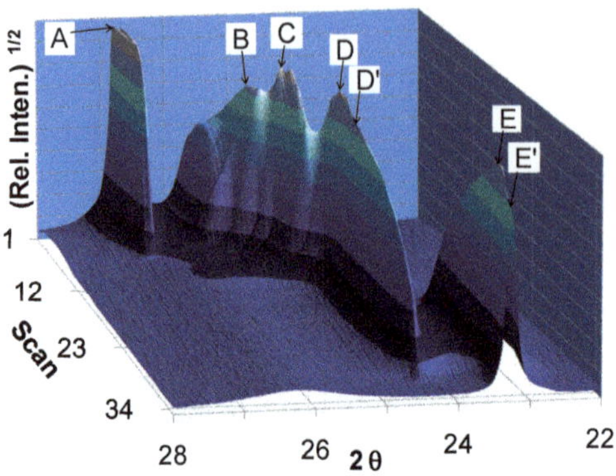

Fig. 4 3-D plot of first charge cycle showing detailed intensity behavior of observed peaks.

The phases were indexed using the relationship between the Bragg angles for the (00n) and (00n+x) peaks in this graphite intercalation compound.[5] The value of n for any particular peak can be calculated from the fact that

$$d_{00n} = \frac{I_c}{n} = \frac{\lambda}{2\sin\theta_{00n}} \quad \text{and} \quad d_{00n+x} = \frac{I_c}{n+x} = \frac{\lambda}{2\sin\theta_{00(n+x)}} \qquad [1]$$

where d_{00n} is the spacing of the (00n) planes, I_c is the unit cell dimension in the direction perpendicular to the graphite planes, n is the Miller indices, x is an integer, λ is the x-ray wavelength, and θ_{00n} is the Bragg angle for the peak (00n). Solving equation 1 for n gives

$$n = \frac{x}{\left[\dfrac{\sin\theta_{00(n+x)}}{\sin\theta_{00(n)}} - 1\right]} \qquad [2]$$

The n value for a particular peak can be calculated from any peak that is at a larger 2θ angle. Normally, a 00n/00n+1 peak pair is used to calculate n values, except when the n+1 peak is not present, and then another (00n+x) peak can be used. By example, phase B was indexed and the Miller indices 004, 005, 006, 010, and 015 were calculated from the peak positions using Equation 2 and x=1, while the Miller indices 007 and 011 were calculated using x=3 and 4 respectively, with the n values calculated for each peak reported in Table S-1. Figure S-2 shows only the indexed phases re-plotted in log[relative intensity] vs. 2θ format in order to show the low angle, low intensity peaks observed in each scan which are not visible in Figure 2. Figure S-2 also shows a broad peak at 26.2°, more prominent in later scans, that remains throughout the charge and most likely represents amorphous carbon present in the original HOPG sample that is not being intercalated.

Analysis of the spectra in Figure 2 using Equation 2 indicates that as the HOPG is initially intercalated with anion, mixed stages are formed that give non-integer Miller indices. It is not until phase B is formed that integer Miller indices appear. After this point in the intercalation process, there are clear transitions between phases with these transitions being either two phase or solid solution in nature. Figure 3 shows that phases C and D and phases D′ and E coexist at the same time that there are clear plateaus in the voltage curve. This is indicative of a two phase transformation where one stage of GIC is transformed to another lower stage, for example a transformation of stage II→I. This is in contrast to the sloping voltage curves seen during the transitions D → D′ and E→E′ where there are no other lower stages forming but the I_c value is decreasing, indicative of a solid solution transition. Figure 3 also shows how the 22.9° peak of phase E shifts to a higher 2θ as the cell is charged from 5 to 5.2V, indicating a contraction of the lattice in the direction perpendicular to the graphite planes. This behavior was reported previously for graphite charged to 5.5V in 2M $LiPF_6$ EMS and it was suggested that a reduction in the molecular vibrations of PF_6 anion as additional anion is packed between graphite sheets[5] might account for this contraction. Figure 4 more clearly shows that as phase D → D′, the intensity of the 003 peak at 24 ° decreases in intensity and that as phase E → E′, the intensity of the 002 peak at 22.9° falls as it shifts to higher angle. The small kink in the voltage curve between phases D and D′ has been observed previously[19,21,23] by a number of researchers and it has been associated with an ordering of the anions on the graphite lattice.[24]

Table 1 Measured phases and Models 1-3 with calculated I_c and percent expansion in c lattice over graphite (Δc).

Measured		Model 1[a]			Model 2[b]			Model 3[c]		
Phase	I_c	Stage	I_c	Δ c	Stage	I_c	Δ c	Stage	I_c	Δ c
B	17.85	V	17.84	6%	IV	17.84	33%	IV	24.09	79%
C	14.48	IV	14.48	8%	III	14.48	44%	III	20.72	106%
D	11.11	III	11.12	10%	II	11.12	65%	II	17.36	158%
D'	11.09	III	11.09	10%	II	11.09	65%	II	17.36	158%
E	7.755	II	7.755	15%	I	7.755	131%	I	14.00	317%
E'	7.725	II	7.725	15%	I	7.725	130%	I	14.00	317%

notes: graphite layer spacing = 3.361 Å, $I_c = d_i + (S-1)\,3.361$Å, S is stage number (1-5)

a) d_i = 4.394 / 4.364Å b) d_i = 7.755 / 7.725Å c) $_{di}$ = 14.0 Å

Table 1 shows all of the indexed phases B-E′ with corresponding I_c axis unit cell dimension calculated as an average of the values for each phase from Table S-1. The calculated values for the three proposed models in the literature with differing intercalate layer spacing (d_i) values, and the expected percent expansion (Δc) in the I_c unit cell dimension with intercalation to the various stages I through IV or V is also given. Two different values for d_i are used in these calculations with phases D' and E' calculated using the smaller d_i values as the lattice appears to contract upon the formation of these phases. The percent expansion is calculated using Equation 3.

$$\Delta c = \left[\left(\frac{I_c}{S \times 3.361}\right) - 1\right] \times 100 \qquad [3]$$

Assignment of the stage for each of the indexed phases is not possible using XRD alone as it is possible to have a stage I GIC with the same I_c unit cell dimension as a stage II GIC but with a different structure, as demonstrated by the identical I_c values for Models 1 and 2. It appears that Model 3 does not fit our observed data and therefore will be disregarded as a possible structure.

In order to determine the stage of each observed phase, in-situ dilatometry was performed on the same cell discussed above after the in-situ XRD and 10 full charge / discharge cycles had been completed. Figure 5 shows several cycles of the dilatometry with the micron change in thickness plotted along with the cell voltage. The large change in crystal thickness (117 - 120 microns) is comparable to the initial thickness (117 microns) and represents a 100% change in thickness upon anion intercalation. This magnitude of change is only consistent with Model 2 (Table 1) where the crystal is cycling between phase B and phase E' as observed during the in-situ XRD. The thickness of the crystal was measured post cycling and registered a final thickness of 272 microns, a 133% increase in thickness from the pristine crystal, in line with the 130-131% change expected from Model 2. Taking this final thickness into account, the dilatometry curve can be replotted to show percent expansion accounting for the thickness at full charge. Figure 6 shows the resulting dilatometry curve with lines showing the percent expansion expected at each of the stages and how they line up as expected with each of the stages from Model 2. The dilatometry supports the assignment of a solid solution transformation when phase D is converted to D′ and E is converted to E′ as the rate of change in the thickness slows during these transitions, especially when D′ is nearly formed, indicating that some packing of anions in the existing stage, instead of new stage formation, may be occurring. The rapid jump in thickness just when phase E is fully formed as determined by XRD, may signal the conversion of the last of the stage II phase to stage I phase. As the in-situ XRD was taken on a fresh cell while the in-situ dilatometry was carried out on a cycled cell, the exact correlation of these two measurements may not be ideal. In order to observe the dilatometry behavior of a fresh cell, a second set of measurements was carried out.

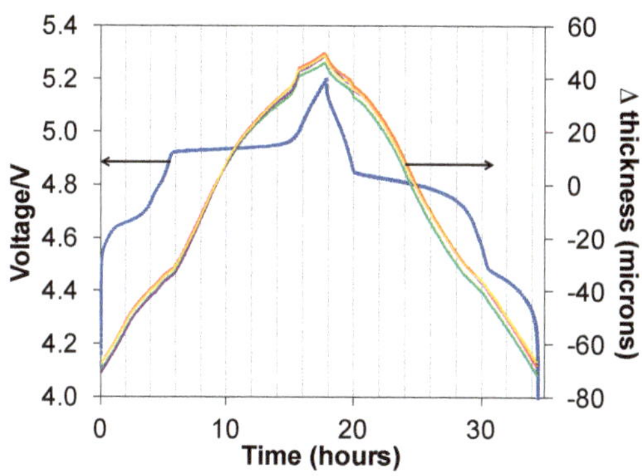

Fig. 5 In-Situ dilatometry on previously cycled cell X4. Voltage profiles for cycles 2-5(■) with the dialatometry data for cycles 2(■), 3(■), 4(■) and 5(■).

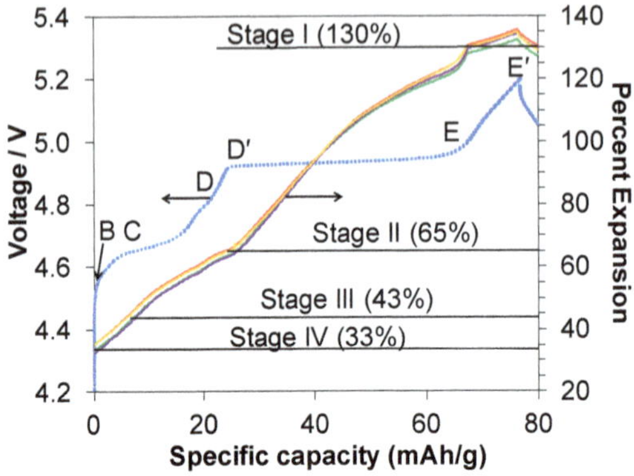

Fig. 6 In-Situ Dilatometry curve on previously cycled cell X4 with percent expansion for each stage I-IV. Voltage profiles for cycles 2-5(■) with the dilatometry data for cycles 2(■), 3(■), 4(■) and 5(■).

Figure 7 shows a plot of the expansion during the first charge cycle plotted along with the voltage-capacity curve on a fresh uncycled cell. The change in thickness is measured to be >90%

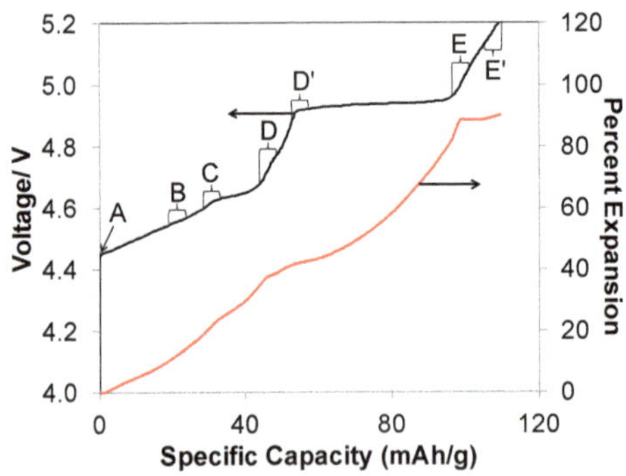

Fig. 7 1st charge curve and In-situ Dilatometry on cell X3.

on the first charge, based on the initial thickness, reaching 98% by the end of 8th cycle, and 120% when the cell was removed from test after a total of 20 charge discharge cycles during which both in-situ dilatometry and in-situ XRD was carried out. Several differences in the dilatometry curve of the fresh cell are obvious as compared to the cycled cell, most notably, the clear plateau when phase E is converted to E′, and the convex nature of the curve when phase D′ is converted to phase E. The plateau from E to E′ would be consistent with a solid solution phase transformation where the lattice is contracting slightly as shown in the XRD. Both sets of dilatometry data support Model 2 as the most likely structure for the anion intercalation compound formed during cycling with the overall expansion and position of the stages being consistent. The lower than expected expansion on the first charge shown in Figure 7 could be the result of incomplete intercalation of the HOPG crystal; there may be random layers that do not initially intercalate on the first cycle but that do so on subsequent cycles, or possibly some change in the cell thickness unrelated to the HOPG crystal, such as relaxation of the separator or change in the lithium thickness on cycling that occurs over the first few cycles. Based on Model 2, we can calculate the size of the PF_6^- anion in the graphite layers from the stage I values in Table 1. Given the graphite layer spacing of 3.361Å and an intercalated layer spacing (d_i) of 7.755-7.725Å, the calculated diameter of the PF_6^- anion would be 4.394-4.364Å. This value is close to the diameter of the PF_6^- anion as calculated from the bond length of PF_6^- and the covalent radius of fluorine[25] (4.36 Å) but smaller than the diameter based on ion volume[26] (5.93 Å). The data would suggest that the PF_6^- anion is in close contact with the graphite sheets when intercalated.

Table 2 Comparison of Phases B-E′ to possible C_nPF_6 compounds.

	Measured (mAh/g)						Theoretical (mAh/g)				
Cell	E′	E	D′	D	C	B	Compound	I	II	III	IV
X4	110	98	54	46	31	24	$C_{18}PF_6$	124	62	41	31
X3	109	98	54	46	32	23	$C_{20}PF_6$	112	56	37	28
G1	114	101	56	49	33	25	$C_{24}PF_6$	93	46	31	23
G2	105	98	54	45	28	22					

Taking a closer look at both the in-situ XRD and dilatometry data with Model 2 established, we can look for evidence of specific compositions, such as $C_{24}PF_6$, $C_{20}PF_6$, or $C_{18}PF_6$ being formed upon cycling. The initial charge capacities for phases B-E′ from

Figures 1 and 7 as well as the in-situ gravimetry experiments are collected in Table 2 along with the theoretical capacities of several anion intercalation compounds. Given that the XRD scans took 45 minutes to complete, each XRD scan has a range of voltage and capacity values associated with it. To determine the capacity value for a particular phase, we determined the point on the voltage-capacity curve where the scan was 50% complete. The exception to this was phase E′ where the value is taken when charging is complete. The capacity values for the in-situ gravimetry experiments were taken from the same points on the voltage-capacity curves determined in the in-situ XRD experiments. Looking at the overall first charge capacity for the stage I phase E′, the most likely composition is $C_{20}PF_6$ (I). The capacity values for phases E, D, C and B, correlate well with the $C_{24}PF_6$ stages I, II, III, and IV respectively, while the stage II phase D′ value is close to the 56 mAh/g value for $C_{20}PF_6$ (II). A sequence of phase formation observed upon charge is therefore proposed as shown below. The transitions between $C_{24}PF_6$ stages are two phase in nature with the average number of graphite layers between anions

$C_{24}PF_6$ (IV) → $C_{24}PF_6$ (III) → $C_{24}PF_6$ (II) → $C_{20}PF_6$ (II) → $C_{24}PF_6$ (I) → $C_{20}PF_6$ (I)

Scheme 1. Proposed stage and composition progression during anion intercalation.

changing over the course of the transformation, while the transitions between $C_{24}PF_6$ and $C_{20}PF_6$ stages can be thought of as solid solution transitions where more anion is simply packed between already intercalated graphite layers. The in-situ XRD shows some additional evidence for this proposed model, specifically the intensity behavior of the 002 peak of the stage I compounds E and E′. The $C_{24}PF_6$ composition has been identified previously as giving the most intense 002 diffraction line in the $C(HSO_4)_x(H_2SO_4)_y$ compound.[24] The reason for this can be seen by looking at the graphite lattice spacing and the relative diameter of the PF_6^- anion as calculated from the bond length of PF_6^- and the covalent radius of fluorine[25] (4.36 Å) or from the diameter based on ion volume[26] (5.93 Å). A single sheet of the anion intercalation compound $C_{24}PF_6$ is shown in Figure S-3, where every 24 carbons is occupied with one PF_6^- anion. The red lines encircle 24 carbon atoms each and form a periodic structure in the graphite layers that cannot be realized with the more anion dense $C_{20}PF_6$ composition. The periodic $C_{24}PF_6$ structure should result in sharper, more intense diffraction lines as compared to the $C_{20}PF_6$ composition. As additional anion is intercalated into the already formed stage I compound, a more disordered structure is likely to be formed leading to a loss in intensity. It has also been suggested that this additional anion increases the coulomb attraction between layers leading to a slight contraction of the lattice.[24] This would be consistent with our XRD data but not with a majority of the dilatometry data, where we continue to see some expansion of the HOPG crystal near the end of the E to E′

transformation. The transition between stage II phases D and D′ may also be examined for evidence of this behavior in the XRD. Figure 4 shows how the intensity of the 003 peak decreases and shifts to higher 2θ as the transformation occurs. The kink in the voltage curve right after phase D is formed is at the same point where a plateau in the dilatometer curve appears. This would support the model of a $C_{24}PF_6$ (II) → $C_{20}PF_6$ (II) transition where the $C_{24}PF_6$ (II) is fully formed and a small increase in anion concentration causes the potential to rise due to coulomb repulsion without a large increase in crystal thickness. Once the $C_{20}PF_6$ (II) phase is fully formed, or once the coulomb repulsion becomes too great, the transition to $C_{24}PF_6$ (I) begins. The 003 peak intensity reaches a maximum at the same point as where the kink begins to appear and would signal the formation of the pure $C_{24}PF_6$ (II) compound. The XRD does not show any superstructure peaks that would be expected for the ordered lattice of PF_6^- anions in the $C_{24}PF_6$ compound so that this structure may just be an average structure that gives rise to stronger diffraction peaks.

Generally, both the XRD and dilatometry curves evolve to some degree upon cycling, where the dilatometry curve for the fresh cell by cycle 8 begins to look similar to the curve in Figure 5 for the cycled cell. The XRD for the two cells at different stages of cycle life appear similar (Figure S-4) with the shift to higher angles for the transitions D to D′ and E to E′ still present. The main difference between the two sets of XRD scans is that the second cell, discharged to 2.0V shows more complete formation of phase B.

The XRD and dilatometry data give a good indication of the overall structure of the anion intercalation compound, but do not give the full composition. It has been found that the weight gain observed in a number of anion intercalation compounds is larger than can be accounted for by anion intercalation alone.[18,20,24,27] This extra weight gain has been determined to be due to the co-intercalation of solvent with the anion. The ratio of solvent to PF_6^- in HOPG intercalated from 0.4M $LiPF_6$ in nitromethane was determined to be in the range of 1.1 to 1.6 nitromethane molecules per anion by NMR.[28] In the chemically intercalated $C_{24}SbF_6$ graphite salt, the nitromethane to anion ratio is close to 1.7 by chemical analysis.[18] We determined the weight gain of fully charged HOPG cathodes in 1.7m $LiPF_6$ FEC:EMC (1:1 w/w) electrolyte for a number of samples used in the paper with the results given in Table 3 along with the weight change expected for 0.6 to 1.0 solvent molecules co-intercalated with each anion. The average increase in weight for the crystal is 92%, corresponding to ~0.7 solvent molecules for each PF_6^- anion with a range from 0.5 to 1 molecule per anion.

An estimate of the available space for solvent co-intercalation can be made looking at Figure S-3 which shows the stage I $C_{24}PF_6$ intercalate to scale where the PF_6 anion is 4.36Å across. The largest remaining space between anions is roughly 5.7Å in diameter with 1 space associated with each anion. Based on molar volume, FEC has a molecular diameter of 4.9Å, while EMC has a molecular diameter of 5.6Å, both within the range where they could occupy the 5.7Å space in the intercalated anion layer. Using bond lengths and covalent radii to estimate size, FEC has an average dimension in the ring plane of 5Å and 2.8Å perpendicular to the ring while EMC is larger at 5.3Å in a planar conformation and 2.1Å in the direction perpendicular to this conformation. It would seem reasonable that one molecule of solvent, either FEC or EMC, can be accommodated in the anion intercalation compound and that the orientation of the co-intercalated molecule has to be such that they fit into the 4.36 Å gap dictated by the anion. FEC would have to be oriented roughly parallel to the graphite sheets to fit into the space while EMC would only need to be parallel if it were in the planar conformation.

Table 3 Percent increase in weight for HOPG crystals recorded at full charge. Calculated percent increase for specific compounds with varying amounts of solvent.

Cell	% increase	Compounds	% increase
X3	109%	$C_{20}PF_6$	60%
X4	88%	$C_{20}PF_6$: 0.6 FEC/EMC	87%
G1	97%	$C_{20}PF_6$: 0.7 FEC/EMC	91%
M10	81%	$C_{20}PF_6$: 0.8 FEC/EMC	95%
M11	91%	$C_{20}PF_6$: 0.9 FEC/EMC	100%
M14	96%	$C_{20}PF_6$: 1.0 FEC/EMC	104%
M15	83%	$C_{24}PF_6$: 1.0 FEC/EMC	87%

The fact that there is a range of weight uptake values might indicate that the solvent is poorly bound to the anion during the intercalation process and therefore a range of values is admitted, or that the crystals are not fully intercalated to the $C_{20}PF_6$ composition in all cases. In order to determine how much of each solvent is present in the intercalated crystals, a number of fully charged samples were analyzed after 10-20 charge/discharge cycles using GC-MS. The crystals were extracted and run on the GC-MS to measure the total ion current for both EMC and FEC in the extract. Calibration curves were then used to determine wt% solvent for both components as shown in Figure S-5. The results on 4 different crystals is given in Table 4 along with the measured ratio of FEC to EMC in the extracted crystals. Table 4 gives an estimate of the total wt% solvent expected during extraction based on the initial HOPG mass and the assumption that at full charge, the $C_{20}PF_6$ composition is the correct one.

Table 4 Weight percent solvent measured using GC-MS compared to weight percent estimated from initial and final crystal mass.

	Measured wt% solvent				
Cell	EMC	FEC	total	Estimated total wt % solvent	ratio FEC/EMC
M10	0.062	0.003	0.065	0.084	0.05
M11	0.081	0.010	0.091	0.129	0.11
M14	0.183	0.013	0.196	0.365	0.07
M15	0.174	0.015	0.189	0.230	0.08

This assumption allows us to calculate the grams of PF_6 in the crystal at full charge. Assuming that the balance of the weight gain in the HOPG crystal is due only to solvent uptake, the wt% solvent in the DMC extract can be calculated with the details given in Table S-2. The results of the analysis show that there is significant solvent in the HOPG crystals, and is on average 70% of that expected based on a $C_{20}PF_6$ composition. An interesting result of the analysis is that the EMC is much preferred over the FEC as co-intercalate for the PF_6^- anion. The ratio of FEC to EMC in the electrolyte is 1.0 while in the HOPG crystal, the ratio is 5-10% of this value. This type of preference has not been observed before in anion intercalation compounds with the intercalation of EMC over FEC possibly resulting from favorable solvation of the anion by EMC or by the availability of EMC in a bulk electrolyte where FEC is coordinated strongly with Li^+ in solution.

The gravimetry and GC-MS data show that solvent is co-intercalated with anion during charge and presumable de-intercalated on discharge. In order to observe this phenomenon and to determine the ratio of solvent to anion during the intercalation process, we performed several in-situ gravimetry experiments. Figure 8 shows the results from a number of charge/discharge cycles with the mAh per mass change recorded vs. cycle number. The three horizontal lines on the graph indicate the theoretical values for the mAh per gram change based on 0, 1 or 2 solvent molecules intercalating per anion. The data shows 3 different charge/discharge rates, C/20 (cycles 1-10), C/5(cycles 13-17) and C/40 (cycles 18-22). The baseline drift in the in-situ measurements accounts for the large error bars on the data and for the fact that a number of cycles are not shown as the drift in the baseline was too large for correction. The data indicates that the weight does change in a consistent manner with ~ 1 solvent molecule co-intercalating/de-intercalating with 1 anion. The data also indicates that this ratio might vary based on the rate or degree of intercalation. Cycles 13-17 are at a higher rate and cycle at roughly half of the capacity of the other cycles in Figure 8. Those cycles tended to show more solvent intercalated per anion than the lower rate cycles. Although it is clear that solvent is co-intercalated with anion between the graphite layers, it appears that the spacing of the graphite sheets is dictated by the anion size and not the solvation sphere of the co-intercalated solvent. Given that the interaction between the anion and the positively charged graphite is similar to an ionic bond while the interaction between solvent and the anion or the positively charged graphite in the GIC is equivalent to an ion-dipole interaction, it would seem reasonable that the stronger anion-graphite lattice interaction would determine structure. Comparing our results to previous studies, it is clear that the interlayer spacing of 7.7-8.0Å is observed in a number of cases for the PF_6^- anion intercalate[18,21] and that where there is disagreement on the structure,[5,20] the XRD data from those studies can support an intercalated layer spacing of 7.7-8.0Å. From our results, there appears to be a strong interaction between the graphite sheets and the PF_6^- anion, as suggested by the intimate contact between the anion and the graphite sheets, and therefore it appears that the anion dictates the structure to a large degree. The fact that in the literature different electrolyte solvents result in a similar structure would suggest that the solvent does not dictate structure and may be only ancillary to the intercalation process. The effect of solvent co-intercalation on cycling efficiency and the effect on the long term cycling of graphite need to be explored further.

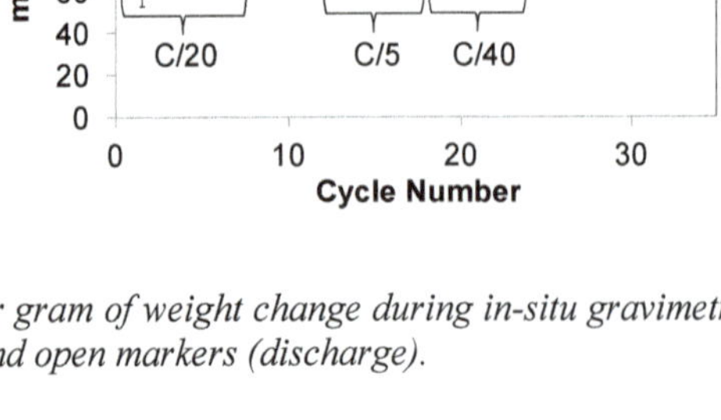

Fig. 8 Capacity per gram of weight change during in-situ gravimetry on cell G2. Closed markers (charge) and open markers (discharge).

4. Conclusions

In-situ X-Ray diffraction in combination with in-situ dilatometry confirmed the formation of a stage I intercalate at full charge with a graphite lattice expansion consistent with PF_6^- anion in close contact with the graphite layers. The size of the intercalated PF_6^- anion is consistent with that calculated from the P-F bond length and the covalent radius of fluorine in PF_6^-. The XRD and dilatometry data also support the idea that specific compositions are formed during the anion intercalation process with the identified phases B-E′ correlated to the compositions given in scheme 1. The $C_{20}PF_6$ (I) composition formed at 5.2V gives an indication of the upper limit on the specific capacity at 112 mAh/g for this anion intercalation system. Gravimetry indicated that the anion intercalation is accompanied by solvent co-intercalation at a ratio of ~0.7 ± 0.2 solvent molecules per anion. Analysis of charged samples removed from cycled cells confirmed this solvent to anion ratio but also indicated that the solvent molecules co-intercalated in a ratio differing from that of the bulk electrolyte with a strong preference for EMC. In-situ gravimetry confirmed that solvent moves with the anion during cycling at roughly a 1:1 ratio but that the ratio may be rate dependent.

Supporting Information Available: Figure of the in-situ gravimetry cell, graph of the indexed single phases for a first charge cycle on cell X4, stage I intercalate diagram, in-situ XRD on cell X3, tables of the indexed phases A-E′ with all identified Miller indices, and table of the raw data used to calculate results in Table 4. This material is available free of charge via the Internet at http://pubs.acs.org.

Acknowledgment

The author would like to acknowledge Dr. Janet Ho for help with dilatometry, Dr. Jan Allen for helpful discussion on XRD indexing, Dr. Kang Xu, and Dr. Arthur Cresce for electrolytes and the US Army Research Laboratory for financial support.

The work has been previously published by ACS Publications:

Jeffrey A. Read, "In-Situ Studies on the Electrochemical Intercalation of Hexafluorophosphate Anion in Graphite with Selective Cointercalation of Solvent", *J. Phys. Chem. C*, 2015, 119 (16), pp 8438–8446. https://doi.org/10.1021/jp5115465

References

[1] Read, J.A.; Cresce, A.V.; Ervin, M.H.; Xu, K., Dual-Graphite Chemistry Enabled by a High Voltage Electrolyte, *Energ. Environ. Sci.* 2014, *7*, 617-620. https://doi.org/10.1039/C3EE43333A

[2] Carlin, R.T.; De Long, H.C.; Fuller, J.; Trulove, P., Dual Intercalating Molten Electrolyte Batteries, *J. Electrochem. Soc.* 1994, *141*, L73-L76. https://doi.org/10.1149/1.2055041

[3] Santhanam, R.; Noel, M., Effect of Solvents on the Intercalation/De-Intercalation Behaviour of Monovalent Ionic Species from Non-Aqueous Solvents on Polypropylene-Graphite Composite Electrode, *J. Power Sources* 1997, *66*, 47-54. https://doi.org/10.1016/S0378-7753(96)02472-X

[4] Santhanam, R.; Noel, M., Electrochemical Intercalation of Cationic and Anionic Species from a Lithium Perchlorate-Propylene Carbonate System-A Rocking Chair Type of Dual-Intercalation System, *J. Power Sources* 1998, *76*, 147-152. https://doi.org/10.1016/S0378-7753(98)00152-9

[5] Seel, J.A.; Dahn, J.R., Electrochemical Intercalation of PF_6 into Graphite, *J. Electrochem. Soc.* 2000, *147*, 892-898. https://doi.org/10.1149/1.1393288

[6] West, W.C.; Whitacre, J.F.; Leifer, N.; Greenbaum, S.; Smart, M.; Bugga R.; Blanco, M.; Narayanan, S.R., Reversible Intercalation of Fluoride-Anion Receptor Complexes in Graphite, *J. Electrochem. Soc.* 2007, *154*, A929-A936. https://doi.org/10.1149/1.2759841

[7] Ishihara, T.; Koga, M.; Matsumoto, H.; Yoshio, M., Electrochemical Intercalation of Hexafluorophosphate Anion into Various Carbons for Cathode of Dual-Carbon Rechargeable Battery, *Electrochem. Solid State Lett.* 2007, *10*, A74-A76. https://doi.org/10.1149/1.2424263

[8] Gunawardhana, N.; Park, G-J; Dimov, N.; Thapa, A.K.; Nakamura, H.; Wang, H.; Ishihara, T.; Yoshio, M., Constructing a Novel and Safer Energy Storing System using a Graphite Cathode and a MoO_3 Anode, *J. Power Sources* 2011, *196*, 7886-7890. https://doi.org/10.1016/j.jpowsour.2011.04.059

[9] Placke, T.; Fromm, O.; Lux, S.F.; Bieker, P.; Rothermel, S.; Meyer, H-W; Passerini, S.; Winter, M., Reversible Intercalation of Bis(trifluoromethanesulfonyl)imide Anions from an Ionic Liquid Electrolyte into Graphite for High Performance Dual-Ion Cells, *J. Electrochem. Soc.* 2012, *159*, A1755-A1765. https://doi.org/10.1149/2.011211jes

[10] McCullough, F.P.; Levine, A.; Snelgrove, R.V. (The Dow Chemical Company), Secondary Battery, U.S. Patent 4,830,938, May 16, 1989.

[11] McCullough, F.P.; Beale, A.F. (The Dow Chemical Company) Secondary Electrical Energy Storage Device and Electrode Therfor, U.S. Patent 4,865,931, Sept. 12, 1989.

[12] McCullough, F.P.; Flexible Carbon Fiber, Carbon Fiber Electrode and Secondary Energy Storage Devices, U.S. Patent 5,518,836, May 21, 1996.

[13] McCullough, F.P.; Flexible Carbon Fiber Electrode with Low Modulus and High Electrical Conductivity, Battery Employing the Carbon Fiber Electrode, and Method of Manufacture, U.S. Patent 5,532,083, Jul. 2, 1996.

[14] Dahn, J.R.; Seel, J.A., Energy and Capacity Projections for Practical Dual-Graphite Cells, *J. Electrochem. Soc.* 2000, *147*, 899-901. https://doi.org/10.1149/1.1393289

[15] Rothermel, S.; Meister, P.; Fromm, O.; Huesker, J.; Meyer, H.W.; Winter, M.; Placke, T., Study of the Electrochemical Behavior of Dual-Graphite Cells using Ionic Liquid-based Electrolytes, *ECS Transactions* 2014, *58*, 15-25. https://doi.org/10.1149/05814.0015ecst

[16] Hu, L.; Zhang, Z.; Amine, K, Fluorinated Electrolytes for Li-ion Battery: An FEC-based Electrolyte for High Voltage $LiNi_{0.5}Mn_{1.5}O_4$/Graphite Couple, *Electrochem. Commun.* 2013, *35*, 76-79. https://doi.org/10.1016/j.elecom.2013.08.009

[17] Borodin, O.; Behl, W.K.; Jow, T.R., Oxidative Stability and Initial Decomposition Reactions of Carbonate, Sulfone, and Alkyl Phosphate-Based Electrolytes, *J. Phys. Chem. C.* 2013, *117*, 8661-8682. https://doi.org/10.1021/jp400527c

[18] Billaud, D.; Pron, A.; Lincoln Vogel, F.; Herold, A., Intercalation of Pyrographite by NO_2^+ and NO^+ Salts, *Mat. Res. Bull.* 1980, *15*, 1627-1634. https://doi.org/10.1016/0025-5408(80)90244-5

[19] Billaud, D.; Chenite, A., Electrochemical Intercalation of Fluorides into Graphite, *J. Power Sources* 1984, *13*, 1-7. https://doi.org/10.1016/0378-7753(84)80049-X

[20] Touzain, A.J.; Bonnetain, L., Insertion Des Ions PF_6^-, AsF_6^- et SbF_6^- dans le Graphite Par Methode Electrochemique, Caracterisation Des Produits Obtenus, *Carbon* 1981, *19*, 193-198. https://doi.org/10.1016/0008-6223(81)90042-7

[21] Zhang, Z.; Lerner, M., Electrochemical Oxidation of Graphite in Organic Electrolytes Containing PF_6^- and ClO_4^-, *J. Electrochem. Soc.* 1993, *140*, 742-746. https://doi.org/10.1149/1.2056151

[22] Daumas, N.; Herold, A., On Relations Between the Stage Concept and the Reaction Mechanisms in the Insertion of Graphite, *COMPTES RENDUS*

HEBDOMADAIRES DES SEANCES DE L ACADEMIE DES SCIENCES SERIE C 1969, *268*, 373.

[23] Yan, W.; Lerner, M., Electrochemical Preparation and Structural Characterization of a Graphite $B[OC(CF_3)_2C(O)O]_2^-$ Intercalation Compound, *J. Electrochem. Soc.* 2004, *151*, J15-J20. https://doi.org/10.1149/1.1636741

[24] Besenhard, J.O.; Wudy, E.; Mohwald, H.; Nickl, J.J.; Biberacher, W.; Foag, W., Anodic Oxidation of Graphite in H_2SO_4 Dilatometry – In Situ X-Ray Diffraction – Impedance Spectroscopy, *Synth. Met.* 1983, *7*, 185-192. https://doi.org/10.1016/0379-6779(83)90046-2

[25] Gillespie, R.; Robinson, E.A., Bond Lengths in Covalent Fluorides. A New Value for the Covalent Radius of Fluorine, *Inorg. Chem.* 1992, *31*, 1960-1963. https://doi.org/10.1021/ic00036a045

[26] Jenkins, H.D.B.; Roobottom, H.K.; Passmore, J.; Glasser, L., Relationships among Ionic Lattice Energies, Molecular (Formula Unit) Volumes, and Thermochemical Radii, *Inorg. Chem.* 1999, *38*, 3609-3620. https://doi.org/10.1021/ic9812961

[27] Miller, G.R.; Resing, H.A.; Vogel, F.L.; Pron, A.; Wu, T.C.; Billaud, D., The Identification of PF_6^- Ions in Intercalated Graphites by ^{19}F and ^{31}P NMR Spectroscopy, *J. Phys. Chem.* 1980, *84*, 3333-3335. https://doi.org/10.1021/j100462a001

[28] Miller, G.R.; Resing, H.A.; Moran, M.J.; Banks, L.; Vogel, F.L.; Pron, A.; Billaud, D., NMR Determination of the Ordering of CH_3NO_2 Co-Intercalated with PF_6^-, AsF_6^-, or SbF_6^- in Graphite, *Synth. Met.* 1983, *8*, 77-82. https://doi.org/10.1016/0379-6779(83)90013-9

Chapter 10

Understanding Transport at the Acid-Alkaline Interface of Bipolar Membranes

Kyle N. Grew[*1], Joshua P. McClure[1], Deryn Chu[1], Paul A. Kohl[2], John M. Ahlfield[2]

[1]U.S. Army Research Laboratory, Sensors & Electron Devices Directorate, 2800 Powder Mill Road, Adelphi, MD 20783

[2]Georgia Institute of Technology, School of Chemical & Bimolecular Engineering, Atlanta, GA 30332-0100

* kyle.n.grew.civ@mail.mil

Abstract

The combination of cation exchange membrane (CEM) and anion exchange membrane (AEM) electrolytes to form of a hybrid, or bipolar membrane (BPM) electrolyte, can have unique advantages for electrochemical systems including fuel cells, electrolyzers, electrodialysis, and solar-to-fuel devices. However, a major challenge for this approach is the development of a stable and active interfacial region (i.e., junction) that adjoins the CEM and AEM layers. Moreover, a fundamental understanding of transport at the CEM-AEM junction is lacking. Therefore, the present study focuses on the theoretical development and analysis of the nature of the BPM interface. A Poisson-Nernst-Planck (PNP) theory is formalized and applied to a representative BPM interface. The findings are reported in terms of bias (i.e., overpotential) in a galvanic device with respect to CEM and AEM material requirements. Specific attention is paid to our interests in the application of the BPM to a fuel cell device with an acidic (CEM) anode and alkaline (AEM) cathode. We demonstrate that a BPM with an acidic CEM anode and alkaline AEM cathode must promote a trap-assisted type of recombination mechanism under forward bias. Without such a mechanism, large overpotentials are needed to drive ionic recombination processes.

Contents

1. Introduction

Low temperature fuel cells have had difficulty in reaching the mass market despite promise as an efficient and scalable power source. Issues associated with costs, reliability, and ease of integration have made it difficult to disrupt established energy storage and conversion technologies. Fuel cell costs are often driven by the use of noble metal catalysts and fluorinated polymeric electrolytes [1]. Numerous research groups explore catalyst materials in search of methods to remove platinum (Pt) and other precious metal electrocatalysts from low temperature fuel cell systems. To their credit, Pt content in polymer electrolyte membrane hydrogen/air fuel cells have dropped from upwards of 5 mg/cm^2 in the 1980s to approximately 0.125-0.3 mg/cm^2 at present [2, 1]. Other portions of the research community have turned their attention from acidic polymer electrolytes, a type form of cation exchange membrane (CEM), to alkaline anion exchange membrane (AEM) materials. While less mature, AEM materials have been steadily improving with notable improvements in metrics such as conductivity and stability [3, 4, 5, 6]. The motivation for the move to AEM technology is the recognition that oxygen reduction reaction (ORR) can be performed without Pt-based electrocatalysts. However, the hydrogen and methanol oxidation reaction at an alkaline ionomer-catalyst interface can also experience a voltage penalty associated with specific adsorption of cationic groups and/or other intermediates [7, 8, 9]. In addition to challenges with catalysts, a fuel cell system's balance of plant (BoP) can turn a simple device into a complex system. The BoP, which is typically comprised of radiators or heat exchangers, humidifiers, flow regulators, and power conditioning components, can also contribute to challenges associated with a fuel cell system's size, cost, and reliability.

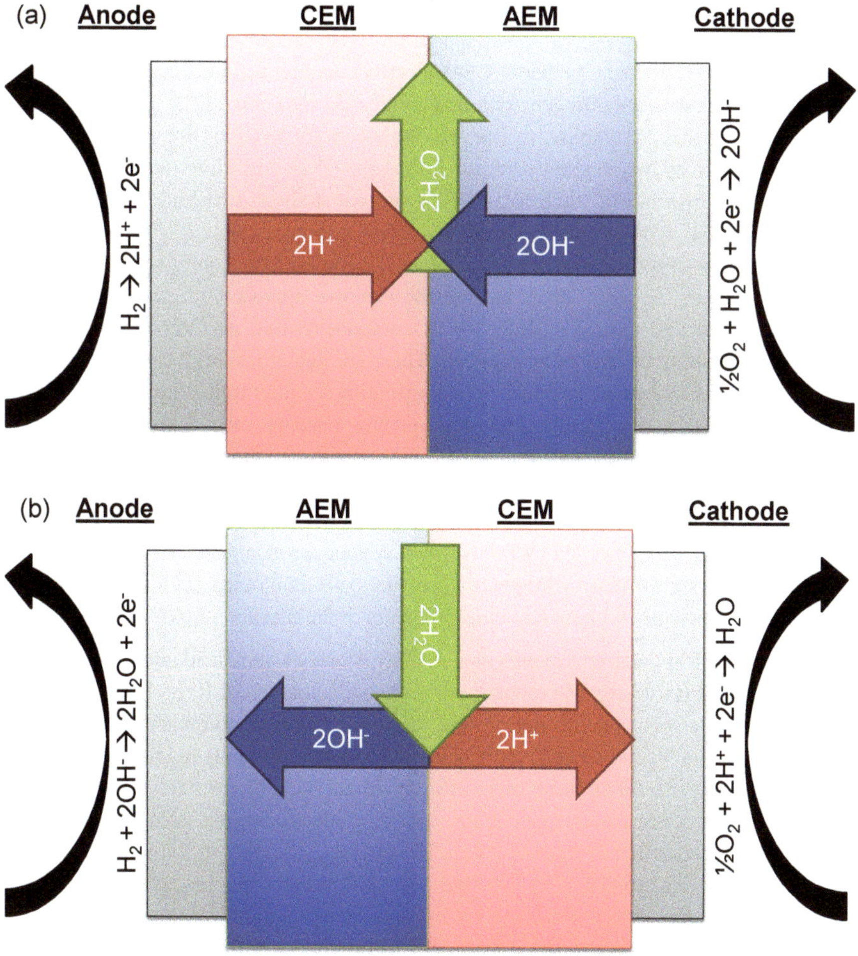

Figure 1. Bipolar membrane fuel cells employing a hybrid acid- and alkaline-membrane are shown. These include: ***(a)*** *an acidic (CEM) anode and alkaline (AEM) cathode with water generated at the bipolar interface and* ***(b)*** *alkaline (AEM) anode and acidic (CEM) cathode with water split at the bipolar interface. Figures are shown for H_2/air.*

The incorporation of a bipolar membrane (BPM) electrolyte may be one way to circumvent many challenges associated with low temperature fuel cells' electrocatalyst costs and BoP. The BPM is a hybrid material, which adjoins a cation conducting CEM with an anion conducting AEM. This is shown for a fuel cell (or electrolyzer) in Fig. 1. In Fig. 1a, oxidation occurs at an acidic CEM electrode and reduction occurs at an alkaline AEM electrode so that hydrogen ions and hydroxide ions migrate to the CEM-AEM interface where they recombine. In Fig. 1b, the electrode reactions are reversed so that water is split into hydrogen and hydroxide ions at the CEM-AEM interface. The intent of the approach shown in Fig. 1a, a focus of the present study, is two-fold. First, an acidic (CEM) anode for hydrogen or methanol oxidation and an alkaline (AEM) cathode for oxygen reduction reactions are electrochemically favorable and require significantly less Pt-based catalysts. Second, there is promise for the approach to simplify the low temperature fuel cell system's BoP, where components that perform water and fuel management operations may be minimized. There are numerous large components to (i) capture the water byproduct and recycle it for diluting a concentrated liquid fuel, such as methanol, or (ii) humidify the fuel and oxidant streams such that the membrane electrolyte is sufficiently hydrated. The BPM approach shown in the Fig. 1a is intrinsically self-hydrating. The creation of a BPM junction (i.e., at CEM-AEM interface) close to the water-consuming cathode can reduce the need for active water management because water is being needed. While further system development efforts are needed, Kohl et al have shown that hydrogen/oxygen fuel cells employing such a configuration operate most efficiently without any humidification of the fuel and/or oxidant [10, 11].

The history of BPM material systems date to discussions on weak acid and base solutions as chemical analogs to impurity-containing semiconductors by H. Reiss [12]. J. Bockris clarified this perspective by noting the electrochemical requirements for such a configuration and demonstrated rectification-like behavior using acidic and alkaline polyelectrolytes [13]. Since these early efforts, bipolar electrolytes have been examined for numerous processes and applications including electrodialysis, acid/base chemical production and purification, chlor-alkali electrolysis, analogs to biological system, macromolecular electronics, and flexible electronics [14, 15, 16, 17, 18, 19, 20]. Analogies to semiconductor theory have been made and utilized to in efforts to understand these systems [21, 22, 15, 21, 23]. However, there have been relatively few attempts to extend this theory to implementation in fuel cell devices [24, 25].

In the present study, we explore the application of the BPM to fuel cells. The theory formalized in this study is applicable to other BPM configurations and applications, making it more general than the present discussions. A consistent framework could be used to study interfacial phenomena in a wide range of electrochemical systems including electro-osmosis, colloids, electrodialysis, acid/base chemical production, chlor-alkali electrolysis, bioelectrochemical systems, and capacitive electrodes to name a few [14, 15, 16, 17, 18, 19, 20, 26, 27, 28]. The strength of the present model is that a consistent framework can be used for acid- and base- solutions with a consistent potential scale; making it straight-forward to implement with other potential- and field-dependent analyses.

Nevertheless, the present effort focusses on the investigation of the BPM fuel cell configuration with an acidic anode and alkaline cathode (Fig. 1a), which has similarities to that of a semiconductor *p-n* junction under forward bias. For completeness, the reverse configuration (Fig. 1b) (i.e., alkaline anode and acidic cathode) is also considered with similarities to a *p-n* junction under reverse bias. Thermodynamic and transport theories are formalized for the examination of the BPM under equilibrium and non-equilibrium conditions. This includes a discussion on the thermodynamics associated with the formation of a bipolar junction and the corresponding built-in junction voltage. The theory governing the space charge region and ionic transport at the BPM interface under bias (overpotential) are discussed. And the reactive processes that result in ionic charge generation and recombination at the BPM interface are examined. We find that trap-like mechanisms are needed to provide a low-loss junction for galvanic currents typical of a fuel cell. Moreover, the BPM system discussed has unique constraints relative to a semiconductor device. This is supported using evidence in the published literature as well as some key figures of merit on what are seen as the most likely (contributing) factors to the rates of the recombination reactions that dictate the magnitude of the losses associated with the bipolar interface. Finally, fabrication approaches to tailor the acid-alkaline interface to further reduce losses associated with the BPM interface are discussed.

2. Theoretical development

2.1 Junction thermodynamics at thermal equilibrium

A thermodynamic approach is used to examine the effect of marrying an acidic CEM with an alkaline AEM, which provides relationships for the "built-in" or "junction potential". This analysis is analogous to the thermodynamic relationships for a *p-n* junction in a semiconductor at thermal equilibrium. For simplicity, we use a hydrogen/oxygen (H_2/O_2) fuel cell, where the global reaction is well known.

$$H_2 + \tfrac{1}{2}O_2 = H_2O \tag{1}$$

The open circuit voltage (OCV), V_{OCV}, may be calculated according to the change in Gibbs free energy, ΔG_f, for the overall reaction.

$$nFV_{OCV} = -\Delta G_f \tag{2}$$

Considering an acidic (CEM) anode and an alkaline (AEM) cathode, the electrochemical hydrogen oxidation reaction (HOR) and ORR are typical.

$$H_2 \rightleftharpoons 2H^+ + 2e^- \tag{3}$$

$$\frac{1}{2}O_2 + H_2O + 2e^- \rightleftharpoons 2OH^- \tag{4}$$

During fuel cell operation, the protons (H^+) in a CEM migrate from the anode to the cathode and the hydroxide (OH^-) in an AEM migrate from the cathode to the anode. Therefore it seems to reason that water will be formed in the vicinity of the junction in the bipolar configuration.

$$2H^+ + 2OH^- \rightleftharpoons 2H_2O \tag{5}$$

The combination of protons and hydroxide at the junction not only makes intuitive sense, its inclusion yields the standard global reaction for the full fuel cell. This observation is important for two reasons. First, we have a stable system at steady-state operation. Second, there are no additional entropic terms from the application of the junction; we have the same global reaction and corresponding changes in enthalpy and free energy. This implies there shouldn't be anomalous exothermic heating effects from the proton and hydroxide recombination (i.e., as would typically accompany the "mixing" of an acid and base in solution). Therefore, as Kohl et al have shown experimentally, we have the same potential at open circuit [29]. It is important to clarify that this does not imply there won't be additional losses associated with the junction as we depart from equilibrium. For completeness, the interfacial reactions for four distinct H_2/O_2 fuel cell configurations are provided in Table 1. These include the standard CEM- and AEM-based polymer electrolyte fuel cells, as well as the two bipolar membrane fuel cell configurations shown in Fig. 1.

Table 1: Standard potentials in acid, alkaline, and bipolar membrane electrolyte fuel cell configurations

Acidic (CEM) Fuel Cell		
Anode (Hydrogen Oxidation)	$H_2 \rightarrow 2H^+ + 2e^-$	0.00 V vs. SHE (25 °C)
Cathode (Oxygen Reduction)	$\frac{1}{2}O_2 + 2H^+ + 2e^- \rightarrow H_2O$	1.23 V vs. SHE (25 °C)
Net:	$H_2 + \frac{1}{2}O_2 \rightarrow H_2O$	1.23 V
Alkaline (AEM) Fuel Cell		
Anode (Hydrogen Oxidation)	$H_2 + 2OH^- \rightarrow 2H_2O + 2e^-$	-0.83 V vs. SHE (25 °C)
Cathode (Oxygen Reduction)	$\frac{1}{2}O_2 + H_2O + 2e^- \rightleftharpoons 2OH^-$	0.40 V. vs. SHE (25 °C)
Net:	$H_2 + \frac{1}{2}O_2 \rightarrow H_2O$	1.23 V
Bipolar Membrane Fuel Cell: Acid (CEM) Anode & Alkaline (AEM) Cathode		
Anode (Hydrogen Oxidation)	$H_2 + 2OH^- \rightarrow 2H_2O + 2e^-$	0.00 V vs. SHE (25 °C)
Junction:	$2H^+ + 2OH^- \rightarrow 2H_2O$	0.83 V vs. SHE (25 °C)
Cathode (Oxygen Reduction)	$\frac{1}{2}O_2 + H_2O + 2e^- \rightleftharpoons 2OH^-$	0.40 V vs. SHE (25 °C)
Net:	$H_2 + \frac{1}{2}O_2 \rightarrow H_2O$	1.23 V
Bipolar Membrane Fuel Cell: Alkaline (AEM) Anode & Acid (CEM) Cathode		
Anode (Hydrogen Oxidation)	$H_2 + 2OH^- \rightarrow 2H_2O + 2e^-$	-0.83 V vs. SHE (25 °C)
Junction:	$2H_2O \rightarrow 2H^+ + 2OH^-$	-0.83 V vs. SHE (25 °C)
Cathode (Oxygen Reduction)	$\frac{1}{2}O_2 + 2H^+ + 2e^- \rightarrow H_2O$	1.23 V vs. SHE (25 °C)
Net:	$H_2 + \frac{1}{2}O_2 \rightarrow H_2O$	1.23 V

To study the thermodynamics associated with the bipolar membrane we introduce a definition for the electrochemical potential $\overline{\mu_i^\alpha}$ of species i in phase α, which is consistent with Bard and Faulkner [30].

$$\overline{\mu_i^\alpha} = \mu_i^o + RTln(a_i^\alpha) + z_i F \psi^\alpha \quad (6)$$

In this definition, the reference chemical potential, μ_i^o, ionic activity a_i^α, and electrical potential of the phase/material in question ψ^α are used. The bar over electrochemical potential designates the difference between a chemical and electrochemical potentials.

To describe the BPM configurations shown in Fig. 1, we must describe the equilibrium process relating the protons and hydroxide ions to water.

$$\overline{\mu_{H^+}} + \overline{\mu_{OH^-}} = \mu_{H_2O} \quad (7)$$

If we expand the electrochemical potentials and note that the hydrogen proton is the predominant carrier in the CEM-phase and the hydroxide anion is the predominant carrier in the AEM-phase, we arrive at an expression for the junction potential.

$$\mu_{H^+}^{o} + RTln\left(a_{H^+}^{CEM}\right) + F\psi^{CEM} + \mu_{OH^-}^{o} + RTln\left(a_{OH^-}^{AEM}\right) - F\psi^{AEM} = \mu_{H_2O}^{o} + RTln\left(a_{H_2O}^{j}\right) \quad (8)$$

Recognizing that $\mu_{H^+}^{o} + \mu_{OH^-}^{o} = \mu_{H_2O}^{o}$, Eq. 8 may be simplified to a Nernst-like expression for the voltage difference across the interface, referred to as a junction potential, V_j, where $a_{H_2O}^{j} = K_W = a_{H^+}a_{OH^-}$.

$$V_j = \left(\psi^{AEM} - \psi^{CEM}\right) = \frac{RT}{F} \cdot ln\left[\frac{a_{H^+}^{CEM} a_{OH^-}^{AEM}}{a_{H_2O}^{j}}\right] = \frac{RT}{F} \cdot ln\left[\frac{a_{H^+}^{CEM} a_{OH^-}^{AEM}}{K_W}\right] = \frac{RT}{F} \cdot ln\left[\frac{a_{H^+}^{CEM}}{a_{H^+}^{AEM}}\right] = \frac{RT}{F} \cdot ln\left[\frac{a_{OH^-}^{AEM}}{a_{OH^-}^{CEM}}\right] \quad (9)$$

It should be noted that in our definition of Eq. 9, the junction potential was taken as $\psi^{AEM} - \psi^{CEM}$. This is a sign specific designation which yields a positive junction potential value for the configuration shown in Fig 1a. The configuration shown in Fig 1b, is simply the negative value of that which is derived in Eq. 9 (i.e. $V_j = \psi^{CEM} - \psi^{AEM}$). These expressions are consistent with the values provided in Table 1.

To confirm this expression and demonstrate its generality, an entire fuel cell may be considered with the junction momentarily treated as a black box. For this exercise, we consider the configuration shown in Fig. 1a. The acidic HOR and alkaline ORR may be written using Eqs. 10 and 11.

$$\mu_{H_2}^{an} - 2\overline{\mu_{H^+}^{an}} + 2F\psi^{an} = 0 \quad (10)$$

$$-\tfrac{1}{2}\mu_{O_2}^{cat} - \mu_{H_2O}^{cat} + 2F\psi^{cat} + 2\overline{\mu_{OH^-}^{cat}} = 0 \quad (11)$$

A fuel cell's voltage is dictated by the difference in electric potential (i.e., $V = [\psi^{cat} - \psi^{an}]$).

$$2FV = 2F[\psi^{cat} - \psi^{an}] = \mu_{H_2}^{an} + \tfrac{1}{2}\mu_{O_2}^{cat} + \mu_{H_2O}^{cat} - 2\overline{\mu_{H^+}^{an}} - 2\overline{\mu_{OH^-}^{cat}} \tag{12}$$

Expanding the electrochemical potential terms and simplifying Eq. 12, an expression is obtained that relates the chemical potential of the fuel and oxidant processes at the junction.

$$2FV = \mu_{H_2}^{an} + \tfrac{1}{2}\mu_{O_2}^{cat} + \mu_{H_2O}^{cat} - 2\left[\mu_{H^+}^{an} + \mu_{OH^-}^{cat}\right] + 2F(\psi^{cat} - \psi^{an}) \tag{13}$$

The last two groups of terms are equivalent to our previous derivation of the relation between potential, protons, hydroxide, and water at the junction.

$$-2\mu_{H_2O}^{j} = -2\left[\mu_{H^+}^{an} + \mu_{OH^-}^{cat}\right] + 2F(\psi^{cat} - \psi^{an}) \tag{14}$$

Substituting Eq. 14 into Eq. 13, an expression for analyzing the voltage in a H_2/O_2 fuel cell is found.

$$2FV = \mu_{H_2}^{an} + \tfrac{1}{2}\mu_{O_2}^{cat} + \mu_{H_2O}^{cat} - 2\mu_{H_2O}^{j} \tag{15}$$

This expression may be analyzed by equating the chemical potentials to the change in free energy within the system; effectively defining a Nernst equation.

$$-2FV = \Delta G_f + RT \cdot \ln\left(\frac{a_{H_2}^{an} \cdot a_{H_2O}^{cat} \cdot a_{O_2}^{cat\frac{1}{2}}}{a_{H_2O}^{j}{}^{2}}\right) \tag{16}$$

Under equilibrium conditions, the chemical potential of water at the cathode can be assumed into the chemical potential of the water produced at the junction at equilibrium, yielding.

$$2FV = \mu_{H_2}^{an} + \tfrac{1}{2}\mu_{O_2}^{cat} - \mu_{H_2O}^{j} \tag{17}$$

and

$$-2FV = \left(G_{H_2} + \frac{1}{2}G_{O_2} - G_{H_2O}\right) + RT \cdot \ln\left(\frac{a_{H_2}^{an} \cdot a_{O_2}^{cat\frac{1}{2}}}{a_{H_2O}^{j}}\right) \tag{18}$$

These expressions are equivalent to those for a H_2/O_2 fuel cell.

$$2FV = \mu_{H_2}^{an} + \frac{1}{2}\mu_{O_2}^{cat} - \mu_{H_2O} \tag{19}$$

Both of the approaches shown provide thermodynamic evidence that the potential difference arising from the bipolar junction satisfies the global reaction. From the perspective of calculating the cell potential, the junction potential may be considered additive to the typical definition, which is consistent for either configuration shown in Fig 1.

$$V = \psi^{cat} - \psi^{an} + V_j \tag{20}$$

It is also worth noting that there are alternative means by which we can show the validity of this expression. For example, the expressions developed in Eq. 8 and Eq. 9 are in effect a simple analysis of the Donnan potential difference between the membranes [31, 32]. Likewise, the Nernst-Planck equation at thermal equilibrium (i.e., zero current) yields a similar expression [29].

2.2 Thermal equilibrium & analogy to semiconductors

To access the characteristics of the mobile ionic species and intrinsic (electrostatic) or junction potentials, an analogy to intrinsic semiconductor materials and the *p-n* junction is utilized. However, there are significant departures between the respective theories that must be addressed. Our present understanding of *p-n* junctions dates to the work from W. Shockley and colleagues, who produced many of the seminal works on the topic [33, 34, 35]. Semiconductor materials are described by the intrinsic carrier concentration, n_i, which defines the number of thermally generated charge carriers at equilibrium. It is a physical property of the material based on the density of available states, N_s, and band gap, E_g.

$$n_i = N_s \cdot exp\left(-\frac{E_g}{2k_bT}\right) \tag{21}$$

The intrinsic carrier concentration scales the product of the mobile electron, n, and hole, p, concentrations.

$$p \cdot n = n_i^2 \tag{22}$$

In an *n*- or *p*-type material, the electron and hole carrier concentrations are defined using the difference between the quasi-Fermi potentials for the electrons φ_n and holes φ_p, as well as the electrostatic potential ψ. The Fermi energies may also be mapped to the electrochemical potential $\mu_F^i(= q\varphi_i)$ [12].

$$n = n_i \cdot exp\left[\frac{q}{kT}(\psi - \varphi_n)\right] \tag{23}$$

$$p = n_i \cdot exp\left[\frac{q}{kT}(\varphi_p - \psi)\right] \tag{24}$$

As shown in Fig. 2, an *n*- or *p*-type material can be created by doping the semiconductor with donors and/or acceptors, respectively. When the semiconductor is doped, it skews the availability of mobile *p*- and *n*-carriers by introducing fixed ionized cores into the material. The doping effectively shifts the quasi-Fermi potential toward the conduction or valence band, and the equilibrium mobile electron and hole concentrations approach the corresponding doping number density (i.e., N_A and N_D).

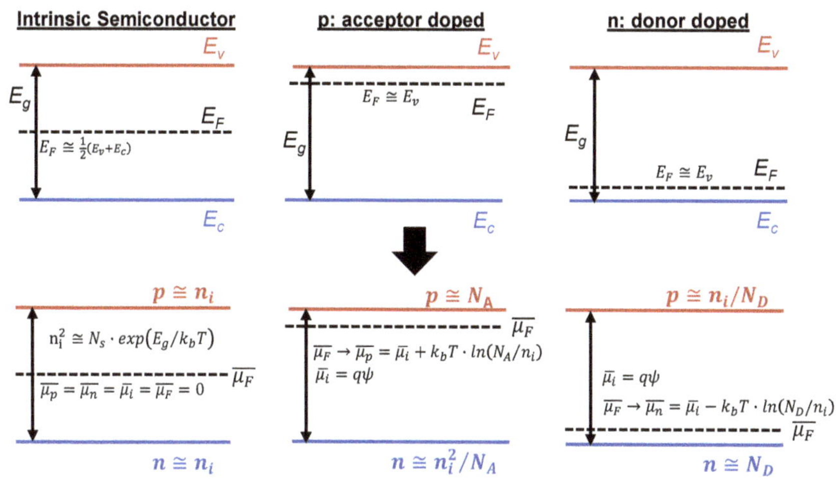

Figure 2. Band energy and carrier considerations of semiconductor materials. The semiconductor baseline serves as an analogy to the polyelectrolyte system.

If a *p*-type semiconductor is placed in intimate contact with an *n*-type semiconductor, as shown in Fig. 3, an equilibration process takes place. An electrochemical potential difference between the (majority) mobile *p*- and *n*-type carriers (i.e., difference in the quasi-Fermi potential) exists when the materials are brought together. This potential difference drives the transport of mobile charge carriers near the interface. Mobile electrons and holes near the *p-n* interface recombine until the quasi-Fermi potential difference is nullified and the system reaches thermal equilibrium. At thermal equilibrium, the quasi-Fermi potentials for the *n*- and *p*-type carriers align to zero potential. However, this equilibration process also results in a local departure from electroneutrality near the *p-n* junction. A depletion, or space-charge, region is formed near the interface and there is compensatory bending of the conduction and valence bands. The intrinsic energy, $E_i(= -q\psi)$, which is related to the electrostatic potential ψ, results from this band bending. Moreover, an electrostatic potential difference and electric field forms across the space-charge region to compensate the departure from electroneutrality. This potential difference is the built-in voltage, V_{BI}, and is dependent upon the band-gap and donor/acceptor concentrations in the respective materials.

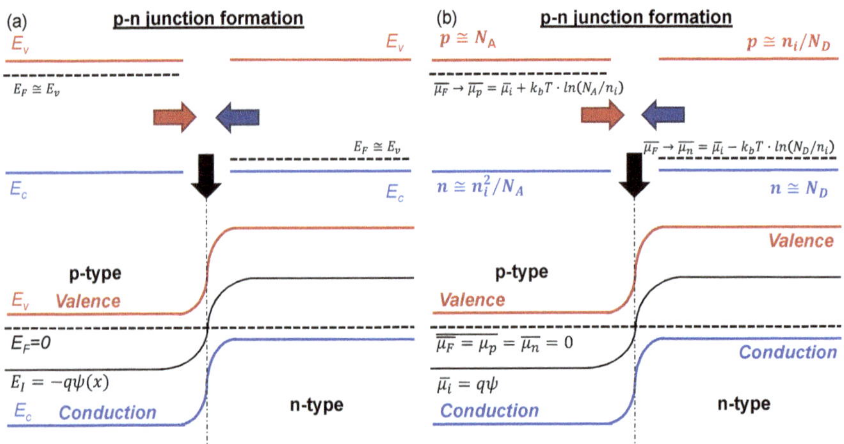

Figure 3. (a) Band energy and (b) considerations of the formation of a semiconductor PN junction.

$$V_{BI} = \frac{k_b T}{q} ln\left(\frac{n_n^o}{n_p^o}\right) = \frac{k_b T}{q} ln\left(\frac{p_p^o}{p_n^o}\right) = \frac{k_b T}{q} ln\left(\frac{N_A N_D}{n_i^2}\right) \tag{25}$$

The variations in electrostatic potential and mobile carrier concentrations can also be calculated by solving Poisson's equation, which requires the dielectric properties of the system, $\epsilon(=\epsilon_o\epsilon_r)$.

$$\frac{d}{dx}\left(\epsilon\frac{d\psi}{dx}\right) = -q\sum z_i n_i$$
$$= q\left(N_d - N_a + n_i \cdot exp\left[\frac{q}{kT}(\phi_p-\psi)\right] - n_i \cdot exp\left[\frac{q}{kT}(\psi-\phi_n)\right]\right) \quad (26)$$

This provides a Poisson-Boltzmann description of the *p-n* interface, and its definition provides a convenient departure point for examining analogy to the electrolytic system.

The extension of the chemical physics associated with a semiconductor to an electrolytic system is not rigorous. The discrete size and mass of ions is different from the elementary charges in a semiconductor. And, the concentration and mobility of the mobile charge carriers present in electrolytes can be drastically different than those in semiconductors [21]. Undoubtedly, certain phenomena which are observed with semiconductors are not likely to extend to electrolytic membranes and vice-versa. Nevertheless, many of the underlying descriptions of the systems are consistent from the perspective of thermodynamics and transport phenomena. The analogy to semiconductors provides a convenient framework for examining a membrane electrolyte system.

For the electrolytic system, we no longer have elementary charges relegated to a band structure. Nonetheless, the H^+ and OH^- concentrations are intrinsically linked via the thermal water auto-ionization constant, K_W, where brackets are used to denote molar concentration.

$$[H^+][OH^-] = K_W(T) \quad (27)$$

The K_W provides an analog to the relationship between mobile hole and electron concentrations in the intrinsic semiconductor. However, unlike semiconductor materials, additional mobile ionic species may be able to participate; a prospect which is discussed in more detail later.

Within this framework, a quasi-Fermi (electrochemical) potential for the mobile ionic carriers may be defined by an excess electrochemical potential, $\overline{\mu_i^{ex}}$.

$$\overline{\mu_i^{ex}} = \overline{\mu_i}(C_i,\psi) - \overline{\mu_i^o}(C_i^o) = RT \cdot ln\left(\frac{C_i}{C_i^o}\right) + z_iF\psi \quad (28)$$

As is typical of thermodynamic systems, the excess electrochemical potential represents the difference between the actual electrochemical potential, $\overline{\mu_\iota}$, and that at defined state (e.g., an ideal system at equilibrium). We define this reference electrochemical potential $\overline{\mu_\iota^o}$ as that of the neutral aqueous solution at thermal equilibrium and spatially separated from any space-charge inducing interfaces (i.e., $\overline{\mu_\iota^o} = f(C_\iota^o = \sqrt{K_W};\ \psi = 0))$. The excess electrochemical potential defines the quasi-Fermi electrochemical potential of the mobile ionic species and therefore goes to zero for a system at thermal equilibrium.

$$[OH^-] = \sqrt{K_W} \cdot exp\left[\frac{1}{RT}(F\psi - \overline{\mu_{OH^-}^{ex}})\right] \quad (29)$$

$$[H^+] = \sqrt{K_W} \cdot exp\left[\frac{1}{RT}(\overline{\mu_{H^+}^{ex}} - F\psi)\right] \quad (30)$$

The definition of the reference concentration and potential are critical to a proper treatment of the bipolar membrane. Other theories developed within the community, which are used for the understanding a variety of electrochemical phenomena at interfaces are built upon electrochemical potential definitions that reference the bulk electrolyte concentration [36, 37, 38, 39]. While the approaches used by these groups lead to more compact forms of Poisson-Boltzmann relationships (i.e., analytically solvable), it is not appropriate for the present study if we are to depart from equilibrium conditions.

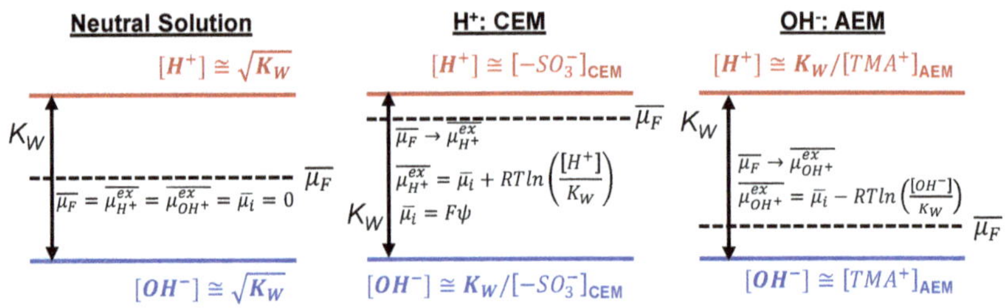

Figure 4. A polyelectrolyte system may be viewed from a similar carrier perspective as the semiconductor materials. Note: Brackets indicate molar concentration.

The relationship between the mobile ionic species are highlighted in Fig. 4 for the application to polymer electrolyte membranes. At the most basic level, parallels between the intrinsic semiconductor in Fig. 2 and electrolyte solutions in Fig. 4 are straightforward. Whereas the bandgap and doping effectively dictate the balance between the available electron and hole carrier concentration in semiconductors, the water auto-ionization constant and does so for polymer electrolytes. A typical CEM has fixed perflurosulfonic acid groups, which provide an acceptor-like behavior (i.e., donates excess hydrogen protons to the solution) whereas the AEM has fixed cationic groups (e.g., trimethylammonium) that provides donor like behavior (i.e., provides excess hydroxide to the solution). The acidic perflurosulfonic acid ($-SO_3^-$) and basic trimethylammonium ($-TMA^+$) groups are used as general proxies for the respective ionic head-groups in this study with the '-' denoting the tether; however, other ionic groups are also prevalent. The tethered head-groups within the AEM and CEM can have a degree of flexibility due to mobility of the tether segment, which does not occur within a crystalline semiconductor. The balance between the H^+ and OH^- carriers in both of these systems are still governed by the water constant K_w (i.e., Eq. 27). In Fig. 4, the system shown assumes that the CEM and AEM polyelectrolyte materials are fully dissociated, which is not always the case [40, 31]. Nonetheless, this donor/acceptor behavior has a similar effect on the polyelectrolyte as it does in the semiconductor. The availability of the respective carriers are skewed and the quasi-Fermi electrochemical potential is pulled toward an acid or alkaline limit.

When CEM and AEM materials are brought into intimate contact, as shown in Fig. 5, a difference in the excess electrochemical potential of the (majority) mobile ionic species (i.e., quasi-Fermi electrochemical potential) results in a recombination process at the junction. During recombination, the excess electrochemical potentials drive toward zero and a depletion, or space charge, region forms. A built-in or junction voltage, V_j, is formed in an analogous manner for the semiconductor interface (and as was also shown from a thermodynamic perspective in Section 2.1).

$$V_j = \frac{RT}{F} ln\left(\frac{[H^+]_{CEM}}{[H^+]_{AEM}}\right) = \frac{RT}{F} ln\left(\frac{[OH^-]_{AEM}[H^+]_{CEM}}{K_W}\right) \cong \frac{RT}{F} ln\left(\frac{a_{H^+}^{CEM} a_{OH^-}^{AEM}}{K_W}\right) \quad (31)$$

With the mobile ionic species defined according to the excess electrochemical potentials, Poisson's equation may also be defined to describe variations in electrostatic potential and mobile ionic species.

$$\frac{d}{dx}\left(\epsilon\frac{d\psi}{dx}\right) = \rho_f - F\left(\sqrt{K_W} \cdot exp\left[\frac{1}{RT}\left(F\psi - \overline{\mu_{OH^-}^{ex.}}\right)\right] - \sqrt{K_W} \cdot exp\left[\frac{1}{RT}\left(\overline{\mu_{H^+}^{ex.}} - F\psi\right)\right]\right) \quad (32)$$

In Eq. 32, $\rho_f(= F[-TMA^+]_{AEM} - F[-SO_3^-]_{CEM})$ is the local (signed) charge density associated with the fixed ionic head-groups in the CEM and AEM. This effectively represents a Poisson-Boltzmann formalism and provides a mean field description of a concentrated electrolytic system. The present study relies on the use of a Boltzmann distribution, which implicitly assumes point-like charges; a noted problem for electrolytic systems which have finite ion size and mass. Several research groups, have addressed this problem for electrochemical interfaces via variational approaches, which re-evaluates the system with additional entropic considerations (e.g., ion size) in the free-energy formulation [36, 37, 38, 39, 41, 42, 43, 44]. From this perspective, ion-specificity and steric effects on phenomena including dielectric decrement, solvation, double-layer structure, and surface tension, among others, continue to be an area of active research.

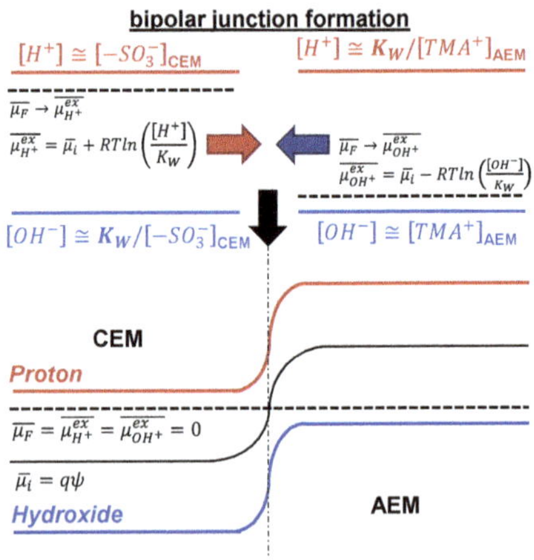

Figure 5. Formation of built-in, or junction potential, at thermal equilibrium in a bipolar membrane.

2.3 Effects of carbon dioxide and carbonates

AEM materials can undergo an exchange from OH^- to (bi)carbonate ($CO_3^=$, HCO_3^-) in the presence of carbon dioxide (CO_2). CO_2 is present at about 400 ppm in ambient air and is also a byproduct of many electrochemical oxidation reactions for carbon-containing fuels like methanol. Because CO_2 is soluble in the water which hydrates AEM materials, it has a direct path into the AEM and fuel cell.

$$H^+ + OH^- = H_2O \tag{33}$$

$$H^+ + HCO_3^- = CO_2 + H_2O \tag{34}$$

$$2H^+ + CO_3^= = CO_2 + H_2O \text{ or } H^+ + CO_3^= = HCO_3^- \tag{35}$$

This phenomena has been studied in some detail [45, 46]. The impact of CO_2 on fuel cells and electrochemical devices is typically two-fold. First, the formation of (bi)carbonates reduces the ionic conductivity of the AEM material. Second, the formation of (bi)carbonates can reduce the pH in the AEM materials. This drop in pH results in a thermodynamic loss; typically on the order of 200 mV at room temperature (i.e., 59 mV/decade in pH decrease) in ambient air. The CO_2 solubility in water decreases with temperature. Elevated temperatures may be used to reduce the magnitude of such losses. Additional methods to separate and/or scrub CO_2 from can also be used. For example, in electrolyzers for the production of dry hydrogen gas, waste hydrated oxygen gas is also produced and it is used to purge the incoming water feed.

In the present study, the presence of CO_2, $CO_3^=$, and HCO_3^- reduces the pH difference (i.e., OH^- activity) across the bipolar junction. This results in a lower equilibrium junction potential, as may be described with Eq. 9. Moreover, additional anions and recombination at the junction changes the problem because CO_2 (or H_2CO_3) is produced in addition to H_2O. Therefore, the BPM system has mixed potentials and multiple mobile carriers (majority), which is a departure from the analogy of a traditional *p-n* junction. However, the framework is sufficiently robust to address this problem. The thermal auto-ionization properties of water remain valid and typical charge conservation principles still apply outside of the bipolar interface's space charge region. Therefore, if we hypothesize that the $CO_3^=$ and HCO_3^- can also directly recombine with the H^+, there is no loss in consistency. The liberation of CO_2 at the junction will require further examination in future studies to ensure that the solvent does not become saturated and generate internal CO_2 pressures (i.e., evolve gaseous CO_2 internal to the membrane).

2.4 Transport equations

Non-equilibrium processes must be considered and understood for the implementation of a BPM into a practical device. Figure 6 highlights the relationships in a BPM fuel cell employing an acidic anode at a potential negative of its potential of zero charge (PZC) and alkaline cathode at a potential positive of its PZC. At equilibrium, each electrochemical interface has a double-layer associated with the distinct phases and corresponding electrochemical potential differences. This double-layer is represented by charge accumulation across the interfaces (i.e., relative to point of zero charge, PZC). The double-layer results in an electric field as well as variations in electrical potential. The magnitude of the electric field and potential variations are dependent upon the polarizability (dielectric) of the media as well as the physical properties. However, as shown in Fig. 6, we are also interested in the response under galvanic conditions where a bias, or overpotential, is necessary for the recombination (or generation) of H^+ and OH^- at the BPM interface.

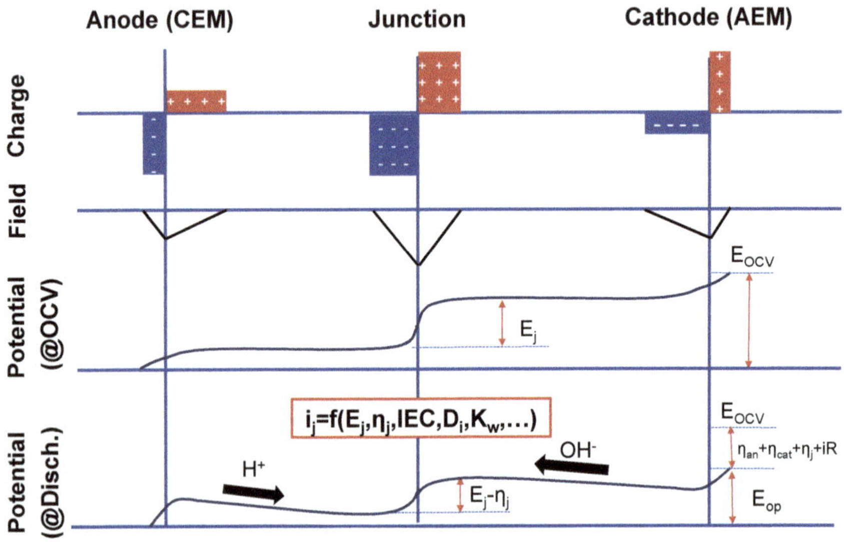

Figure 6. Interactions between the space-charge (relative to point of zero charge, PZC), electric-field, and electric potential in a bipolar membrane of the kind shown in ***Fig. 1a****. The potential is shown at both thermal equilibrium (i.e., open circuit, or OCV), as well as with a galvanic current (e.g., in a fuel cell with Faradaic current being produced).*

Transport and continuity expressions are necessary to analyze the system under bias. In semiconductor materials, the drift-diffusion equations are used in conjunction with Poisson's equation (Eq. 26). There are numerous treatises on this subject, which are not repeated here [33]. In the electrolytic framework, a Nernst-Planck based framework exists for describing transport in the CEM and AEM materials, which must be coupled with the solution to Poisson's equation (Eq. 32).

$$\overrightarrow{\iota_{H^+}} = -FD_{H^+}\nabla[H^+] + \left(\frac{F^2 D_{H^+}}{RT}\right)[H^+]\nabla\psi = -\left(\frac{FD_{H^+}}{RT}\right)[H^+]\nabla\overline{\mu_{H^+}^{ex.}} \qquad (36)$$

$$\overrightarrow{\iota_{OH^-}} = FD_{OH^-}\nabla[OH^-] + \left(\frac{F^2 D_{OH^-}}{RT}\right)[OH^-]\nabla\psi = -\left(\frac{FD_{OH^-}}{RT}\right)[OH^-]\nabla\overline{\mu_{OH^-}^{ex.}} \tag{37}$$

The Nernst-Planck transport equations provided in Eqs. 36-37 relate the mobile ionic current densities (shown for H^+ & OH^- here), $\vec{\iota_i}$, to the gradients in ionic concentration and electrostatic potential. The diffusivity of the mobile carrier D_i and standard physical constants are used. The equations can be reduced to that of the gradient of the quasi-Fermi or electrochemical potential with the total current density being conserved. It is found by summing that of the respective mobile ionic carriers.

$$\vec{\iota} = \overrightarrow{\iota_{H^+}} + \overrightarrow{\iota_{OH^-}} = -\frac{F}{RT}\left(D_{H^+}[H^+]\nabla\overline{\mu_{H^+}^{ex.}} + D_{OH^-}[OH^-]\nabla\overline{\mu_{OH^-}^{ex.}}\right) \tag{38}$$

The continuity equations for the Nernst-Planck equations are determined by taking the divergence of the respective current densities, which are equated to the net volumetric rate of (mobile) ionic charge recombination.

$$\nabla\cdot\overrightarrow{\iota_{H^+}} = -\left(\frac{FD_{H^+}}{RT}\right)\nabla\cdot\left([H^+]\nabla\overline{\mu_{H^+}^{ex.}}\right) = \dot{R}_{rg} \tag{39}$$

$$\nabla\cdot\overrightarrow{\iota_{OH^-}} = -\left(\frac{FD_{OH^-}}{RT}\right)\nabla\cdot\left([OH^-]\nabla\overline{\mu_{OH^-}^{ex.}}\right) = \dot{R}_{rg} \tag{40}$$

Because the total current density is conserved, the overall volumetric rate $\dot{R}_{rg}$ is a summation of the competing recombination and generation processes. The volumetric rate is consistent between the respective mobile carriers. The mechanisms and rates are discussed in more detail in the subsequent section; however, it suffices to say that they are a function of the local mobile ionic concentrations and quasi-Fermi electrochemical potentials in the vicinity of the space charge region.

An interesting observation is made by introducing a bias (overpotential) and tracking its influence, whereby the product of the mobile carriers is scaled by the overpotential in the vicinity of the junction.

$$[H^+][OH^-] = K_W\cdot\exp\left(\frac{F\eta}{RT}\right) \tag{41}$$

This type of expression is shown analytically using Eqs. 29-30, and holds when the difference between the H^+ and OH^- quasi-Fermi electrochemical potentials is equal to $F\eta$.

Accordingly, Eq. 41 holds identically at the metallurgical junction (i.e., position of maximum charge recombination rate) and generally through much of the space charge regions. It provides a useful point of verification for the implementation of the numeric model.

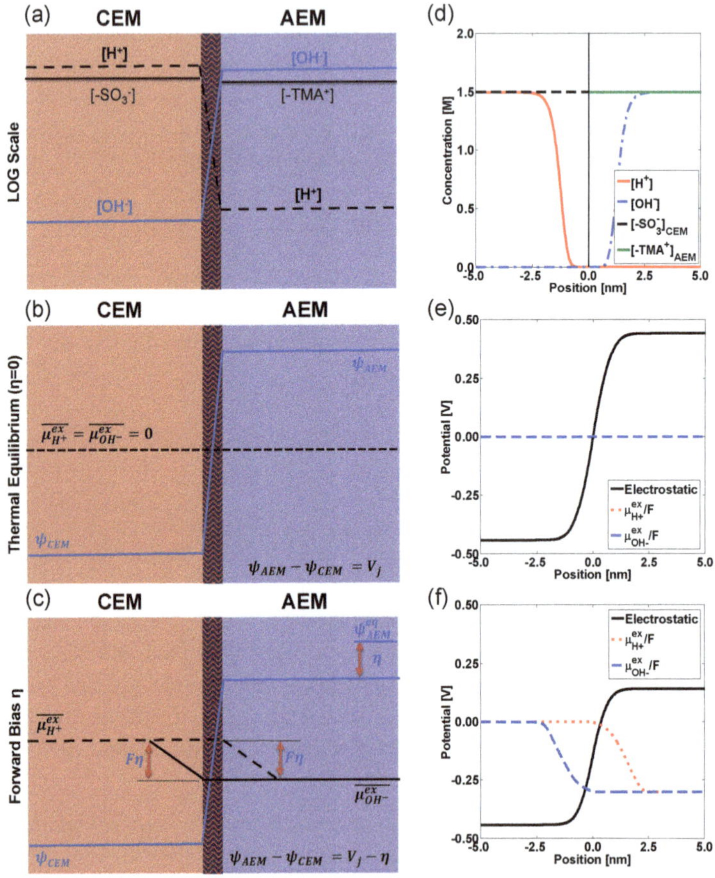

Figure 7. Notional representation of the anticipated (a) charge distribution and (b) intrinsic (electrostatic) and quasi-Fermi potential in a bipolar membrane at thermal equilibrium. Under (c) forward bias η, the intrinsic and quasi-Fermi (electrochemical) potentials are lowered resulting in differences between the H^+ and OH^- quasi-Fermi electrochemical potentials. Predicted trends for equilibrium (d) charge distribution and (e) potentials, as well as (f) system under bias are also provided for reference.

Numeric solutions are needed to study the effects of bias on the BPM interface response. A bias, or overpotential η, must be applied to the coupled system of the transport, continuity, and Poisson's equations. This is handled by applying the bias to the intrinsic (electrostatic) potential and the corresponding quasi-Fermi electrochemical potentials. It should occur at the boundary of the BPM junction at a location that is sufficiently far from the depletion region that the gradients of the electrostatic potential and mobile carrier concentrations effectively go to zero. Figure 7 highlights concentration, quasi-Fermi electrochemical potentials and intrinsic potentials in a BPM fuel cell of the type represented in Fig 1a. Specifically, the potentials of a system at thermal equilibrium in Fig. 7b is contrasted with those in a system with a forward bias (overpotential) η applied across the membrane in Fig. 7c. The bias reduces the electrostatic potential difference across the space charge region and introduces a difference between the quasi-Fermi electrochemical potentials. Far from the junction, these quasi-Fermi electrochemical potentials take on the same value. However, in the vicinity of the space-charge region, there is a difference between the quasi-Fermi electrochemical potentials which drives ionic recombination (or generation in a reverse bias scenario). To this effect, corresponding results predicted by the model for a symmetric bipolar membrane are also shown in Fig. 7. Disregarding the mechanism associated with ionic recombination, Fig. 7d provides model predictions for the fixed and mobile (carrier) ionic concentrations and Fig. 7e provides the corresponding potentials at thermal equilibrium for a bipolar membrane with symmetric fixed charge concentrations of 1.5 M and assuming comparable ionic diffusivities. These assumptions are not necessary within the model construct, but provide a simpler demonstration of the physical phenomena being described. Under forward bias, the model results in Fig 7f shows a decrease of the electrostatic (intrinsic) potential within the material as well as bending of the H^+ and OH^- quasi-Fermi potentials (shown on voltage scale).

The theoretical formalism presented in this study is complementary to other Nernst-Planck based studies dealing with multicomponent transport in electrochemical systems, for example [26, 28, 47, 45, 46]. In the present model, the coupling of Eq. 32 with Eqs. 39-40 provides additional fidelity via the (i) explicit coupling of the Nernst-Planck model to Poisson's equation to account for space charge at interfaces, and (ii) use of the intrinsic concentration scale K_W to consistently account for potential-based effects in acidic- and alkaline-media. As this present model transitions from the charged interfaces, local electroneutrality regains its validity. In this limit, when coupled with typical conservation principles, the model becomes equivalent with the noted studies (and numerous others) which utilize PNP formalism and/or a Nernst-Planck model in conjunction with a thermodynamic treatment (e.g., Donnan equilibrium, Henry's law, etc.) of the interfaces.

2.5 Mechanisms of charge generation & recombination in the space-charge region

To address the characteristics of the bipolar membrane interface, it is necessary to identify the mechanisms which participate in the ionic recombination and/or generation. There are significant differences between the polymer electrolytes forming the BPM and the semiconductor materials in the *p-n* junction. Four of the most significant include the (i) type and mass of the mobile carriers, (ii) doping and fixed charge concentrations, (iii) diffusivity or mobility of the mobile carriers, and (iv) chain segment mobility of the head-groups in the AEM and CEM. There are severe implications of these deviations. The transition from elementary charge carriers to ions suggest that factors including the differences in momentum (mass), thermal motion (Brownian), steric (ionic radii), and chemical (solvation) aspects of the mobile charge carriers uniquely influence the respective systems. Moreover, polymer electrolytes support large fixed charge concentrations; much large than the typical doping concentrations in semiconductors. However, ions can have different degrees of hydration and their mobility (or diffusivity) tend to be much smaller than the elementary charges in semiconductors.

In an ideal *p-n* junction, minority charge injection dominates the charge transfer process (i.e., via low level injection), which is typically characterized by small band gaps as well as relatively large minority carrier lifetimes and diffusions lengths (i.e., relative to the depletion region widths). This scenario does not seem to hold for a BPM. The acid-base reactions $H^+ + OH^- \rightarrow H_2O$ are among the fastest known reactions in aqueous chemistry with a recombination (forward) rate constant of $k_{recomb.}(=1.3\text{x}10^{11}\ M^{-1}s^{-1})$ and an equilibrium constant of K_W [15, 48]. The fast recombination process yields carrier lifetimes and diffusion lengths on the order of $\tau(=7\text{x}10^{-12}\text{ s})$ and $L_i\left(=\sqrt{D_i\tau_i}=0.3\text{ nm}\right)$ for conditions that are typical of the polymer electrolytes in the BPM. The depletion layer widths may also be calculated for our system.

$$W_T = W_{CEM} + W_{AEM} = \sqrt{\frac{2\epsilon V_j}{F}\left(\frac{1}{[-SO_3^-]_{CEM}} + \frac{1}{[-TMA^+]_{AEM}}\right)} \quad (42)$$

$$[-SO_3^-]_{CEM} \cdot W_{CEM} = [-TMA^+]_{AEM} \cdot W_{AEM} \quad (43)$$

Eq. 42-43 approximate the depletion layer widths by assuming that the space charge region is effectively depleted of mobile ionic charge carriers. It is an approximation, but provides a useful characteristic length. Our detailed transport model does not utilize such an approximation, but must be solved numerically. However, because the concentrations of the mobile ionic species is significantly depressed in the space charge region, the detailed model yields nearly identical values for the depletion layer width as those in Eq. 42-43. Values for the overall and relative depletion layer widths for a BPM at thermal equilibrium (i.e., $\eta \rightarrow 0$) are provided in Fig. 8. The fixed charge concentration scales reported in Fig. 8 are analogous to a volume-based ion exchange capacity. These values can be converted to the traditional mass-specific ion exchange capacity using corresponding membrane electrolyte density. Regardless of the scale used, it is apparent that the space-charge region is compact for the BPM as is typical of double-layers in condensed phases. This is due in large part to the head-group density within the AEM or CEM (i.e., fixed charge concentration).

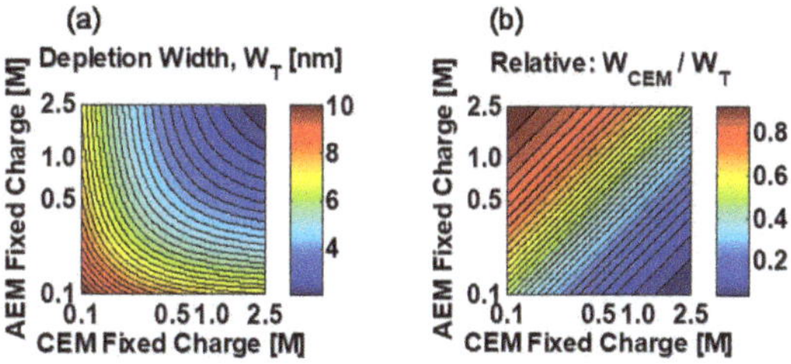

Figure 8. Predicted (a) absolute and (b) relative depletion layer widths for the acid-alkaline interface as a function of the relative ion-exchange capacity (fixed charge) values for the CEM and AEM materials. Analysis assumes the interface is fully depleted and complete dissociation (ionization) of the respective fixed ionic groups. Surface plots are shown with the fixed charge concentrations spaced with a log distribution.

It may be noted that the characteristic diffusion lengths for the mobile ionic charges is roughly an order of magnitude smaller than the calculated depletion layer widths. That is to say that the diffusive minority carrier (low-level) injection that is characteristic of an ideal *p-n* junction is not expected to be appreciable for our BPM interface. This is actually an encouraging observation for the use of BPM materials in electrochemical devices. Low-level injection at an ideal *p-n* junction is typically characterized by large biases to drive diffusion limited current. The small characteristic ionic diffusion length for the BPM (relative to the depletion layer width) is indicative that that there are faster recombination processes at play which have lower losses.

As a result, our attention turns to an analogue of band-to-band recombination. From an intuitive standpoint, this mechanism is the most satisfying. It is characterized by the direct interaction of the mobile species as they traverse the space charge region. Effectively, this is a bimolecular acid-base reaction and may be viewed from the perspective of mass-action kinetics. Bimolecular reaction processes (Langevin) have been recognized as being significant in organic semiconducting materials when the characteristic carrier transport length is smaller than critical distance $r_c(= q^2/4\pi\epsilon k_b T)$

$$\dot{R}_{rg} \cong F k_{recomb}\{[H^+][OH^-] - K_W\}$$
$$= F k_{recomb} K_W \left\{ exp\left(\frac{\left[\overline{\mu_{H^+}^{ex.}} - \overline{\mu_{OH^-}^{ex.}}\right]}{RT}\right) - 1\right\} \tag{45}$$

[49]. The characteristic mobile carrier diffusion length is roughly an order of magnitude smaller than r_c for our BPM, substantiating its presence. The overall rate of recombination for a Langevin mechanism has been noted as being proportional to the transport coefficients for the mobile carriers [49].

$$\dot{R}_{rg} \cong \frac{F^2(D_{H^+} + D_{OH^-})}{RT\epsilon} \cdot \{[H^+][OH^-] - K_W\}$$
$$= \frac{F^2(D_{H^+} + D_{OH^-})}{RT\epsilon} K_W \left\{ exp\left(\frac{\left[\overline{\mu_{H^+}^{ex.}} - \overline{\mu_{OH^-}^{ex.}}\right]}{RT}\right) - 1\right\} \tag{44}$$

It is interesting that an alternative expression for the bimolecular mechanism may be calculated according to the noted mass-action kinetics approach (i.e., via using the acid-base recombination rate constant, k_{recomb})

Evaluation of these rate expressions reveals that the rate coefficient terms are identical; a product of the acid-base type reactions being transport limited. This is a characteristic of many acid-base reactions and is a well explored theoretical limit [48].

An alternative mechanism involves the use of so-called trap-centers, which is often found at *p-n* junctions that have large band gaps, short lifetimes and low resistivity. This mechanism is commonly known as a Shockley-Read-Hall (SRH) trap-assisted recombination-generation mechanism [34, 35]. Recombination-generation centers that fall within the semiconductor material's forbidden energy gap are responsible for this mechanism. The trap-centers provide an energy well that can capture and/or emit electrons and/or holes. Practically speaking, the recombination-generation centers may be dislocations, impurities, defects, surfaces, and/or other heterogeneities within the semiconducting materials. While the rules are different, these traps behave in a manner analogous to chemical catalysts in that they provide an alternative pathway (i.e., via

Coulombic interactions) with multiple steps having a lower overall energy barrier and yield a higher probability/rate of interaction. A trapping-type rate expression may be developed for the BPM system.

$$\dot{R}_{rg} = F\left\{\frac{[H^+][OH^-] - K_W}{([H^+] + [H^+]_1)\tau_{OH^-} + ([OH^-] + [OH^-]_1)\tau_{H^+}}\right\} \tag{46}$$

Equation 46, is a direct ionic analog to the well-known SRH recombination rates [35, 34]. In Eq. 46, $[H^+]_1$ and $[OH^-]_1$ are the concentrations of the protons and hydroxide when the quasi-Fermi potential corresponds to that of the trap. We assume that the traps fall at the intrinsic energy level for this study (i.e., follow the electrostatic potential). Therefore, these concentrations take on that of the intrinsic system (i.e., $[H^+]_1 = [OH^-]_1 = \sqrt{K_W}$). The minority mobile carrier life time values, τ_{OH^-} and τ_{H^+}, also appear in Eq. 46. These values represent the mean lifetime of the respective mobile ions in the opposite membrane material (i.e., OH^- ions in the CEM). Accordingly, the lifetime is implicitly linked to the ion exchange capacity (IEC) via the concentration of the fixed ionic groups which we will hypothesize can promote a trap-like behavior. We take the lifetime values to be that the diffusion-limited acid-base reaction, $\tau_{OH^-} \cong (k_{recomb} \cdot [-SO_3^-])^{-1}$ and $\tau_{H^+} \cong (k_{recomb} \cdot [-TMA^+])^{-1}$. As a point of reference, lifetimes c.a. $\tau_{OH^-} = \tau_{H^+} = 7\text{x}10^{-12}$ s correspond to a one molar system; however, we note that these lifetime values could be different for other systems including weak acids/bases and/or with other counterions. Implementing the discussed assumptions, and substituting Eqs. 29-30 into Eq. 46, we arrive at a trap-assisted recombination rate which corresponds to our dependent variables (i.e., $\overline{\mu_{H^+}^{ex.}}, \overline{\mu_{OH^-}^{ex.}}$, and ψ).

$$\dot{R}_{rg} = F\sqrt{K_W}\left\{\frac{exp\left(\frac{\left[\overline{\mu_{H^+}^{ex.}} - \overline{\mu_{OH^-}^{ex.}}\right]}{RT}\right) - 1}{\left(exp\left[\frac{1}{RT}\left(\overline{\mu_{H^+}^{ex.}} - F\psi\right)\right] + 1\right)\tau_{OH^-} + \left(exp\left[\frac{1}{RT}\left(F\psi - \overline{\mu_{OH^-}^{ex.}}\right)\right] + 1\right)\tau_{H^+}}\right\} \tag{47}$$

Prior to implementing such a mechanism, a critical question arises. Can such a mechanism may be present in a system comprised of polymer electrolytes, whereby the discrepancies in the mass and thermal motion associated with the ionic species (i.e., relative to elementary charges) become significant? Specifically, are there centers within the BPM's space charge region which can serve as de-facto traps for mobile ions? We speculate that dopants (i.e., fixed ionic charge groups) are persistent throughout the material and there are numerous additional heterogeneities due to the polymeric nature of the materials, phase segregation, and other material interfaces. However, is counterion condensation and/or the presence of other heterogeneities' sufficient for promoting trap-like behavior? This issue will be discussed in further detail within our results.

The reverse bias case is not examined in detail within this study; however, some comments are merited. Under the conditions of reverse bias, as would be the case for an electronic device where water splitting occurs at the CEM-AEM interface, Fig. 1b, the issue is no longer recombination of H^+ and OH^- but their generation (i.e., water splitting). Such a configuration has been employed in variety of electrochemical devices and the potential difference across the junction increases with the applied (reverse) bias. As is characteristics of a semiconductor *p-n* junction, the reverse bias case is characterized by relatively small currents until some critical voltage threshold is reached. In the semiconductor, this is typical of a breakdown due to impact ionization and/or tunneling processes, such as a Zenner diode. While it is not the primary focus of the present study, several groups have associated this behavior in BPM materials with electric-field assisted water splitting and Onsager's Second Wien effect [21, 50, 51, 22]. Further, the redox stability window for water is approximately 1.23 V at 300 K. A potential difference across the bipolar junction that is greater than 1.23 is reached at relatively large reverse biases (c.a. -0.4 V). These biases tend to coincide with onsets in galvanic current in numerous experimental efforts with no and/or dilute external electrolyte/salt concentrations [52, 53, 29]. The other mechanisms described may still be participating in either the generation and/or recombination process; however, they are typically suppressed and do not significantly contribute to the reverse current.

2.6 Numeric methods & solution procedures

The described equations represent a set of coupled, non-linear ordinary differential equations, which take the form of a boundary value problem. Two methods were used to obtain solutions. The first method uses the boundary value problem solver in Matlab, which utilizes a colocation methodology. Second, the identical equations were also programed into and solved with COMSOL v.5.1, which uses a finite element methodology. The results were examined to ensure mesh independence and compared for verification purposes. All Matlab results were completed on a personal desktop computer (Dell T7500 with two Intel Xeon 2.67 GHz processors and 12 GB of RAM) while COMSOL results were generated on a single node of a Cray XC40. Results presented in this study were consistently taken from COMSOL.

3. Results & discussion

The analysis of the BPM for fuel cell applications begins with predictions of the junction potential and depletion layer widths at thermal equilibrium. As may be inferred from Eq. 9 and described in Eqs. 42-43, both of these factors are dependent upon the fixed charge concentration of the respective polymer electrolytes. As noted in Table 1, the nominal junction potential is on the order of 830 mV for typical acidic CEM and alkaline AEM materials at room temperature, when the activity of all species equals unity. This corresponds to the standard potential difference between the HOR in acidic and alkaline media (vs. SHE, 25 °C). Given the Nernst-like nature of Eq. 9, reduction in either or both electrolyte concentration(s) (i.e., IEC) lowers the junction potential. This is a result of a reduction in ionic activity, but the effect is small considering typical IEC values for AEM and CEM materials. One of the items noted in the thermodynamic analysis of the CEM-AEM interface, Fig. 1, is that the interface is stable and there are no new or additional entropic terms relative to a standard fuel cell. This observation has also been reported experimentally, where stable open circuit voltage values were demonstrated in BPM fuel cells in both of the configurations highlighted in Fig. 1 [29, 10].

The depletion layer widths shown in Fig. 8 were calculated with Eqs. 42 and 43. The depletion layer widths are shown as contour plots, with the CEM and AEM IEC values independently varied and shown with log-spacing in Fig. 8. It is immediately apparent that the depletion layer width is extremely compact at typical IEC values, i.e. 4 to 10 nm. This is important because it highlights the relatively small region of the BPM which is responsible for ionic recombination or water splitting. By comparison, the size of conductive channels in a typical AEM or CEM are on the order of 2 to 20 nm and the length of the tether between the ionic head-group and the polymer backbone is on the order of 1 nm. The present study assumes that the CEM and AEM employ strong acid and base head-groups, respectively. They were treated as being fully dissociated with a hydrated head-group and hydrated mobile conductive ion, but this need not be the case. A polymer electrolyte material where a weak acid or base head-group were used would result in a partially dissociated fixed ion head-group. Such a scenario would reduce the corresponding ionic activity (i.e., junction potential) as well as the unscreened head-group concentration (i.e., effective doping level) that establishes the depletion layer width.

The ion concentrations within the CEM-AEM depletion region are shown for two cases in Fig. 9, which provides the electrostatic potential Fig. 9a, electric-field Fig. 9b, space charge Fig. 9c, and mobile ionic concentrations for a BPM which has been exposed to 400 ppm CO_2 Fig. 9d as well as a CO_2-free junction Fig. 9e. In Fig. 9, the red vertical lines depict the width of the depletion region according to Eqs. 42 and 43. They were calculated for the condition with ambient CO_2 present, which are more compact than the CO_2-free case (i.e., by approximately 10% because the pH difference between the two sides of the junction is lower with CO_2). The agreement between the predictions from Eqs. 42 and 43 are apparent in Fig. 9c, which provides the space charge accumulation in the depletion region. The baseline properties used throughout this study consist of a system at standard ambient conditions (i.e., T = 300 K, P = 1 atm) with a relative dielectric permittivity $\varepsilon_r (= 35)$ as is typical of a strong electrolyte [29]. More robust treatments which account for the large electric fields are available, such as Booth's approach, but have minimal impact on the observed response [54]. The fixed charge concentration of the CEM $[-SO_3^-]_{CEM} (= 1.5\ M)$ and AEM $[-TMA^+]_{AEM} (= 1.5\ \mathrm{M})$ as well as the proton $D_{H^+} (= 6.0x10^{-9}\ \mathrm{m^2/s})$ and hydroxide ionic diffusivity $D_{OH^-} (= 6.0x10^{-9}\ \mathrm{m^2/s})$ values were used so as to provide symmetric properties across the BPM interface. These values deviate slightly from those which are expected for a real system; however, they simplify the interpretation of the results. This issue will be discussed in additional detail momentarily.

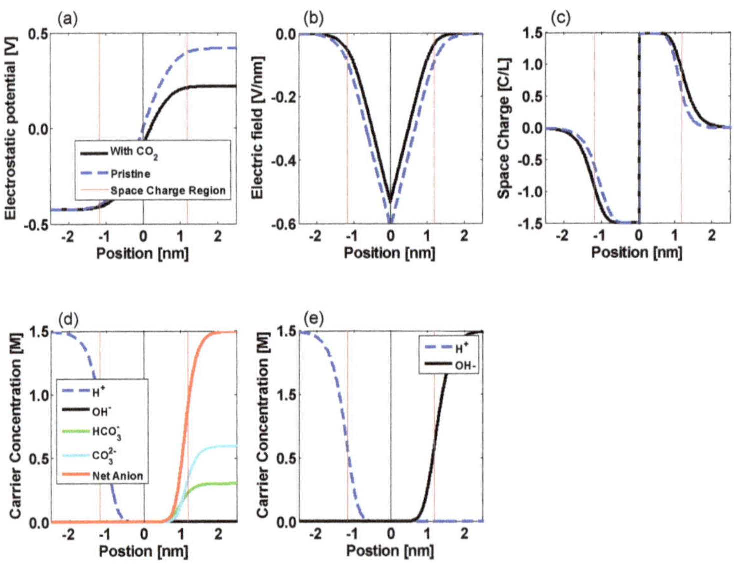

Figure 9. Bipolar junction characteristics at thermal equilibrium with 400 ppm ambient CO_2 present as well as in its pristine form, including the predicted (a) electrostatic potentials, (b) electric field, (c) space-charge, (d) carrier concentrations with CO_2 present, and (e) carrier concentrations in pristine form.

Two additional observations can be made regarding Fig. 9. The first is that the CO_2-free system exhibits an electrical potential difference, or junction potential, in excess of 800 mV. This junction potential is the result of the large electric field created by the depletion of mobile ionic charge and resulting charge separation for the head-groups within the space charge region. The theory is sufficiently robust to treat this scenario where ambient CO_2 (400 ppm) has equilibrated with the membrane and equilibrium between CO_2, HCO_3^-, CO_3^{2-}, and OH^- has been achieved. As was previously noted, the difference in the junction potential (c.a. 200 mV) and electric field relative to the pristine case is a direct result of the suppression of the OH^- concentration and activity in the AEM materials. The second observation in regard to Fig 9 is the magnitude of the space charge and electric fields that are present within the depletion region. These extreme conditions are not unprecedented for electrochemical interfaces, such as electrode/electrolyte interfaces. However, they do play important role and their influence extends beyond the voltage-current response of the BPM materials.

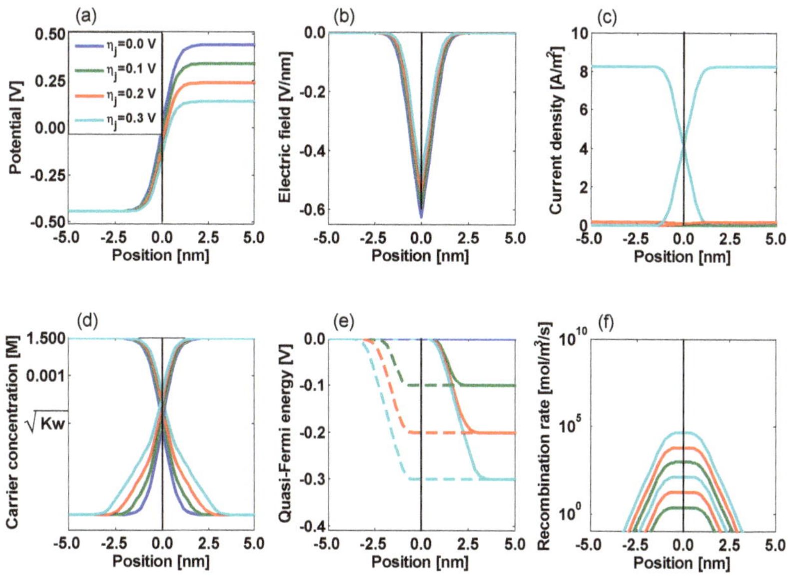

Figure 10. Predicted BPM response as a function of overpotential (i.e., forward bias) for the bimolecular, or mass-action-kinetic, recombination mechanism. Results include the (a) electrical potential, (b) electric field, (c) current density, (d) mobile carrier concentrations, (e) quasi-Fermi potentials on the voltage scale (e), and (f) volumetric ionic recombination rates.

With an appreciation for the nature of the BPM system at thermal equilibrium, it is prudent to move to galvanic conditions. The primary focus of the present study is for a BPM fuel cell which employs an acidic anode and alkaline cathode as is shown in Fig. 1a. This is representative of forward-bias condition and implies that the bias, or overpotential, reduces the electrostatic potential across the junction. This decreases the electric field at the junction and causes the depletion region to narrow. The additional mobile ions enter the space charge region (Eq. 41) where they can recombine. The opposite is true of the reverse configuration, which would be representative of a galvanic device in the configuration shown in Fig. 1b. Operation in this configuration would involve a reverse bias condition, where the bias would increase the potential across the BPM junction, increase the electric field, and force water splitting to occur.

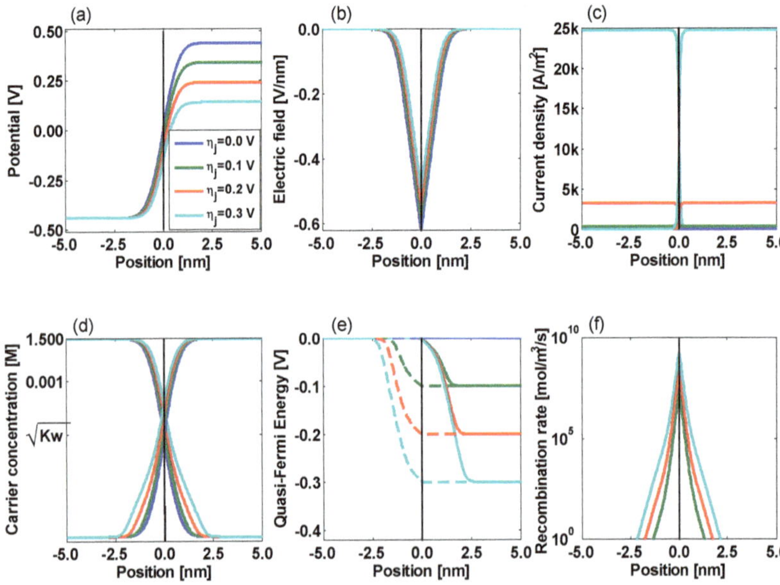

Figure 11. Predicted BPM response as a function of overpotential (i.e., forward bias) for the combined bimolecular and trap-assisted recombination mechanism. Results include the (a) electrical potential, (b) electric field, (c) current density, (d) mobile carrier concentrations, (e) quasi-Fermi potentials on the voltage scale, and (f) volumetric ionic recombination rates.

Focusing on the configuration in Fig. 1a, one of the interesting aspects of the theoretical approach is that it may be used to explore the effects of different mechanisms. We take this approach by exploring the differences between a system with bimolecular recombination processes relative to those involving a combined bimolecular and trap-assisted recombination processes. A sampling of results at different forward biases are shown in Fig. 10 and Fig. 11, respectively.

First, the bias reduces the potential difference across the bipolar junction, which in turn reduces the electric field and forms a more compact space charge region. As shown in Fig. 10a-b and Fig. 11a-b, these processes are nearly identical regardless of the recombination mechanism being considered. However, there is a dramatic difference in current density (c.a. three orders of magnitude) achieved for the BPM with the bimolecular, or mass action, based mechanism in Fig. 10c relative to that of the combined bimolecular and trap-assisted mechanism in Fig. 11c. The higher current density is the result of the differences in the net recombination rates between the mechanisms. This manifestation is especially apparent in the tails of the minority carrier concentration in Fig. 10d and Fig. 11d as well as the area between the H^+ and OH^- quasi-Fermi electrochemical potentials in Fig. 10e and Fig. 11e. Specifically, for a given bias, this area is much smaller for the system which includes the trap-assisted mechanism in Fig. 11e. This is an indication that ionic recombination is much more effective in this system. In semiconductor *p-n* junctions, Shockley has noted that identified that the area between the quasi-Fermi potential approaches zero and the system effectively becomes (series) resistive in nature when the recombination rates become sufficient fast [33]. The increased recombination rate when the trap-assisted mechanism is considered may be directly examined by comparing Fig. 10f and Fig. 11f. The maximum recombination rate for the combined bimolecular and trap-assisted mechanisms are nearly five orders of magnitude larger than those for the bimolecular reaction on their own. It is these rates which have led to the higher current densities and more compact quasi-fermi electrochemical potential windows.

Both Fig. 10 and Fig 11 considered the symmetric bipolar conditions noted at the beginning of the results and discussion. However, we expect both the ionic diffusivity and fixed change concentrations to be different in the CEM and AEM materials for a real system. From a heuristic perspective, we the hydroxide and proton diffusivity values have a modest impact on the trap-assisted recombination rates. For example, a five-fold decrease in D_{OH^-} only changes the current response by a few of percent. An asymmetric fixed charge concentrations, (e.g., considering $[-SO_3^-]_{CEM}(= 1.5\ M)$ and $[-TMA^+]_{AEM}(= 1.0\ M)$) increases the current by about 20% at a given overpotential. However, a lower ionic conductivity typically accompanies a lower fixed charge concentration. If we consider a concurrent decrease in D_i (i.e., for the majority ionic species in the membrane with the reduced fixed charge concentration) the current response falls back to within 10% of the baseline (symmetric) case noted above. Both of these parameters will necessarily influence the membrane resistance and associated transport processes, which are not explicitly discussed. Different characteristics are likely to occur if the bimolecular recombination process is rate limiting. Reduced fixed charge concentrations can increase concentration of minority carriers within the space charge region and therefore the current response; however, care needs to be taken as the diffusivity values are also coefficients to the diffusion-limited bimolecular rate expression in Eq. 44.

In addition to the details shown in Fig. 10 and Fig 11, simplified representations of the voltage-current characteristics are informative. For diodes, saturation equations are used to describe the behavior and can be used to simplify the representation of our system. An analogous expression may be used for the BPM using an exchange current density $i_{o,j}$ in place of a saturation current.

$$i = i_{o,j}\left[exp\left(\frac{\alpha_j F}{RT}\eta\right) - 1\right] \tag{48}$$

Because we are dealing with an ionic system a reaction symmetry factor α_j is included. This expression is typically valid for any forward bias as well as reverse biases up to some critical value. For example, it is not valid when dielectric breakdown occurs in semiconductors under extreme reverse bias. The same is true of the spontaneous splitting of water under large electric fields for galvanic devices. The current-voltage result at conditions consistent with Fig. 10 and Fig. 11 is shown in Fig. 12a, where the symbols represent the detailed PNP analysis and the lines correspond to Eq. 48. These results are shown on a semi-log and standard scale in Fig. 12a and Fig. 12b, respectively. While not shown, the expression also satisfactorily describes the reverse bias up to the aforementioned critical potential, which is characterized by very small reverse current densities that are typical of a diffusion limited system. The exchange current density and symmetry factors for the respective mechanisms are noted. The trap-assisted recombination mechanism has an exchange current density roughly 5 orders of magnitude larger than that of the bimolecular recombination. The difference in the losses associated with a BPM interface in a fuel cell for the respective mechanisms is readily apparent in Fig 12b. The trapping mechanism yields reasonable overpotential values whereas those associated with the bimolecular mechanism are over-burdensome and will result in a low-current or inefficient device.

Both of the exchange current characteristics used in Fig. 12 may be improved by tailoring the BPM interface making the real surface area greater than the superficial area. For example, sub-micrometer electrospun polymer electrolyte fibers can provide acid-alkaline interfaces with significantly higher interfacial area than the nominal cell area. However, given the length scales of the depletion layer widths, the full interfacial area may not be active. This suggests that a realistic increase in CEM-AEM junction area needs to have a period which is greater than the depletion layer thickness.

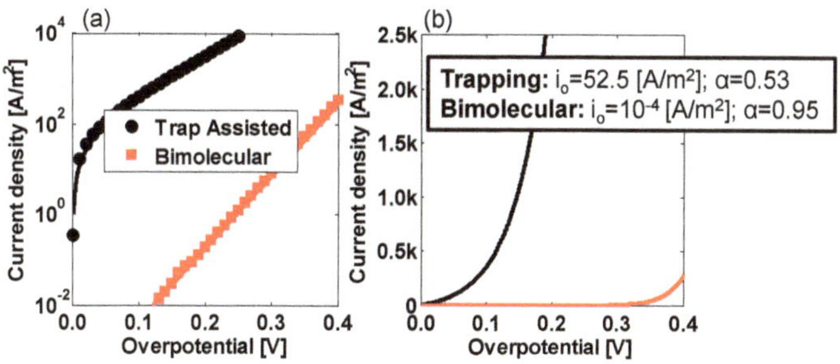

Figure 12. Analysis of the BPM current-voltage characteristics as a function of recombination mechanism, which compares the detailed PNP results (symbols) to the simple saturation model (lines) provided in Eq. 48. Results include a ***(a)*** *semi-log plot as well as a* ***(b)*** *linear plot.*

This observation leads to the obvious question of what other materials-based modifications and/or structures may be used to further increase the current density at modest bias in a BPM-based device under galvanic conditions. Results from two concepts are provided in Figs. 13 with graphical representation of these materials configurations in Fig. 14. The exchange current density and symmetry factors for these studies are provided in Table 2. The first concept is to reduce the doping level (i.e., lower the AEM and/or CEM IECs), as shown in Fig 14a-c. The second approach is to intermix regions of the CEM and AEM material, as shown in Fig. 14d-e. This may be achieved through attaching both AEM and CEM head-groups to the polymer in the depletion region or co-electrospinning CEM and AEM materials. On the basis of these approaches, results for the combined trapping and bimolecular recombination processes are shown in Fig. 13a while those for the isolated bimolecular recombination processes are shown in Fig. 13b.

Table 2: Calculated junction properties for the simple saturation model.

Conditions	Trapping + Bimolecular		Bimolecular Only	
	i_o [A/m^2]	α	i_o [A/m^2]	α
Abrupt: 1.5P/1.5A [M]	52.5	0.53	7.33x10^{-5}	1.00
Abrupt: 0.75P/0.75A [M]	33.5	0.54	2.0x10^{-4}	0.97
Abrupt: 0.15P/0.15A [M]	15.9	0.54	2.0x10^{-3}	0.97
Mixed: 0.75P & 0.75A (20 nm) [M]	33.5	0.54	5.33x10^{-4}	1.02
Mixed: 0.15P & 0.15A (20 nm) [M]	15.5	0.54	4.02x10^{-4}	0.94

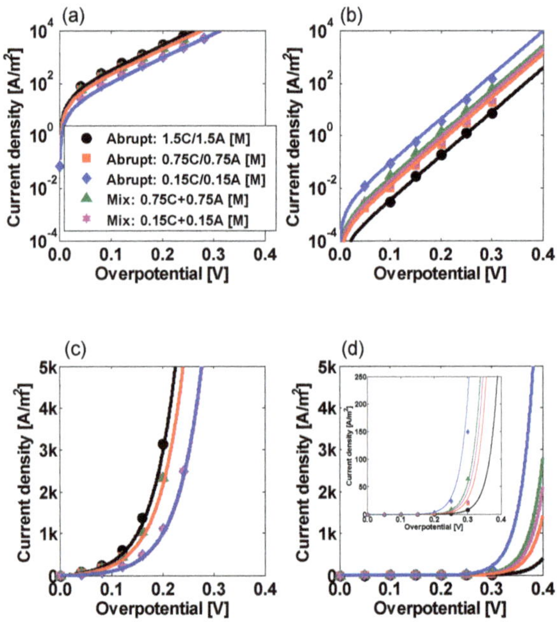

Figure 13. Model results (symbols) versus simple saturation model (lines) with values provided in Table 2 for different bipolar interface conditions. Results are shown on a log scale for a (a) combined bimolecular and trap-assisted and (b) bimolecular recombination mechanism. The results are also shown on a linear scale for the (c) combined bimolecular and trap-assisted and (d) bimolecular recombination mechanisms with the insert showing the low current density regime with higher fidelity. Graphical representation of the fixed charge concentration profiles are provided in Fig. 14.

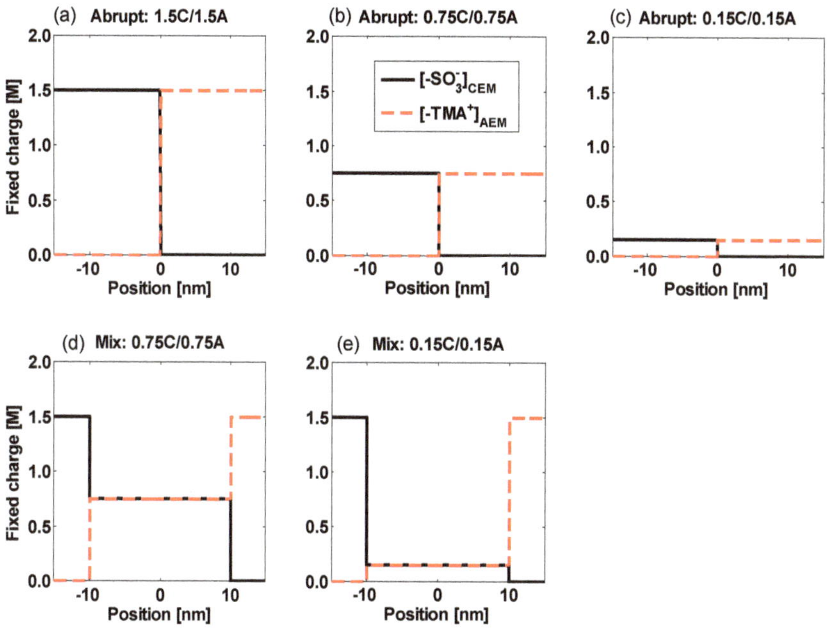

Figure 14. Graphical representation of the fixed ion doping concentrations for the studies in Fig. 13. These include an abrupt CEM-AEM junction with fixed charge concentrations of (a) $[\text{-}SO_3^-]_{CEM}$ = 1.500 M and $[\text{-}TMA^+]_{AEM}$ = 1.5 M (baseline case), (b) $[\text{-}SO_3^-]_{CEM}$ = 0.75 M and $[\text{-}TMA^+]_{AEM}$ = 0.75 M, and (c) $[\text{-}SO_3^-]_{CEM}$ = 0.15 M and $[\text{-}TMA^+]_{AEM}$ = 0.150 M. In addition to the abrupt junctions, two cases of mixed CEM and AEM materials over 20 nm of the junction are considered. This includes cases where the CEM and AEM fixed charge concentrations within the 20 nm overlap region are (d) $[\text{-}SO_3^-]_{CEM}$ = 0.75 M and $[\text{-}TMA^+]_{AEM}$ = 0.75 M and (e) $[\text{-}SO_3^-]_{CEM}$ = 0.15 M and $[\text{-}TMA^+]_{AEM}$ = 0.15 M.

Interestingly, Fig. 13 shows that a BPM interface employing the lowest fixed charge concentrations (i.e., 1/10 the baseline case), have the highest bimolecular recombination rate and current density at a given bias. The lower fixed ion concentrations will inherently influence the junction potential as well as the ionic conductivity, water transport, and accessibility of the fixed charges within these polymeric materials. It is beyond the scope of this work to explore the trade-offs between all of these dependencies. Instead, we focus on their effect on the recombination processes. The lower fixed change concentrations effectively increase the concentration of minority mobile ionic species, which is leading to the higher exchange current density values. One can envision a system with a thin BPM interface region employing polymer electrolyte materials with low IECs, sandwiched by more typical CEM and AEM materials. The question which arises is whether sufficient rates of ionic and water transport through a BPM interface with low IEC materials can be accomplished in a stable manner. Other interfacial catalysts and modifications can also be considered. Weak acid and base materials have been used at the BPM interface in commercial applications to produce acids and bases from salt solutions [16]. These systems increase the minority carrier concentration and can improve interactions within the junction region. As noted in Table 2, these types of approaches yield modest variations in the exchange current density for the trap-assisted recombination mechanism. However, the exchange current density for the bimolecular recombination mechanism increases by nearly an order of magnitude.

The results of the analysis provided in Fig. 13 are two-fold. First, this suggests that if the bimolecular recombination process is rate limiting, these types of approaches are needed in order to achieve a high-current, BPM fuel cell device. Such an approach is less prudent if the trap-assisted mechanism is at play. The second is that we need to better understand which recombination mechanism(s) are rate-limiting for the BPM materials. The immediate question is whether a trap-assisted mechanism is practical for an ionic system. There are several important pieces of evidence which may be pointed to in order to suggest that such a mechanism is possible and rate limiting for this system.

One of the few existing efforts associated with understanding the recombination kinetics in a BPM fuel cell within the literature, was recently presented by Xiang and coworkers who utilized transition state theory (TST) [24]. Their study used TST with published experimental data [29] to fit rate constants and examine transport in a BPM fuel cell. Several implicit assumptions were made in this study, including the use of Butler-Volmer kinetics to form the saturation current-voltage curve (i.e. similar to Eq. 48) and that the minority ionic concentrations in the entire space charge region are governed by Eq. 41, which has its own limitations. Nevertheless, examination of the exchange current densities predicted with this approach (c.a. $20\ A/m^2$) support the values we predict with the use of the trap-assisted mechanism within the BPM.

Having already hypothesized that Coulombic interactions for the fixed charges and/or a counterion condensation process could promote a trap-assisted recombination mechanism, a comparison between the nature of the space charge region and bulk polymer electrolyte materials can be made. The characteristic lengths of physical processes within the vicinity of the bipolar acid-alkaline junction are compared to those in the bulk membranes (i.e., away from any space charge effects). The Bjerrum length $l_{Bjerrum}$ is often used is often used to calculate ionic association in chemical systems. It is attributed to Coulombic interactions between ions and defined related to the thermal energy k_bT. Its values typically range which from about 0.7 nm at room temperature in pure liquid water to 1.6 nm for a typical CEM or AEM.

$$l_{Bjerrum} = \frac{q^2}{4\pi k_B T \epsilon} \tag{49}$$

$$l_E = \frac{k_B T}{q|\vec{\boldsymbol{E}}|} \tag{50}$$

A characteristic electric field length l_E may also be defined.

Within the bulk electrolyte, electric fields with magnitudes on the order of 10^3 V/m are typical which yield electric field length values on the order of 10^4 nm. However, as shown in the preceding figures, the magnitude of the electric fields can significantly exceed 10^8 V/m within the space charge region at the acid-alkaline BPM interface. This yields characteristic electric field length values on the order of 10^{-2} nm. By comparing the ratio of Eq. 49 and Eq. 50, it becomes apparent that different physical effects apply. The ratio $\chi(= l_{Bjerrum}/l_E)$ tells us that the local Coulombic interactions play different role in the space charge region than in bulk polymer electrolytes. Inspection shows that χ ranges from 10 to 100 in the BPM's space charge region and 10^{-4} to 10^{-5} in the bulk electrolyte. This highlights the different role that local Coulombic interactions are playing within the space charge region. It is these unique characteristics which could facilitate a trap-like recombination mechanism as a direct result of a physical process similar to counterion condensation or local screening the mobile ions.

With evidence of unique physical interactions in the space charge region, we next turn our attention to evidence within the published literature. The first is Nyquist plots from *in situ* electrochemical impedance spectroscopy (EIS) performed on BPM fuel cells at a variety of conditions. Several publications have shown consistent trends associated with the high frequency resistance (HFR) of BPM fuel cell membrane electrode assemblies (MEAs), which is typically associated with MEA resistance, across a range of humidification, stoichiometry, and operational conditions [29, 10]. The fact that the HFR is consistent is indicative that the resistance within the MEA is constant/stable and that there is not an additional condition-dependent interfacial junction resistance. This does not preclude an intrinsic interfacial resistance, such as may occur from phase segregation or the presence of an interfacial film (e.g., from processing). The same Nyquist plots which show consistent HFR values, exhibit a charge transfer resistance that is on the order of 1 to 4 Ωcm^2 depending upon the cell potential, humidification, and fuel/oxidant flow rate [10, 29, 11, 55]. The reported *in-situ* AC impedance spectra exhibited a single loop characteristic in these studies and do not show any diffusion limited tails, characteristic of a Warburg impedance element. While small cells were used in those studies (c.a. 2 cm^2), the overall charge transfer resistance can be compared to those which would be caused by the recombination mechanisms discussed here. The exchange current densities for the trap-assisted recombination mechanism correspond to charge transfer resistances on the order of 1 Ωcm^2 for an MEA of comparable size. Those which are characteristic of the bimolecular reaction process would be 5 orders of magnitude larger. Given the magnitude of the charge transfer resistance predicted for the respective recombination mechanisms, as well as the clear dependence of the experimentally reported data on operational conditions, it is somewhat surprising that the reported EIS spectra do not exhibit distinct features for the electrode and junction processes. Undoubtedly, additional EIS study of these systems including a broader frequency range as well as the development of an impedance model for the BPM fuel cell are needed. However, the fact that the EIS spectra exhibit clear charge-transfer behavior and that the magnitude of the charge transfer resistances for a trap-assisted mechanisms are of the appropriate order of magnitude for these spectra (along with the electrode processes) provides compelling evidence that such a mechanism is rate limiting.

The final experimental evidence which suggests that a mechanism similar to the trap-assisted recombination mechanism is occurring within the BPM is the corresponding fuel cell voltage-current density characteristics which have been achieved [55, 25]. The current-voltage characteristics calculated for a BPM which is rate limited by a bimolecular (mass-action kinetics) recombination mechanism are simply not sufficient to generate the fuel cell current-voltage characteristics shown in these studies.

4. Conclusions

In conclusion, we have introduced a theoretical framework for addressing transport in BPM materials for fuel cell applications. The theoretical framework was developed in an analogous manner to the *p-n* junction in semiconductor devices, with major discrepancies between the systems being directly addressed. The most critical step in the development of this analogy-based model was the development of the Poisson's and the transport equations relative to the water auto ionization constant. Emphasis was placed on identification of a detailed mechanism associated with ionic recombination under forward bias (overpotential). The configuration is representative of a fuel cell device with an acidic anode, alkaline cathode, and water generated at the bipolar interface. Both a bimolecular mechanism, which is analogous to band-to-band recombination or mass-action kinetics, and a trap-assisted mechanism were identified as possible. Based on the analyses and associated rates, a trap-assisted recombination mechanism would be needed to produce a practical galvanic device such as a fuel cell device. On the basis of characteristic forces in the bulk electrolyte relative to the bipolar space charge region, it was argued that the Coulombic interactions between the mobile ionic species and the fixed ionic charges would promote such a mechanism through counterion condensation and/or screening processes. The presence of a low-loss mechanism is supported by a range of published experimental data, and based on the present findings, the response characteristics could only be explained by a mechanism similar to the trap-assisted process. Nonetheless, additional efforts are needed to better understand and explain these processes. Finally, several materials-based methods are needed to further reduce the losses associated with ionic recombination at the BPM interface. Increasing the interfacial area through corrugation or the use of polymer electrolytes provides a tractable approach to further reduce losses associated with the acid-alkaline polymer electrolyte interface.

Acknowledgments

KNG, JPM, and DC gratefully acknowledge the U.S. Department of the Army and U.S. Army Material Command (AMC) for supporting this work. PAK and JMA gratefully acknowledge financial support from the from the U.S. Office of the Deputy Assistant Secretary of the Army for Defense Exports and Cooperation (DASA-DE&C) and the U.S. Army Research Laboratory, W911NF-14-2-0008. The authors would also like to thank Dr. David Baker & Dr. Oleg Borodin of the U.S. Army Research Laboratory and Ms. Lisha Liu of Georgia Tech for helpful discussions.

Symbols & Nomenclature

Greek:

α	Electrochemical symmetry factor [#]
ε_o	Vacuum permittivity = 8.8542×10^{-12} [F/m]
ε_r	Medium specific relative permittivity[#]
ϵ	Total dielectric permittivity ϵ (= $\varepsilon_o\varepsilon_r$) [F/m]
η	Bias or overpotential [V]
μ_i	Chemical potential of species i [J]
$\bar{\mu}_i$	Electrochemical potential of species i [J]

$\bar{\mu}_i^{ex}$	Excess electrochemical potential of species i [J]
ρ	Space charge density [C/m^3]
σ	Conductivity [S/m]
τ_i	Lifetime of carrier species i [s]
φ_i	Quasi-Fermi potential of species i [V]
χ	Ratio of characteristic lengths [#]
ψ	Electric potential [V]

Roman:

a_i	Chemical activity of species i [#]
A	Area [m^2]
C_i	Molar concentration of species i [mol/m^3]
D_i	Diffusivity of species i [m^2/s]
E	Energy [eV]
$\vec{\boldsymbol{E}}$	Electric field [V/m]
F	Faraday's constant = 96485.333 [C/mol]
G	Gibbs free energy [J/mol]
$\vec{\boldsymbol{\imath}}$	Current density [A/m^2]
i_o	Exchange current density [A/m^2]
k_{recomb}	Acid-base reaction rate [m^3/mol/s]
K_W	Water auto-ionization constant [mol^2/m^6]
k_b	Boltzmann constant [J/K]
l	Length [m]
L	Characteristic diffusion length [m]
n	Electron number density [#/m^3]
n	Intrinsic carrier number density [#/m^3]
N_i	Number density of donor, acceptor, or density of states [#/m^3]
N_A	Avogadro's number = 6.022141x10^{23} [#/mol]
p	Hole number density [#/m^3]
q	Elementary charge = 1.602x10^{-19} [C]
R	Universal gas constant = 8.3145 [J/mol/K]
R_i	Volumetric rate [mol/m^3/s]
T	Temperature [K]
V	Voltage [V]
W	Depletion layer width [m]
x	Position [m]
z	Charge number [#]

References

[1] J. Marcinkoski, J. Spendelow, A. Wilson and D. Papageorgopoulos, "Fuel Cell System Cost - 2015," U.S. Department of Energy (2015).

[2] J. Spendelow and D. Papageorgopoulos, "Platinum Gorup Metal Loading in PEMFC Stacks," U.S. Department of Energy (2011).

[3] J. Varcoe and R. Slade, *Fuel Cells,* 5(2), 187 (2004). https://doi.org/10.1002/fuce.200400045

[4] G. Merle, M. Wessling and K. Nijmeijer, *J. Memb. Sci.,* 377 (1-2), 1 (2011). https://doi.org/10.1016/j.memsci.2011.04.043

[5] M. Hickner, A. Herring and E. Coughlin, *J. Poly. Sci. Part B: Poly. Phys.,* 51, 1727 (2013). https://doi.org/10.1002/polb.23395

[6] J. Varcoe, P. Atanassov, D. Dekel, A. Herring, M. Hickner, P. Kohl, A. Kucernak, W. Mustain, K. Nijmeijer, K. Scott, T. Xu and L. Zhuang, *Envergy Environ. Sci.,* 7, 3135 (2014). https://doi.org/10.1039/C4EE01303D

[7] M. Unlu, D. Abbott, N. Ramaswamy, X. Ren, S. Mukerjee and P. Kohl, *J. Electrochem. Soc.,* 158 (11), B1423 (2011). https://doi.org/10.1149/2.075111jes

[8] S.-D. Yim, H. Chung, J. Chlistunoff, D.-S. Kim, C. Fujimoto, T.-H. Yang and Y. Kim, *J. Electrochem. Soc.,* 162 (6), F499 (2015). https://doi.org/10.1149/2.0151506jes

[9] W. Sheng, H. Gasteiger and Y. Shao-Horn *J. Electrochem. Soc.,* 157 (11), B1529 (2010). https://doi.org/10.1149/1.3483106

[10] M. Unlu, J. Zhou and P. Kohl, *Fuel Cells,* 10 (1), 54(2010). https://doi.org/10.1149/1.3468700

[11] M. Unlu, J. Zhou and P. Kohl, *J. Electrochem. Soc.,* 157 (10), B1391 (2010). https://doi.org/10.1149/1.3468700

[12] H. Reiss, *J. Chem. Phys.,* 21 (7), 1209 (1953). https://doi.org/10.1063/1.1699165

[13] B. Lovrecek, A. Despic and J. O. Bockris, *J. Phys. Chem.,* 63 (5), 750 (1959). https://doi.org/10.1021/j150575a030

[14] R. Buck, N. Surridge and R. Murray, *J. Electrochem. Soc.,* 139 (1), 136 (1992). https://doi.org/10.1149/1.2069158

[15] L. Hegedus, N. Kirschner, M. Wittmann and Z. Nosticzius, *J. Phys. Chem. A,* 102(32), 6491 (1998). https://doi.org/10.1021/jp981488h

[16] J. Balster, S. Srinkantharajah, R. Sumbharaju, I. Punt, R. Lammertink, D. Stamatialis and M. Wessling, *J. Memb. Sci,* 365, 389 (2010). https://doi.org/10.1016/j.memsci.2010.09.034

[17] O. Cayre, S. Chang and O. Velev, *J. Am. Chem. Soc.,* 129 (35), 10801 (2007). https://doi.org/10.1021/ja072449z

[18] N. Leventis, M. Schloh, M. Natan, J. Hickman and M. Wrighton, *Chem. Mater.,* 2(5), 568 (1990). https://doi.org/10.1021/cm00011a020

[19] A. Mauro, *Biophyhys. J.,* 2, 179 (1962). https://doi.org/10.1016/S0006-3495(62)86848-9

[20] C. Chidsey and R. Murray, " *Science,* 231 (4733), 25 (1986). https://doi.org/10.1126/science.231.4733.25

[21] I. Bassignana and H. Reiss, *J. Mem. Sci., 15*, 27 (1983). https://doi.org/10.1016/S0376-7388(00)81360-5

[22] S. Mafe, J. Manzanares and P. Ramirez, *Phys. Rev. A,* 42 (10), 6245 (1990). https://doi.org/10.1103/PhysRevA.42.6245

[23] H. Coster, *Biophys. Journal,* 5, 669 (1965). https://doi.org/10.1016/S0006-3495(65)86745-5

[24] S. Peng, S. Lu, J. Zhang, P.-C. Sui and Y. Xiang, *Phys. Chem. Chem. Phys.,* 15, 11217 (2013). https://doi.org/10.1039/c3cp52150h

[25] S. Peng, X. Xu., S. Lu, P.-C. Sui, H. Djilali and Y. Xiang, *J. Power Sources,* 299, 273 (2015). https://doi.org/10.1016/j.jpowsour.2015.08.104

[26] J. Dykstra, P. Biesheuvel, H. Bruning and A. Ter Heijne, *Phs. Rev. E,* 90, 013302 (2014). https://doi.org/10.1103/PhysRevE.90.013302

[27] M. Bazant, M. Sabri Kilic, B. Storey and A. Ajdari, *Adv. Colloid Interface Sci.,* 152, 48 (2009). https://doi.org/10.1016/j.cis.2009.10.001

[28] J. Paz-Garcia, J. Dykstra, P. Biesheuvel and H. Hamelers, *J. Colloid & Int. Sci.,* 442, 103 (2015). https://doi.org/10.1016/j.jcis.2014.11.045

[29] M. Unlu, J. Zhou and P. Kohl, *J. Phys. Chem. C,* 113 (26), 11416 (2009). https://doi.org/10.1021/jp903252u

[30] A. Bard and L. Fauklner, Electrochemical Methods Fundamentals and Applicaitons, 2'nd Ed., John Wiley & Sons, Inc., (2001)

[31] F. Helfferich, Ion Exchange, New York: McGraw-Hil Book Co., Inc. (1962).

[32] A. Galama, J. Post, M. Cohen Stuart and P. Biesheuvel, *J. Memb. Sci,* 442, 131 (2013). https://doi.org/10.1016/j.memsci.2013.04.022

[33] W. Shockley, *Bell Systems Technical Journal,* 28 (3), 435 (1949). https://doi.org/10.1002/j.1538-7305.1949.tb03645.x

[34] W. Shockley and W. Read Jr., *Phys. Rev.,* 87 (5), 835 (1952). https://doi.org/10.1103/PhysRev.87.835

[35] C.-T. Sah, R. Noyce and W. Shockley, *Proceedings of the IRE,* 45 (9), 1228 (1957). https://doi.org/10.1109/JRPROC.1957.278528

[36] M. Kilic, M. Bazant and A. Ajdari, *Phys. Rev. E,* 75, p. 021502, (2007). https://doi.org/10.1103/PhysRevE.75.021502

[37] M. Kilic, M. Bazant and A. Ajdari, *Phys. Rev. E,* 75, p. 021503, (2007). https://doi.org/10.1103/PhysRevE.75.021503

[38] P. Biesheuvel, A. Franco and M. Bazant, *J. Electrochem. Soc.,* 156 (2), B225 (2009). https://doi.org/10.1149/1.3021035

[39] P. Biesheuvel, M. van Soestbergen and M. Bazant, *Electrochimica Acta,* 54, 4857 (2009). https://doi.org/10.1016/j.electacta.2009.03.073

[40] K. Grew, D. Chu and W. Chiu, *J. Electrochem. Soc.,* 157 (7), B1024 (2010). https://doi.org/10.1149/1.3368728

[41] D. Ben-Yaakov, D. Andelman, D. Harries and R. Podgornik, *J. Phys.: Condens. Matter,* 21, 424106 (2009). https://doi.org/10.1088/0953-8984/21/42/424106

[42] D. Ben-Yaakov, D. Andelman and R. Podgornik, *J. Chem. Phys.,* 134, 074705 (2011). https://doi.org/10.1063/1.3549915

[43] D. Ben-Yaakov, D. Andelman, R. Podgornik and D. Harries, *Curr. Opin. Colloid Interface Sci.,* 16, 242 (2011). https://doi.org/10.1016/j.cocis.2011.04.012

[44] D. Ben-Yaakov, Y. Burak, D. Andelman and S. Safran, *EPL,* 79, 48002 (2007). https://doi.org/10.1209/0295-5075/79/48002

[45] K. Grew, X. Ren and D. Chu, *Electrochem. Solid-State Lett.,* 14, B127 (2011). https://doi.org/10.1149/2.011112esl

[46] T. Myles, K. Grew, A. Peracchio and W. Chiu, *J. Power Sources,* 296, 225 (2015). https://doi.org/10.1016/j.jpowsour.2015.07.044

[47] M. Tedesco, H. Hamelers and P. Biesheuvel, *J. Memb. Sci.,* 510, 370 (2016). https://doi.org/10.1016/j.memsci.2016.03.012

[48] J. P. Warrick, J. Auborn and E. Eyring, " *J. Phys. Chem.,* 76 (8), 1184 (1972). https://doi.org/10.1021/j100652a017

[49] M. Kuik, L. Koster, G. Wetzelaer and P. Blom, *Phys. Rev. Lett.,*. 107 (25), 256805, (2011). https://doi.org/10.1103/PhysRevLett.107.256805

[50] L. Onsager, *J. Chem. Phys.,* 2, 599 (1934). https://doi.org/10.1063/1.1749541

[51] V. Kaiser, S. Bramwell, P. Holdsworth and R. Moessner, *Nat. Mater.* 12, 1033 (2013). https://doi.org/10.1038/nmat3729

[52] S. Mafe, P. Ramirez and A. Alcaraz, *Chem. Phys. Lett.,* 294, 406 (1998). https://doi.org/10.1016/S0009-2614(98)00877-X

[53] S. Mafe and P. Ramierz, *Acta Polymer,* 48, 234 (1997). https://doi.org/10.1002/actp.1997.010480702

[54] F. Booth, *J. Chem. Phys.,* 19 (4), 391 (1951). https://doi.org/10.1063/1.1748233

[55] M. Unlu, J. Zhou, A.-R. I., H. Kim and P. Kohl, *Electrochimica Acta,* 56, 4439 (2011). https://doi.org/10.1016/j.electacta.2011.02.017

Chapter 11

Stabilizing High-Voltage $LiCoO_2$ Cathode in Aqueous Electrolyte with Interphase-forming Additive

Fei Wang[1], Yuxiao Lin[2], Liumin Suo[3], Xiulin Fan[1], Tao Gao[1], Chongyin Yang[1], Fudong Han[1], Wei Sun[1], Yue Qi[1], Kang Xu[4] and Chunsheng Wang[1*]

[1]Department of Chemical and Biomolecular Engineering, University of Maryland, College Park , MD 20742 , USA

[2]Department of Chemical Engineering and Materials Science, Michigan State University, East Lansing, MI 48824, USA

[3]Department of Nuclear Science and Engineering, Massachusetts Institute of Technology, Cambridge, MA 02138, USA

[4]Electrochemistry Branch, Sensor and Electron Devices Directorate, Power and Energy Division, U.S. Army Research Laboratory, Adelphi, MD 20783, USA.

[*]cswang@umd.edu

Abstract

Aqueous lithium ion batteries (ALIBs) receive increasing attention due to their intrinsic non-flammable nature. Their practical applications, however, have been prevented by the poor electrochemical stability of aqueous electrolytes, which severely restricts the choice of anode and cathode materials for such batteries and lead to low energy densities. It has been demonstrated that "water-in-salt" electrolytes can provide an expanded voltage window of ~3.0 V, which can accommodate electrode materials that were otherwise forbidden in conventional aqueous electrolytes. In this work, we showed that the presence of additive could further stabilize the aqueous electrolyte at high voltage cathode surface. By forming a cathode electrolyte interphase (CEI) through electrochemical oxidation of tris(trimethylsilyl) borate (TMSB), we successfully stabilized $LiCoO_2$ in water-in-salt electrolyte at a high cut-off voltage that corresponds to 0.7e electron charge transfer. To our knowledge, this is also the first electrolyte additive known to form interphase in aqueous electrolytes. The formed CEI significantly suppressed electrolyte oxidation as well as cobalt dissolution from $LiCoO_2$ into electrolytes, allowing $LiCoO_2$ to be stably charged/discharged at a higher cut-off voltage with a high capacity utilization of 170 mAh/g. When paired with Mo_6S_8 anode, the 2.5 V aqueous full cell delivered a high energy density of 120 Wh/kg for 1000 cycles with an extreme low capacity decay rate of 0.013% per-cycle.

Keywords

Aqueous Lithium-ion Battery, $LiCoO_2$, High Voltage, High Energy Density, Water-in-salt Electrolyte

Contents

1. Introduction

Aqueous lithium ion batteries (ALIBs) attract intense attention due to the intrinsic non-flammable nature introduced by the aqueous electrolytes therein[-4], which also make it possible to rid of the rigorous moisture-free manufacturing environment and heavy reliance on the battery management systems at module or pack levels[5-7]. Despite these advantages, the practical applications of ALIBs remains remote, primarily limited by the inferior energy densities that has been imposed by the narrow electrochemical stability window of water (1.23 V). Recently, the electrochemical stability window of aqueous electrolyte was shown to be able to expand significantly at high salt concentrations via the formation of interphase on anode surface and suppression of the electrochemical activity of water [12]. Such stability window of 3.0V (1.9-4.9 V vs. Li) provides unprecedented flexibility in selecting electrochemical couples, allowing many electrode materials that are otherwise forbidden in conventional aqueous electrolytes. Hence, identifying electrodes with suitable lithiation/delithiation potentials as well as stability for this new class of aqueous electrolytes becomes new challenge in maximizing the deliverable energy density of a full aqueous Li-ion battery.

While spinel-structured $LiMn_2O_4$[13-19] suffers from poor cycling stability, and olivine-structured $LiFePO_4$[20-22] is characterized by typically low potential/low energy density, layered $LiCoO_2$ has been a reliable cathode material since the birth of Li-ion batteries, delivering a usable reversible capacity of ~ 140 mAh/g when charged to 4.2 V vs. Li[23], corresponding to 50% lithium removal to $Li_{0.5}CoO_2$. This low voltage $LiCoO_2$ (LVLCO) has maintained the status of flagship chemistry due to the excellent cycling stability[24-27], however, when efforts were made to remove more than 0.5 fraction of lithium by charging up to 4.5 V vs. Li, the layered structure becomes destabilized, and Co-dissolution accompanied by extensive electrolyte decomposition results in deteriorated cycling stabilities.

Although the lithiation/delithiation potential of LVLCO rests within the electrochemical stability window of the normal aqueous electrolyte, the LVLCO still suffers from poor cycling stability due to side reactions including the intercalation of proton into LVLCO [28, 29], the reactions between the LVLCO and dissolved oxygen in the electrolyte[2], the dissolution of Co into electrolyte[30] and oxygen-evolution due to oxidation of water. On the other hand, the full electrochemical stability window of water-in-salt electrolyte, with anodic limit at 4.9 V vs. Li, is actually not fully used by LVLCO that is charged to only 4.2 V vs. Li. Recent efforts were made to charge $LiCoO_2$ to 4.5 V vs. Li in order to utilize more capacity from this cathode, and a capacity of ~180 mAh/g (corresponding to 0.7 fraction of Li removed, defined as HVLCO) was reported in organic electrolytes without structural destruction [23, 31-35]. However, the attained capacity of HVLCO gradually decays during charge/discharge cycles due to the growth of surface impedance[23]. It was believed that both Co-dissolution into electrolytes and other side reactions between $Li_{0.3}CoO_2$ and electrolytes are responsible for the poor cycling stability[23,30,31,35]. Surface coating or formation of cathode electrolyte interphases (CEI) on $LiCoO_2$ can effectively suppress these parasitic processes and significantly enhance the capacity retention of HVLCO[23,31,35]. It was believed that the formation of CEI was supported by the oxidative decomposition of non-aqueous solvents, which provides chemical building blocks in a smilar manner as they did in the formation of solid electrolyte interphase (SEI) on anode [36, 37]. Such chemical source, however, is absent in aqueous electrolytes, as the oxidation of neither solvent (water) nor salt anion leads to dense solid deposition on electrode surface. Therefore, a third component responsible for interphase should be introduced.

While the use of additives have been very popular in tailoring SEI on anode surfaces, recently, additives designed for cathode protection were also reported [32, 38-43]. However, to our best knowledge, there has never been any attempt to use additive approach in aqueous electrolyte systems.

In this work we identified such an additive for water-in-salt electrolyte. By electrochemical oxidation of tris(trimethylsilyl) borate (TMSB), we successfully formed a CEI on the HVLCO surface, which significantly suppressed Co-dissolution from the cathode even when charged to 4.5 V, allowing $LiCoO_2$ to deliver reversible capacity corresponding to 0.7 fraction of Li removal in water-in-salt electrolytes. When coupled with Mo_6S_8 anode, a 2.5V aqueous full cell HVLCO/Mo_6S_8 prodcues a high energy density of 120 Wh/kg for 1000 cycles at an extremely low capacity decay rate of 0.013% per-cycle.

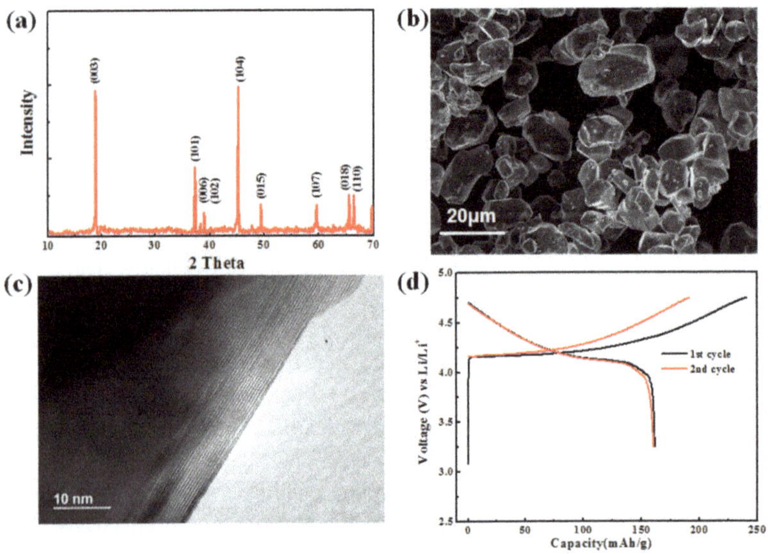

Fig. 1 (a) X-ray Diffraction (XRD), (b) Scanning electron microscope (SEM) and (c) Transmission electron microscopy (TEM) image of $LiCoO_2$ powder. (d) The typical voltage profile of $LiCoO_2$ between 4.7V to 3.2V (vs Li^+/Li) in water-in-salt electrolyte (21m LiTFSI in H_2O) at 0.2 C measured in a three-electrode cell using Pt as counter and Ag/AgCl as reference electrodes.

The XRD patterns of pristine $LiCoO_2$ are shown in Fig. 1a, where all peaks can be indexed to the hexagonally packed O_3 layered structure with an R3m space group. SEM image (Fig. 1b) reveals that all particles range between several to tens of microns in size. TEM image (Fig. 1c) confirms the typical layer structure of $LiCoO_2$ with clean and well-defined surface. This cathode can be reversibly charged/discharged between 3.0V to 4.5V in typical non-aqueous electrolytes without severe surface structure changes[35]. In water-in-salt electrolytes, the potential of redox lithiation/delithiation will positively shift by ~0.2V due to the high salt concentration [12], the corresponding upper cut-off voltage of HVLCO should be 4.7 V for the same fraction of Li removal. The galvanostatic lithiation/delithiation behavior of HVLCO in water-in-salt electrolytes was evaluated in a three-electrode cell with Pt as counter and Ag/AgCl as reference electrodes. Thus, a reversible discharge capacity of 170 $mAhg^{-1}$ was delivered by HVLCO when cycled between 3.2V and 4.7V (vs Li^+/Li) (Fig. 1d). Comparing the voltage profiles, one can easily tell that the lithiation/delithiation behavior of HVLCO in water-in-salt electrolytes is rather similar to its behavior in non-aqueous electrolytes, except for the positive potential upshift by 0.2V (Fig. S1).

The lithiation/de-lithiation potential of chevrel phase Mo_6S_8 (~2.2 V) has been shown to rest safely within the electrochemical stability window of water-in-salt electrolytes, where it can be charged/discharged for thousands of cycles with negligible fading [12]. When coupling it with HVLCO, we produce a 2.5 V full aqueous Li-ion cell. The HVLCO/Mo_6S_8 mass ratio was set to 1:1 in order to compensate the irreversible capacity incurred by the formation of SEI at the Mo_6S_8 anode during the initial cycles. Fig. 2a shows the charge/discharge curves of HVLCO/Mo_6S_8 full cell between 2.5 V (4.7 V vs Li for $LiCoO_2$ cathode) and 1.0 V. Two distinct voltage plateaus were observed at 0.5 C for the full cell, at about 2.0 V and 1.5 V, respectively, because of the two reversible redox couples of Mo_6S_8, which is de-lithiated at 2.24 V and 2.67 V in water-in-salt electrolyte, respectively (Fig. S2). A discharge capacity of 60 mAh/(g of total electrode mass) and an energy density of 120 Wh/kg were obtained from such a full cell for 20 cycles, after which a rapid capacity fading occurred (Fig. 2b). However, a very stable capacity lasted for over 50 cycles when the the same full cell was charged/discharged between 2.1 V and 1.0 V (Fig. S3). The electrochemical impedance spectra (EIS) at different cycle numbers (Fig. 2c) revealed that the interfacial resistances remain stable between 100-150 ohm during the first 20 cycles, followed by a quick increases to >400 ohm at the 21st cycle.

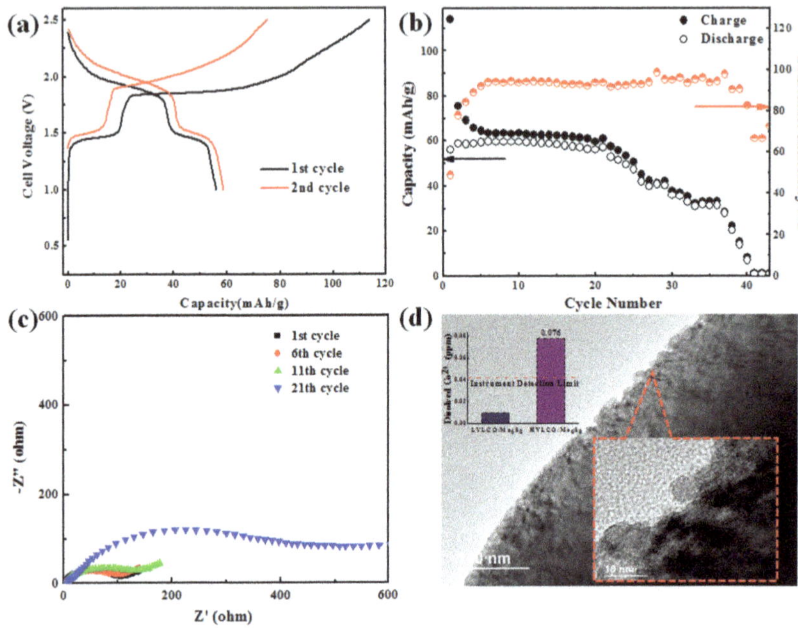

Fig. 2 The electrochemical performance of HVLCO/Mo_6S_8 full cells (a) The typical voltage profile in water-in-salt electrolytes between 1.0 V and 2.5 V at 0.5 C; (b) The cycling stability and Coulombic efficiencies at 0.5 C rate; (c) Electrochemical impedance spectroscopy (EIS) at the 1st, 6th, 11th, 21th cycles; (d) The TEM image for the cycled HVLCO electrodes in 21m LiTFSI electrolyte (inset is the amount of dissolved Co^{2+} in electrolyte measured using ICP).

Since Mo_6S_8 has been known to stably cycle in water-in-salt electrolyte [12], this rapid capacity fading should very likely arise from the cathode side. Thus the surface structure of the cycled HVLCO was examined using HRTEM (Fig. 2d), which reveals that the smooth edges of pristine material (Fig. 1a) became rough with small notches after 50 cycles in water-in-salt electrolytes, while the layer structure of $LiCoO_2$ remain intact as shown in the high-resolution TEM (inset of Fig. 4d &S4). The notches on the surface of HVLCO shows disordered nanocrystals with smaller d-spacing than that of $LiCoO_2$, which could be indexed as cobalt oxides (Fig. S4). We believe that the formation of cobalt oxide layer resulted from the reaction between the electrolyte and HVLCO surface, which became highly oxidative upon more than 0.5 fraction of Li removal. In addition, Co-dissolution also occurred simultaneously, as evidenced by the inductively coupled plasma (ICP) spectroscopy analysis conducted on the electrolyte recovered from the cycled full cell (Fig. 2d inset), which shows that the concentration of the dissolved Co in electrolyte recovered from 50 cycles between 2.5 and 1.0V is 0.0776 ppm, while none detected if the cell was cycled at lower voltage limits of 2.1 and 1.0 V. Hence we conclude that the overall effects of both surface defects presence, Co dissolution and formation of cobalt oxide at high high voltages are responsible for the rapid capacity fading of HVLCO in water-in-salt electrolyte. A similar correlation among surface defect, the capacity fading, and resistance growth was also established previously for LVLCO cycled in aqueous electrolyte between 4.1V and 2.7V (vs Li^+/Li) [27].

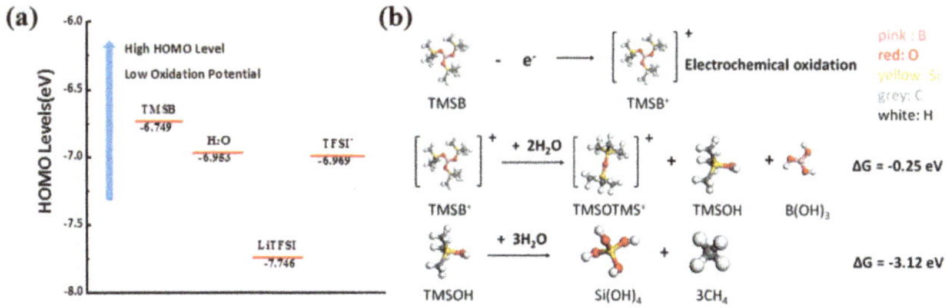

Fig. 3 (a) HOMO levels of TMSB, H_2O, LiTFSI, EC and $TFSI^-$; (b) Schematic representation of possible mechanisms for electrochemical oxidative decomposition of TMSB.

The most effective and economical strategy in suppressing the damage of HVLCO surface structure and Co dissolution is to introduce extra surface protection through electrolyte additives, whose electrochemical oxidation forms a solid interphase in a similar manner as SEI is formed on anode surface. Although additive approach has been extensively used in non-aqueous electrolytes (cite: Xu, Chem Rev. 2004, 104, 4303), its application in aqueous electrolyte is hitherto unknown, and more stringent requirements have to be met in this particular case: (1) the additive itself must possesses lower oxidation potential than the water in the water-in-salt electrolyte, which is ca. 4.9 V vs. Li; and (2) the additive itself must remain chemically stable against water (non-hydrolysable). After extensive screening, we identified a borate ester, tris(trimethylsilyl) borate (TMSB), as an qualifying cathode additive, which does not hydrolyze in the water-in-salt electrolyte (Fig. S5). This additive has also been used in non-aqueous electrolytes to form CEI on cathode [38, 41, 44]. First principles calculations was carried out on TMSB, whose highest occupied molecular orbital (HOMO) level, a critical element determining oxidation reaction[45], was compared with those of H_2O and TFSI anion in water-in-salt electrolyte.

As shown in Fig. 3a, the HOMO level of TMSB is the highest among all the molecules or anions, indicating that it will be the first to be oxidized when the cell voltage increases, before water or salt anion. [46] If the decomposition product is densely-packed solid and possesses electrolyte nature, i.e., ionically conductive but electronically insulating, then TMSB would serve as an excellent CEI additive in aqueous electrolyte.

To understand what TMSB will end up with, the oxidative pathway of TMSB in water was explored by using the first principles calculations, whose energetically favorable oxidation pathway in water-in-salt electrolyte was shown in Fig. 3b. According to the simulation, the continuous hydrolytic reactions of oxidized TMSB ($TMSB^+$) will lead to the formation of boric acid and orthosilicic acid together with $TMSOTMS^+$ and TMSOH, which is an intermediate product. Although boric acid and orthosilicic acid dissolve in water, they are insoluble in the water-in-salt electrolyte. Thus, they could co-precipitate on cathode surface and serve as a robust CEI. The protection mechanism of TMSB additive in aqueous electrolyte should differ from TMSB in non-aqueous electrolytes. In the later, the improved cycling performance was ascribed to the removal of HF impurity and the suppression of LiF formation [47, 48]. However, no LiF or HF should exist, at least on the cathode side, in our water-in-salt battery. On the other hand, oxygen evolution from aqueous electrolyte should become the main parasitic process that CEI prevents.

Based on the theoretical calculation, the electrochemical performance of HVLCO/Mo_6S_8 full cell was evaluated in 21m LiTFSI electrolyte with 0.1 wt% TMSB additive. Fig. 4a shows the charge/discharge behavior of the full cell at 0.5 C rate. In the first charge, an additional plateau at about 2.25 V (highlighted by circle) was observed, which could be ascribed to the TMSB oxidation . The 2.25 V plateau disappeared in the following cycles, indicates that the oxidation of TMSB is an irreversible process that only occurred in the first cycle, and once CEI is formed, TMSB should be completely consumed from the electrolyte. After the first charging, the following voltage profiles becomes identical to those of HVLCO/Mo_6S_8 full cell in additive-free electrolytes, suggesting that TMSB does not affect the eelctrochemical lithiation/delithiation of both HVLCO cathode or Mo_6S_8 anode except for the 1st cycle where CEI forms on HVLCO surface.

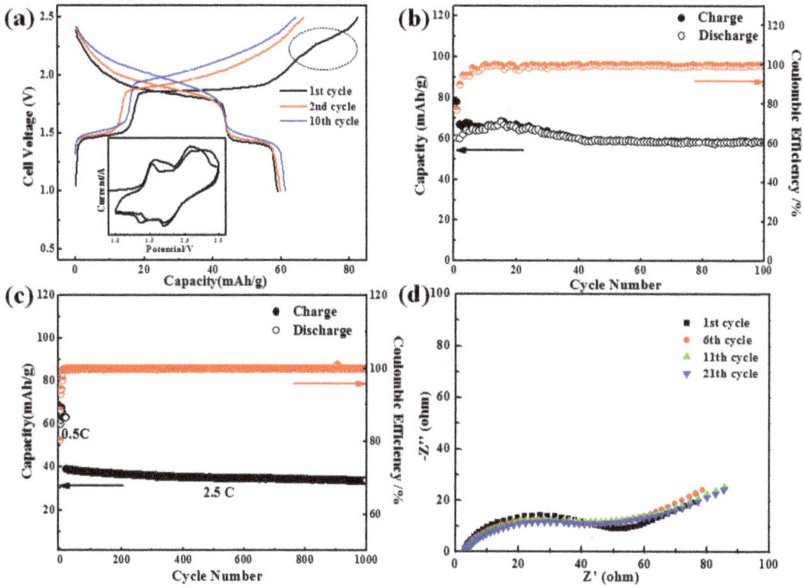

Fig. 4 The electrochemical performance of HVLCO/Mo_6S_8 full cell in 21m LiTFSI electrolyte + 0.1% TMSB (a) The typical voltage profile at constant current of 0.5 C between 1.0 V and 2.5 V (inset: CV curves); (b) The cycling stability and Coulombic at a low rate of 0.5 C rate; (c) The cycling stability and coulombic efficiencies at a high rate of 2.5 C rate after first 20 cycles at 0.5C; (d) Electrochemical impedance spectroscopy (EIS) at the 1st, 6th, 11th, 21th cycles.

The oxidation of TMSB must have formed a stable surface interphasial phase,which significantly suppressed parasitic reactions on the high oxidizing HVLCO surface, because the columbic efficiency increased from 94% of TMSB-free electrolyte (Fig. 2b) to ~100% in 21m LiTFSI with 0.1% TMSB electrolyte (Fig. 4b) after just first 8 cycles. Moreover, the irreversible capacity induced by formation of CEI on HVLCO cathode in the first charge, which should produce extra Li^+, was cancelled out by the negative irreversible capacity from formation of a SEI on Mo_6S_8 anode, which consumes Li^+. The net result is thus the columbic efficiency in the first cycle improved from 50% of TMSB-free electrolyte (Fig. 2b) to 75% in 21mLiTFSI+0.1% TMSB (Fig. 4b). Remarkably a 92% capacity retention was achieved for the full cell after 100 cycles at 0.5 C rate. The long term cycling stability of the HVLCO/Mo_6S_8 full cell was examined at a high rate of 2.5 C (Fig. 4c), where extremely high capacity retention, with minimum capacity decay rate of 0.013% per-cycle, was observed for 1000 cycles in 21m LiTFSI with 0.1% TMSB electrolytes. The stable interfacial impedance as detected by EIS at different cycles (Fig. 4d) confirmed the effective protection of CEI on HVLCO surface. At 0.5C HVLCO/Mo_6S_8 full cell using 21m LiTFSI-0.1% TMSB electrolyte provides an energy density of 120 Wh kg^{-1}, which is among the highest energy density in all the aqueous batteries reported to date (Fig. S6).

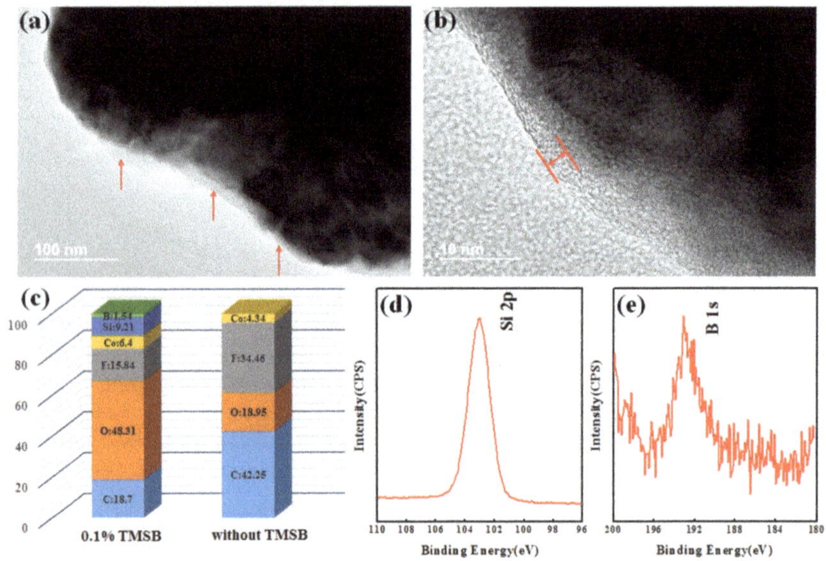

Fig. 5 (a) and (b) The TEM image for the HVLCO electrodes after cycling in 0.1% TMSB-containing 21m LiTFSI electrolyte; (c) Atomic compositions of the surface films on HVLCO electrodes after cycling; XPS spectra for (d) Si 2p and (e) B 1s on the surface of the HVLCO electrode after cycling in 21m LiTFSI-0.1% TMSB electrolyte.

The chemical composition of CEI on HVLCO surface formed by the oxidation of TMSB additive was analyzed by transmission electron microscopy (TEM). Fig. 5a and 5b shows the surface morphology of HVLCO electrode after 100 charge/discharge cycles in the electrolytes with TMSB additive. A ~10 nm thick surface layer is coated on a

HVLCO particle, which should be the electrochemical oxidation product of TMSB additive that generates a conformal coating on cathode particle during the initial charging. The HR-TEM image (Fig. 5b) further reveals the amorphous nature of the surface coating layer. In contrast with Fig. 2d, no obvious surface defects were observed on HVLCO that is now protected by CEI , suggesting that the surface layer eliminates Co-dissolution and hence improves the cycle performance of HV-LCO. The suppression of Co dissolution was also confirmed by the ICP analysis (Fig. S7), where the concentration of dissolved Co in electrolytes after 100 cycles was negligible in the presence of TMSB as additive in water-in-salt electrolyte.

The composition of CEI was determined using XPS by examining the surfaces of the cycled HVLCO cathode with and without TMSB additive. The atomic compositions of the surface layer are listed in Fig. 5c. Compared with the surface composition of HVLCO cycled in additive-free water-in-salt electrolyte, additional signals corresponding to Si (Si2p) and B (B1s) elements are clearly detected on the surface of HVLCO cycled in presence of 0.1% TMSB (Fig. 5d and Fig. 5e) . The binding energy of Si2p located at ~103eV, between the Si^{4+} (103.4eV) and Si^{3+} (102.1eV).[49] Therefore, we assign it to species of $Si(OH)_4$ containing minor fractions of $Si(OH)_3$. Similarly, the binding energy peak of B 1s at 193eV is attributed to the $B(OH)_3$ with minor Boron subhydroxides[50], which is in agreement with the simulation results in Fig. 3b. All these electrochemical, surface analysis and first principle calculation converge to suggest that TMSB additive in water-in-salt electrolytes can form protective CEI on HVLCO surface, thus allowing HVLCO to be stably charged/discharged at a high cut-off voltage, delivering a high capacity of 170 mAh g-1 with excellent stability for 1000 cycles, which is hitherto the highest reported for any aqueous Li-ion batteries. What more important than the avhieved energy density and cycling stability is the fact that the interphasial additive approach works in aqueous electrolyte. This new finding opens up a very critical pathway in future direction of aqueous electrolytes of better electrochemical stability and lower cost.

2. Conclusions

We have demonstrated that the high voltage $LiCoO_2$ can provide a high capacity of 170 mAh g^{-1} with exceptional cycling stability in water-in-salt electrolyte when its surface is protected by an interphase formed by additive. Paired with Mo_6S_8 anode, the 2.5 V HVLCO/Mo_6S_8 aqueous full cell delivered a high energy density of 120 Wh/kg for 1000 cycles with an extremely low capacity decay rate of 0.013% per-cycle, which is the highest energy density and longest cycle stability in all aqueous Li-ion batteries reported to date. The XPS and the first principle calculation demonstrated that TMSB additive was electrochemically oxidized and formed a densely-packed CEI on the surface of the

HVLCO, preventing Co dissolution and other parasitic reactions between HVLCO and electrolytes. The success of additive approach in aqueous electrolytes will certainly provide a new strategy of widening voltage windows of aqueous electrolytes at low cost, eventually making it bpossible to develop low cost high energy density and long cycle life aqueous Li-ion batteries with intrinsic safety.

3. Experimental Section

3.1 Materials

Lithium bis(trifluoromethane sulfonyl) imide ($LiN(SO_2CF_3)_2$, LiTFSI) (>98%) and water (HPLC grade) were purchased from Tokyo Chemical Industry and Sigma-Aldrich, respectively. The water-in-salt electrolyte (21m LiTFSI) is prepared by dissolving LiTFSI in water according to molality (21mol salt in 1 kg water). The tris(trimethylsilyl) borate(TMSB) was purchased from Sigma-Aldrich. 0.1% mass ration of TMSB additive was added to the water-in-salt electrolyte to form the 21m LiTFSI-0.1%TMSB electrolyte.

The $LiCoO_2$ materials were purchased from MTI Corporation. Chevrel phase Mo_6S_8 was prepared according to the previous reports.

3.2 Materials Characterizations

The morphology of the sample was investigated by scanning electron microscopy (SEM, Hitachi SU-70) and transmission electron microscopy (TEM, JEM 2100 FEG, 200 keV). All the samples for ex situ TEM were recovered from full aqueous Li-ion battery in 2032 coin cell configuration after electrochemical cycling. The samples were washed by DME three times and then dried under vacuum for two hours. X-ray diffraction (XRD) patterns were obtained on Bruker Smart 1000 (Bruker AXS, Inc.) using Cu Kα radiation with an airtight holder from Bruker. X-ray photoelectron spectroscopy (XPS) was conducted on a high sensitivity Kratos AXIS 165 X-ray photoelectron spectrometer with Mg Kα radiation. For ICP measurements, the electrolyte for the full aqueous Li-ion battery with 3mg cathode assembled in a bottle was diluted in 2% HNO_3 solution to form 25mL. ICP-OES measurements were performed using a Shimadzu ICPE-9820 Dual View Spectrometer.

3.3 Electrochemical Measurements

The $LiCoO_2$ and Mo_6S_8 electrodes were fabricated by compressing active materials, carbon black, and Polytetrafluoroethylene (PTFE) at weight ratio of 8:1:1 onto the stainless steel grid, respectively. The three-electrode devices for cathode consists of

$LiCoO_2$ composite (about 2mg) as working, platinum foil as the counter and Ag/AgCl as the reference electrode. Cyclic voltammetry (CV) was carried out using CHI 600E electrochemical work station at scanning rate of 0.1 mV/s for these composite working electrodes. CV was also applied to determinate the electrochemical stability window at 10 mV/s using 316 stainless steel grid (200-mesh sieve) as both working and counter electrodes, which were thoroughly cleaned ultrasonically in high purity alcohol, and then washed three times with high purity water and dried before measurement. The potentials vs. Ag/AgCl were converted to those vs. standard Li^+/Li, supposing that the potential of Ag/AgCl electrode was 3.239 V vs Li+/Li.

The full ALIB cell was assembled in CR2032-type coin cell using $LiCoO_2$ cathode (about 10 mg/cm^2), Mo_6S_8 anode (about 8 mg/cm^2) and glass fiber as separator. The charge-discharge experiments were performed on a Land BT2000 battery test system (Wuhan, China) at room-temperature.

3.4 DFT Calculation

All electron DFT calculation with unrestricted spin is conducted under Dmol3 modules in Materials Studio within GGA PW91 functional. 3.5 version Double Numerical plus *d*-functions (DND) basis set was chosen. Conductor-like screening model (COSMO), in which solvent is approximated by a dielectric continuum surrounding the solute molecules outside of a molecular cavity, is used to investigate the influence of water solvent and the dielectric constant of water, 78.54 was applied. To ensure the configuration of lowest energy, geometry optimization was conducted with several different initial configuration and the optimized configuration with lowest energy is selected.

Acknowledgements

The authors gratefully acknowledge funding support from DOE ARPA-E (DEAR0000389). We also acknowledge the support of the Maryland Nano Center and its NispLab. The NispLab is supported in part by the NSF as a MRSEC Shared Experimental Facility.

Supplementary Information: Stabilization of High-Voltage $LiCoO_2$ Cathode in Water-in-Salt Electrolytes

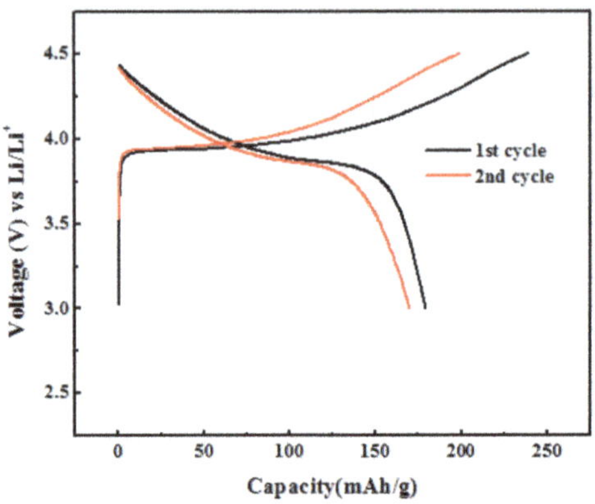

Fig. S1 The typical voltage profile of $LiCoO_2$ between 3.0V and 4.5V (vs Li^+/Li) in organic electrolyte (1M $LiPF_6$ in EC: DMC, 1:1 by volume) at constant current of 0.2 C with lithium as counter electrode and reference electrode.

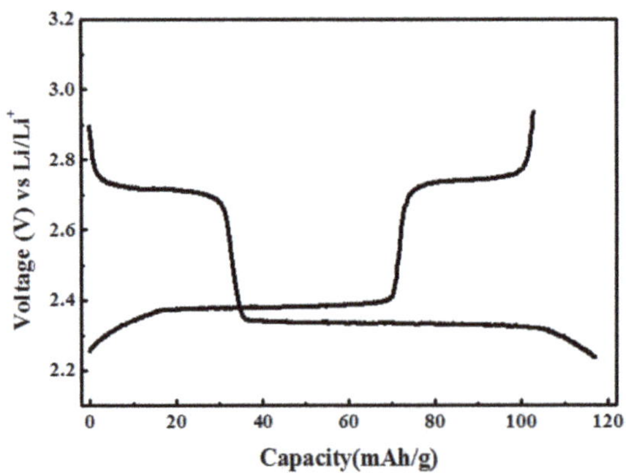

Fig. S2 The typical voltage profile of Mo_6S_8 in water-in-salt electrolyte at constant current of 0.2 C with Pt as counter electrode and the Ag/AgCl as the reference electrode.

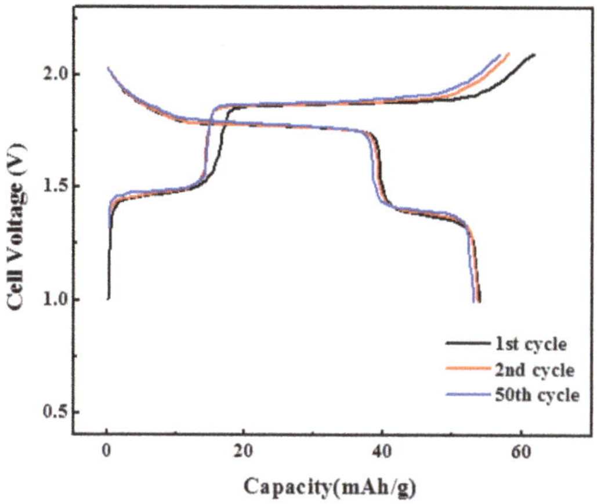

Fig. S3 The typical voltage profile of the full aqueous Li-ion cell employing Mo_6S_8 and LVLCO cycled between 1.0V and 2.1V in 21m LiTFSI electrolyte at constant current of 0.5 C.

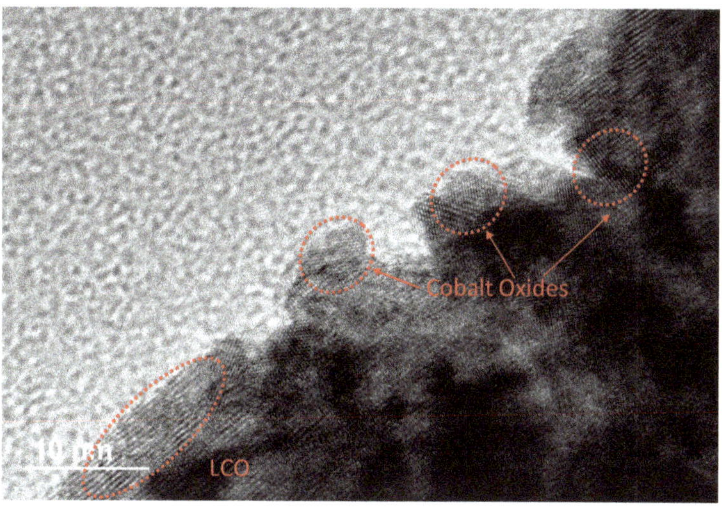

Fig. S4 The TEM image for the HVLCO electrodes after cycling in 21m LiTFSI electrolyte.

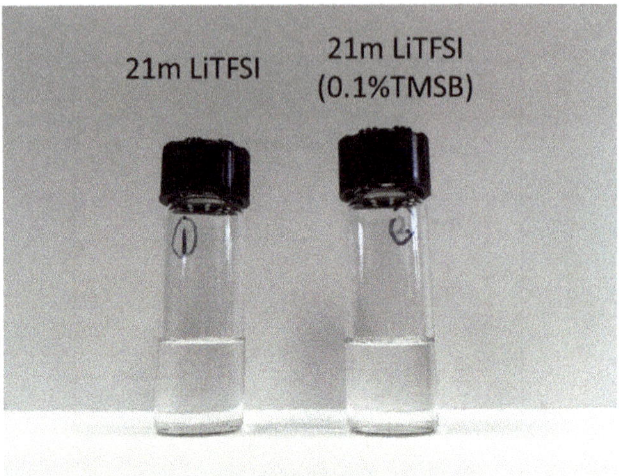

Fig. S5 The photograph of the 21m LiTFSI electrolyte and 21mLiTFSI-0.1wt% TMSB electrolyte.

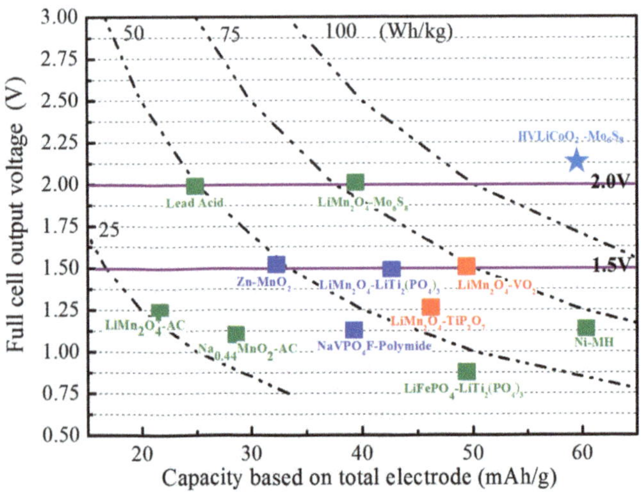

Fig. S6 Performances comparison for aqueous batteries based on various electrochemical couples

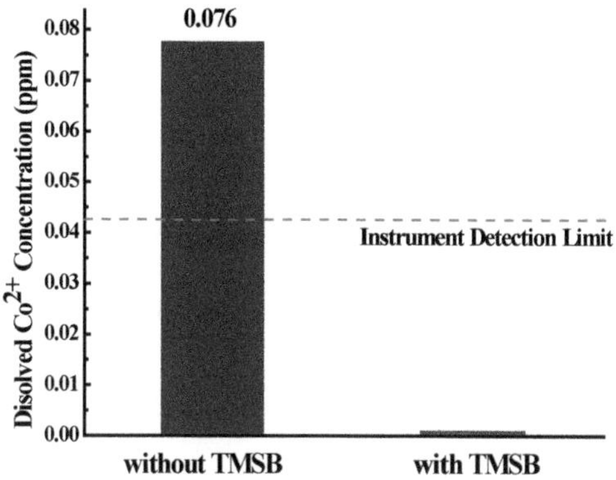

Fig. S7 The comparison for the dissolved Co concentration in TMSB-added and TMSB-free electrolyte.

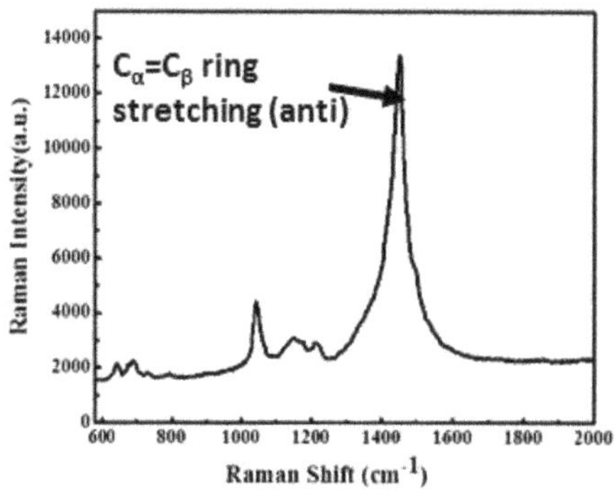

Fig. S8 Raman spectroscopy of high voltage $LiCoO_2$ in 21m LiTFSI with thiophene additive. The electrochemical polymerization of thiophene was achieved during the charge process.

Reference

[1] W. Li, J. R. Dahn and D. S. Wainwright, *Science*, 1994, 264, 1115-1118. https://doi.org/10.1126/science.264.5162.1115

[2] J.-Y. Luo, W.-J. Cui, P. He and Y.-Y. Xia, *Nat. Chem.*, 2010, 2, 760-765. https://doi.org/10.1038/nchem.763

[3] H. Kim, J. Hong, K.-Y. Park, H. Kim, S.-W. Kim and K. Kang, *Chem. Rev.* , 2014, 114, 11788-11827. https://doi.org/10.1021/cr500232y

[4] W. Tang, Y. S. Zhu, Y. Y. Hou, L. L. Liu, Y. P. Wu, K. P. Loh, H. P. Zhang and K. Zhu, *Energy Environ. Sci.* , 2013, 6, 2093-2104. https://doi.org/10.1039/c3ee24249h

[5] B. Dunn, H. Kamath and J.-M. Tarascon, *Science*, 2011, 334, 928-935. https://doi.org/10.1126/science.1212741

[6] J. B. Goodenough and Y. Kim, *Chem. Mater.* , 2010, 22, 587-603. https://doi.org/10.1021/cm901452z

[7] M. Armand and J. M. Tarascon, *Nature*, 2008, 451, 652-657. https://doi.org/10.1038/451652a

[8] S. F. Lux, L. Terborg, O. Hachmoeller, T. Placke, H. W. Meyer, S. Passerini, M. Winter and S. Nowak, *J. Electrochem. Soc.*, 2013, 160, A1694-A1700. https://doi.org/10.1149/2.039310jes

[9] M. S. Hong, S. H. Lee and S. W. Kim, *Electrochem. Solid-State Lett.* , 2002, 5, A227. https://doi.org/10.1149/2.039310jes

[10] D. Gordon, M. Y. Wu, A. Ramanujapuram, J. Benson, J. T. Lee, A. Magasinski, N. Nitta, C. Huang and G. Yushin, *Adv.Energy.Mater*, 2016, 6, 1501805. https://doi.org/10.1002/aenm.201501805

[11] D. W. McOwen, D. M. Seo, O. Borodin, J. Vatamanu, P. D. Boyle and W. A. Henderson, *Energy Environ. Sci.* , 2014, 7, 416-426. https://doi.org/10.1039/C3EE42351D

[12] L. Suo, O. Borodin, T. Gao, M. Olguin, J. Ho, X. Fan, C. Luo, C. Wang and K. Xu, *Science*, 2015, 350, 938-943. https://doi.org/10.1126/science.aab1595

[13] W. Tang, L. L. Liu, Y. S. Zhu, H. Sun, Y. P. Wu and K. Zhu, *Energy Environ. Sci.* , 2012, 5, 6909-6913. https://doi.org/10.1039/c2ee21294c

[14] J.-Y. Luo and Y.-Y. Xia, *Adv. Funct. Mater.* , 2007, 17, 3877-3884. https://doi.org/10.1002/adfm.200700638

[15] J. Yan, J. Wang, H. Liu, Z. Bakenov, D. Gosselink and P. Chen, *J. Power Sources* 2012, 216, 222-226. https://doi.org/10.1016/j.jpowsour.2012.05.063

[16] S. Liu, S. H. Ye, C. Z. Li, G. L. Pan and X. P. Gao, *J. Electrochem. Soc.*, 2011, 158, A1490. https://doi.org/10.1149/2.094112jes

[17] Y. G. Wang and Y. Y. Xia, *Electrochem. Commun.* , 2005, 7, 1138-1142. https://doi.org/10.1016/j.elecom.2005.08.017

[18] H. Wang, Y. Zeng, K. Huang, S. Liu and L. Chen, *Electrochim. Acta* 2007, 52, 5102-5107. https://doi.org/10.1016/j.electacta.2007.02.004

[19] Q. T. Qu, L. J. Fu, X. Y. Zhan, D. Samuelis, J. Maier, L. Li, S. Tian, Z. H. Li and Y. P. Wu, *Energy Environ. Sci.* , 2011, 4, 3985-3990. https://doi.org/10.1039/c0ee00673d

[20] M. Manickam, P. Singh, S. Thurgate and K. Prince, *J. Power Sources* 2006, 158, 646-649. https://doi.org/10.1016/j.jpowsour.2005.08.059

[21] P. He, J.-L. Liu, W.-J. Cui, J.-Y. Luo and Y.-Y. Xia, *Electrochim. Acta* 2011, 56, 2351-2357. https://doi.org/10.1016/j.electacta.2010.11.027

[22] Y. Y. Hou, X. J. Wang, Y. S. Zhu, C. L. Hu, Z. Chang, Y. P. Wu and R. Holze, *J. Mater. Chem. A*, 2013, 1, 14713-14718. https://doi.org/10.1039/c3ta13472e

[23] Z. Chen and J. R. Dahn, *Electrochim. Acta* 2004, 49, 1079-1090. https://doi.org/10.1016/j.electacta.2003.10.019

[24] R. Ruffo, F. La Mantia, C. Wessells, R. A. Huggins and Y. Cui, *Solid State Ionics* 2011, 192, 289-292. https://doi.org/10.1016/j.ssi.2010.05.043

[25] G. J. Wang, Q. T. Qu, B. Wang, Y. Shi, S. Tian, Y. P. Wu and R. Holze, *Electrochim. Acta* 2009, 54, 1199-1203. https://doi.org/10.1016/j.electacta.2008.08.047

[26] R. Ruffo, C. Wessells, R. A. Huggins and Y. Cui, *Electrochem. Commun.* , 2009, 11, 247-249. https://doi.org/10.1016/j.elecom.2008.11.015

[27] A. Ramanujapuram, D. Gordon, A. Magasinski, B. Ward, N. Nitta, C. Huang and G. Yushin, *Energy Environ. Sci.* , 2016, 9, 1841-1848. https://doi.org/10.1039/C6EE00093B

[28] X. Gu, J.-l. Liu, J.-h. Yang, H.-j. Xiang, X.-g. Gong and Y.-y. Xia, *J. Phys. Chem. C* 2011, 115, 12672-12676. https://doi.org/10.1021/jp202846p

[29] J. Choi, E. Alvarez, T. A. Arunkumar and A. Manthiram, *Electrochem. Solid State Lett.*, 2006, 9, A241-A244. https://doi.org/10.1149/1.2184495

[30] D. Aurbach, B. Markovsky, A. Rodkin, E. Levi, Y. S. Cohen, H. J. Kim and M. Schmidt, *Electrochim. Acta* 2002, 47, 4291-4306. https://doi.org/10.1016/S0013-4686(02)00417-6

[31] Y. Sun, J. Han, S. Myung, S. Lee and K. Amine, *Electrochem. Commun.* , 2006, 8, 821-826. https://doi.org/10.1016/j.elecom.2006.03.040

[32] J.-N. Lee, G.-B. Han, M.-H. Ryou, D. J. Lee, S. Jongchan, J. W. Choi and J.-K. Park, *Electrochim. Acta* 2011, 56, 5195-5200. https://doi.org/10.1016/j.electacta.2011.03.041

[33] F. Zhao, Y. Tang, J. Wang, J. Tian, H. Ge and B. Wang, *Electrochim. Acta* 2015, 174, 384-390. https://doi.org/10.1016/j.electacta.2015.05.180

[34] X. Dai, A. Zhou, J. Xu, Y. Lu, L. Wang, C. Fan and J. Li, *The Journal of Physical Chemistry C*, 2016, 120, 422-430. https://doi.org/10.1021/acs.jpcc.5b10677

[35] H. Xia, L. Lu, Y. S. Meng and G. Ceder, *J. Electrochem. Soc.*, 2007, 154, A337. https://doi.org/10.1149/1.2509021

[36] M. Gauthier, T. J. Carney, A. Grimaud, L. Giordano, N. Pour, H. H. Chang, D. P. Fenning, S. F. Lux, O. Paschos, C. Bauer, F. Maglia, S. Lupart, P. Lamp and Y. Shao-Horn, *J. Phys. Chem. Lett.* , 2015, 6, 4653-4672. https://doi.org/10.1021/acs.jpclett.5b01727

[37] K. Xu, *Chem. Rev.* , 2014, 114, 11503-11618. https://doi.org/10.1021/cr500003w

[38] X. Zuo, C. Fan, J. Liu, X. Xiao, J. Wu and J. Nan, *J. Power Sources* 2013, 229, 308-312. https://doi.org/10.1016/j.jpowsour.2012.12.056

[39] A. von Cresce and K. Xu, *J. Electrochem. Soc.*, 2011, 158, A337. https://doi.org/10.1149/1.3532047

[40] L. Xia, Y. Xia and Z. Liu, *Electrochim. Acta* 2015, 151, 429-436. https://doi.org/10.1016/j.electacta.2014.11.062

[41] H. Rong, M. Xu, B. Xie, X. Liao, W. Huang, L. Xing and W. Li, *Electrochim. Acta* 2014, 147, 31-39. https://doi.org/10.1016/j.electacta.2014.09.105

[42] S. Tan, Z. Zhang, Y. Li, Y. Li, J. Zheng, Z. Zhou and Y. Yang, *J. Electrochem. Soc.*, 2012, 160, A285-A292. https://doi.org/10.1149/2.066302jes

[43] Y. Liu, L. Tan and L. Li, *J. Power Sources* 2013, 221, 90-96. https://doi.org/10.1016/j.jpowsour.2012.08.028

[44] Z. Cai, Y. Liu, J. Zhao, L. Li, Y. Zhang and J. Zhang, *J. Power Sources* 2012, 202, 341-346. https://doi.org/10.1016/j.jpowsour.2011.10.101

[45] Y. K. Han, J. Yoo and T. Yim, *J. Mater. Chem. A*, 2015, 3, 10900-10909. https://doi.org/10.1039/C5TA01253H

[46] X. L. Liao, P. Y. Sun, M. Q. Xu, L. D. Xing, Y. H. Liao, L. P. Zhang, L. Yu, W. Z. Fan and W. S. Li, *Applied Energy*, 2016, 175, 505-511. https://doi.org/10.1016/j.apenergy.2016.03.114

[47] J. H. Li, L. D. Xing, R. Q. Zhang, M. Chen, Z. S. Wang, M. Q. Xu and W. S. Li, *J. Power Sources* 2015, 285, 360-366. https://doi.org/10.1016/j.jpowsour.2015.03.113

[48] Y. M. Song, J. G. Han, S. Park, K. T. Lee and N. S. Choi, *J. Mater. Chem. A*, 2014, 2, 9506-9513. https://doi.org/10.1039/C4TA01129E

[49] J. W. He, X. Xu, J. S. Corneille and D. W. Goodman, *Surf. Sci.* , 1992, 279, 119-126. https://doi.org/10.1016/0039-6028(92)90748-U

[50] Y. J. Wang, J. F. Fan and M. Trenary, *Chem. Mater.* , 1993, 5, 192-198. https://doi.org/10.1021/cm00026a007

SECTION II

The Power Sciences

Paul Barnes
U.S. Army Research Laboratory
Email: paul.n.barnes.civ@mail.mil

1. Introduction

The research being conducted for Power Sciences is arranged into 6 primary interactional areas. These are:

Compact power – such as micro-combustion and energy conversion

Thermal sciences and engineering – such as phase change and thermal interfaces

Wide band-gap materials – such as (Al)GaN high-power electronics

Microscale-fabricated power electronics – such as stretchable electronics and wireless power

Alternative energy – such as long-lived power from radionuclides/isomers

Catalytic reactions – such as fuel combustion and processing

Generally the research performed in Power Sciences is more fundamental in nature as opposed to applied. It falls into two categories [1]. The one is a systematic study to gain the basic knowledge necessary to determine the means by which a recognized need may be met. The other is the systematic study of the power related sciences directed toward a greater understanding of the fundamental aspects of phenomena without specific applications in mind. The intent is for the farsighted high-payoff research to provide the basis for technological prowess and progress toward an end application for the warfighter. Typically, the research being conducted in science and engineering will result in published reports and shared broadly within the scientific community. An introductory overview is given.

2. Overview

2.1 Compact Power

This developmental area is focused on compact fueled power sources. By studying hybrid heterogeneous and homogeneous micro-reactors along with integrated heat recuperation efforts it is expected that efficient, small-scale heat generation is possible. The intent is to develop hybrid micro-reactors that will convert JP-8 and other fuels into portable electric power, with a multi-fuel capability and instant refueling. A current topic of emphasis is combustion-based thermophotovoltaics. Two key parts are small scale combustion of logistics fuel and thermophotovoltaic energy conversion. These thermophotovoltaics can reduce the size and weight of power sources to be used for small systems and Soldier power. One goal is a high energy density system than existing rechargeable batteries.

2.2 Thermal Science and Engineering

Efforts on thermal management are geared to improve compact temperature control and heat removal from high power systems and devices. For example, device thermal management aims to improve of electronics performance to enable increased power densities. This includes not only electronics packaging but a consideration of multi-domain electrical/thermal/mechanical co-design. Another example of research in this area is phase change science. In this case the intent is to obtain enhanced solid-to-liquid thermal energy storage and liquid-to-vapor heat removal to protect high power devices and systems. In addition to power electronics, research is also conducted for solid-state thermal-power conversion. The development of improved thermoelectric and elastocaloric materials can provide enhanced compact solid-state heat pumping and thermal control. Advances in 3D electronics packaging, phase change cooling, and high figure of merit solid-state materials can reduce the size, weight, and thermal overdesign of power systems.

2.3 Wide Bandgap Materials and Devices

(Al)GaN will provide improved efficiencies and performance over state-of-the-art and near-term materials used in electronics, especially for high-power applications. GaN promises higher carrier mobility which results in less loss with smaller on-resistance which then leads to higher switching speed. The importance of higher switching speeds is to enable smaller transformers and hence a reduction in weight. Indeed a few GaN devices are reaching the market albeit they are more expensive since they are new. The cost will decrease over time as is often the case with new electronic devices and having GaN on Si as the device, if appropriate, is low cost in and of itself. It is expected that pulsed power applications with their associated need for higher voltage devices will push for materials with even wider energy gaps. This will drive the introduction of Al into the material as AlGaN which can exceed the energy gap of diamond.

2.4 Photovoltaics

Energy harvesting can be of key importance for on-site power generation without the need to bring the fuel along. Photovoltaics is an area that has been gaining traction commercially, but it is still important to seek novel ways to not only improve the efficiency of the photovoltaics, but also in their application. Mobile, flexible, light-weight, modular solar sheets with integrated optically active nanostructures that can maximize absorption and reduce glint/glare over a broad range of incident angles is one example. Developing novel ways to materially improve the efficiency of the device will also need to consider a new way to provide power. The goals are to achieve a high power per unit area and weight, provide enhanced power output over an entire day, and reduce the physical load. Technologically, this can be achieved by integrating nanostructures into high efficiency materials to enhance light trapping and absorption while simultaneously minimizing optical, thermalization, recombination and below-bandgap losses.

2.5 Thermal to Electric Energy Conversion

The thermal to electric energy conversion research is a good example where multiple approaches are being considered for a research thrust where each show promise in different application space: thermophotovoltaics, thermoelectrics, and pyroelectrics. The thermophotovoltaics were already discussed under the compact power section. Here is a summary for the other two. This also shows how the areas are interactional as mentioned earlier.

Thermoelectrics are being investigated to improve the conversion efficiency at a smaller temperature differential. The intent is to develop next-generation technologies for improved energy efficient devices. New materials and device technologies can electronically improve system efficiency following a variety of research directions for example via heterostructures to improve zT. There is also the potential of new phenomena that can impact the state-or-the-art such as the Rashba effect. The benefit to the warfighter by using electronically converted heat to electrical energy includes a reduction in overall system energy needs, improvement in fuel utilization and power redundancy, and a reduction in the logistics trail for fuel/energy resupply.

The research on pyroelectrics was initiated with an aim toward reducing parasitic losses by a few orders of magnitude and developing new thermodynamic cycles with increased power and energy density. Expected directions include the development of new measurements and materials towards engineering "Giant" pyroelectric coefficients. This will provide the basis and solid footing needed to truly advance the state-of-the-art. The direction of our research is guided by the expected relevance to the Army such as energy beaming, remote activation, armored tank cabin cooling, and so on.

2.6 Energy-Efficient Electronics

The applications areas that are currently emphasized as the main objectives for the development of more energy efficient electronics are: 'energy harvesting' sensors, low-power power electronics for long-lived sources, and wireless power technologies. Broader considerations expand this to passively powered sensors and electronics and low-power power management in general. It should be noted that the intent of the energy harvesting sensors is to extend the sensing capabilities of the relevant devices.

The development of energy efficient electronics for energy and power applications is expected to center around three major components related to sources, distribution, and loads. A key issue to consider is that battery life limits mission time with additional challenges associated with battery recharge and/or replacement. These activities are expected to directly contribute to Soldier-portable electronics, distributed sensors, and communications for sensors/radios development. All of these will benefit from the improvement of persistent sensors, extended battery life, and reduced maintenance.

Stretchable electronics are also being explored as highly deformable power components and systems based on non-traditional materials, i.e. liquid metals, for Soldier wearable electronics. These stretchable electronics are expected to offer orders of magnitude lower resistivity than current alternative stretchable devices. Initially the work will center on inductors based on the non-toxic liquid metal galinstan in microfluidic channels. The plan is to demonstrate functionality of up to 100% mechanical strain. Success in achieving these electronics with improved power efficiency should provide a lower temperature increase and thereby less discomfort for the user and lead to wireless power coupling in stretchable systems.

2.7 Alternative Energy

In general, the Army's access to energy storage is restricted to its reliance on chemicals, whether in the form of batteries, fuels, explosives, propellants, and so on. The intrinsic energy density of chemicals is limited by purely physical properties, to an upper limit of about 10^4 J/g. Reaching beyond this rather hard limit is possible with nuclear reactors, where fissionable materials have energy densities around 10^{11} J/g. However, the scale of nuclear reactors, with their accompanying shielding, control and waste-management systems, would seem to make their use for Army applications infeasible. The only other known materials that could provide energy storage far in excess of that allowed by chemicals would be radioisotopes, whether of the traditional type, e. g. tritium, or of a new class of isomeric materials. For traditional radioisotopes, more applied research is underway to develop prototypes for long-duration power sources. At the same time, fundamental physics research into isomeric materials seeks to understand mechanisms that control the release of energy from those non-traditional radioisotopes, with potential applications involving energy storage and directed gamma emission. In all, the possibility of reaching intrinsic energy storage of 10^9 J/g and extremely-long-duration serves to motivate work in this area.

3. Summary

A brief overview of some of our developmental work for Power Sciences is given via a few examples. The area is more comprehensive, but the sampling will serve as an orientation into the research and as an introduction for the next several chapters in this book.

References

[1] P.N. Barnes, Making Department of Defense Basic Research Purple (Joint), but NOT the Department of Defense Laboratories, Tech Report No. ARL-TR-6762, Dec 2013.

Chapter 12

Nuclear Metastables for Energy and Power: Status and Challenges

James J. Carroll

U. S. Army Research Laboratory, Adelphi, MD 20783, USA

Abstract

The Army's ever-increasing demands for energy are confronted with a fundamental limitation on the intrinsic energy density of chemical fuels and batteries: a "chemical limit". Radioisotopes and, in particular. nuclear metastables (isomers) may provide the only feasible means of surpassing this limit, and by many orders of magnitude, for future applications. This paper summarizes the present state-of-the-art in basic research on nuclear isomers with a focus on their potential value as switchable ultra-energy-density power sources.

Keywords

Energy Storage and Release, Chemical Limit, Radioisotopes, Nuclear Isomers, Ultra-Energy-Density Materials

Contents

1. Introduction

Energy, and the ability to deliver it as required, underpins everything the Army does. Great effort is being expended to develop new tools for the Army of the future, whether in enhancing the ability to understand the battlespace, project greater lethality, move more quickly and efficiently, etc. However, it is easy to forget that without sufficient energy and power, none of these improvements are feasible. Thus, there is clearly a strong motivation to expand the Army's access to energy and this drives a search for materials offering more energy density and longer-lived power sources, most obviously in research and development into new battery technology. That effort ultimately faces a fundamental "chemical limit" as depicted in Fig. 1.

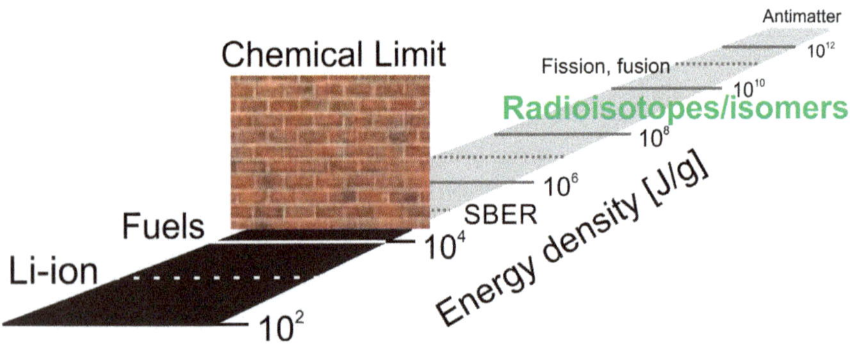

Figure 1: Schematic of intrinsic energy density (J/g in log scale) for different materials, where those below the "chemical limit" are presently in use by the Army. Of the materials with energy densities far beyond the chemical limit, radioisotopes and nuclear isomers are the most likely to support Army applications in the deep future.

The intrinsic energy-storage density of any material is determined by a characteristic energy corresponding to some basic building block. For chemicals, that building block is the atom and each electron within the atomic volume is bound with 10 – 100 eV, with binding energies between atoms in molecules of a few eV. This translates into a macroscopic intrinsic density for stored energy of less than 10^5 J/g, e. g. gasoline is about 46,000 J/g.[1] Chemical batteries store less, on the order of about 10^3 J/g for a lithium ion battery, while food is surprisingly energetic: nearly half the energy density of fuels like

[1] Energy densities are taken from [1] unless otherwise noted.

gasoline. No matter the form, this type of energy storage will be restricted by fundamental physics to values below a chemical limit of about 5×10^4 J/g.

There is clearly great value in attempting to achieve improvements in chemical energy storage, as one cannot envision any alternative to the chemical battery for many applications. Still, is there any way to access energy storage beyond the chemical limit? To do so requires a willingness to explore entirely new classes of materials. Far beyond the chemical limit lies nuclear energy storage in fissionable materials, near 10^{11} J/g. Again, the intrinsic energy density is determined by the fundamental building block, in this case the atomic nucleus, with constituent nucleons (neutrons and protons), each bound with a mean energy of about 8 MeV. Nuclear (fission) reactors exist, but are unwieldly and impractical for the Army. Tritium-deuterium fusion is characterized by an intrinsic energy density of the order of 3×10^{11} J/g, but "hot" fusion reactors are likely to be very large and have not yet demonstrated break-even. On the other hand, "cold" fusion, sometimes called Low Energy Nuclear Reactions (LENR), has yet to be demonstrated to the satisfaction of the scientific community. Antimatter has by far the highest intrinsic energy density, about 10^{14} J/g, but there is currently a very limited ability to store antimatter for long times. Closer to the chemical limit lies energy storage in strongly-compressed matter, such as metallic hydrogen (about 2×10^5 J/g [2]), but with confirmed production in only microscopic amounts [3]. Similarly, energy storage via structural bond energy release (SBER) has been explored for nanodiamonds. Moving significantly past the chemical limit will be challenging for Army applications.

While there are a number of different possible energy-storage media beyond the chemical limit, the most practical are likely those based on radioisotopes, which store on the order of 10^9 J/g. Some examples are tritium (^3H), with a half-life of 12.32 years and storing about 6×10^8 J/g; and ^{63}Ni, with a half-life of 101.2 y and storing 10^8 J/g. These materials are commodities and are available in needed quantities. The energy release mechanism for these examples is spontaneous radioactive decay and the decay emission is solely in the form of high-energy electrons (β^- particles). The long half-lives of these and some other radioisotopes are key qualities, being much longer than the typical useful life of most chemicals, which degrade over shorter times: gasoline over a period of several months to a year, and chemical batteries over a few years. Thus, radioisotopes may serve an important role in niche applications like drop-and-forget sensors for battlespace awareness, where replacement of chemical batteries is impractical and where the lifetime of power supplies is more important than even a much higher energy density.

The utilization of radioisotopes in practical applications will rely on the development of reliable means of converting the released energy into electrical current. The general methods fall into two broad categories: direct conversion, where the emitted radiation creates charge carriers in a semiconductor; and indirect conversion, where the emitted radiation first creates photons in a phosphor and then those photons create charge carriers in a photovoltaic. Both methods have their advantages and disadvantages, as described elsewhere [4-7], but considerable effort is underway to exploit different schemes to maximize the efficiency and radiation tolerance of the energy-conversion materials. The goal would be to provide the greatest longevity of a power source, based on the long lifetimes of selected radioisotopes like tritium, ^{63}Ni, etc. At the same time, the slow natural decay rate of those radioisotopes strongly inhibits the continuous power output to low levels.[2] A non-standard type of radioisotope, exhibiting more than one lifetime of significant duration, may provide a way to benefit from the storage of a large energy density for long times, yet enable higher output power on demand via a controlled energy release. This would give a disruptive form of energy storage far beyond the chemical limit in a new class of ultra-energy material.

2. Nuclear isomers

The atomic nucleus is a dynamic system in which the constituent particles (nucleons: neutrons and protons) follow complex quantized motions.[3] In some cases, this motion can be described collectively, as when the protons oscillate together with respect to the neutrons (the giant dipole resonance), or when waves move along the surface of the nucleus (collective vibrations). In other instances, the motion can be described effectively using a single-particle model, where individual nucleons move in orbitals of definite energy, angular momentum,[4] and parity within a mean potential provided by the other nucleons. These behaviors can exist in the same nucleus, at different energies, giving a manifold of nuclear levels that can be relatively simple (see Fig. 2(a) for ^{208}Pb) or quite rich (see Fig. 2(b) for ^{180}Ta). Electromagnetic transitions are possible between some levels, with rates that are dependent on the differences in energies, angular momenta, and parities of those levels. These transitions typically have energies on the order of tens to hundreds of keV.

Most nuclear excited states have half-lives much shorter than 10 ns. However, there are certain notable exceptions called isomers, first discovered almost 100 years ago. These are metastable excited states having half-lives longer than about 10 ns, and in the most

[2] A low level of continuous power could be accumulated in a coupled conventional battery for higher power availability on an occasional basis.

[3] A detailed discussion of collective and single-particle nuclear models may be found in Ref. [8].

[4] The magnitude of the angular momentum of a quantum state is $\hbar\sqrt{J(J+1)}$, where J is the angular momentum quantum number. Since the angular momentum is determined by the quantum number, in the text "angular momentum" will be used to refer to J.

interesting cases half-lives of years or greater. The longest-lived isomer is an excited state in ^{180}Ta (one of the many levels in Fig. 2(b)), that lies 77.2 keV[5] above the ground state: This isomer, designated as ^{180m}Ta ("m" for metastable), has never been seen to decay and has a half-life of at least 4.5×10^{16} years [10]. The ground state[6] of this nuclide is unstable, with readily-observed decays by β^- emission or electron capture, and a half-life of 8.154 h. With such a long half-life, ^{180m}Ta represents the only naturally-occurring isomer, produced in and expelled from stars, with 0.012% of terrestrial tantalum comprised of this isotope. A more extensive discussion of isomers and the physical causes for their long lifetimes may be found in Refs. [11-13].

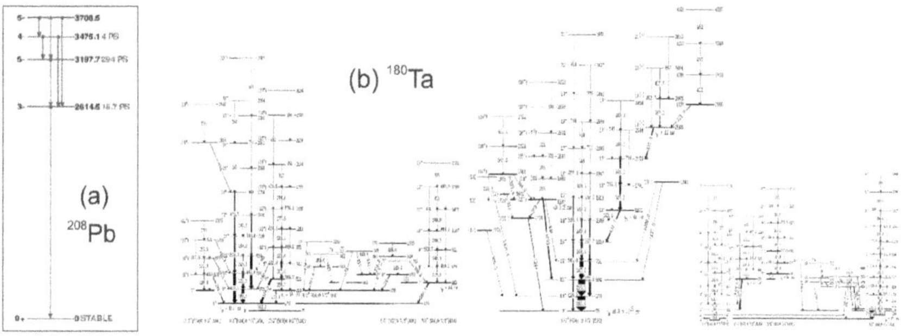

Figure 2: Known nuclear levels up to about 3700 keV for (a) ^{208}Pb (derived from [9]) and ^{180}Ta (adapted from [14]). Electromagnetic transitions between levels are shown by the arrows. Energies are in keV. The important takeaway is that there is great variability in the complexity of nuclear behavior and that "every nucleus is different" [15].

One may immediately consider that the isomer of ^{180}Ta permits the storage of a large amount of energy (4.1×10^7 J/g) in that metastable state for an exceptionally long time. The ground state of this nuclide is non-existent within a sample,[7] unless nuclei were transferred from the isomer to the ground state via some mechanism. In that case, decay of the ^{180}Ta ground state would release an additional 702 keV on a much faster time scale. The total energy stored in ^{180}Ta is then 4.2×10^8 J/g. The question then becomes: Are there mechanisms by which nuclei of ^{180}Ta (or isomers of other nuclides) can be

[5] All nuclear data taken from Ref. [9] unless otherwise indicated.
[6] The ground state is the lowest-energy state of that nucleus, with its characteristic constituents (specific numbers of neutrons and protons), even if that state is unstable and can decay (transmute) to another nuclide (different numbers of neutrons or protons).
[7] In the parlance of laser physics, the collection of ^{180}Ta nuclei comprises a fully-inverted medium.

transferred from the long-lived (energy-storing) state to the shorter-lived (energy-releasing) state?

3. Isomer depletion

Motivated by both scientific issues and potential practical applications, ^{180}Ta has been investigated using a variety of theoretical approaches and experimental methods. For example, the most detailed gamma-ray spectroscopic study to-date [14] provided the level scheme shown in Fig. 2(b). In that work accelerated beams of charged particles were used to produce ^{180}Ta by reactions in which the beam particles were fused into targets of ^{176}Yb. The fusion and subsequent "evaporation" of a few neutrons from the hot nuclei left ^{180}Ta in high-energy (much greater than the 3700 keV shown in Fig. 2(b)), high-angular-momentum states. Subsequent rapid decay of these states led to various sequences of gamma rays emitted by the allowed electromagnetic transitions following different pathways. These gamma rays were observed by an array of High Purity Germanium (HPGe) detectors, and their energies and arrival times were recorded. After the multi-day experiment, the data were sorted to extract those events where on a nucleus-by-nucleus basis gamma rays were emitted in close time coincidence, defining particular decay cascades for that nucleus. Evaluation of all the collected nucleus-by-nucleus events allowed the development of the level scheme shown in Fig. 2(b).

Separately, experiments were performed to directly test if exposure of a tantalum sample to high-energy photons could cause a transfer of population from the isomer to the ground state. Electron accelerators were used to produce bremsstrahlung for these tests, where the continuous spectrum of photons was used to irradiate tantalum samples containing ^{180m}Ta. It was expected that some states lying higher than the isomer would be found that could serve as intermediaries between the isomer and ground state, following the process depicted in Fig. 3. The figure shows a single mediating state, also referred to as a depletion level, above the isomer, which is in turn higher in energy than the ground state. The downward arrows signify electromagnetic decays from higher to lower states: the solid arrow is a single transition while the dashed arrow could be a single transition or a cascade of multiple transitions. The upward grey arrow signifies the absorption of a photon to excite the depletion level from the isomer. Thus, a transfer of population from the isomer to the ground state (a depletion of the isomer population) would require 1) absorption of a photon in the transition from isomer to depletion level, and 2) subsequent decay of the depletion level by single transition or cascade to the ground state. The ground state, if unstable, could decay to release additional energy. The amount of energy released directly and promptly in the depletion process is just that stored in the isomer.

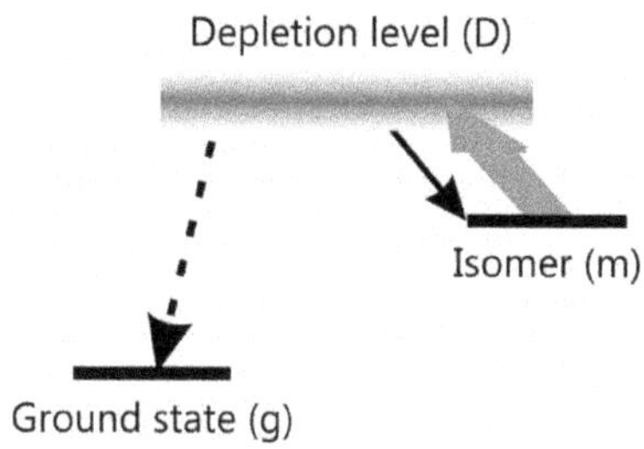

Figure 3: Schematic showing the process of isomer depletion via a higher-lying intermediate state (depletion level).

Depletion of ^{180m}Ta was indeed demonstrated experimentally, first reported in 1988 [16] and the process was later elucidated in more detail [17, 18]. In all, eight depletion levels were identified in the later results.

An examination of the levels shown in Fig. 2(b) [14] and the depletion measurements [18] made possible the first connection between results of these two different types of experiments. The first three depletion levels, previously identified only by the energies required to reach them from the isomer, were found [19] to coincide with three specific levels in the scheme of Fig. 2(b). This defined the entire depletion pathway from isomer to ground state, although a mystery remains. Excitation of a depletion level should sometimes be followed by a "back decay," returning the nucleus to the isomer, as depicted in Fig. 3 by the solid down arrow. However, none of the expected back decays have been observed for ^{180}Ta. This is likely due to the sensitivity limitations of previous experiments, leaving a motivation for future measurements.

Depletion of ^{180m}Ta has been experimentally demonstrated, showing that it is possible to release on demand radioisotope energy stored at densities far beyond the chemical limit. While proving the concept, the minimum energy required to cause a release for this nuclide was around 1000 keV, greater than the amount of energy stored in the isomer (77.2 keV) and the resulting ground state decay (702 keV). Fortunately, there are another 10 known isomers with half-lives greater than 1 year, as listed in Table 1. Investigations are underway to search for depletion mechanisms for many of those nuclides.

Table 1 Nuclides having isomers with half-lives ($T_{1/2}$) greater than one year. Isomer energies (E_m) are given, as well as angular momenta (J), parities (π), and isomer and ground-state decays. Decay modes are: IT = isomer transition (electromagnetic decay toward the ground state); β^- and β^+ = electron and positron decay; ε = atomic electron capture by the nucleus; and α = alpha decay. Uncertainties in the values are given in Ref. [9]

Nuclide	Isomer				Ground state			
	$T_{1/2}$ [y]	E_m [keV]	J^π	Decay modes	$T_{1/2}$	J^π	Decay modes	Decay energy [keV]
^{93}Nb	16.12	30.77	$1/2^-$	IT: 100%	Stable	$9/2^+$	NA	NA
^{102}Rh	3.742	140.73	$6^{(+)}$	IT: 0.2% $\varepsilon+\beta^+$: 99.8	207.3 d	(1-, 2-)	β^-: 22% $\varepsilon+\beta^+$: 78%	1151
^{108}Ag	438	109.466	6^+	IT: 9% $\varepsilon+\beta^+$: 91%	2.382 min	1^+	β^-: 97% $\varepsilon+\beta^+$: 3%	1646
^{113}Cd	14.1	263.54	$11/2^-$	IT: 0.1% β^-: 99.9%	8.04 × 10^{15} y	$1/2^+$	β^-: 100%	323
^{121}Sn	43.9	6.31	$11/2^-$	IT: 78% β^-: 22%	27.03 h	$3/2^+$	β^-: 100%	401
^{166}Ho	1200	5.969	7^-	β^-: 100%	26.824 h	0^-	β^-: 100%	1855
^{178}Hf	31	2446.09	16^+	IT: 100%	Stable	0^+	NA	NA
^{180}Ta	> 4.5 × 10^{16}	77.2	9^-	NA	8.154 h	1^+	β^-: 15% ε: 85%	702
^{186}Re	2 × 10^5	149	(8^+)	IT: 100%	3.7183 d	1^-	β^-: 93% ε: 7%	1072
^{192}Ir	241	168.14	(11^-)	IT: 100%	73.829 d	4^+	β^-: 95% ε = 5%	1455
^{242}Am	141	48.60	5^-	IT: 99.5% α = 0.5%	16.02 h	1^-	β^-: 83% ε = 17%	665

4. Isomer depletion for ^{108m}Ag and ^{186m}Re

The Army Research Laboratory (ARL) has conducted studies of a number of the long-lived isomers in Table 1, and has instituted major research campaigns focused on several of these: ^{108m}Ag (Silver Campaign, begun 2011), ^{186m}Re (Rhenium Campaign, begun 2013), and ^{192}Ir (Iridium Campaign, begun 2017). Major results from the Silver and Rhenium Campaigns, to-date, will be discussed here to illustrate the status and challenges to this research topic.

The nuclide ^{108}Ag is of interest in several ways: the isomer half-life is long (438 y), the ground-state half-life is much shorter (2.4 min), and the energy released as a result of ground-state decay is relatively large (1646 keV). The intrinsic energy density of this material is 1.5×10^9 J/g and for those nuclei shifted from the isomer to ground states, the power output is increased by a factor of nearly 100,000,000 (from the ratio of the half-lives). Scientifically, ^{108}Ag is of interest as it is a "transitional" nucleus proposed to exhibit a rare behavior of possessing nearly-degenerate bands of states exhibiting a chiral nature [21]. With these motivations, a gamma spectroscopic experiment was designed and performed as a collaboration between ARL; the University of Surrey, UK; the Joint Institute for Nuclear Research, Russia; and the Tata Institute for Fundamental Research (TIFR), India. The experiment was performed at TIFR using an accelerated ion beam and the Indian National Gamma Array. Similar to the case of the ^{180}Ta spectroscopic experiment described above, gamma rays emitted from excited ^{108}Ag nuclei were measured and their energies and arrival times recorded, so that gamma-ray coincidences could be analyzed. Nearly-degenerate bands of states were indeed found, but were identified as arising from different orbital configurations rather than from the same configuration, but with left-handed or right-handed summing of nucleon angular momenta [21].

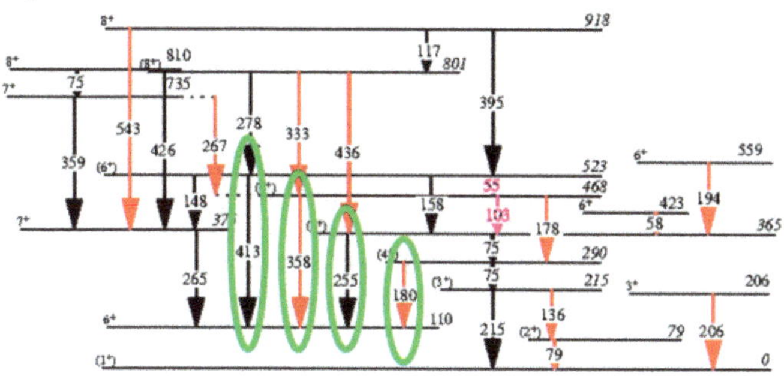

Figure 4: Partial level scheme of ^{108}Ag, adapted from Ref. [20]. Four transitions were identified from the gamma spectroscopic data, circled in green, which represented measured gamma decay from higher-lying states to the 438-y isomer at about 110 keV.

The data were further analyzed to determine if there were states above the isomer which could serve as depletion levels as depicted in Fig. 3. Indeed, four such levels were located, as shown in Fig. 4. For example, in the figure a decay transition at 413 keV is circled, representing measured gamma decay from a state at 523 keV to the isomer at 110 keV. The state at 523 keV also exhibits other decay branches at 158 keV, 148 keV and 55 keV. These gamma rays were measured in the experiment and their placement in the level scheme inferred from spectra like that in Fig. 5. Since the highlighted decay transitions from higher-lying states to the isomer exist, those same transitions would allow absorption of photons from the isomer to the higher-lying states to initiate isomer depletion. Thus, the states at 290, 365, 468 and 523 keV can be identified as depletion levels.

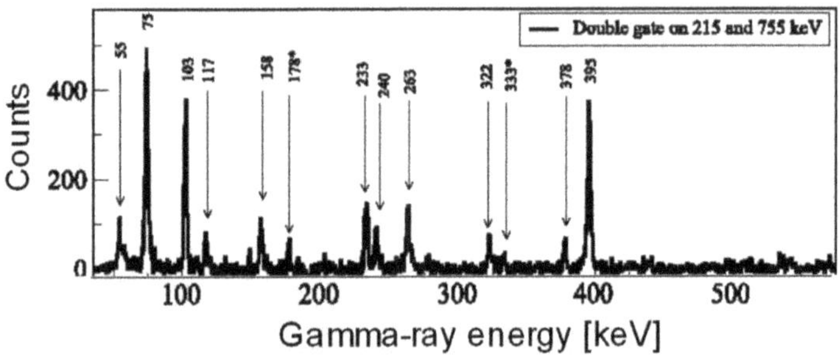

Figure 5: Spectrum of gamma rays coincident simultaneously with gamma rays at 755 keV (not shown in Fig. 4, but decaying into the 918 keV state) and 215 keV. This spectrum (adapted from Ref. [20]) establishes that, for example, the 55 keV transition and the 158 keV transition are both part of cascades that include both the 755 keV and 215 keV transitions. However, this by itself does not mean that the 55 keV and 158 keV transitions are in coincidence with each other and part of the same cascade (they are not – see Fig. 4 – since one nucleus cannot emit both a 55 keV gamma ray and a 158 keV gamma ray as they are in parallel paths). Comparisons between this spectrum and other coincidence spectra were needed to correctly place the gamma rays and states in the level scheme.

The probability for isomer depletion via a particular intermediate state is given by the energy-integrated cross section (*ICS*):

$$ICS = \left(\frac{2J_D+1}{2J_m+1}\right)\left(\frac{\pi\hbar c}{E_{m\to D}}\right)^2 b^{\gamma}_{m\to D} b_{D\to g} \Gamma_D \quad , \tag{1}$$

where J_D and J_m are the angular momenta of the depletion level and isomer, respectively, and $E_{m\to D}$ is the energy of the transition between those levels. The $b^{\gamma}_{m\to D} = b^{\gamma}_{D\to m}$ is the gamma branching ratio (without inclusion of electron conversion[8]) for the transition from isomer to depletion level, while $b_{D \to g}$ is the total branching ratio for the fraction of decays of the depletion that lead toward the ground state, whether directly or by one or more cascades. All the factors on the right side of Eq. (1) are known except for the energy width Γ_D of the depletion level. This can only be determined from the half-life of the depletion level, which could not be measured in that particular experiment. Thus, the contributions of each of the four depletion levels to energy release from the isomer are unknown.

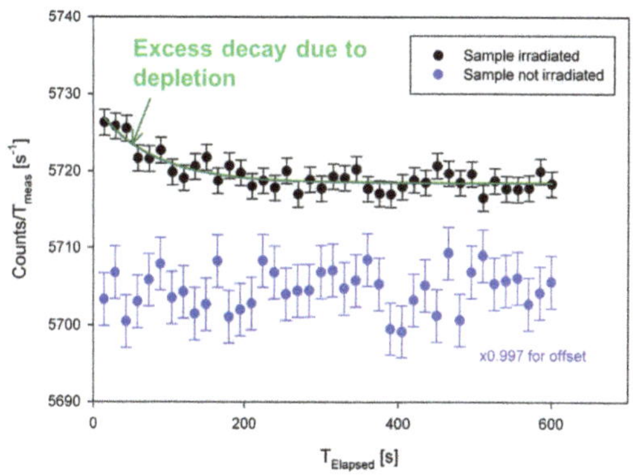

Figure 6: Count rates of radiation emitted from the ^{108m}Ag sample after exposure to 6 MeV bremsstrahlung (black points) and without exposure (blue points, offset downward for clarity), adapted from Ref. [13]. Multiple repetitions of the exposure, transport, counting sequence were combined to improve the statistical accuracy of the measurement.

As another experiment in the Silver Campaign, a direct test of isomer depletion was performed at ARL, as described in detail in Ref. [13]. A sample was obtained that contained ^{108m}Ag and that sample was exposed to 6 MeV bremsstrahlung produced with an electron linear accelerator: the irradiating photon spectrum extended up to a maximum of 6 MeV, the energy of the accelerated electrons. After irradiation, the

[8] Electron conversion occurs when an electromagnetic decay transition in a nucleus transfers energy to an atomic electron that is then ejected from the atom. This decay mode increases the overall transition rate, sometimes negligibly for higher-energy nuclear transition energies and sometimes significantly for lower-energy transitions. However, only the transition probability due to gamma emission corresponds in the inverse to photon absorption, so only that part is included for the excitation step from isomer to depletion level. For decays from the depletion level toward the ground state, any mode contributes.

sample was transferred by remote control to a detector and radiation emitted from the sample was measured as a function of time. This sequence was repeated by an automated system and the results of each separate measurement were combined to improve statistical accuracy of the results, which are shown in Fig. 6. Without exposure to the bremsstrahlung, the rate of radiation emission from the sample is constant within statistics, due to steady decay of the ^{108}Ag ground state which is being fed at an essentially constant rate by the decay of the 438 year isomer. However, after exposure to the bremsstrahlung the sample emits excess β^- radiation with the characteristic 2.4 min half-life of the ground state, due to population shifted from the isomer to the ground state during the exposure. After several ground-state half-lives have elapsed, the emitted radiation returns to the steady level seen without exposure. Without the offset in the figure, the baselines of the with- and without-exposure measurements coincide.

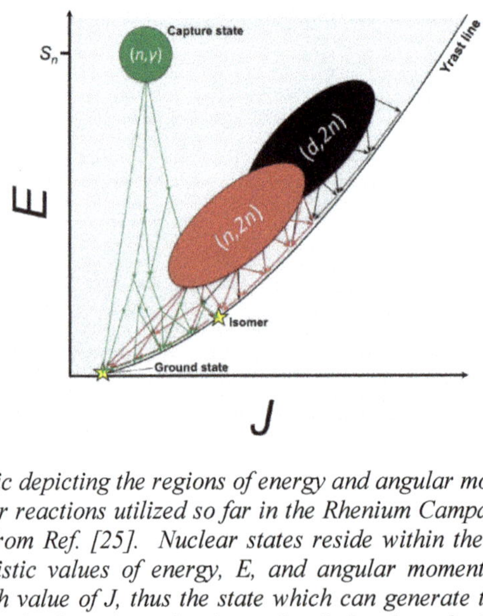

Figure 7: Schematic depicting the regions of energy and angular momentum accessed by the different nuclear reactions utilized so far in the Rhenium Campaign for investigation of ^{186}Re, adapted from Ref. [25]. Nuclear states reside within the grey region, having different characteristic values of energy, E, and angular momentum, J. The lowest-energy state at each value of J, thus the state which can generate that value of angular momentum most easily, lies along the "yrast line" that bounds the region of nuclear states. The ground state and 200,000 year isomer of ^{186}Re lie on the yrast line.

The experimental results in Fig. 6 demonstrate depletion of the ^{108m}Ag isomer and an overall *ICS* could be obtained. However, this integral cross section would be the cumulative effect of all the possible depletion levels, weighted by the numbers of photons at each separate energy $E_{m \to D}$. At present a full characterization of the depletion levels and their roles in enabling an energy release from ^{108m}Ag is not available, but further experiments utilizing new techniques are planned in the Silver Campaign to provide this information.

The Rhenium Campaign has thus far included multiple experiments conducted across four continents, focused on the nuclide ^{186}Re with its 200,000 year isomer. The half-lives of isomer and ground state, and the stored energy, are attractive for potential applications, and this nucleus is fertile ground for scientific study as relatively little confirmed level information can be found in the nuclear database of Ref. [9]. No isomer sample is currently available for direct depletion tests, but gamma spectroscopic measurements have been performed, with excited ^{186}Re nuclei produced by thermal neutrons [22], fast neutrons [23], and charged particles [24]. Each reaction provides access to a different energy and angular momentum region in the level scheme, as depicted in Fig. 7. Beginning the Rhenium Campaign, there was no information on the location of potential depletion states in the level scheme. Thus, it was necessary to initiate a series of experiments which each provided access to states in different regions of the energy and angular momentum space: the availability of research facilities that lie outside DoD was also a critical factor, based on scientific interest in this nuclide and the ability to form ad hoc research teams for different studies.

One such study was performed at the Budapest Research Reactor, from which thermal neutrons were utilized to produce excited states of ^{186}Re due to neutron capture on an enriched ^{185}Re target.[9] The ground state of ^{185}Re has an angular momentum of 5/2, so that capture of a thermal neutron (zero orbital angular momentum, 1/2 spin angular momentum) could only produce ^{186}Re in capture states of angular momentum 2 or 3. The small locus of these states is shown in Fig. 7 at the neutron separation energy S_n and depicted by "(n, γ)". Equipartition of energy takes place rapidly (on the scale of 10^{-19} s) so there is very little chance for a neutron to "evaporate" from the ^{186}Re nucleus. Instead, the nuclei deexcite by cascades of "statistical" gamma-ray transitions (green in Fig. 7) of dominant electric dipole character that do not change the mean angular momentum of the nucleus (but do spread the distribution of angular momentum) until reaching states near the yrast line. There are also high-energy ("primary") transitions that reach directly from the capture states to near the yrast line, even to the ground state. Figure 8 shows the measured in-beam spectrum obtained with a single large HPGe detector. The Continuum region contains a high density of unresolved peaks due to the many statistical transitions, while primary transitions directly from the capture states to those near the ground state lie

9 The natural abundance of ^{185}Re is 37.4%. The sample was prepared to enhance the amount of this isotope to 96.74%.

at the highest energies. Resolved lower-energy gamma-ray peaks are seen and correspond to transitions between states nearer the yrast line. Some new information was obtained from this experiment, but the use of a single detector rather than an array precluded coincidence measurements necessary for full development of a level scheme.

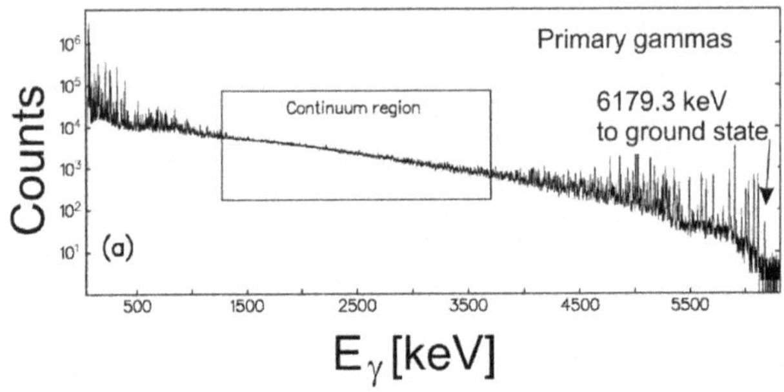

Figure 8: Gamma-ray spectrum obtained from ^{186}Re after thermal neutron capture, adapted from Ref. [22]. The "Continuum region" contains a very high density of unresolved gamma-ray peaks, while the higher-energy region contains "Primary gammas" due to transitions directly from the capture states to levels near the yrast line.

Figure 9 shows a partial level scheme deduced from (*n, 2n*) and (*d, 2n*) reaction studies [23, 24], which populate initial states in ^{186}Re at lower energies, but higher angular momentum (see Fig. 7). Among the other information, the energy of the 200,000 year isomer was determined with much greater precision at 148.2(5) keV [23], compared to the previous value of 149(7) keV. In addition, a number of new levels and transitions were added to the level scheme shown in Fig. 9, including several transitions to the isomer from higher-lying states. Such transitions would be a feature of depletion levels, as shown in Fig. 3. However, it would also be necessary for there to exist a cascade of transitions from a potential depletion level toward the ground state in order to release the isomer energy: These transitions were not yet observed, so that the states connected to the isomer in the right side of the level scheme of Fig. 9 have not exhibited any connection to those states feeding the ground state on the left side of the level scheme. This is likely just a matter of the sensitivity of the experiments, due to the detectors and reactions employed.

Future experiments are planned in the Rhenium Campaign, including coincidence measurements from (*n, γγ*) experiments using cold neutrons in collaboration with scientists at NIST. An isomeric sample is anticipated to become available during 2018 and first direct depletion tests are planned with that sample using bremsstrahlung. In addition, an initial study has been performed on ^{93}Nb and a full-scale Iridium Campaign was initiated. Finally, an entirely new experimental method is under development that would benefit from a laser-based source of pseudo-monochromatic photons in the hundreds of keV to MeV range.

Summary

So far, induced depletion of isomers has been experimentally demonstrated for six nuclides via different mechanisms: ^{180}Ta by bremsstrahlung [16-18, 26] and Coulomb excitation [27]; ^{68}Cu by Coulomb excitation [28]; ^{177}Lu [29] and ^{178}Hf[10] [31] by thermal neutrons; ^{108}Ag by bremsstrahlung [13]; and ^{93}Mo by the newly-observed process of nuclear excitation by electron capture [32]. The Army Research Laboratory and its subject matter experts have been instrumental in achieving four of these discoveries. All these measurements have shown that it is indeed possible to cause, upon demand, a release of stored nuclear energy and to greatly increase the power output for the affected nuclei. Spectroscopic measurements as described above are challenging, and even if performed over just a few days it typically takes a year or more to complete the mapping of states and transitions that reveals level schemes like that of Figs. 2b, 4 and 9. Plus mappings by different reactions are often required to provide the best characterization of potential depletion levels. New experimental techniques may also be required to provide all the needed information from which to understand the depletion mechanisms.

At the present, samples prepared to contain energy-storing isomeric nuclei are only available in a few cases. Thus, an additional challenge is to obtain such samples for the most interesting cases, in at least quantities sufficient for testing, meanwhile learning more about the process of isomer production. Overall, all these studies have steadily advanced the state-of-the-art in nuclear physics and the ability to assess the potential of radioisotopes and isomers for future applications ranging from power sources to directed gamma energy. The vision of isomers for power and energy applications is high-risk, but high-payoff, and research remains strictly at the fundamental level. At the same time, research into improved methods for conversion from radioisotope decay emission into electrical power is progressing at a steady pace and those technologies will be ready in the nearer term, not only enabling the use of standard radioisotopes but also eventually isomers for Army applications.

---

[10] Reports of depletion of ^{178m2}Hf induced by low-energy photons near 10 keV were not substantiated by detailed independent tests (see Ref. [30] and references therein).

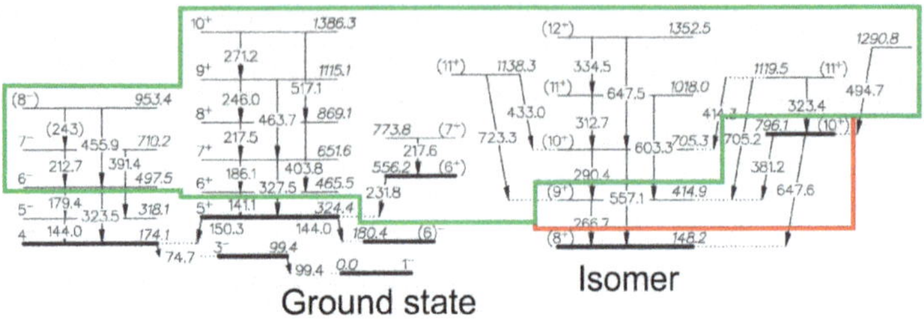

Figure 9: Partial level scheme for ^{186}Re, deduced from (n, 2n) and (d, 2n) reaction experiments, adapted from Ref. [24]. Level and transition energies are in keV. Those states boxed in red were first determined in the work of Ref. [23] while those boxed in green were first determined in the work of Ref. [24].

Apart from the research challenges, a strategic perspective might well spur questions like: "Why bother?" or "Why should the US Army perform research, and even provide scientific leadership, in this area?" To the first question is offered the response: Because radioisotopes represent the most likely way to access energy densities far beyond the chemical limit for Army applications, and isomers may provide a way to exert control over the release of this stored energy. The need to avoid technological surprise is also an important consideration. To the second question is offered the response: There is a wide range of studies in academia and DoE into interesting and esoteric properties of nuclei, e. g. nuclear halos and shape coexistence, plus industry and DoE clearly expend considerable efforts related to nuclear reactors. However, only the unique US Army expertise permits a focus on learning about energy-release mechanisms from isomers while leveraging expertise and facilities outside DoD to great advantage. In this context, these questions are really: "If not radioisotopes and isomers, then how can one access energy beyond the chemical limit?" and "If not the US Army, then who?"

Acknowledgments

The author wishes to thank Dr. Christopher J. Chiara, ORAU/ARL for valuable suggestions on the manuscript. The author also wishes to acknowledge the core ARL team (Drs. Marc Litz and Christophe Chiara, and John Russo and William Ray) and that many scientists have contributed to ARL research in this field from many institutions around the world, including: Air Force Institute of Technology, Argonne National Laboratory, Australian National University, Centre for Energy Research (Hungary),

Florida State University, Joint Institute for Nuclear Research (Russia), Lawrence Berkeley National Laboratory, Los Alamos National Laboratory, Michigan State University, National Centre for Nuclear Research (Poland), Nicolaus Copernicus University (Poland), Tata Institute for Fundamental Research (India), University of Osaka (Japan), University of Surrey (UK), US Naval Academy, and others.

References

[1] Energy density, https://en.wikipedia.org/wiki/Energy_density, 2017.

[2] N. E. Galushkin, N. N. Yazvinskaya, and D. N. Galushkin, arXiv:1312.6851v1 [cond-mat.mtrl-sci] (2013).

[3] R. P. Dias and I. F. Silvera, Science 355, 715 (2017). https://doi.org/10.1126/science.aal1579

[4] M. R. Khan, et al., Solid-State Elec. 136, 24 (2017). https://doi.org/10.1016/j.sse.2017.06.010

[5] J. Russo, W. Ray, and M. S. Litz, Appl. Energy 191, 10 (2017). https://doi.org/10.1016/j.apenergy.2017.01.029

[6] L. Olsen, P. Cabauy, and B. Elkind, Physics Today 65, 12 (2012). https://doi.org/10.1063/PT.3.1820

[7] Isotope beta-battery approaches for long-lived sensors: Technology review, M. Litz, (Army Research Laboratory, 2014), ARL-TR-7048.

[8] A. deShalit and H. Feshbach, Theoretical Nuclear Physics, Volume 1: Nuclear Structure, Ed. (Wiley, New York, 1990).

[9] Evaluated Nuclear Structure Data File (ENSDF), www.nndc.bnl.gov, National Nuclear Data Center, Brookhaven National Laboratory, 2018.

[10] B. Lehnert, et al., Phys. Rev. C 95, 044306 (2017). https://doi.org/10.1103/PhysRevC.95.044306

[11] P. M. Walker and G. D. Dracoulis, Nature 399, 35 (1999). https://doi.org/10.1038/19911

[12] P. M. Walker and J. J. Carroll, Nucl. Phys. News 17, 11 (2007). https://doi.org/10.1080/10506890701404206

[13] J. J. Carroll, et al., AIP Conf. Proc. 1525, 586 (2013). https://doi.org/10.1063/1.4802396

[14] G. D. Dracoulis, et al., Phys. Rev. C 58, 1444 (1998). https://doi.org/10.1103/PhysRevC.58.1444

[15] "Summarizing the nucleus", G. D. Dracoulis, in The Nucleus: New Physics for the New Millennium, edited by F. D. Smit, R. Lindsay and S. V. Förtsch (Kluwer, New York, 1999). https://doi.org/10.1007/978-1-4615-4257-5

[16] C. B. Collins, et al., Phys. Rev. C 37, 2267 (1988). https://doi.org/10.1103/PhysRevC.37.2267

[17] D. Belic, et al., Phys. Rev. Lett. 83, 5242 (1999). https://doi.org/10.1103/PhysRevLett.83.5242

[18] D. Belic, et al., Phys. Rev. C 65, 035801 (2002). https://doi.org/10.1103/PhysRevC.65.035801

[19] P. M. Walker, G. D. Dracoulis, and J. J. Carroll, Phys. Rev C 64, 061302 (2001). https://doi.org/10.1103/PhysRevC.64.061302

[20] J. Sethi, et al., J. Phys. G 43, 015103 (2015). https://doi.org/10.1088/0954-3899/43/1/015103

[21] J. Sethi, et al., Phys. Lett. B 725, 85 (2013). https://doi.org/10.1016/j.physletb.2013.06.054

[22] D. A. Matters, et al., Phys. Rev. C 93, 054319 (2016). https://doi.org/10.1103/PhysRevC.93.054319

[23] D. A. Matters, et al., Phys. Rev. C 92, 054304 (2015). https://doi.org/10.1103/PhysRevC.92.054304

[24] D. A. Matters, et al., Phys. Rev. C 96, 014318 (2017). https://doi.org/10.1103/PhysRevC.96.014318

[25] Nuclear structure of 186Re, D. A. Matters, Ph. D. Thesis, Air Force Institute of Technology (2017) AFIT-ENP-DS-16-D-017, www.dtic.mil/dtic/tr/fulltext/u2/1032026.pdf.

[26] C. B. Collins, et al., Phys. Rev. C 42, 1813 (1990). https://doi.org/10.1103/PhysRevC.42.R1813

[27] M. Schumann, et al., Phys. Rev. C 58, 1790 (1998). https://doi.org/10.1103/PhysRevC.58.1790

[28] G. Georgiev, et al., Intl. J. Modern Physics E 15, 1505 (2006). https://doi.org/10.1142/S0218301306005095

[29] O. Roig, et al., Phys. Rev. C 83, 064617 (2011). https://doi.org/10.1103/PhysRevC.83.064617

[30] J. J. Carroll, et al., Phys. Lett. B 679, 203 (2009). https://doi.org/10.1016/j.physletb.2009.07.025

[31] S. A. Karamian and J. J. Carroll, Phys. Rev. C 83, 024604 (2011). https://doi.org/10.1103/PhysRevC.83.024604

[32] C. J. Chiara, et al., Nature 554, 216 (2018). https://doi.org/10.1038/nature25483

Chapter 13

Microcombustion of Heavy Fuels for Multifuel Portable Power Generation

C.M. Waits[1]*, S. Karnani[2]
[1]US Army Research Laboratory, Adelphi, MD
[2]General Technical Services
*christopher.m.waits.civ@mail.mil

Abstract

Microreactors that combine heterogeneous and homogeneous reaction for combustion of heavy fuels are studied as efficient heat sources in thermal to electric converters for high specific energy Soldier wearable power sources. Confinement and fuel decomposition are found to dominate reactor operation in hybrid microreactors of *n*-dodecane. Specifically, confinement and *n*-dodecane decompoisition into lighter species increases heat and mass transport to the catalyst wall and overall reactor temperature. Conditions are described to acheive complete conversion at high surface temperatures and heat loss conditions relevant to thermophotovoltaic energy conversion. Reactor design, modeling, and materials are discussed as well as necessary next steps to extend Soldier's mission durations and reduced their physical burden due to power.

Keywords

Microcombustion, Homogeneous-Heterogeneous Combustion, Portable Power Conversion, Soldier Power

Contents

1. Introduction

Compact power sources, capable of storing and delivering large amounts of energy, are critical for numerous military applications: from providing Soldier power for dismounted operations – with limited to no resupply for multiple days – to unmanned air vehicles (UAVs) operating for hours at a time. Today, batteries are the only energy source used on the battlefield that can deliver power in the 10-100 Watt range. The weight burden associated with Soldier power and UAV run-time is directly tied to a power source's energy density, which, for state-of-the-art rechargeable lithium-ion batteries [1], is limited to <150 W·h/kg. A dismounted Infantryman or scout deployed in Afghanistan in 2013, carried an average of 9.7 lbs. of batteries with an expectation of increasing to more than 14 lbs. for a 72-hour mission if every Soldier is connected to the communications network [2].

An alternative strategy might be realized through the use of high energy density liquid fuels. In terms of storage, on a per-weight basis, JP-8 (11.9×10^3 W·h/kg) greatly exceeds batteries. A JP-8 fueled power source, with a 10% system efficiency, weighing 1 kg without fuel, and providing 20-W electric for 72-hours would be capable of providing >600 W·h/kg, thereby reducing a Soldier's weight burden for power by 4x compared to a 150 W·h/kg battery. Fuel cell technologies are capable of high conversion efficiencies and power density, however, they require special fuels like hydrogen or methanol. Thermal-to-electrical conversion (TEC) technologies that rely on fuel flexible burners as a heat source are prime candidates albeit at lower conversion efficiencies. Thermophotovoltaics (TPVs), a class of TEC where photons emitted from a heated surface are converted directly to electrons – via a photovoltaic semi-conductor – in practice, have demonstrated total system efficiencies exceeding 20% [3-7] with theoretical efficiencies >50% for thermal emitter temperatures exceeding 1273K [8]. In addition to logistical fuels, such as the Army's use of JP-8, TECs open the possibility of multi-fuel operation, which would provide rapid re-charging and greatly expand the sourcing capability of the fuel.

A combustion-based heat source for a TEC would have the following components: (1) a liquid fuel vaporizer or atomizer; (2) a fuel and air mixer; (3) a micro- and meso-scale combustion reactor (microreactor) to convert the fuel's stored chemical energy into heat delivered to a TEC element; and (4) a heat exchanger to recuperate exhausted sensible heat from gas products to heat cold incoming reactants. A large body of knowledge exists on heat exchangers, reactant mixing, and approaches to vaporize/atomize fuel. For portable power applications, a combustion-based TEC system must be lightweight and highly compact – to achieve high specific energy density – and capable of complete fuel conversion. These requirements pose a challenge for microreactors due to high heat losses, high flow velocities, and short residence times that can cause flame extinction, quenching, and blowout.

A microreactor is a complex component that couples transport, thermal management, and temperature sensitive chemistry capable of operating within a small stability window. Our studies, described in the subsequent sections of this chapter, utilize pre-heated air, vaporization, and mixing to fill existing knowledge gaps associated with highly compact reactors in order to accomplish the following aim: To determine the performance limits, material suitability, fuel flexibility, and required operating conditions to efficiently convert the immense chemical energy found in a liquid hydrocarbon to usable electrical energy, in a lightweight, compact microreactor to support the dismounted Soldier.

2. Major Concepts

The most effective means of harvesting thermal energy for TEC, while achieving stable and complete combustion, with modest temperature gradients at high temperatures and flow velocities, is not clear. The reason is due to a competition between supporting endothermic chain initiating, propagating, and branching reactions with the removal of energy from the reactor in order to produce electricity. Although liquid hydrocarbon fuels are energy dense, significant activation energy barriers must be overcome to initiate and sustain the radical pool responsible for combustion's characteristic exothermicity. The simultaneous release and abstraction of energy combined with the high-activation energies required during gas-phase combustion can lead to extinction as nearby heat transfer surfaces, both, thermally and radically quench the system chemistry [9-14]. High flow velocities – exceeding laminar flame speeds by up to an order of magnitude – and small reactor dimensions – to maximize heat loss, which is an important requirement in any TEC application – can also lead to blowout (30-40 cm/s is typical for simple hydrocarbons at stoichiometric conditions [15]). Accommodating both issues underscores the need to consider catalyst support, to encourage chemistry, albeit, at a lower activation energy, or recirculating heat from combusted products to increase reactivity. The activation energy of a stoichiometric propane-air mixture in the gas phase is more than 5 times greater than when platinum, a common combustion catalyst, is present [16]. However, it is unlikely that a fuel can fully react at high flow velocities and short residence times via surface chemistry alone.

As shown in Figure 1 peak temperatures are much lower for heterogeneous operation versus homogeneous operation, thus global reaction rates based on heterogeneous chemistry are slower than gas-phase rates, a reasonable conclusion if we assume an Arrhenius-type form can be used to describe global rates:

$$k = Ae^{-\frac{E_a}{k_B T}} \tag{1}$$

where A is kinetic pre-exponential factor, E_a is the activation energy of the reaction, k_B is Boltzmann's constant, and T is temperature. The reaction rate is largely determined by the activation energy and the temperature. For a given temperature, generally, homogeneous reactions rates are much faster than a heterogeneous equivalent; however, heterogeneous reactions can be sustained at lower temperatures.

Figure 1 demonstrates the small operating window for homogeneous chemistry when using a similar reactor geometry fueled with propane and air. When comparing the range of inlet velocities for a system that relies solely on gas-phase chemistry to one that includes surface (heterogeneous) support, we find a noticeable extension of the blowout limit and a lowering of the extinction limit.

In addition to slower reaction rates at the surface, the overall conversion becomes limited by transport time scales at high flow velocities and globally fast reaction rates:

$$\tau_D \propto \frac{h^2}{D} \tag{2}$$

where h is the minimum dimension of the combustion volume and D is the reactant diffusivity, which is a strong function of the reactant species and temperature. If diffusion times, τ_D, exceed reactant residence times, τ_R, then fuel breakthrough and/or high CO concentration in the exhaust is possible. Surface area to volume enhancements, such as high aspect ratio channels can improve transport to surfaces; however, the increased resistance to the flow would affect the required pumping power and potentially compromise the total system performance.

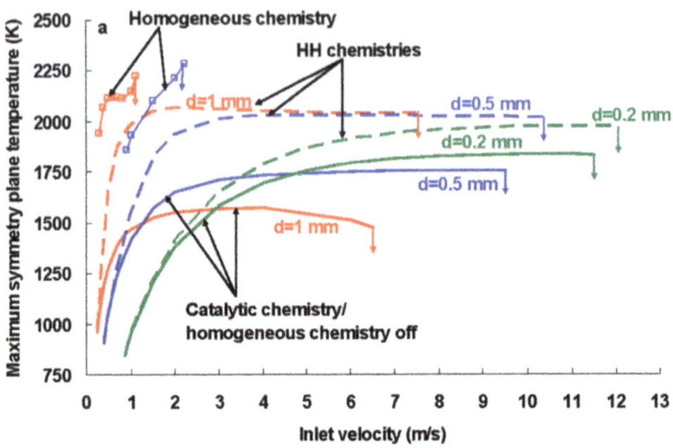

Figure 1. Comparison of extinction and blowout limits of propane for homogenous, heterogeneous, and combined homogeneous and heterogeneous reactions. Adapted from [17].

Heat loss from the reactor surfaces is a major factor in sustaining stable and complete combustion. The heat loss due to the TEC can be 5 to 15 W/cm^2 which acts to lower overall reaction temperatures thereby affecting both reaction rate and species transport. Thermal management is critical to maintaining operating temperatures that can sustain stable and complete combustion. Heat recirculation, or excess enthalpy reactors, and catalyst assisted gas-phase combustion are two approaches to overcome the limitations of heterogeneous dominated or homogeneous only combustion. ARL has examined catalytic supported gas-phase combustion extensively as well as the addition of heat recirculation in micro-channel reactors.

2.1 Hybrid homogeneous-homogeneous combustion

In the context of microreactors, catalytic assisted combustion or hybrid Heterogeneous-Homogeneous (HH) operation may be achieved by adding a catalyst layer to a reactor wall with a level of confinement comparable to the flame quenching distance (~1 mm). Catalyst-supported heterogeneous chemistry offers low temperature ignition and inherent stability against extinction while extending the blowout limits of the gas-phase reactions. Conversely, the homogeneous reactions increase conversion rates, thereby preventing incomplete combustion [17-20]. By coupling the best features of each combustion mode, the reactor can achieve stable operation over a wide range of heat loss conditions and inlet velocities (Figure 1). Furthermore, wall temperatures in HH combustion are lower than those encountered in purely homogeneous combustion. By avoiding the extreme temperatures characteristic of purely homogenous combustion, NO_x production is greatly reduced and devices need not rely on costly exotic materials such as rare earth oxides [5].

However, with catalyst present, homogeneous reactions may be weakened as radicals are adsorbed on the catalyst surface [21]. Fortunately, vigorous preheating of the gas from surface reactions acts to intensify homogeneous chemistry. In short contact time reactors operating near atmospheric pressure, homogeneous reactions play an important role at moderate temperatures (~1000 K) by decomposing fuel into lighter intermediates, which can diffuse quickly to catalyst walls, where surface chemistry provides flame stability. Ideally, the net effect for HH reactors is that surface reactions act to stabilize a weaker flame [21-23].

To date, research on small scale (gap height, $g \approx 0.3$-2 mm) HH combustors have focused mainly on hydrogen, methane and propane; however, the future Army will require current and future power systems to operate using energy dense liquid fuels as mission times extend and baseline power demands increase [24]. Diffusive transport plays a critical role in determining stability in HH reactors, especially at high inlet velocities. Either confinement must increase, or, heavy fuels must break into smaller, faster diffusing, species in order to overcome transport limitations at low residence times. Our efforts have examined how the use of heavy fuels, specifically *n*-dodecane – a sulfur free JP-8 surrogate – in HH combustors can act to lengthen the stable window of operation in a way that is amenable to next generation thermal-to-electric energy conversion.

2.2 Heat recirculation

Excess enthalpy burners, or heat-recirculating burners are an important paradigm to consider for promoting stability in microburners losing significant amounts of heat [25]. By utilizing the enthalpy of the burner's exhaust stream to pre-heat incoming reactants, excess enthalpy burners produce reaction temperatures in excess of the incoming mixture's adiabatic flame temperature [25-28]. Both HH operation and heat recirculating concepts are important for stabilizing combustion and improving efficiency by driving reactions toward completion and maximizing energy extraction.

3. Approach

Experimental reactors and models suitable for combustion with liquid fuels were developed to evaluate materials, inlet conditions, and geometry on HH operation. As previously described, reactor performance is governed by the time and temperature history of reactants passing through the reactor. Heat loss from the reactor to the TEC element (desired) and to the environment (undesired), both, have a large effect on the temperature, which thereby influences chemical kinetics, heat release rates and, finally, the residence time required for complete heat release within the system. Heat transfer and species transport between the gaseous species and the wall surfaces are determined, primarily, by the level of confinement, or gap height between adjacent surfaces. Therefore, in the presented researches, we present the findings of two joint experimental/numerical efforts that focused on (1) Evaluating the interaction between surface support and gas-phase chemistry in an insulated reactor with varying levels of confinement, and (2) Examining the role of heat loss on combustion products and temperature. We focus on radiative heat loss due to our interest in TPV energy conversion; however, this approach can be readily used to design reactors for any TEC element by accounting for the relevant heat loss mechanisms.

The following performance parameters are tracked during experiments: chemical conversion of reactants into CO_2, water, CO, and surface temperature as a function of fuel input. In addition to combustion performance, other parameters and properties, including temperature uniformity, total energy balance, physical limitations of the burner and catalyst materials, and, finally, the chemical stability of the catalyst against different fuels, especially those containing sulfurous compounds affect system performance and reactor lifetime. The latter has been a focus of parallel efforts within ARL.

Recently, our efforts have culminated in the development and refinement of a sophisticated design tool that allows the evaluation of multiple configurations, materials, heat recirculation, and the impact of simultaneous heat release and abstraction on combustion.

3.1 Experiments

3.1.1 Fully insulated Parallel Plate Reactor

The fully insulated reactor consists of two thin stainless steel plates (0.9 mm thick) separated by an interchangeable stainless steel gasket and follows the approach be Norton *et al.* [29]. A schematic and photograph of the reactor are shown in Figure 2 and Figure 3, respectively. The reaction zone consists of catalyst-coated alumina inserts separated by thin strips of compressed alumina insulation. The reaction zone is 5 cm long and 1 cm wide; its height is governed by the gasket thickness, which allows different gap heights to be evaluated. A more complete description is available in the literature [30].

Figure 2. (Top) a lengthwise view of the reactor. (Bottom, not to scale) a cross-section

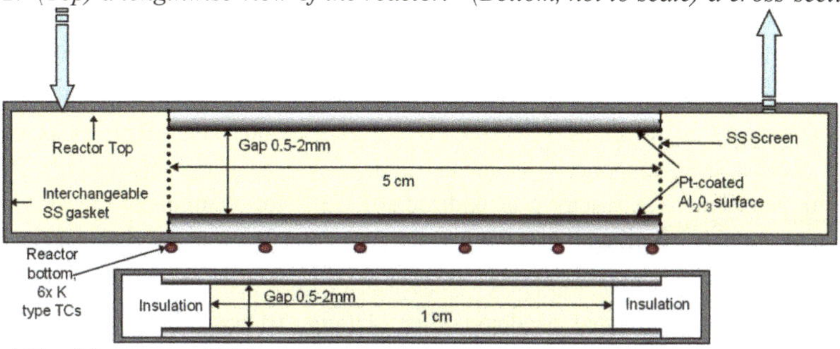

of the width of the reactor [30].

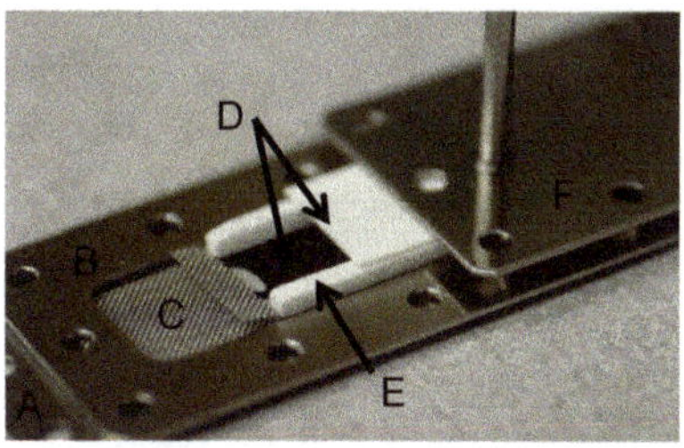

Figure 3. An exploded view of the reactor. A. Reactor bottom. B. Interchangeable height-controlling gasket. C. Stainless steel mesh. D. Catalytic inserts. E. Alumina foam insulation. F. Reactor top [30].

The temperature along the length of the reactor is monitored by six K-type thermocouples welded, in 10 mm intervals, along the center of the bottom reactor plate. The reactor assembly is sandwiched between two 5 cm thick needled silica mats (AVSil AB100S) wrapped in aluminum tape for insulation.

Three equivalence ratios (φ=0.5, 0.7, 0.9) and three gap heights (g =0.5, 1.0, 2.0 mm) were tested at various volumetric flow rates (from 1.5 to 3.0 SLPM). This corresponds to chemical enthalpy input rates from 43 to 156 W and inlet velocities ranging from 180 to 1410 cm/s (calculated at T_{in}=423 K). Additional details can be found in Table 1 of [30].

3.1.2 High Heat Loss Slot Reactor

The high heat loss reactor has a square slot configuration (20 mm x 20 mm) and was constructed using 1 mm thick plates of Hastelloy X (Figure 4). The internal reactor walls are coated with a layer of 5% wt. Pt on alumina (Aldrich). Fiberglass insulation jackets the inlet portion of the reactor in order to minimize heat loss and prevent fuel condensation. A more complete description is available in the literature [31].

Thermocouples at locations TC1 and TC2 (Figure 4) measure temperatures along the inlet section, and an optical pyrometer (Fluke MI3) measures temperature along the reactor centerline. The reactor was oxidized at 1373 K for 30 minutes in order to develop a stable oxide coating, ensuring a uniform surface emissivity for pyrometer measurements.

A camera with a mounted bandpass filter (430nm, 10 nm FWHM, Edmund Optics) placed above the reactor captures CH* chemiluminescence images within the reactor slot. A gas chromatograph measuring batch samples at the center of the reactor exit reports the presence and concentration of permanent gas species. The equivalence ratio is φ=0.9 for all tested conditions. The total fuel/air flow rates reported are 2.0, 3.0 and 4.0 SLPM, which correspond to chemical energy input rates of 104, 156 and 208 W, respectively.

Both the parallel plate reactor and slot reactor experiments use the same fluid controls, vaporization, pre-heat temperatures, and ignition. Thermal mass flow controllers maintain, both, dry air and propane flow rates. A syringe pump delivers *n*-dodecane (Aldrich) into a heated air stream to vaporize the fuel. The temperature of the fuel/air mixture is fixed at 423 K. To ignite the system, the reactor is heated with a butane torch until self-sustaining operation is possible.

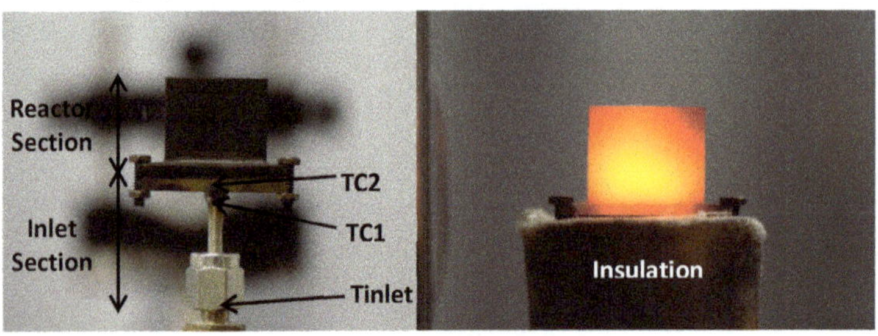

Figure 4. Photograph of the combustor not operating (left) and operating (2 SLPM, 104 W input) with insulation covering the inlet (right) adapted from [31].

4. Major findings

4.1 Fully Insulated Parallel Plate Reactor

4.1.1 Experiments

The temperatures measured along the length of the reactor as a function of equivalence, flow rate, and increasing levels of confinement are presented in Figure 5 and Figure 6. Self-sustained operation was achieved with catalyst applied to the surface while self-sustained operation with catalyst-free inserts was unsuccessful.

Under the least confined conditions tested (g=2 mm, Figure 5), the mass transport time is dominated by the diffusion of *n*-dodecane, which is much larger than the residence time of this reactor, $\tau_D/\tau_{residence}$~10. Wall temperatures are nearly independent of flow rate suggesting reactor operation is dominated by surface reactions. Wall temperatures increase with increasing equivalence ratio, i.e., increasing fuel concentration and power input. The temperature conditions correspond to the low temperature oxidation (LTO) regime (~500-700 K based on the behavior of similar n-alkanes at 1 atm); however, given the short residence times, both fuel conversion and heat release from LTO is small [32, 33], thus, homogeneous reaction rates when g=2 mm are negligible.

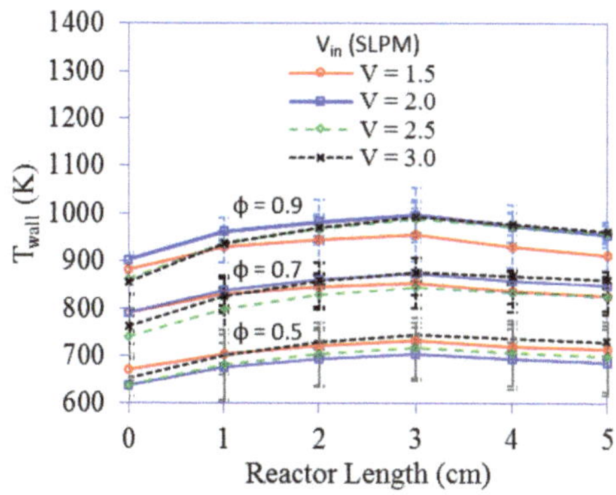

Figure 5. Location dependent wall temperature for the reactor g=2 mm for ϕ=0.5, 0.7, and 0.9 and $\dot{V}$=1.5, 2.0, 2.5, and 3.0 SLPM. Adapted from [30].

As confinement increases (g=1 mm, Figure 6a), diffusion times, τ_D, decrease. Reactor wall temperatures are nearly independent of flow at low power inputs (ϕ=0.5), suggesting the reactor operation is dominated by surface reactions similar to what was overserved in the g=2 mm reactor. Wall temperatures increase significantly at ϕ=0.7. The reactor reached two steady states for flow rates of 2.0 and 2.5 SLPM (the steady state temperature approached from a higher steady state temperature operation is plotted in Figure 6a). The reactor wall temperatures are significantly lower at the highest flow rate (3.0 SLPM) suggesting a blowout of the homogeneous reactions. The multiple steady states observed and the apparent blowout of homogeneous reactions at 3.0 SLPM suggest that homogeneous chemistry plays an important role in this reactor configuration. Furthermore, the jump in wall temperatures from ~1000-1100 K is in agreement with initiation of gas-phase reactions. At ϕ=0.9, the reactor wall temperatures are high and a consistent increase with flow rates and power input.

At the highest level of confinement evaluated (g=0.5 mm, Figure 6b), the centerline-to-wall diffusion time, τ_D, is the shortest. The reactor wall temperature increases as the fuel input increases (increasing ϕ) and constant flow rate, thus increasing input power. Over the first 1 cm, the reactor wall temperatures are flow rate independent for ϕ=0.5, suggesting that homogeneous reactions are limited in this region. At flow rates of 2.5 and 3.0 SLPM, when ϕ=0.7, temperatures increase significantly (> 1100 K) into a range where homogeneous reactions are expected to be more important and reactor operation is no longer dominated by only surface reactions. At ϕ=0.9, maximum temperatures and temperature gradients are near the reactor inlet.

Propane was compared to *n*-dodecane for ϕ = 0.9 to observe differences that might be attributed to fuel properties (Figure 7) [34]. Generally, our findings show that surface chemistry plays a smaller role and favors lighter species as confinement decreases, which is consistent with the literature [18]. The drop in reactor temperature at the largest flow rates tested (3 SLPM), however, suggests that homogeneous reactions are susceptible to blowout at high velocities. In the case of intermediate confinement, g=1 mm, reactor temperatures are in good agreement for both *n*-dodecane and propane – a desired behavior for fuel flexible operation.

In the case of g=0.5 mm, it is remarkable that, both, the peak and mean temperatures, are greater when the reactor is fueled by *n*-dodecane than with propane. Since propane is a smaller molecule and has a greater diffusivity than *n*-dodecane, it would be expected that propane is more readily transported to the reactor walls and, therefore, leads to higher reactor temperatures. Numerical modelling suggests that a transport-limited reaction involving ethylene, the most abundant intermediate of *n*-dodecane decomposition [35], can account for this rapid temperature rise near the reactor inlet. The rapid coupling between surface and gas-phase chemistry, and temperature gradients, seen in both modelling and experiments suggests that there may be a lower limit to the level of confinement for a HH reactor to act fuel flexibly.

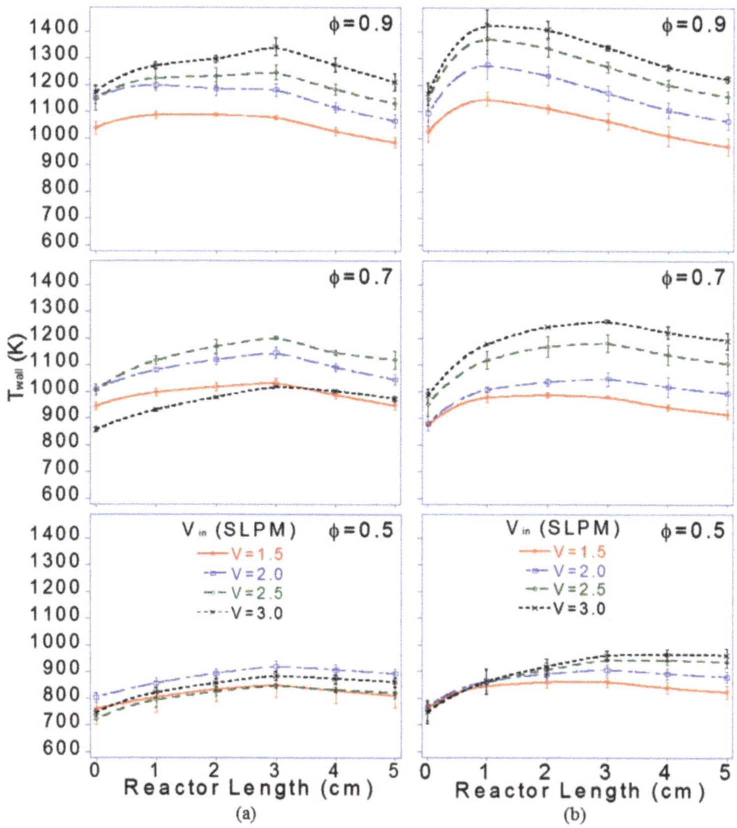

Figure 6. Location dependent wall temperature for $\phi = 0.5$, 0.7, and 0.9 and $\dot{V}$=1.5, 2.0, 2.5, and 3.0 SLPM for the reactor (a) g = 1 mm and (b) g = 0.5 mm. Adapted from [30].

4.1.2 Numerical modeling

The thermal characteristics of the parallel plate reactor under transport limited reactions were modeled using the commercial CFD software, FLUENT [36]. A fast global reaction at the catalytic surface is used in the software to achieve transport limited operation. Full details of the model can be found in [30]. The reactor walls is 0.65 mm thick with a constant thermal conductivity k=20 W/m/K consistent with the thermal resistance of the stainless steel/alumina reactor walls over the range of temperatures of interest. A heat loss coefficient of h=40 W/m2/K is applied to the outer walls and the ends of the reactor walls are assumed to be well insulated. A uniform velocity boundary condition is prescribed at the reactor inlet and the temperature is set to 423 K as in the experiment. At the outlet, pressure is set to 1 atm.

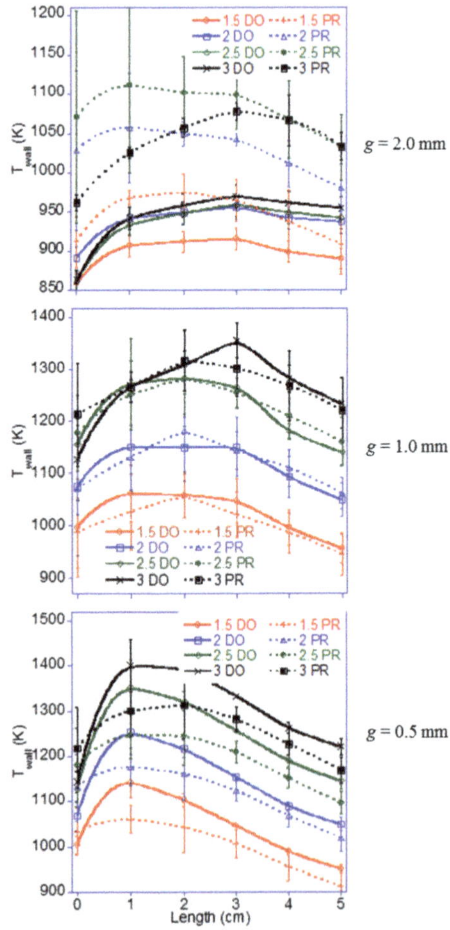

Figure 7. Results for g=2.0 mm, 1.0 mm, and 0.5 mm from top to bottom respectively for propane (PR) and n-dodecane (DO) with ϕ = 0.9 and flow rates of 1.5, 2.0, 2.5, and 3.0 SLPM [34].

Figure 8 shows numerical results for two cases that are intended to qualitatively illustrate the two observed operating regimes for *n*-dodecane: (1) at higher temperatures, which are governed by the mass transport of ethylene $\tau_D/\tau_{residence}$~0.1 and (2) at lower temperatures where the mass transport of *n*-dodecane $\tau_D/\tau_{residence}$~10 is a limiting factor. Under low temperature conditions (2), gas-phase reaction rates are negligible and the reactor is governed either by the mass transport of fuel to the catalyst surface, or the subsequent reactions at the catalyst surface; in the higher temperature case, (1), alkanes are expected to readily decompose into smaller fragments through homogeneous reactions, with ethylene being the most abundant. The characteristic diffusion time in the low temperature case (2) is larger than the available residence time, hence fuel conversion is low and reactor temperatures are nearly indistinguishable. At higher temperatures (1), the residence time is sufficient to fully oxidize the fuel, and wall temperatures are dominated by fast transport-limited reactions near the reactor inlet. Qualitatively, both regimes exhibit similar behaviors controlled by transport rates, temperatures, and heat transfer between the walls and the gas. Condition 2 was observed for a g = 2.0 mm channel gap, while it was only observed for the lowest thermal inputs when g = 1.0 mm. Condition 1 was observed for the larger thermal inputs of g = 1.0 mm and all of the input conditions when g = 0.5 mm.

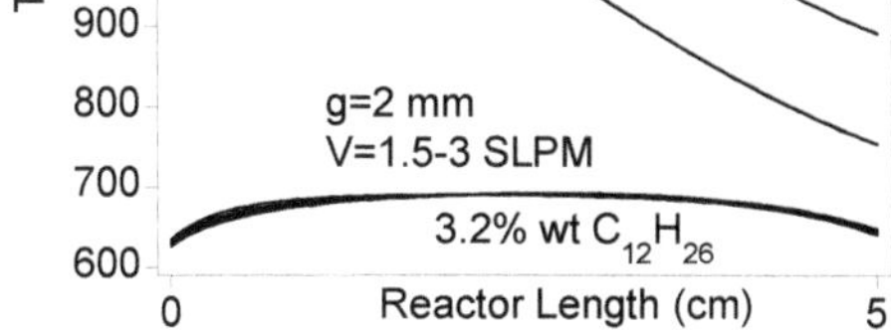

Figure 8. Numerical results for two transport limited reactor operation: (nominal 1) a high temperature reactor, governed by the mass transport of ethylene $\tau_D/\tau_{residence}$~0.1 and (nominal 2) a low temperature reactor limited by mass transport of n-dodecane $\tau_D/\tau_{residence}$~10 [30].

The influence of confinement and gas-phase reactions on the reactor's temperature and temperature uniformity is investigated using a boundary layer model of the fully insulated parallel plate reactor for select conditions (g = 0.5 and 1 mm). The model incorporates Cantera [37], an open-source chemical kinetics library, to evaluate *n*-dodecane oxidation kinetic rates and transport [35, 38]. Due to an incomplete understanding of surface chemistry coupled with gas-phase kinetics, we do not attempt a complete model of the chemistry occurring in the reactor. However, some understanding of gas-phase processes, specifically fuel decomposition near the inlet, can be gained from a homogeneous model. For example, rapid gas-phase decomposition of *n*-dodecane, at the high wall temperatures examined, results in low *n*-dodecane mass fractions. Therefore, we suspect the homogeneous model can provide a reasonable approximation of *n*-dodecane decomposition. It should be noted that, however, radical absorption/desorption at the catalytic surfaces may have a large effect on gas-phase intermediates and products. We restricted our analysis to conditions where unburnt gas is rapidly heated (O[ms]) from 423 K to 1100 K—similar to premixed flame conditions, which our selected mechanism is designed to model.

The extreme temperature gradients, seen in the highest temperature case in Figure 6, ($\dot{V}$=3.0 SLPM, g=0.5 mm, ϕ=0.9) suggests very intense chemistry near the reactor inlet; whereas milder, more desirable gradients are observed in the less confined, but identical power input ($\dot{V}$=2.5, 3.0 SLPM, g=1 mm, ϕ=0.9), cases suggest that fuel consumption is more gradual. Indeed, modeling results (Figure 9) display similar trends: the rapid decomposition of *n*-dodecane near the inlet of the reactor for more confined conditions ($\dot{V}$=3.0 SLPM, g=0.5 mm, ϕ=0.9); and a more gradual fuel decomposition, extending well into the reactor, when less confined ($\dot{V}$=3.0 SLPM, g=1 mm, ϕ=0.9). In each case, ethylene is the dominant intermediate. Furthermore, the homogeneous model does not predict vigorous exothermicity until near the middle of the reactor where ethylene is rapidly consumed. Therefore, we can expect heat release near the reactor inlet, as suggested by CFD, to depend on catalytic surface reactions.

The contrast between Figure 5 and Figure 6 and the modeling results of Figure 8 and Figure 9 support that reactor confinement is a key factor in determining performance: where a highly confined reactor enhances thermal transport (in the axial and transverse directions) and mass transfer, which intensifies gas-phase reactions, leading to extreme temperatures and gradients.

4.2 High Heat Loss Slot Reactor

Complete oxidation was achieved in a 0.5-mm gap slot reactor, while incomplete oxidation was observed for a 1-mm gap slot reactor. Temperatures along the centerline of a 0.5 mm gap slot reactor for three different *n*-dodecane/air (ϕ=0.9) volumetric flow rates are shown in Figure 10. Along the inlet section (denoted as the negative values along the x-axis in Figure 10), temperature is nearly independent of flow rate because of limited gas-phase heat release and negligible back conduction. Temperatures in the inlet section are also much lower when compared to the fully insulated 2.0-mm gap parallel plate reactor. In addition, unoxidized fuel fragments were observed in a 1.0-mm gap slot reactor exhaust, suggesting an increase in confinement is needed to account for the higher heat losses and shorter residence times of the slot reactor design . Model results suggest and experimental chemiluminescence photographs confirm (Figure 10b) that gas-phase reaction intensity increases with the system's input chemical energy.

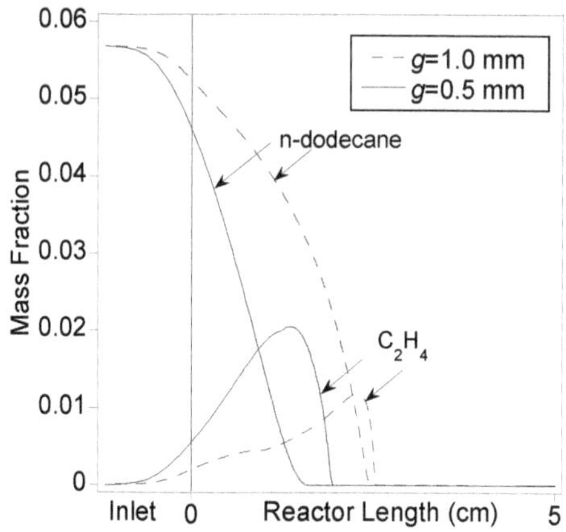

Figure 9. Mass fractions (averaged in the transverse direction) of n-dodecane and ethylene for $\dot{V}$=3.0 SLPM, g=0.5 mm, ϕ=0.9 and $\dot{V}$=3.0 SLPM, g=1 mm, ϕ=0.9 [30].

Experimental analysis showed that it would be possible to operate the catalyst supported reactor under large heat losses with near-complete oxidation of heavy hydrocarbons without the presence of a vigorous flame. Since large alkane fuels rapidly decompose at high temperatures (>1100 K), appreciable amounts of these small hydrocarbon intermediates — with ethylene being among the most abundant — and CO would be expected when oxidation is incomplete. Based on the results of the GCMS analysis, we conclude that oxidation is nearly complete in the 0.5-mm gap slot reactor.

A boundary layer fluid model, similar to the one used to model the insulated reactor, combined with gas-phase chemistry and a modified surface reaction model provides a detailed understanding of the high heat loss slot reactor. The gas phase chemical mechanism is capable of capturing low and high temperature chemistry [39].

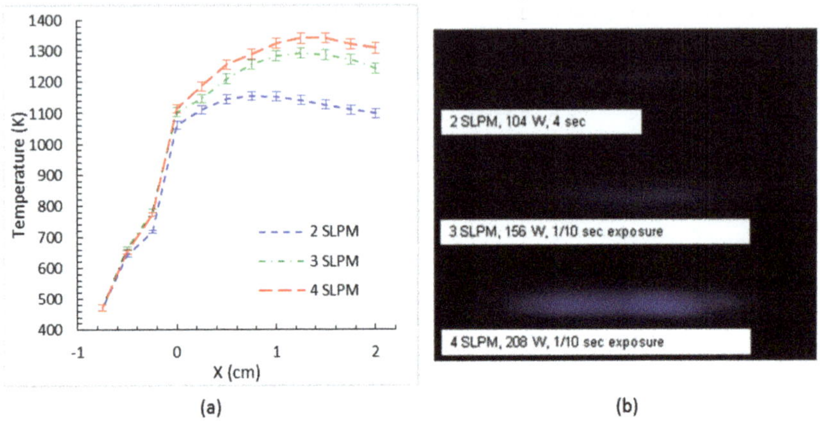

Figure 10. (a) Experimentally measured wall temperatures along the inlet section and the combustor centerline in a 0.5 mm gap slot reactor and (b) chemiluminescence images taken from above the reactor, looking into the reactor channel. Adapted from [31].

A surface chemistry model was developed and combined with the gas phase chemistry to obtain reasonable estimates of reactant consumption, and, therefore, heat release, of major species at the catalytic surface. The model uses a hierarchical approach to describe chemistry at the catalyst surface: using detailed rates to describe CH_4 oxidation on Pt and treating larger, carbon containing, species as lumped irreversible reactions with kinetic rates described by a relatively large sticking coefficient and first order dependence on [Pt(s)]. The CH_4 chemistry is a subset of the Zerkle mechanism [40] with CO reaction rates updated using oxidation on Pt rates by Deutschmann [41]. Lumped reactions for C_2 and larger carbon containing species are described by:

$$C_xH_yO_z + (x+y+z)[Pt] \rightarrow x[C] + y[H] + z[O] \tag{3}$$

and the reaction rates are described by:

$$\dot{s}_k = \frac{\gamma}{\Gamma}\sqrt{\frac{RT}{2\pi W_k}}[Pt][k], \gamma = 0.01 \tag{4}$$

where γ is the sticking coefficient, Γ is the total surface site concentration (2.72×10^{-9} mol/cm^2) and k represents the gas phase species. The surface model was tested numerically using a lumped parameter model (e.g., [42]) with mass transport coefficients [43] empirically determined by a series of catalytic foam experiments (e.g., [44]), which provided temperature profiles and exhaust gas compositions from lean *n*-dodecane combustion. The order of magnitude of the sticking coefficient was adjusted until the numerical model agreed with experimental results. Numerically determined species profiles suggest that surface reactions in the foam catalytic reactor quickly become transport limited at high temperatures (> 920 K). Details of the model and the catalytic foam experiment can be found in the supplementary data of [31].

Figure 11 shows predicted heat release rates (HRR) along a reactor length with varying degrees of confinement. Both, surface heat release rates and gas phase fuel consumption rates (shown in [31]), increase due to increased confinement and through heat transfer from exothermic surface heat release to the gas phase. This combination of effects results in more complete oxidation and therefore, greater combustor efficiency. Rapid gas phase consumption of fuel occurs at temperatures greater than 1200 K (Figure 12). At lower temperatures, the reduced fuel consumption rates in the gas phase suggest significant fuel breakthrough for fast reactant flow rates.

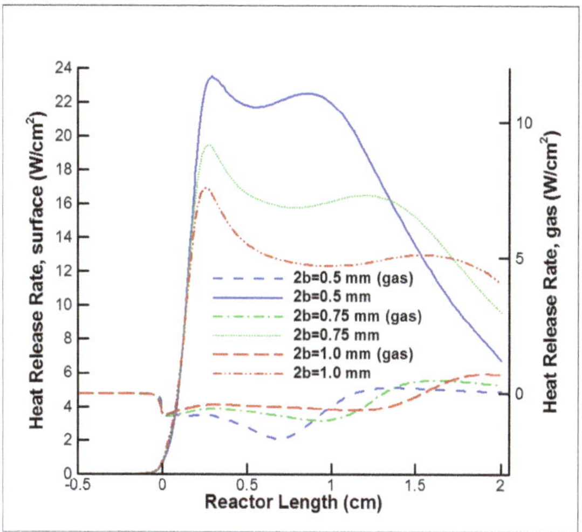

Figure 11. Predicted HRR from surface and gas phase reactions along the length of the reactor as a function of gap height ($\dot{V}$ =3 SLPM, Twall = 1300 K)[31].

In all cases, rapid heat release, due to surface chemistry, is observed near the reactor inlet. As surface kinetics deplete any available fuel, heat release decreases. Subsequent high temperatures and high flow rates in the gas-phase promote the rapid decomposition of reactants, which fuels a second heat release zone near the surface. Gas-phase reactions are primarily endothermic for all cases examined, aiding fuel decomposition and radical formation. Small positive heat release rates downstream in the reactor suggest mild gas-phase oxidation.

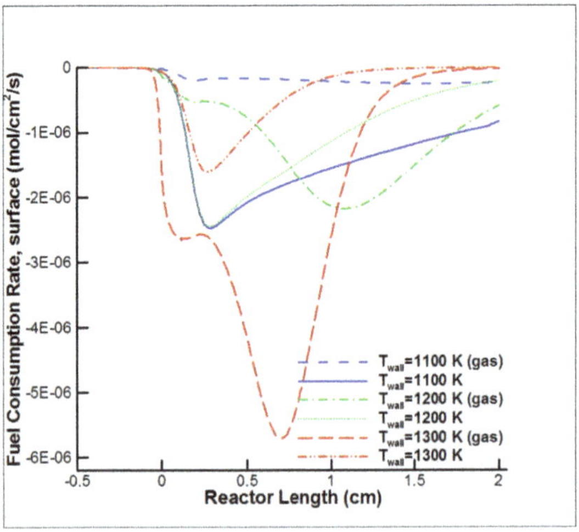

Figure 12. Predicted surface and gas phase fuel consumption along the length of the reactor as a function of wall temperature ($\dot{V}$ =3 SLPM, 2b=0.5 mm) [31].

5. Discussion

We have extended the understanding of liquid-fueled microreactors that rely on coupled surface and gas-phase chemistry for stability. Previous experimental and numerical studies focused on propane and lighter hydrocarbon fuels. In those studies, fuel transport and surface chemistry played a larger role than fuel reactivity at short residence times [19]. That is, fuels with higher diffusivities (e.g. methane versus propane) have a larger stability window near blowout. With heavier fuels, the intermediates produced during fuel decomposition play a large role in determining reactor stability. The highest confinement case examined in Figure 7, which showed higher reactor wall temperatures for *n*-dodecane instead of propane indicates heavy fuels may have a larger stability window with respect to blowout, owing to the diffusivity of intermediate fuel fragments.

Further studies with heavy hydrocarbons and the role of fuel intermediates is a necessary next step to optimize microreactors thermal characteristics.

Confinement greatly impacts thermal and mass transport and, therefore, the intensity of heterogeneous and gas-phase reactions. Highly confined conditions lead to rapid gas-phase reaction rates fueled by heterogeneous reactions and, thus extreme temperatures and temperature gradients, which have been observed in propane-fueled systems [18, 45, 46]. For heavier hydrocarbons, however, intermediate combustion products govern stability limits and temperatures. Reducing confinement, on the other hand, leads to insignificant gas-phase reaction rates. In this case, *n*-dodecane transport and reactivity are the governing properties. This is demonstrated by the lower wall temperatures in Figure 7 for *n*-dodecane when compared to propane for the 2-mm gap.

Confinement is a design parameter that can be used to achieve stable and complete combustion under high heat loss conditions or minimize the differences different fuels have on wall temperature. The high heat loss reactor experimentally shows the role of confinement to extend the stability limit for complete conversion. This was observed in the case that a 1.0-mm channel resulted in low temperature and fuel breakthrough while the 0.5-mm channel achieved high temperatures and complete conversion. This was also evident in the models showing that only a 0.5-mm channel results in nearly complete fuel oxidation (Figure 12). We observed under a specific level of confinement, 1-mm channel gap in a fully insulted reactor, the surface temperature, and therefore thermal performance of the reactor, is similar for propane and *n*-dodecane.

HH reactors for TEC would, ideally, have an optimum confinement to achieve both complete conversion for multiple fuels as well as uniform temperatures. In the fully insulated parallel plate reactors, a more uniform temperature profile was observed for the 1.0-mm gap while the 0.5-mm gap reactor exhibited higher peak temperatures and sharper gradients. At the largest gap, 2.0 mm, the experiments showed high uniformity, but low temperatures. In addition, the reactors were unable to fully react the input fuel due to high heat loss.

A practical reactor for TEC applications must recirculate the exhaust or recuperate the sensible heat exhausted. The enthalpy lost in the exhaust of the slot reactor is estimated to vary between ~35 W for 104 $W_{chemical}$ input to ~90 W for 208 $W_{chemical}$ input. Therefore about 35-45% of heat is lost to the exhaust and the remainder is lost to convection and radiation. Convective losses can be reduced by having a vacuum gap or the cavity between the emitter and photovoltaic cell filled with thermally insulating gas. Heat recirculation and external heat recuperation are required to address the appreciable amounts of energy lost through the exhaust. Initial results using a design similar to that reported in [47] show the heat lost to the exhaust can be reduced to 20% by recirculating the exhaust of the slot reactor over the surface by wrapping the flow around the outside surface in two additional channels.

A detailed reactor design and optimization will require capturing the physics that our previous model did not account for, including multidimensional diffusion in the gas phase as well as heat conduction along reactor walls, radiation losses to the environment, and the three dimensional nature of the flow field near the reactor's side walls. Initial efforts have used ANSYS Fluent to model 2-D steady state, laminar flow with simplified gas and surface chemistry to model the performance of a heat recirculating reactor aimed at enhancing the thermal efficiency [48] . Thermal efficiencies, η_{th}, up to 60% are predicted in the simulations when evaluating radiation from all surfaces. If these can be harnessed, then > 18% overall chemical to electric conversion efficiency is feasible when combined with the best reported TPV efficiencies [3, 5, 6]. In a planar reactor configuration, however, not all radiation can be harvested, for example from the sides and tip of the reactor [47]. Thermal management will be critical in the reactor design to achieve the high thermal efficiencies necessary for TEC applications.

In addition to comprehensive numerical models, a parallel effort, underway, focuses on reduced-order numerics to enable parametric comparisons of materials, geometries, input chemical energy rates, and heat exchangers on total system performance. Our aim is to narrow the design space for more detailed numerical work. Initial results suggest that reduced length scales in microreactors, which are necessary to effectively couple surface and gas-phase chemistry, allow us to reduce the expense of modeling transport in exchange for detailed chemistry and heat transfer. The computational requirements when using this method are 100x less (hours versus hundreds of hours), without a loss in accuracy when measured to experimentally observable parameters.

Reactor material thermal properties should be analyzed to determine the performance limitations and design requirements beyond just the combustion. For example, high temperature operation combined with the seemingly trivial stresses of operating in vacuum – due to the pressure difference between inside and outside the channel – can easily reach the yield strength of materials for wide channels with thin walls [47]. Furthermore, a similar deformation is expected when coupling thermal emitters, necessary for enhanced radiative efficiency in TPV converters, to the reactor wall, because of a thermal expansion (CTE) mismatch. Chan et al. have shown that small channels can result in stable geometry during high temperature operation in vacuum and during the vacuum brazing process [49], but the overall thermal efficiency limit may be governed by the upper temperature limit of the Inconel reactor.

Additional considerations for practical use with heavy hydrocarbons include fuel vaporization and thermal stability. Thermal breakdown of the fuel can cause fouling on the inner walls and in some cases degrade tubing and reactor materials [50]. Fuel decomposition and catalyst surface fouling can occur especially during the ignition period where the temperature of the fuel and air mixture is moderate. If the catalyst sites are not replenished, ignition times and, thus, the energy required to ignite will increase over multiple start/stops [51] .

As mentioned previously, catalyst supported combustion has remaining challenges associated with deposition and stability. The nature of the catalyzed surface in the experimental reactors is largely unknown and can greatly affect performance. The overall effect of sulfur when moving to JP-8 and high temperature operation over long durations is largely unknown for microreactor applicable to TEC. Additional studies on the catalyst directly deposited on Inconel-600 and the impact of Inconel oxidation on catalyst reactivity and stability is necessary. The modeling capability discussed to predict reactor performance will also provide insights into the catalyst chemistry and high temperature exposure. This information can be used to inform catalyst research to evaluate stability at conditions appropriate to extended operation.

6. Summary

Hybrid Heterogeneous-Homogeneous (HH) reactors operating with *n*-dodecane and propane were studied under insulating and extreme heat loss conditions. Experiments show that HH combustion enables complete conversion in sub-quenching distance channels and under high heat loss conditions relevant to TEC applications. Models were developed that provided insight into the coupling between surface and gas-phase chemistry and the differences in operation for heavy fuels (*n*-dodecane) when compared to lighter species (propane).

Confinement was shown to greatly affect combustion within the reactor. Low confinement led to overall lower temperatures, where *n*-dodecane conversion was primarily through reactions with the catalyzed wall. The operation was limited by the diffusion time of the reactants to the wall. At high levels of confinement, the overall temperatures are high and the reactant decomposition is primarily a gas-phase process. Heat release, however, was shown through simulation to be due to surface reactions in the high heat loss reactor. In high confinement conditions, and at high temperature (>1100 K), fuel decomposition into intermediates was shown to play a large role in reactor operation.

Microreactors using the HH approach show great promise as hydrocarbon fueled heat sources for TEC applications. However, additional research and investigation of the engineering trade space is necessary to enable high specific energy wearable power sources for extending the Soldier's mission or reducing the physical burden of carrying energy on mission. The following is a list of key areas for further exploration:

- Military relevant, complex hydrocarbon mixtures including JP-8 and fuel flexibility;
- Simplified modelling approaches are necessary with short computation times to adequately explore materials, flow conditions, geometry, and thermal management as applied in TEC applications;
- Catalyst durability against fuel chemistry and high temperature operation;
- Fuel thermal stability and the impact of the ignition process on reactor operation;
- Integration with TEC elements such as selective emitters in TPV applications.

References

[1] Ultralife, *Li-145 Ultralife battery datasheet*, N.N.U. Corporation, Editor. 2010.

[2] Meredith, M.S.P. and M.D. Bergmann, *Soldier Power: A Growing Operational Concern*, in *Infantry*. 2013, U.S. Army Infantry School: Fort Benning, GA. p. 5-8.

[3] Wernsman, B., et al., Greater Than 20% Radiant Heat Conversion Efficiency of a Thermophotovoltaic Radiator/Module System Using Reflective Spectral Control. IEEE Transactions on Electron Devices, 2004. 51(3): p. 512-515. https://doi.org/10.1109/TED.2003.823247

[4] Waits, C.M., *Thermophotovoltaic Energy Conversion for Personal Power Sources*, U.S.A.R. Laboratory, Editor. 2012, U.S. Army Research Laboratory: Adelphi, MD.

[5] Bitnar, B., W. Durisch, and R. Holzner, *Thermophotovoltaics on the move to applications.* Applied Energy, 2013. 105: p. 430-438. https://doi.org/10.1016/j.apenergy.2012.12.067

[6] Bermel, P., et al., Design and global optimization of high-efficiency thermophotovoltaic systems. Optics Express, 2010. 18(S3): p. A314-A334. https://doi.org/10.1364/OE.18.00A314

[7] Fraas, L., et al. Design and Thermal Modeling of a Portable Fuel Fired cylindrical TPV Battery Replacement. in 9th World Conference on Thermophotovoltaics (TPV-9). 2010. Valencia, Spain.

[8] Scranton, G., et al. Highly efficient thermophotovoltaics enabled by photon re-use. in 2016 IEEE 43rd Photovoltaic Specialists Conference (PVSC). 2016. Portland, OR.

[9] Maruta, K., et al., *Characteristics of Microscale Combustion in a Narrow Heated Channel.* Combustion, Explosion and Shock Waves, 2004. 40(5): p. 516-523. https://doi.org/10.1023/B:CESW.0000041403.16095.a8

[10] Richecoeur, F. and D.C. Kyritsis, *Experimental study of flame stabilization in low Reynolds and Dean number flows in curved mesoscale ducts.* Proceedings of the Combustion Institute, 2005. 30(2): p. 2419-2427. https://doi.org/10.1016/j.proci.2004.08.015

[11] Pizza, G., et al., *Dynamics of premixed hydrogen/air flames in mesoscale channels.* Combustion and Flame, 2008. 155(1): p. 2-20. https://doi.org/10.1016/j.combustflame.2008.08.006

[12] Pizza, G., et al., *Dynamics of premixed hydrogen/air flames in microchannels.* Combustion and Flame, 2008. 152(3): p. 433-450. https://doi.org/10.1016/j.combustflame.2007.07.013

[13] Maruta, K., et al., *Characteristics of combustion in a narrow channel with a temperature gradient.* Proceedings of the Combustion Institute, 2005. 30(2): p. 2429-2436. https://doi.org/10.1016/j.proci.2004.08.245

[14] Pizza, G., C.E. Frouzakis, and J. Mantzaras, *Chaotic dynamics in premixed hydrogen/air channel flow combustion.* Combustion Theory and Modelling, 2012. 16(2): p. 275-299. https://doi.org/10.1080/13647830.2011.620174

[15] Dirrenberger, P., et al., *Measurements of Laminar Flame Velocity for Components of Natural Gas.* Energy & Fuels, 2011. 25(9): p. 3875-3884. https://doi.org/10.1021/ef200707h

[16] Zhao, Z., et al., The initial temperature and N2 dilution effect on the laminar flame speed of propane/air. Combustion Science and Technology, 2004. 176(10): p. 1705-1723. https://doi.org/10.1080/00102200490487553

[17] Stefanidis, G.D. and D.G. Vlachos, *Controlling Homogeneous Chemistry in Homogeneous−Heterogeneous Reactors: Application to Propane Combustion.*

Industrial & Engineering Chemistry Research, 2009. 48(13): p. 5962-5968. https://doi.org/10.1021/ie801480m

[18] Karagiannidis, S., J. Mantzaras, and K. Boulouchos, Stability of hetero-/homogeneous combustion in propane- and methane-fueled catalytic microreactors: Channel confinement and molecular transport effects. Proceedings of the Combustin Institute, 2011. 33: p. 3241-3249. https://doi.org/10.1016/j.proci.2010.05.107

[19] Karagiannidis, S., et al., *Hetero-/homogeneous combustion and stability maps in methane-fueled catalytic microreactors.* Proceedings of the Combustion Institute, 2007. 31(2): p. 3309-3317. https://doi.org/10.1016/j.proci.2006.07.121

[20] Di Benedetto, A., V. Di Sarli, and G. Russo, *A novel catalytic-homogenous micro-combustor.* Catalysis Today, 2009. 147(Supplement): p. S156-S161. https://doi.org/10.1016/j.cattod.2009.07.030

[21] Coward, H.F. and P.G. Guest, *IGNITION OF NATURAL GAS-AIR MIXTURES BY HEATED METAL BARS1.* Journal of the American Chemical Society, 1927. 49(10): p. 2479-2486. https://doi.org/10.1021/ja01409a017

[22] Mantzaras, J. and C. Appel, Effects of Finite Rate Heterogeneous Kinetics on Homogeneous Ignition in Catalytically Stabilized Channel *Flow Combustion.* Combustion and Flame, 2002. 130: p. 336-351. https://doi.org/10.1016/S0010-2180(02)00384-X

[23] Mantzaras, J. and P. Benz, *An Asymptotic and Numerical Investigation of Homogeneous.* Combustion and Flame, 1999. 119: p. 455-472. https://doi.org/10.1016/S0010-2180(99)00071-1

[24] TRADOC, The U.S. Army Functional Concept for Movement and Maneuver 2020-2040. 2017, TRADOC Pamphlet 525-3-6.

[25] Lloyd, S.A. and F.J. Weinberg, *Limits to energy release and utilisation from chemical fuels.* Nature, 1975. 257(5525): p. 367-370. https://doi.org/10.1038/257367a0

[26] Lloyd, S.A. and F.J. Weinberg, *A burner for mixtures of very low heat content.* Nature, 1974. 251(5470): p. 47-49. https://doi.org/10.1038/251047a0

[27] Ronney, P.D., *Analysis of non-adiabatic heat-recirculating combustors.* Combustion and Flame, 2003. 135(4): p. 421-439. https://doi.org/10.1016/j.combustflame.2003.07.003

[28] Ahn, J., et al., *Gas-phase and catalytic combustion in heat-recirculating burners.* Proceedings of the Combustion Institute, 2005. 30(2): p. 2463-2472. https://doi.org/10.1016/j.proci.2004.08.265

[29] Norton, D.G., E.D. Wetzel, and D.G. Vlachos, *Thermal Management in Catalytic Microreactors.* Industrial & Engineering Chemistry Research, 2006. 45(1): p. 76-84. https://doi.org/10.1021/ie050674o

[30] Tolmachoff, E.D., W. Allmon, and C.M. Waits, Analysis of a high throughput n-dodecane fueled heterogeneous/homogeneous parallel plate microreactor for portable power conversion. Applied Energy, 2014. 128(Supplement C): p. 111-118. https://doi.org/10.1016/j.apenergy.2014.04.057

[31] Tolmachoff, E.D., et al., Modeling and experimental analysis of n-dodecane oxidation in platinum-coated channels. Combustion and Flame, 2015. 162(10): p. 3674-3680. https://doi.org/10.1016/j.combustflame.2015.07.006

[32] Biet, J., et al., Experimental and Modeling Study of the Low-Temperature Oxidation of Large Alkanes. Energy & Fuels, 2008. 22(4): p. 2258-2269. https://doi.org/10.1021/ef8000746

[33] Suzuki, S., et al., Study on cetane number dependence of diesel surrogates/air weak flames in a micro flow reactor with a controlled temperature profile. Proceedings of the Combustion Institute, 2013. 34(2): p. 3411-3417. https://doi.org/10.1016/j.proci.2012.06.162

[34] Tolmachoff, E.D., W.R. Allmon, and C.M. Waits, *A Method for Microscale Combustion of Near Stoichiometric Energy Dense Liquid Fuel Mixtures.* Journal of Physics: Conference Series, 2013. 476(1): p. 012074. https://doi.org/10.1088/1742-6596/476/1/012074

[35] You, X., F.N. Egolfopoulos, and H. Wang, Detailed and simplified kinetic models of n-dodecane oxidation: The role of fuel cracking in aliphatic hydrocarbon combustion. Proceedings of the Combustion Institute, 2009. 32(1): p. 403-410. https://doi.org/10.1016/j.proci.2008.06.041

[36] *ANSYS Fluent 14.0 Documentation.* 2012, ANSYS Inc.: Lebanon, NH.

[37] Goodwin, D., et al., *Cantera: An object-oriented software toolkit for chemical kinetics, thermodynamics, and transport processes.* 2013: https://code.google.com/p/cantera.

[38] Wang, H., et al., *Simplified n-dodecane oxidation model.* 2007: http://ignis.usc.edu/Jet_Fuel_MechI.html.

[39] Narayanaswamy, K., P. Pepiot, and H. Pitsch, A chemical mechanism for low to high temperature oxidation of n-dodecane as a component of transportation fuel surrogates. Combustion and Flame, 2014. 161(4): p. 866-884. https://doi.org/10.1016/j.combustflame.2013.10.012

[40] Zerkle, D.K., et al., Understanding Homogeneous and Heterogeneous Contributions to the Platinum-Catalyzed Partial Oxidation of Ethane in a Short-Contact-Time Reactor. Journal of Catalysis, 2000. 196(1): p. 18-39. https://doi.org/10.1006/jcat.2000.3009

[41] Deutschmann, O., et al., *Hydrogen assisted catalytic combustion of methane on platinum.* Catalysis Today, 2000. 59(1): p. 141-150. https://doi.org/10.1016/S0920-5861(00)00279-0

[42] Pfefferle, L.D., Heterogeneous/homogeneous reactions and transport coupling for catalytic combustion systems: a review of model alternatives. Catalysis Today, 1995. 26(3): p. 255-265. https://doi.org/10.1016/0920-5861(95)00147-7

[43] Degenstein, N.J., Spatially Resolved Species and Temperature Profiles in the Catalytic Partial Oxidation of Methane and Ethane, in Graduate School. 2007, University of Minnesota. p. 178.

[44] Lee, I.C., J.G. St. Clair, and A.S. Gamson, *Catalytic partial oxidation of isobutanol for the production of hydrogen.* International Journal of Hydrogen Energy, 2012. 37(2): p. 1399-1408. https://doi.org/10.1016/j.ijhydene.2011.09.121

[45] Norton, D.G., E.D. Wetzel, and D.G. Vlachos, *Fabrication of Single-Channel Catalytic Microburners: Effect of Confinement on the Oxidation of Hydrogen/Air Mixtures.* Industrial & Engineering Chemistry Research, 2004. 43(16): p. 4833-4840. https://doi.org/10.1021/ie049798b

[46] Niket S. Kaisare, S.R.D., Dionisios G. Vlachos, *Stability and performance of catalytic microreactors: Simulations of propane catalytic combustion on Pt.* Chemical Engineering Science, 2008. 63: p. 1098-1116. https://doi.org/10.1016/j.ces.2007.11.014

[47] Waits, C.M., et al., Energy Analysis of n -Dodecane Combustion in a Hetero/Homogeneous Heat-Recirculating Microreactor for Portable Power Applications. Journal of Physics: Conference Series, 2016. 773(1): p. 012060. https://doi.org/10.1088/1742-6596/773/1/012060

[48] Tolmachoff, E.D. and C.M. Waits, Analysis of catalyst placement strategies for efficient heat harvesting in aradiant Heterogeneous/Homogeneous heat-

recirculating microcombustor, in Fall Technical Meeting of theWestern States Section of theCombustion Institute (WSS/CI 2015 Fall Meeting). 2015: Provo, Utah.

[49] Chan, W.R., et al., Enabling efficient heat-to-electricity generation at the mesoscale. Energy & Environmental Science, 2017. https://doi.org/10.1039/C7EE00366H

[50] Altin, O. and S. Eser. Carbon deposition from thermal stressing of petroleum fuels. in The 228th Amearical Chemical Society (ACS) National Meeting. 2004. Philadelphia, PA.

[51] O'Brien, C.P., et al., Deactivation of Pt/Al_2O_3 during propane oxidation at low temperatures: Kinetic regimes and platinum oxide formation. Journal of Catalysis, 2016. 337: p. 122-132. https://doi.org/10.1016/j.jcat.2016.02.012

Chapter 14

Catalytic Oxidation of Hydrocarbons and Army Jet Fuels for Small Scale Combustion

Ivan C. Lee

U.S. Army Research Laboratory, Sensors and Electron Devices Directorate, 2800 Powder Mill Road, Adelphi, MD 20783, USA

ivan.c.lee2.civ@mail.mil

Abstract

In this Chapter we summarize recent ARL efforts in catalytic oxidation of propane, liquid hydrocarbon surrogates and jet fuel samples for soldier power generation. A previous review discussed the progress of small portal combustors in the literature [1]. About a decade ago, ARL initiated some fundamental studies to understand how hydrocarbon oxidation occurs on solid catalysts including combustion of military logistics jet fuel JP-8 [2]. This Chapter reviews our recent efforts in this subject in past 5 years [3-12]. From these efforts, we have learned that we could achieve complete microscale combustion of complex liquid hydrocarbon fuels in presence of a catalyst. We have also gained better insights on how propane oxidation occurs with quantifications of key parameters such as activation energy barriers and diffusion barriers. Finally, we have demonstrated complete catalytic oxidation of jet fuels with high sulfur content (~700 parts per million by weight) in short contact time of millisecond scale. Such small contact time operation space allows future designs of compact soldier power generation from military logistics jet fuels.

Keywords

Jet Fuel, Hydrocarbon, Combustion, Microcombustion, Mesoscale Combustor, Soldier Power

Contents

1. Background

Catalytic combustion of jet fuels is an enabling technology for small scale combustion in soldier and squad compact power generation [1]. Typical mechanisms of hydrocarbon combustion involve initiation, propagation, chain-branching, and termination reactions with free radicals as intermediates. Once the combustion is initiated, chain reactions produce free radicals in the propagation or chain-branching reaction. The free radicals recombine in the termination reactions. For example, H_2-O_2 combustion consists of as many as 40 reactions and 8 species: H_2, O_2, H_2O, OH, O, H, HO_2 and H_2O_2 [13]. Some of the elementary reactions are

$$H_2 + M \rightarrow H + H + M \qquad \text{Initiation}$$

$$H + O_2 \rightarrow O + OH \qquad \text{Chain-Branching}$$

$$H_2 + OH \rightarrow H_2O + H \qquad \text{Propagation}$$

$$H + OH + M \rightarrow H_2O + M \qquad \text{Termination}$$

where M represents a collision partner.

1.1 Surface effects (interfacial phenomena)

Microscale combustion differs from the mesoscale, or large-scale combustion when the dimensions of combustor change in order of magnitude. As the size decreases, surface to volume ratio (S/V) increases. The chance of combustion species hitting the combustor wall (surface effect) become more frequent. Therefore, interfacial phenomena such as flame quenching (thermal quenching and radical quenching) becomes a problem [14,15]. The consequence is incomplete combustion of hydrocarbon fuels. The implication is poor fuel efficiency in microcombustion of hydrocarbon fuels.

In thermal quenching, flame extinction occurs when the heat of combustion cannot compensate the heat loss to the surrounding environment through the wall. The wall acts as an enthalpy sink, and eventually flame extinction occurs. In the literature, a series of microcombustion experiments were performed to determine quenching distance with various wall materials (quartz, stainless steel, copper, yttria-stabilized zironia (YSZ), and alumina) at different inside wall temperatures [14]. The investigators concluded that thermal quenching dominates at low temperature (500 K), whereas the radical quenching plays a dominant role at higher temperature (1000 K). In addition to the heat loss type extinction mechanism, blowout was the second thermal quenching mechanism [16,17]. Blowout occurs when a flame gets swept out of the reactor at low residence time (i.e. high flow rates).

In radical quenching, the free radicals were destroyed to form stable products by undergoing termination reactions at the reactor wall. These wall reactions prevent the buildup of free radicals that would lead to explosions. However, the combustion is terminated if all the radicals recombined to form stable species on the walls of a combustor. When the size of the combustor becomes small, the removal of these radical intermediates is more likely in microscale combustion and disturbs the combustion reactions. In another study, other investigators studied the relative significance of heat loss (thermal quenching) and removal of active radicals at the wall (radical quenching) in stoichiometric combustion of a methane-air mixture [18]. They measured quenching distance with virgin and annealed plates of stainless steel, alumina, and quartz at wall temperature from 100–800 °C. Their experimental results indicated that these two quenching phenomena were temperature-dependent. At low temperatures, quenching distances are independent of the surface characteristics such as oxygen vacancy, grain boundary, or impurities. At high temperatures, surface characteristics strongly affect the quenching distance. These results suggest that radical quenching plays a significant role in the quenching process at high temperatures. Since the typical adiabatic flame temperature of hydrocarbon is over 2200 K, one would expect that minimizing radical quenching is very important for sustaining gas phase microcombustion.

1.2 Time scaling effects

In addition, time scaling effects also play a critical role in microscale combustion. Theoretically, stable combustion can be achieved when reactant residence time is larger than chemical reaction time and diffusion time [14,15]. Otherwise, incomplete combustion occurs. As a result, lower combustion efficiency will lead to insufficient heat generation to achieve sustained combustion.

At a given fluid (fuel and air) velocity, the reactant residence time decreases with the size of combustor. The conversion (chemical reaction time) is a challenge to maintain stable combustion in microscale. This fundamental time constraint can be quantified in terms of the Damkohler number, which is the ratio of gas residence time to the characteristic chemical reaction time. The Damkohler number must be greater than unity to ensure complete combustion. Superadiabatic combustion has been proposed to increase the Damkohler number and to enhance heat transfer [19]. It is noted that superadiabatic combustion offers higher burning rates, increased flame stability, and prolonged residence time because of combined exceptional effects of conduction, radiation, and surface convection. Superadiabatic combustion refers to the strong overheating of the reaction zone, where the temperature far exceeds the thermodynamic combustion temperature [20]

In parallel to this, Reynolds number (Re) scales as Re~L, where L represents the characteristic length of the combustor. Re, therefore, decreases when the combustor becomes smaller. Re in a microscale combustor is usually much smaller than 2000, at which only laminar flow is possible. Without turbulent mixing, the fuel-air mixing for non-premixed streams would rely on molecular diffusion.

1.3 Flame stability and extinction characteristics in microchannel

Microcombustion of hydrocarbon in microchannel reactors have been achieved in past decades with new understanding of flame stability and extinction characteristics [21-23]. Models [24,25] were developed to explain the experimental observations of flame stabilization, non-linear flame oscillations, and flames with repetitive extinction and ignition [26].

For instance, Maruta et al. [21] performed experimental studies to investigate the characteristics of microscale combustion in a microchannel heated by an external source. Oscillatory combustion was observed for low velocities of the flow. This can be explained by repetitive extinction and ignition due to the stable-to-unstable transition of the flame.

In another study, Miesse et al. [22] studied the microcombustion of methane and observed a sequence of isolated reaction zone structures (flame cells) along the gas flow direction in the microchannel. They summarized the flame stability observation in a flame structure regime map for various initial mixture strength (0.29<equivalent ratio <3.5) and total gas flow rate ($34 \leq Re \leq 82$). Prakash et al. [23] modified the previous microcombustor by Miesse et al. [22] such that chemiluminescence imaging and acoustic emissions can be used to characterize the ignition to transient dynamics and, finally, to formation of a stable edge-like flame and distinct cellular structures. They concluded that the flame cell structure was initially dominated by a hydrodynamic instability, which then transitioned to a steady-state flame cell structure. This stabilization was introduced by a strong coupling between the thermal and fluid fields [23].

Furthermore, Minaev et al proposed a one-dimensional nonlinear evolutionary equation of the flame front model [25], which took into account the inertial effects such as flame front acceleration and rate of flame temperature variation. These inertial effects were found to be important to describe oscillatory modes of flame propagation. Kessler and Short [24] described an unsteady ignition sequence in a simple two-dimensional model of a single-pass conductive microburner. Their model predicted the unsteady microcombustion modes, as seen in previous experimental studies [23,26], in which a reaction wave was driven rapidly down the channel towards the inlet via a sequence of oscillatory ignition and quenching transient extinction and repetitive ignition. This model [24] also showed how the initial axial wall temperature gradient was found to be central to determining combustion evolution to a variety of sustained combustion modes.

In a previous review [1], we identified three common approaches to achieve micro-scale combustion with improved flame stability and extinction characters: (1) heat recirculation in microchannels and Swiss roll combustor, (2) combustion in inert porous media, and (3) addition of catalyst coating in either of two aforementioned approach. We recently published another literature review [8] on the use of catalytic materials for hydrocarbon oxidation. In this Chapter we summarize our most recent effort on gaining deeper understanding on the catalytic oxidation chemistry of hydrocarbons with increasing complexity of fuel composition, starting from propane, dodecane, then dodecane-xylene mixtures, and finally various kinds of jet fuel samples.

2. Defense relevance

This Chapter describes a potential path for lightening the physical load of individual warfighters and for extending the range and the endurance of the RAS by identifying new materials and processes to convert energy dense fuels such as jet fuels (40 kJ/mol) to thermal energy. Then thermal energy could be converted to electrical energy by using thermoelectric or thermophotovolatic power generation. Success development of the new materials and processes will reduce the use of heavy batteries. For instance, a 50W-96 hour mission currently requires an equivalent of nineteen BB2590 rechargeable batteries with 53L in volume and 59 lbs in weight. Alternatively, a 5% efficient JP-8 fueled-rechargeable system with a compact catalytic fuel combustor, thermal-to-electrical converter and two BB 2590 batteries would be 17 L in volume and 27 lbs in weight. This illustrate the potential impact of catalytic materials to the size-weight and power (SWaP) for combat effectiveness with enhanced logistics efficiency.

3. Major concepts

3.1 Use of a descriptor in volcano-type relationship for catalyst design

To design new generation of non-precious metal combustion catalyst, one major concept is to predict optimal values of a materials descriptor by correlating the experimental measurement of the turnover frequency with the calculated value of such descriptor from density functional theory [3]. The turnover frequency (TOF), defined as molecules of alkane reacted per catalyst site per second, was calculated from experimental kinetic data under differential conditions and normalized with the number of surface sites. The materials descriptor could be the atomic binding energy of a heteratom on a metal catalyst composite. Such correlation often appears as a volcano-type relationship giving an optimal value of the material descriptor at the peak of the "volcano" curve. Similar volcano-type relationships have been reported for ammonia decomposition [30,31,32], isopropanol dehydrogenation [33] and desulfurization [34,35]. Our recent work demonstrates the usefulness of this major concept, leading to identification of optimal values of descriptor [3]. For instance, the propane combustion is correlated with carbon binding energy of various transition metal as shown in Figure 1. Consequently, we were able to identify and to confirm experimentally that Ag-Pd bimetallics catalyst is superior to monometallic Pd, Ag, and Pt catalysts [3]. This discovery of a less expensive combustion catalysts such as Ag-Pd bimetallic catalyst is beneficial for design new metal catalyst.

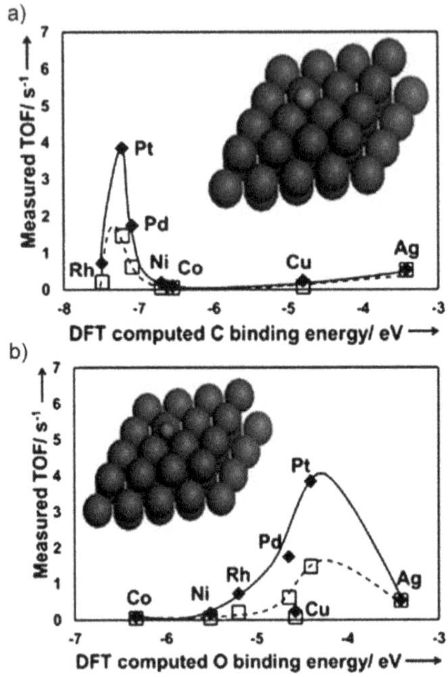

Fig. 1 Volcano curves of propane oxidation at 673K and 613K using C binding energy or O binding energy as a descriptor [3].

3.2 Use of a surrogate compound in a homologous series

Another major concept is the use of homologous series for simplify the complexity of logistics military fuels such as jet fuels. Jet fuel is a complex mixture of different classes of hydrocarbons including paraffins (also known as alkanes), cycloalkanes, alkyl benzenes, and napthalenes (Figure 2). According to military specifications MIL-DTL-5624 and MIL-DTL-83133E, these military jet fuels contain as high as 3000 part-per-million by weight (pppmw) of sulfur including derivatives of benzothiophene compounds [27]. This complexity of fuel composition poses a great challenge of formulating our fundamental research plan to understand hydrocarbon fuel microcombustion when interfacial phenomena and time scale effect become increasing important as the size of combustor decrease to soldier portable scale or so called mesoscale combustor.

To address the fuel complexity challenge, one idea is to identify a surrogate chemical compound to represent each homologous series of compounds in the jet fuel mixture. For instance, a recent work aims to find a surrogate compound to represent the alkane homologous series by generating a volcano curves with experimentally measured turnover frequency and descriptor like binding energy as similar to Figure 1. Briefly, we compared the volcano curves for methane, ethane, propane and isobutane oxidation, and we determined that propane is an excellent surrogate of the homologous alkane series [28]. An implication is that we could potentially apply the same concept to identify a surrogate compound each class hydrocarbon in Figure 2. The significance is that we could use propane as a surrogate so that we can perform some fundamental studies with in-situ FTIR studies to understand alkane oxidation on molecular level.

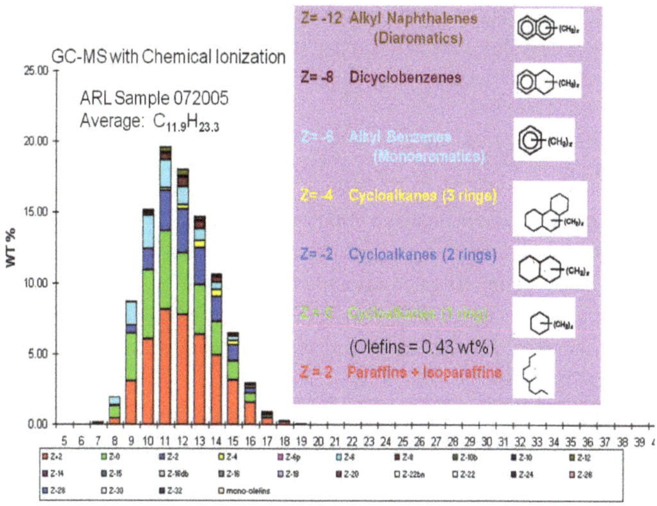

Fig. 2 Chemical analysis of a typical military jet fuel JP-8 sample (x-axis showing carbon chain length of the hydrocarbon species and y-axis showing the weight percent of each hydrocarbon class)

4. Approach

Our approach is first to simplify the complexity of fuel composition by applying the concept of homologous series so that we use propane as the representative compound for alkanes [6,9,10,11]. Then we add the complexity by including aromatics such as xylene

[7]. Finally, we study more complex fuel mixtures such as synthetic jet fuels and petroleum-based jet fuels [12]. Second, we also utilize a catalytic approach to improve fuel microcombustion performance in a meso-scale combustion system [7]. Since catalyst combustion relies on surface reactions, the high surface area of a meso-scale combustor provides a larger number of active sites upon which the reactant molecules can adsorb and react. Additionally, chemical reactions will theoretically only occur on catalyst surfaces, so the location of the heat source is fixed, unlike in homogenous combustion, where the flame may anchor at different locations depending on the operating conditions [30]. This thermal stability makes a catalytic system much more attractive for integration with a thermoelectric/thermophotovoltaic device than a comparable non-catalytic system [30]. Several studies have shown that catalysts can help stabilize liquid fuel combustion at very short residence times, even for petroleum-based jet fuels [36-38]. Most of the existing literature investigates the role of the catalyst in fuel reformation (i.e. partial oxidation), particularly for fuel cell applications [39-46]. Very few studies have been conducted on complete oxidation of liquid fuels with catalysts at the meso-scale combustor [41].

5. Methodology

We employ both experimental and computational chemistry techniques in concert in order to elucidate important kinetics parameters of the catalytic oxidation of jet fuel and its surrogates. First, we investigated catalytic oxidation of propane, jet fuel surrogates in heat-recirculated mesoscale combustor to understand combustion performance operation range, and we also used kinetic analysis to extract global parameters such as activation energy barriers (Figure 3).

A molecular level understanding of mechanism of surface reactions in catalytic oxidation requires careful experimental designs with in-situ spectroscopic methodology. Recently, we utilize diffuse reflectance infrared Fourier-transform spectroscopy (DRIFTS) with online gas analysis to monitor whether sintering of catalyst nanoparticles, metal oxide formation or accumulation of poisoning species reduce the reactivity of propane combustion (Figure 4). Then, we also analyzed with theoretical calculations with density functional theory to identify the unknowns in the infrared spectra [9,10]. Kinetics modeling of the temperature programmed oxidation profiles of the oxy-carbon surface species allowed us to determine their activation energy barriers and surface diffusion parameters [11].

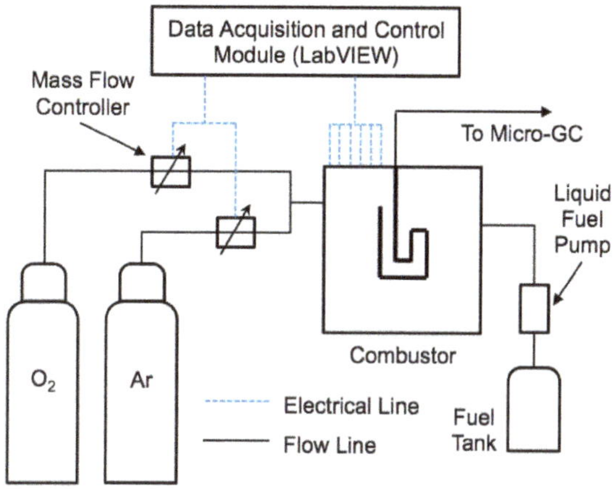

Fig. 3 Meso-scale heat recirculated combustor at ARL [5,6,7].

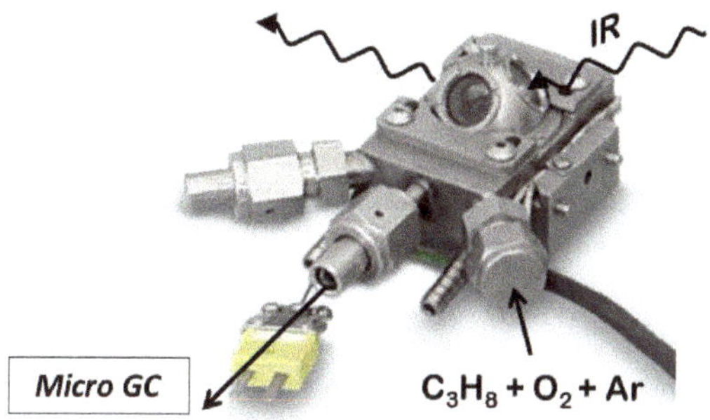

Fig. 4 Experimental setup for catalytic propane oxidation with DRIFTS.

Finally, we use a short contact time reactor with porous catalytic foams to investigate military jet fuel combustion (Figure 5). Catalytic porous foam provides a combination of homogenous oxidation reaction with heterogeneous oxidation reaction.

6. Major findings/results and discussion

6.1 Propane combustion

We have studied the combustion behavior of propane over platinum and rhodium catalysts in a meso-scale heat recirculating combustor in an effort to predict their performance using a liquid fuel. (Figure 1) [6]. The extinction limits of these two catalysts were compared to those of non-catalytic combustion in the same combustor (Figure 6). The results showed that the use of a catalyst greatly expanded the range of stable operating conditions, in terms of both extinction limits and flow rates supported. Moreover, conversion, product selectivity/ yield, and activation energy using these two catalysts were compared. The Rh catalyst was found to exhibit a higher propane conversion rate, reaching a maximum of 90.4% at stoichiometric conditions (as compared to only 61.4% offered by the Pt catalyst under lean conditions), but the Pt catalyst had superior CO_2 selectivity for most of the examined conditions, indicating more of the heat released being used for product formation as opposed to being lost to the environment. However, despite having a higher rate of heat loss, the propane combustion with the Rh catalyst produced an overall higher amount of enthalpy than the Pt due to its superior fuel conversion. The Pt catalyst also had a significantly smaller activation energy (13.8 kJ/mol) than the Rh catalyst (74.7 kJ/mol), except at equivalence ratios richer than ϕ = 1.75 (corresponding to catalyst temperatures below 500 °C), where it abruptly changed to 211.4 kJ/mol, signifying a transition from diffusion-limited reactions to kinetically limited reactions at this point. The propane combustion results suggest that Rh would be a more suitable catalyst for use in liquid-fueled meso-scale combustors, as fuel conversion has been found to be a limiting factor for combustion stability in these systems, and as its higher output energy allows for greater flexibility of use.

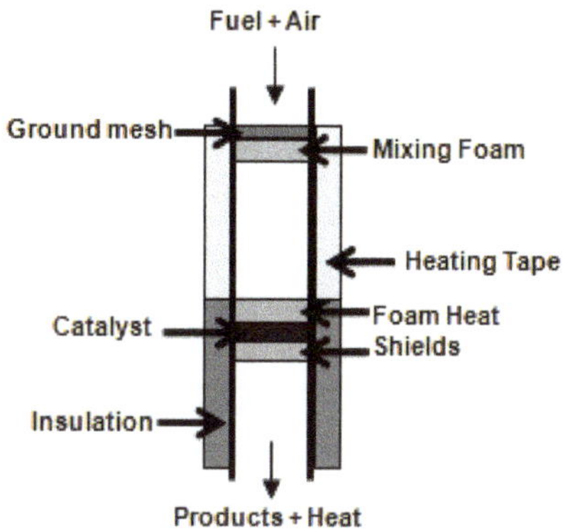

Fig. 5 Short contact time reactor with porous foam catalyst for jet fuel combustion studies.

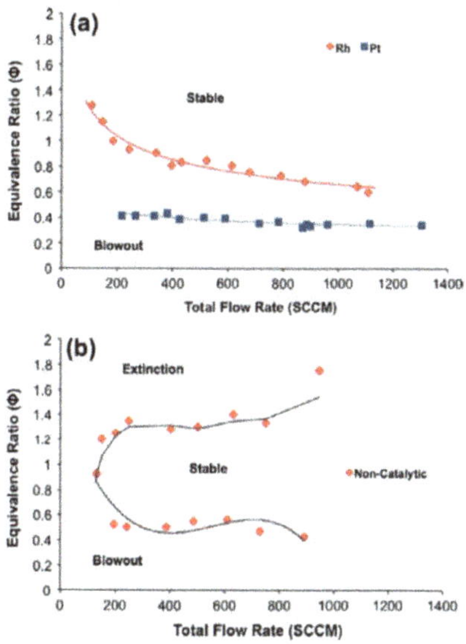

Fig. 6 Pt and Rh expands the stable operation regime for propane combustion [6].

In our in-situ spectroscopic study on catalytic propane oxidation [9,10,11], we studied the kinetics of propane combustion over Pt/Al_2O_3 as a function of the O_2/C_3H_8 ratio in the 150-300 °C temperature range. We found that the chemisorbed oxygen coverage on the platinum nanoparticles controlled the propane oxidation reaction rate in a large extent with 3 kinetic regimes [9]. At high O_2/C_3H_8 ratios, the platinum nanoparticles are saturated with oxygen and the reaction rate is zero-order with respect to the oxygen partial pressures in this regime. As the oxygen coverage decreases with decreasing O_2/C_3H_8 ratio, the reaction rate increases and the reaction order changes from zero-order to negative-order in the oxygen partial pressure (Figure 7).

The reaction rate is controlled to a large extent by the oxygen coverage on the platinum nanoparticles. However, at lower temperatures and higher oxygen pressures there is a slow deactivation of the catalyst that cannot be explained by a slow change in the oxygen coverage. We also observed a slow deactivation of the CO_2 production rate at O_2/C_3H_8 ratio of 20 and below 250 °C over hundreds of minutes that is not consistent with a slow change in the chemisorbed oxygen coverage, and must be related to deactivation of the catalyst. There are 3 potential deactivation mechanisms including (1) sintering of Pt nanoparticles causing reduction of number of catalytic active sites, (2) surface oxy-carbon species formation blocking the active sites, and (3) surface reconstruction of Pt nanoparticles in oxygen environment. We utilize the diffuse reflectance infrared Fourier-transform spectroscopy (DRIFTS) of adsorbed CO was performed to track the evolution of the nanoparticle structure over the course of the propane oxidation reaction and to determine which of the three potential deactivation mechanism is responsible for the deactivation. The details of the investigation of deactivation mechanism is published recently [9]. Briefly speaking, by comparing our CO-DRIFTS results with density functional density calculations, we demonstrated that the slow deactivation was caused by reconstruction of the platinum nanoparticles. We found that the platinum nanoparticles are significantly reconstructed during the course of the reaction, including the formation of a platinum oxide (PtO) which has a characteristic CO-DRIFTS band at 2123 cm^{-1}(Figure 8).

We also observed that the extent of PtO formation decreased with increasing temperature and, as a result, deactivation of the catalyst was less severe at higher temperatures. Unexpectedly, increasing the oxygen partial pressure resulted in less PtO formation. We believe that a different platinum oxide phase such PtO_2 or Pt_3O_4 was formed at higher oxygen pressures, which was reduced to metallic platinum during CO exposure at 25 °C, and therefore was not detectable by CO-DRIFTS [9].

In parallel to our observation of surface reconstruction of Pt nanoparticles, we use in-situ DRITS to study the growth of oxygenated carbonaceous (oxy-carbon) species on the surface of Pt/Al_2O_3 during total oxidation of propane and their subsequent burn-off during temperature program oxidation [10,11]. We found that propane transformed to a variety of surface oxy-carbon species as shown in Figure 9. Although these species are spectators in the propane oxidation mechanism, we found that there was an equilibrium for these species to diffuse away from Pt to the Al_2O_3 support and to oxidize to CO_2 under oxygen rich environment. We observed that the growth of oxy-carbon species on Al_2O_3 support was consistent with a diffusion-limited process, and we modeled their spillover from the platinum nanoparticles and their growth on the Al_2O_3 with the Fick's second law for two-dimensional, radial diffusion with Equation 1 [11].

$$\frac{\partial c}{\partial t} = \frac{1}{r}\frac{\partial}{\partial r}\left(rD\frac{\partial c}{\partial r}\right) \tag{1}$$

Then we estimated a pre-exponential factor of $7.9 \cdot 10^{-14}$ cm^2/s and an activation barrier of 24 kJ/mol by performing the kinetic analysis with an Arrhenius expression describing the rate of oxy-carbon species diffusion on the Al_2O_3 support.

We also observed the oxidation of these oxy-carbon species to CO_2 during the burn-off process with temperature-programmed oxidation (TPO) [10,11]. During TPO experiments, the oxy-carbon species diffuse quickly to the platinum nanoparticles where they are oxidized to CO_2. This burn-off process is reaction-limited. We estimated the activation barriers of 147 kJ/mol for oxidation of acetate species and 112 kJ/mol for oxidation of higher reactivity enolate, aliphatic ester, and acetone species using 2nd-order kinetic rate expressions.

6.2 Dodecane and dodecane-xylene combustion

We have investigated effect of aromatic content on combustion performance by studying the oxidation behavior of dodecane and two mixtures of dodecane and m-xylene (90/10 wt. % and 80/20 wt. %) over an Rh catalyst in a meso-scale heat recirculating combustor [7]. The fuel conversion, product selectivities, and reaction kinetics were calculated, and the global combustion behavior observed. The results showed that increasing the amount of m-xylene in the fuel increased the fuel conversion from 85% (pure dodecane) to 92% (90/10) and further to 98% (80/20). The presence of xylene also significantly increased CO_2/H_2O selectivity and decreased CO/H_2 selectivity. Global activation energy increased linearly with increase in xylene content to support that addition of aromatic species to fuel lowers the overall reactivity. The non-catalytic reaction was also simulated using Chemkin software to determine the effect of the Rh catalyst on the combustor performance and to analyze the difference in chemical mechanisms. The results revealed that the catalyst promotes total oxidation over partial oxidation, and lowers the global activation energy by up to 70% from 196 kJ/mol to 60.2 kJ/mol.

Table 1 Activation energy Ea for various dodecane-xylene mixtures in non-catalytic and catalytic (Rh) oxidation [7].

Dodecane :Xylene Composition	Non-Catalytic, kJ/mol	Catalytic, kJ/mol
100:0	196.0	60.2
90:10	227.3	86.7
80:20	262.1	116.0

6.3 Jet fuels combustion

We have evaluated the combustion characteristics of two synthetic aviation fuels (from Air Force Research Laboratory) a conventional petroleum-based jet fuel (from Fort Belvior, VA) in a meso-scale heat recirculating combustor [5]. One of the alternative jet fuels was synthesized via a Fischer–Tropsch (F–T fuel) process, while the other was produced from tallow (bio-jet fuel). The petroleum-based fuel used in this study was JP-8 with sulfur content of about 700 ppmw. The combustion and extinction behavior of the above fuels and their mixtures (50% synthetic fuel and 50% JP-8 by volume in JP-8) in the meso-scale combustor using oxygen under fuel-rich and fuel-lean non-premixed combustion conditions was examined. The synthetic fuels exhibited stable combustion over a range of equivalence ratios at each fuel flow rate; however, stable combustion was not achieved for JP-8 under any of the examined conditions. The mixtures also exhibited somewhat unstable combustion phenomena as those seen with JP-8, but dampened enough such that mostly stable combustion could occur. Fuel characterization analysis was performed for each fuel, and their respective thermal performances evaluated. Both the F–T and bio-jet fuels reached a maximum thermal efficiency of about 95% near their respective rich extinction limits. The mixtures exhibited somewhat poor thermal performances, with a maximum thermal efficiency of about 75%. The results reveal that composition of the fuel plays a prominent role in the flame stability and thermal performance in meso-scale combustors, as more complex species (such as aromatics, found in JP-8) have a slower reaction rate than simple species. The short residence time available in the combustion zone of the micro-combustion chamber does not allow to fully combust the complex fuel species, resulting in flame instability and formation of soot.

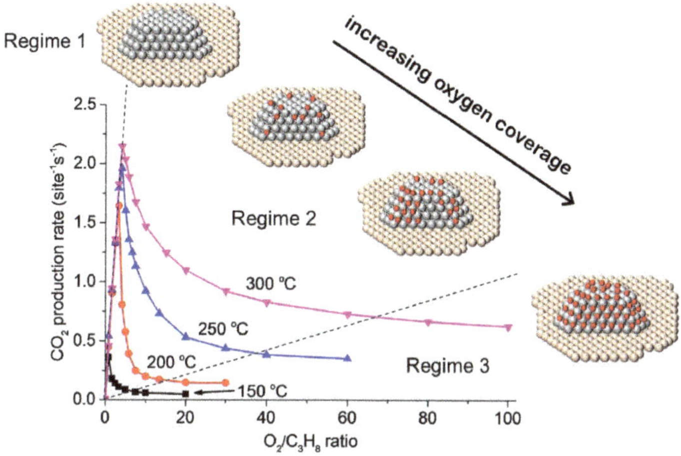

Fig. 7 Propane combustion on Pt/Al_2O_3 shows the presence of 3 kinetic regimes [9].

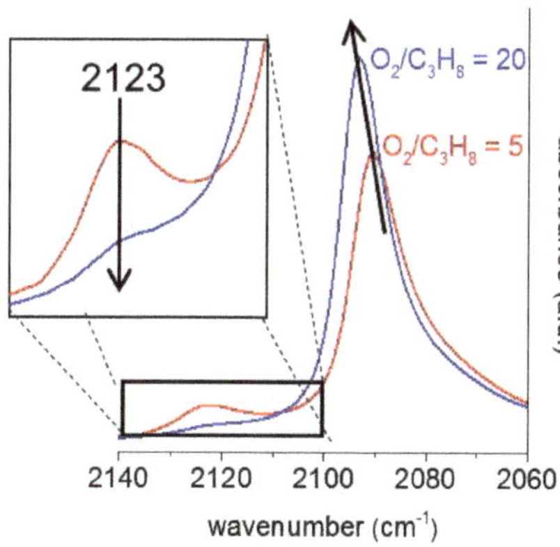

Fig. 8 CO-DRITS spectra showing PtO formation after 1000 minutes of propane oxidation on Pt/Al_2O_3 at 200 °C with O_2/C_3H_8 = 5 or 20 [9].

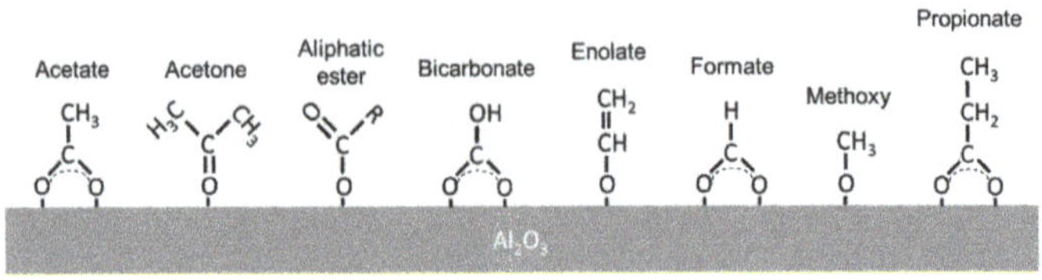

Fig. 9 A variety of oxy-carbon surface spectator species forms as observed in DRITS spectra during propane oxidation on Pt/Al_2O_3 [10].

Furthermore, we have compared the catalytic combustion of JP-8 with 700ppmw sulfur over a porous catalytic Pt or Rh foam in a short contact time reactor [12]. As aforementioned, presence of sulfur in petroleum-based jet fuels often deteriorate the catalyst performance over time. Recently, we have demonstrated that Rh/Al_2O_3 catalyst is more stable than Pt/Al_2O_3 catalysts (Figure 10). The ignition time delay for Pt-catalyzed JP-8 combustion is getting longer and longer after 8-hour time-on-stream daily. Time to reach steady-state increases with increasing number of startup cycles over Pt, but does not change with number of cycles over Rh. It is apparent that Rh can tolerate the presence of sulfur in JP-8 without much deterioration in ignition and steady state combustion performance.

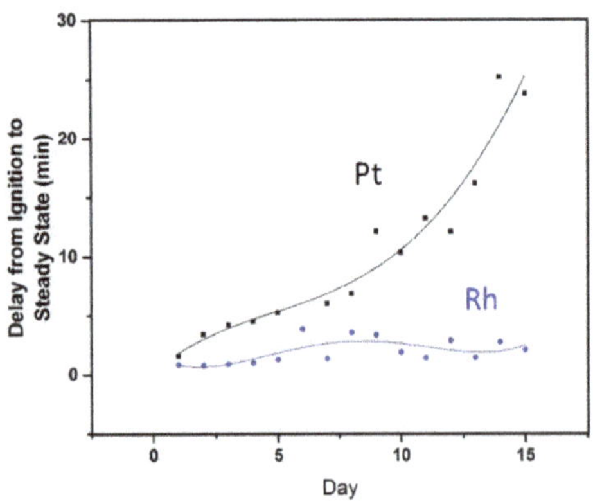

Fig. 10 Catalytic combustion of JP-8 over Rh catalyst maintains a better combustion performance than Pt catalyst.

7. Summary

In this Chapter, we have first put forward the importance of studying hydrocarbon combustion for solider and squad power generation. We have illustrated our approach and methodology on how we study the fundamental and applied aspects of catalytic combustion in small scale. Our work reveals quantitative mechanistic insights into both the carbon growth and catalyst regeneration processes, which are important for designing efficient hydrocarbon conversion processes. For instance, we would like to design cost-effective and highly active combustion catalyst for multiple fuels, including propane, sulfur-free synthetic jet fuels and high sulfur traditional petroleum-based jet fuels. We would also like to develop the startup ignition and shut-down protocols so that we could avoid formation of certain inactive metal oxides.

Acknowledgement

The author would acknowledge the contributions from the former team members Alex Booth, Teresa Wierzbicki and Casey O'Brien (now an assistant professor at Notre Dame) for their work in combustion in porous catalytic foams, catalytic heat recirculating combustor and in-situ FTIR studies, respectively. They spent hours to collect and analyze the data shown here in this Chapter. We thank Professor Dion Vlachos (University of Delaware) and Professor Ashwani Gupta (University of Maryland) as collaborators in the joint publications cited this Chapter.

References

[1] I.C. Lee, Small scale burner review, Army Research Laboratory Technical Report (July 2009) ARL-TR-4881.

[2] I.C., Lee Rhodium supported on thermally enhanced zeolite as catalysts for fuel reformation of jet fuels, Catalysis Today 136 (2008) 258. https://doi.org/10.1016/j.cattod.2008.03.027

[3] N. Rao Peela, W. Zheng, I.C. Lee, A.M. Karim, D.G. Vlachos, Core–shell nanocatalyst design by combining high-throughput experiments and first-principles simulations, ChemCatChem 5 (2013) 3712-3718. https://doi.org/10.1002/cctc.201300553

[4] G. Lee, W. Zheng, IC. Lee, DG Vlachos, Volcano curves for homologous series reactions: Small alkane-total oxidation, (submitted).

[5] T.A. Wierzbicki, I.C. Lee, A.K. Gupta, Performance of synthetic jet fuels in a meso-scale heat recirculating combustor, Applied Energy 118 (2014) 41-47. https://doi.org/10.1016/j.apenergy.2013.12.021

[6] T.A. Wierzbicki, I.C. Lee, A.K. Gupta, Combustion of propane with Pt and Rh catalysts in a meso-scale heat recirculating combustor, Applied Energy 130 (2014) 350-356. https://doi.org/10.1016/j.apenergy.2014.05.069

[7] T.A. Wierzbicki, I.C. Lee, A.K. Gupta, Rh assisted catalytic oxidation of fuel surrogates on Rh in a meso-scale combustor, Applied Energy 145 (2015) 1-7. https://doi.org/10.1016/j.apenergy.2015.01.097

[8] T.A. Wierzbicki, I.C. Lee, A.K. Gupta, Recent advances in catalytic oxidation and reformation of jet fuels, Applied Energy 145 (2015) 1-7. https://doi.org/10.1016/j.apenergy.2015.01.097

[9] C.P. O'Brien, G.R. Jenness, H. Dong, D.G. Vlachos, I.C. Lee, IC, Deactivation of $Pt/A_{l2}O_3$ during propane oxidation at low temperatures: kinetic regimes and platinum oxide formation, Journal of Catalysis 337 (2016) 122-132. https://doi.org/10.1016/j.jcat.2016.02.012

[10] C.P. O'Brien, I.C. Lee, A detailed spectroscopic analysis of the growth of oxy-carbon species on the surface of Pt/Al_2O_3 during propane oxidation, Journal of Catalysis 347 (2017) 1-8. https://doi.org/10.1016/j.jcat.2016.12.021

[11] C.P. O'Brien, I.C. Lee, "Kinetic modeling of spillover and temperature-programmed oxidation of oxy-carbon surface species on Pt/Al2O3," Journal of Physical Chemistry C 121 (2017) 12329-12336. https://doi.org/10.1021/acs.jpcc.7b03858

[12] I.C. Lee, A. Booth, Catalytic combustion of dodecane and jet fuels with Pt and Rh, in preparation.

[13] Turns, S. R. *An Introduction to Combustion*, McGraw Hill, 2nd Edition (2000).

[14] Miesse, C. M., Masel, R. I., Jensen, C. D., Shannon, M. A., Short, M., "Submillimeter-scale Combustion," AICHE J. 50 (2004) 3206. https://doi.org/10.1002/aic.10271

[15] Hua, J. S.; Wu, M.; Kumar, K., "Numerical Simulation of the Combustion of Hydrogen-air Mixture in Micro-scaled Chambers. Part I: Fundamental Study," Chemical Engineering Science 60 (2005), 3497. https://doi.org/10.1016/j.ces.2005.01.041

[16] Kaisare, N. S.; Vlachos, D. G., "Optimal Reactor Dimensions for Homogeneous Combustion in Small Channels," Catalysis Today 201 (2007), 96. https://doi.org/10.1016/j.cattod.2006.07.036

[17] Maruta, K.; Takeda, K.; Ahn, J.; Borer, K.; Sitzki, L.; Ronney, P. D.; Deutschmann, O., "Extinction Limits of Catalytic Combustion in Microchannels,"

Proceedings of the Combustion Institute 29 (2002) 957. https://doi.org/10.1016/S1540-7489(02)80121-3

[18] Kim, K. T.; Dae, H.L.B.; Kwon, S., "Effects of Thermal and Chemical Surface-flame Interaction on Flame Quenching," Combustion & Flame 146 (2006), 19. https://doi.org/10.1016/j.combustflame.2006.04.012

[19] Xue, H.; Yang, W. M.; Chou, S. K.; Shu, C.; Li, Z. W., "Microthermophotovoltaics power system for portable mems devices," Microscale Thermophysical Engineering 9 (2005), 85. https://doi.org/10.1080/10893950590913431

[20] Aldushin, A. P.; Rumanov, I. E.; Matkowsky, B. J.., "Maximal Energy Accumulation in a Superadiabatic Filtration Combustion Wave," Combustion & Flame 118 (1999), 76. https://doi.org/10.1016/S0010-2180(98)00163-1

[21] Maruta, K.; Parc, J. K.; Oh, K. C.; Fujimori, T.; Minaev, S. S.; Fursenko, R. V., "Characteristics of Microscale Combustion in a Narrow Heated Channel," Combustion Explosion and Shock Waves 40 (2004), 516. https://doi.org/10.1023/B:CESW.0000041403.16095.a8

[22] Miesse, C.; Masel, R. I.; Short, M.; Shannon, M. A., "Experimental Observations of Methane-oxygen Diffusion Flame Structure in a Sub-millimetre Microburner," Combustion Theory and Modelling 9 (2005), 77. https://doi.org/10.1080/13647830500051661

[23] Prakash, S.; Armijo, A. D.; Masel, R. I.; Shannon, M. A., "Flame Dynamics and Structure Within Sub-millimeter Combustors," AICHE J 53 (2007), 1568. https://doi.org/10.1002/aic.11180

[24] Kessler, D. A.; Short, M., "Ignition and Transient Dynamics of Sub-limit Premixed Flames in Microchannels," Combustion Theory and Modelling 12 (2008), 809. https://doi.org/10.1080/13647830801956295

[25] Minaev, S.; Maruta, K.; Fursenko, R., "Nonlinear Dynamics of Flame in a Narrow Channel with a Temperature Gradient," Combustion Theory and Modelling 11 (2007), 187. https://doi.org/10.1080/13647830600649364

[26] Maruta, K.; Kataoka, T.; Kim, N. I.; Minaev, S.; Fursenko, R., "Characteristics of Combustion in a Narrow Channel with a Temperature Gradient," Proceedings of the Combustion Institute 30 (2005), 2429. https://doi.org/10.1016/j.proci.2004.08.245

[27] Lee, I.C.; Ubanyionwu, H., "Sulfur Contaminants in Military Jet Fuels," Fuel 87 (2008) 312. https://doi.org/10.1016/j.fuel.2007.05.010

[28] Lee, G., Zheng, W., Lee, IC., Vlachos, DG, "Volcano curves for homologous series reactions: Small alkane-total oxidation," (submitted).

[29] Ahn J, Eastwood C, Sitzki L, Ronney P, "Gas-phase and catalytic combustion in heat-recirculating burners," Proc Combust Inst 30 (2005) 2463–2472. https://doi.org/10.1016/j.proci.2004.08.265

[30] Balandin, A.A., *Modern State of the Multiplet Theor of Heterogeneous Catalysis1*, in *Advances in Catalysis*, H.P. D.D. Eley and B.W. Paul, Editors. 1969, Academic Press. p. 1-210.

[31] Jacobsen, C.J.H., et al., "Catalyst Design by Interpolation in the Periodic Table: Bimetallic Ammonia Synthesis Catalysts," Journal of the American Chemical Society, 123 (2001) 8404-8405. https://doi.org/10.1021/ja010963d

[32] Logadottir, A., et al., "The Brønsted–Evans–Polanyi Relation and the Volcano Plot for Ammonia Synthesis over Transition Metal Catalysts" Journal of Catalysis, 197 (2001) 229-231. https://doi.org/10.1006/jcat.2000.3087

[33] Tamura, M., et al., "Volcano-Curves for Dehydrogenation of 2-Propanol and Hydrogenation of Nitrobenzene by SiO2-Supported Metal Nanoparticles Catalysts As Described in Terms of a d-Band Model," ACS Catalysis 2 (2012) 1904-1909. https://doi.org/10.1021/cs300376u

[34] Chianelli, R.R., et al., "Transition metal sulfide catalysis: Relation of the synergic systems to the periodic trends in hydrodesulfurization," Journal of Catalysis 86 (1984) 226-230. https://doi.org/10.1016/0021-9517(84)90366-X

[35] Pecoraro, T.A. and R.R. Chianelli, "Hydrodesulfurization catalysis by transition metal sulfides," Journal of Catalysis 67 (1981) 430-445. https://doi.org/10.1016/0021-9517(81)90303-1

[36] Kaisare NS, Vlachos DG, "A review on microcombustion: fundamentals, devices and applications," Prog Energy Combust Sci 38 (2012)321–359. https://doi.org/10.1016/j.pecs.2012.01.001

[37] Kyritsis DC, Roychoudhury S, McEnally CS, Pfefferle LD, Gomez A, "Mesoscale combustion: a first step towards liquid fueled batteries," Exp Thermal Fluid Sci 28 (2004)763–770. https://doi.org/10.1016/j.expthermflusci.2003.12.014

[38] Kyritsis DC, Coriton B, Faure F, Roychoudhury S, Gomez A, "Optimization of a catalytic combustor using electrosprayed liquid hydrocarbons for mesoscale power generation," Combust Flame 139 (2004) 77–89. https://doi.org/10.1016/j.combustflame.2004.06.010

[39] Lee IC, "Rhodium supported on thermally enhanced zeolite as catalysts for fuel reformation of jet fuels," Catal Today 136 (2008) 258–265. https://doi.org/10.1016/j.cattod.2008.03.027

[40] Krummenacher JJ, West KN, Schmidt LD, "Catalytic partial oxidation of higher hydrocarbons at millisecond contact times: decane, hexadecane, and diesel fuel," J Catal 215 (2003) 332–343. https://doi.org/10.1016/S0021-9517(03)00011-3

[41] Dreyer BJ, Lee IC, Krummenacher JJ, Schmidt LD, "Autothermal steam reforming of higher hydrocarbons: n-decane, n-hexadecane, and JP-8," Appl Catal A 307 (2006) 184–194. https://doi.org/10.1016/j.apcata.2006.03.039

[42] Strohm JJ, Zheng J, Song C., "Low-temperature steam reforming of jet fuel in the absence and presence of sulfur over Rh and Rh-Ni catalysts for fuel cells," J Catal 238 (2006) 309–320. https://doi.org/10.1016/j.jcat.2005.12.010

[43] Zheng J, Strohm JJ, Song C., "Steam reforming of liquid hydrocarbon fuels for micro-fuel cells. Pre-reforming of model jet fuels over supported metal catalysts," Fuel Process Technol 89 (2008) 440–448. https://doi.org/10.1016/j.fuproc.2007.11.024

[44] DuBois TG, Nieh S., "Selection and performance comparison of jet fuel surrogates for autothermal reforming," Fuel 90 (2011)1439–448. https://doi.org/10.1016/j.fuel.2010.12.040

[45] Cheekatamarla PK, Finnerty CM, "Synthesis gas production via catalytic partial oxidation reforming of liquid fuels," Int J Hydrogen Energy 33 (2008) 5012–5019. https://doi.org/10.1016/j.ijhydene.2008.07.003

[46] Kim DH, Ryu JW, Choi EH, Gong GT, Lee H, Lee BG, et al. "Production of synthesis gas by autothermal reforming of iso-octane and toluene over metal modified Ni-based catalyst," Catal Today 136 (2008) 266–272. https://doi.org/10.1016/j.cattod.2008.02.021

[47] Tolmachoff, E., Booth, AD, Lee IC, Allmon, WR, Waits, CM, "Modeling and Experimental Analysis of n-dodecane Oxidation in a Platinum-coated Channels," Combustion and Flame 162 (2015) 3674-3680. https://doi.org/10.1016/j.combustflame.2015.07.006

Chapter 15

Pyroelectric Energy Conversion for Army Applications

Yomery Espinal[1,2], Cameron Nevile[1], Felisa Sze[1], Andrew N. Smith[3], Nikolay Ter-Gabrielyn, Nicholas R. Jankowski[1], C Mike Waits[1], and Brendan M. Hanrahan[1]

[1]US Army Research Laboratory, 2800 Powder Mill Road, Adelphi MD, 20783-1197, USA

[2]Department of Material Science and Engineering, University of Connecticut, Storrs, CT, 06269, USA

[3]US Naval Academy, 121 Blake Road, Annapolis, MD, 21402

Abstract

Satisfying energy requirements during extended Army missions in remote locations remains a challenge. Wireless thermal energy conversion, storage and harvesting provide a possible solution to persistent energy needs, however improving efficiency remains an area of opportunity. One alternative to deliver power to far-off locations is to wirelessly transmit power nodes. This can be accomplished via inductive or capacitive coupling for short distances (<10cm) or a laser driven source for long distances (>10cm) transmission. Pyroelectric thin films are attractive for energy conversion applications because they have the potential for high efficiency at low temperature fluctuations and high power density conversion under rapid thermal cycling. Ferroelectric capacitors are materials of choice for PEC due to well-established fabrication methods and well characterized material properties. In addition, classical thermodynamic analysis provide guidance for selecting regimes of operating temperatures and electric fields that will optimize performance and maximize work output.

This chapter explores the thermodynamics of the energy conversion cycle and performance of the Brayton cycle as compared to the conventional Ericsson pyroelectric cycle. Cycle performance is examined using coefficients extracted from measured isothermal polarization hysteresis loops. It is found that the Brayton cycle is slightly more efficient than the Ericsson cycle over the range of temperatures tested and has significant efficiency improvements with increasing pyroelectric coefficients. The results from actual energy conversion cycles differed slightly from simulated performance, confirming the known challenges with synchronizing pyroelectric cycles with realistic thermal excitation. Finally, we also demonstrate the first realistic pyroelectric conversion cycle at frequencies with improved power densities. A modulated laser heat source is used to rapidly cycle temperature and illustrate the potentially high impact of efficient pyroelectric energy conversion as an alternative energy harvesting or power generation technology.

Keywords

Pyroelectrics, Energy Harvesting, Wireless Power, Thin Films

Contents

1. Introduction

The majority of worldwide electrical energy production originates from thermal-to-electric conversion. Thermal-to-electric energy conversion can be accomplished by thermoelectric (TE), Thermophotovoltaic (TPV), thermionic (TI), and pyroelectric mechanisms. Among these mechanisms, pyroelectric energy conversion (PEC) shows the most promise as a thermal energy receiver in a wireless power scheme. PEC requires a time-variant temperature, easily obtained by a modulated laser source, whereas a TE needs a tuned spatial temperature gradient, which can be challenging to establish and maintain over constrained receiver dimensions. Temperatures needed for efficient energy conversion can be as low as room temperature for PEC[1], while TI and TPV both require temperatures in excess of 500°C for operation. When comparing wireless PEC to using photovoltaics, the pyroelectric will not have the efficiency of a band-gap matched diode, but will have the advantages of wavelength-agnostic operation and operability at longer wavelengths where diode-based conversion is unavailable such as above >3 μm.

Certain materials such as thermoelectrics or pyroelectrics can intrinsically convert thermal-to-electrical energy by utilizing asymmetrical electrical or crystalline material properties and could one day provide significant electrically energy production. In pyroelectric materials, a temperature dependence on dielectric displacement creates a 2nd order relationship between thermal and electrical properties. Early studies on pyroelectric energy conversion (PEC) yielded disappointingly low potential conversion efficiencies of <1% [2]. These conversion efficiencies were increased by an order of magnitude in the 1980's when Olsen and others devised thermodynamic conversion cycles using pyroelectric properties and engineered thermal and electrical processes [3]. Pyroelectric thin films have garnered attention for their superior pyroelectric coefficients from strain engineering [4-6], graded/multilayer structures [7-9], and potential for significant energy conversion by sustaining large electric fields [10, 11]. There has not been a subsequent re-analysis of PEC cycles to properly take advantage of the high breakdown strength and low thermal mass of thin films.

One of the first studies on pyroelectric conversion cycles was performed by Fatuzzo *et al.*, who related the entropy and Gibbs free energy to the dielectric displacement, temperature, and specific heat. This study concluded that the conversion efficiency could not exceed 1% efficiency based on the ratio of stored thermal energy to "destroyed polarization" [2]. Drummond was the first to propose a thermodielectric energy conversion method using cascaded materials with properties optimized to specific temperature ranges and built a system based on this concept [12]. Margosian examined a Stirling-type PEC (then termed 'thermoelectrostatic') cycle for space applications [13]. Following this work, Olsen published a series of studies on PEC cycles [3, 14, 15], concluding the Ericsson cycle both provided the most work output and was the most practical to implement considering the bulk-ceramic or polymer-based conversion systems which were being targeted. In 2008, Sebald *et al.*, provided a thorough analysis of the thermodynamic processes of the Ericsson cycle showing that there were two regimes of conversion, one with linear pyroelectric response and a non-linear response regime that can be modeled using the electrocaloric response of the material [16]. It is worth noting that within the electrocaloric community, which is making use of similar materials/physical phenomena, the Carnot cycle is the most widely considered conversion cycle because those systems are focused on maximizing the adiabatically induced temperature difference, although other cycles including the Brayton cycle have been explored [17, 18]. Sebald and Olsen separately showed that the Carnot cycle was impractical for bulk pyroelectrics [3, 19]. The advantages of thin films over bulk pyroelectrics can be inferred from the PEC energy and power equations. Pyroelectric energy conversion takes place between different temperature dependent dielectric displacements, *D,* and applied electric fields, *E*, where the work output can be calculated from Eqn. 1

$$W = \oint E dD \qquad [1]$$

Therefore work output can be improved by increasing ΔE and ΔD. Thin films can withstand high electric fields relative to bulk materials which leads to larger values of ΔE and the application of such fields can arise from relatively low voltages, e.g., 1 V over a 1 μm film results in a 10 kV/cm electric field. Previous demonstrations with bulk materials had been limited by breakdown of surrounding media [20]. The power output, *P,* from the pyroelectric conversion cycle is the product of the cycle frequency, ω, and work per cycle, *W*,

$$P = W\omega \qquad [2]$$

Cycle frequency can be much higher with thin films because of their lower thermal mass. Because pyroelectric energy density scales with surface area, rather than volume, the thermal mass reduction occurs without a proportional loss. Bhatia *et al.* recently took this to an extreme and realized a four orders-of-magnitude increase in pyroelectric power density over bulk materials using 150 nm-thick thin films and temperature cycles at 1 kHz with thin film barium titanate [21]. Laser-heated pyroelectrics, enabling wireless power schemes, have also been demonstrated to over 100 Hz cycle frequencies [22].

Considering this, a thermodynamic PEC process can be tailored to thin-film systems in order to provide both efficiency and power density benefits that exceed previously explored conversion schemes. To support this we have developed both thermodynamic and heat transfer models based on measured material properties. In this system we assume the use of a pulsed heat source, such as from a laser, and the transient thermal conditions are sufficiently controllable to drive the material through the desired thermodynamic cycle. These assumptions allow the implications of increased cycle rates to be explored. Using these models we also demonstrate the first realistic energy harvesting concept using PEC at the frequencies where power densities become attractive. The modulated laser heat source used in this work provides the benefits of both high frequency thermal cycles and long-distance energy transfer. We measure the intrinsic power density of laser-stimulated energy harvesting of a thin film PZT receiver coated with a low-reflectivity, nanostructured IrO_2 top electrode as a new wireless power technology.

2. Pyroelectric Materials

2.1 Sample Preparation

Thin film lead zirconate titanate (PZT) samples were prepared by chemical solution deposition with a Zr:Ti ratio of 60:40. 150 mm Silicon substrates were coated with 500 nm of thermally grown SiO_2. Next, a 30 nm layer of Ti was sputter-deposited (Unaxis Clusterline 200) then oxidized at 750 °C in a furnace with an O_2-rich environment. The TiO_2 then acts as template for a {111}-textured Pt bottom electrode [23], which is sputter-deposited to a thickness of 100 nm. Chemical solution deposition using 2-methoxyethanol solvent was used to grow the PZT thin films. A distilled solution of lead (III) acetate trihydrate was mixed with zirconium (IV) n-proproxide and titanium (IV) isopropoxide in an appropriate ratio to achieve 60:40 Zr:Ti cation ratios. The solution is spun on to the substrate at a speed of 2500 rpm and pyrolized at 350 °C for 2 min to remove a majority of the organic species. Following pyrolysis, the films were annealed (AG Associates Heatpulse 610 RTA) at 700 °C for 60s in an O_2 environment, resulting in a ~60 nm film, measured using ellipsometry. This process is repeated until the film measures 500 nm and confirmed crystalline (*Figure 1*) using x-ray diffraction (Rigaku Ultima III Diffractometer). IrO_2 top electrodes are reactively sputtered (Unaxis Clusterline 200) at 30 °C in a 3:1 $Ar:O_2$ plasma from an Ir target to a thickness of 100 nm, measured using cross-section SEM (*Figure 1*b). Further details can be found in reference [24]. Top electrode structures are defined using photolithography and ion-milling (4wave 4W-PSIBE).

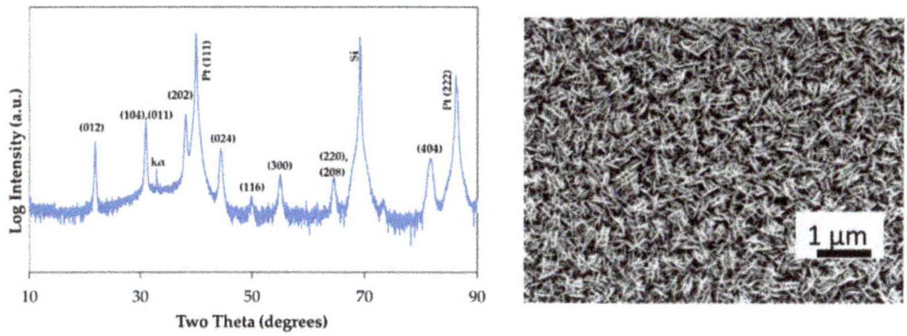

(a) (b)

Figure 1. (a) X-ray diffraction pattern for PZT pyroelectric thin film with IrO_2 *top electrode. (b)SEM image of the nanostructured* IrO_2 *top electrode Electrical characterization of the PZT materials was measured using isothermal bipolar dielectric hysteresis measurements. Dielectric hysteresis loops were measured at 10 kHz over ±20 V using the Radiant Technologies Precision Premier II ferroelectric testing unit in conjunction with a heated probe station with a temperature resolution of ±0.1 °C.*

2.2 Determination of Pyroelectric Coefficient and Dielectric Constant

Once the isothermal relationship between the polarization and electric field was determined, the pyroelectric and dielectric constants of the material were determined using the following relationships.

$$dD = \left(\frac{dD}{dT}\right)_E dT + \left(\frac{dD}{dE}\right)_T dE = pdT + \epsilon dE \qquad [3]$$

Where,

$$p = \left(\frac{dD}{dT}\right)_E \qquad [4]$$

$$\epsilon = \left(\frac{dD}{dE}\right)_T \qquad [5]$$

Each isothermal D-E measurement was curvefit using a cubic spline interpolation to model the relationship between polarization and electric field. The isothermal polarization curvefits were then evaluated at even increments of the electric field in order to construct isoelectric field relationships. The isoelectric field relationships were then curvefit using a 3rd order polynomial fit and evaluated at even increments of temperature. The curvefit isotherm polarization curves are shown over a temperature range of 30 to 120 C and 5 to 200 kV/cm (*Figure 2*).

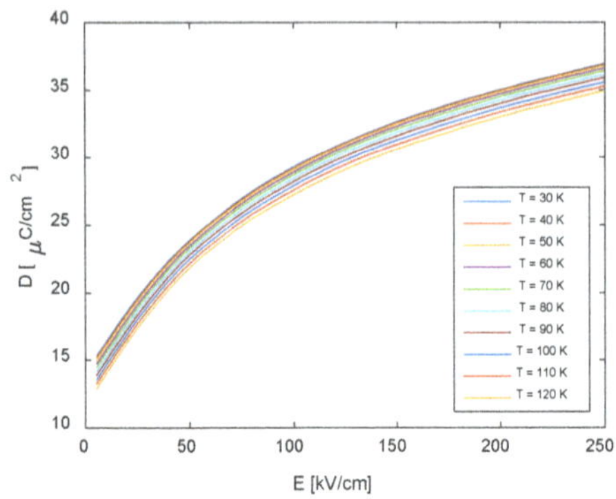

Figure 2. Curvefit results for Polarization versus Electric Field at Constant Temperature..

The dielectric constant, defined as the change in polarization with respect to electric field at constant temperature, was determined from the numerical derivative of the isothermal polarization curvefits with respect to electric field using a center difference approach (*Figure 3*). The resulting dielectric constant was a strong function of electric field but relatively independent of temperature over the tested range which is far from the Curie temperature of the material (~350 °C).

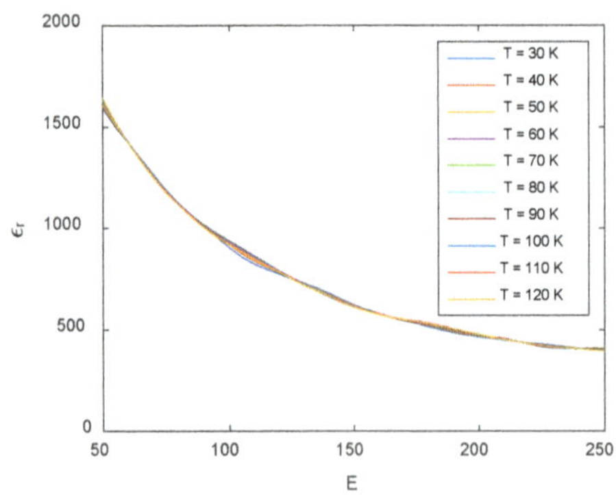

Figure 3. Relative permittivity of thin film 6040 PZT determined from Isothermal Hysteresis Loops.

The pyroelectric coefficient, defined as the change in polarization with temperature at constant electric field, was then determined from the numerical derivative of the isoelectric field polarization curvefits with respect to temperature using a center difference approach (*Figure 4*). The resulting pyroelectric constant was a strong function of temperature but relatively independent of electric field over the range of applied fields. The deviation of the pyroelectric coefficient relationships from predictions based on Landau-Ginzberg-Devonshire theory highlight the challenge of using isothermal dielectric hysteresis loops to interpolate pyroelectric properties. An in-depth discussion of the source of the deviation is beyond the scope of this chapter, but the origins likely arise from thermal strains coupling to the piezoelectric response of the film, as well as the susceptibility of dielectric hysteresis measurements to leakage current.

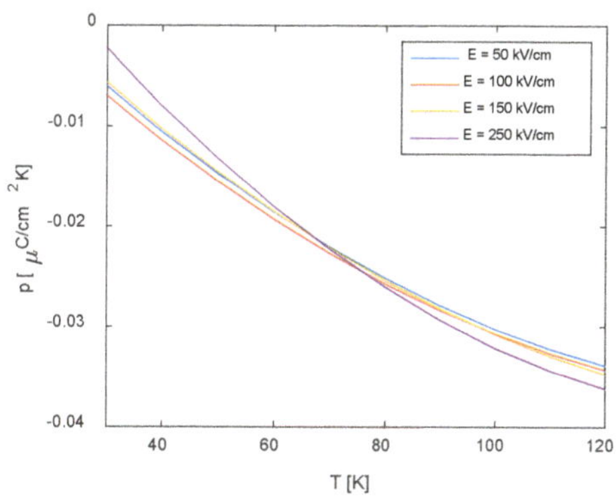

Figure 4. Pyroelectric constant of thin film 6040 PZT determined from Isothermal Hysteresis Loops.

3. Thermodynamic Theory

Olsen's seminal pyroelectric cycling study demonstrated that using the Ericsson cycle equivalent for pyroelectrics produced 30% more energy than other compared cycles [3]. More recently, the authors derived analytical relationships exploring the work and power relationships for the pyroelectric Ericsson cycle and compared it to the Brayton equivalent cycle [25]. Representative phase diagrams are shown in *Figure 5* for both the Ericsson and Brayton Cycle. As can be seen, the primary differences between the two cycles are that the field changes occur at constant temperature in the Ericsson cycle (dashed red lines), whereas they occur isentropically in the Brayton cycle. The net work for pyroelectrics cycles can be determined by performing the cyclic integral of the polarization-electric field state diagrams; and cyclic integral on a temperature-entropy diagram for internally reversible cycles:

$$W_{net} = \oint DdE = \oint TdS \qquad [6]$$

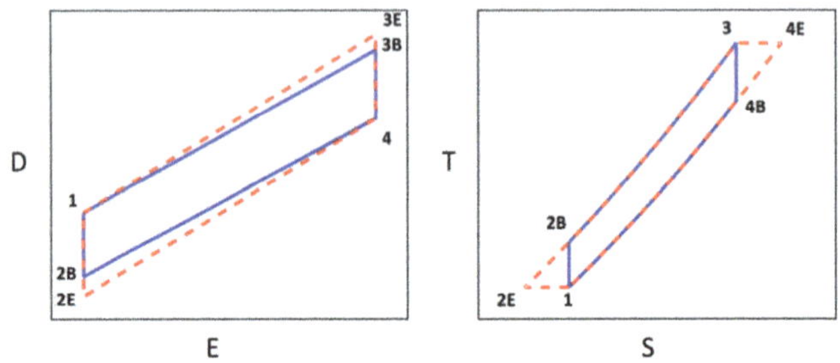

Figure 5. Polarization (D) vs Electric field (E) and Temperature (T) vs Entropy (S) diagrams for the pyroelectric implementation of the ideal Brayton cycle (solid lines), where isentropic charging and discharging replace the Ericsson isothermal conditions (shown with dashed lines).

By inspection of *Figure 5* it can be seen that the Ericsson cycle produces more work per cycle when the two cycles are operating between the same temperature and electric field strength. However, thermal analysis showed that there is strong coupling between the transient operating conditions and the driving thermal reservoirs for both cycles. This results in significant differences between the frequency dependent behaviors of the two cycles. Full thermodynamic and thermal derivations of those relationships including a detailed analysis comparing the cycles can be found in [25]. A summary of the cycle work and efficiency expressions for Ericsson and Brayton cycles are included in Table 1.

Table 1. Heat relationships for processes within the Brayton and Ericsson cycles from [25]

	Brayton	Ericsson
1→2 Charging	Adiabatic $Q_{12} = 0$	Isothermal $Q_{12} = p(E_H - E_L)T_L$
2→3 Heat Addition	Isoelectric Field $Q_{23} = C_E\left(T_H - T_L e^{-p(E_H-E_L)/C_E}\right)$	Isoelectric Field $Q_{23} = C_E(T_H - T_L)$
3→4 Discharging	Adiabatic $Q_{34} = 0$	Isothermal $Q_{34} = -p(E_H - E_L)T_H$
4→1 Heat Rejection	Isoelectric Field $Q_{41} = C_E\left(T_L - T_H e^{p(E_H-E_L)/C_E}\right)$	Isoelectric Field $Q_{41} = C_E(T_L - T_H)$
Net Work	$W_B = C_E\left[T_H\left(1 - e^{p\Delta E/C_E}\right) + T_L\left(1 - e^{-p\Delta E/C_E}\right)\right]$	$W_E = -C_E(T_H - T_L)\left(p\Delta E/C_E\right)$
Output efficiency	$\eta_B = 1 - e^{\frac{p\Delta E}{C_E}}$	$\eta_E = \left(\frac{T_H}{\Delta T} - \frac{C_E}{p\Delta E}\right)^{-1}$

3.1 Constant Pyroelectric Coefficient Work Relationships

In [25] it was identified that pyroelectric thermal cycle output is strongly influenced by the dimensionless ratio of the electrocaloric effect to the material thermal capacity:

$$r_{ec} = {-p\Delta E}/{C_E} \qquad [7]$$

Using the cycle work expressions in Table 1, we can further examine the dependence of work and efficiency for the two cycles here. Because of the isothermal and isoelectric switching conditions, the Ericsson work output is simply, with increasing work output in proportion to the electrocaloric ratio:

$$W_E = -C_E(T_H - T_L)\left(\frac{p\Delta E}{C_E}\right) \qquad [8]$$

This is contrasted with the Brayton work output relationship, which is made more complex by the non-isothermal condition during charge and discharge:

$$W_B = C_E(T_H + T_L) - C_E T_L e^{-p\Delta E/C_E} - C_E T_H e^{p\Delta E/C_E} \qquad [9]$$

Expanding Eqn. 9 using the first 3 terms in a Taylor series for the exponential function with the electrocaloric ratio helps illustrate the comparison of the Brayton cycle net work relative to the Ericsson cycle:

$$W_B = -C_E(T_H - T_L)\frac{p\Delta E}{C_E} - C_E\frac{T_H + T_L}{2!}\left(\frac{p\Delta E}{C_E}\right)^2 - C_E\frac{T_H - T_L}{3!}\left(\frac{p\Delta E}{C_E}\right)^3 + \cdots \qquad [10]$$

As can be seen, work output for the two cycles converges when the magnitude of $p\Delta E/C_E$ is small, and the Brayton work output per cycle relative to Ericsson decreases as the magnitudes of $p\Delta E/C_E$ increases.

A similar examination can be made of cycle efficiency, defined as:

$$\eta_{th} = W_{net}/Q_{in} \qquad [11]$$

Eqn. 11 shows that the Brayton cycle is increasingly more efficient than the Ericsson cycle for growing $p\Delta E/C_E$ ratios. Applying a Taylor expansion with respect to $p\Delta E/C_E$ of the cycle efficiency terms from Table 1:

$$\eta_B = 1 - e^{\frac{p\Delta E}{C_E}} = -\frac{p\Delta E}{C_E} - \frac{1}{2!}\left(\frac{p\Delta E}{C_E}\right)^2 - \frac{1}{3!}\left(\frac{p\Delta E}{C_E}\right)^3 + \cdots \qquad [12]$$

$$\eta_E = \left(\frac{T_H}{\Delta T} - \frac{C_E}{p\Delta E}\right)^{-1} = -\frac{p\Delta E}{C_E} - \frac{T_H}{\Delta T}\left(\frac{p\Delta E}{C_E}\right)^2 - \left(\frac{T_H}{\Delta T}\right)^2\left(\frac{p\Delta E}{C_E}\right)^3 + \cdots \qquad [13]$$

We can see from Eqns. 12 and 13 that while the two cycles have the same efficiency for small values of $p\Delta E/C_E$, the Ericsson cycle efficiency will always be lower than that of Brayton due to its dependence on the reservoir temperatures, the ratio of which can never exceed one. Physically, Brayton's increased efficiency can be attributed to the replacement of an isothermal process, requiring heat addition, with an adiabatic one. These work and efficiency relationship comparisons can be seen in *Figure 6*.

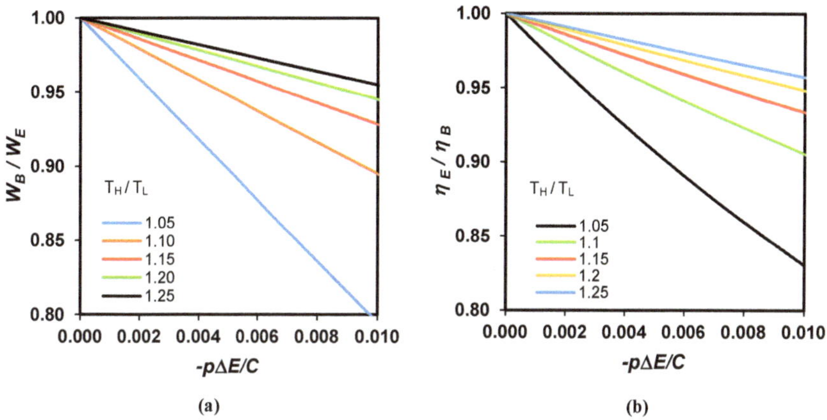

Figure 6. Ratio of (a) net work and (b) cycle efficiency between the Brayton and Ericsson cycles over a range of material and temperature ratios.

In summary, the Brayton cycle's increased efficiency comes at the expense of net work per cycle. These results are consistent with *Figure 5* where the area of the polarization – electric field loop is diminished for the Brayton cycle compared to the Ericsson cycle. However, because power is proportional to both work output and cycle frequency, a thermal transient analysis is required to understand the full implications of operation the two cycles.

3.2 Frequency Impacts on Cycle Temperature

It was shown in [25] that PEC power output is not simply linearly proportional to frequency as Eqn. 2 suggests. This occurs primarily because the net work shows a frequency dependence as non-ideal cooling conditions limit achievable temperature changes. Here we further examine the pyroelectric thin-film transient response assuming the thin film pyroelectric device has a lumped thermal capacitance and can be represented by a single temperature. Additionally, a constant thermal resistance to the ambient environment is assumed to be primarily due to convective cooling using an effective convection heat transfer coefficient h_{eff}, where a value of h_{eff} = 100 W/mK has been chosen as a reasonable value for either forced air cooling or area enhanced passive air cooling using a heat sink. With these assumptions and the representative values for a thin film PZT material shown in Table 1, we can predict the temperature of the thin film PZT as a function of time [25]. Note that for this discussion a pyroelectric coefficient value of $p = -0.0025$ C/m^{2}K was arbitrarily chosen to make cycle differences more distinct.

Table 2. Transient model parameters.

T_H	250°C	p	-0.0025 C/m^{2}K
T_L	*varied*	C_E	3.15 J/cm^{3}K
T_∞	25°C	h_{eff}	100 W/m^{2}K
E_H	1000 kV/cm	P_p	10 W/cm^2
E_L	100 kV/cm	l	500 nm

The electric charging and discharge were assumed to be nearly instantaneous for the Brayton Cycle, while for Ericsson thee process time depended on the charge and discharge rates that were set to maintain process isothermality. The maximum cycle temperature was held fixed at 250 °C and the heater power density was held constant at a rate of 10 W/m^2. For the Ericsson cycle these values were then used to determine the duration of the isoelectric heat addition process and the rate and duration of the isothermal discharge process. The remainder of the cycle time is taken up by cooling governed by the overall thermal resistance, $R_{th} = 1/h_{eff}$. This cooling period generally dominates the cycle time, and increasing cycle frequency generally results in raising the process lower temperature, T_L, by shortening this cooling time.

Figure 7 shows both the Brayton and Ericsson cycles for a frequency of 10 Hz. The low temperature, T_L, approaches the sink temperature, T_∞, in both cases. However, the Ericsson cycle must reject heat during the isothermal charging which keeps the T_L value for Ericsson slightly above the Brayton cycle.

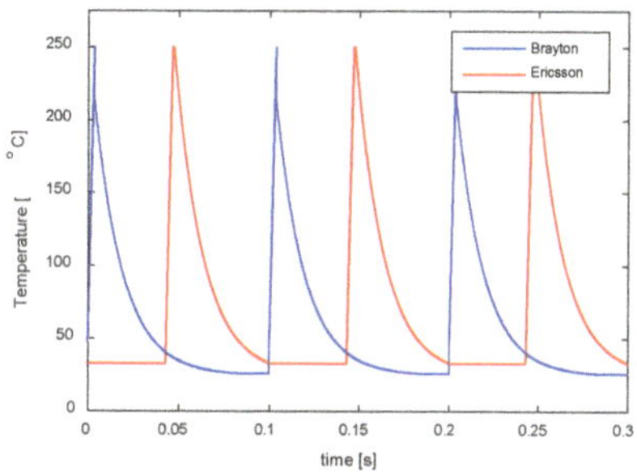

Figure 7. Transient Brayton cycle vs Ericsson Cycle for $f = 10Hz$, $T_H = 250°C$, $h_{eff} = 100W/m^2K$, $p = -0.0025\ C/m^2K$.

Figure 8 shows the same results for a frequency of 50 Hz. The time required for the isothermal heat addition is limited by the rate of heat input and can become significant at higher frequencies. The charge and discharge are nearly instantaneous for the Brayton cycle which leaves additional time for cooling. The Brayton cycle therefore reaches lower temperatures, T_L, especially when operating at higher frequencies.

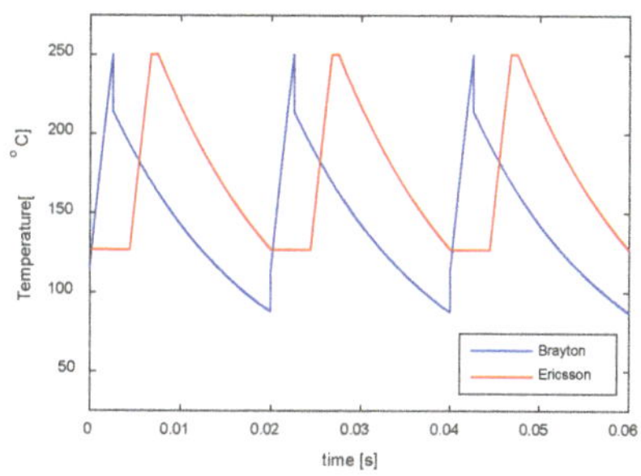

Figure 8. Comparison of transient Brayton cycle versus Ericsson Cycle for $f = 50Hz$, $T_H = 250°C$, $h_{eff} = 100W/m^2K$, $p = -0.0025\ C/m^2K$.

The phase diagram for the transient results shown in Figure 7 and Figure 8 are shown in Figure 9. The temperature-entropy diagram at 10 Hz, Fig. 9(a), and 50 Hz, Fig. 9(b), show that the Ericsson cycle produces more net work at low frequencies, while the Brayton cycle produces more work at high frequencies. This is largely due to the fact that the Brayton cycle maintains a larger temperature difference across the cycle. The same conclusion can be drawn from the polarization-electric field diagrams. The hot temperature, T_H, is held constant, so as T_L decreases for both cycles the isoelectric heat addition and rejection leg diminish and there is less work produced per cycle. Thus, for both cycles work per cycle becomes frequency dependent as shorter cycle periods shrink the cycle temperature band, with a larger impact on Ericsson due to the time taken by the isothermal portions of the cycle.

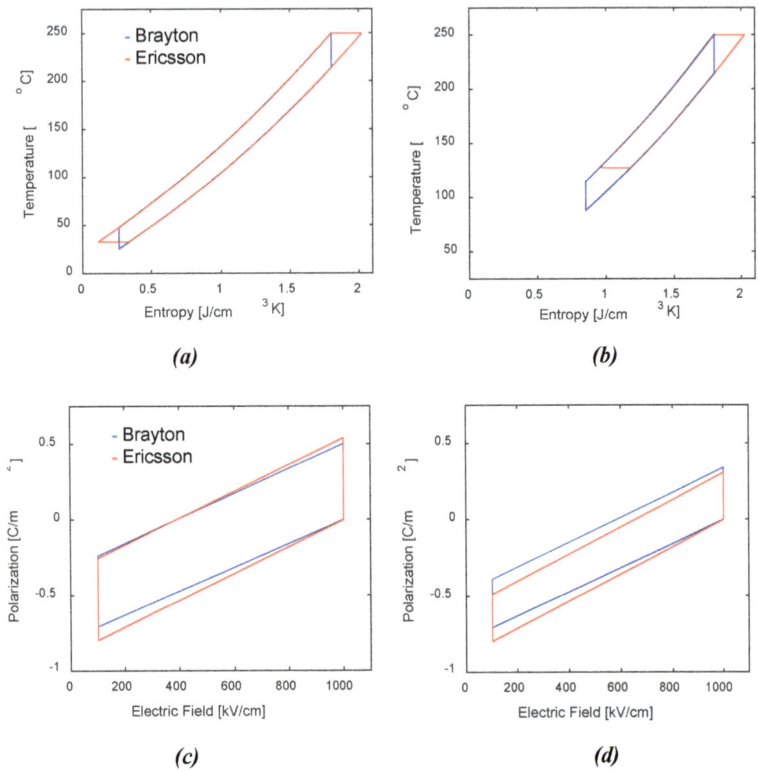

Figure 9. Temperature-Entropy diagrams that at (a) 10Hz and (b) 50 Hz; Polarization-Electric Field diagrams at (c) 10 Hz and (d) 50 Hz, $T_H = 250°C$, $h_{eff} = 100 W/m^2K$, $p = -0.0025\ C/m^2K$.

3.3 Frequency Dependent Cycle Work and Power Output

Figure 10 shows both the calculated cycle net work and output power as a function of frequency. At very low frequencies there is not much change in T_L or W, producing a steady increase in power for both cycles as cycle rate increases. Eventually, as was shown in *Figure 8*, T_L begins to increase more with frequency and new work starts to roll off for both cycles. Ericsson's larger time requirement for the same operating temperatures causes a much steeper decline in cycle work than for Brayton. This causes both net work and power to eventually drive to zero. Brayton on the other hand has a much more gradual decrease in work output such that increasing cycle rate continues to increase power output. This continues until at high frequencies the cycle work decreases in direct proportion to the increase in frequency and output power saturates.

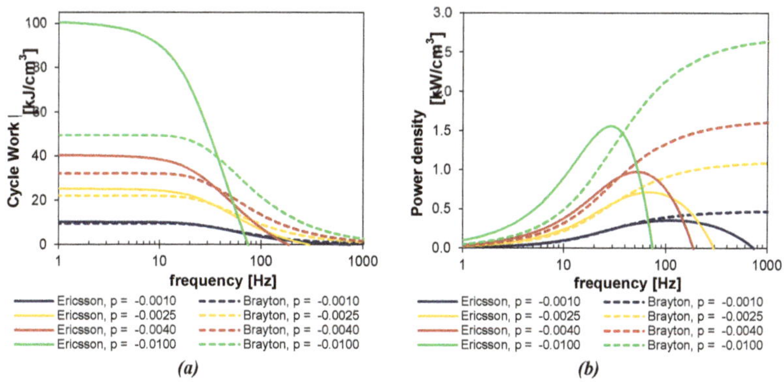

Figure 10. Output power density for Ericsson and Brayton cycles as a function of cycle frequency for several values of p.

3.4 Frequency Impact on Thermodynamic Efficiency

Because frequency has an indirect effect on cycle work output, the previously discussed thermodynamic efficiencies will be affected as well. *Figure 11* shows that the absolute efficiency of the ideal Brayton cycle is independent of frequency as both the net work and heat addition decrease at an equal rate with increasing switching rate. However, the ideal Ericsson efficiency decreases with increasing frequency because of the dependence of the efficiency on the operating temperatures, T_H and T_L, as seen in Eqn. 13. In our model, T_L increases with increasing frequency due to limited cooling, which causes the Ericsson cycle efficiency to decrease and approach zero when T_L approaches T_H. *Figure*

11 also shows the cycle's thermodynamic efficiency, which is the cycle efficiency relative to an ideal Carnot cycle, defined as

$$\eta_{Carnot} = 1 - T_{\infty}/T_H \qquad [14]$$

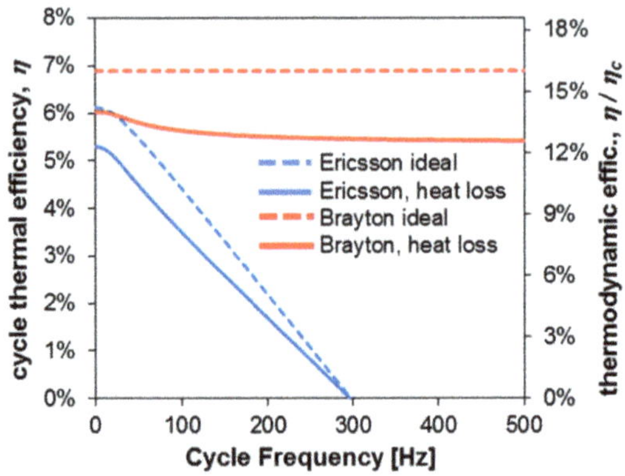

Figure 11. Thermal efficiency of the Brayton and Ericsson cycles. Heat transfer from the system during the heat addition processes lowers the thermal efficiency.

The ideal cycle only accounts for the heat added to the system during the heat addition process. However, since the system is always losing heat to the substrate additional heat must be added which reduces the actual thermal efficiency. The Brayton cycle only requires additional heat during the isoelectric field heating. The total heat added becomes the following, where the pulsed heater power is P_{P} and t_{23} is the duration of the isoelectric field heating.

$$Q_{in,\mathrm{B}} = P_{\mathrm{P}} t_{23} \qquad [15]$$

The additional heat for the Ericsson cycle is required for both the isoelectric field and the isothermal heating processes. The new heating term is shown below where $t_{34}$is the duration of the isothermal heating process.

$$Q_{in,\mathrm{E}} = P_{\mathrm{P}} t_{23} - T_H p \Delta E + t_{34}\,(T_H - T_{\infty})/R_{th} \qquad [16]$$

Complete expressions for each of time segment can be found in [25]. The thermal efficiency is reduced in both cases simply because more heat is required to produce the same amount of net work. This represents another advantage for the Brayton cycle as there is less time spent at higher temperatures, where one of primary loss mechanisms for PEC is temperature-driven leakage current [26]. The isothermal discharge process represents additional time during the Ericsson cycle where heat must be added to the cycle.

4. Brayton Cycle Experiment

4.1 Energy Conversion Set-up

PEC cycles were performed within a modified cryogenic probe station. The large thermal mass of the cryogenic probe station prevented fast temperature cycling, which results in low pyroelectric current, I_p , because of the relationship,

$$I_p = pA\frac{dT}{dt} \qquad [17]$$

where p is the pyroelectric coefficient, A is the capacitor area, and dT/dt is the heating/cooling rate. To increase dT/dt, a small ceramic heating chip was placed in thermal contact with the cold plate, which could be powered against the cold plate, and free-cooled by the cryogenic cold plate. Temperature was measured using a thin-film platinum resistance temperature detector (RTD). A maximum dT/dt of 10 °C/s with a span of 25 °C was measured. The system is described in *Figure 12*.

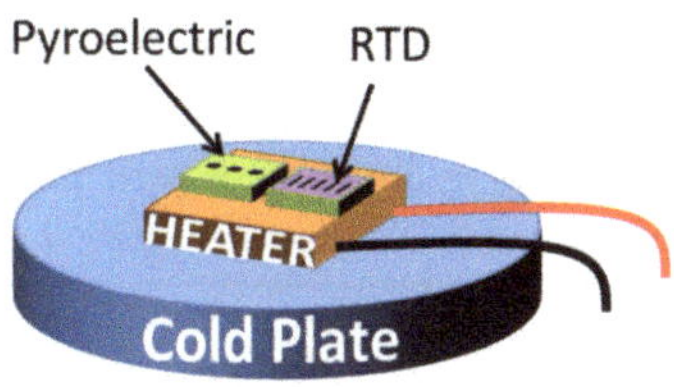

Figure 12. Schematic of energy conversion testing set-up.

Direct temperature measurements of the pyroelectric chip were not performed. Instead, the RTD chip was used as a surrogate temperature sensor. Pyroelectric characterization of the chip is shown in *Figure 13* where the out of phase current arises from the pyroelectric effect. A pyroelectric coefficient of 1.2 x 10-4 C/m^2K can be calculated from this data and Eqn. 17.

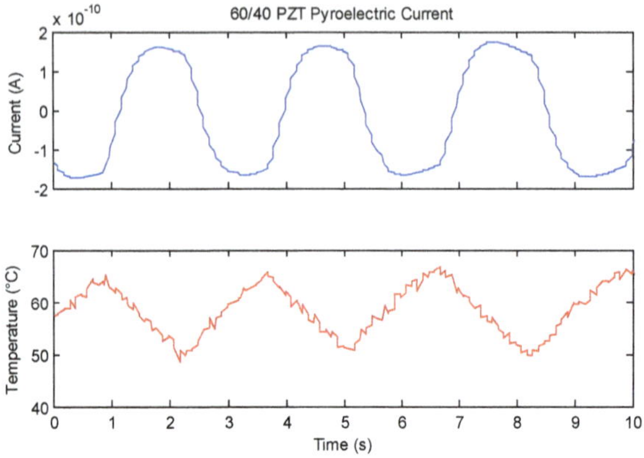

Figure 13. Pyroelectric current measurement with heating cycles.

LabVIEW software was used to synchronize the thermal and electrical cycles needed to perform energy conversion. Electric field changes were triggered based on measured surrogate temperatures and pyroelectric current was measured throughout the cycles.

4.2 Results and Discussion

Synchronized temperature and electric field cycles were applied to a thin film 60:40 PZT sample in a pyroelectric Brayton cycle. Temperatures were driven at a frequency of 0.2 Hz between 46-72 °C and charging/discharging took place in line with heater power, triggered by the temperature. A cycle is shown in *Figure 14*.

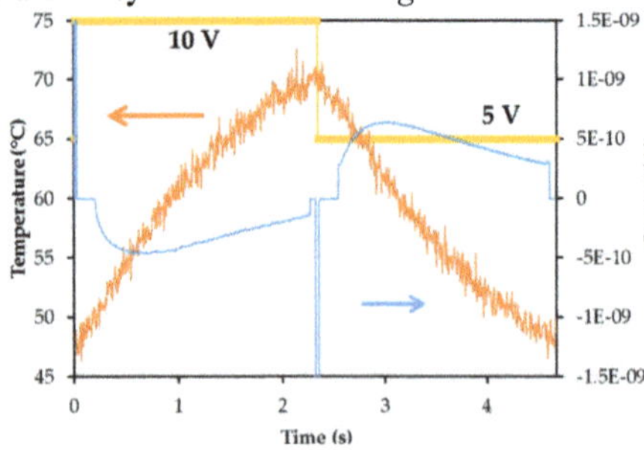

Figure 14. Pyroelectric Brayton cycle showing temperature, current, and applied voltage. Charging and discharging currents are clipped in this graph.

The thermally- and electrically-induced current were 4-orders of magnitude in difference which presented some measurement challenges. To measure each process, a blanking and gain-switching period was introduced on the current amplifier before and after the charge-discharge processes, which results in an under-estimate of the pyroelectric current [27].

It is helpful to view the cycle in terms of dielectric displacement, which is calculated by integrating the cycle current, and applied electric field then comparing to the isothermal hysteresis loops. *Figure 15*shows the modelled and measured D-E conversion loops.

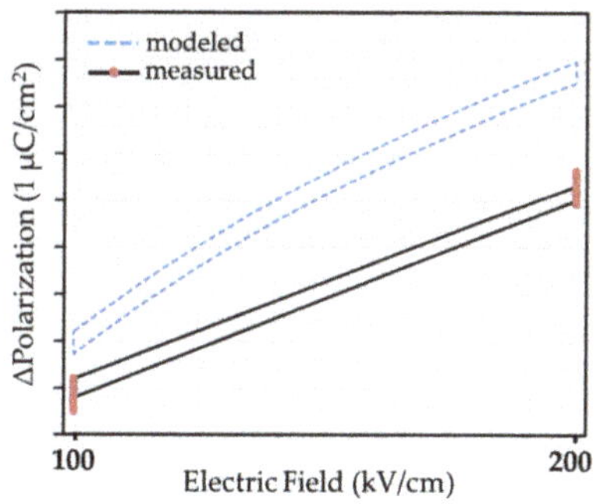

Figure 15. Brayton pyroelectric energy conversion cycles, modeled from isothermal hysteresis loops and measured during cycle implementation. Polarization offset introduced for clarity.

An energy conversion density of 41 mJ/cm^3 per cycle can be calculated from Eqn. 1 using the integrated current. For comparison, Bhatia *et al.* measured 3 mJ/cm^3 conversion in a thin film pyroelectric from an electric field span of 15 kV/cm, or about 7% the energy conversion density of this work using 15% the applied electric field. The authors in this study also tested a range of electric field ramping rates which has in effect testing more Brayton-like, or Ericsson-like conversion cycles [21]. They found that there was no major difference in energy density which was likely from unintended miss-timing of their conversion cycles where for example heating and discharging occur simultaneously [28].

The measured energy conversion density can also be compared with zero applied field, resistive cycles, like those used by Yu et al. on bulk ceramics [29]. The work per cycle can be calculated for a resistive pyroelectric cycle from ref. [19] where,

$$W_{resist} = \frac{\pi p^2}{4\varepsilon}(\Delta T)^2 \qquad [18]$$

Using the temperature range in *Figure 14*, an energy density of 721 μJ/cm^3 is calculated using this material in a resistive cycle, which is over 50X less than the measured energy conversion using the Brayton pyroelectric cycle. One of the significant benefits of the thin film regime is the ability to withstand large electric fields, which is not taken advantage of in this type of cycle.

An ideal Brayton cycle, modeled from measured isothermal hysteresis loops, would have an energy conversion density of 48 mJ/cm^3 or about 17% greater than the measured cycle over the same temperature and electric field range. The main source of this difference is from non-ideal synchronization which is a persistent challenge for PEC [28]. Electric field changes were triggered on the heater power signal, where it's assumed the electrical time constants are much greater than the thermal time constants of the system. For example, in a typical cycle the cooling continued for around 20 ms after heater power was initiated, meaning that cooling was happening against the applied electric field and the work per cycle is decreased.

The cycle power can be calculated by the product of the work and cycle frequency or by product of the applied voltage and measured current, where a negative value represents current delivered to the circuit. Using the calculated energy conversion density and a cycle frequency of 0.21 Hz, the cycle power is 8.6 mW/cm^3.

5. Wireless Power Transmission via Modulated Laser Irradiation

These experiments are the first to report a realistic energy harvesting concept using PEC at the frequencies where power densities become attractive. The modulated laser heat source used in this work provides the benefits of both high frequency thermal cycles and long-distance energy transfer. In these experiments we measure the intrinsic power density of laser-stimulated energy harvesting of a thin film PZT receiver coated with a low-reflectivity, nanostructured IrO_2 top electrode as a new wireless power technology.

5.1 Fabrication and Characterization of Pyroelectric Receivers

Fabrication of pyroelectric receivers: Thin film pyroelectric receivers are formed from a Pt/PZT/IrO_2 capacitor structure with an area of 2.21 x 10^{-3} cm^2. The capacitors are prepared by chemical solution deposition with a Zr:Ti ratio of 60:40 in the same manner described in section 2.1 of this manuscript. A Zr-rich PZT was chosen because it was found to have an increased pyroelectric activity versus morphotropic 52:48 PZT (*Figure 16a*). The 52:48 PZT composition has peak piezoelectric properties, that can deleteriously couple polarization to temperature via piezoelectricity and thermal expansion strains, known as the secondary pyroelectric effect [30]. An IrO_2 top electrode is chosen because the nanostructured top electrode serves three purposes for this system: enable electrical probing of the pyroelectric thin film, reduce leakage current through an oxygen vacancy reduction mechanism [26], and act as a low reflectivity IR absorber (*Figure 16b*).

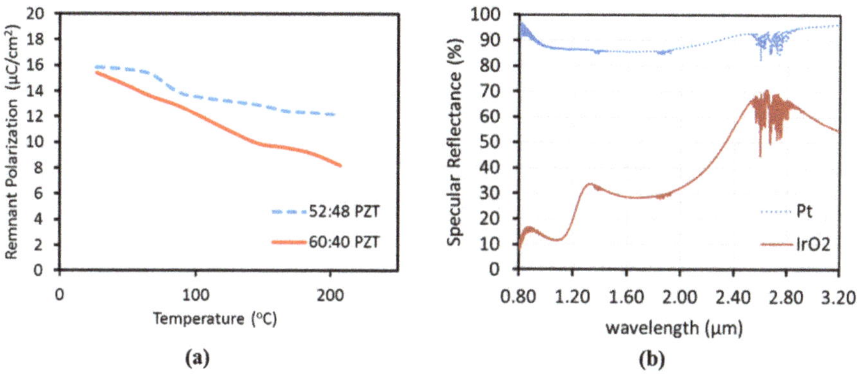

Figure 16. (a) Remnant polarization versus temperature extracted from adiabatic hysteresis loops for a 500 nm thick PZT with IrO_2 top electrodes at either the morphotropic 52:48 or slightly Zr-rich 60:40 composition, where pyroelectric coefficient is dP/dT, (b) Specular reflectance of pyroelectric capacitor structures with two different top electrodes, either nanostructured IrO_2 or Pt.

Material Characterization: Isothermal dielectric hysteresis loops were measured using a Radiant Precision Premier II tester in conjunction with a Lakeshore Model TTPX heated probe station. Bipolar triangular waveforms were applied to the pyroelectric film at +/-35 V amplitudes and 1 kHz at 25 ±0.1 K temperature intervals. The dielectric constant and pyroelectric coefficients were determined by taking numerical derivatives of the polarization versus electric field data measured at specific temperatures.

The pyroelectric constant was also determined by measuring the out-of-phase current from a triangle wave temperature from 46-65 °C imposed on the sample by a ceramic heater in an environmentally-controlled chamber, similar to the technique described in [31]. The chip temperature was determined from an adjacent thin film pt resistive temperature detector chip located on the same ceramic heater, thus experiencing the same thermal cycling.

Energy Conversion Model: The measured transient temperature response was compared to a lumped capacitance model to determine the thermal decay time constant of the device. The lumped capacitance model was then used to predict the temperature response as a function of laser power, phase and duty cycle. Assuming constant properties, the polarization was calculated as a function of time from the temperature and applied electric field and used to predict the power density. The ideal power density was calculated using the same methodology with an instantaneous rise time on the electric field in phase with the heat source.

Wireless power testing: The test circuit consisted of an arbitrary waveform generator (Keithley 3390), a pyroelectric receiver, and a current amplifier (Stanford Research Systems SR570) connected in series (*Figure 17*). A data acquisition module (National Instruments 9201 DAQ) was used to record current and applied voltage data simultaneously at sampling frequencies ranging from 200 S/cycle at 100 Hz to 10 kS/cycle at 0.1 Hz. A LDX-36070 laser driver operating in quasi-constant waveform mode powered a 3.2 W CW laser with a wavelength of 1.5 μm at 20% duty cycle at various frequencies. Electrical power was calculated from the product of the *in operando* voltage and current averaged over 3-5 cycles. Power density is calculated using the pyroelectric thin film dimensions.

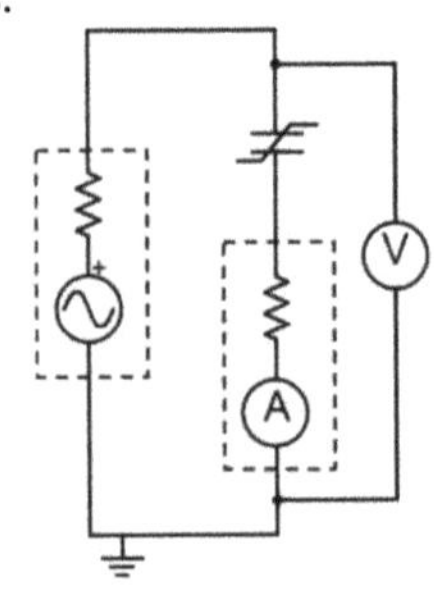

Figure 17. Circuit diagram for the wireless power tests. An electric field is applied to the pyroelectric/ferroelectric capacitor using a function generator, and voltage and current are simultaneously measured across the capacitor.

5.2 Wireless Power Experiments

Isothermal dielectric hysteresis loops were measured to confirm pyroelectric activity of the film (*Figure 18b*). A pyroelectric constant of -1.3 x 10^{-4} C/m^{2}K between 27-207 °C was determined from these indirect measurements. The pyroelectric coefficient was also measured from current measured under temperature cycling with a *dT/dt* of 10 K/s and a span of 25 K, extracting a -1.2 x 10^{-4} C/m^{2}K pyroelectric coefficient, in agreement with the indirect coefficient.

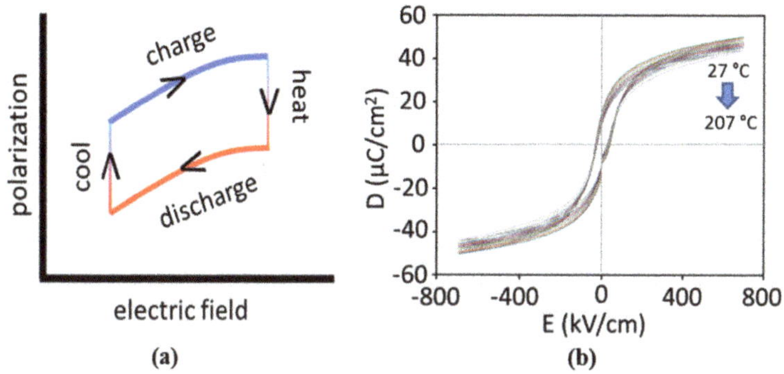

Figure 18. (a) Schematic of the PEC cycle, (b) isothermal hysteresis loops measured from 27-207 °C.

Temperature oscillations necessary for energy conversion were induced in the thin film pyroelectric receiver by a modulated fiber-coupled diode laser with a wavelength of 1.5 μm focused on the sample to a spot size of 1.5 mm at frequencies from 0.1-100 Hz. An electric field ranging from 2-12 kV/cm was triggered on the laser pulse in order to recreate the energy conversion cycle described in *Figure 18a*. Higher density electric fields could not be realized due to current limits within the measurement set-up. A 20% laser duty cycle, the minimum of the laser power supply, was used for all tests. The high temperature, high electric field stage of the PEC cycle will have the highest leakage current-driven losses, so time in the regime should be limited [26]. A current amplifier was used to measure charging/discharging (~μA) and the pyroelectric heating/cooling (~nA) currents. The high impedance of this measurement resulted in power losses for charging and discharging steps, scaling with the square of frequency and applied field. PEC cycles at 1 Hz are shown in *Figure 19(a)* with arbitrary starting polarizations.

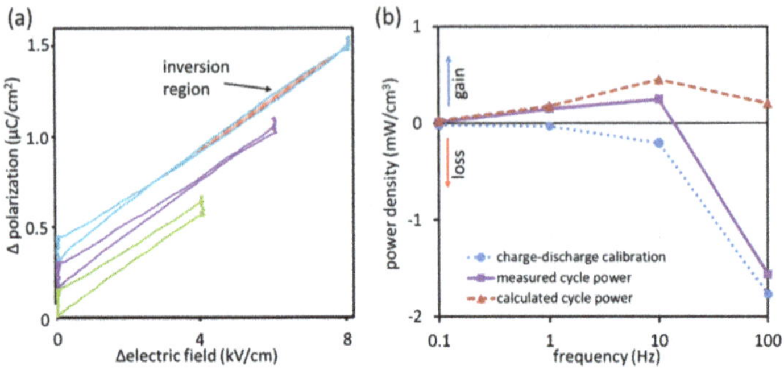

Figure 19. (a) Polarization versus electric field for ΔE = 4, 6, and 8 kV/cm cycles at 1 Hz. (b) Power density comparison for 0.1-100 Hz cycle frequencies, between charge cycle loss, measured cycle power, and calculated cycle power without charge cycle losses for ΔE = 1kV/cm.

Increasing electric fields appear to narrow the loop and change the charge/discharge curvature up to ΔE = 8 kV/cm, where the loop inverts in the region highlighted. At higher applied fields the charging and discharging current is increasingly impeded by the input resistance of the current amplifier, creating an apparent energy loss mechanism that is not intrinsic to the energy conversion. For example, once the charging step is completed at ΔE = 1 kV/cm, 1.55 nC/m^2 of apparent charge continues into the pyroelectric. When ΔE = 4 kV/cm, 24 nC/m^2 continues to charge the capacitor, up to 209 nC/m^2 at ΔE = 10 kV/cm, obscuring pyroelectric polarization change and resulting in apparent energy loss. Therefore the calculated energy conversion was determined using calibrated charge-discharge losses at each applied electric field and frequency (ex. *Figure 20*). Energy conversion data were taken to the point where the measured conversion began to show apparent loss. Future realizations of high frequency PEC should seek to limit the impedance of the energy harvesting circuit and reduce the dielectric constant of the active pyroelectric material, allowing for higher electric fields.

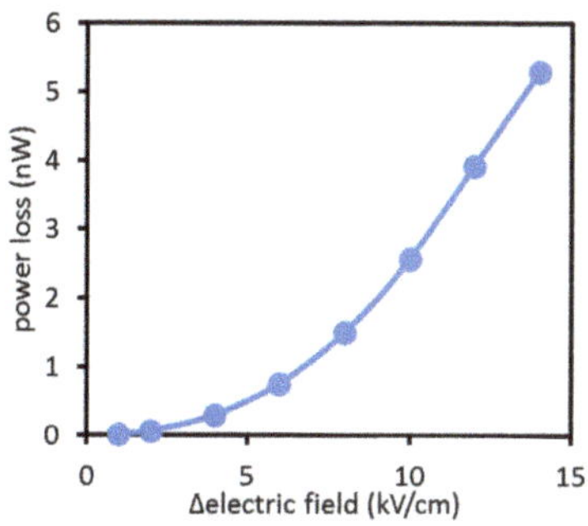

Figure 20. Measured charge/discharge power loss at 10Hz cycle frequencies from 1-14 kV/cm.

Figure 19(b) compares the measured, charge-discharge, and calculated cycle power of the pyroelectric. Power was determined by multiplying the current and voltage waveforms measured simultaneously using the circuit in *Figure 17*, similar to techniques used for previous high frequency PEC demonstrations [21]. This method allows for the application of an electric field and the direct observation of the pyroelectric cycle, in contrast to more harvesting-focused circuits utilizing a load [32, 33]. Calibration measurements were performed at all frequencies and applied fields used for energy conversion tests.

The temperature swing at each excitation frequency was calculated from Eqn. 17 by integrating measured pyroelectric current and using a pyroelectric coefficient, *p*, of -1.2×10^{-4} C/m^2k from previous calibrated temperature measurements. Under a constant laser power, ΔT ranged from 26.5 °C at 0.1 Hz to 0.16 °C at 100 Hz. For a given absorbed radiation power it is expected that there will be a specific frequency that maximizes converted power.

5.3 Wireless Power Transmission

A pyroelectric cycle consisting of charging followed by isoelectric heat addition, discharging, and isoelectric heat rejection was used for PEC, similar to the Olsen or Brayton cycle [25]. To maximize extracted work, charging would be nearly instantaneous so that there is no temperature dependent polarization change working against the applied field and heating would commence immediately after the pyroelectric capacitor is charged so there is no cooling, reducing the ΔT. Practically, the temperature never reaches steady-state, gradual charging and discharging are necessary to minimize measurement impedance, and there is some delay between the initiation and realization of heating. Figure 23a shows the pyroelectric cycle current, electric field, and laser power at 10 Hz with various phase lags between the initiation of charging and heating.

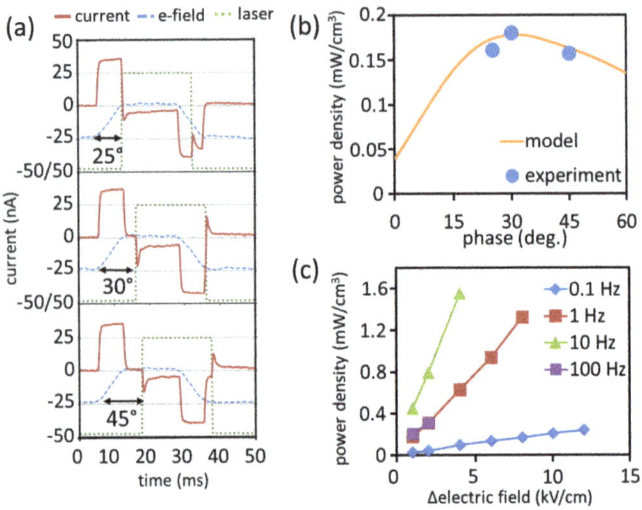

Figure 21. (a) 10 Hz pyroelectric energy conversion cycles with various phase lag between charging/discharging and heating/cooling cycles controlled by manipulating the phase lag between the electric field and laser. (b) Simulated power density from measured voltage and current waveforms compared to measurements(c) Power density for energy conversion cycles at 0.1, 1, 10, and 100 Hz.

A 30° phase lag was found to be superior, producing the maximum measured energy density of 46 μJ/cm^3. A phase-lag of 25° results in cooling before the system fully discharges. The re-polarization from cooling works to oppose the de-polarization from discharging, reducing the amount of electrical work completed by polarization to 44 μJ/cm^3. A large phase lag of 45° resulted in a reduced energy density of 36 μJ/cm^3 due to the delayed initiation of the heating and therefore a lower temperature increase while charged. The temperature and applied electric field waveforms were used to model a continuum of phase lag effects on energy conversion. The modelled and measured power densities are shown to be in good agreement in *Figure 21(b)*.

The frequency and electric field influence on power density is shown in *Figure 21(c)*. The 10 Hz cycle frequency resulted in the highest measured power density of 1.6

mW/cm^3, balancing frequency and ΔT. At a given input laser power, ΔT reduces with increasing frequency since the heating duration is reduced per cycle.

The pyroelectric receiver is being periodically laser-heated then passively cooled, necessitating that charging and discharging take place against transient temperatures even at the optimum phase. *Figure 22(a)* shows the thermal and measured electric field signals for the first half of a 1 Hz cycle, highlighting the region where both heating and discharging are reducing the polarization. *Figure 22(b)* compares an ideal adiabatic (instantaneous) charge/discharge to the realistic charging rate, which was limited by the measurement impedance for this demonstration, showing the effect of the superimposed processes as a reduction in cycle area, thus energy density.

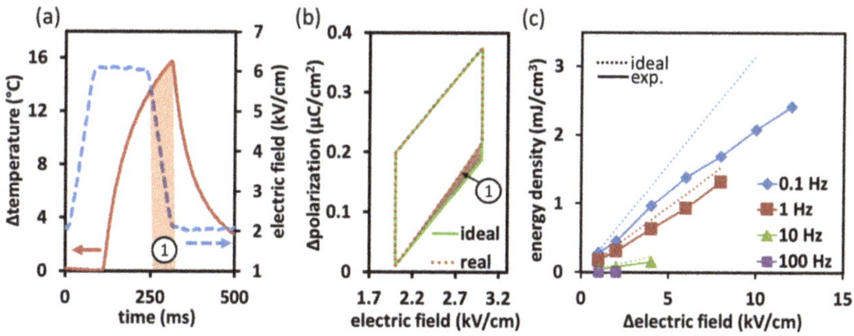

Figure 22. (a) The first half of a 1 Hz pyroelectric energy conversion cycle showing electrical and thermal cycles. Region (1) is highlighted as processes with significant deviation from the ideal cycle operation. (b) Modelled energy conversion cycle with ideal instant charge/discharged and the realized signal, continuing to highlight the main loss region. (c) Energy density for energy conversion cycles at 0.1, 1, 10, and 100 Hz comparing ideal and experimental energy density.

A maximum energy conversion of 2.4 mJ/cm^3 was obtained at a ΔE of 12 kV/cm and 0.1 Hz, due to the largest ΔT induced on the sample. The ideal conversion can be estimated using Eqn. 19 which represents the net work determined using Eqn. 1 assuming constant pyroelectric coefficients. Eqn. 19 was derived for the Olsen cycle but can also be used for the Brayton cycle when $p{\cdot}\Delta E$ is much less than the heat capacity of the material.

$$W = p\Delta T\Delta E \quad [19]$$

The converted energy densities presented in *Figure 22(c)* range from 72-92% of the ideal value, resulting from the transient thermal waveform and charge/discharge rate.

5.4 Wireless Power Calculation

The measured power (P_{meas}) is the product of the circuit voltage and current. The voltage is the sum of the voltage across the PEC (V_{pyro}) and the voltage induced in the series resistance of the current amplifier. The current is the sum of the current induced by the pyroelectric effect (I_{pyro}) and that used to charge/discharge ($I_{c/d}$) the pyroelectric receiver, where:

$$P_{meas}(t) = \left(V_{pyro}(t) + \left(I_{pyro}(t) + I_{c/d}(t)\right) * R\right) * \left(I_{pyro}(t) + I_{c/d}(t)\right) \quad [20]$$

We define the pyroelectric power out as the product of the pyroelectric current and the voltage across the PEC. That is, it's the power from the pyroelectric if charging and discharging required no current,

$$P_{pyro}(t) = V_{pyro}(t) * I_{pyro}(t) \quad [21]$$

In order to isolate the pyroelectric power, every cycle was also run without laser pulses, for which $I_{pyro}(t) = 0$. These cycles gave the power lost across series resistance while charging/discharging, $P_{c/d}$,

$$P_{c/d}(t) = \left(V_{pyro}(t) + I_{c/d}(t) * R\right) * I_{c/d}(t) \quad [22]$$

The charge/discharge power was then subtracted from the measured power to give the calculated power, P_{calc},

$$\begin{aligned} P_{calc}(t) &= P_{meas}(t) - P_{c/d}(t) \\ &= V_{pyro}(t) * I_{pyro}(t) + I_{pyro}^2(t) * R + I_{pyro}(t) * I_{c/d}(t) \\ &\quad * R \end{aligned} \quad [23]$$

$$P_{calc}(t) = P_{pyro}(t) + I_{pyro}^2(t) * R + I_{pyro}(t) * I_{c/d}(t) * R \quad [24]$$

Two error terms remain, but from our experimental data, these terms are several orders of magnitude smaller than the calculated power. Thus:

$$I_{pyro}^2(t) * R + I_{pyro}(t) * I_{c/d}(t) * R \ll P_{calc}(t) \qquad [25]$$

$$P_{calc}(t) \approx P_{pyro}(t) \qquad [26]$$

Average power was then calculated by integrating over 3-5 cycles.

$$P_{avg} = \frac{1}{n\tau} \int_{t_0}^{t_0+n\tau} P_{calc}(t)\delta t \qquad [27]$$

In order for the current amplifier to accurately measure the current/polarization in the pyroelectric, it must be connected in series with the pyroelectric. Connecting a voltmeter to their shared node would create a parallel path for current to flow through instead of the current amplifier, potentially invalidating current results. Instead, the power across the pyroelectric is inferred using the voltage across both the pyroelectric and current amplifier, and retroactively removing the power loss introduced by the current amplifier input impedance. The measurement technique represents an ideal harvester that allows us to understand PEC under optimal conditions. Practical harvesting of pyroelectric power would require a circuit to efficiently charge and discharge, but methods have been suggested previously as in ref [34].

Conclusions

Pyroelectric energy conversion is most efficient when performed within a thermodynamic cycle that alternates between thermal and electrical stimuli. Analogous to large-scale turbines and engines, the specific thermodynamic cycle needs to be chosen to optimize power density based on the system and practicalities of implementation. In this work we have analyzed the Ericsson cycle, which has remained the standard for over 30 years, in comparison to a pyroelectric Brayton cycle. This analysis is specific to thin film-based PEC systems, where it's assumed that high electric fields will be used and fast cycling is enabled by low thermal mass. The Brayton cycle, which replaces isothermal charging with adiabatic charging, is found to be more efficient for all cases due to the lack of heat addition during the charging cycle at the expense of energy density. A power density of 8 mW/cm^3 was obtained using the Brayton cycle for a thin film system with stimulation frequency of 0.2 Hz. At high cycle frequencies (>100 Hz), the isothermal charging process in the Ericsson cycle becomes a rate-limiting process, and therefore reduces the temperature span reached by a given input power, reducing the power density of the process. The practical implementation of a pyroelectric Brayton cycle is expected to be a challenge because of losses associated high currents resulting from instantaneous charging, and will be the subject of future work.

In wireless power experiments power transfer was demonstrated by inducing temperature oscillations on a pyroelectric receiver provided by a modulated laser, resulting in a measured maximum energy density per cycle and power density of 2.4 mJ/cm^3 and 1.6 mW/cm^3 respectively. The energy density was evaluated for cycle frequencies from 0.1 Hz to 100 Hz with a constant duty cycle. The maximum energy density was achieved at 0.1 Hz where the highest temperature change occurred, while the 10 Hz cycle resulted in the highest power density, balancing temperature change and cycle frequency. Compared to existing research in this field, this study is the first to address the practicality of fast thermodynamic cycles including the timing of the electrical and thermal waveforms as well as the implications of attempting a thermodynamic cycle on a transient thermal profile. Measurement impedance losses resulting from either high frequency or high applied fields were found to obscure the energy conversion intrinsic to the sample. Future work will focus on improving the electrical energy dissipation in the testing system and further optimizing the thermodynamic cycle for transient temperatures.

References

[1] G. Zhang, X. Zhang, T. Yang, Q. Li, L.-Q. Chen, S. Jiang, *et al.*, "Colossal Room-Temperature Electrocaloric Effect in Ferroelectric Polymer Nanocomposites Using Nanostructured Barium Strontium Titanates," *ACS Nano,* vol. 9, pp. 7164-7174, 2015/07/28 2015.

[2] E. Fatuzzo, H. Kiess, and R. Nitsche, "Theoretical Efficiency of Pyroelectric Power Converters," *Journal of Applied Physics,* vol. 37, pp. 510-516, 1966. https://doi.org/10.1063/1.1708205

[3] R. B. Olsen, D. A. Bruno, and J. M. Briscoe, "Pyroelectric conversion cycles," *Journal of Applied Physics,* vol. 58, p. 4709, 1985.

[4] J. Karthik, J. C. Agar, A. R. Damodaran, and L. W. Martin, "Effect of 90° Domain Walls and Thermal Expansion Mismatch on the Pyroelectric Properties of Epitaxial PbZr0.2Ti0.8O3 Thin Films," *Physical Review Letters,* vol. 109, p. 257602, 12/21/ 2012.

[5] J. Karthik and L. W. Martin, "Pyroelectric properties of polydomain epitaxial Pb(Zr1−x,Tix)O3 thin films," *Physical Review B,* vol. 84, p. 024102, 07/01/ 2011.

[6] Z.-G. Ban and S. P. Alpay, "Dependence of the pyroelectric response on internal stresses in ferroelectric thin films," *Applied Physics Letters,* vol. 82, pp. 3499-3501, 2003. https://doi.org/10.1063/1.1576503

[7] M. T. Kesim, J. Zhang, S. P. Alpay, and L. W. Martin, "Enhanced electrocaloric and pyroelectric response from ferroelectric multilayers," *Applied Physics Letters,* vol. 105, p. 052901, 2014. https://doi.org/10.1063/1.4892455

[8] R. V. K. Mangalam, J. C. Agar, A. R. Damodaran, J. Karthik, and L. W. Martin, "Improved Pyroelectric Figures of Merit in Compositionally Graded PbZr1–xTixO3 Thin Films," *ACS Applied Materials & Interfaces,* vol. 5, pp. 13235-13241, 2013/12/26 2013.

[9] F. Jin, G. W. Auner, R. Naik, N. W. Schubring, J. V. Mantese, A. B. Catalan, *et al.*, "Giant effective pyroelectric coefficients from graded ferroelectric devices," *Applied Physics Letters,* vol. 73, pp. 2838-2840, 1998. https://doi.org/10.1063/1.122607

[10] M. Valant, A.-K. Axelsson, F. Le Goupil, and N. M. Alford, "Electrocaloric temperature change constrained by the dielectric strength," *Materials Chemistry and Physics,* vol. 136, pp. 277-280, 10/15/ 2012.

[11] A. S. Mischenko, Q. Zhang, J. F. Scott, R. W. Whatmore, and N. D. Mathur, "Giant electrocaloric effect in thin-film PbZr(0.95)Ti(0.05)O3," *Science,* vol. 311, pp. 1270-1, Mar 03 2006. https://doi.org/10.1126/science.1123811

[12] J. E. Drummond, V. Fargo, J. Ream, J. M. Briscoe, and D. Brown, "Experimental Demonstration of Heat to Electricity Conversion within a Dielectric," in *10 Intersociety Energy Conversion Engineering Conference*, 1979, p. 1819.

[13] P. M. Margosian, "Thermal electrostatic generator variable capacitance device for converting thermal energy to electric energy," NASA NASA-TN-D-2763, 1965.

[14] R. B. Olsen, J. M. Briscoe, D. A. Bruno, and W. F. Butler, "A pyroelectric energy converter which employs regeneration," *Ferroelectrics,* vol. 38, pp. 975-978, 1981/10/01 1981.

[15] R. B. Olsen, D. A. Bruno, J. M. Briscoe, and E. W. Jacobs, "Pyroelectric conversion cycle of vinylidene fluoride-trifluoroethylene copolymer," *Journal of Applied Physics,* vol. 57, pp. 5036-5042, 1985. https://doi.org/10.1063/1.335280

[16] S. Gael, P. Sebastien, and G. Daniel, "Energy harvesting based on Ericsson pyroelectric cycles in a relaxor ferroelectric ceramic," *Smart Materials and*

Structures, vol. 17, p. 015012, 2008. https://doi.org/10.1088/0964-1726/17/01/015012

[17] S. J. Smullin, Y. Wang, and D. E. Schwartz, "System optimization of a heat-switch-based electrocaloric heat pump," *Applied Physics Letters,* vol. 107, p. 093903, 2015. https://doi.org/10.1063/1.4928716

[18] Y. Jia and Y. Sungtaek Ju, "A solid-state refrigerator based on the electrocaloric effect," *Applied Physics Letters,* vol. 100, p. 242901, 2012. https://doi.org/10.1063/1.4729038

[19] G. Sebald, E. Lefeuvre, and D. Guyomar, "Pyroelectric energy conversion: Optimization principles," *IEEE Transactions on Ultrasonics, Ferroelectrics, and Frequency Control,* vol. 55, pp. 538-551, 2008. https://doi.org/10.1109/TUFFC.2008.680

[20] Y. L. Felix, J. Hwan Ryul, S. L. Christopher, and P. Laurent, "Pyroelectric energy conversion using PLZT ceramics and the ferroelectric–ergodic relaxor phase transition," *Smart Materials and Structures,* vol. 22, p. 025038, 2013. https://doi.org/10.1088/0964-1726/22/2/025038

[21] B. Bhatia, A. R. Damodaran, H. Cho, L. W. Martin, and W. P. King, "High-frequency thermal-electrical cycles for pyroelectric energy conversion," *Journal of Applied Physics,* vol. 116, p. 194509, 2014. https://doi.org/10.1063/1.4901993

[22] B. Hanrahan, C. Neville, A. Smith, N. Ter-Gabrielyan, N. Jankowski, and C. M. Waits, "Wireless Power Transmission via Modulated Laser Irradiation of Pyroelectric Thin Films," *Advanced Materials Technologies,* p. 1600178, 2016.

[23] L. M. Sanchez, D. M. Potrepka, G. R. Fox, I. Takeuchi, K. Wang, L. A. Bendersky, *et al.*, "Optimization of PbTiO3 seed layers and Pt metallization for PZT-based piezoMEMS actuators," *Journal of Materials Research,* vol. 28, pp. 1920-1931, 2013.

[24] Daniel Potrepka, Glen Fox, and R. Polcawich, "IrO2 Top Electrodes for PbZr0.52Ti0.48O3 Functionality," in *IEEE International Symposium on Applications of Ferroelectrics (ISAF)*, Singapore, Singapore, 2015.

[25] N. Jankowski, A. Smith, and B. Hanrahan, "Thermal Modeling of Pyroelectric Energy Conversion from Pulsed Heat Sources," in *ASME SHTC 2016*, Washington, D.C. , 2016.

[26] B. Hanrahan, L. Sanchez, C. M. Waits, and R. G. Polcawich, "Improved pyroelectric performance for thin film lead zirconate titanate (PZT) capacitors with

IrO 2 electrodes," *Smart Materials and Structures,* vol. 25, p. 015025, 2016. https://doi.org/10.1088/0964-1726/25/1/015025

[27] F. Sze and B. Hanrahan, "Pyroelectric Energy Harvesting: Model and Experiments," 2016.

[28] A.N. Smith, B.M. Hanrahan, C.J. Neville, and N. R. Jankowski, "Timing A Pulsed Thin Film Pyroelectric Generator For Maximum Power Denisty," in *PowerMEMS 2016*, Paris, France, 2016, p. 4.

[29] T. Yu, G. Zhang, Y. Yu, Y. Zeng, and S. Jiang, "Pyroelectric energy harvesting devices based-on Pb[(MnxNb1−x)1/2(MnxSb1−x)1/2]y(ZrzTi1−z)1−yO3 ceramics," *Sensors and Actuators A: Physical,* vol. 223, pp. 159-166, 3/1/ 2015.

[30] T. Tong, J. Karthik, L. W. Martin, and D. G. Cahill, "Secondary effects in wide frequency range measurements of the pyroelectric coefficient of Ba0.6Sr0.4TiO3 and PbZr0.2Ti0.8O3 pitaxial layers," *Physical Review B,* vol. 90, p. 155423, 10/14/ 2014.

[31] B. Bhatia, J. Karthik, T. Tong, D. G. Cahill, L. W. Martin, and W. P. King, "Pyroelectric current measurements on PbZr0.2Ti0.8O3 epitaxial layers," *Journal of Applied Physics,* vol. 112, p. 104106, 2012. https://doi.org/10.1063/1.4766271

[32] C. R. Bowen, J. Taylor, E. LeBoulbar, D. Zabek, A. Chauhan, and R. Vaish, "Pyroelectric materials and devices for energy harvesting applications," *Energy & Environmental Science,* vol. 7, pp. 3836-3856, 2014. https://doi.org/10.1039/C4EE01759E

[33] D. Zabek, J. Taylor, V. Ayel, Y. Bertin, C. Romestant, and C. R. Bowen, "A novel pyroelectric generator utilising naturally driven temperature fluctuations from oscillating heat pipes for waste heat recovery and thermal energy harvesting," *Journal of Applied Physics,* vol. 120, p. 024505, 2016. https://doi.org/10.1063/1.4958338

[34] R. B. Olsen, "Pyroelectric energy converter and method," ed: Google Patents, 1987.

Chapter 16

Novel Measurement Methods for Thermoelectric Power Generator Materials and Devices

Patrick J. Taylor[1], Adam Wilson[1,2], Jay R. Maddux[1,3], Theodorian Borca-Tascuic[2], Samuel P. Moran[2], Eduardo Castillo[2], Diana Borca-Tascuic[2]

[1]US Army Research Laboratory, Sensors and Electron Devices Directorate, 2800 Powder Mill Road, Adelphi, MD 20783

[2]Department of Mechanical, Aerospace and Nuclear Engineering, Rensselaer Polytechnic Institute 110 8th Street, Troy, NY 12180-3590

[3]General Technical Services, LLC, 3100 Route 138 West, Wall Township, NJ 07719

Abstract

Thermoelectric measurements are notoriously challenging. In this work, we outline new thermoelectric characterization methods that are experimentally more straightforward and provide much higher accuracy, reducing error by at least a factor of two. Specifically, three novel measurement methodologies for thermal conductivity are detailed: steady-state isothermal measurements, scanning hot probe, and lock-in transient Harman technique. These three new measurement methodologies are validated using experimental measurement results from standards, as well as, candidate materials for thermoelectric power generation. We review thermal conductivity measurement results from new half-Heusler (ZrNiSn-based) materials, as well as, commercial $(Bi,Sb)_2(Te,Se)_3$ and mature PbTe samples. For devices, we show characterization of commercial $(Bi,Sb)_2(Te,Se)_3$ modules, pre-commercial PbTe/TAGS modules, and new high accuracy numerical device simulation of Skutterudite devices. Measurements are validated by comparison to well-established standard reference materials, as well as evaluation of device performance, and comparing to theoretical prediction obtained using measurements of individual properties. The new measurement methodologies presented here provide a new, compelling, simple and more accurate means of materials characterization, providing better agreement with theory.

The work has been previously published by IntechOpen:
Patrick J. Taylor, Adam Wilson, Jay R. Maddux, Theodorian Borca-Tasciuc, Samuel P. Moran, Eduardo Castillo and Diana Borca-Tasciuc, Novel Measurement Methods for Thermoelectric Power Generator Materials and Devices, in Eds. Sergey Skipidarov and Mikhail Nikitin, Thermoelectrics for Power Generation - A Look at Trends in the Technology, IntechOpen, Chapter 16, 2016. https://dx.doi.org/10.5772/65443

Keywords

Thermal Conductivity, Seebeck Coefficient, Electrical Resistivity, ZT, Device Efficiency.

Contents

1. Introduction

The efficiency with which a TE power generator can convert heat energy to electricity is determined, in part, by thermal conductivity, κ, of the materials used for fabricating TE devices. Experimental measurement of that property usually results in surprisingly significant error, > ±10 % [1]. There are many causes of that error, and in this chapter, we provide new solutions which are experimentally faster, yield results which are more consistent with physical devices, and address several sources of experimental error, which may reduce uncertainty by a factor of two or more.

In this work, we describe several new, more accurate techniques to measure thermal conductivity, κ, of thermoelectric materials. The first is based on detailed control of the heat flows within a sample under steady-state conditions; so-called *steady-state isothermal technique*. The second is non-destructive micro-scale analysis technique called *scanning hot-probe (or scanning thermal microscopy)*. And the third is *lock-in transient Harman* method, which is a comprehensive modification of transient Harman technique, employing a lock-in procedure and considers detailed contact effects. A new interesting follow-on is frequency-dependent Nyquist analysis, which presages a different perspective on the materials analysis.

The truest test of the accuracy of measurement is comparison with fabricated devices. To support the validation of measurements of individual material properties, we outline new device metrics, which allows comparison between theoretical and measured device efficiency. We outline a new *slope-efficiency method*, which can be used to determine informative index $ZT_{maximum}$ of any device. The second method of device evaluation is a numerical device model called the *discretized heat balance model*, which considers a piecewise continuous collection of discrete layers within a device, where boundary heat flows have energy and current continuity relationships, and enable incredibly easy determination of device efficiency.

2. Novel measurements of thermal conductivity

2.1 Steady-state isothermal technique

This new measurement of κ leverages Peltier heat, ± Q_Π, an electronically-controlled internal heat source unique to thermoelectric materials. Peltier heat causes either heating (+Q_Π) or cooling (-Q_Π) at the junction between a thermoelectric material and a metal by passing the proper polarity of electric current, ± I.

Q_Π was first employed to roughly estimate κ by Putley [2]. In Putley's experiment, convective, parasitic and non-symmetric heat flows required correction factors larger than 20 %. Harman dramatically improved upon Putley's demonstration by performing the measurements in vacuum, and by reducing other parasitic heat flows. Despite these improvements, error is still obtained, because parasitic heat flows are non-zero [3].

In these past studies, principal parasitic heat flows causing error include: conduction along lead wires, conduction along thermocouples, Joule heating within lead wires, and radiation. The magnitudes of these parasitic heat flows can be as large as 30 % of Peltier heat. Penn quantified the significance of parasitic heat flows in Harman's technique, and showed, that they induced error of more than 10 % [4]. Bowley and Goldsmid [5], as well as, Buist [6] reported that parasitic heat flows cause error, usually larger than 20 %.

The focus of the present work is description of a new, correctionless method to measure κ by balancing two independently controlled heat sources: Q_{Π}, and a radiatively-coupled input heat as per the Stefan-Boltzmann law, Q_{SB}. In this new method, Q_{Π} and Q_{SB} can be independently balanced. A finite temperature difference across the sample imposed by Q_{SB} can be cancelled and even inverted by application of Q_{Π}. When exactly cancelled, there is no temperature difference across the sample (*i.e.* $\Delta T \equiv 0$) and a steady-state isothermal condition is obtained leaving the steady-state temperature of the sample exactly equal to that of the surroundings, T_e. Because there is no ΔT, parasitic heat flows, such as those along lead-wires/thermocouples (Q_{wires}) and radiative heat loss, ($Q_{radiation-error}$), which would otherwise cause significant error [7], converge exactly to zero. Analysis of this technique using a Peltier cooler demonstrates error of less than ±1 %, an improvement of over an order of magnitude [8]. When considering thermoelectric power generators, other considerations must be taken into account to determine experimental uncertainty (such as view angle for radiative heat flows from the environment, material emissivity, etc.) which are more complicated and beyond the scope of the work presented here. However, as is demonstrated qualitatively here, experimental uncertainty is reduced using the steady-state isothermal technique by significantly more than a factor of two.

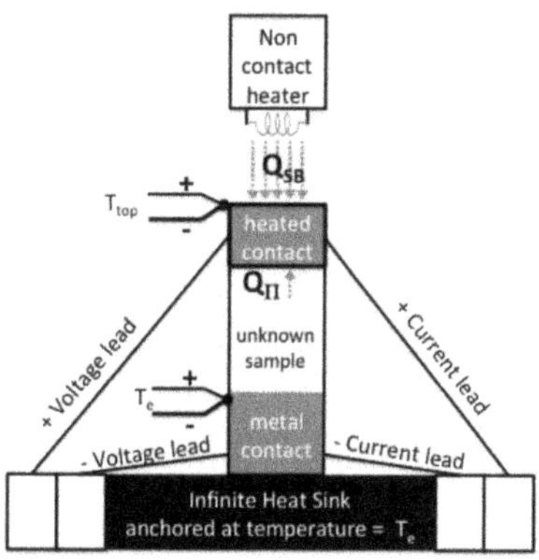

Fig. 1 Experimental schematic of steady-state isothermal technique.

Assume a sample having the temperature of one end anchored to the temperature of the environment, T_e, by a large heat-sink, and the opposite end, T_{top}, having small thermal mass and capable of temperature diversion by the application of Q_{SB}. Vacuum is used to obviate convective heat flow, and Q_{SB} is applied by small heater having approximately the same subtended area as that of the cross-sectional area (A) of the sample, such that Q_{SB} is localized to the top and there is no direct line-of-sight along the sample's length (ℓ). Thermocouples are attached to each end to determine the temperature difference across the sample, and electrical leads are used for passing ± I to control Q_{Π}. Figure 1 depicts this experimental setup.

When Q_{SB} is applied, the temperature of the heated contact will increase with respect to T_e and some magnitude of Q_{κ} will be conducted through the sample. At the heated contact, contributing heat flows include Q_{SB}, Q_{κ}, $Q_{radiation\text{-}error}$ and Q_{wires}. Radiation-error flow $Q_{radiation\text{-}error}$ is governed by the emissivity (ε), Stefan-Boltzmann constant (σ), sidewall temperature ($T_{sidewall}$) and sidewall area ($A_{sidewall}$) of the sample. Q_{wires} follows the usual Fourier's law description where, for simplicity, aspect ratio (A_{wires}/ℓ_{wires}) and thermal conductivity (κ_{wires}) of all wires and thermocouples are combined into one lumped parasitic term. When electrical current flows through the sample, Q_{Π} is absorbed at the heated contact, where the first Kelvin relation gives $Q_{\Pi} = (\alpha_{sum}IT)$ and α_{sum} is the sum of Seebeck coefficients of the sample and the contact metal, and T is the temperature of the heated contact. An equal and opposite value of Q_{Π} is liberated at the contact between the sample and large heat-sink, but is too small to cause any measurable temperature change of the heat-sink, and is therefore negligible. Including Q_{Π} at the heated contact yields equation (1), which, under steady-state conditions sums to zero:

$$\Sigma Q = Q_{SB} - Q_{\Pi} - Q_{\kappa} - Q_{radiation\text{-}error} - Q_{wires} = 0 \qquad (1)$$

To quantify the magnitude of Q_{κ}, a range of electrical currents can be passed, which enables Q_{Π} to absorb a corresponding range of Q_{SB} at the contact, so we get

$$Q_{SB} = (\alpha_{sum}\, IT) + [\kappa\, (\tfrac{A}{\ell})\Delta T] + \varepsilon\sigma A_{sidewall}\left(T^4_{sidewall} - T_e^{\,4}\right) + \left[\kappa_{wires}\left(\tfrac{A_{wires}}{\ell_{wires}}\right)\Delta T\right] \qquad (2)$$

For progressively larger I, Q_{Π} absorbs increasingly more of Q_{SB} and temperature of the heated contact begins to converge to T_e, such that the overall ΔT across the sample goes to zero. As ΔT becomes smaller with increasing I, the only relevant heat flows are Q_{SB}, Q_{Π}, and Q_{κ} because $Q_{radiation\text{-}error}$, and Q_{wires} are only statistically significant [7] for larger ΔT, say >10 K, and all parasitic heat flows converge to zero. Therefore, under these conditions at any given I, ΔT across the sample is required to satisfy equation (3):

$$Q_{SB} = (\alpha_{sum}IT) + \left[\kappa\, (\tfrac{A}{\ell})\,\Delta T\right] \qquad (3)$$

From the requirement imposed by equation (3), a new method for measuring κ is obtained. Equation (3) is solved to show the dependence of ΔT on electrical current. By taking the derivative of (3), prior knowledge of Q_{SB} isn't required, because it is a constant, and the analysis yields the following:

$$\frac{\partial \Delta T}{\partial I} = -\left[\frac{\alpha_{sum} T}{\kappa\left(\frac{A}{\ell}\right)}\right] \tag{4}$$

To determine κ, equation (4) is solved using the slope at ΔT =0 of steady-state ΔT as function of I and that slope is combined with α_{sum}, as well as, geometrical aspect ratio (A/ℓ) of the sample [8]. Figure 2 shows dependence of temperature difference between ends of the thermoelement sample on value of passing current (left panel) and real picture measurement configuration (right panel).

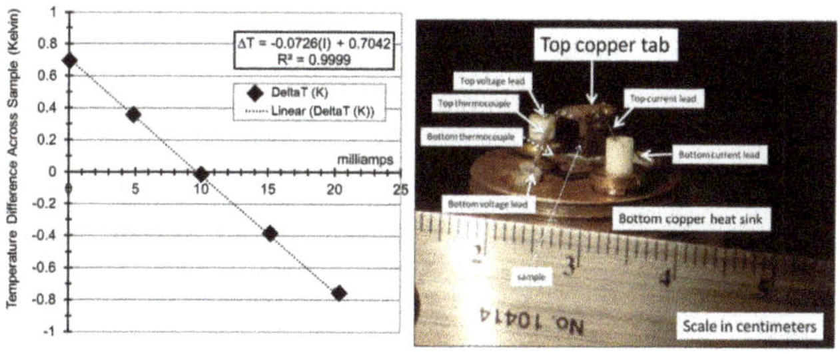

Fig. 2 (Left) Linear ΔT decrease across sample by application of milliamps of current (I) for Peltier cooling and (Right) Picture showing a fully connected sample.

2.1.1 Thermal conductivity of n-type half-Heusler

Thermal conductivity of a half-Heusler alloy (Figure. 3) was collected (triangles) and is presented with respect to previously published data (squares, circles, diamonds) measured by laser-flash thermal diffusivity technique and reported by researchers from GMZ corporation [9]. Because of the speed, ease, and simplicity of the new technique presented, there is opportunity for significantly more collected data. One data point can be collected in seconds. However, as can be seen in Figure 3, it is consistent to within experimental error and falls within the bounds of the published laser-flash data.

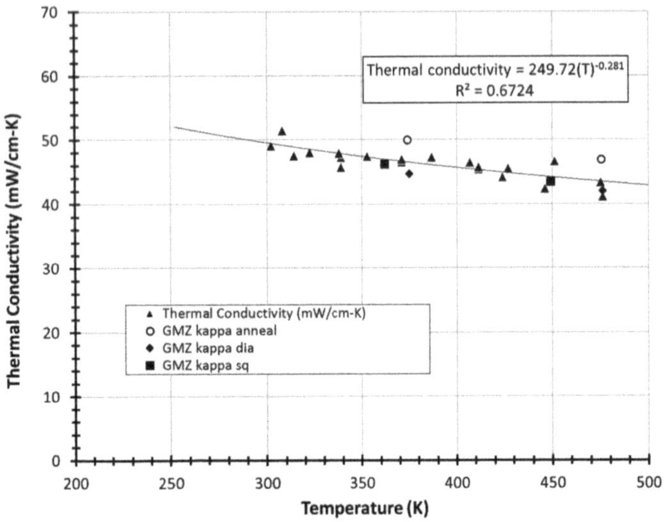

Fig. 3 Thermal conductivity of n-type half-Heusler measured by steady-state isothermal technique as compared to published data [9].

2.1.2 Thermal conductivity of PbTe

One category of high performance thermoelectric materials are near-degenerate semiconductors. Such materials do not directly obey the Wiedemann-Franz relationship between electrical resistivity and thermal conductivity, due to the significant contribution of lattice thermal conduction to the total thermal conductivity. However, utilizing a modified Wiedemann-Franz relationship to find the thermal conductivity due to electron flow allows direct, real-time deconvolution of lattice thermal conductivity ($\kappa_{lattice}$) from electronic contribution ($\kappa_{electronic}$). For charge carriers' concentrations near degeneracy, and random scattering of charge carriers, Rosi et al. [10] describe how $\kappa_{electronic}$ can be determined using electrical resistivity, ρ, by:

$$\kappa_{electronic} = \left(\frac{\pi^2}{3}\right)\left(\frac{k_B}{q}\right)^2\left(\frac{T}{\rho}\right) \quad (5)$$

If thermal conductivity and electrical resistivity are measured, then $\kappa_{lattice}$ can be determined by $\kappa_{lattice} = [\kappa - \kappa_{electronic}]$.

Figure 4 shows temperature dependences of PbTe thermal conductivity (including measured data) and deconvolution of electronic and lattice contributions to total thermal conductivity.

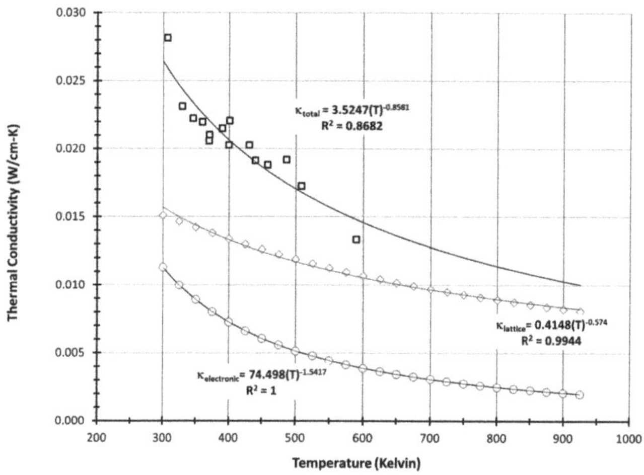

Fig. 4 Measurement of thermal conductivity of PbTe, and deconvolution of electronic and lattice contributions to total thermal conductivity.

2.2 Scanning hot probe

Scanning hot probe technique provides measurement of local thermal conductivity and Seebeck coefficient of a sample by measuring average probe temperature when probe tip and sample are in “thermal contact”, i.e., when probe tip is in physical contact with the sample or is at known distance near enough to the sample to induce measurable heat exchange between probe and sample. Average probe temperature is also measured far from the sample, to account for the amount of heat lost to the surroundings and through the probe contacts. Difference in average probe temperature between these two cases is due to the heat transferred to the sample, which may be quantified through the following analytical derivation. Figure 5 depicts the thermal exchange between probe, sample, and surroundings, as well as the series thermal resistance network between probe and sample.

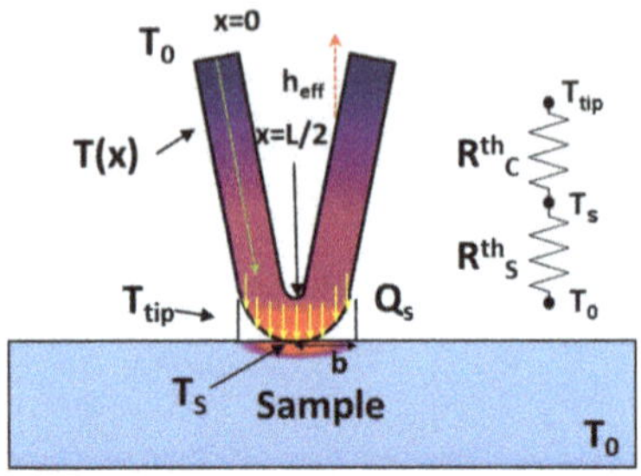

Fig. 5 Diagram showing thermal phenomenology around probe tip. Reproduced from [39] with permission from The Royal Society of Chemistry.

For steady-state probe heating using DC current, (or AC current at low frequency, when the heat capacity effects are negligible, and temperature rise amplitude is frequency independent and equivalent to DC temperature rise to good approximation) the governing equation describing amplitude of temperature profile of the probe shown in Figure 5 is given by [11]:

$$\frac{d^2T^*}{dx^2} - \left(\frac{2h_{eff}}{k_P r} - \frac{I^2\rho_0 TCR}{k_P \pi^2 r^4}\right)T^* + \frac{I^2\rho_0}{k_P \pi^2 r^4} = 0, \tag{6}$$

where $T^* = T(x) - T_0$, $h_{eff} = h + 4\epsilon\sigma T_0^3$ (here, h is the convective heat transfer coefficient, ϵ is the probe's emissivity, T_0^3 is an approximation for the exact (T^4 - T_o^4) term and σ is Stefan-Boltzmann constant. ρ_0 and k_P are the probe's electrical resistivity and thermal conductivity, respectively, TCR is the probe's temperature coefficient of resistance, I is root-mean-square electrical current passed through the probe, and r is the radius of the probe. Contribution from radiation is negligible for $\Delta T<100$ K [12]. To obtain an analytical solution to the second order differential equation, two boundary conditions are employed. The first assumption is that the ends of the probe are at ambient temperature (i.e. $T(0) = T_0$). The second assumption is that the tip region of the probe of length 2b is of uniform temperature, and by energy balance at the probe tip region, we get:

$$-k_P A \frac{dT^*}{dx}\Big|_{x=L/2-b} + I^2\rho_0\left(1 + TCR \times T^*|_{x=L/2-b}\right)\frac{b}{A} = \frac{Q_S}{2}, \tag{7}$$

where the left-hand side is heat conduction and Joule heating of the probe, $A = \pi r^2$ is the probe's cross-sectional area, and L is the length of the probe, and where the right-hand side is heat transfer through one leg of the probe (thus half the total heat transfer to the sample, by symmetry). Finally, heat transfer rate between probe and sample, Q_s, is

$$Q_s = \frac{\Delta T_{tip}}{R_C^{th} + R_S^{th}} = \frac{\Delta T_S}{R_S^{th}}, \tag{8}$$

where $\Delta T_{tip} = T_{tip} - T_0$ and $\Delta T_S = T_S - T_0$ are temperature of the probe and sample, respectively, in the tip region, and R_S^{th} is sample's thermal resistance. Solving equation (6) to obtain temperature profile along the probe for a given value of Q_s yields the following expression:

$$\Delta T_P(x) = C_1 e^{\lambda x} - C_2 e^{-\lambda x} + \frac{\Gamma}{\lambda^2}, \tag{9}$$

where $\lambda = \frac{I^2 \rho_0}{k_P \pi^2 r^4}$, $\Gamma = \frac{2h_{eff}}{k_P r} - \frac{I^2 \rho_0 TCR}{k_P \pi^2 r^4}$, and constants C_1 and C_2 are easily obtained by applying boundary condition $T(0) = T_0$.

If the sample is bulk, or has bulk-like thickness, thermal conductivity is found from R_S^{th} by employing semi-infinite medium assumption and 2D bulk sample assumption [30]:

$$R_S^{th} = \frac{1}{4\kappa_{sample} b} \tag{10}$$

If the sample is thin film of thickness l on substrate, and is thin enough, that there is negligible heat spreading in in-plane directions of the sample, then thermal conductivity is found by solving the expression for the series thermal resistance across substrate and film, with 1D heat transfer across the thickness of the film [30]:

$$R_S^{th} = \frac{1}{4\kappa_{substrate} b} + \frac{l}{\pi \kappa_{film} b^2} \tag{11}$$

When heat transfer may be multidimensional and anisotropic, models developed by Son et. al. [13] for laser heating may be used to predict thermal resistance of the sample, based on the respective values of thermal conductivity for the film and substrate.

Data collected from scanning hot thermoelectric probe experiment are probe voltage, voltage across a reference resistor, Seebeck voltage, and photodetector voltage for position sensing. Value of current passing through the system is obtained by dividing voltage across the reference resistor by known electrical resistance of that resistor. Probe resistance is then found by dividing probe voltage by that value of current. Often, instead of single reference resistor, Wheatstone bridge is utilized. Figure 6 depicts the difference in circuit between measurement taken a) with and b) without Wheatstone bridge.

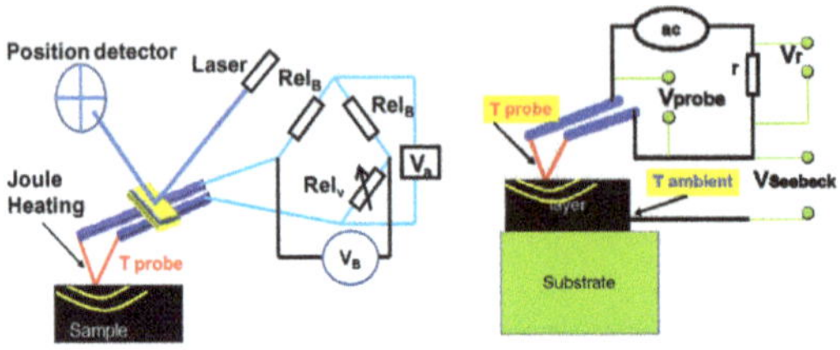

Fig. 6 Schematic scanning thermal microscopy circuits (left) with Wheatstone bridge and (right) with a reference resistor.

In DC mode, with resistor wired in series with the probe, measured probe voltage, V_P, may be expressed in terms of voltage across reference resistor (V_r) as:

$$V_P = R_P I = \frac{R_P V_r}{R_r}, \tag{12}$$

where I is DC current passing through the circuit, R_P is electrical resistance of the probe, and R_r is known electrical resistance of reference resistor. Probe's electrical resistance is proportional to temperature rise above ambient, when the probe undergoes Joule heating. It may be expressed as:

$$R_P = \Delta T_P (R_0 TCR) + R_0 + R_c, \tag{13}$$

where ΔT_P is average probe temperature rise, R_0 is nominal probe electrical resistance at 19.9 °C, (not including electrical contacts to the circuit), when the probe is not being heated, R_c is electrical resistance arise from contacts' and circuit's wiring, and TCR is probe's temperature coefficient of resistance, in terms of 1/°C.

Defining average probe thermal resistance as average probe temperature rise divided by Joule heating power, we can write average probe thermal resistance as:

$$R_P^{th} = \frac{\Delta T_P}{I^2 R_0 (1 + TCR * \Delta T_P)}. \tag{14}$$

Equation (14) allows determination of R_P^{th} by the slope of probe temperature rise with power applied, reducing the overall experimental uncertainty compared with a single value of temperature at a given power. If the circuit uses Wheatstone bridge, equations differ only by method of finding electrical resistance of the probe. In this case, R_P reduces to:

$$R_P = R_0 + \frac{V_B}{V_A} \frac{(R_B + R_0)^2}{R_B} + R_c, \tag{15}$$

where V_B and $V_A$are voltages across bridge side and probe side respectively. R_B is the total resistance of bridge side of the circuit, and R_0 is no heating resistance of the probe side of the bridge. With the probe resistance obtained, remaining equations are left unchanged. When AC current of amplitude I_0 is passed through the circuit, then measured probe resistance can be expressed as:

$$R_P = \underbrace{R_0\left(1 + TCR * \Delta T_{P,DC}\right)}_{DC\ Component} + \underbrace{R_0 TCR * \Delta T_{P,2\omega}\cos(2\omega t + \phi)}_{AC\ Component} \quad (16)$$

The probe tip voltage is expressed as:

$$V = I_0 R_P(1 + \mathrm{TCR} * \Delta \mathrm{T}_{P,\mathrm{DC}})\cos(\omega t) + \frac{I_0 R_P TCR * \Delta T_{P,2\omega}}{2}[\cos(3\omega t + \phi) + \cos(\omega t + \phi) \quad (17)$$

Thus, temperature amplitude is determined to be:

$$\Delta T_{ave} = \frac{2V_{3\omega}}{TCR * V_{1\omega}} \quad (18)$$

To obtain sample thermal conductivity, the probe must be calibrated. Quantities in Equations (6 - 11), which are not determined directly from experimental measurement are: h_{eff}, TCR, k_P, A, L, ρ_0, b, and R_C^{th}. To be fully calibrated, these quantities must be known. The probe manufacturer specifies values for TCR, k_P and ρ_0, and these values are used in this work. Values A and L may be found by determining probe's geometry (typically from SEM or microscope images, but may also be determined by measuring R_P^{th} in vacuum and in air). h_{eff} is determined by measuring R_P^{th} far from contact, and matching the value predicted by analytical model by adjusting h_{eff} and integrating equation (9) from x = 0 to L and dividing by L, with $Q_s = 0$ to obtain the average probe temperature when no heat is transferred to the sample. Finally, probe-to-sample thermal exchange parameters, b and R_C^{th}, must be determined. Typically, these values have been assumed to be sample-independent for given probe-to-sample contact force or probe-to-sample distance. As such, calibration strategies utilize measurements on two samples. However, these parameters are now shown to change with sample thermal conductivity. Figure 7 demonstrates change in b and R_C^{th} with sample thermal conductivity. Alternatively, if the sample is electrically grounded and probe tip is capable of making good electrical contact with the sample, then sample with known thermal conductivity and Seebeck coefficient may be used to determine both b and R_C^{th} simultaneously. Figure 7 demonstrates this calibration strategy, together with the typical "intersection method" using two or more samples. Care must be taken to calibrate in the correct range of thermal conductivity, as samples with thermal conductivity of higher than 1.1 W/mK yield a different pair of b and RthC values compared with samples with thermal conductivity of 1.1 W/mK and lower. Table 1 gives results of measurements taken with properly calibrated Wollaston probe tips, showing good agreement with independent measurements.

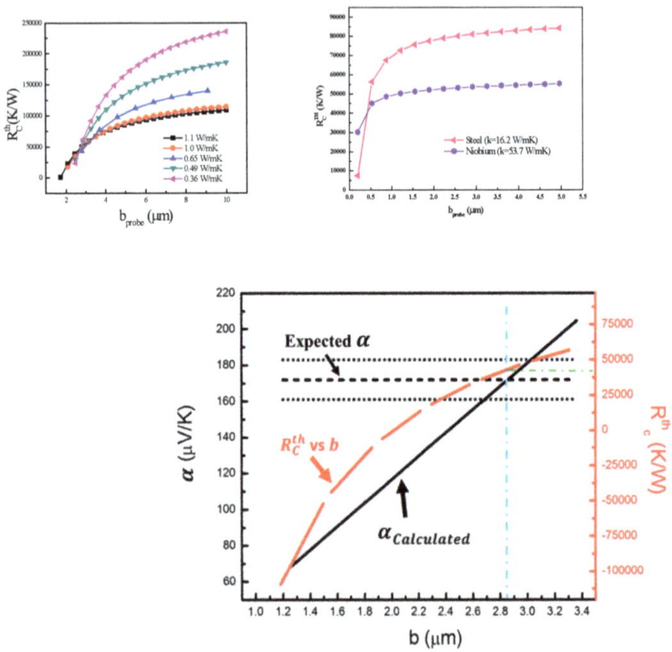

Fig. 7 Empirical Probe Calibration Strategies for Determining b and R_C^{th} using (top) Intersections of Curves taken from Samples with known κ for (left) low thermal conductivity and (right) high thermal conductivity and (bottom) determining b and R_C^{th} using a single sample with known κ and α. Adapted from [39] with permission from The Royal Society of Chemistry.

Table 1 Tabular results for a range of materials [Refs. 13-20]. Reproduced from [39] with permission from The Royal Society of Chemistry.

Sample	***l***	***b***	R_P^{th}	R_C^{th}	R_S^{th}	κ_{film}, **this work //** κ_{film}, **expected**
SiGe film on glass substrate	1.8 µm	2.8±0.3 µm	14,948± 54 K/W	44,927± 7820 K/W	76,134± 9494 K/W	1.22±0.21 W/K·m // 1.23±0.12 W/K·m
Fe-doped PCDTBT (1:1 doping concentration)	3.0 µm	2.8±0.3 µm	15,220± 155 K/W	44,927± 7820 K/W	87,022± 14,631 K/W	1.03±0.15 W/K·m [28]
PCDTBT (non-doped)	3.0 µm	2.8±0.3 µm	17,866± 204 K/W	44,927± 7820 K/W	358,859± 66,204 K/W	0.25±0.04 W/K·m //0.20±0.02 W/K·m[28]
Tellurium Film	2.74 µm	2.8±0.3 µm	15,749± 75.5 K/W	44,927± 7820 K/W	112,476± 6,480 K/W	0.79±0.04 W/K·m //0.78±0.08 W/K·m[29]
Au film on silicon substrate	150 nm	428±24nm	11,624± 157 K/W	40,191± 1532 K/W	5505± 253 K/W	104.2±67.4W/ K·m //110±2 W/ K·m[30]
PEDOT CAL	Bulk	2.8±0.3 µm	17,429± 217 K/W	44,927± 7820 K/W	241,732± 37,672 K/W	0.37±0.05 W/K·m // 0.36 W/K·m
PANI-5 % GNP CAL	Bulk	2.8±0.3 µm	17,018± 115 K/W	44,927± 7820 K/W	188,595± 27,836 K/W	0.47±0.06 W/K·m // 0.49 W/K·m[25]
PANI-7 % GNP CAL	Bulk	2.8±0.3 µm	16,314± 118 K/W	44,927± 7820 K/W	131,760± 25,913 K/W	0.68±0.08 W/K·m //0.65 W/ K·m[25]
p-type Bi_2Te_3 CAL	Bulk	2.8±0.3 µm	15,700± 145 K/W	44,927± 7820 K/W	92,113± 11,911 K/W	0.97±0.11 W/K·m // 1.0 W/K·m
Borosilicate Glass CAL	Bulk	2.8±0.3 µm	15,516± 134 K/W	44,927± 7820 K/W	82,313± 9787 K/W	1.08±0.11 W/K·m // 1.1 W/K·m
AISI 304 Steel CAL	Bulk	428±24 nm	13,811± 119 K/W	40,191± 1532 K/W	37,511± 3511 K/W	15.6±2.2 W/K·m // 16.2 W/K·m
Goodfellow® 99.9 % pure Niobium CAL	Bulk	428±24 nm	12,194± 140 K/W	40,191± 1532 K/W	10,632± 2329 K/W	54.9±8.9 W/K·m // 53.7 W/K·m

2.3 Transient and lock-in harman techniques to decouple material ZT and thermoelectric properties

Finally, a new method of measuring material thermal conductivity by simultaneously measuring thermoelectric figure of merit (ZT), electrical resistivity (ρ), and Seebeck coefficient (α) is proposed. ZT is dimensionless measure of the efficiency of material at converting thermal into electrical energy, or vice-versa, at a given temperature, T, and may be expressed as:

$$ZT = \alpha^2 T/\rho\kappa \quad (19)$$

Thus, thermal conductivity is obtained if the other terms in equation (19) are known. This new method also allows for measuring intrinsic ZT with reduced experimental error by accounting for losses through non-ideal contacts and geometry.

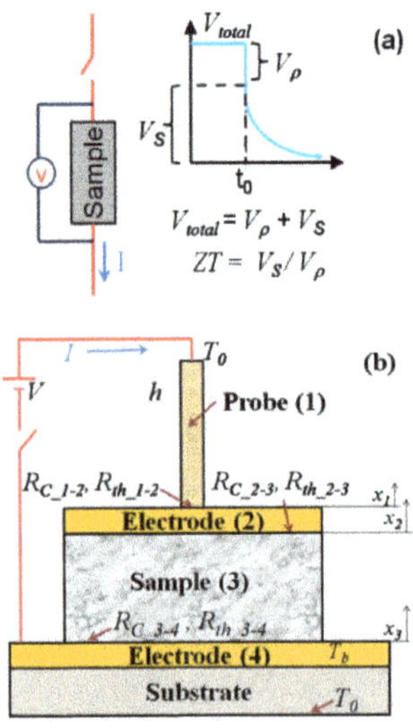

Fig. 8 Schematic of lock-in Harman technique, under assumptions of (top) ideal contacts and (bottom) considering non-negligible thermal and electrical resistances arise from contacts. Reprinted from [23] with the permission of AIP Publishing.

Conventional application of Harman method uses four probes – two to pass current, and two to measure the voltage response of the sample. Harman demonstrated that, while electrical response of the sample was nearly instantaneous, voltage generated by Seebeck effect, which is thermally driven, is much slower. By taking advantage of this fact, thermal signal could be determined from voltage response of the sample to a sudden change in voltage over time, or response to an AC current passed, locking into thermally driven signal. It was shown, by letting $\alpha = V_\alpha/\Delta T$, $\kappa = -\Delta T * A/(\alpha * T * I)$, and $\rho = V_\rho * A/(L * I)$, that ZT could be reduced to the ratio of resistive voltage to Seebeck voltage (*i.e.* $ZT = V_S/V_\rho$). This assumes ideal contacts (negligible thermal and electrical losses through the contact leads) and that temperature rise is due only to Peltier heating, neglecting effects of Joule heating. However, these effects are often difficult to mitigate, and may be accounted for by appropriate modeling (see Figure 8).

From resistive voltage, one may be able to determine electrical resistivity of the material; however, Seebeck coefficient and thermal conductivity remain coupled in equation for ZT. To decouple them, Seebeck coefficient may be simultaneously determined by adding a pair of thermocouple wires at the top and bottom surfaces of the sample as per Figure 9. If we label electric potential in each corner of the sample E_1-E_4, respectively, using Ivory technique [21], we may find Seebeck coefficient from taking voltage measurements across the sample. If voltage is measured at opposite corners, then voltage values measured are E_{13} and E_{24}. Value α is determined from expression below, where m is the slope of E_{13}vs E_{24}:

$$\alpha_{sample} = \lim_{E_{13}\to E_{24}} \frac{1}{1-m}\alpha_{ba} + \alpha_b \qquad (20)$$

This technique for measuring Seebeck coefficient reduces the required number of voltage measurements to determine α from three to two and mitigates mismatch in thermocouples, since DC offsets are removed by using a slope. It also allows for AC measurements of total voltage, and determination of ZT from Nyquist diagrams.

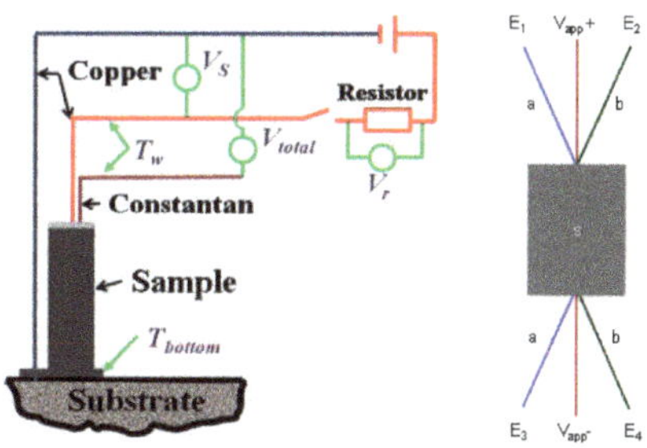

Fig. 9 Schematic of sample measurement, including thermocouple contacts (Ivory technique). (Left) sectional view (reprinted from [23] with the permission of AIP Publishing) and (right) view of sample from above.

2.3.1 Transient harman technique – analytical model

Several experimental setups used by the research community for thermoelectric characterization of thin films employ clean room micro-fabrication techniques to pattern a metallic electrode on top of the sample, while others use bonded wires or micro-manipulated probes to make electrical contact with the top surface of the film sample [21-23, 31, 32]. Configuration modeled in this work is similar to these situations, as shown in Figure 9. Thermoelectric film (3) with cross-sectional area, A_3, is deposited on substrate; metallic electrode (2) covers the top film's surface; and electrically conductive probe wire (1) of diameter, d_1, is brought in contact with the top surface of the sample. Substrate electrode (4) situated at the interface between thermoelectric film and substrate is used to close the loop and pass current into the film. The substrate electrode is assumed to have negligible electrical and thermal resistance. Its contribution to thermoelectric transport is therefore neglected, with exception of Peltier effect. Substrate electrode temperature is assumed to be the same as the top surface of the substrate (T_b). Electrical and thermal contact resistances expressed as specific values $R_{C\_i\text{-}j}$ and $R_{th\_i\text{-}j}$ are assumed at interfaces between adjacent layers indexed by i and j, with ($j=i+1$). Classical thermoelectric transport model, which neglects electron-phonon non-equilibrium effects, is developed by assuming that the thickness of thermoelectric film is much larger than phonon-electron thermalization length [33]. Under these conditions, thermoelectric transport in the probe, electrode, and sample is considered one dimensional. In each layer i, x is the spatial coordinate; h, k and ρ are convection heat transfer coefficient, thermal conductivity and electrical resistivity, respectively; P is perimeter, T is absolute temperature, α is Seebeck coefficient, A_3 is area perpendicular to thermoelectric transport direction, and J is current density.

Equation (21) represents steady state energy balance in layers 1-3; the first term on the left side represents heat conduction, the second represents lateral convection and the third represents Joule heating. Thermal radiation and temperature dependence of thermoelectric properties have been neglected as small temperature differences are assumed to occur during the experiments.

$$\frac{d^2T_i}{dx_i^2} - \frac{h_i P_i}{k_i A_i}(T_i - T_0) + \frac{J_i^2 \rho_i}{k_i} = 0; \quad i = 1,2,3 \tag{21}$$

General solution of equation (21) for temperature profile in the layer i is of the form,

$$T_i(x_i) = T_0 + c_{1,i}\, e^{\sqrt{\frac{h_i P_i}{k_i A_i}} x_i} + c_{2,i}\, e^{-\sqrt{\frac{h_i P_i}{k_i A_i}} x_i} + \frac{J_i^2 \rho_i A_i}{k_i h_i}; \quad i = 1-3 \tag{22}$$

Integration constants $c_{1,\,i}$ and $c_{2,\,i}$ are determined using six boundary conditions. First, the temperature at free end of the probe of length l_1 is assumed to be ambient temperature as stated by:

$$T_1 = T_0\,;\quad x_1 = l_1 \tag{23}$$

Second, the temperature of the end of the probe in contact with the surface is assumed to be constant temperature (see expression in the ensuing discussion, equation 31).

The third boundary condition considers energy conservation at the interface between the film and the substrate,

$$-k_3 A_3 \left.\frac{dT_3}{dx_3}\right|_{x_3=0} + J_3 \alpha_3 A_3 T_3\big|_{x_3=0} = \frac{T_b\big| - T_3\big|_{x_3=0}}{R_{th\_3\text{-}4} / A_3} + \frac{1}{2} J_4^2 A_4 R_{C\_3\text{-}4} =$$
$$\frac{T_0 - T_b}{\Theta_{subst}} + J_4^2 A_4 R_{C\_3\text{-}4} + J_4 \alpha_4 A_4 T_b \tag{24}$$

In equation (24), the left side represents heat transfer rate out of the interface. It includes heat conduction and Peltier terms in layer 3, respectively. The middle section of equation (24) represents one way to express heat transfer rate entering the interface. It is written as the sum between heat conduction across the interface thermal contact resistance and contact Joule heating term deposited at the interface in layer 3. It is assumed, that the total contact Joule heating is split equally on both sides of the interface. The right side of equation (24) represents the second way to express heat transfer rate entering the interface and includes: 1) substrate heat conduction transfer rate written as the temperature difference across the substrate divided by substrate's thermal conduction resistance; 2) total Joule heating due to electrical contact resistance; and 3) Peltier contribution due to electric current flowing through the substrate electrode. It is assumed, that the bottom surface of the substrate is at ambient temperature. Thermal conduction resistance of the substrate, Θ_{subst}, can be determined by conduction shape factor. For instance, for sample of diameter d_3 on semi-infinite substrate with thermal conductivity k_3 shape factor is 0.5 $k_s^{-1} d_3^{-1}$.

Similarly to equation (23), the fourth boundary condition is energy balance at the interface between probe and electrode,

$$-k_1 A_1 \left.\frac{dT_1}{dx_1}\right|_{x_1=0} + J_1 \alpha_1 A_1 T_1\big|_{x_1=0} + h_2 (A_2 - A_1)\left(T_2\big|_{x_2=l_2} - T_0\right) =$$
$$\frac{T_2\big|_{x_2=l_2} - T_1\big|_{x_3=0}}{R_{th\_1\text{-}2} / A_1} + \frac{1}{2} J_2^2 A_2 R_{C\_1\text{-}2} = -k_2 A_2 \left.\frac{dT_2}{dx_2}\right|_{x_2=l_2} + J_2 \alpha_2 A_2 T_2\big|_{x_2=l_2} + J_2^2 A_2 R_{C\_1\text{-}2} \tag{25}$$

Here heat transfer rate exiting the interface (the left side of the equation) also includes convection from the top surface of the electrode to the ambient. Similar analysis is performed at the electrode-sample interface, as stated in (26):

$$-k_2A_2\left.\frac{dT_2}{dx_2}\right|_{x_2=0}+J_2\alpha_2A_2T_2\big|_{x_2=0}=\frac{T_3\big|_{x_3=l_3}-T_2\big|_{x_2=0}}{R_{th\_1\text{-}2}/A_1}+\frac{1}{2}J_3^2A_3R_{C\_2\text{-}3}=$$
$$-k_3A_3\left.\frac{dT_3}{dx_3}\right|_{x_3=l_3}+J_3\alpha_3A_3T_3\big|_{x_3=l_3}+J_3^2A_3R_{C\_2\text{-}3} \tag{26}$$

Finally, continuity of electrical current in the layers of the sample requires:

$$J_1A_1=J_2A_2=J_3A_3=J_4A_4 \tag{27}$$

Modeling approach discussed above can be used to study in detail effects on temperature profile due to thermal and electrical properties of individual layers and contacts.

Thermal conductivity of thermoelectric film is typically determined from relationship between temperature rise (usually the measured surface temperature) and dissipated power. In addition, difference between surface and substrate temperature together with Seebeck voltage developed across the film is used to calculate Seebeck coefficient of the film. Practitioners in thermoelectric field need a way to evaluate steady state surface temperature before electrical current is switched off for transient Harman method under non-ideal boundary conditions.

The main strategy pursued here is to use the superposition principle to calculate total temperature rise by solving separately for temperature solutions under Joule heating and Peltier effects. Rather than using full set of equations (21 - 27), several assumptions are made in this section in order to arrive at an easy to use expression for surface temperature, which still reflects the main thermoelectric transport mechanisms in many practical situations. These assumptions are: 1) electrode's contributions (layer 2) to thermoelectric transport are neglected because metallic electrode layers typically have low Seebeck coefficient similar to the probe and much lower electrical and thermal resistances compared to thermoelectric films; 2) Seebeck coefficient of the current probe is neglected; 3) convection terms on the film surfaces are neglected; 4) substrate thermal resistance and film-substrate electrode thermal contact resistances are neglected when compared to film thermal resistance, since thermoelectric film samples are typically low thermal conductivity films on high thermal conductivity substrates; and 5) Joule heating at film-substrate electrode contact is neglected because under assumption 4) the substrate acts as a heat sink.

Total temperature, T, in thermoelectric film is then divided into two components as: T=T'+T*, where T' is linear temperature component, LTC, that is independent of Joule heating terms and includes Peltier effects, while T* is nonlinear component, NLTC, and takes into account Joule heating effects, including electrical contact resistance heating. Then the set of equations (21 - 27) for T′ becomes:

$$-k_3 A_3 \frac{d^2 T_3^{'}}{dx_3^2} = 0 \tag{28}$$

$$T_3^{'} = T_0 ; \quad x_3 = 0 \tag{29}$$

$$-k_3 A_3 \left.\frac{dT_3^{'}}{dx_3}\right|_{x_3 = l_3} + J_3 A_3 \alpha_3 T_S^{'} = \frac{T_S^{'} - T_w^{'}}{R_{th\_1\text{-}3}/A_1} = q_w^{'} ; \qquad x_3 = l_3 \tag{30}$$

where T_s' and q_w' are temperature of the top surface of the sample and heat transfer rate through the probe, respectively. Heat transfer rate through the probe is calculated using a fin model with ambient temperature at free end and constant temperature T_w' at its base [23]:

$$q_w^{'} = a(T_w^{'}\text{-}T_0) \tag{31}$$

where the constant *a* is defined as:

$$a = \frac{\sqrt{h_1 P_1 k_1 A_1}}{\tanh\left(\frac{l_1^2 h_1 P_1}{k_1 A_1}\right)^{0.5}} \tag{.32}$$

Then, the solution for LTC of the top surface of the sample can be calculated as:

$$T_S^{'} = \frac{(a_{th} l_3/A_3 + k_3) T_0}{a_{th} l_3/A_3 + k_3 - J_3 l_3 \alpha_3} \tag{33}$$

where a_{th} is total heat conductance through the contact and the probe defined as:

$$a_{th} = \frac{a}{1 + a R_{th\_1\text{-}3}/A_1} \tag{34}$$

Next, T* is calculated from the following equations:

$$\frac{d^2 T_3^*}{dx_3^2} + \frac{J_3^2 \rho_3}{k_3} = 0 \tag{35}$$

$$T_3^* = 0 ; \quad x_3 = 0 \tag{36}$$

$$-k_3 A_3 \left.\frac{dT_3^*}{dx_3}\right|_{x_3=l_3} + J_3 A_3 \alpha_3 T_3^* \Big|_{x_3=l_3} + J_3^2 A_3 R_{C\_1\text{-}3} =$$
$$\frac{T_S^* - T_w^*}{R_{th\_1\text{-}3}/A_1} + \frac{1}{2} J_1^2 A_1 R_{C\_1\text{-}2} = q_w^* \;; \qquad x_3 = l_3 \quad . \tag{37}$$

Heat transfer along the probe is calculated by solving the fin model with volumetric Joule heating and a temperature rise equal to zero (relative to the ambient) at free end of the fin (away from the sample). The equation for the probe heat transfer is:

$$q_w^* = aT_w^* - b \,, \tag{38}$$

where the constant b is expressed as:

$$b = -a \frac{J_3^2 A_3^2 \rho_1}{h_1 P_1 A_1} \left[1 - \frac{1}{\cosh[(l_1^2 h_1 P_1 / k_1 A_1)^{0.5}} \right]. \tag{39}$$

Then, the solution for NLTC is given by:

$$T_S^* = \frac{J_3^2 l_3 (R_{C\_1\text{-}3} \frac{A_3}{A_1} + \frac{\rho_3 l_3}{2}) - \frac{l_3 b_{th} b_{cont}}{A_3}}{a_{th} l_3 / A_3 + k_3 - J_3 l_3 \alpha_3} \,, \tag{40}$$

where,

$$b_{th} = \frac{b}{1 + aR_{th\_1\text{-}3}/A_1} \qquad b_{cont} = 1 + \frac{\frac{1}{2} J_3^2 \frac{A_3^2}{A_1} R_{C\_1\text{-}3}}{b} \frac{aR_{th\_1\text{-}3}}{A_1} \,. \tag{41}$$

Finally, total temperature of the top surface of the sample, T_s, is:

$$T_S = \frac{1}{a_{th} l_3 / A_3 + k_3 - J_3 l_3 \alpha_3} \begin{bmatrix} (a_{th} l_3 / A_3 + k_3) T_0 + \\ J_3^2 l_3 (R_{C\_1\text{-}3} \frac{A_3}{A_1} + \frac{\rho_3 l_3}{2}) - b_{th} b_{cont} \frac{l_3}{A_3} \end{bmatrix} \tag{42}$$

where the first term contains Peltier effect's induced contributions to the surface temperature, the second term includes Joule heating effects from the sample and contact, and the third term includes Joule heating contribution from the probe wire.

Temperature of the probe at junction with the sample surface is then calculated as:

$$T_w = \frac{T_S + \frac{R_{th\_1-3}}{A_1}\left(\frac{1}{2}J_3^2\frac{A_3^2}{A_1}R_{C\_1-3} + aT_0 - b\right)}{1 + aR_{th\_1-3}/A_1} \quad . \tag{43}$$

Understanding how to eliminate or reduce the effects due to heat loss and electrical and thermal contact resistances is critical in designing test structures amenable for accurate thermoelectric transport measurements. Parasitic effects are expected to be different for macroscale vs microscale samples, and this section focuses on microscale samples. To illustrate these effects, surface temperature predictions as a function of current density are discussed for thermoelectric sample of 10×10×10 μm^3 in contact with copper probe of 5 μm diameter and 1.3 mm length. Thermoelectric properties of thermoelectric film are similar to n-type $Bi_2Te_{2.7}Se_{0.3}$ and are listed in Table 2.

Figure 10 shows rise of surface temperature with respect to ambient temperature, calculated from equation (42) for a range of specific thermal contact resistances. Electrical resistance of contact was assumed to be equals to theoretical limit predicted for electrical boundary resistance between Bi_2Te_3 and metal electrode [33]. Direction of electrical current was chosen such, that sample surface undergoes Peltier cooling. At low current densities, Peltier cooling term dominates over Joule heating terms and temperature of the top surface of the sample decreases linearly as electrical current density increases. After reaching the maximum cooling temperature at optimum current density, Joule heating terms start to dominate over Peltier terms. Parasitic conduction heat transfer effect is apparent even at very low current densities, as shown by inset in Figure 10. It leads to reduction of temperature difference across the sample as compared to predictions of ideal Harman model. On the other hand, as thermal contact resistance increases, the sample cooling is stronger because the thermal barrier created at the contact reduces heat transfer rate with the probe. Importance of thermal barrier effect is gauged by comparison between thermal resistances of the probe, probe-sample contact, and the sample itself. The modeled probe has thermal resistance of ~5×10^4 K/W, which is similar to 6×10^4 K/W thermal resistance of the sample; therefore, a significant heat transfer occurs through the probe. As the thermal contact resistance increases, the probe heat transfer is reduced, particularly after contact thermal resistance becomes of the same order as thermal resistance of the probe. Alternative way to minimize the probe heat transfer rate could be realized by reducing diameter of the probe. However, besides practical challenges, this may have a negative impact associated with increase in resistance of electrical contact as shown in figure 10.

Table 2 Sample parameters and thermal/thermoelectric properties. Reprinted from [23] with the permission of AIP Publishing.

	Prob	Sample	Electro
Material	Cu	$Bi_2Te_{2.7}S$	In
Lateral dimension	76.2[a]	1014[b]	1014[b]
Length (mm)	20	1.8	0.05
Seebeck coefficient (μV/K)	1.84	-212	1.68
Thermal conductivity	401	1.5	82
Convection heat transfer coefficient	200[c]	10	10
Electrical resistivity (nOhm	17.1	10500	84

a = Diameter, b = Width, c = The convection heat transfer coefficient for 5 μm probe is 3000 (W/m2K)

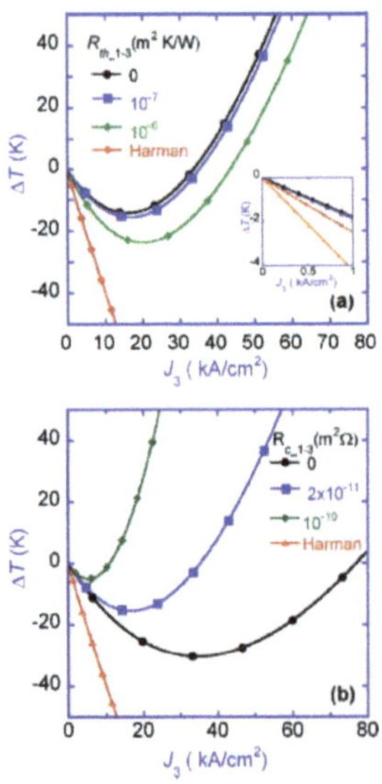

Fig. 10 Calculated temperature response for specified R_{th} and R_c. Reprinted from [23] with the permission of AIP Publishing.

Heat transfer through the substrate could also play a major role in establishing the surface temperature. For large current densities, the strength of heat transfer through the substrate is indicated by large positive temperature difference measured across the sample. This difference is in contrast with the predictions of ideal Harman model, where Joule heating effects never generate a temperature difference across the sample.

The effect of electrical contact resistance in absence of thermal contact resistance is investigated in Figure 10. The modeled probe has electrical resistance of ~1 Ohm, which is similar to electrical resistance of the sample, therefore Joule heating effects can occur simultaneously in the sample and the probe. In addition, even for very low specific electrical contact resistance of 1×10^{-11} Ohm×m^2, electrical contact resistance for the modeled probe is considerable (0.5 Ohm). Electrical contact resistance increases by two orders of magnitude if the probe diameter is reduced by factor of 10. This illustrates strong requirements to control the probe and contact electrical resistances in transient Harman experiments performed on film on substrate samples. This is because good thermoelectric samples have low electrical resistances, so Joule heating effect in contacts and probe can easily become dominant. Inspection of the probe and wire Joule heating terms suggests mitigation of the electrical contact resistance problem may be achieved by preparing thick film samples (large l_3) for measurements, but with small cross-sectional area (A_3) in order to increase the relative importance of electrical resistance of the sample. Since this strategy leads to an increase in thermal resistance of the sample, one must simultaneously address the need to mitigate parasitic probe heat conduction effects by choosing a probe with thermal resistance larger than that of the sample. The probe's thermal resistance may be calculated by knowing the material properties and diameter of the probe.

One proposed strategy for determining the properties under non-ideal conditions is the bipolar method, where transient Harman experiments are performed using direct and reversed current directions and where measured Seebeck and resistive voltages across the sample are averaged. This is believed to eliminate Joule heating effects and reveal the intrinsic Peltier effects in the sample [23, 31]. However, as demonstrated in this work, when non-ideal boundary conditions are present, parasitic effects cannot be always completely eliminated by this strategy. Nevertheless, the analysis below demonstrates the ability to exploit this behavior to determine both thermal and electrical transport properties of the samples and their contacts.

Bipolar resistive voltage difference measured across the sample $\Delta V_{\rho\pm}$ is related to total electrical resistance through the expression:

$$\frac{V_{\rho+}-V_{\rho-}}{J_3A_3}=\frac{\Delta V_{\rho\pm}}{J_3A_3}=2*\left(R_{C\_1-3}/A_1+R_{C\_3-4}/A_3+\frac{\rho_3 l_3}{A_3}\right). \quad (44)$$

To find expression for bipolar Seebeck voltage, first Seebeck coefficient, α_s, is expressed as measured V_S as function of surface temperature of the sample as:

$$\alpha_S=\frac{-V_S}{T_S-T_0}, \quad (45)$$

where V_s is experimental Seebeck voltage measured between the probe and electrode situated at the bottom of the sample. If the probe temperature is measured, then:

$$\alpha_S=\frac{-V_S}{(T_w-T_0)\left(1+\frac{aR_{th\_1-3}}{A_1}\right)+\frac{R_{th\_1-3}}{A_1}\left(b-\frac{1}{2}J_3^2A_3^2\frac{R_{C\_1-3}}{A_1}\right)}. \quad (46)$$

When bipolar method is used, then α_s can be extracted from bipolar Seebeck voltage difference $\Delta V_{S\pm}$ and temperature difference $\Delta T_{w\pm}$ using expression:

$$\alpha_S=\frac{-\Delta V_{S\pm}}{\Delta T_{S\pm}}=\frac{-\Delta V_{S\pm}}{\Delta T_{w\pm}\left(1+\frac{aR_{th\_1-3}}{A_1}\right)}. \quad (47)$$

Next, under small currents approximations:

$$k_3+a_{th}\frac{l_3}{A_3}>>J_3l_3\alpha_3 \quad \& \quad T_0\left(k_3+a_{th}\frac{l_3}{A_3}\right)>>J_3^2l_3(R_{C\_1-3}\frac{A_3}{A_1}+\frac{\rho_3l_3}{2})-\frac{l_3b_{th}b_{cont}}{A_3} \quad (48)$$

a simplified expression for bipolar surface temperature difference is obtained as:

$$\Delta T_{S\pm}=\frac{2T_0}{k_3+a_{th}l_3/A_3}J_3l_3\alpha_3 \quad (49)$$

Expression for ZT obtained through bipolar technique is found as:

$$ZT_0=\frac{-\Delta V_{S\pm}}{\Delta V_{\rho\pm}}\left(1+\frac{a_{th}l_3}{k_3A_3}\right)*\left(1+\frac{R_{C\_1-3}l_3}{\rho_3A_3A_1}+\frac{R_{C\_3-4}l_3}{\rho_3A_3^2}\right) \quad (50)$$

Another strategy is to perform experiments over a range of currents and use differential changes in V_S and $V_{\square}$ with current I. Under small current conditions, the following expression is then obtained for the figure of merit:

$$ZT_0 = \frac{-dV_S / dI}{dV_\rho / dI}\left(1 + \frac{a_{th} l_3}{k_3 A_3}\right) * \left(1 + \frac{R_{C\_1-3} l_3}{\rho_3 A_3 A_1} + \frac{R_{C\_3-4} l_3}{\rho_3 {A_3}^2}\right). \tag{51}$$

Neither bipolar nor differential current methods alone are able to account for all parasitic effects. As a result, these effects must be considered in data reduction or otherwise minimized. A variable thickness method [23, 31, 32] is used to account for electrical contact resistance effects, while heat losses and thermal resistance effects are neglected. A different method to determine all thermoelectric properties without the need for extensive sample preparations is outlined below.

The strategy explored here is to use bipolar experiments performed over a wide range of currents rather, than small current regime required by above methods. It is expected, that at large currents experimental Seebeck voltage and temperature signals become sensitive to electrical transport properties of the sample and contacts and could be used to determine the sample and contact thermoelectric properties. In addition to Seebeck and resistive voltage drops, method requires measurement of the sample surface temperature or the probe temperature (at the contact with the sample's surface).

Proposed strategy takes into consideration selective sensitivity of thermal signals to Peltier and combined Peltier and Joule heating effects under low and high current regimes, respectively. Under small current approximations, temperatures of the probe and sample surface are linear with current, and thermal conductivity can be expressed as a function of experimentally measured slope of the probe temperature as:

$$k_3 = \frac{l_3}{A_3}\left[\frac{\alpha_3 T_0}{\left(1 + \frac{a R_{th\_1-3}}{A_1}\right) dT_w / dI} - a_{th}\right]. \tag{52}$$

In equation (52), value of Seebeck coefficient is substituted from equation (47), which is valid at any current. Next, equation (44), equation (47) and equation (52) are substituted into equation (43). For the sake of discussion it is assumed, that specific electrical contact resistance is similar at the top and bottom contacts (other assumptions are discussed in section 3). After the above substitutions, predicted probe temperature and Seebeck coefficient become a function of two unknowns, specific electrical and thermal contact resistances, which are then used to fit experimental signals under large current regime for both direct and reverse currents. This strategy allows the unique determination of all thermoelectric properties of the sample and electrical and thermal contact resistances. Details of the fitting procedure are presented in experimental validation section.

2.3.2 Lock-in harman technique – analytical model

To find ***frequency-dependent*** temperature solution in the sample, governing equation and boundary conditions for the problem were first expressed as a function of time and then transformed to frequency domain. Governing thermal transport equation for the sample with attached wires was derived by balancing the energy in infinitesimal length dx of wire or sample domain treated as Joule heated fin and neglecting Thompson effects. Approach is similar to the steady state model [33]. The governing equation is:

$$\frac{\partial^2 T_i(x,t)}{\partial x_i^2} - m_i^2(T_i - T_0) + \frac{J^2 \rho_i}{2k_i} = \frac{1}{\Delta_i}\frac{\partial T_i(x,t)}{\partial t} \tag{53}$$

The first term on the left side accounts for conduction through the wire/sample, where T_i is temperature along length of the wire/sample as a function of position, x, and time, t. The second term represents convective heating or cooling from the environment, where $m_i^2 = h \cdot p/(k \cdot A)$, with h being heat transfer coefficient, p wire/sample's circumference, A its cross-sectional area, and T_0 ambient temperature. The third term accounts for Joule heating, and the right side of equation is transient heat storage term. J is electrical current density, and Δ is thermal diffusivity.

Figure 11 represents the same configuration of sample and wires as considered before, but here shows heat transfer domains used in the model. 1D heat transfer was modeled in each of seven domains, one each for the six wires plus another for the sample. Details of the boundary conditions are given below.

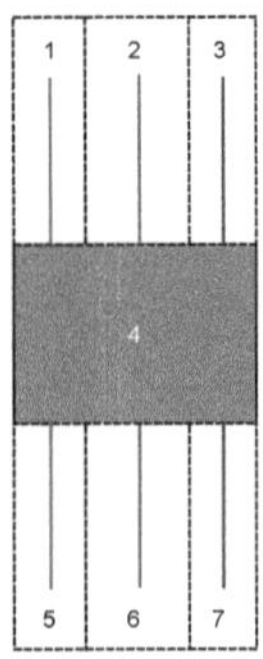

Fig. 11 Schematic of the sample showing heat transfer domains.

Temperature solution requires a total of fourteen boundary conditions, two for the sample and four for each set of wires. The wire boundary conditions are as follows: two boundary conditions per wire (or six total) were defined by assuming that the end of each wire was at room temperature, since the wires in the experiment were relatively long compared to their width and measurements were conducted in ambient conditions. The remaining two boundary conditions per wire (summing to twelve in total) are that the ends of each wire in contact with the sample are at a fixed temperature. The two boundary conditions across the sample come from the fact that heat transfer at the interfaces must be balanced. For the interface between the first set of wires (domains 1-3) and the sample (domain 4), the energy balance yields equation (54):

$$q_{1-3} + J_4^2 A_4 R_{14} + J_4 \alpha_{34} A_4 T_4(0,t) = -k_4 A_4 \frac{\partial T_4(0,t)}{\partial x_4} \quad (54)$$

The first term is the rate of heat conduction to and from the wires and is a function of temperature gradient at the wire-sample interface and thermal conductivity of the wires. This is calculated using the pin fin equation below for the experimental results. The second term accounts for Joule heating due to electrical contact resistivity between the current lead and the sample, R_{14}. The third term is Peltier heating at the interface, where $\alpha_{34} = \alpha_3 - \alpha_4$, relative Seebeck coefficient between the wire (3) and the sample (4). The right side of the equation is the heat conducted through the sample. Form of boundary condition for the other wire-sample interface is identical. To transform the problem from a partial differential equation in time to an ordinary differential equation in frequency, ω, we used Fourier transform. Before applying Fourier transformation, it was convenient to represent the sample temperature as a Fourier series. This was possible since temperature is function of periodic excitation signal and is therefore itself periodic. Temperature as the sum of its DC component (η=0) and all of the harmonics (all other values of η) of fundamental frequency, ω_0, is given by equation (55):

$$T_i(x,t) = \sum_{-\infty}^{\infty} T_i(x, n\omega_0) e^{i\eta\omega_0 t} \quad (55)$$

Substituting this into governing equation and applying Fourier transform gives transformed governing equation:

$$\frac{\partial^2 T_i(x,\omega)}{\partial x_i^2} - m_i^2 T_i(x,\omega) + m_i^2 T_0 \sqrt{2\pi}\delta(\omega) + \frac{J^2 \rho_i}{8k_i}\sqrt{2\pi}(2\delta(\omega) + \delta(\omega - 2\omega_0) + \delta(\omega + 2\omega_0)) = \frac{i\omega}{\Delta_i} T_i(x,\omega) \quad (56)$$

Dirac delta function, δ, is employed because signals of constant frequency in the time domain become delta functions in frequency domain. Conduction, convection, and heat storage terms are present at each harmonic with additional convection term present at ω = 0. Joule heating occurs only at ω=0 and ω =2·ω_0. The transformation was next applied to the boundary conditions. The transformed boundary condition for the first interface is given by equation (57):

$$q_{1-3} + \frac{J_4 A_4 R_{14}}{2}\sqrt{2\pi}\left(2\delta(\omega) + \delta(\omega - 2\omega_0) + \delta(\omega + 2\omega_0)\right) + \frac{J_4 S_{34} A_4}{2}\sum_{-\infty}^{\infty} T_4(0, n\omega_0)\sqrt{2\pi}\left(\delta(\omega - \omega_0(n+1)) - \delta(\omega - \omega_0(n-1))\right) = -k_4 A_4 \frac{\partial T_4(0,\omega)}{\partial x_4} \tag{57}$$

Joule heating term is again present at $\omega = 0$ and $2\cdot\omega_0$, while Peltier heating term occurs only at the fundamental frequency. This demonstrates mathematically how Joule and Peltier components of heat transfer are separated by measuring the harmonics of temperature. All measurements in this work use Peltier component. Joule component is not used in measurements described here. Since this work focuses on the first harmonic measurements, the solution for the first harmonic and its derivative are given by equation (58) and equation (59), where R is the root of homogenous form of governing equation and is given by equation (60):

$$T_i(x_i, \omega) = c_{i1,1\omega} e^{R_{1\omega} x} + c_{i2,1\omega} e^{-R_{1\omega} x} \tag{58}$$

$$\frac{\partial T_i(x_i, \omega)}{\partial x_i} = c_{i1,1\omega} R_{1\omega} e^{R_{1\omega} x} - c_{i2,1\omega} R_{1\omega} e^{-R_{1\omega} x} \tag{59}$$

$$R = \sqrt{m^2 + \frac{i\omega}{\Delta}} \tag{60}$$

The undetermined coefficients were solved numerically using transformed boundary conditions.

2.3.3 Experimental results – transient harman

Figure 12 shows experimentally measured total and Seebeck voltages as a function of current.

Resistive voltage drop obtained after subtracting Seebeck voltage from total voltage includes contributions from the sample, probe-sample contact, and sample-substrate electrode contact. Inset in Figure 12 shows an example of measured voltage as a function of time during an experiment at 163 mA. The figure of merit calculated according to classical Harman method yields an average value of 0.11, much smaller, than the manufacturer value of 0.85. This discrepancy is due to parasitic effects neglected in classical technique.

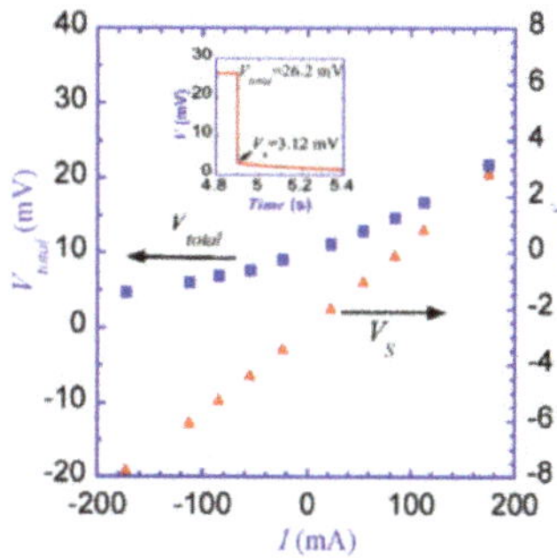

Fig. 12 The deconvolution of resistive and Seebeck voltage contributions. Reprinted from [23] with the permission of AIP Publishing.

Measured probe temperature is linear with current for the smallest two bipolar currents and yields a slope of -55.1 K/A and was substituted in equation (52). Originally, only one probe was modeled in contact with the sample, the energy loss across constantan wire was evaluated and found to be an order of magnitude smaller, than for copper wire, therefore its effect is expected to be negligible.

Transient Harman experiments performed under the highest direct and reverse current conditions were used in the fitting of sample's thermoelectric properties and contact resistances. In the fitting procedure, thermal contact resistance $R_{th\_1\text{-}3}$ was varied over a wide range, typically between 1.0×10^{-8} m$^2\times$K/W and 1.0×10^{-5} m$^2\times$K/W with step of 1.0×10^{-8} m^2 K/W. For each $R_{th\_1\text{-}3}$ value, sample's Seebeck coefficient was determined from equation (47) and sample's thermal conductivity was determined from equation (52). In addition, two sets of solutions for electrical contact resistance as function of $R_{th\_1\text{-}3}$ are generated by fitting experimental Seebeck coefficient and the probe temperatures. The first set of solutions is obtained by minimizing the mean square deviation between Seebeck coefficient values calculated from equations (46 - 47). The second set of solutions is obtained by minimizing the mean square deviation between experimental temperatures of the probe and the predictions of equation (43). In equation (43), T_s was substituted from equation (42), thermal conductivity was substituted from equation (52), Seebeck coefficient from equation (47) and resistivity of the sample from equation (41). It was found, that the first set of solutions is more sensitive to thermal contact resistance, while the second set was more sensitive to electrical contact resistance. The intersection of the two sets of solutions leads to the unique solution for thermal and electrical contact resistances and thermoelectric properties of the sample.

Since the original sample-substrate interface from commercial Peltier device has a negligible electrical contact resistance compared with the sample, $R_{c,3-4}$ was taken as zero in equation (41). It was also assumed that $R_{c,1-3}$ and $R_{th,1-3}$ represent respectively lump electrical and lump thermal contact resistance contributions due to the probe-indium interface, indium layer (negligible contribution), and indium-sample interfaces. However, measured resistive voltage drop through the constantan probe does not include electrical resistance contribution due to the probe-indium interface. Therefore, the fitting procedure was repeated several times under four different assumptions regarding contact electrical resistance. The assumptions and the fitting results are presented and discussed below.

Table 3 Summary of cases simulated to explore effect of contacts in transient Harman measurements.

Case #	$R_{th,1-3}$ (K/W)	k_3 (W/mK)	α_3 (μV/K)	$R_{c,1-3}$ (m² Ω)	ρ_3 (Ωm)
1	1.1×10^{-6}	1.38	-212	2.9×10^{-10}	5.7×10^{-5}
2	6.6×10^{-7}	1.48	-218	6.2×10^{-10}	Non-physical (negative)
3	9.4×10^{-7}	1.43	-215	3.9×10^{-10}	3.3×10^{-5}
4	7.8×10^{-7}	1.46	-217	4.9×10^{-10}	1×10^{-5} (manufacturer specified)

The thermoelectric properties arise from each of the following four cases are summarized above in table 2.3.2. In case 1, electrical contact resistance R_{c_1-3} was assumed to originate only from the probe-indium interface. In case 2, electrical contact resistance R_{c_1-3} was assumed to originate only from the indium-sample interface. In case 3 electrical contact resistance R_{c_1-3} was assumed to be split equally between probe-indium and indium-sample interfaces. Finally, in case 4 manufacturer's value for sample resistivity $\rho_3 = 1\times10^{-5}$ Ohm×m, was used in the fitting process, which allowed the exact determination of the split of probe-sample electrical contact resistance between two contributions. In this case, $R_{c,1\text{-}2}$=4.9×10^{-10} Ohm×m^2 may be split as 1.1×10^{-10} Ohm×m^2 due to the probe-indium interface and 3.8×10^{-10} Ohm×m^2 due to indium-sample interface. Relatively large values of measured electrical and thermal contact resistances are due to the imperfections of the mechanical contact under the small contact load used in this proof-of-concept experiment. The indium-pellet interface dominates the contact resistance.

The highest deviations between measured sample's Seebeck and thermal conductivity as compared with manufacture's values are respectively 6 μV/K (3 %) and 0.12 W/(m×K) (8 %). These deviations are smaller, than the experimental uncertainty. The uncertainty in thermal conductivity due to propagation of the uncertainty in temperature and voltage measurements was calculated to be 0.26 W/(m×K). Similarly, for Seebeck coefficient, the uncertainty is equal to 9.9 μV/K. To accurately determine the sample resistivity, resistive voltage drop should be measured through the copper probe, which is used to pass electrical current, at the same time as through the constantan wire, so total electrical contact resistance and sample resistivity can be accurately determined. When correct resistivity of the sample was employed in the fitting, the sample thermal conductivity was within 3 % of manufacture's values.

Figure 13 shows comparison between measured and calculated temperature of the probe as a function of electrical current passed through it.

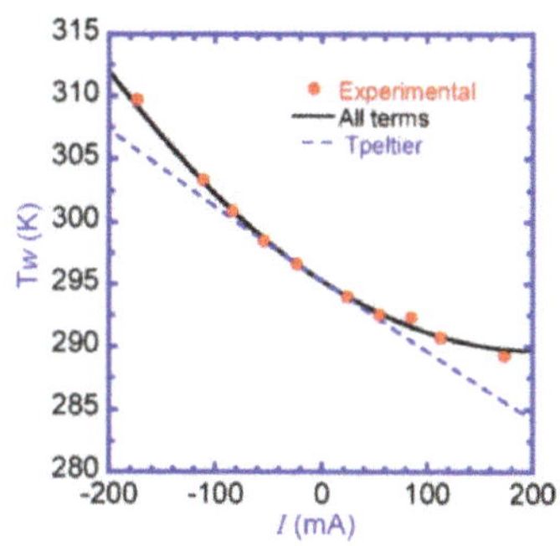

Fig. 13 Measured versus calculated probe temperature. Reprinted from [23] with the permission of AIP Publishing.

The theoretical predictions use the fitted thermoelectric properties and employ equation (43) with either all terms or only Peltier terms. The theoretical predictions were performed for all cases 1-4 and, since they superpose along the same line, they are not individually distinguishable in Figure 13. Joule heating effects are important at large currents in tested sample, as demonstrated by the discrepancy between Peltier heating only predictions and combined Peltier and Joule heating model. Predictions based on solving equation (20 - 27), that also include the convection on the sample surface and the contributions from indium electrode, show no significant difference with prediction based on equation (43). There is an excellent agreement between experimental and modeling data over entire electrical current range. Data sets for intermediate current values (~85 mA) show also excellent agreement, although they have not been used in the fitting.

2.3.4 Experimental results – lock-in Harman

ZT and individual thermoelectric properties may also be characterized experimentally using a **Nyquist** plot (plotting imaginary vs. real parts of the complex voltage signal or sample temperature rise). Nyquist analysis of voltage measurements across the sample allows for direct calculation of the slope m, which, in turn, yields α. ZT is obtained from finding each of V_R and V_S from the different regimes represented in the plot (see Figure 14). V_R is obtained from the value of the real part of the voltage response when the imaginary part is equal to zero, and V_S is the radius of the circular portion of the Nyquist plot.

The samples measured were bulk bismuth telluride alloys with dimensions of $4.5\times3.8\times3.8$ mm^3. Thin layer of gold was deposited on either end to improve adhesion and current spreading between the sample and lead wires. One lead wire and one thermocouple were soldered to either end of the sample. Current was applied through un-insulated 50.8 µm diameter copper wire, and voltage was measured using 50.8 µm E-type thermocouples. Two sets of voltage measurements were made across the sample using each set of thermocouple wires for excitation frequencies between 10 mHz and 10 Hz. Amplitudes of resulting voltages, E_{13} and E_{24}, are plotted in Figure 14. When the signal is applied at low frequencies, then measured voltage is sum of total voltage across the sample and Seebeck voltage in thermocouple's wires. As frequency is increased, thermal component in the sample and wires decays and voltage approaches resistive voltage of the sample, and E_{13} approaches E_{24}.

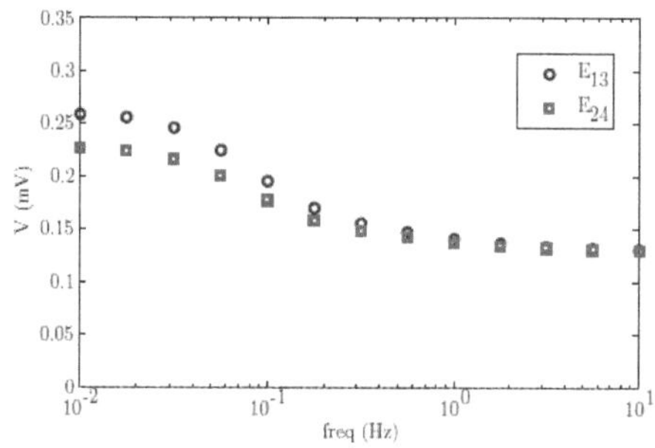

Fig. 14 E_{13} and E_{24} voltages as a function of frequency.

These two voltages were used to find Seebeck coefficient by Ivory's technique [21] using equation (21). Non-imaginary values of two voltages are plotted against each other in Figure 15 and resulting Seebeck coefficient α_{sample}=202.6±1.4 μV/K. Real parts of signals are used, as these are components, that are in phase with excitation signal and as a result are in phase with each other. Amplitudes may be out of phase with each other and imaginary part is much smaller. Advantage of Ivory technique is that the magnitudes of measured voltages are greater, than in traditional technique, if α_{sample} is larger than average value of α_a and α_b, which is often the case, when measuring thermoelectric materials. This is because voltage measured across the sample is equal to $\Delta T(\alpha_a-\alpha_{sample})$, whereas that measured in the thermocouples is $\Delta T(\alpha_a-\alpha_b)/2$. This assumes, that the sample is symmetric and that ΔT is total temperature gradient across the sample. Thus, temperature gradient measured by one set of thermocouples is $\Delta T/2$. Since temperatures on the two sides of the sample are 180° out of phase, the total temperature difference is twice the temperature amplitude registered on one side. The larger signal results in better signal-to-noise ratios and less error in the final calculation.

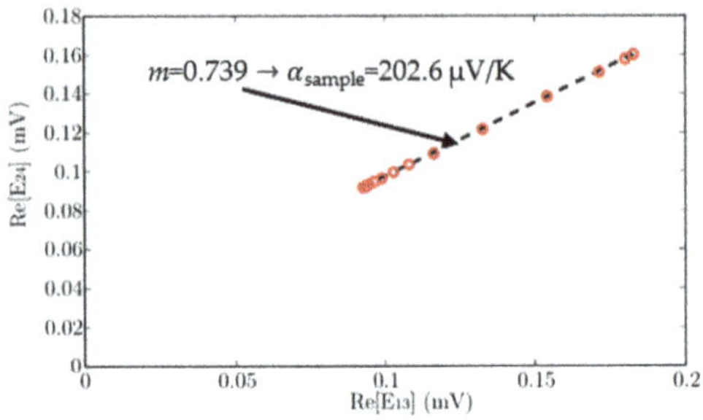

Fig. 15 Real parts of E_{24} and E_{13} to find the slope, m, and Seebeck coefficient.

Total voltage in the sample V_{sample} was calculated using equation (61) and then plotted on Nyquist diagram, shown in Figure 16.

$$V_{sample} = \frac{\alpha_b E_{13} - \alpha_a E_{24}}{\alpha_b - \alpha_a} \quad (61)$$

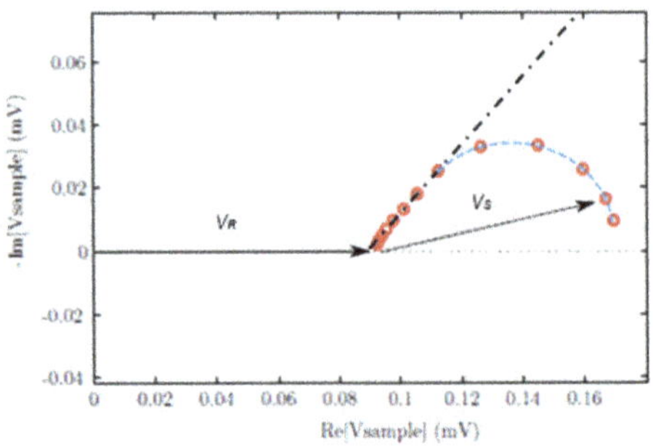

Fig. 16 Voltage measurements plotted on Nyquist diagram.

Obtained data is again shown as superposition of resistive V_R and Seebeck V_S voltages, and V_{sample} and V_R can be found from extrapolating the data to the real axis as described in introduction. High frequency behavior of real devices may not obey the -45° assumption, if contacts have significant heat capacity [20]. As seen in Figure 16, behavior of the sample deviates somewhat from -45°, which can be attributed to heat capacity of the solder between the wires and the sample. From Figure 16, V_{sample} is 0.17 mV and V_R is 0.088 mV. Extrinsic ZT of the device, is 0.93 for this measurement. To find intrinsic ZT of the material, then non-idealities in measurement system must be accounted for.

To find equation for intrinsic ZT, the derivation of ZT performed by Harman can be repeated to include terms for heat loss and contact resistance. This adds two corrections factors as shown in equation (62). The first is the ratio of heat lost from the sample to Peltier heat generation at the wire-sample interface. This heat loss may be due to either convection or radiation from the end of the sample or conduction through the contacts. The second correction factor is the ratio of voltage drop across the contacts to that in the sample, which is equal to resistance of the contacts divided by that of the sample.

$$ZT = \frac{V_S}{V_R}\left[\left(1 - \frac{q_{loss}}{q_{Peltier}}\right)\left(1 - \frac{V_C}{V_R}\right)\right]^{-1} \tag{62}$$

For this experiment, radiation and convection from the sample itself were negligible compared to heat loss through the contacts and only the latter was considered. Since ZT was calculated using voltages approximating DC, where heat transfer is steady state, and high frequency AC, which is not affected by heat losses, steady-state equation can be used to account for heat loss. Conduction through the wires is described by equation (63), which is the fin equation, where the base temperature is equal to that of the wire-sample interface, T_s, and T_0 is the ambient temperature.

$$q_w = \sqrt{hpkA}(T_S - T_0)tanh\left(\sqrt{\frac{hP}{kA}}L\right) \tag{63}$$

For long thin wires, the hyperbolic tangent goes to one and may be neglected. We considered losses by convection through the wire and temperature at the end of the wire away from the sample was assumed to have reached ambient (since wires were long and thin). Even though the sample temperature was not measured directly, the temperature gradient across the sample was found from measured Seebeck coefficient value, α_{sample}, and Seebeck voltage, V_S. Once T_s was found this way and substituted in equation (63), q_w was determined for each of two wires, summed up and the value used in equation (62). In the equation, $q_{Peltier} = \alpha IT$. After all these substitutions, intrinsic ZT of the material was calculated as 1.04.

Sample's resistivity was determined as 7.0×10^{-6} Ohm×m based on resistive voltage, V_R, and neglecting the contribution of the contacts. This was confirmed in independent measurement probing resistive voltage profile along length of the sample, which resulted in value of 7.25×10^{-6} Ohm×m. Using intrinsic ZT, measured Seebeck coefficient and the first value of resistivity, thermal conductivity of 1.55 W/(m×K) is obtained, while if the second resistivity value is used k is determined as 1.6 W/(m×K).

The same material properties were also found by fitting the predictions based on the numerical model described above in equation (58 - 60) to experimental data. The fitting is shown in Figure 17, where Seebeck voltage data was converted to temperature amplitude using measured Seebeck coefficient. Adjusting thermal conductivity in the model and using a least squares fit, thermal conductivity was 1.55 W/(m×K) and thermal diffusivity was 9.5×10^{-3} cm^2/s.

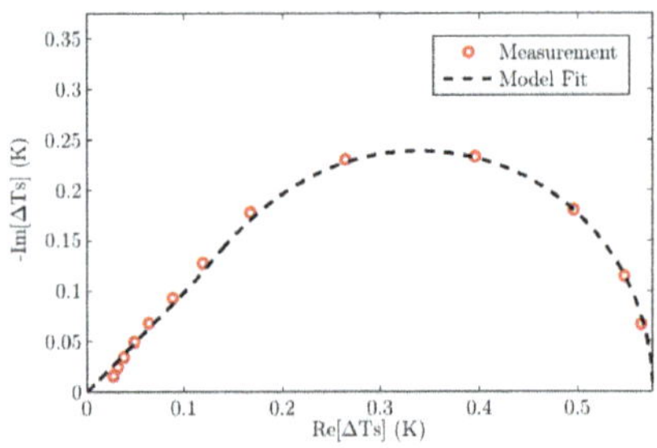

Fig. 17 Sample temperature and fitted prediction plotted on Nyquist diagram.

Overall, calculated uncertainty in this experiment was small with that in Seebeck coefficient and extrinsic ZT measurements being lower, than that of calculated intrinsic ZT. The latter is due to the uncertainty in calculating the heat loss through the wires, specifically calculating the heat transfer coefficient between the wires and the air. The heat transfer coefficient was calculated assuming horizontal cylinder in air. While the uncertainty for the heat transfer coefficient was determined to be about 2 %, its uncertainty was assumed to be closer to 10 %. This was done because there is some additional uncertainty surrounding the assumptions of free convection and due to its dependence on lab conditions. The uncertainty could be improved by testing the sample in an evacuated chamber, eliminating entirely the need to calculate heat transfer coefficient. The uncertainty of extrinsic ZT is due to that of voltage measurements, and was assumed to be 1 % of the measured value, and temperature, assumed to be 296±2 K. The uncertainty in voltage measurement was assumed 1 % as conservative estimate. The error of the device was lower, but noise in the system and variation between measurements was closer to 1 %. Some error in Seebeck coefficient measurement will be present due to the assumption, that temperature gradients across all the wires in each thermocouple are identical. Since the junctions of thermocouples were somewhat embedded in solder, there may be slight temperature gradient between two wire-solder interfaces. However, this difference was assumed to be negligible compared to temperature gradient along lengths of wires, and the uncertainty in Seebeck coefficient measurement was calculated as less than 1 %. The uncertainty in V_{sample} was similarly low, while that in determining V_R from the sample voltage was calculated to be about 2-3 %. These were calculated using error propagation from the uncertainty of the measurements. For extrinsic value of ZT, the uncertainty was calculated as 2.7 %. With addition of the uncertainty in the heat loss calculation, that for intrinsic ZT increased to 5.9 %.

3. Verification strategies for measurements

3.1 Slope-efficiency method: rapid measurement of device $ZT_{maximum}$.

Maximum electrical power output, P_{max}, of any thermoelectric generator (TEG) depends on open-circuit voltage, V_{oc}, and occurs when internal device resistance, R_{int}, exactly equals to resistance of external load. When $R_{int}=R_{load}$, then total system resistance = $2R_{int}$, and V_{oc} drops by exactly half leading to:

$$P_{max} = \frac{{V_{oc}}^2}{4R_{int}}. \tag{64}$$

For TEG consisting of some number "i" of individual "thermocouples" connected in series and each having n-type thermoelement and p-type thermoelement, Seebeck effect relates V_{oc} to the temperature difference, ΔT, induced by the heat source as described:

$$V_{oc} = \Sigma_0^i(\alpha_n + \alpha_p)_i\, \Delta T, \tag{65}$$

where α_n and α_p are values of n-type and p-type Seebeck coefficients from each individual thermoelement, respectively. Thus, the sum of Seebeck coefficients from i thermocouples is ensemble-average proportionality between V_{oc} and ΔT. Likewise, R_{int} is the sum of resistances from i thermocouples, and it is the ensemble-average electrical resistivity of n-type (ρ_n) and p-type (ρ_p) thermoelements times their respective area (A)-to-length (ℓ) values:

$$R_{int} = \Sigma_0^i\left(\rho_n \frac{\ell}{A} + \rho_p \frac{\ell}{A}\right)_i. \tag{66}$$

P_{max} can be expressed in terms of Seebeck coefficients:

$$P_{max} = \frac{\left(\Sigma_0^i(\alpha_n+\alpha_p)_i\right)^2 \Delta T^2}{4R_{int}}. \tag{67}$$

This expression highlights first important point: P_{max} increases as ΔT^2. So, for large electrical power output, the largest possible ΔT is desired.

The efficiency, Φ, with which TEG can convert heat flow, Q, to electrical power is also important because the most electrical power possible from a given amount of heat flow is desirable. A new expression for the efficiency of a TEG can be obtained starting with expression for P_{max}. The ratio of electrical energy generated per given amount of input heat energy is the definition of efficiency:

$$\Phi = \frac{P_{max}}{Q}. \tag{68}$$

Equation (68) can be rewritten, assuming for simplicity a unicouple (i=1), as:

$$\Phi = \frac{(\alpha_n+\alpha_p)^2 \Delta T^2}{4R_{int}Q}. \tag{69}$$

The flow of heat is dominated by thermal conductivity of the materials from which TEG is constructed, so Fourier's law can be used to express Q:

$$\Phi = \frac{(\alpha_n+\alpha_p)^2 \Delta T^2}{4R_{int}\left([\kappa_n+\kappa_p]\frac{A}{\ell}\Delta T\right)} \tag{70}$$

Then expressing R_{int} as described earlier:

$$\Phi = \frac{(\alpha_n+\alpha_p)^2 \Delta T^2}{4\left(\rho_n\frac{\ell}{A}+\rho_p\frac{\ell}{A}\right)\left([\kappa_n+\kappa_p]\frac{A}{\ell}\Delta T\right)}. \tag{71}$$

For planar TEG devices, the values of ℓ of both n-type and p-type thermoelements are equal; however, cross-sectional areas of n-type and p-type may be quite different. Identifying cross-sectional area of n-type as A_n and that of p-type as A_p allows a simplification, yielding Φ in terms of measurable materials properties and temperature difference.

$$\Phi = \frac{1}{4}\left(\frac{(\alpha_n+\alpha_p)^2}{\left(\frac{\rho_n}{A_n}+\frac{\rho_p}{A_p}\right)(\kappa_n A_n+\kappa_p A_p)}\right)\Delta T \tag{72}$$

The proportionality between Φ and ΔT will be termed "Z_{device}".

$$Z_{device} = \left(\frac{(\alpha_n+\alpha_p)^2}{\left(\frac{\rho_n}{A_n}+\frac{\rho_p}{A_p}\right)(\kappa_n A_n+\kappa_p A_p)}\right). \tag{73}$$

Note that when area-to-length ratios are optimized for maximum efficiency, this relationship reduces to the common, well-known expression for device ZT:

$$Z_{max} = \left(\frac{(\alpha_n+\alpha_p)}{\left(\sqrt{\kappa_n\rho_p}+\sqrt{\kappa_p\rho_n}\right)}\right)^2 \tag{74}$$

TEG efficiency can be measured as function of ΔT, and the slope of that data should equal to:

$$\frac{\partial \Phi}{\partial \Delta T} = \frac{1}{4}\left(\frac{(\alpha_n+\alpha_p)^2}{\left(\frac{\rho_n}{A_n}+\frac{\rho_p}{A_p}\right)(\kappa_n A_n+\kappa_p A_p)}\right) = \frac{Z_{device}}{4}. \qquad (75)$$

This expression highlights a second important point, that Φ should linearly increase as a function of ΔT according to the slope indicated by ¼ of the quantity in parentheses. This makes sense, because Φ increases linearly with ΔT, and P_{max} increases as ΔT^2. Taking the ratio yields a simple linear dependence on ΔT. Note that the materials properties are all temperature dependent, so taking the derivative would necessarily yield higher-order terms. However, we make use of the following assumptions: 1.) the temperature dependence of the electrical component of thermal conductivity depends on the mobility of charge carriers, and the electrical resistivity depends on the inverse of that mobility, so these dependencies can be assumed to first-order to completely cancel. 2.) The Seebeck coefficient does have a relatively small, but finite temperature dependence; however the derivative of the Seebeck coefficient should yield a temperature dependence of T^{-1} which approximately cancels with the temperature dependence of lattice thermal conductivity in the denominator. So, to first order the linearity of the slope would be expected, and is in fact experimentally observed as will be shown.

A new index to determine maximum ZT of TEG device can be obtained by measuring the slope of TEG efficiency. To calculate maximum ZT, four times the slope of TEG efficiency multiplied by maximum temperature, under which TEG displays linear behavior with respect to ΔT. Outside the linear regime of TEGs, the basic properties properties can no longer be described by these functions. Therefore, measure of maximum ZT of TEG device can be determined by

$$\text{(Maximum)}\ \ Z_{device}T = 4\left(\frac{(\alpha_n+\alpha_p)^2}{\left(\frac{\rho_n}{A_n}+\frac{\rho_p}{A_p}\right)(\kappa_n A_n+\kappa_p A_p)}\right)T_{maximum} \qquad (76)$$

The significance of this analysis is it allows unique means to rapidly obtain $ZT_{maximum}$ and confirm properties and individual measurements. Measurements can be confirmed by measuring slope of efficiency as function of ΔT and ZT can be obtained and compared to theoretical ZT as calculated by individual measurements.

3.1.1 Analysis of commercial $(Bi,Sb)_2(Te,Se)_3$ module

Efficiency of commercial $(Bi,Sb)_2(Te,Se)_3$ device is presented in Figure 18. This device is designed for high thermal impedance and has optimum performance window from nearly room-temperature to roughly 425 K. Slope of efficiency was determined and is shown in inset in Figure 18, and is given as 0.0004/K. As expected, the slope is highly linear function of ΔT until deviation from non-linearity begins at 405 K. The maximum ZT can be obtained by the simple relationship, and observed maximum temperature of roughly 405 K:

$$Z_{device}T_{maximum} = 4\left(\frac{\partial\Phi}{\partial\Delta T}\right)T_{maximum}. \tag{77}$$

Obtained value of $ZT_{maximum}$ is equal to 0.7, which is consistent with established values for commercial devices.

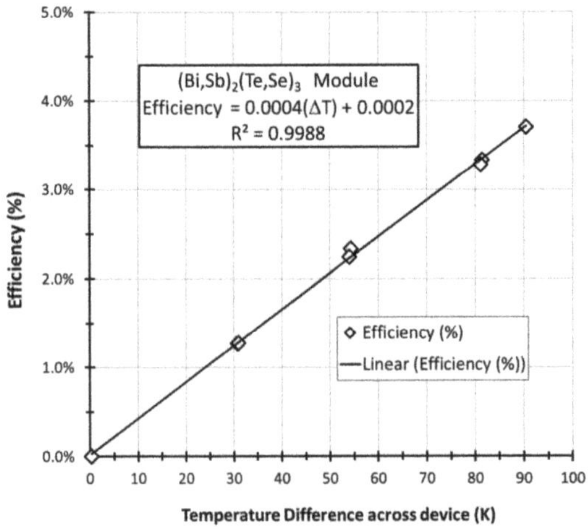

Fig. 18 Slope of efficiency from $(Bi,Sb)_2(Te,Se)_3$ to determine ZT_{max}.

3.1.2 Analysis of PbTe/TAGS module

Efficiency of PbTe/TAGS device is presented below. Figure 19 shows temperature dependence of PbTe/TAGS module efficiency. At low temperature, the slope is somewhat non-linear function of ΔT because, it is well-known, that properties of these materials are uninteresting at low-temperature, but optimum at elevated temperature. So, slope of efficiency is measured in temperature range, where properties are linear, >500 K. The maximum ZT can therefore be obtained by equation (77), and observed maximum temperature for linear device behavior, which for the device being measured is equal to 873 K. Obtained slope is 0.0002/K resulting in value $ZT_{maximum}$ = 0.7, which is consistent with established values for well-known PbTe/TAGS modules.

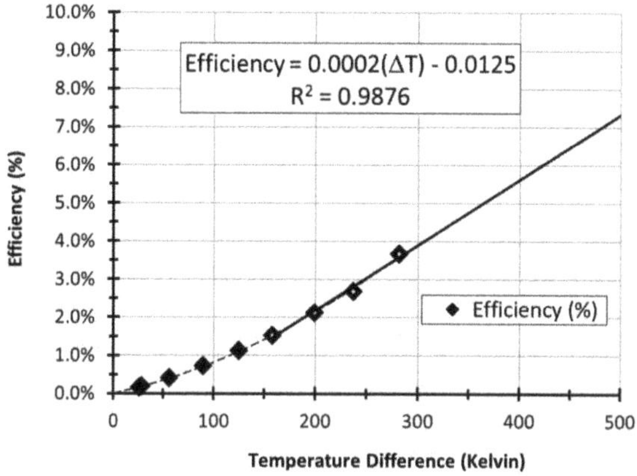

Fig. 19 Slope of efficiency from pre-commercial PbTe/TAGS to determine ZT_{max}.

3.2 Discretized heat-balance model and analysis

More detailed device analysis and performance modeling including effects of temperature-dependent material properties may be accomplished through the use of numerical methods. One technique for performing numerical analysis on TEG was reported by Lau and Buist [34] and later confirmed and expanded upon by Hogan and Shih [35]. It involves partitioning the legs of TEG into virtual segments for computational purposes, where each segment is taken to be isothermal. Neighboring segments then vary in temperature such, that governing thermoelectric heat balance equations based on constant parameter theory are satisfied [36, 37]. This process is illustrated in Figure 20.

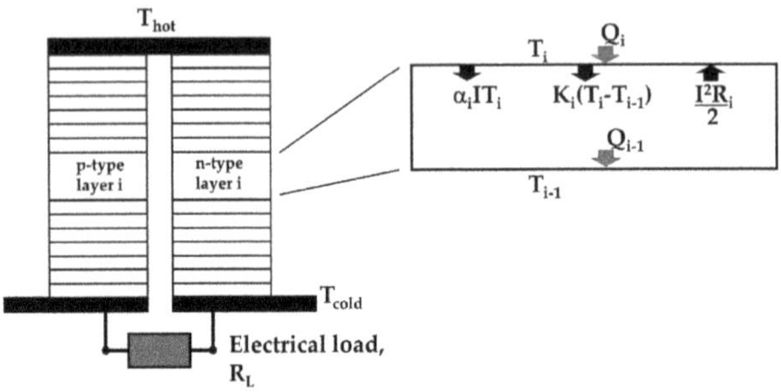

Fig. 20 Discrete communicating layers having thermal and electrical flux continuity. Expanded view of ith segment explicitly showing the heat flows that must be balanced to maintain continuity through the bulk of the TEG device.

Based on constant-parameter theory [37] and heat balance at the top surface of the *ith* segment, we have:

$$Q_i = \alpha_i IT_i - I^2R_i/2 + K_i(T_i\text{-}T_{i-1}) \tag{78}$$

where the total heat flux Q_i into the segment is the sum of the individual components indicated. Here α_i is the Seebeck coefficient of the ith segment at temperature T, R_i is the electrical resistance of the ith segment at temperature T, K_i = thermal conductance of the ith segment at temperature T. Last, I and T_i are the electrical current and temperature of the ith segment respectively. The electrical power, P, is determined by

$$Q_{i-1} = Q_i - P \tag{79}$$

Which is also the P delivered to the external load resistor, R_L.

$$P = I^2R_L \tag{80}$$

The discrete heat balance equations (Equations 78 and 79) derived in this manner can be easily solved for single leg of TEG with an iterative technique [34, 35]. For a given T_C, an initial estimate is made for the heat delivered to the cold junction, Q_C, and temperature and heat flow in each segment is determined sequentially, ending with a numerical solution for heat absorbed at the hot junction Q_H and the hot-side temperature T_H. If calculated T_H is not equal to the desired T_H boundary condition, then Q_C is adjusted accordingly and the process repeated until the desired T_H is achieved.

The initial hot-side temperature of segment is taken as a uniform temperature for the entire segment and its thermoelectric properties are then determined from a curve fit to measured data. Adjacent segments attain different temperatures as the system is solved according to the energy balance requirements. Thus, temperature-dependent effects are fully incorporated into the model. In fact, Hogan and Shih [35] were able to demonstrate excellent agreement using the discrete approximation as compared with an exact analysis of temperature-dependent TEG performance by Sherman et.al. [38].

Figure 21 shows temperature profiles calculated for n-type Skutterudite material with temperature-dependent properties, operating at the indicated boundary conditions. For simplicity, electrical current is treated here as though there were an external load resistance matched to the internal resistance of the thermoelectric material leg, thus producing maximum output power.

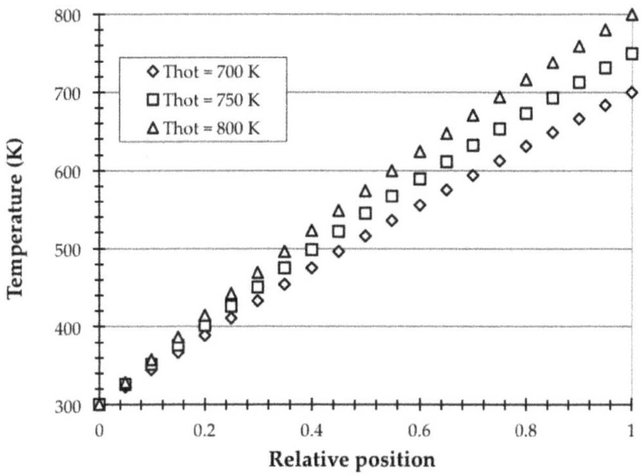

Fig. 21 Temperature profile in n-type Skutterudite at maximum power. $T_C = 300$ *K.*

It is also instructive to examine calculated heat flow through the leg as this helps illustrate thermal-to-electrical conversion process. Figure 22 shows heat flow corresponding to temperature profiles depicted in Figure 21. Looking from the hot-side to the cold-side of the leg (or right to left on the plot), heat flow is reduced as thermal energy is converted to electrical power and delivered to the load. Examining the specific case of T_H = 750 K and T_C = 300 K, there is approximately 5.7 W of thermal power incident on the hot-side and 5.0 W rejected at the cold-side, leaving 0.7 W which is delivered as electrical power to the load. So the conversion efficiency within just one of the legs of the TEG itself is simply 0.7 W/5.7 W = 12.3 %. The same methodology may be applied to the companion p-type leg to complete the analysis of a full TEG at a given set of temperature boundary conditions. Proper temperature profiles of n-type and p-type legs together are shown in Figure 23. Taking T_H = 800 K and T_C = 300 K, the combined incident thermal power is equal to 6.42 $\text{Watts}_{\text{thermal}}$ (n-type) + 7.85 $\text{Watts}_{\text{thermal}}$ (p-type) = 14.27 $\text{Watts}_{\text{thermal}}$ (TEG). And that rejected to the cold-side is equal to 5.58 $\text{Watts}_{\text{thermal}}$ (n-type) + 7.16 $\text{Watts}_{\text{thermal}}$ (p-type) = 12.74 $\text{Watts}_{\text{thermal}}$ (TEG). This calculation finds the overall efficiency = 1.53 W/14.27 W = 10.7 %.

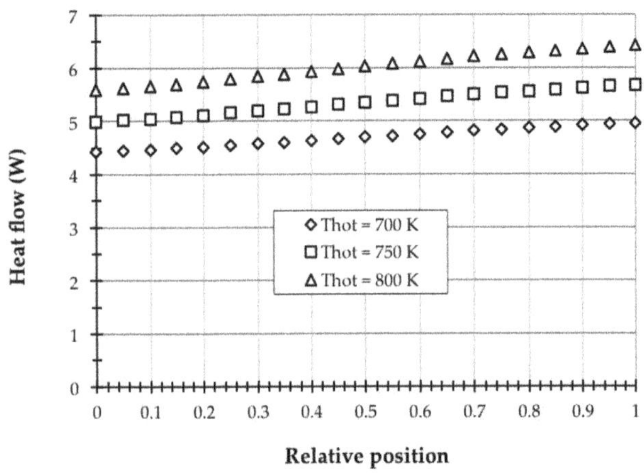

Fig. 22 Heat flow profile in n-type Skutterudite sample at maximum output power, corresponding to temperature profiles depicted in Figure 21. $T_C = 300$ *K.*

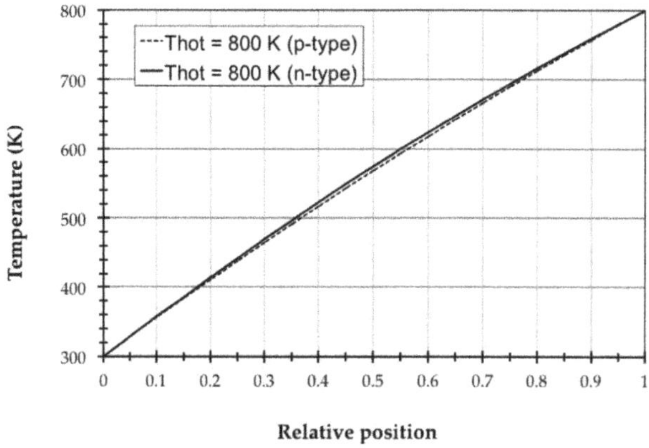

Fig. 23 Temperature profiles in both n- and p-type legs of TEG. $T_C = 300$ K.

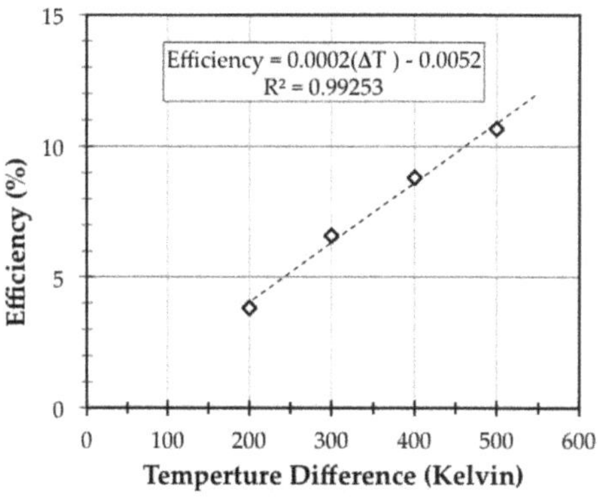

Fig. 24 Calculated efficiency of single p-n couple TEG. T_C=300 K and $T_h(max)$=800K.

This discussion has highlighted a simple, but powerful temperature-dependent phenomenological model for precisely calculating temperature profiles, heat flows, power outputs and efficiencies in a single leg of TEG. Full TEG device modelling is accomplished by simultaneously solving the discrete heat balances as described for each leg (subject to hot- and cold-side boundary conditions) along with the simultaneous energy balance relationship required for electrical power being delivered to the load.

Figure 24 shows calculated efficiency using the *discretized heat balance* theory of idealized Skutterudite n-type Skutterudite p-type device.

Electrical and thermal contact resistivities are defined to be zero, but could easily be included as finite quantities, which would add penalties to the efficiency. Slope of efficiency identified in the best-fit is quantified first using unitless efficiency data. The equation is then re-included on the plot after converting to percent. This is so that $ZT_{maximum}$ can be calculated using **slope/efficiency method** described in section 3.1, and for overall clarity in the final plot. Slope of efficiency is 0.0002/K, and the upper-limit temperature is 800 K, so, following equation (77), so value $ZT_{maximum}$= 0.64 is obtained for the device.

Therefore, to confirm measurements for device fabricated using materials, from which measurements were collected, it could be assembled and the efficiency measured. If measurements are accurate and not overestimated, then performance should be consistent with $ZT_{maximum}$ value equals to 0.64. The slope of the data should be roughly 0.0002/K.

Conclusions

In conclusion, we have presented several novel approaches to the significant challenge of accurately determining the thermal conductivity of thermoelectric materials. The new solutions can be much faster experimentally, and they successfully address several sources of experimental error. The overall result is significantly reduced error, which may reduce uncertainty by a factor of two or more. Further, we introduce new approaches to compare device performance with physical property measurements as a novel means of confirming measurements. Using this approach, the new measurements can be clearly seen to yield physical property measurements which are more consistent with physical device performance.

The first new thermal conductivity measurement method, *steady-state isothermal technique*, improves accuracy by collecting data under conditions where thermal losses and errors are unimportant. The validity was confirmed by comparing the thermal conductivity extracted from a Peltier cooling device with the lab measured value: the error was ~2%. [8] The second is non-destructive micro-scale analysis technique called the *scanning hot-probe*, and the third is *lock-in transient Harman* method, which is a comprehensive modification of transient Harman technique. The second and third methods reduce error by highly detailed treatment of interfacial contact effects including electrical contact resistance and thermal contact effects. The high accuracy for both of these methods is obtained by comparison to established standard reference materials whose properties are well-known and accepted. A new interesting follow-on is frequency-dependent Nyquist analysis, which presages a different perspective on the materials analysis, and even further simplified measurement.

The truest test of the accuracy of measurement is comparison with fabricated devices. To support the validation of measurements of individual material properties, we have outlined new device metrics, which allows comparison between theoretical and measured device efficiency. We outline a new *slope-efficiency method*, which can be used to determine informative index $ZT_{maximum}$ of any device. The second method of device evaluation is a numerical device model called the *discretized heat balance model*. Using this modelling approach, we showed that a a piecewise continuous collection of discrete layers within a device, where boundary heat flows have energy and current continuity relationships, can yield an incredibly easy theoretical determination of device efficiency, which can be compared with experimental values.

References

[1] Wang H, Bai S, Böttner S, Chen L, Harris F, Kiss L, Kleinke H, Konig J, Lo J, Mayolet A, Porter W, Sharp J, Smith C, Tritt T. Transport Properties of Bulk Thermoelectrics-An International Round-Robin Study, Part I: Seebeck Coefficient and Electrical Resistivity. Journal of Electric Materials. 2013:42:4 p.654-664. https://doi.org/10.1007/s11664-012-2396-8

[2] Putley E. Thermoelectric and Galvanomagnetic effects in Lead Selenide and Telluride. Proceedings of the Physical Society. Section B. 1955:68 p.35. https://doi.org/10.1088/0370-1301/68/1/306

[3] Harman T, Cahn J, Logan M. Measurement of Thermal Conductivity by Utilization of the Peltier Effect. Journal of Applied Physics. 1959:30:9 p.1351. https://doi.org/10.1063/1.1735334

[4] Penn A. The correction used in the adiabatic measurement of thermal conductivity using the Peltier effect. Journal of Scientific Instruments. 1964:41:10 p.626. https://doi.org/10.1088/0950-7671/41/10/311

[5] Bowley A, Cowles L, Williams G, Goldsmid H. Measurement of the figure of merit of a thermoelectric material. Journal of Scientific Instruments. 1961:38:11 p.433. https://doi.org/10.1088/0950-7671/38/11/309

[6] Buist R. A new method for Testing Thermoelectric materials and Devices. In: Proceedings of the 11th International Conference on Thermoelectrics (ICT '92); 7-9 October 1992; Arlington, Texas. Available from: https://www.tetech.com/wp-content/uploads/2013/10/ICT92RJB.pdf [accessed 2016-08-22]

[7] Taylor P, Jesser W, Rosi F, Derzko Z. A model for the non-steady-state temperature behaviour of thermoelectric cooling semiconductor devices. Semiconductor Science and Technolology. 1997:12:4 p.443. https://doi.org/10.1088/0268-1242/12/4/018

[8] Taylor P, Maddux J, Uppal P. Measurement of Thermal Conductivity Using Steady-State Isothermal Conditions and Validation by Comparison with Thermoelectric Device Performance. Journal of Electronic Materials. 2012:41:9 p.2307. https://doi.org/10.1007/s11664-012-2178-3

[9] Joshi G, Yan X, Wang H, Liu W, Chen G, Ren. Enhancement in Thermoelectric Figure-Of-Merit of an N-Type Half-Heusler Compound by the Nanocomposite Approach. Advanced Energy Materials. 2011:1:4 p.643 https://doi.org/10.1002/aenm.201100126

[10] Rosi F, Hockings E, Lindenblad N. Semiconducting Materials for Thermoelectric Power Generation. RCA (Radio Corporation of America) Review. 1961:22. OSTI: 4838510

[11] Borca-Tasciuc T. Scanning Probe Methods for Thermal and Thermoelectric Property Measurements. Annual Review of Heat Transfer. 2013:16:80 p.211-258. https://doi.org/10.1615/AnnualRevHeatTransfer.v16.80

[12] Lefèvre S, Volz S, Chapuis P. Nanoscale heat transfer at contact between a hot tip and a substrate. Internation Journal Heat and Mass Transfer. 2006:49:1–2 p.251. https://doi.org/10.1016/j.ijheatmasstransfer.2005.07.010

[13] Son Y, Pal S, Borca-Tasciuc T, Ajayan P, Siegel R. Thermal resistance of the native interface between vertically aligned multiwalled carbon nanotube arrays and their SiO2/Si substrate. Journal Applied Physics. 2008:103:2 p.024911. https://doi.org/10.1063/1.2832405

[14] Zhang Y, Castillo E, Mehta R, Ramanath G, Borca-Tasciuc T. A noncontact thermal microprobe for local thermal conductivity measurement. Review of Scientific Instruments. 2011:82:2 p.024902. https://doi.org/10.1063/1.3545823

[15] Puyoo E, Grauby S, Rampnoux J, Rouvìre E, Dilhaire S. Thermal exchange radius measurement: Application to nanowire thermal imaging. Review of Scientific Instruments. 2010:81:7 p.073701. https://doi.org/10.1063/1.3455214

[16] Lefèvre S, Saulnier J, Fuentes C, Volz S. Probe calibration of the scanning thermal microscope in the AC mode. Superlattices and Microstructures. 2004:35:3–6 p.283. https://doi.org/10.1016/j.spmi.2003.11.004

[17] Maiz J, Rojo M, Abad B, Wilson A, Nogales A, Borca-Tasciuc D, Borca-Tasciuc T, Martín-González M. Enhancement of thermoelectric efficiency of doped PCDTBT polymer films. RSC Advances. 2015:82:5 p.66687-66694. https://doi.org/10.1039/C5RA13452H

[18] Abad B, Rull-Bravo M, Hodson S, Xu X, Martin-Gonzalez M. Thermoelectric properties of electrodeposited tellurium films and the sodium lignosulfonate effect. Electrochimica Acta. 2015:169:37. https://doi.org/10.1016/j.electacta.2015.04.063

[19] Langer G, Hartmann J, Reichling M. Thermal conductivity of thin metallic films measured by photothermal profile analysis. Review of Scientific Instruments. 1997:68:1510. https://doi.org/10.1063/1.1147638

[20] Abad B, Alda I, Díaz-Chao P, Kawakami H, Almarza A, Amantia D, Gutierrez D, Aubouy L, Martín-González M. Improved power factor of polyaniline nanocomposites with exfoliated graphene nanoplatelets (GNPs). Journal of Materials Chemistry A. 2013:1:35 p.10450. https://doi.org/10.1039/C3TA12105D

[21] Ivory J. Rapid Method for Measuring Seebeck Coefficient as ΔT Approaches Zero. Review of Scientific Instruments. 1962:33:9 p.992. https://doi.org/10.1063/1.1718048

[22] Harman T, Cahn J, Logan M. Measurement of Thermal Conductivity by Utilization of the Peltier Effect. Journal of Applied Physics. 1959:30:9 p.1351. https://doi.org/10/1063/1.1735334

[23] Castillo E, Hapenciuc C, Borca-Tasciuc T. Thermoelectric characterization by transient Harman method under non-ideal contact and boundary conditions. Review of Scientific Instruments. 2010:81:4 p.044902. https://doi.org/10.1063/1.3374120

[24] Smith G, Wolfe R. Thermoelectric Properties of Bismuth-Antimony Alloys. Journal of Applied Physics. 1962:33:3 p.841. https://doi.org/10.1063/1.1777178

[25] Labounty C, Shakouri A, Bowers J. Design and characterization of thin film microcoolers. Journal of Applied Physics 2001:89:4059. https://doi.org/10.1063/1.1353810

[26] Bian Z, Zhang Y, Schmidt H, Shakouri A. Thin film ZT characterization using transient Harman technique. In: Proceedings of the 24th International Conference on Thermoelectrics (ICT '05); 19-23 June 2005; Clemson, South Carolina. New York: IEEE; 2005. p. 76-78. https://doi.org/10.1109/ICT.2005.1519868

[27] Singh R, Bian Z, Zeng G, Zide J, Christofferson J, Chou H, Gossard A, Bowers J, Shakouri A. Transient Harman measurement of the cross-plane ZT of InGaAs/InGaAlAs superlattices with embedded ErAs nanoparticles. In: Proceedings of the MRS Fall Meeting (MRS Fall '05); 27 November – 2 December 2005; Boston, Massachusetts. New York: Cambridge University Press; 2005. p. 886.

[28] Venkatasubramanian R, Siivola E, Colpitts T, O'Quinn B. Thin-film thermoelectric devices with high room-temperature figures of merit. Nature. 2001: 413:597. https://doi.org/10.1038/35098012.

[29] da Silva L, Kaviany M. Micro-thermoelectric cooler: interfacial effects on thermal and electrical transport. International Journal of Heat and Mass Transfer. 2004:47:2417. https://doi.org/10.1016/j.ijheatmasstransfer.2003.11.024

[30] Incropera F, DeWitt D. Fundamentals of heat and mass transfer. 5th ed. New York: Wiley; 2002. ISBN: 0471386502

[31] Sharp J, Bierschenk J, Lyon Jr. H. Overview of solid-state thermoelectric refrigerators and possible applications to on-chip thermal management. Proceedings of the IEEE. 2006:94:8 p.1602. https://doi.org/10.1109/JPROC.2006.879795

[32] De Marchi A, Giaretto V. An accurate new method to measure the dimensionless figure of merit of thermoelectric devices based on the complex impedance porcupine diagram. Review of Scientific Instruments. 2011:82:10 p. https://doi.org/104904. 10.1063/1.3656074

[33] Efremov M, Olson E, Zhang M, Lai S, Schiettekatte F, Zhang Z, Allen L. Thin-film differential scanning nanocalorimetry: heat capacity analysis. Thermochimica Acta. 2004:412:1 p. 13. https://doi.org/10.1016/j.tca.2003.08.019

[34] Lau P, Buist R. Calculation of thermoelectric power generation performance using finite element analysis. In: Proceedings of the 16th International Conference on Thermoelectrics (ICT '97); 26-29 August 1997; Dresden, Germany. New York: IEEE; 1997. p.563

[35] Hogan T, Shih T. Modeling and Characterization of Power Generation Modules Based on Bulk Materials. In: Rowe D, editor. Thermoelectrics Handbook: Macro to Nano. 1st ed. Washington DC: CRC Press; 2005. p.12-1. https://doi.org/10.1201/9781420038903.ch12

[36] Buist R. Calculation of Peltier Device Performance. In: Rowe D, editor. CRC Handbook of Thermoelectrics. 1st ed. Washington DC: CRC Press; 1995. p. 143-156. https://doi.org/10.1201/9781420049718.ch14

[37] Cobble M. Calculation of Generator Performance. In: Rowe D, editor. CRC Handbook of Thermoelectrics. 1st ed. Washington DC: CRC Press; 1995. p.489. https://doi.org/10.1201/9781420049718.ch39

[38] Sherman B, Heikes R and Ure R. Calculation of Efficiency of Thermoelectric Devices. J. Appl. Phys. 1960:31:1. https://doi.org/10.1063/1.1735380

[39] Wilson A, Munoz M, Abad B, Perez J, Maiz J, Schomacker J, Martin-Gonzalez M, Borca-Tasciuc D, Borca-Tasciuc T. Thermal conductivity measurements of high and low thermal conductivity films using a scanning hot probe method in the 3ω mode and novel calibration strategies. Nanoscale. 2015:7:15404-15412. https://doi.org/10.1039/C5NR03274A

Chapter 17

Charged Quantum Dots for High-Efficiency Photovoltaics and IR Sensing

A. Sergeev[1*], V. Mitin[1], N. Vagidov[2] and K. Sablon[3]

[1]Dept. of Electrical Engineering, State University of New York, Buffalo, Buffalo, NY 14260-1920, U.S.A.

[2]Optoelectronic Nanodevices LLC, Amherst, NY 14226-5126, U.S.A.

[3]U.S. Army Research Laboratory, Adelphi, MD 20783, U.S.A.

Abstract

Charged quantum dots (QDs) create nanoscale potential profile, which in turn provides effective control of critical optoelectronic characteristics, such as photoelectron lifetime, photon absorption, its spectral characteristics, responsivity to electromagnetic radiation, generation-recombination noise, etc. In this way, optimization of optoelectronic properties via selective doping of QD nanomaterials substantially improves performance of QD solar cells and IR photodetectors. This chapter presents the concept of nanoscale optoelectronic engineering via QD charging and review experimental studies of solar cells and detectors with charged dots.

Keywords

Optoelectronic Nanomaterials, Charged Quantum Dots, Photoelectron Lifetime

Contents

The work has been previously published by John Wiley & Sons, Inc.:
A. Sergeev, V. Mitin, N. Vagidov and K. Sablon, Charged Quantum Dots for High-Efficiency Photovoltaics and IR Sensing, in Eds. Serge Luryi, Jimmy Xu, Alex Zaslavsky, Future Trends in Microelectronics: Frontiers and Innovations, John Wiley & Sons, Inc., Chapter 18, 2013. https://doi.org/10.1002/9781118678107.ch18

1. Introduction

For many years significant efforts of scientific and engineering communities have been devoted to IR solar energy conversion which could drastically increase the output of the usable electric energy. As shown in Fig. 1, in a single-junction solar cell (SC) the conversion efficiency is limited by the phonon-mediated relaxation of photoelectrons to band-edges during the process of conversion of photons with energies above the bandgap as well as by the cut-off of IR photons with energies below the bandgap. The maximum theoretical efficiency for conversion of unconcentrated solar energy, known as the Shockley-Queisser limit, is 31% for AM0 spectrum.[1] In single-junction GaAs SCs more than 30% of solar energy falls into the IR sub-bandgap spectral region and more than 20% of solar energy is lost during the thermalization of high-energy photocarriers.

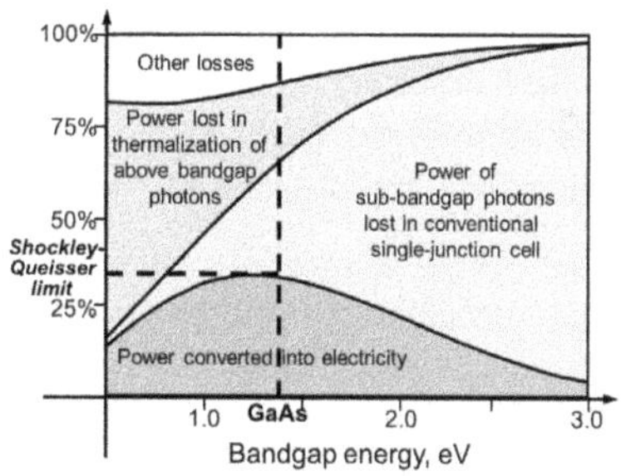

Fig. 1 Usable power, thermalization losses, and losses of sub-bandgap photons as a function of the bandgap in a single-junction solar cell.

Quantum-dot (QD) optoelectronic materials and structures have attracted much attention[2-4] due to their ability to adjust the electron transitions to specific IR photons and to increase electron coupling to IR radiation. The strong coupling enables high quantum efficiencies crucial for conversion of IR radiation and sensing. However, together with strong coupling to radiation, the optoelectronic materials for IR conversion and sensing should also provide long photocarrier lifetime and small recombination losses.

Initial hopes for long lifetime of photoexcited carriers were pinned on the theoretically predicted suppression of the electron-phonon relaxation in QDs which is known as a "phonon bottleneck". Over several years, the results were controversial. Some data supported and some data contradicted the phonon bottleneck concept.[5] Further investigations have shown that while the relaxation in QDs is slowed down compared with the bulk semiconductors, this effect is weak and insufficient for sensing and photovoltaic conversion.[6]

While the electron-phonon interaction is weakly suppressed in QDs, strong confinement of electrons in QDs drastically enhances inter-electron interaction. Therefore, in the relaxation of high energy photocarriers the electron energy is redistributed in the electron subsystem and is not wasted in the form of heat via electron-phonon interaction. For the first time, this effect was experimentally observed by Schaller and Klimov.[7] They found that the inverse Auger processes lead to very effective multiplication of electron-hole pairs. The multiple exciton generation (MEG) by absorbed high energy photons is especially favorable for photovoltaic conversion. Theoretical calculations[6] show that in a single-junction SCs, MEG with maximum carrier multiplication increases the limiting efficiency by one third, to ~44%. However, for tandem SCs the MEG-related improvement of photovoltaic efficiency is limited to 2% only. Recent experiments[8] demonstrated that MEG in PbSe QDSC increases the internal quantum efficiency up to 130% and the total photocurrent by ~1 mA cm^{-2} which is still small for practical applications.

Quantum dots also provide numerous possibilities for 3D band engineering. The most investigated QDoSC design is the intermediate-band solar cell where the intermediate band (IB) is formed by electron tunneling between QDs.[3,9] In such a photovoltaic device, the electron–hole pairs may be generated in a two-step process. Electron is excited from the valence band to IB by the first photon and then it is excited from IB to a conduction band by a second photon. Such solar cell with optimized position of IB is expected to have a limiting efficiency of 62%. To obtain IB, QDs of similar shapes and sizes should have regular positions in QD material or structure. In solid-state structures with QD layers, the correlated positions of QDs are realized via a local stress transferred from one layer to another. Accumulation of strain results in various defects which in turn enhance recombination processes. While various technologies for reduction of the stress accumulation are proposed and employed, recombination losses in QD structures are still strong and prevail over the IR converted power. This is one of the main reasons why the QD band engineering shows a very limited progress in photovoltaic and sensing applications.

Looking for other ideas of employing of QDs, we have recently proposed[10,11] and experimentally investigated[12-14] possibilities for engineering of 3D nanoscale potential by charged quantum dots in order to increase the photoelectron lifetime and reduce the recombination losses. In this technology the nanoscale potential profile is created by **q**uantum dots with **b**uilt-**in** **c**harge (Q-BIC) which are charged via selective doping of the inter-dot space.

2. 3D nanoscale potential profile in Q-BIC structures

Operation of semiconductor devices is essentially based on manageable potential barriers for electrons and holes. In the majority of modern devices the potential profile changes only in one direction whereas 3D potential profile can provide more functionality and scalability to optoelectronic materials and structures. Manageable nanoscale potential is an effective tool for engineering of electron processes for specific applications. Effective control of electron processes via manageable nanoscale profile opens wide perspectives for the development of adaptive optoelectronic materials.

Charged quantum dots provide unique possibilities for creation of specific 3D potential profiles favorable for photovoltaic and sensing applications. Potential barriers around QDs are always created when electrons from dopants outside QDs fill the dots. In the simplest case, uniformly distributed charged quantum dots create local potential barriers. Charged quantum dots with correlated positions, such as QD clusters, planes, rows etc. may be used to create higher collective potential barriers which separate the groups of QDs from conducting channels.

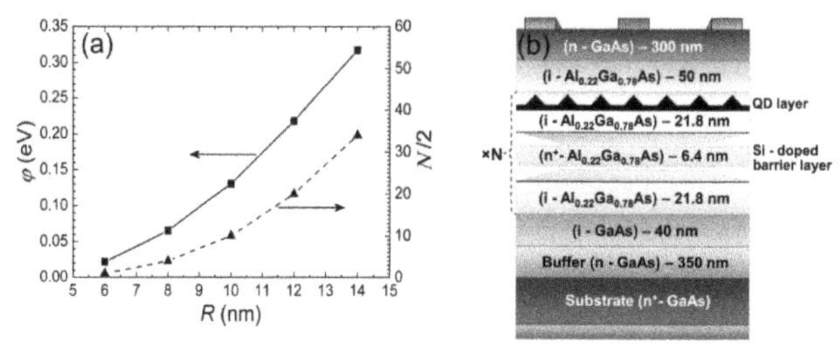

Fig. 2 (a) Electric potential, φ, of half-filled spherical InAs quantum dot in GaAs matrix (left) and a number of electron states, N, localized in a quantum dot (right) versus the quantum dot radius, R. (b) Schematics of Q-BIC solar cell structure with δ-doping of the spacer.

It is well-understood that the QDs are effective centers of recombination. To suppress the photocarrier capture into QDs, the potential barrier around QD should be two-three times larger than k_BT. Therefore, at room temperature, the local barriers should be ~ 0.05 eV. To evaluate and illustrate the potential created by a single dot, in Fig. 2(a) we present the electric potential of the half-filled spherical

InAs quantum dot in GaAs matrix as a function of a QD radius, R. As seen in Fig. 2, for spherical dots even a weak QD charge of 2 – 3 electrons per dot creates substantial potential barriers that prevent capture of thermally excited electrons.

Now let us discuss real structures and nanoscale potential profiles in these structures. In the outgoing research we employ Q-BIC structures with random positions of dots in QD planes. The Q-BIC structures have been fabricated using molecular beam epitaxy with typical growth temperatures of 500 ± 10 °C. The InAs QDs were grown on GaAs or AlGaAs surfaces by deposition of approximately 2.1 monolayers of InAs. To charge quantum dots, the structures have been doped in two different ways: with intra-dot doping, i.e. direct doping of QD layer, and with inter-dot (selective) doping, i.e. doping of the spacer that separates QD layers. The typical structure of the solar cell with δ-doping in the middle of the spacer is shown in Fig. 2(b). The dopant sheet concentration was chosen to provide up to 6 electrons per dot.

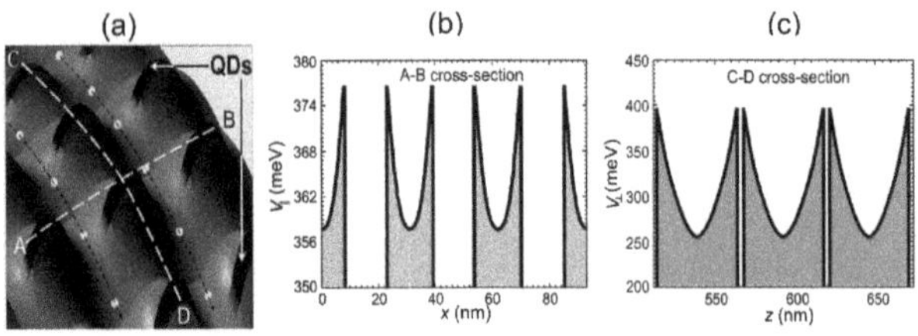

Fig. 3 (a) 3D potential profile in Q-BIC structure with QDs in the form of truncated pyramids. (b) Potential profile of A-B cross-section. (c) Potential profile of C-D cross-section.

The potential barriers in real structures are strongly anisotropic. To calculate the potential profile in such devices we employed nextnano3 software.[15] As seen from Fig. 3(a), the charged QDs in the form of truncated pyramids with a large base and small height create strongly asymmetric potential barriers. The barriers in the QD planes, i.e. in the direction perpendicular to the current, are substantially smaller than the barriers in the direction of the current. Therefore, the electron trapping is most effective for electrons moving along the A-B direction shown in Fig. 3(a). The fraction of such electrons decreases in structures with vertically correlated positions of QDs where electrons move mainly in the channels between dots – see the white circles in Fig. 3(a). Let us note that the potential barrier in the A-B direction increases by ~2.5 meV per electron, which is approximately one order of magnitude smaller than the increase of potential per electron in C-D direction, compare the scales in Fig. 3(b) and (c). The potential in the C-D direction is similar in magnitude to the spherical dots of Fig. 2(a).

3. Photoelectron kinetics in Q-BIC structures

Processes in quantum dots strongly depend on the electron/hole level structure. As it will be shown, the huge difference between electron and hole trapping by QDs are mainly determined by structures of electron and hole energy levels. The level structure in InAs/GaAs QDs was investigated in photoluminescence measurements. All data show practically equidistant level positions of electrons and holes,[16-18] as it is shown in Fig. 4(a). The total level spacing, $\Delta E = \Delta E_e + \Delta E_h$, which was directly determined in photoluminescence experiments, found to be 60 – 80 meV. The ratio, $\Delta E_e/\Delta E_h$, evaluated in the papers[16-19] changes from 2 to 8. By associating equidistant positions of energy levels with the quasi-parabolic confinement potential in InAs/GaAs QDs, we accept the spacing ratio, $\Delta E_e/\Delta E_h$, to be $(m_h/m_e)^{1/2} \approx 4$, which is in agreement with the scope of data. This leads to $\Delta E_e \approx 55$ meV and $\Delta E_h \approx 14$ meV. Thus, the electron spacing in QDs significantly exceeds the thermal energy and therefore electron transitions cannot be induced by thermal phonons, while the hole transitions are easily induced by thermal phonons. So we conclude that to stimulate electron escape from QD by IR radiation, the n-doping is strongly preferable.

Figure 4 shows the IR-assisted photogeneration of electron and hole pairs in undoped (Fig. 4(b)) and n-doped InAs/GaAs QD structures (Figs. 4(c), 4(d), and 4(e)). Figure 4(c) describes a two-step process induced by n-doping with IR transition of electron from the localized to the conducting state. Figures 4(d) and 4(e) show other n-doping induced processes that involve inter-electron interaction in QDs. In Fig. 4(d) the radiation excites two electrons to QD excited states, then one of these electrons transfers to the conducting state and the other transfers to a low-energy state. In Fig. 4(e) the relaxation of electron to a low-energy state leads to the escape of a hole from the QD. We would like to highlight that, because of the small hole level spacing, the hole escape from QDs can be easily generated by hot (excited) electrons. Thus, doping should stimulate the radiation-induced electron escape from QDs. Therefore, contrary to the lasing applications of QDs, which are strongly enhanced by p-doping,[19] the photovoltaic applications require n-doping.

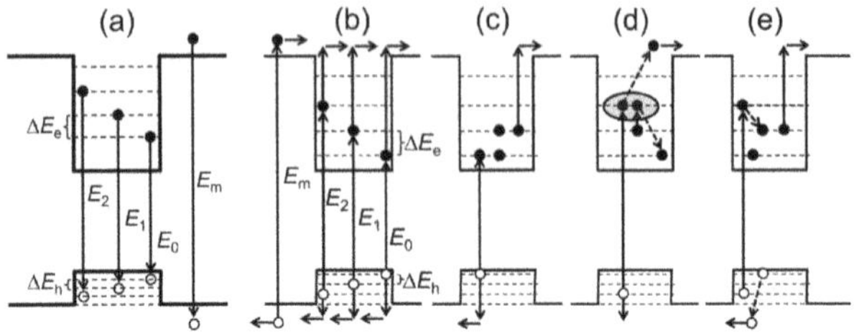

Fig. 4 Level structure in InAs/GaAs quantum dots and transitions measured in photoluminescence measurements (a); IR-assisted photogeneration of electron-hole pairs in undoped QD structures (b) and n-doped Q-BIC structures (c, d, e).

The photoelectron capture by repulsive n-charged QDs may be realized via tunneling through the barrier or thermal excitation above the barrier. Calculations show[10] that at room temperature for QDs with the radius larger 5 - 10 nm the thermoexcitation processes dominate over tunneling in the photoelectron capture. In this case, the temperature dependence of the capture rate is given by[10]

$$\frac{1}{\tau^{e}_{capt}} = \pi N_d R^3 \tau_e^{-1} \exp\left[-\frac{Ne^2}{k_B T \kappa R}\right], \tag{1}$$

where N_d is the dot concentration, τ_e is the electron-phonon inelastic scattering time which corresponds to transitions with characteristic electron level spacing in QD, eN is the charge of the quantum dot, and κ is the permittivity.

The hole capture by the repulsive n-charged QDs is given by[20]

$$\frac{1}{\tau^{h}_{capt}} = \frac{4}{3}\pi N_d (R^*)^3 \tau_h^{-1}, \tag{2}$$

where τ_h is the hole cascade relaxation time ($\tau_h \approx \tau_0 (3k_B T/\Delta E_h)$ and τ_0 is the relaxation time between adjacent hole levels); R^* is the dot radius, R, or the Thomson's radius, $R_{Th} = e^2N/(\kappa k_B T)$, depending on which of these parameters is the largest one. According to Fig. 2(a), for the suitable for Q-BIC technology parameters, the characteristic capture radius, R^*, is equal to the dot radius. Thus, according to Eqs. (1) and (2), the n-charging of QDs exponentially suppresses the electron capture processes, but does not practically affect the hole capture.

Suppression of photoelectron capture by n-charging of QDs exponentially increases the photoelectron lifetime in all QD unipolar devices. In QD photo-detectors the long photoelectron lifetime increases the photoresponse, decreases the generation-recombination noise, and improves the sensitivity.[13] The n-charging of QDs is favourable for photovoltaic applications also. In undoped or weakly doped InAs/GaAs QD structures the capture of electrons is substantially faster than the capture of holes, which has a cascade mechanism with up and down motion through hole energy levels in QD. Suppression of the fastest capture processes by *n*-charging of QDs decreases the recombination via QDs and recombination losses.

We would like to highlight that under sunlight the electron and hole capture rates are equated by changing of QD charge. In the "underdoped" structures DQs will be filled by photoelectrons, in the "overdoped" structures electrons will leave QDs. Therefore, to optimize the potential profile in QD solar cell, one should choose the optimal doping that provides equal electron and hole capture rates, i.e. $\tau^{e}_{capt}(N) = \tau^{h}_{capt}(N)$, where the filling of QDs, N, should correspond to the number of dopants per dot. As we discussed above, the hole capture rate is practically independent on QD charge and, therefore, at the optimal *n*-doping the capture rate of electrons and holes under the sunlight is equal to the slow hole capture rate.

4. Q-BIC solar cells and IR detectors

We have experimentally investigated quantum dot infrared photodetectors (QDIPs) and QDoSCs based on Q-BIC structures with random positions of dots in QD planes. The main parameters of our Q-BIC solar cells are summarized in Table 1.

Table 1 Parameters of Q-BIC solar cell devices.

N (electrons)	J_{SC} (mA/cm^2)	$\Delta_{IR}J_{SC}$ (mA/cm^2)	V_{OC} (V)	Fill Factor (%)	Efficiency (%)
Ref cell (no dots)	14.6	0	0.81	80	9.07
0	15.1	4.1	0.77	77	9.31
2	17.3	7.2	0.74	76	9.73
3	18.5	8.1	0.79	75	12.1
6	24.3	9.3	0.78	72	14.0

The Q-BIC solar cell with maximal n-doping (six electrons per dot) demonstrates strong increase in the short-circuit current, J_{SC}, from 15.1 mA/cm^2 to 24.3 mA/cm^2, with negligible degradation of the open-circuit voltage, V_{OC}. The third column in Table 1 shows the IR contribution to the short-circuit current, J_{SC}. These data have been directly obtained in experiments, where the above-bandgap part of solar spectrum was completely suppressed by the corresponding filter (so the reference cell without QDs did not show any photoresponse). As seen from Table 1, the conversion of IR radiation of Q-BIC solar cells increases with the increase of the built-in-dot charge. At the *n*-doping which provides six electrons per dot, the conversion of solely IR radiation via QD transitions gives ~9 mA/cm^2 contribution to the short-circuit current. Thus, in our current Q-BIC devices the harvesting and conversion of IR part of solar spectrum give additional 5% to the photovoltaic efficiency.

In Table 2 we have summarized parameters of our QDIP devices. In devices B44 and B52, charging of QDs has been realized via doping of QD layers (intra-dot doping); whereas in devices B45 and B53, the QDs were charged by doping of spacers (inter-dot doping). The effective average charge of QDs and the corresponding potential barrier in the QD plane (see Fig. 3) were found by modeling, described in Section 2. The measured spectral dependences show the maximum photoresponse at ~ 329 meV with the half maximum width ~40 meV.

Table 2 Parameters of Q-BIC IR photodetectors.

Device	Doping	Donor concentration (10^{11}cm^{-3})	Effective QD charge (e)	Potential barrier (meV)
B44	Intra-dot	2.7	1.8	4.5
B45	Inter-dot	2.7	2.8	7
B52	Intra-dot	5.4	3,45	8.6
B53	Inter-dot	5.4	6.1	15.3

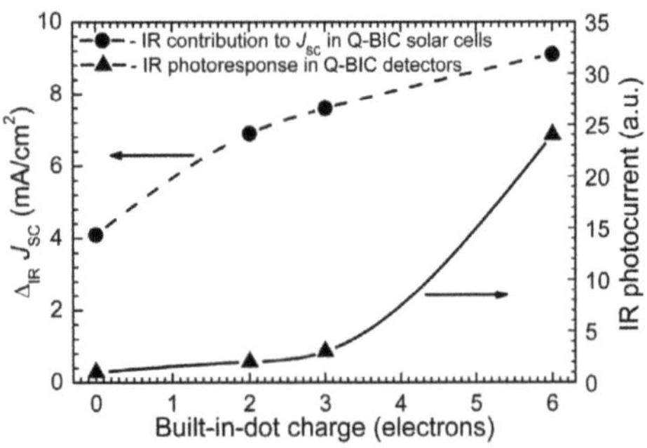

Fig. 5 IR contribution to the short-circuit current of Q-BIC InAs/GaAs solar cell (left) and photoresponse of Q-BIC IR photodetector as a function of the built-in-dot charge (right).

Fig. 5 summarizes the performance of our Q-BIC solar cells and photodetectors as a function of the built-in-dot charge. As it is seen from Fig. 5 (right), the photoresponse of Q-BIC IR photodetectors increases ~25 times, if the QD filling increases to 6 electrons per dot.

Our data are limited by the value of built-in-dot charge of six electrons per dot. However, the obtained results show that the improvements of solar cell and photodetector performances due to the built-in-dot charge are still far from saturation. Further doping-related improvements and their limitations are under investigation. We also started investigations of QD structures with correlated positions of QDs (Fig. 3(a)) that promise further improvements in optoelectronic characteristics.

5. Conclusions

Charging of quantum dots provides an effective way for managing 3D nanoscale potential profile, which in turn control photoelectron processes. Charging of dots creates local potential barriers around single dots, if dots are arbitrarily distributed in the medium, and collective barriers around dot clusters, rows etc. if quantum dots form specific structures. Manipulations with potential barriers provide an effective tool for suppression of fast capture processes of photocarriers by quantum dots. This allows one to increase the photocarrier lifetime and to reduce the recombination losses. The charging of dots also enhances the electron coupling to infrared radiation and multi-step absorption of sub-bandgap photons. Nanomaterials with charged QDs have a number of features attractive for photovoltaic and sensing applications.

Acknowledgments

This work was supported by the National Science Foundation (award ECCS-1236459) and the Air Force Office of Scientific Research (AFOSR).

References

[1] W. Shockley and H. J. Queisser, "Detailed balance limit of efficiency of p-n junction solar cells," *J. Appl. Phys.* **32**, 510 (1961). https://doi.org/10.1063/1.1736034

[2] V. Ryzhii, "The theory of quantum-dot infrared phototransistors," *Semicond. Sci. Technol.* **11**, 759 (1996). https://doi.org/10.1088/0268-1242/11/5/018

[3] A. V. Barve, S. J. Lee, S. K. Noh, and S. Krishna, "Review of current progress in quantum dot infrared photodetectors," *Laser & Photonics Reviews*, June 17, 2009.

[4] A. Luque and A. Martı, "Increasing the efficiency of ideal solar cells by photon induced transition at intermediate levels," *Phys. Rev. Lett.* **78**, 5014 (1997). https://doi.org/10.1103/PhysRevLett.78.5014

[5] A. J. Nozik, "Spectroscopy and hot electron relaxation dynamics in semiconductor quantum wells and quantum dots," *Annu. Rev. Phys. Chem.* **52**, 193 (2001). https://doi.org/10.1146/annurev.physchem.52.1.193

[6] A. J. Nozik, "Fundamentals and applications of quantum confined structures," Chapter 3 in *Nanostructured and Photoelectrochemical Systems for Solar Photon Conversion,* ed. M. D. Archer and A. J. Nozik, Imperial College Press, 2008. https://doi.org/10.1142/9781848161542_0003

[7] R. Schaller and V. Klimov, "High efficiency carrier multiplication in PbSe nanocrystals: implications for solar energy conversion," *Phys. Rev. Lett.* **92**, 186601 (2004). https://doi.org/10.1103/PhysRevLett.92.186601

[8] O. E. Semonin, J. M. Luther, S. Choi, H. Y. Chen, J. Gao, A. J. Nozik, and M. C. Beard, "Peak external photocurrent quantum efficiency exceeding 100% via MEG in a quantum dot solar cell," *Science*, **334**, 1530 (2011). https://doi.org/10.1126/science.1209845

[9] S. Tomić, "Intermediate-band solar cells: Influence of band formation on dynamical processes in InAs/GaAs quantum dot arrays," *Phys. Rev. B* **82**, 195321 (2010). https://doi.org/10.1103/PhysRevB.82.195321

[10] A. Sergeev, V. Mitin, and M. Stroscio, "Quantum-dot photodetector operating at room temperatures: diffusion-limited capture," *Physica B* **316-317**, 369, (2002). https://doi.org/10.1016/S0921-4526(02)00510-0

[11] A. Sergeev, L-H. Chien, N. Vagidov, and V. Mitin, "Quantum-dot infrared photodetectors: In search of right design for room-temperature operation," in S. Luryi, J. Xu, and A. Zaslavsky, eds., *Future Trends in Microelectronics: From Nanophotonics to Sensors and Energy*, New Jersey, Wiley-IEEE Press, 2010, pp. 385-393. https://doi.org/10.1002/9780470649343.ch32

[12] K. A. Sablon, J. W. Little, V. Mitin, A. Sergeev, N. Vagidov, and K. Reinhardt, "Strong enhancement of solar cell efficiency due to quantum dots with built-in charge," *Nano Letters* **11**, 2311 (2011). https://doi.org/10.1021/nl200543v

[13] V. Mitin, A. Antipov, A. Sergeev, N. Vagidov, D. Eason, and G. Strasser, "Quantum dot infrared photodetectors: Photoresponse enhancement due to potential barriers," *Nanoscale Research Letters,* **6**, 21 (2011).

[14] K. A. Sablon, A. Sergeev, N. Vagidov, A. Antipov, J. Little, and V. Mitin, "Effective harvesting, detection, and conversion of IR radiation due to quantum dots with built-in charge," Nanoscale Research Letters **6**, 584 (2011). https://doi.org/10.1186/1556-276X-6-584

[15] http://www.nextnano.de/nextnano3/

[16] F. Adler, M. Geiger, A. Bauknecht, F. Scholz, H. Schweizer, M. H. Pilkuhn, B. Ohnesorge, and A. Forchel, "Optical transitions and carrier relaxation in self assembled InAs/GaAs quantum dots," *J. Appl. Phys.* **80**, 4019 (1996). https://doi.org/10.1063/1.363361

[17] J. Siegert, S. Marcinkevičius, and Q. X. Zhao, "Carrier dynamics in modulation-doped InAs/GaAs quantum dots," *Phys. Rev. B* **72**, 085316 (2005). https://doi.org/10.1103/PhysRevB.72.085316

[18] H.-Y. Liua, Q.-F. Daia, Li-J. Wua, S. Lana, V. A. Trofimov, and S. A. Varentsova, "Effects of p-type doping on the optical properties of InAs/GaAs quantum dots," *Solid State Comm.* **152**, 435 (2012). https://doi.org/10.1016/j.ssc.2011.11.044

[19] D. G. Deppe, H. Huang, and O. B. Shchekin, "Modulation characteristics of quantum-dot lasers: The influence of p-type doping and the electronic density of states on obtaining high speed," IEEE J. Quant. Electronics **38**, 1587 (2002). https://doi.org/10.1109/JQE.2002.805246

[20] K. A. Sablon, J. W. Little, A. Sergeev, N. Vagidov, and V. Mitin, "Effects of quantum dot charging on photoelectron processes and solar cell characteristics," to appear in *Solar Energy Materials and Solar Cells* (2013).

Chapter 18

IVT Measurements of GaN Power Schottky Diodes with Drift Layers Grown by HVPE on HVPE GaN Substrates

R.P. Tompkins[1*], M.R. Khan[1], R. Green[1], K.A. Jones[1], J.H. Leach[2]

[1] U.S. Army Research Laboratory, 2800 Powder Mill Road, Adelphi, Maryland 20783, USA

[2] Kyma Technologies, 8829 Midway West Road, Raleigh, North Carolina 27617, USA

[*] randy.tompkins.civ@mail.mil

Abstract

To date, a majority of epitaxial layers for vertical gallium nitride (GaN) power Schottky diodes have been grown by metalorganic chemical vapor deposition (MOCVD). In this work, we investigate the electrical properties of vertical GaN Schottky diodes with drift layers grown by hydride vapor phase epitaxy (HVPE) on moderately-doped freestanding HVPE GaN substrates. Room temperature IV data is presented for devices tested where results for diode characteristics such as the breakdown voltage, V_b, specific on-resistance, $R_{on\text{-}sp}$, ideality factor, n, and barrier height, Φ_b, are measured for devices across the 1 cm x 1 cm sample. The smallest diodes, which are 30 μm in diameter, show the smallest specific on-resistance, whereas the breakdown voltage (defined as the voltage corresponding to a current of 10 mA/cm^2) is independent of device size across the wafer. IV data show an average value of 1.06 ± 0.06 for n and 0.80 eV ± 0.04 eV for Φ_b with little variation across the wafer, suggesting a reasonable metal-semiconductor interface across the entire sample. For one of the 300 μm devices tested, we also examine the IV properties as function of temperature from 25 °C to 250 °C in increments of 25 °C and extract a zero temperature Φ_b of 0.908 eV and Richardson's constant of 4.44 A cm^{-2} K^{-2} which is significantly less than the theoretical value of 26.9 A cm^{-2} K^{-2}.

Keywords

Gallium Nitride, Hydride Vapor Phase Epitaxy, Schottky Diode

Reprinted with permission from Springer Nature: J. Mater. Electron. "IVT measurements of GaN power Schottky diodes with drift layers grown by HVPE on HVPE GaN substrates," R.P. Tompkins, M.R. Khan, R. Green, K.A. Jones, J.H. Leach, (2016).
https://doi.org/10.1007/s10854-016-4536-z

Contents

1. Introduction

With its wide band gap, and thus high critical field, GaN is of great interest for high power electronics applications. The Schottky junction is fundamental to these applications, where Schottky diodes serve in power circuits and Schottky junctions are the gate contact for AlGaN/GaN high electron mobility transistors (HEMTs). Growth and fabrication of GaN high power Schottky diodes has many challenges including: lack of low cost, readily available bulk GaN substrates, high background n-type carrier concentrations (> 10^{16} cm^{-3}) in unintentionally doped drift layers, and management of high electric fields at the device periphery. Having a low n-type carrier concentration in the drift layer is important to achieve a high breakdown voltage, V_b, device, as the maximum V_b is inversely proportional to the doping level [1]. Few groups can controllably dope n-type GaN below 10^{16} cm^{-3} due to the large number of intrinsic defects, background impurities such as Si and O, and compensating carbon impurities. In our previous work, we have shown that C, presumably acting as a deep level acceptor, compensates the background n-type dopants and increasing the specific on-resistance, $R_{on\text{-}sp}$, of Schottky diodes [2]. While controllable doping levels < 10^{16} cm^{-3} can be achieved using metalorganic chemical vapor deposition (MOCVD) [3, 4], it is difficult to do so without the presence of compensating carbon impurities, where carbon is an intrinsic component of the trimethylgallium (TMGa) source, and growth occurs at temperatures ≥ 1050 °C.

While MOCVD is the most common technique for growing epitaxial device layers for power Schottky diodes, few investigations have examined the use of hydride vapor phase epitaxy (HVPE) for device layer growth. HVPE growth is based on the reaction of HCl with liquid gallium to form GaCl, which subsequently reacts with NH_3 to form GaN. HVPE can realize a much higher growth rate compared to MOCVD, as high as ~ 200 μm/hr compared to ~ 2 - 3 μm/hr for MOCVD growth, and thus HVPE growth is much more common in producing thick freestanding substrates, typically produced on lattice mismatched substrates such as sapphire or GaAs [5, 6]. Subsequent removal of the backside substrate and polishing produces a freestanding pseudo-bulk GaN substrate. Thick layers reduce the overall threading dislocation density from ~ 10^9 cm^{-2} for heteroepitaxial films to 10^5 - 10^6 cm^{-2} for films grown on freestanding HVPE substrates [5, 6]. HVPE substrates have been used for both power [3, 4] and optoelectronic [7] devices. HVPE provides an alternative to MOCVD, where HVPE does not have any direct sources of carbon. In addition to the advantage of reducing the background C, the ability to grow both film and substrate in the same reactor, as well as the higher growth rate, should lower the overall cost of GaN power Schottky diodes. In our previous work, we measured carrier concentrations as low as 10^{15} cm^{-3} in freestanding HVPE substrates [8]. Diodes fabricated on such structures have breakdown voltages > 900 V, but at the cost of a high specific on-resistance due to the thick low-doped substrate [8]. An ideal Schottky diode structure consists of a highly doped substrate with a lightly doped drift layer, where the substrate contribution to the series resistance is as small as possible (≤ 1%). However, growth of highly doped n-type HVPE GaN substrates with good crystallinity is challenging due to an increase in tensile stress with increased Si doping during substrate growth [9 - 11]. Growth of the drift layer by HVPE also presents challenges. Previous results have shown that unintentionally doped (UID) layers grown by HVPE have large concentrations (> 10^{17} cm^{-3}) of silicon [5], the most common n-type dopant in GaN, where the Si atom sits on the Ga site. The origin of the Si is unknown, but a prime candidate is the etching of the quartz reactor by the HCl. This challenge must be solved for growth of drift layers for Schottky devices where doping layers ≤ 10^{16} cm^{-3} are desired.

In this work, we report our initial results of vertical GaN Schottky diodes with drift layers grown by HVPE on freestanding HVPE substrates of moderate doping (7.0 x 10^{17} cm^{-3}). We report IV properties both at room temperature and as a function of temperature (IVT).

2. Experimental

Unintentionally doped HVPE GaN films with free carrier concentrations of 3.0 x 10^{16} cm^{-3} ± 4.29 x 10^{15} cm^{-3} were grown by HVPE on freestanding double-sided polished HVPE (1 cm x 1 cm area) n-type substrates doped 7.0 x 10^{17} cm^{-3}. Carrier concentrations were measured at room temperature in a probe station by capacitance-voltage (CV) at a frequency of 200 kHz and DC bias of 25 mV on the fabricated device and reference samples for the substrate. A frequency of 200 kHz, as opposed to the common 1 MHz frequency, was used because of the high resistivity of the drift layer [12]. Typical full width half maximum values from x-ray diffraction (XRD) rocking curves were 90 arcsec for the (0002) symmetric reflection and 200 arcsec for the (10-12) asymmetric reflection. The substrate thickness was approximately 450 μm and the drift layer thickness is estimated to be between 6.6 - 12.3 μm.

Before fabrication, we degreased the substrates in acetone followed by methanol for a period of 5 minutes with sonication in each solvent to remove organic contaminants on the front and back surfaces. We fabricated vertical Schottky diodes by initially depositing ohmic contacts on the entire backside of the sample using electron beam evaporation. Prior to ohmic contact deposition, samples were etched in HCl for 30 s to remove the oxide layer and organic contaminants. The ohmic contact consisted of a Ti/Al/Ni/Au (250/2200/650/500 Å) metal stack that was annealed at 750 °C in N_2 gas for 30 s using rapid thermal annealing. A 1 μm plasma enhanced chemical vapor deposition (PECVD) non-stoichiometric silicon nitride (SiN_x) film was deposited for a field plate structure to reduce the electric field around the device periphery. We patterned the SiN_x layer using standard contact photolithography techniques and subsequent etching in a 10:1 buffered oxide etch (BOE) solution. Standard contact photolithography and lift-off techniques were then used to fabricate the Schottky contacts. Prior to electron beam deposition of the Schottky contact, samples were again etched in HCl to remove the oxide layer and organic contaminants on the sample surface. The Schottky contact consisted of a Ni/Au (500/1500 Å) un-annealed metal stack. Schottky contacts were circular with a diameter of 30 μm, 100 μm, 200 μm, or 300 μm.

After fabrication, 36 devices were IV tested at room temperature using a probe station in air (no florinert) with an Agilent 4155B semiconductor parameter analyzer. From the room temperature IV curves, $R_{on\text{-}sp}$, Φ_b, and n were determined using conventional methods for thermionic emission current transport. The reverse IV was also measured to determine the leakage current, I_s, where V_b was defined as the reverse voltage at which I_s reaches a value of 10 mA/cm^2.

After initial room temperature measurements, we packaged the sample in a 40-pin dual inline IC package (DIP) to accommodate our IVT test apparatus consisting of a computer control of an Agilent 4155C semiconductor parameter analyzer and an Omega temperature controller. Individual devices were wire bonded using gold wire bonds. The packaged devices were then loaded into the IVT setup equipped with standard source measurement units (SMUs) and aluminum nitride ceramic heaters with integrated K-type thermocouples as the heat source. After allowing for temperature stabilization, we performed IVT measurements in both the forward and reverse directions. The device temperature was increased in increments of 25 °C starting at room temperature up to a maximum temperature of 250 °C. We then analyzed the forward IV curves using the thermionic emission model with series and shunt resistances, assuming negligible contributions of the tunneling current and generation recombination current to the total current. We determined *n*, Φ_b at T=0, and estimated Richardson's constant, A^*, from the data.

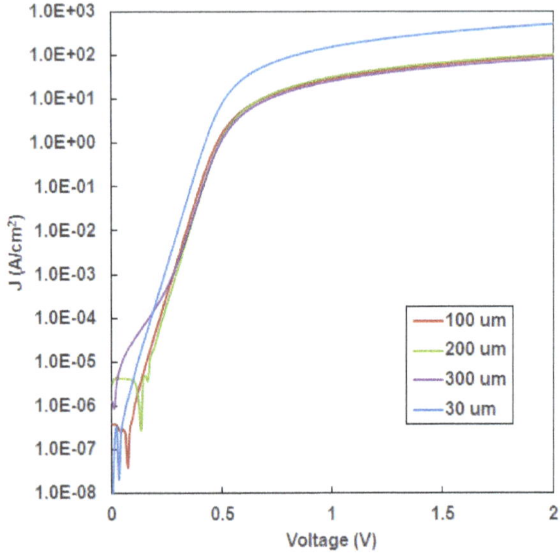

Fig. 1 Room temperature forward IV curves of diodes with minimum $R_{on\text{-}sp}$ for diode size ranging from 30 μm to 300 μm.

3. Results and discussion

Figure 1 shows forward IV curves for each size device that have the smallest measured $R_{on\text{-}sp}$ at room temperature. Performing a linear fit to the forward IV curves, the minimum values for $R_{on\text{-}sp}$ were 2.7 mΩ-cm², 13.98 mΩ-cm², 11.6 mΩ-cm², and 11.8 mΩ-cm² for 30, 100, 200 and 300 μm diameter diodes, respectively. While these numbers for $R_{on\text{-}sp}$ are high compared to the ideal case, they are comparable to what others have measured for vertical diodes on freestanding HVPE substrates [13 -15]. Because of large variation in $R_{on\text{-}sp}$ across the sample, we will express results for $R_{on\text{-}sp}$ in terms of the median. Median values of $R_{on\text{-}sp}$ with increasing size were 3.0, 19.8, 24.6, and 24.5 mΩ-cm². Increasing $R_{on\text{-}sp}$ with diode size has been observed in previous studies and has been attributed to the lower change in ratio of Schottky contact resistance and the spreading resistance for the smaller devices [11]. Note that the doping of the substrate is 7.0 x 10^{17}

cm^{-3}, an increase in doping of the HVPE substrate would lower $R_{on\text{-}sp}$. Figure 2 is a plot as a function of the barrier height. Average values of the *n* were 1.02 ± 0.01, 1.02 ± 0.01, 1.09 ± 0.06, and 1.10 ± 0.10 for 30, 100, 200, and 300 μm diameter devices, respectively. The average Φ_b correspond to 0.77 ± 0.01 eV, 0.81 ± 0.01 eV, 0.79 ± 0.07 eV, and 0.80 ± 0.03 eV with increasing diode size. For all devices tested irrespective of size, *n* was 1.06 ± 0.06 and Φ_b was 0.80 ± 0.04 eV. An ideality factor of 1.06 suggests an ideal diode with thermionic emission as the dominant current transport mechanism. The ideal value for Φ_b is approximately 1.1 eV. Our measured value of 0.80 eV for the barrier height is a reasonable result for Ni/Au contacts to GaN, where others have measured similar numbers [16]. Our results for *n* and Φ_b suggest a metal-semiconductor interface of reasonable quality uniformly across the entire sample.

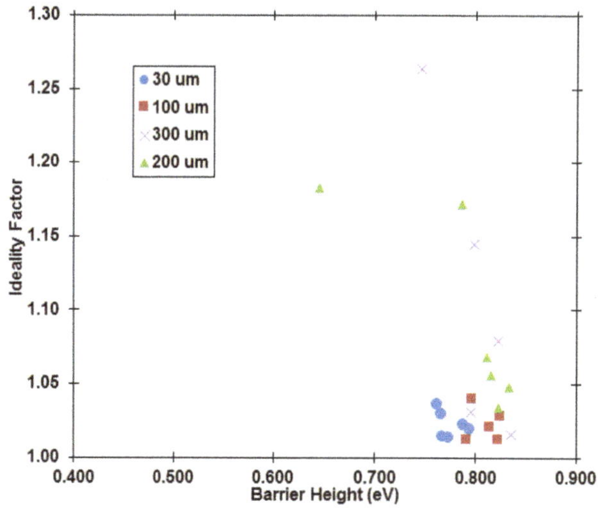

Fig. 2 Room temperature ideality factor vs. barrier height data for each diode tested.

The breakdown voltages for diodes with the various diameters are shown in Figure 3. A current density of 10 mA/cm^2 was to define V_b to prevent destructive breakdown of the devices and be able to perform future measurements. Select devices pushed beyond a 10 mA/cm^2 current density limit destructively broke down where damage was observed around the device periphery, as shown in Figure 4. Reverse voltages corresponding to such breakdown were between 100 – 200 V. The breakdown voltage corresponding to a current density of 10 mA/cm^2 is plotted for each individual device as a function of diode diameter. The average values for each size are -37.3 ± 3.0 V, -41.8 ± 5.1 V, -41.4 ± 5.1 V, -30.3 ± 5.6 V for 30, 100, 200 and 300 μm devices, respectively. Note that for the 300 μm devices, a single device with an outlier value of -2.1 V was omitted before computing the average and uncertainty. The measured values for V_b are low compared to their theoretical maximum for each size tested. With the exception of the 300 μm diodes, the values of V_b fall within one standard deviation of each other, with values for the 300 μm devices falling just outside the range. In previous studies we have observed the lowest V_b for the largest 300 μm devices [8, 17]. Those studies attributed this observation to the increased probability of defects, specifically threading dislocations, intersecting the larger pads of the largest devices compared to the smaller ones [18, 19]. While the numbers for V_b were lower than their theoretical maximum, in this study V_b was uniform across the sample. We expect that an optimized combination of field plates and ion implantation isolation should improve their reverse bias characteristics.

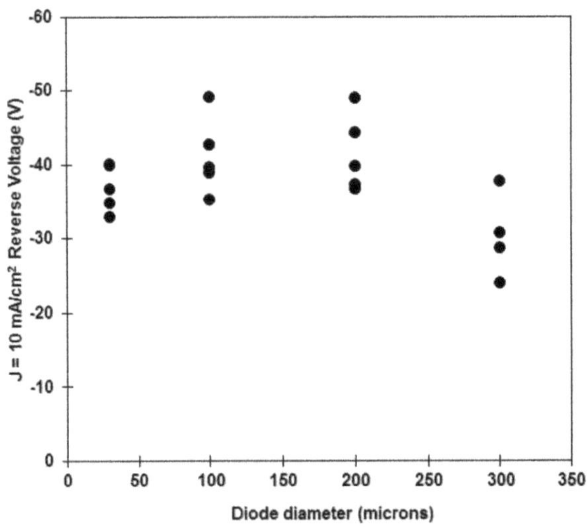

Fig. 3 Room temperature reverse IV data showing reverse voltage corresponding to a current density of 10.

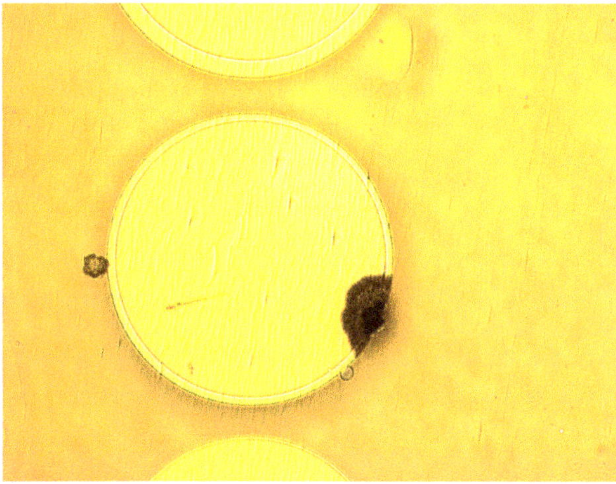

Fig. 4 Optical microscopy image showing device with destructive breakdown at the Schottky contact periphery. Destructive breakdown for this device occurred at a reverse voltage of ~ 200 V. An approximate scale is shown in the diagram.

After room temperature IV characterization, we performed IVT measurements on select devices. The total current (I) is composed of thermionic emission current (I_{te}), generation and recombination current (I_{gr}), tunneling current (I_{tun}), and series and shunt resistance components (I_{sh}) given by the equation [20]:

$$I = I_{te} + I_{gr} + I_{tun} + I_{sh} \quad \text{(Eq. 1)}$$

Where each individual current component is given by the following equations [20]:

$$I_{te} = I_{te0}\left[exp\left\{\frac{q(V-IR_s)}{k_bT}\right\} - 1\right] \quad \text{(Eq. 2)}$$

$$I_{gr} = I_{gr0}\left[exp\left\{\frac{q(V-IR_s)}{2k_bT}\right\} - 1\right] \quad \text{(Eq. 3)}$$

$$I_{tun} = I_{tun0}\left[exp\left\{\frac{q(V-IR_s)}{E_0}\right\} - 1\right] \quad \text{(Eq. 4)}$$

$$I_{sh} = \frac{V-IR_s}{R_{sh}} \quad \text{(Eq. 5)}$$

With I_{te0}, I_{gr0} and I_{tun0} corresponding to the saturation current for each of the current components. R_s and R_{sh} are the series and shunt resistances respectively. E_0 is the tunneling energy which is equal to:

$$E_0 = E_{00} \coth\left(\frac{E_{00}}{k_b T}\right) \qquad \text{(Eq. 6)}$$

Where E_{00} is the characteristic energy given by:

$$E_{00} = \frac{qh}{4\pi}\left(\frac{N_d}{m^* \varepsilon_s}\right) \qquad \text{(Eq. 7)}$$

Based on the values of the ideality factors measured at room temperature, thermionic emission should be the dominant current transport mechanism. Assuming a value of $0.22m_e$ for the electron effective mass [21] and our measured doping concentration of 3.0 x 10^{16} cm^{-3} for N_d, we calculate a characteristic energy of 1.90 meV; considering that thermionic emission dominates over field emission where $E_{00} << kT$, and that the current from generation-recombination processes (I_{gr}) are negligible in such a wide bandgap semiconductor, the contributions to the forward conduction from I_{tu} and I_{gr} are negligible.

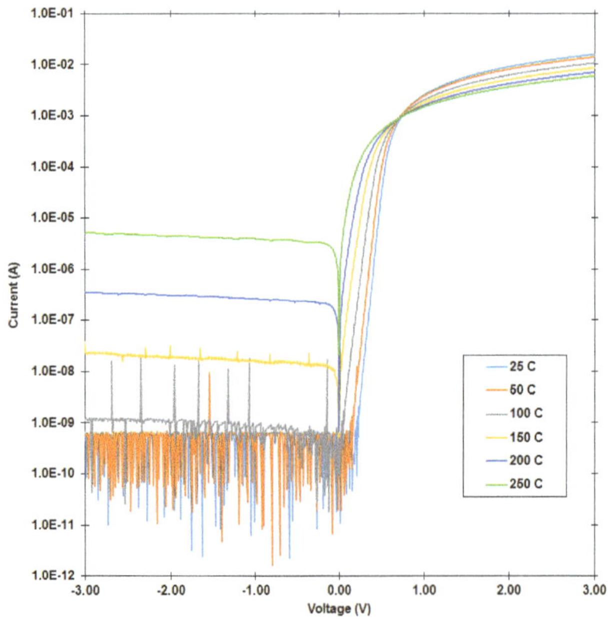

Fig. 5 Forward and reverse IVT plot for a 300 µm diameter Schottky diode.

Figure 5 shows forward and reverse IV curves for the temperature range of 298 to 523 K with a step size of 25 K for a 300 μm device. Measurements were taken from -3 V to 3 V with a step size of 0.02 V and the current compliance was set to 85 mA to reach a current density of 100 A/cm^2 commonly defined as the turn-on voltage. The forward IV shows a decrease in resistance with temperature for low voltages that is attributed to the decrease in barrier height. As expected, at high forward bias, the series resistance component dominates the transport and thus resistance increases with increasing temperature. The reverse IV shows an increase in leakage current with increasing temperature. With an increase in temperature, electrons have higher energies which promotes leakage, whereas the barrier height increases which reduces leakage; the trend that is observed depends on which component dominates [22]. We observe an increase in leakage current with temperature which suggests the electron energy component dominates.

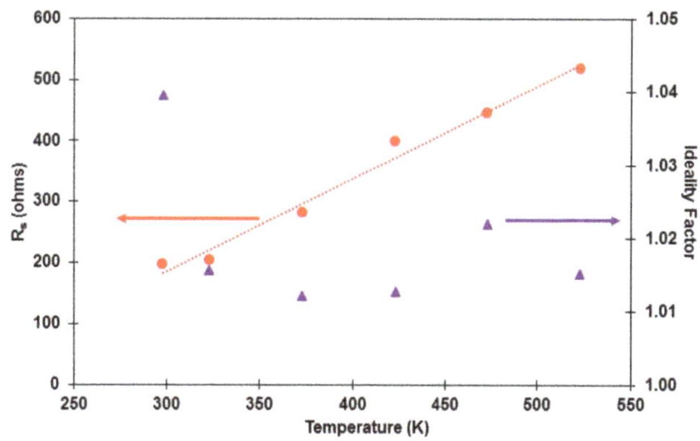

Fig. 6 Series resistance (R_s) and ideality factor (n) plotted as a function of temperature for a 300 μm device.

The forward IV curves shown in Fig. 5 were fitted using equations 1 - 5 and proprietary software developed by Semetrol LLC [23]. Fixed negligible values of 10^{-55} A for I_{tun} and I_{gr} were used in all fits, based on the values of E_{00}. Thus, only the thermionic emission and resistance contribution to the forward conduction were examined. Four terms were extracted from the fitting process n, R_s, R_{sh}, and I_{te} based on initial guessed values of 1.00, 100 Ω, 10^{20} Ω, and 10^{-9} A. The best fit to the data was determined by minimizing the mean squared error. The mean squared error for fits to the individual curves were between 8.11 x 10^{-4} and 6.68 x 10^{-6}. Figure 6 shows a plot of ideality factor and series resistance as a function of temperature. As expected for thermionic emission across a barrier, n was between 1.01 and 1.04. The series resistance was found to increase linearly as expected from a value of 196 Ω to 518 Ω. For measurements at each temperature, the shunt resistance was ≥ 9.04 x 10^{18} Ω.

The GaN Richardson's constant can be estimated from the IVT data. The saturation current density for thermionic emission is given by:

$$J_0 = A^* T^2 exp\left[-\frac{q\phi_b}{kT}\right] \quad \text{(Eq. 8)}$$

which can also be written as:

$$ln\left(\frac{J}{T^2}\right) = ln(A^*) - \frac{q\phi_b}{kT} \quad \text{(Eq. 9)}$$

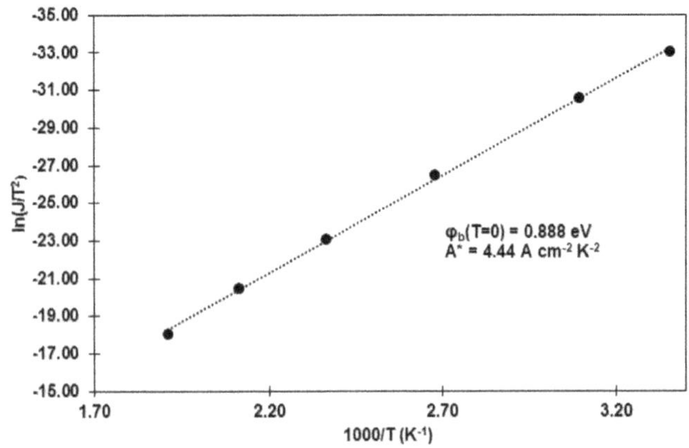

Fig. 7 Richardson's plot for the 300 μm Schottky diode tested.

Figure 7 is the common Richardson's plot, $\ln(J_{te0}/T^2)$ vs. 1000/kT. From this data we extract the T = 0 thermionic emission barrier of 0.908 eV. From the slope intercept we extract a Richardson's constant of 4.44 A cm^{-2} K^{-2} which is significantly smaller than the theoretical value of 26.9 A cm^{-2} K^{-2} [24] for GaN. Previous experimental works have also underestimated the value of Richardson's constant from the Richardson's plot for Ni Schottky contacts on n-type device layers grown by MOCVD [25, 26], Ni/Au contacts to n-type MBE material [28] as well as Au contacts to n-type HVPE material [27]. Such observations have been attributed to chemical reactions at the interface [25], enhanced tunneling current through surface states [26], presence of an oxide layer [26, 27], and Schottky barrier height inhomogeneity [28]. Because our contacts were not annealed, it is unlikely that chemical reactions at the interface are the reason for the discrepancy between the measured and theoretical value of A*. Also, it is unlikely that the presence of an oxide layer is a reason for the discrepancy because both the ideality factor and barrier height would have been affected as well, where the values we measured were reasonable. However, any of the other explanations remain plausible. Future work will investigate such possibilities for Ni/Au Schottky contacts to HVPE GaN material.

4. Conclusion

Vertical GaN power Schottky diodes were fabricated on HVPE epitaxial films grown on freestanding HVPE GaN substrates, where we present our initial results of room temperature IV and IVT measurements. Doping in the epitaxial film was measured to be 3.00×10^{16} cm^{-3} $\pm$ 4.29×10^{15} cm^{-3} using CV measurements. Room temperature IV data showed uniformity of barrier height and ideality factor across the wafer suggesting a uniform metal-semiconductor interface. Forward IV measurements showed that the smallest devices had the lowest specific on-resistance. Ideally, this observation would not be the case; however, a possible reason is the lower change in ratio of Schottky contact resistance and the spreading resistance for the smaller devices. Reverse IV showed a breakdown voltage less then theory for the given drift layer thickness and doping, where the breakdown voltage was defined as 10 mA/cm^2 to prevent destructive breakdown of the devices. The destructive nature of device breakdown suggests that the electric field around the device periphery is too high causing the destructive breakdown, not material defects. We expect that an optimized combination of both field plate structures and ion implantation isolation will improve V_b. IVT results for the largest device showed an underestimation of the Richardson's constant, which is commonly observed. The most likely reasons for this observation are enhanced tunneling current through surface states and/or Schottky barrier height inhomogeneity and will be a subject of further investigation.

References

[1] B.J. Baliga: Fundamentals of Power Semiconductor Devices (Springer Science, New York, NY, 2008) p. 15.

[2] R.P. Tompkins, T.A. Walsh, M.A. Derenge, K.W. Kirchner, C.B. Nguyen, K.A. Jones, P. Suvarna, M. Tungare, N. Tripahi, F. Shahedipour-Sandvik, *J. Mater. Res,*. 26 (23) 2895 (2011). https://doi.org/10.1557/jmr.2011.360

[3] Y. Saitoh, K. Sumiyoshi, M. Okada, T. Horii, T. Miyazaki, H. Shiomi, M. Ueno, K. Katayama, M. Kiyama, and T. Nakamura, *Appl. Phys. Exp.* 3, 081001 (2010). https://doi.org/10.1143/APEX.3.081001

[4] I.C. Kizilyalli, A.P. Edwards, H. Nie, D. Disney, and D. Bour, IEEE Trans. On Electron Devices, 60 3067 (2013). https://doi.org/10.1109/TED.2013.2266664

[5] K. Fujito, S. Kubo, H. Nagaoka, T. Mochizuki, H. Namita, S. Nagao, *J Crystal Growth*, 311 3011 (2009). https://doi.org/10.1016/j.jcrysgro.2009.01.046

[6] S.K. Mathis, A.E. Romanov, L.F. Chen, G.E. Beltz, W. Pompe, and J.S. Speck, *Phys. Status. Solidi.* A., 179, 125-145 (2000). https://doi.org/10.1002/1521-396X(200005)179:1%3C125::AID-PSSA125%3E3.0.CO;2-2

[7] M. Funato, M, Ueda, Y. Kawakami, Y. Narukawa, T, Kosugi, M. Takahashi and Takashi Muka, *J. Journal of Appl. Phy.*, 45, L659 (2006). https://doi.org/10.1143/JJAP.45.L659

[8] R.P. Tompkins, T.A. Walsh, M.A. Derenge, K.W. Kirchner, S. Zhou, C.B. Nguyen, K.A. Jones, G. Mulholland, R. Metzger, J.H. Leach, P. Suvarna, M. Tungare, F. Shahedipour-Sandvik, Solid State Electron., 79 238 (2013).

[9] G.S. Sudhir, Y. Peyrot, J. Kru¨ ger, Y. Kim, R. Klockenbrink, C. Kiesielowski, M.D. Rubin, E.R. Weber, W. Kriegseis, and B.K. Meyer, Mater. Res. Soc. Symp. Proc. 482, 525 (1998).

[10] A. Cremades, L. Go¨rgens, O. Ambacher, M. Stutzmann, andF. Scholz, Phys. Rev. B 61, 2812 (2000). https://doi.org/10.1103/PhysRevB.61.2812

[11] L.T. Romano, C.G. Van de Walle, J.W. Ager III, W. Go¨tz, and R.S. Kern, J. Appl. Phys. 87, 7745 (2000). https://doi.org/10.1063/1.373529

[12] J.D. Wiley, G.L. Miller, *IEEE Trans. Electron. Dev.* 22 265 (1975). https://doi.org/10.1109/T-ED.1975.18117

[13] Y. Wang, S. Alur, Y. Sharma, F. Tong, R. Thapa, P. Gartland, T. Issacs-Smith, C. Ahyi, J. Williams, M. Park, M. Johnson, T. Paskova, E. Preble, and K. Evans, *Semicond. Sci. Technol.* 26 022002 (2011). https://doi.org/10.1088/0268-1242/26/2/022002

[14] Y. Wang, H. Xu, S. Alur, Y. Sharma, F. Tong, P. Gartland, T. Issacs-Smith, C. Ahyi, J. Williams, M. Park, G. Wheeler, M. Johnson, A. Allerman, A. Hanser, T. Paskova, E.A. Preble, and K. Evans, Phys. Status Solidi C 8 2430 (2011). https://doi.org/10.1002/pssc.201001158

[15] Y. Zhou, D. Wang, C. Ahyi, C. Tin, J. Williams, M. Park, N. Williams, and A. Hanser, Solid State Electon. 50 1744 (2006). https://doi.org/10.1016/j.sse.2006.09.009

[16] B.J. Zhang, T. Egawa, G.Y. Zhao, H. Ishikawa, M. Umeno, and T. Jimbo, *Appl. Phys. Lett.*, 79 2567 (2001). https://doi.org/10.1063/1.1410355

[17] R.P. Tompkins, J.R. Smith, K.W. Kirchner, K.A. Jones, J.H. Leach, K. Udwary, E. Preble, P. Suvarna, J.M. Leathersich, and F. Shahedipour-Sandvik, *J. Electron. Mat.* 43 850 (2014). https://doi.org/10.1007/s11664-014-3021-9

[18] J.W. Johnson, F.R. LaRoch, F. Ren, B.P. Gila, M.E. Overberg, C.R. Abernathy J.I. Chyi, C.C. Chuo, T.E. Nee, C.M. Lee, K.P. Lee, S.S. Park, Y.J. Park, and S.J. Pearton,Solid State Electron. 45, 405 (2001). https://doi.org/10.1016/S0038-1101(01)00059-4

[19] J.W. Johnson, A.P. Zhang, W.B. Luo, F. Ren, S.J.Pearton and S.S. Park et al, IEEE Transactions Electon Devices 49, 504 (2002).

[20] K. Suzue, S.N. Mohammad, Z.F. Fan, W. Kim, O. Aktas, A.E. Botchkarev, H. Morkoc, J. Appl. Phys. 80(8), 4467 (1996). https://doi.org/10.1063/1.363408

[21] A.M. Witowski, K. Pakula, J.M. Baranowski, M.L. Sadowski, and P. Wyder, Appl. Phys. Lett. 75, 4154 (1999). https://doi.org/10.1063/1.125567

[22] Z. Xiaoling, L. Fei, Lu Changzhi, X. Xuesong, L. Yang, and M. Sn, J. of Semiconductors, 30 034001-1 (2009). https://doi.org/10.1088/1674-4926/30/3/034001

[23] http://www.semetrol.com/

[24] A.M. Witowski, K. Pakula, J.M. Baranowski, M.L. Sadowski, and P. Wyder, Appl. Phys. Lett. 75 4145 (1999). https://doi.org/10.1063/1.125567

[25] J.D. Guo, F.M. Pan, M.S. Feng, R.J. Guo, P.F. Chou and C.Y. Chang, J. Appl. Phys. 80 1623 (1996). https://doi.org/10.1063/1.363822

[26] L.S. Yu, Q.Z. Liu, J. Xing, D.J. Qiao, S.S. Lau, and J. Redwing, J. Appl. Phys. 84 2099 (1998). https://doi.org/10.1063/1.368270

[27] P. Hacke, T. Detchprohm, K. Hiramatsu, and N. Sawaki, Appl. Phys. Lett., 63 2676 (1993). https://doi.org/10.1063/1.110417

[28] S. Dogan, S. Duman, B. Gurbulak, S. Tuzemen and H. Morkoc, Phys. E. 41, 646 (2009). https://doi.org/10.1016/j.physe.2008.10.020

Chapter 19

MEMS-Based and Switched-Capacitor Approaches for Miniature Power Supply Applications

Sarah S. Bedair[a], Christopher D. Meyer[a], Nathan Lazarus[a], Christopher Dougherty[b], Jeffrey S. Pulskamp[a], Brian Morgan[a], Ronald Polcawich[a], Xue Lin[b], Rizwan Bashirullah[b], Iain Kierzewski[a], Joel Martin[a], Brian Power[a]

[a]Army Research Laboratory, 2800 Powder Mill Rd. Adelphi, MD 20783, USA

[b]Dept. of Electrical & Computer Engineering, Univ. of Florida, Gainesville, FL 32611, USA

Abstract

This chapter ummarizes work towards creating mm-scale power converters using high-frequency CMOS as well as MEMS and micro-machined passives. Reducing the converter size is largely motivated by creating power supplies for micro-robotic platforms (with millimeter and milligram scales) without negatively impacting robotic system functionality. MEMS power passives are first presented where thin-film piezoelectric transformers and resonators are shown as an electromechanical approach to achieve ultra-miniature passives at the chip scale. Piezoelectric transformers fabricated with thin-film lead zirconate titanate (PZT) on silicon are measured and show ~60% efficiencies with 240 and 75 Ω loads. These transformers have resonant frequencies ranging between 14 and 20 MHz. Work towards creating transmission lines fabricated with air-core inductors and ferroelectric capacitors is also presented. Finally, a fully-integrated bi-directional converter in CMOS is shown driving mm-scale robotic wings made with PZT. The converter's maximum efficiency is 77% at ~800μW load with 9V output and demonstrates >3x voltage boost in 0.13-μm triple-well CMOS.

Keywords

Autonomous Systems, mm-scale Power Supplies, Power Conversion, MEMS Power Passives, Micro-machined Power passives, Thin-film Piezoelectric Transformers, Switched Capacitor Converters, Transmission-line Converters

The work has been previously published by SPIE:
Sarah S. Bedair, Christopher D. Meyer, Nathan Lazarus, Christopher Dougherty, Jeffrey S. Pulskamp, Brian Morgan, Ronald Polcawich, Xue Lin, Rizwan Bashirullah, Iain Kierzewski, Joel Martin, Brian Power, "MEMS-based and switched-capacitor approaches for miniature power supply applications". In Proc. SPIE-DSS, 2014, pp. 90831T- 1-12. https://doi.org/ 10.1117/12.2051775

Contents

1. Introduction

There is a growing interest in the research on power supplies on chip [1] Waldron, F., Foley, R., Slowey, J., Alderman, A. N., Narveson, B. C. and OMathuna, S. C., "Technology roadmapping for power supply in package (PSiP) and Power Supply on Chip (PwrSoC)," IEEE Trans. on Power Electronics 28 (9), 4137-4145 (2013). [2], which have a number of applications including portable electronics, microprocessors, and miniature sensors. High-power-density power supplies which can adapt to a variety of sources and loads for future Micro-Autonomous Systems Technologies (MAST) are in need to maintain the lifetime of the battery or power source. Some of these MAST type systems are ultra-small with cm scale system sizes [3] [4] and even systems that are envisioned to be mm-scale [5]. An analysis of the power requirements for flight at the mm-scale shows maintained system mobility for > 1 hour and km range of flight based on currently available thin-film batteries and power supplies occupying > 90% of the total system mass [6]. This analysis, however, assumes mm-scale power converters with near-perfect efficiencies. Further power analysis and feasibility of power approaches at these scales are also detailed in [7].

Future mobile MAST platforms motivate both ultra-small power management units and high power-density sources. The total load power is expected to be low for these systems (<1 W), but there are varying load requirements and they range from 100's μW (kΩ) to 10–100 mW (10–100 Ω) [7]. With a large range of load requirements and the additional stringent size requirements, alternatives to commercial-off-the-shelf components must be explored. Ultra-miniature form factors with high-efficiency motivate miniature, high quality passives operating in the 10–100's MHz. With multiple passive types tightly integrated with high performance active circuitry, either through monolithic or hybrid integration [8], a wide variety of loads could be accommodated with such fine scale integration of heterogeneous components. Figure 1 illustrates this vision of multiple passive types and active components, all integrated either monolithically or through hybrid approaches. Current research on chip-scale power supplies includes topologies with thin-film inductors and magnetic transformers [9]. Magnetic materials-based inductors and transformers have shown moderate performances [10] with inductor quality factors ranging between 1 and 30 in the 10-100 MHz frequency range and between 0.8 and 9 in the 100-1000 MHz range. In contrast, air-core inductor approaches [11] have shown inductance densities similar to magnetic-material based devices (> 100 nH / mm^2) but with higher quality factors (>10) which render devices with practical use in the 100-1000 MHz frequencies.

Alternant approaches to traditional magnetic-materials-based passives are explored in this paper. The vision for this work is to leverage fabrication processes, or similar spin-off processes [12], used to make piezoelectric mm-scale robotic platforms [5] to implement the passive components needed in a converter. In other words, the approach is to leverage passives that can be available in a process common with the mm-scale robotics fabrication process. These passives include piezoelectric resonators, which can be used in resonant converter topologies, piezoelectric transformers, and ferroelectric capacitors, which offer high-dielectric-constant capacitors on chip. Active components could be integrated in various ways including heterogeneous integration [13] or through flip-chip bonding approaches.

This paper provides a summary of the authors' works toward chip-scale power-supply applications. First, chip-scale piezoelectric transformers and resonators, which are electromechanical devices with high equivalent electromechanical inductance densities (> 10,000 nH / mm^2 [14]), are introduced. Other electromechanical approaches offer great potential for chip-scale applications [15]. Although different circuit topologies are needed for the use of piezoelectric resonators and transformers, the inductance density comparison with traditional magnetic inductor and transformer approaches illustrates the ultra-miniature size that can be achieved with such electromechanical passives.

Secondly, the monolithic integration of air-core inductors with either PZT or air-dielectric capacitors is considered for the implementation of transmission-line converters. Such approaches can introduce conversion schemes allowing multiple output voltages which can be generated from a single transmission line, offering potential savings in terms of component count and overall area. Finally, switched-capacitor converter approaches are explored where a fully-integrated CMOS switched-capacitor, bi-directional converter driving a 2 mm piezoelectric wing is summarized and presented. To the best of the author's knowledge, this is the first fully integrated converter used to drive mm-scale robotic wings.

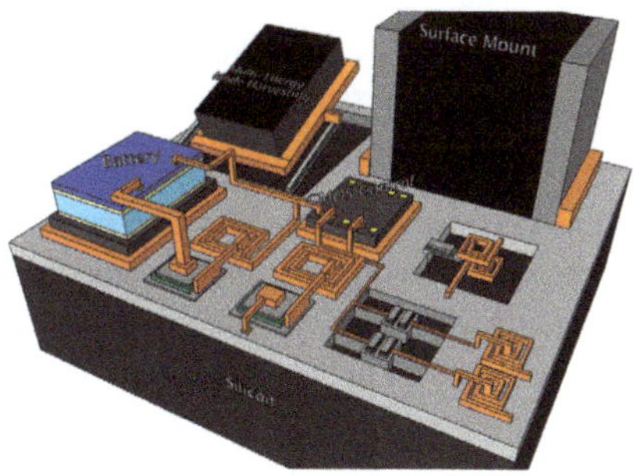

Fig. 1 Envisioned power supply on chip concept. Miniature passives including piezoelectric transformers, inductors and capacitors are tightly integrated along with active CMOS switching circuitry and control. Thin-film batteries are envisioned to be integrated as well.

2. MEMS power passives

2.1 Piezoelectric RLC resonators and transformers

Bulk piezoelectric transformers have been implemented in the past due to its attractive features which include compact size, high voltage isolation, high conversion ratios and very high power densities [16]. Piezoelectric resonators and transformers in their thin-film form could offer an ultra-miniature passives approach, in contrast to their larger, magnetic counterparts. When assessing the power densities of piezoelectric transformers or piezoelectric resonant passives compared with magnetic-materials-based approaches, the electromechanical approach offers ~3 orders of magnitude higher energy and power densities [17]. This can be arrived at when considering the theoretical electromechanical strain versus magnetic saturation limits. Piezoelectric transformers and piezoelectric resonant approaches are pursued in this work for this reason, where the authors are motivated by ultra-miniature passives solutions, especially for systems with total size on the order of mm-cm.

There has been a significant amount of work on the utility of thin-film piezoelectric resonators for RF filtering applications. With these applications, piezoelectric resonant filters have been implement using ZnO [18] and AlN [19] [20] materials as the piezoelectric transducers with some work using thin-film PZT [21-23]. In contrast to the filtering efforts, there has been minimal work on the utility of these electromechanical resonators as power passives for voltage and current transformers in supply-on-chip applications [14], [24-26]. These works include the use of both thin-film Aluminum nitride and lead zirconate titanate (PZT) resonators either as one-port resonators replacing the passives in a resonant converter [27] or as a two-port device implemented in a piezoelectric transformer operation.

Table 1 [28] lists the relevant material properties for the typical thin-film piezoelectric materials seen in prior art. When comparing the various piezoelectric materials, PZT is an attractive choice due to its high e_{31} piezoelectric stress constant and electromechanical coupling factors, which is relevant for the contour-modes typically reported in the MEMS resonator field. PZT has also been the choice material for bulk piezoelectric transformer for this particular reason.

Table 1 Material properties of piezoelectric materials typically used in thin-film resonator applications [28]. Dielectric constants, $e_{33,f}$, electromechanical coupling factors, $k^2_{p,f2}$, and piezoelectric stress constant, e_{31}, are listed for these materials.

	ZnO	AlN	PZT 52/48
$\varepsilon_{33,f}$	8 - 12	10.4	400 - 1300
$e_{31,f}$ (C/m^2)	-0.43 to -0.8	-0.9 to -1.0	-10 to -18
$k^2_{p,f}$	0.06 -0.085	0.065 - 0.11	0.3-0.40

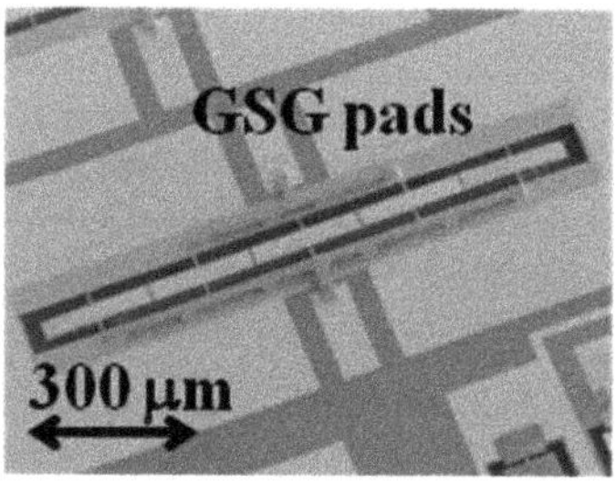

Fig. 2 Scanning electron micrograph of a PZT piezoelectric transformer with ~14 MHz resonant frequency. For this particular device, the fifth harmonic of the length-extensional mode is used. There is asymmetry in the number of input and output pads used to access the device. For voltage boost, the number of input pads is 3, while the number of output pads is 2. To buck the voltage the input and output pads would be swapped.

A two-port piezoelectric transformer is considered here where a higher-order (order n = 5), length-extensional mode is used for operation as shown in Figure 2, where an asymmetry in the number of input and output electrodes is designed to achieve voltage and current transformation between the input and output ports. The most basic understanding of the piezoelectric transformer operation can be gained by first looking at the power balance between the input and output ports. Assuming a low loss resonator, the power transferred into the input port is equivalent to the power emerging from the output port, $P_{in} = P_{out}$, and the current flowing into the device is proportional to the input electrode area, A_{in}, and number, N_{in}. Similarly, the current flowing out is proportional to the output electrode area, A_{out}, and number, N_{out}. The resulting conversion ratios can be estimated by

$$I_{in}V_{in} \cong I_{out}V_{out}$$
$$\frac{V_{out}}{V_{in}} \cong \frac{A_{in}N_{in}}{A_{out}N_{out}} \qquad \text{(1a), (1b)}$$

Following this above analysis, one might erroneously conclude that the omission of electrodes in a higher-order-mode construct, for example, could lead to higher conversion ratios. This is not the case since the low-loss resonator assumption would not be valid. This becomes evident when evaluating the motional resistance of the resonator for a high-order length-extensional mode resonator [14]

$$R_{m,n} = \frac{n\pi^2 w_{tot} t_{tot} \sqrt{Y_c \rho_c}}{16 N_{in} N_{out} Q e_{31,f}^{\ 2} w_{in} w_{out}} \tag{2}$$

where $e_{31,f}$ is the effective transverse piezoelectric stress constant, Q is the quality factor, t_{tot} the total device thickness, w_{tot} the total device width, w_{in} is the input electrode width, and w_{out} is the output electrode width. Equation (2) reveals a larger $R_{m,n}$ resulting with omission of an output electrode while keeping the same higher order harmonic design. For example, if one considers an $n = 9$ extensional mode resonant design, under the low loss assumption, the conversion ratio should be 4:5 if $N_{in} = 5$ and $N_{out} = 4$. Keeping N_{in} equivalent to 5 and reducing N_{out} to 1 does not result in a 1:5 conversion ratio since $R_{m,n}$ will be reduced by a factor of 4, which negatively and directly impacts the conversion ratio. With single-layer piezoelectric transformers, the conversion ratio has generally been limited (<10); higher conversion ratios will result with multi-layer stacking of the piezoelectric material as previously shown with bulk piezoelectric transformers [29].

The equivalent circuit for electromechanical transformers is shown in Figure 3(a). The mass, m, stiffness, k, and motional impedance, c, can be considered in its electrical equivalent circuit form in Figure 3(b) by conversion through its transduction coefficients at its input, Γ_{in}, and output, Γ_{out}, ports. The relevant piezoelectric transformer characteristics can be further derived [14] resulting in geometry and material dependent efficiency and open circuit voltage gains described by

$$\eta_{peak} \cong \frac{1}{1 + \dfrac{1}{8Qk_{31}^{\ 2}} \dfrac{n\pi^3}{N_{in}} \left(1 + \dfrac{t_{Si}}{t_{PZT}} \dfrac{Y_{Si}}{Y_{PZT}}\right)} = \frac{1}{1 + \dfrac{1}{\text{FOM}}} \tag{3}$$

$$\left| \frac{V_{out}(j\omega_{\max})}{V_{in}(j\omega_{\max})}, {}_{oc} \right| \cong \frac{8Qk_{31}^{\ 2}}{\dfrac{n(n-1)}{n^2 - 1} \pi^3 \left(1 + \dfrac{t_{Si}}{t_{PZT}} \dfrac{Y_{Si}}{Y_{PZT}}\right)} = \text{FOM} \cdot (n+1) \tag{4}$$

These are dependent on the mode number, n, silicon thickness, t_{si}, PZT thickness, t_{pzt}, silicon modulus, Y_{si}, and PZT modulus, Y_{pzt}. The figure of merit, FOM, is dependent on the effective electromechanical coupling factor, k_{eff}^2, and Q

$$\text{FOM} = k_{eff}^{\ 2} Q \tag{5}$$

$$k_{eff}^{\ 2} = \frac{8k_{31}^{\ 2} N_{in}}{n\pi^3 \left(1 + \dfrac{t_{Si}}{t_{PZT}} \dfrac{Y_{Si}}{Y_{PZT}}\right)} \tag{6}$$

PZT-on-silicon piezoelectric transformers were fabricated with 2, 4, and 10 μm silicon-on-insulator (SOI) substrates. All substrates were prepared with 0.5 μm PZT and follow the process sequence detailed in [21]. Different device silicon thicknesses were fabricated to experimentally assess and verify the dependence of FOM and k_{eff}^2 on device Si:PZT thickness ratios. Higher order harmonics, n, of the length extensional modes (200 μm lengths and 40 μm widths) were also considered for the three thickness ratios. The results are shown in Figure 4. For each extensional mode considered the corresponding k_{eff}^2 does degrade, as expected, with increasing silicon thicknesses. The FOMs exhibit a more complicated dependence on the thickness ratios since the device Qs also has a dependence on silicon thickness; for example, volumetric Qs are expected to increase with increasing thicknesses. A zero silicon thickness is included as a reference and illustrates that although a high $k_{eff}^2 = 4.6\%$ results with PZT only devices, the PZT material quality factor limits the device to a measured $Q = 38$. This results in a FOM = 1.75 and low corresponding peak efficiency of 40%. The incorporation of silicon is advantageous and results in higher Qs and FOMs despite the reduction in k_{eff}^2. This is true for the 2, 4 and 10 μm silicon devices fabricated and tested. For most devices, there is a slight degradation in FOM between the 2 and 4 μm silicon devices but this is followed by a further improvement in FOM with thicker silicon. The device impedances were extracted from two-port scattering parameters and the peak efficiency, corresponding to the resistive load matched to the equivalent output impedance of the devices, was estimated from the measured impedances of the devices. The resulting peak efficiency and the measured FOM are shown in Figure 5. Efficiencies as high as ~60% are expected for the $n = 3$ devices with 2 and 4 μm device silicon. Future work is needed to further understand the complicated dependence the device FOM has with silicon thickness and its implications in power converter applications.

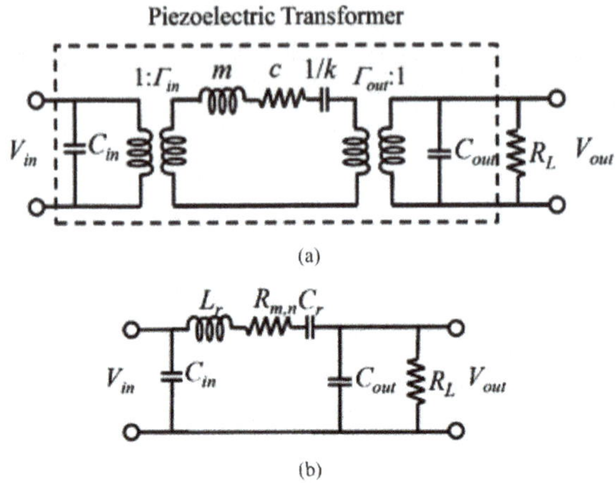

Figure 3. Electromechanical equivalent circuit for piezoelectric resonators used in a two-port piezoelectric transformer configuration.

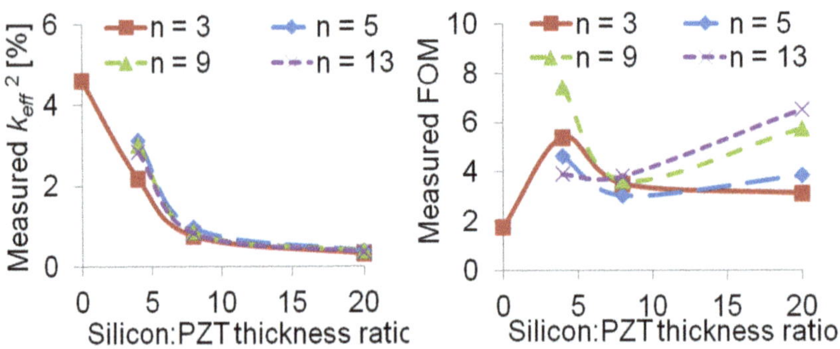

Fig. 4 Measured electromechanical coupling factors (left) and FOM (right) with varying silicon to PZT thickness ratios. High order harmonics of the length-extensional mode are considered.

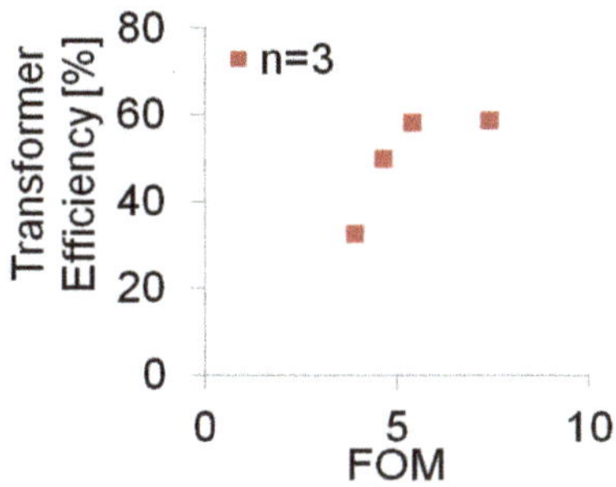

Fig. 5 Efficiency versus FOM for the varying silicon to PZT thickness ratio devices. The zero-silicon-thickness value is also included. This is shown for the 3rd harmonic of the length-extensional mode.

2.2 Power transfer and handling considerations

It is important to address the distinction between the utility of these piezoelectric resonators in power converters versus RF filtering applications. Unlike RF systems where maintaining power levels is of the utmost importance given the limited amount of available power, power converters are intended to operate at the highest efficiency as possible since maximizing efficiency increases battery life and subsequent mission time, and minimizes excessive heating due to I^2R losses. For maximum efficiency, the source resistance (largely dominated by the power switch being used) should be at a minimum. For example, similar to the analysis in [14], the following is true if the current through the source resistance, R_s, is the same as the current through $R_{m,n}$ and R_{Ls}, which is the equivalent series resistance of the load.

$$\eta = \frac{I_{rms}{}^2 R_{Ls}}{I_{rms}{}^2 R_{Ls} + I_{rms}{}^2 R_{m,n} + I_{rms}{}^2 R_s} = \frac{R_{Ls}}{R_{Ls} + R_{m,n} + R_s} = \frac{1}{1 + \frac{R_{m,n}}{R_{Ls}} + \frac{R_s}{R_{Ls}}} \tag{7}$$

In power conversion systems, R_s should be as close to zero as possible, but is typically 50 Ω in RF systems. Therefore, there is an important distinction for R_L load conditions for maximum efficiency versus conditions for maximum power delivery. Details on the load conditions for maximum efficiency and the load conditions for peak power delivery (where the efficiency with peak power delivery is 50%) for piezoelectric transformers can be found in the following references [26] [30] [31].

The incorporation of silicon not only improves device Q and FOM, as detailed previously, it also improves device power handling. Consider a critical vibration amplitude, x_{cr}, corresponding to the material in the resonator stack limiting the critical strain and resulting displacement, the energy stored per unit cycle, E_m, is

$$E_m = \frac{1}{2} k_o x_{cr}^{\ 2} \propto E_c A_c x_{cr}^{\ 2} \tag{8}$$

where k_c is the stiffness, E_c the modulus, and A_c the cross-sectional area of the composite. With thicker silicon both E_c and A_c increase, resulting in larger energy stored per unit cycle and power handling, P, which is proportional to both the maximum energy stored and the operating frequency, $P \sim E_m f_r$. Two similar devices, one with and the other without silicon, were measured with various power levels incident on the device. The device without silicon and 1 μm of PZT (30 μm wide x 150 μm long x 1 μm thick) had a maximum power handling of 0.5 W / mm^3. In contrast, a device constructed with 1 μm PZT on 3 μm of device silicon (30 μm wide x 150 μm long x 4 μm thick) was able to handle at least 100 μW of power, which translates to a device power density > 10 W / mm^3. This power measurement was limited to the maximum power output for the network analyzer used. Further improvements in power densities can be made with either increasing the frequency or increasing the device silicon, but this must be balanced with the FOM to ensure high device efficiencies.

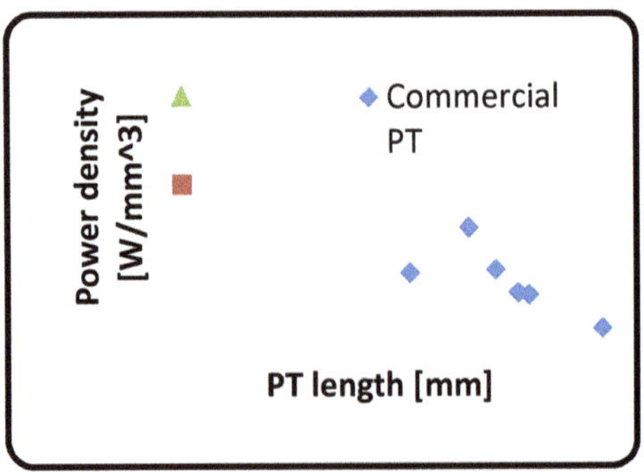

Fig. 6 Power density comparison for thin-film PZT only and PZT on silicon resonators. Commercially available, bulk piezoelectric transformers (PT) are included as a reference [32].

3. Micromachined power passives

3.1 Transmission-line-based power converters

A process where air-core power inductors are monolithically integrated with PZT and air-dielectric capacitors (copper-on-PZT process [12]) is also being explored for the purposes of creating transmission lines for high frequency power conversion. With the operating frequencies of high performance power switches increasing in recent years, transmission lines have been investigated as possible replacements for conventional power inductors. Prior work at Ericsson Power Solutions has demonstrated a technique for performing DC-AC and DC-DC power conversion by allowing current to accumulate for multiple transits of a transmission line [33]. The transmission line is used as a substitute for the inductor in traditional switching-mode power converter topologies; Figure 7 (a) shows a transmission-line-based variant of a buck-boost power converter. Unlike in an inductor, energy in a transmission line is distributed along the length of the line. By controlling the line input to time multiplex the signal, the same line can be used by multiple power converters simultaneously, where separate waves travel along the length of the line. This method allows inductance to be quickly re-distributed between different converters based on the timing of the control switch, reducing the need for excess inductance for each converter intended only for peak operation. With multiple power converters sharing the line, switches can also be combined in some cases, allowing for possible reductions in the system size.

The transmission line converter circuit in Figure 7 (a) is operated in two distinct phases. During the first phase, current is allowed to build in the transmission line over several one-way propagation times after switch S_A is turned on; during this period, switch S_B is turned off and the voltage across the load is zero. During the second phase, switch S_A is turned off and S_B is turned on, resulting in a controlled rectangular pulse across the load of width twice the propagation time. The controlled discharge of a transmission line (a "pulse-forming line" or PFL) is a well known method for creating a rectangular pulse of known duration [34]. The amount of time the current is allowed to accumulate sets the energy that is released in the controlled output interval, allowing the effective gain to be changed by controlling the switch timing. The cycle is then repeated, resulting in a rectangular pulse train.

3.2 PZT-copper implementation

Two different approaches for creating low-loss transmission lines in the aforementioned copper-on-PZT process were investigated. In the first, a coplanar waveguide (CPW) in thick copper layers, is periodically loaded with a thin capacitor with 0.5 μm PZT as the dielectric layer (Figure 7(b)). Since PZT has a very high dielectric constant, with values greater than 1000 [35], the line resonance is relatively low, occurring at 100 MHz. With subsampling, allowing current to build up for more than two propagation times for a higher magnitude pulse, this line can be operated at 10 MHz or less, within the range of currently demonstrated power switch technology. Higher frequency variants, in the hundreds of megahertz range, based on MEMS capacitors and inductors (Figure 7 (c)) have also been fabricated and are in the process of being tested. Since air is a low-loss dielectric, using an air-gap capacitor could give better power conversion performance when compared with a transmission line loaded with PZT capacitors with higher loss tangents.

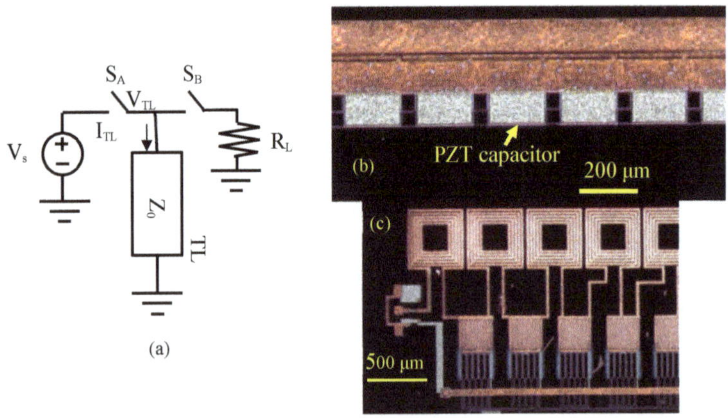

Fig. 7 (a) Circuit schematic of a buck-boost variant of a transmission line power converter and implementations of (b) an on-chip copper transmission line loaded with PZT capacitors and (c) using micromachined capacitors and inductors

4. 10V fully-integrated bidirectional switched capacitor ladder converter

The mobility of autonomous micro-systems is ultimately limited by the availability of a suitable power system. For smaller robots, efficiency and power density considerations have lead to significant interest in piezoelectric actuators for transforming power into locomotion [6]. Driving these actuators, however, requires specialized high voltage converters. Herein we present a brief summary of a fully integrated (zero external components) bidirectional switched-capacitor (SC) converter implemented in 0.13μm triple-well CMOS [36]. The SC converter is capable of actively charging/discharging a 5nF load while following 500 Hz arbitrary waveforms with ~3x step-up conversion (n_{conv}=3) and maximum output voltage (V_{OUT}) of 10V, limited only by well-substrate breakdown.

4.1 High-voltage SC ladder

Shown in Figure 8a is the 3-stage SC ladder architecture along with ideal floating clock drivers (in gray). Transmission gate pull-up switches are employed for complete signal range as V_{IN} varies and all switches are implemented with 3.3V I/O devices. SC ladder drive signals are also shown, using Φ_1 as an example, and are separated into 3 clock voltage domains: the static "low" (L) and the dynamic "medium" (M) and "high" (H) domains. Alternating between V_{IN} or V_{IN}+V_{DD}, the M domain clocks peak at 6.6V for max V_{IN}. Similarly, the H domain signals clock between $2V_{IN}$ or $2V_{IN}$+V_{DD} reaching 9.9V. All capacitors are MIM structures chosen for minimal bottom-plate parasitic capacitance and acceptable density (1fF/μm2 or equivalently 1nF/mm^2) with C1=500pF, C2=C3=250pF. The size of the SC ladder structure was optimized using a charge vector representation [37] while also accounting for dynamic loss of the MOS switches. Practical gate drive signals, which must limit each $|V_{GS}|$ to ~3.3V while following the boosted input signal, are generated using a feedforward approach by bootstrapping the M and H voltage domains to V_{IN} and $2V_{IN}$, respectively. Each switch actuates properly as long as its respective source voltage approaches its unloaded ideal value (V_{IN}, $2V_{IN}$ or $3V_{IN}$) with error much less than the device threshold voltage V_{TH}.

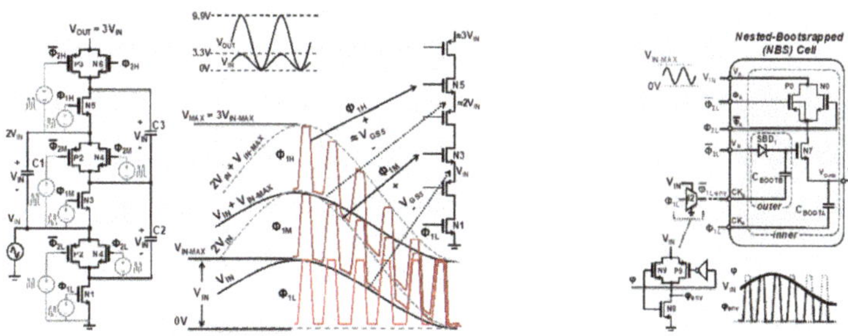

(a) (b)

Fig. 8 Simplified schematic of SC ladder converter: (a) Bidirectional SC ladder converter with gate-drive signals and (b) Nested-bootstrapped switch (NBS) cells and simulated operation.

4.2 NBS cell drivers

A nested-bootstrapped switch is shown in Figure 8(b), comprising outer and inner bootstrapped switch (BTS) subsections. Analog MUX I2 creates a clock signal with a V_{IN} envelope and drives the outer BTS formed by C_{BOOTB} and 10V Schottky barrier diode SBD_1 [38]. The inner and main BTS is formed by a compound stacked switch of N7 and T-gate P0/N0 that biases C_{BOOTA} to very nearly V_{IN}. As it follows Φ_{1L}, N7 is deactivated by making its V_{GS} 0V or a negative value (accumulation), safely leaving $|V_{GS}|\leq 3.3$V but allowing a 6.6V swing. The T-gate serves as a high-impedance to stop C_{BOOTA} from discharging when Φ_1 is logic high and also holds N7's drain at $\sim V_{IN}$ using the bulk-substrate capacitance of P0. Fundamentally, the top-plate voltage of C_{BOOTA} is Φ_{1M}, approaching its ideal value with both rising/falling V_{IN} values and without the voltage drop associated with a diode based implementation. Aggregate error when using multiple NBS cells in series is thus minimized. Most importantly, $|V_{GS}|$ and $|V_{GD}|$ of all devices are limited to 3.3V during "on" and "off" states.

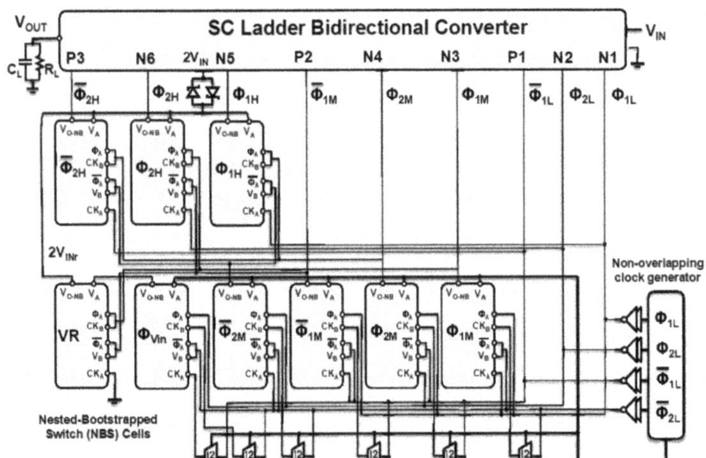

Fig. 9 System block diagram detailing connections of NBS cells and creation of clock signals.

Figure 9 depicts the complete system composed of 9 NBS cells and the SC ladder itself. The 3.3V supply voltage, not shown, powers the non-overlapping clock generator and inverter buffers for the L domain. Aside from the varying V_{IN} connection and Φ_{IN} clock reference, it is the only input. MN3, MN4 and MP2 are driven by M domain cells directly. H domain signals are developed using a combination of those in the M domain and a replica voltage ($2V_{INr}$) that is twice the value of V_{IN}, created separately from the similar valued SC ladder node. Anti-parallel SBDs tie together $2V_{IN}$ and the replica during momentary loading.

4.3 Measurements

Plots of overall efficiency vs. V_{OUT} and P_{OUT} are shown in Figure 10 for resistive loads with various values of f_{SW}. Maximum efficiency is 77% at ~800μW load with 9V output. Near 100μW output levels, however, comparably high efficiency is possible by decreasing f_{SW}. Additionally, at 0.25MHz the converter can drive exceedingly light loads, slightly above 10μW, with 50% efficiency.

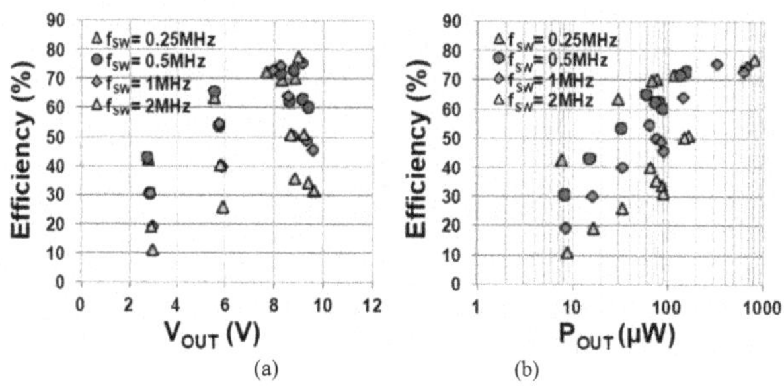

Fig. 10 Measured efficiency vs. (a) V_{OUT} and (b) P_{OUT} for switching frequency f_{SW}=0.25MHz, 0.5MHz, 1MHz and 2MHz [36].

5. Conclusion

A summary of work towards chip-scale power supplies has been presented. Different electrical passives approaches include chip-scale piezoelectric transformers and transmission lines created with air-core inductors monolithically integrated with capacitors. Thin-film piezoelectric transformers fabricated with PZT on silicon were presented and show promising results with ~60% device efficiencies for first generation devices. Such an approach is very attractive since it offers a passives solution with more than two orders of magnitude of reduced size when compared with magnetic approaches. Additionally, work towards passives creating on-chip transmission lines was introduced. Such an approach offers a way to create highly tailored voltage pulses delivered to a load and a means for using the same energy storage element to deliver power to varying loads with reduced switch count. Finally, a fully integrated bidirectional converter implemented using a switched capacitor topology in 0.13 μm CMOS was presented. The converter demonstrates a 1:3 conversion ratio while driving a mm-scale piezoelectrically-actuated wing with nF load capacitance. 50% efficiency is demonstrated driving an extremely light load of 10 μW. A peak efficiency of 77% is demonstrated with ~800 μW load. Future work includes further understanding and generation designs with the different MEMS and microfabricated passives approach presented here. Full-integration of the CMOS converter with a thin-film battery directly integrated onto microfabricated piezoelectric wings is also underway.

References

[1] Waldron, F., Foley, R., Slowey, J., Alderman, A. N., Narveson, B. C. and OMathuna, S. C., "Technology roadmapping for power supply in package (PSiP) and Power Supply on Chip (PwrSoC)," IEEE Trans. on Power Electronics 28 (9), 4137-4145 (2013). https://doi.org/10.1109/TPEL.2012.2227821

[2] Lee, F. C. and Li, Q., "High-frequency integrated point-of-load converters: overview," IEEE Trans. on Power Electronics 28 (9), 4127-4136. https://doi.org/10.1109/TPEL.2013.2238954

[3] Wood, R. J., "The first takeoff of a biologically inspired at-scale robotic insect," IEEE Trans. on Robotics 24 (2), 341-347 (2008). https://doi.org/10.1109/TRO.2008.916997

[4] Bergbreiter, S. and Pister, K., "Design of an autonomous jumping microrobot," Proc. 2007 IEEE Conference on Robotics and Automation, 7–12 (2007). https://doi.org/10.1109/ROBOT.2007.363827

[5] Pulskamp, J., Smith, G., Polcawich, R., Kroninger, C. and Wetzel, E., "Two degree of freedom PZT MEMS actuated flapping wings with integrated force sensing", Proc. of 2010 Solid State Sensor, Actuator and Microsystems Workshop, 390-391 (2010).

[6] Pulskamp, J. S., "Millimeter-scale MEMS enabled autonomous systems: system feasibility and mobility" Proc. Of SPIE Vol. 8873, 83731G-1 – 83731G-10 (2012).

[7] Morgan, B., Bedair, S.S., Pulskamp, J. S., Polcawich, R. G., Meyer, C., Dougherty, C., Line, X., Arnold, D., Bashirullah, R., "Power Considerations for MAST Platforms," Proc. SPIE 7679, 1-12 (2010). https://doi.org/10.1117/12.849515

[8] Meyer, C. D., Bedair, S. S., Morgan, B. C. and Arnold, D., "Micromachined wiring board with integrated microinductor for chip-scale power conversion," IEEE Trans. on Power Electronics, published online (2014).

[9] Mathuna, S. C. O., O'Donnell, T., Wang, N. and K. Rinne, "Magnetics on silicon: an enabling technology for power supply on chip," IEEE Trans. on Power Electronics 20 (3), 585-592, (2005). https://doi.org/10.1109/TPEL.2005.846537

[10] Gardner, D. S., Schrom, G., Paillet, F., Jamieson, B., Karnik, T. and Borkar, S., "Review of on-chip inductor structures with magnetic films," IEEE Trans. on Magnetics 45 (10), 4760-4766, (2009). https://doi.org/10.1109/TMAG.2009.2030590

[11] Meyer, C. D., Bedair, S. S., Morgan, B. C. and Arnold, D. P., "High-inductance-density, air-core, power inductors, and transformers designed for operation at 100-500 MHz," IEEE Trans. on Magnetics 46 (6), 2236-2239, (2010). https://doi.org/10.1109/TMAG.2010.2045742

[12] Bedair, S. S., Pulskamp, J. S., Meyer, C. D., Mirabelli, M., Polcawich, R. G. and Morgan, B., "High-performance micromachined inductors tunable by lead zirconate titanate actuators," IEEE Electron Device Letters 33 (10), 1483-1485 (2012). https://doi.org/10.1109/LED.2012.2207700

[13] Meyer, C. D., Bedair, S. S., Trocchia, S. M., Mirabelli, M. A., Benard, W. L., Ivanov, T. G. andBoteler, L. M., "Heterogeneous chip integration into silicon templates by through-wafer copper electroplating," ECS 2012 Transactions, 163-169, (2012).

[14] Bedair, S. S., Pulskamp, J. S., Polcawich, R. G., Morgan, B., Martin, J. L. and Power, B., "Thin-film piezoelectric-on-silicon resonant transformers," IEEE J. of MEMS 22 (6), 1383-1394 (2013). https://doi.org/10.1109/JMEMS.2013.2262585

[15] Cheng, S. and Arnold, D. P., "Microfabricated electrodynamic transformers for electromechanical power conversion," J. Micromech. Microeng. 23, 114002, 1-10 (2013). https://doi.org/10.1088/0960-1317/23/11/114002

[16] Carazo, A. V., "50 years of piezoelectric transformers. Trends in the technology," *Proc. Materials Res. Soc. Symp.* 785, D1.7.1-D1.7.12 (2004).

[17] Flynn, A. M. and Sanders, S. R., "Fundamental limits on energy transfer and circuit considerations for piezoelectric transformers," IEEE Trans. on Power Electronics 17, 8-14 (2002). https://doi.org/10.1109/63.988662

[18] Ho, G. K., Abdolvand, R., Sivapurapu, A., Humad, S. and Ayazi, F., "Piezoelectric-on-silicon lateral bulk acoustic wave micromechanical resonators," IEEE J. of MEMS 17 (2), 512-520 (2008). https://doi.org/10.1109/JMEMS.2007.906758

[19] Piazza, G., Stephanou, P. L. and Pisano, A. P., "Piezoelectric aluminum nitride vibrating countour-mode MEMS resonators," IEEE J. of MEMS 16 (6), 1406–1418, (2007).

[20] Olsson, R. H., Fleming, J. G., Wojciechowski, K. E., Baker, M. S. and Tuck, M. R., "Post-CMOS compatible aluminum nitride MEMS filters and resonant sensors," in Proc. IEEE Freq. Contr. Symp., 412–419 (2007). https://doi.org/10.1109/freq.2007.4319108

[21] Pulskamp, J. S., Bedair, S. S., Polcawich, R. G., Smith, G. L., Martin, J., Power, B. and Bhave, S. A., "Piezoelectric electrode-shaping for the excitation and detection of arbitrary modes in arbitrary geometries," IEEE Trans. Ultrason. Ferroelectr. Freq. Contr. 59 (5), pp. 1043–1060, May 2012.

[22] Pulskamp, J. S., Bedair, S. S., Polcawich, R. G., Judy, D. and Bhave, S. A., "Ferroelectric PZT RF MEMS resonators," in Proc. IEEE Freq. Contr. Symp., 353–358 (2011). https://doi.org/10.1109/fcs.2011.5977870

[23] Bedair, S. S., Pulskamp, J. S., Polcawich, R. G., Judy, D., Gillon, A., Bhave, S. and Morgan, B., "Low loss micromachined lead zirconate titanate, contour mode resonator with 50 Ω termination," in Proc. IEEE MEMS, 708–712.3 (2012). https://doi.org/10.1109/memsys.2012.6170285

[24] Vasic, D., Sarraute, E., Costa, F., Sangouard, P. and Cattan, E., "Piezoelectric micro-transformer based on SOI structure," Sens. Actuators A 117 (2), 317–324 (2005). https://doi.org/10.1016/j.sna.2004.06.009

[25] Kim, S. K. and Seo, Y. H., "Fabrication and characterization of the piezoelectric microtransformer based on microelectromechanical systems," Appl. Phys. Lett. 88 (26), 263510–263513 (2006). https://doi.org/10.1063/1.2218056

[26] Bedair, S. S., Pulskamp, J., Morgan, B. and Polcawich, R., "Performance model of electrode tailored thin film piezoelectric transformers for high frequency switched mode power supplies," in Proc. Power MEMS, 436–438 (2009).

[27] Imtiaz, A. M. and Khan, F. H., "Film bulk acoustic resonator (FBAR) based power converters: a new trend featuring EMI reduction and high power density," in Proc.

IEEE Applied Power Electronics Conf. & Exp. (APEC), 1485-1491 (2013). https://doi.org/10.1109/apec.2013.6520494

[28] Polcawich, R. G. and Pulskamp, J. S., "Additive processes for piezoelectric materials: Piezoelectric MEMS," in MEMS Materials and Processes Handbook, 1st ed., R. Ghodssi and P. Lin, Ed., Springer, New York, NY, 273–344 (2011). https://doi.org/10.1007/978-0-387-47318-5_5

[29] Carazo, A. V., "Multilayer piezoelectric transformer," U.S. Patent 6 614144, Sept. 2003.

[30] Ivensky, G., Zafrany, I., Ben-Yaakov, S., "Generic operational characteristics of piezoelectric transformers," IEEE Trans. on Power Electronics, 17 (6), 1049-1057 (2002). https://doi.org/10.1109/TPEL.2002.805602

[31] Kartashev, I., Vontz, T. and Florian, H., "Regimes of piezoelectric transformer operation," Meas. Sci. Technol. 17 (8), 2150–2158 (2006). https://doi.org/10.1088/0957-0233/17/8/014

[32] http://www.mmech.com/transformers

[33] Sander, S., "Buck and boost converters with transmission lines," IEEE Trans. on Power Electronics 27(9), 4013-4020 (2012). https://doi.org/10.1109/TPEL.2012.2188044

[34] Rim, G., Pavlov, E. P., Lee, H., Kim, J. S. and Choi, Y. W., "Pulse forming lines for square pulse generators," IEEE Trans. on Plasma Sciences 31(2), 196-200 (2003). https://doi.org/10.1109/TPS.2003.810185

[35] Bar-Chaim, N., Brunstein, M., Grunberg, J. and Seidman, A., "Electric field dependence of the dielectric constant of PZT ferroelectric ceramics," J. of Applied Physics 45(6), 2398-2405 (1974). https://doi.org/10.1063/1.1663605

[36] Dougherty, C.M.; Lin Xue; Pulskamp, J.; Bedair, S.; Polcawich, R.; Morgan, B.; Bashirullah, R., "A 10V fully-integrated bidirectional SC ladder converter in 0.13μm CMOS using nested-bootstrapped switch cells," VLSI Circuits (VLSIC), 2013 Symposium on , vol., no., pp.C172,C173, 12-14 June 2013

[37] Seeman, M.D.; Sanders, S.R., "Analysis and Optimization of Switched-Capacitor DC–DC Converters," Power Electronics, IEEE Transactions on , vol.23, no.2, pp.841,851, March 2008

[38] Li, P.; Xue, L.; Bhatia, D.; Bashirullah, R., "Digitally assisted discontinuous conduction mode 5V/100MHz and 10V/45MHz DC-DC boost converters with integrated Schottky diodes" ISSCC, 2010 pp.206,207, 7-11 Feb. 2010

Chapter 20

Power Management for Small Scale Systems

Christopher D. Meyer*[a], Sarah S. Bedair[a], Brian C. Morgan[a], Xue Lin[b], Rizwan Bashirullah[b], David P. Arnold[b], Iain M. Kierzewski[a], and Nathan S. Lazarus[a]

[a]US Army Research Laboratory, 2800 Powder Mill Road, Adelphi, MD, 20783-1197, USA

[b]Dept. of Electrical and Computer Engineering, University of Florida, P.O. Box 116200, Gainesville, FL 32611-6200, USA

Abstract

Contemporary electronic systems often contain power circuits to support the unique power conversion or conditioning needs of each of the various subsystems. Each of these power circuits is generally implemented with discrete passive and active electronic components soldered next to the load devices on the printed circuit board. As greater levels of functionality are demanded within diminishing size and weight allowances, power management solutions will increasingly demand highly miniaturized power converters that are more tightly integrated into single-package solutions or even directly integrated onto the points of source and load. Experimental converters have demonstrated great potential in switching at very high frequencies (100+ MHz) to reduce the size of the requisite passive storage elements (inductors, transformers, and capacitors) to values that may be suitable for in-package or on-chip integration. However, integrating the passives into the same package as the active switching and control circuitry remains a significant fabrication challenge due to material incompatibility and inadequate performance of the passives.

This paper discusses progress towards a fully integrated power converter module with a focus on microfabrication processes for both passive component development and wafer-level packaging. The passive components have been optimized for high performance at hundreds of MHz through the use of thick copper traces, intricate three-dimensional winding patterns. The capability of detaching the passives from the fabrication wafer produces a passives substrate that can serve directly as a routing platform for full integration of all components into a single-package solution.

The work has been previously published by SPIE:
CD Meyer, SS Bedair, BC Morgan, L Xue, R Bashirullah, DP Arnold, IMKierzewski, NS Lazarus, "(Invited) Power management for small scale systems," in Proc. SPIE 9083, Baltimore, MD, June 2014.
https://doi.org/10.1117/12.2050626

Keywords

Power Management, Microfabrication, Packaging, Inductors, Transformers

Contents

1. Introduction

Embedding power conversion circuitry directly into the individual points of load of multifunctional electronic systems promises not only reduced overall system size but also enables highly granular power management. Researchers across industry, government, and academia are actively developing power system-on-chip and system-in-package solutions to satisfy a variety of needs in handheld-size-and-smaller platforms [1]. In the mobile phone sector, for instance, it is desirable to replace the multitude of low drop out (LDO) linear regulators that provide individualized voltage levels for each subsystem component with higher efficiency switch mode power supplies [2]. Microprocessor manufacturers desire voltage regulator and converter modules that are monolithically integrated directly on the integrated circuit (IC) die in order to accommodate very large and very fast current transients [3] and further could enable fast, per-core dynamic voltage and frequency scaling in many-core processors [4]. For micro-autonomous robotics, power converters are required to mesh the divergent needs of sensors, actuators, microprocessors, radio frequency (RF) devices, and power sources, which can span several orders of magnitude in both voltage and power levels, while minimizing system weight to meet mobility requirements [5].

However, the components of a typical power circuit—particularly the passives (inductors, transformers, and capacitors)—have tended to be quite bulky and have proven challenging to integrate. Increasing the switching frequency of the converters up from the typical hundreds of kilohertz and into the tens to hundreds of megahertz has been shown to be a promising route to reducing the size of passives by reducing the required values of inductance and capacitance [6]. Very high frequency switching also provides the faster transient response that is desired for highly granular power management. Correspondingly, new microfabrication methods have been developed in order to produce reduced-size passives that maintain high power and energy densities [7], [8], [9]. Beyond component development, integration challenges also need to be addressed, such as temperature and chemical incompatibilities and cost. A variety of integration options have been explored with varied degrees of integration.

Our group is developing a wafer-level packaging technology for embedding bare die and surface-mount parts within a wiring board that includes built-in, high-density magnetic passives to realize a system-in-package solution with fully integrated power management. The technology is based on a multilayer metallization process that is capable of fine feature resolution (down to ~5 μm) in thick copper films (up to 30 μm per layer with up to 4 layers), which makes it ideally suited to producing air-core microinductors and microtransformers with high inductance densities, low resistances, and high quality factors—particularly at the hundreds of megahertz frequencies that are desirable for chip-scale power converters. From a packaging perspective, the key advancement is the detachment of the thick film copper wiring from the fabrication substrate to create a standalone wiring board, such as shown in Figure 1, which contains high performance air-core microinductors and microtransformers. This wiring board can be combined with the power controller IC and the load devices in different ways with varying degrees of integration. Some options are illustrated in Figure 2. A step up in integration level from the non-integrated case is mounting the power IC on top of the passives wiring board to form a single-chip power module, such as the one pictured in Figure 3 and presented in a previous publication [10]. If the load takes the form of a relatively compact surface-mount chip, it too could be mounted on the wiring board, so that the wiring board would act as a carrier for an integrated, power-managed subsystem. Alternatively, the wiring board can be used as an interposer layer, providing the power passives within a chip stack.

Fig. 1 Detached thick film copper wiring board with integrated microinductors and a microtransformer.

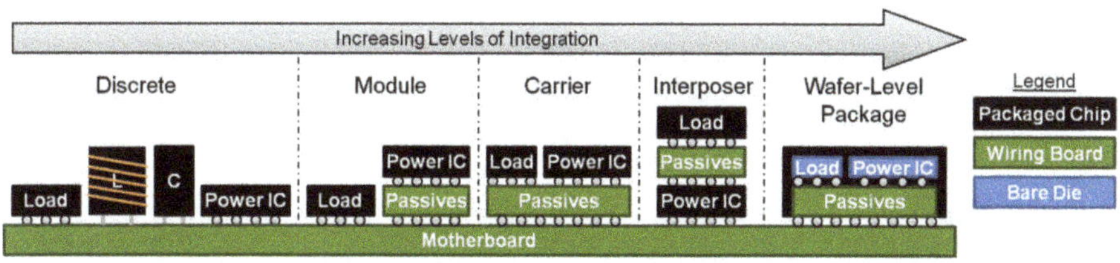

Fig. 2 Options for utilizing wafer-detached wiring board to integrate passive components with power IC.

Fig. 3 Thick-film copper wiring board used to realize integrated power converter module [10].

Finally, the concept that is presented in this work is the wafer-level package, shown in Figure 4. Bare dies and accessory surface-mount components are bonded to the thick-film multilayer copper wiring on a silicon wafer. The whole structure is then covered with a protective, structural encapsulant. Finally, the underlying silicon fabrication wafer is removed to produce a standalone chip that has exposed copper landing pads on the bottom of the chip in the style of a classical no-lead surface-mount package. This paper will outline our group's progress in fabricating these wafer-level packaged power modules, discuss considerations and options for component bonding, and provide preliminary characterization of the microfabricated passives. The results are intended to be a demonstration of the fabrication and packaging technology, although testing of functional devices remains as future work.

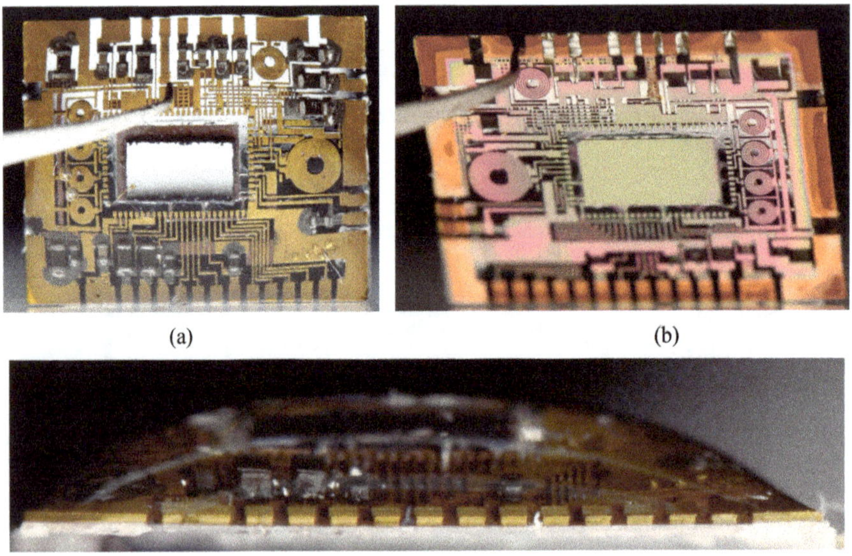

(a) (b)

(c)

Fig. 4 Photographs of wafer level package with built-in microinductors and microtransformer and embedded, soldered surface-mount capacitors and a wire bonded bare silicon die (non-functional demonstration).

2. Fabrication process

2.1 Metallization

The integrated power modules are realized through a wafer-level microfabrication process, an outline of which is illustrated in Figure 5. The process starts with plasma enhanced chemical vapor deposition of a silicon dioxide dielectric layer over the surface of a blank silicon handle wafer. This oxide is intended to serve as a passivation layer in the final module chip after removal of the silicon handle wafer to electrically and environmentally insulate regions where no external connection in desired. The oxide is visible as the areas of pink or green on the underside of the chip in Figure 4(b). Reactive ion etching (RIE) is used to create openings in the oxide that will expose the copper pads on the underside of the chip as needed for soldering.

The copper traces are then deposited through a previously described [10] [11] wafer-level, thick-film, multilayer metallization process. Thick traces of copper are electroplated layer-by-layer through patterned, 30-μm-thick photoresist molds. A thin film of sputtered copper is required underneath the photoresist at the start of each layer to provide a conductive surface on which the much thicker film of copper can be electroplated.

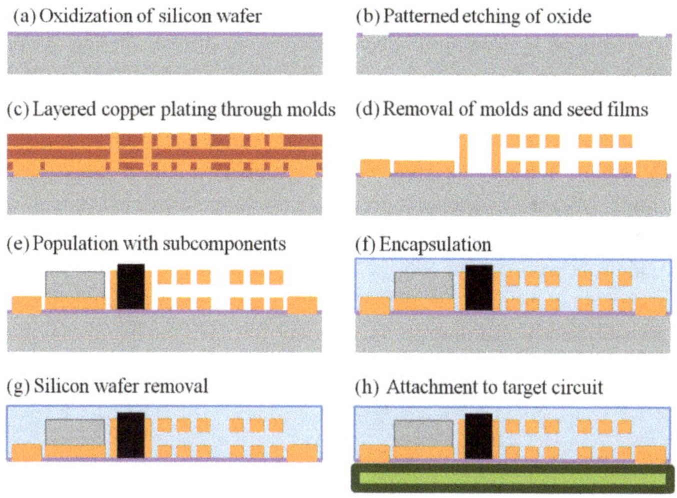

Fig. 5 Outline of fabricaton process for realizing wafer-level packaged power module chips.

The uppermost copper surface may be capped with alternative metal films (e.g. nickel and gold) to assist in later flip-chip or wire bonding as discussed in Section 2.2. Two suitable options for depositing the capping metals are electroplating and evaporation. When electroplating, the upper copper trace is electroplated short of the top of the surrounding photoresist mold, so that subsequent electroplating of the capping metals will then fill to the top of the mold. The wafer is first rinsed in diluted hydrochloric acid to brighten the electroplated copper. A diffusion barrier of nickel is electroplated in a nickel sulfamate bath. The wafer is rinsed again in dilute hydrochloric acid, and the final gold coating is electroplated in a gold sulfite bath. When evaporating the capping film, the uppermost copper layer is again plated a couple microns short of the top of the surrounding photoresist mold to provide a steep discontinuity. Titanium and then gold are deposited by electron beam evaporation. Because of the directionality of the deposition, the subsequently evaporated film will have poor step coverage between the electroplated copper trace and the surrounding photoresist mold. Removing the photoresist mold will then simultaneously remove the evaporated capping metal in between traces while leaving the film deposited on top of the copper.

The photoresist molds and copper seed films are left in place until all layers have been deposited, at which point the molds and seeds are removed by ultrasonic agitation in photoresist stripper with mild heating. Any residual copper seed film on the silicon is subsequently removed by etching in a ceric ammonium nitrate solution. At this stage in the fabrication process the resulting copper structures are freestanding on the silicon wafer, as illustrated in Figure 5(d) and demonstrated in the scanning electron micrograph (SEM) image in Figure 6(a). It should be noted that earlier iterations of this process [12], in which the copper traces were 10 μm thick, required structural supports between the upper and lower traces. The supports prevented the long suspended runs of traces from snapping down when drying the devices after removal of the photoresist moldings in the stripper solution, a problem that can be seen in Figure 6(b). The 30 μm thick traces, however, require no interlayer structural support for the devices presented in this work, as the traces are sufficiently stiff to resist the capillary forces of the liquid as it dries [11].

Fig. 6 SEM images of transformer with (a) 30 um thick copper and (b) 10 um thick copper traces. The upper layer copper traces in (b) can be seen to have snapped down in contact and shorting with the lower layer trace.

2.2 Surface-mount and bare die subcomponent integration

Surface-mount and bare die components can be integrated with the freestanding copper windings at the wafer level by pick-and-place, flip-chip, or wire bonding. Here we report on manual placement of surface-mount capacitors and wire bonding of a silicon die.

Press-fit sockets formed in the copper wiring assist with placement of the smallest surface-mount components. These sockets feature teeth in the uppermost copper layer that deform plastically when a part is inserted to lock the part in place while accommodating variations in part size. The sockets were tested with a 01005 (imperial)-sized 47 Ω resistor and a 47 pF capacitor, as depicted in Figure 7, having nominal footprints of 400 μm × 200 μm. The sockets were sized 420 μm × 220 μm and the teeth were each sized 10 μm wide × 20 μm long. As shown in Figure 7(b), the sockets enabled 01005-sized chips to be placed manually using tweezers while achieving a high degree of placement accuracy. The images in Figure 8 depict the deformation of the teeth and show that the teeth may deflect either vertically or horizontally. While the teeth are intended primarily to hold the chip in place during subsequent soldering, a resistance of 46.9 Ω was measured by four-point probing between the opposing ends of the copper socket containing the 47 Ω resistor using only the teeth to make electrical contact to the resistor.

(a) (b)

Fig. 7 SEM images of a surface-mount resistor (left) and capacitor (right) inserted into toothed sockets formed in multilayer copper wiring, as viewed (a) off-axis and (b) from above.

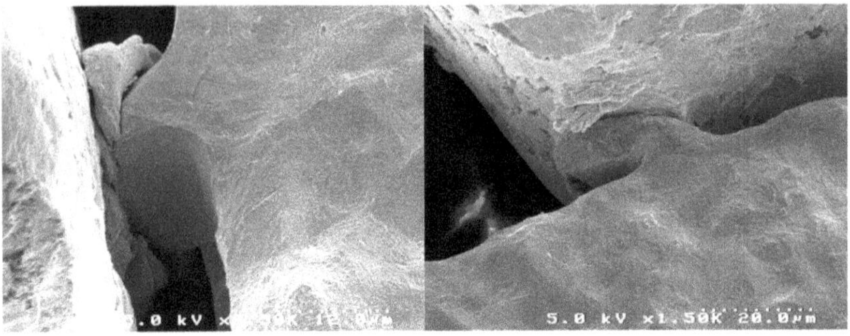

Fig. 8 SEM images showing deformation of teeth in socket following insertion of surface mount part.

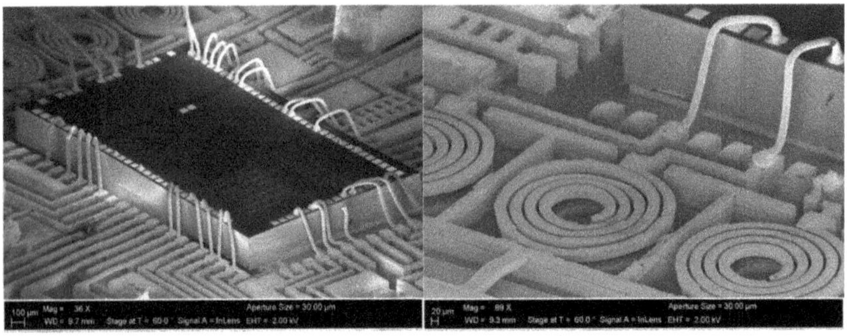

Fig. 9 SEM of CMOS die with gold wire bonds to evaporated-gold capped copper wiring.

Because of the difficulty in gold wire bonding to copper, both electroplated and evaporated capping layers of gold were tested for providing a top surface that would be better suited for gold wire bonding. The processes of depositing each are described in Section 2.1. In testing, the electroplated gold cap was not found to provide any benefit for wire bonding. In contrast, the evaporated gold film was found to greatly improve the bondability of the surface.

2.3 Encapsulation and wafer substrate removal

After all of the subcomponents are attached and bonded to the copper wiring, the chip is then encapsulated to structurally secure all of the parts. The encapsulant should have the properties of low viscosity to fill between the narrow features, good adhesion to metals and glass, dimensional stability during curing, and chemical compatibility with silicon removal chemicals. In this report, the ultraviolet-light curing Norland optical adhesive NOA 88 was used as the encapsulant.

Removal of the silicon fabrication wafer was achieved in xenon difluoride vapor, which isotropically etches silicon with no discernible effect on silicon dioxide, copper, or the encapsulant. The encapsulated silicon chip was also not visibly affected by the vapor, though further testing will need to ensure that no electrical change has occurred. In commercial application, it may be preferable to achieve faster removal by wet chemical etching or by lifting off with an intermediate layer between the chip and the wafer. These alternate methods were not tested and would need to be checked for chemical compatibility against silicon dioxide and the encapsulating compound.

3. Microinductor and transformer characterization

The microinductors and microtransformer have been preliminarily characterized by separate fabrication on a borosilicate glass substrate. The glass substrate enables the devices to be characterized while affixed to a wafer for ease of connection while eliminating the impact of the presence of a semiconducting silicon substrate. Connection was made using RF probes in ground-signal-ground configuration. Scattering parameters were measured using a Rohde & Schwarz ZVA8 vector network analyzer and converted to impedance parameters. The fabricated module included two inductor designs—one 45 nH inductor with an outer diameter of 714 μm and four 18 nH inductors each with an outer diameter of 555 μm. The chip also included one microfabricated 45 nH : 45 nH transformer with an outer diameter of 1171 μm. The trace widths and inter-trace spacings are 30 μm and 15 μm, respectively, for the inductors, and 32 μm and 16 μm for the transformer.

The plots in Figure 10 of the measured inductance, resistance, and quality factor for the 45 nH inductor typify the overall frequency-dependent impedance characteristics of air core microinductors. The inductance is relatively constant over a range of frequencies from dc before increasing near the self-resonant frequency. This self resonance occurs primarily due to the interlayer and interwinding capacitance present in the inductor. The inductance falls sharply at self-resonance, where the reactive part of impedance is zero by definition. Above this frequency, the capacitance dominates and the impedance appears capacitive (negative reactance). The resistance is also flat at lower frequencies, though rises at frequencies well below self-resonance due to skin and proximity effects [13]. The sharp rise in resistance near the self-resonant frequency is a result of the combination of inductance and capacitance shifting most of the impedance into the real part. The quality factor generally increases with frequency up to the self-resonant frequency. Since the quality factor is interpreted in this work as the ratio of reactive to real part of impedance, the quality factor decreases to zero at the self-resonant frequency as the reactive part is by definition zero.

The 45 nH inductor measured a low frequency resistance of 0.32 Ω, peak quality factor of ~12 over the range of 130–300 MHz, and a self-resonant frequency of 787 MHz. The 18 nH inductor measured a low frequency resistance of 0.19 Ω, peak quality factor of ~21 over the range of 1.26–2.24 GHz, and a self-resonant frequency of 4.43 GHz.

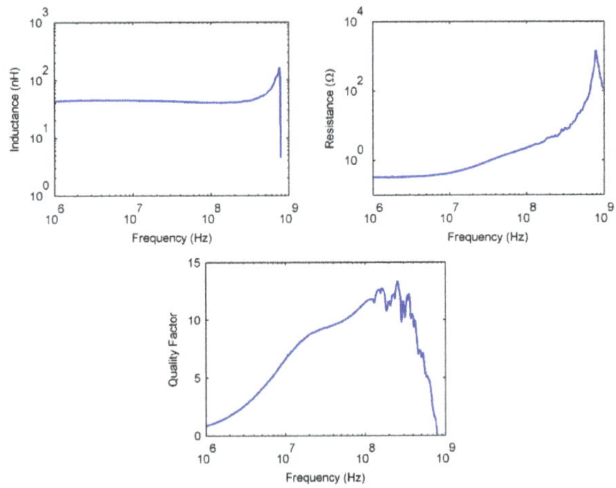

Fig. 10 Measured impedance of microfabricated inductor to be used in packaged power converter module.

The transformer consists of two 45 nH coils that are interleaved for mutual magnetic coupling between the coils. The coils each measured a low frequency resistance of 0.36 Ω, peak quality factor of ~18 over the range of 390–570 MHz, and self-resonant frequency 1.13 GHz. The absolute performance of these individual coils is better than the similar-valued inductors (higher quality factor and self-resonant frequency) due primarily to the larger diameter of the transformer coils. From the measured two-port impedance parameters, the mutual inductance was determined to be 43 nH, equating to a coupling coefficient k of 0.95.

The performance of encapsulated devices is expected to deviate somewhat from these results. Particularly, the presence of the dielectric encapsulant between coil traces will introduce a significant additional shunt capacitance. The added capacitance will have the effect of reducing the self-resonant frequency of the devices.

4. Conclusions and future work

This paper has provided an overview of progress towards a wafer-level packaged power module with integrated microinductors and microtransformer as well as surface mount capacitors. Microfabrication processes were used to deposit self-supported, multilayer, thick-film copper structures that were subsequently detached from the silicon fabrication wafer. These thick-film copper structures were used to form high performance integrated magnetic passives as well as all necessary interconnects for a fully integrated power converter. Sockets with deformable teeth were shown to ease the placement and connection of the smallest surface-mount components. The next steps in the development of this technology include testing the module with a functional power converter control IC and testing the attachment of the module to a PCB for practical use.

Acknowledgements

Wire bonding services were provided by Quik-Pak (a division of Delphon), San Diego, CA.

References

[1] Liu, Y.-F. and O'Mathuna, C., "Editorial: Special Issue on Power Supply on Chip, 2013," IEEE Trans. Power Electron. 28 (9), 4125–4126 (2013). https://doi.org/10.1109/TPEL.2013.2242713

[2] Hayashi, Z., Katayama, Y., Edo, M., and Nishio, H., "High-Efficiency DC–DC Converter Chip Size Module With Integrated Soft Ferrite," IEEE Trans. Magn. 39 (5), 3068–3072 (2003). https://doi.org/10.1109/TMAG.2003.815895

[3] Schrom, G., et al., "Feasibility of monolithic and 3D-stacked DC-DC converters for Microprocessors in 90nm Technology Generation," Proc. ISLPED, 263-268 (2004). https://doi.org/10.1145/1013235.1013302

[4] Kim, W., Gupta, M. S., Wei, G.-Y., and Brooks, D., "System Level Analysis of Fast, Per-Core DVFS using On-Chip Switching Regulators," Proc. Intl. Symp. On High Performance Computer Architecture, 123-134 (2008).

[5] Morgan, B., et al., "Power Considerations for MAST Platforms," Proc. SPIE 7679, 76790V (2010). https://doi.org/10.1117/12.849515

[6] Xue, L., et al., "Towards Miniature Step-Up Power Converters for Mobile Microsystems," Proc. Applied Power Electronics Conf. and Expo., 1451–1455 (2011). https://doi.org/10.1109/apec.2011.5744783

[7] Tseng, V. F. G. and Xie, H, "Design and fabrication of a high-density multilayer metal–insulator–metal capacitor based on selective etching," J. Micromech. Microeng. 23 (3), 035025 (2013). https://doi.org/10.1088/0960-1317/23/3/035025

[8] Kim, J., et al., "Microfabrication of air core power inductors with metal-encapsulated polymer vias," J. Micromech. Microeng. 23(3), 035006 (2013). https://doi.org/10.1088/0960-1317/23/3/035006

[9] Moazenzadeh, A., et al., "Wire bonded 3D coils render air core microtransformers competitive," J. Micromech. Microeng. 23(11), 114020 (2013). https://doi.org/10.1088/0960-1317/23/11/114020

[10] Meyer, C. D., Bedair, S. S., Morgan, B. C., and Arnold, D. P., "Micromachined Wiring Board with Integrated Microinductor for Chip-Scale Power Conversion," IEEE Trans. Power Electron., In Press (2014). https://doi.org/10.1109/TPEL.2013.2296507

[11] Meyer, C. D., Bedair, S. S., Morgan, B. C., and Arnold, D. P., "Influence of Layer Thickness on the Performance of Stacked Thick-Film Copper Air-Core Power Inductors," IEEE Trans. Magn. 48 (11), 4436–4439 (2012). https://doi.org/10.1109/TMAG.2012.2197185

[12] Meyer, C. D., Bedair, S. S., Morgan, B. C., and Arnold, D. P., "High-Inductance-Density, Air-Core, Power Inductors and Transformers Designed for Operation at 100–500 MHz," IEEE Trans. Magn. 46 (6), 2236–2239 (2010). https://doi.org/10.1109/TMAG.2010.2045742

[13] Wheeler, H. A., "Formulas for the Skin Effect," Proc. Of the I.R.E. 30 (9), 412–424 (1942).

Chapter 21

Power Considerations for MAST Platforms

Brian Morgan*[a], Sarah Bedair[a], Jeffrey S. Pulskamp[a], Ronald G. Polcawich[a], Christopher Meyer[a,b], Christopher Dougherty[b], Xue Lin[b], David Arnold[b], Rizwan Bashirullah[b], Ryan Miller[c], Mark Roosz[c]

[a]Army Research Laboratory, 2800 Powder Mill Rd. Adelphi, MD 20783 USA;

[b]Dept. of Electrical & Computer Engineering, Univ. of Florida, Gainesville, FL 32611 USA;

[c]Air Force Research Laboratory, Wright-Patterson AFB, OH USA

*brian.morgan6@arl.army.mil

Abstract

Scaling down autonomous robotic systems introduces numerous challenges in mechanical design, electrical/sensor subsystems, and autonomous control. One particularly daunting task is the design of the power system, since this will ultimately limit all microrobot or micro-UAV's operations. Power sources like lithium polymer batteries possess sufficient power density for basic mobility (walking, fixed wing flight, flapping/hovering), but improved power sources are needed that offer increased energy density in order to extend mission lifetimes – preferably pushing from minutes to multiple hours or days. Additionally, the source power must be efficiently converted and distributed to the various microrobot subsystems. Each system may require a different voltage, current, and duty cycle. This paper will review some of the power-specific challenges related to developing small, mobile autonomous systems.

Keywords

Autonomous Systems, UAV, Mobility, Power Sources, Power Conversion, MEMS Passives

Contents

The work has been previously published by SPIE:
B. Morgan, S.S. Bedair, J. S. Pulskamp, R. G. Polcawich, C. Meyer, C. Dougherty, Xue Lin, D. Arnold, R. Bashirullah, R. M., M. Roosz, "Power Considerations for MAST Platforms," SPIE Defense, Security, & Sensing 2010 Workshop, Orlando, FL, April 5-9, 2010 (invited). https://doi.org/10.1117/12.849515

1. Introduction

In recent years, unmanned aerial vehicles (UAV) have been critical in providing real time intelligence, surveillance and reconnaissance (ISR) to Warfighters on the ground. As the demand for such assets increases, the prospect of supplying small units with the capability to autonomously provide their own ISR in real time becomes highly appealing. Existing UAV, however, can be bulky and difficult to transport for dismounted troops, particularly when considering that supporting equipment may dwarf the UAV – for example, the 430g WASP (from Aerovironment) requires a >5kg ground control station [1]. In addition, mission capabilities for current large UAV are hampered by the need to maintain line of sight during their manual operation, and the size / speed of such fixed wing vehicles restricts access to numerous locations of interest, such as alleyways or inside buildings. Thus, a pressing need exists to develop smaller systems that alleviate the logistics burden of operating large systems while also providing access to this new range of environments that previously remained unreachable with unmanned systems.

The U.S. Army Research Laboratory's Micro Autonomous Systems & Technology (MAST) program is focused on developing autonomous mobile microsystems to enhance tactical awareness in urban and complex terrains. These systems are projected to be palm size or smaller to provide a combination of stealth and accessibility to restricted areas, and whose autonomous operation will enable missions beyond line of sight from a controller. The resulting systems will not only provide better situational awareness for the warfighter, but will simultaneously reduce the cost and logistics footprint necessary to support larger systems for the same missions. However, scaling mobile systems to the desired level introduces numerous challenges due to the size / weight constraints imposed. While there is no lack of imagination for how these devices could be used or features that could be integrated into such a platform, the ultimate capability of any such system is severely constrained by the available power systems, of which there are few at this scale.

This paper will address the critical issue of power for MAST platforms by discussing both general and specific power requirements for mobile microsystems and potential power sources for small autonomous systems. We will then review internal efforts to develop novel power conversion systems at the cubic-millimeter, milligram scale that addresses the significant voltage & power gap that often exists between the anticipated power sources and loads.

2. Mobility power requirements

Since the emphasis of MAST is the development of *mobile* microsystems, one must start by identifying fundamental requirements for power/weight ratio's that should be targeted for small-scale systems. This depends heavily on the locomotion modality being used. For example, Autumn et al. report that "mobile crawling" (i.e. like a gecko) requires a system operating up to 10 W/kg [2] – a relatively low power density as far as power sources are concerned. However, hovering flapping flight can require fully an order of magnitude more at ~100 W/kg [3-5]. Fixed wing flight lies between those two extremes, and depends on the vehicle size and speed, as shown in Figure 1. Hopping also has some interest for future locomotion [6], but it remains a relatively immature technology, so is not discussed here.

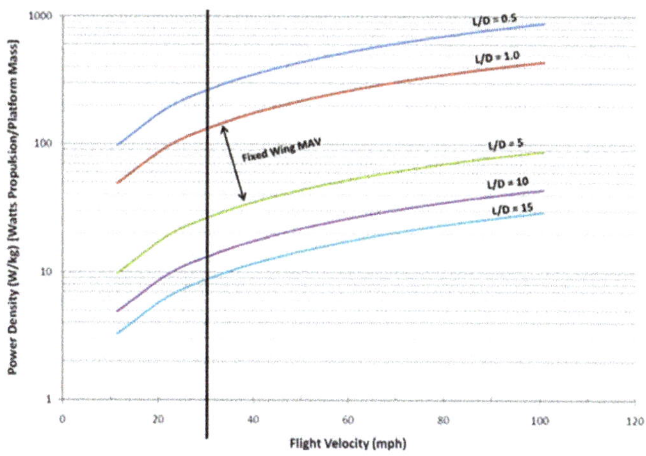

Fig. 1 Required system power density for fixed wing flight given different Lift to Drag (L/D) ratios.

The power densities quoted above are for the entire system, however the power source is only a fraction (often 25-50%) of the total system weight, so a power source must perform at 2-4X higher power density than the mechanical power required for locomotion. Furthermore, inefficient delivery of said power into mechanical motion through mechanical transmissions (or electrical converters), a particular challenge at reduced scales, requires yet another ~2X increase in system power density [5]. Figure 2 shows graphically the envisioned power density relationship for "real" power sources as a function of locomotion modality and fundamental system mobility requirements (precise numbers vary widely depending on system and application).

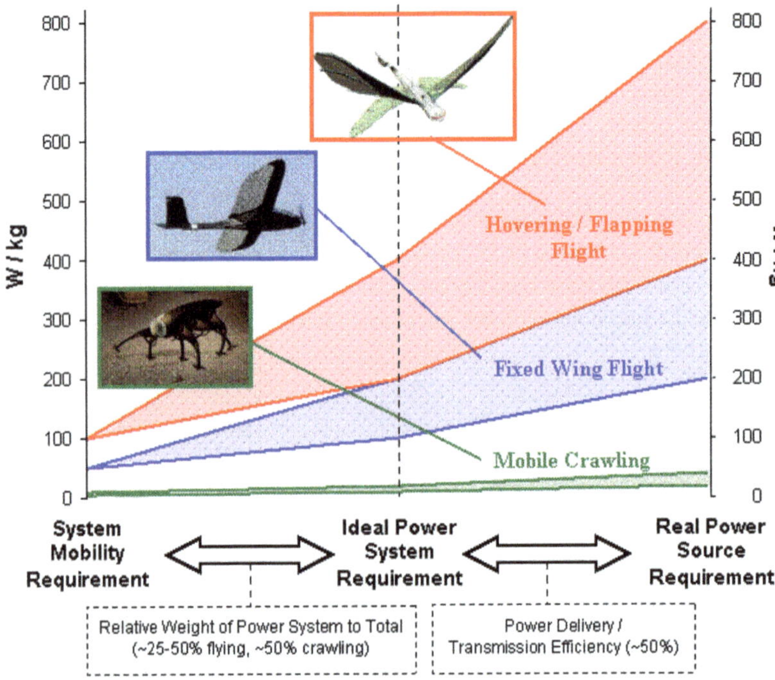

Fig. 2 Specific Power (Avg) required for Mobile Microsystems (<10W).

Power densities in this range (<1000 W/kg) are easy to achieve using commercially available lithium ion batteries, which have been used to power numerous UAV demonstrations in the few hundred gram range. However, when operated at such high power densities, the energy density of Lithium ion or lithium polymer batteries (and other technologies) typically decreases [7], resulting in short mission times. Furthermore, as systems scale down, the specific energy and specific power of a particular power technology tends to decrease as packaging and balance of plant penalties become significant. In Figure 3, we have plotted the expected endurance (locomotion only) as a function of specific energy density for power sources operating at the required power density. One must note that this plot of anticipated locomotion "endurance" should not be confused with "mission time" which may include periods of dormancy or low power stationary sensing.

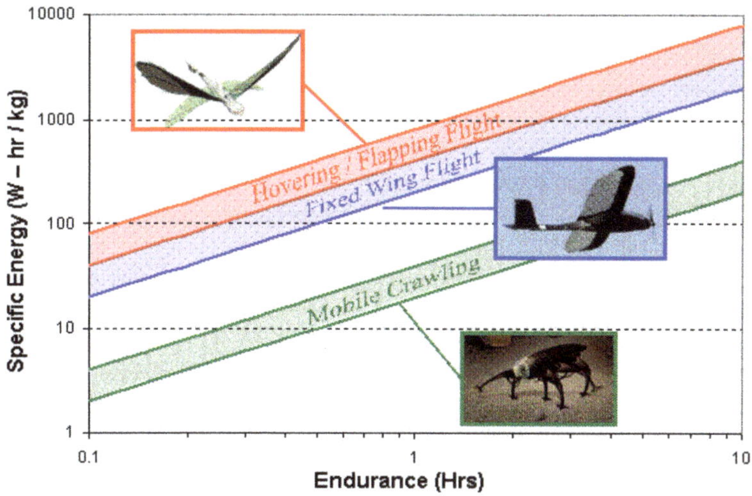

Fig. 3 Specific energy required for mobile Microsystems locomotion.

Some applications will require the system to locomote continuously (while presumably sensing or mapping its surroundings). In that case, Figure 3 is a good starting point for estimating mission lengths for different power sources (usually batteries), under the assumption that power for mobility dominates over power for sensing, navigation, etc. The opposite extreme would be a mission where a sensing mote could be 'dropped' somewhere to "perch-and-stare" indefinitely. In such a case, a low average power is required over a long time, making an energy dense source like a fuel cell or an environmental energy harvester attractive. However, the vast majority of operations will contain periods of both locomotion and "perch-and-stare," significantly complicating the power landscape.

The anticipated duty cycle for locomotion/perching is therefore a good starting point to down select which technologies may be appropriate for a MAST platform, with the understanding that the net power system needs to meet a basic power density threshold to enable mobility. One must also note that although these power constraints may limit the functionality or intelligence of an individual small autonomous system, higher fidelity information should be attainable through the use of group behavior.

3. Small (<50 gram) power sources

The approximate ranges being targeted for average power and energy density have been overlaid with the Ragone plot in Figure 4 for the three locomotion modalities identified. As discussed earlier, the inherent performance (and simplicity) of lithium polymer batteries has made them particularly attractive as prototyping power sources for platforms at the MAST-scale, yet their energy density remains modest. A nice comparison of sub-5g lithium-polymer batteries was performed by the Naval Research Lab (in collaboration with DARPA), which found that a 4.7g Lithium-Polymer battery was capable of outputting up to 130 Whr/kg at 7.7W (>1600 W/kg), versus 160 Whr/kg at low discharge rates [7]. For a 0.8g battery, the energy density dropped even further to <90 Whr/kg when operated at similar power densities. As an alternative in the sub-gram regime, thin film batteries [8] offer tremendous specific power and specific energy. Their solid-state construction enables minimal packaging, high cycle life, and excellent self discharge characteristics – all potentially important characteristics for a MAST system. However, the absolute battery size (only 10's of μm thick due to the vacuum deposition process) limits the amount of stored energy, limiting their direct applicability to only the smallest of MAST systems. While incremental improvements in LiPo technology are being pursued (new electrodes, electrolytes, etc), no large leaps in rechargeable battery technology appear to be imminent. With that in mind, below we discuss some alternative small power sources with applicability to some MAST platforms and scenarios.

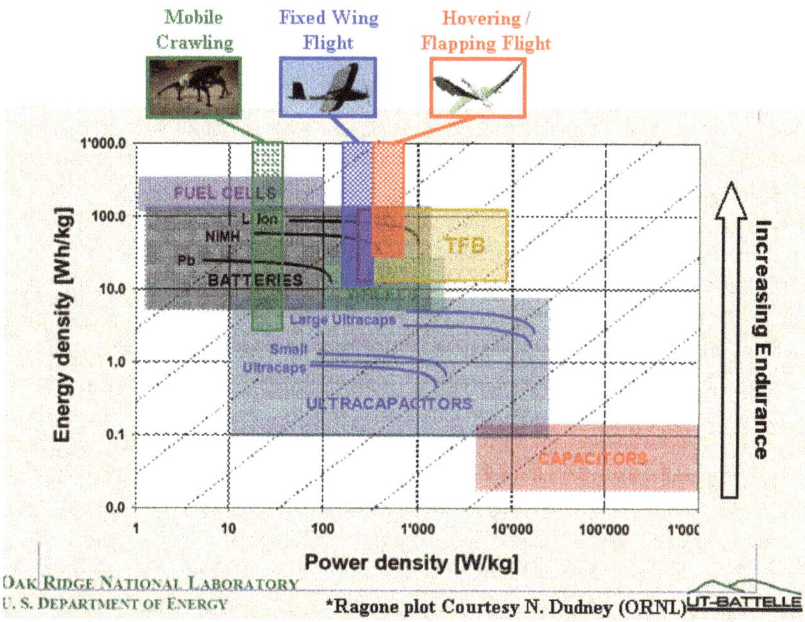

Fig. 4 Average Power & Energy required for Mobile Microsystems (<10W). Note: requirements are for continuous (average) power consumption. Short periods of high Power-on-Demand are likely necessary to perform functions like burst communications.

The most straight forward approach to achieving longer locomotion times is the use of a primary (non-rechargeable) battery. The Lithium Carbon Monoflouride (LiCFx) battery chemistry has been in used for many years as a high energy, low power solution and are commonly found in devices such as pacemakers, drug pumps and various other medical equipment. LiCFx has demonstrated energy densities > 450 Whr/kg [9] at low discharge rates however suffers from reduced energy density and thermal management issues when subjected to the moderate power draws necessary for sustained flight on current MAV's. While additional work on the cell chemistry is needed to increase the power density of this battery, the thermal issues can be mitigated through intelligent placement of a MAV power system in a well ventilated area such as the surface of a wing or the body of the MAV. Depending on the severity of the penalty to energy density incurred by producing a more power-dense battery cell, a hybrid system with a secondary power-dense source maybe the preferred approach. Such hybrid approaches require significant analysis of engineering tradeoffs, particularly at these reduced scales. For example, even gluing 2 types of batteries (a high power LiPo and a high energy LiCFx) together requires additional power management electronics that reduce efficiency and add weight.

Another popular option is fuel cell/battery hybrid – where the high energy density of a fuel cell is augmented during take-off or 'dash' operations by the high power battery. This approach has proven quite successful at the man-portable scale (few kg's), where fuel cell-powered UAV flights conducted by both the Air Force and Naval Research Laboratories have demonstrated system energy densities in excess of 600 Whr/kg [10] resulting in flight time increases of 3-4X versus battery-only solutions [11,12]. Fueled systems also benefit from the fact that they become progressively lighter as the mission progresses and fuel is consumed (whereas a dead battery weighs the same as a full one). However, a fuel cell approach is extremely challenging for MAV power systems, since they rely heavily on a sufficient amount of high energy dense fuel to make up for the dead weight associated with the fuel cell stack and balance of plant (BOP) which consists of fuel pumps, air compressors, electronics, etc. Due to the extremely small nature of a MAV power system and the severe non-linearities involved with scaling such a system, the BOP often dominates the overall system weight, driving down energy densities levels to below that of a conventional lithium battery. Fuel cells may also be sensitive to the surrounding environment, requiring fresh air, certain humidity levels, etc, further limiting the MAST-type missions in which they could be used. A significant amount of research has been conducted on passive BOP systems to reduce system weight by relying on methods such as wicking and capillary action to replace conventional pumps [13]. Fuel cells with a passive BOP have inherently low power output due to the slow rate at which fuel is delivered and are not currently able to provide sufficient power for sustained flight. Passive fuel cells do present an interesting prospect for trickle charging a secondary power source which is capable of meeting high peak power requirements such as a lithium battery or capacitor.

Energy harvesters are attractive for some scenarios, but for realistic scenarios, the payoff for energy harvesting is only apparent in extended missions lasting for several days. Harvesters also typically need some intelligence to couple efficiently to the surrounding (e.g solar tracking), and one must weigh the fact that when not in use, an energy harvester is dead weight, which increases the overall power/energy required for locomotion. Solar cells are the widest known harvesters, and offer efficiencies of ~5% for organic versions [14] and demonstrated efficiencies near 30% for more complicated multi-junction cells (with hopes for 50% [15]). When in direct sunlight, 30% efficiency can equate to 30 mW/cm^2, but that power is reduced by ~100X or more when a system is moved indoors. Furthermore, sunlight is is only available for ~8 hours per day. Alternative harvesters like vibration or thermoelectric devices can operate in dark locations and throughout the entire day. Venkatasubramanian et al reported >1 mW/cm^2 for a 2 C temperature gradient using Bi_2Te_3 superlattice structures [16]. Creating this temperature gradient requires extra design for thermal management that may add weight, but it could potentially be mitigated by using the entire MAST vehicle as a heat sink. Vibration harvesting is extremely tricky to evaluate due to the wide range of vibration amplitudes and frequencies found in different environments. For example, the model developed by Roundy [17] predicts anywhere from 0.5 mW/cm^3 to 100 mW/cm^3 depending on the vibration amplitude (1-10 m/s^2) and frequency (50-350 Hz). The power drops off dramatically when the resonant frequency of the converter is poorly matched to the vibration source, so when considering use of a vibration harvester in a MAST application, one would prefer to know the vibration spectrum of the target environment *a priori* to maximize harvested power. Radio frequency (RF) harvesting requires a specifically tuned antenna to capture background electromagnetic radiation but can operate regardless of time of day and weather. A drawback of RF harvesting however is that it requires antennas to be tuned to a small frequency range prior to the mission. The power output varies throughout the mission depending on the MAV's proximity to and strength of an RF source such as a radio/tv tower, cell phone tower or even a wireless computer router. Companies such as RCA [18] and Intel [19] have demonstrated the feasibility of RF harvesting in urban environments, but power draws were in the milliwatt and microwatt ranges, respectively. Since RF harvesting is a relatively low weight energy harvesting method, requiring only an antenna and electronics, it is feasible that this power solution would be hybridized with a rechargeable source to provide trickle charging to supplement other power sources for extended duration missions. Depending on the area of operation, it may also be feasible to recharge at high rates by relocating the MAV closer to the RF transmitter for increased performance.

4. MAST sub-system needs

As discussed earlier, the MAST program is not specifically aimed at one platform, rather it is developing numerous capabilities and sub-systems that could later be integrated in some combination to fulfill operational needs. However, the breadth of research being conducted quickly illustrates some of the challenges regarding development of a power sub-system to meet these needs.

Figure 5 shows the results of an informal survey (conducted by ARL) of the MAST consortium members concerning the anticipated power and voltage needs of each particular sub-system. The sub-systems are categorized into sensors / processing, actuators / locomotion, and sources. In general, the high voltages for locomotion are required to run piezoelectric [20] or dielectric elastomer actuators (>100 V/μm [21]) envisioned as necessary for mobility towards the few grams & below range, while lower voltage / higher current locomotion mechanisms (like DC motors) are only applicable to larger systems (approx 10g and up). For sensing, the voltages required tend to stay much closer to typical battery voltages (within 2-3X of a 3.7V LiPo battery), but also dip as low as 0.2V for very efficient processing approaches.

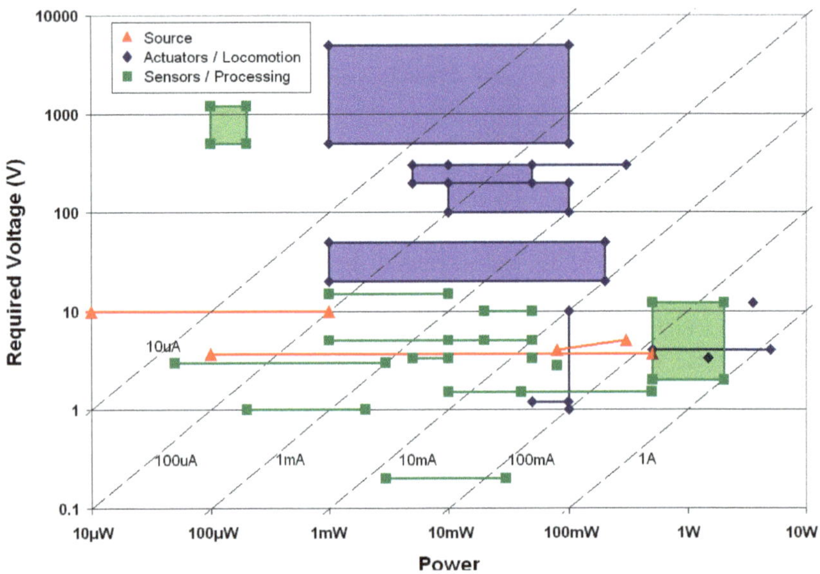

Fig. 5 Power & voltage requirements for sub-systems under development in MAST (as of Feb 2010).

While the data plotted in Figure 5 are primarily estimations, and one particular autonomous system will not contain all these components, it is readily apparent that both power and voltage requirements for autonomous robots will span a wide range (potentially multiple orders of magnitude). Thus, not only must a power source contain sufficient energy to operate each system, but the power distribution must be capable of efficiently delivering wide ranges of power and voltage to various parts of the bot. Such a power distribution system adds both weight and inefficiencies to the overall system, further exacerbating the problem. Naturally, precise power requirements depend heavily on application, but the wide range of power specifications from the survey gives pause to even the most optimistic power engineers.

5. mm-scale power delivery

Given our survey of MAST sub-systems power needs, it became apparent that inherent battery or power source performance is only part of the solution; one must also efficiently deliver the available power into the desired format for use by the actuators. For example, approximately 15% of the total weight of Fearing's 2.4g walking robot was dedicated to drive electronics & regulators operating at <80% efficiency [22]. Thus, a mobile Microsystem must carry both a tiny power source with high power & energy density, but also a vanishingly small, ultra efficient power converter capable of delivering said power across multiple orders of magnitude in voltage. Since inefficiencies in power conversion require oversizing of batteries or other power sources, research on efficient miniaturization of power converters is just as critical as improvements in battery (or other power source) technology. As an extreme example, Prof. Robert Wood's group at Harvard University is developing ~100mg-class flapping wing systems [23,24]. Yet the miniature (80mg) autotransformer-based power converters developed by Prof Wood's group operate at ~50% efficiency when boosting the 100-200mW of power up to the desired 200V [20]. Not only does this converter comprise a majority of the total system mass, but its low efficiency essentially wastes half the energy contained in the battery. Therefore, the following sections outline research being led by the Army Research lab to address this unique problem by creating mm^3/mg-scale power converters using high frequency CMOS circuits and MEMS passives.

5.1 High frequency CMOS converters

In its simplest form, a basic switched inductor voltage converter can be assembled using one inductor, one switch, one rectifier and a filter capacitor. Since CMOS logic and switches are extremely small, the passive components (inductors & capacitors) tend to dominate the size of most converters. Ideally, shrinking the size of passive components is as easy as increasing the switching frequency – if each passive is thought of as an energy reservoir, the same power can be transferred by using a small passive many times instead of a large passive less often. In fact, under fixed ripple constraints, component size decreases as a function of 1/frequency squared, so a significant reduction in overall volume and size should be possible. However, at high switching frequencies, particularly >20MHz, CMOS switches exhibit far from ideal behavior, and processing the desired voltages (10's to 100's of V) becomes extremely challenging.

To address the increased switching losses at higher frequencies, it is necessary to relate device and passive component characteristics to power conversion requirements (i.e. efficiency, conversion ratio, power density, etc) for efficient switch-mode power converter topologies (i.e. switched capacitor (SC) and switch inductor (SI)). The impact of the numerous architectural power conversion choices must also be evaluated in the context of power delivery strategies. Thus, we developed a simplified model (Fig 6 (a)) for a generalized boost converter consisting of a transformer, representing the ideal conversion ratio (n) of the topology, equivalent output resistance R_{OUT}, load resistance R_{LOAD}, and current source I_{DYN} representing dynamic power loss. R_{OUT} is the key performance metric of the converter [25], as it incorporates the conduction losses of the switches, inductors, and, in the case of switched capacitor (SC) converters, the loss associated with charging and discharging the capacitors. R_{OUT} was used to minimize converter component sizes using a Lagrange optimization approach. The inductor size for the SI boost (Fig 6 (b)) and the total capacitance for the SC Dickson (Fig 6(c)) topologies can be estimated by partitioning the dynamic and conduction losses within the converter and using MATLAB to calculate optimum size given these power constraints. The two topologies are compared graphically in Fig 7 to estimate the required passive component sizes for each, the SI (Fig 7(a)) and SC (Fig 7(b)), in terms of conversion ratio, output power, and switching frequency for a given efficiency. For example, for an SI boost converter with conversion ratio of 10 and output power P_{out}=0.35W, from Fig 7(a), an inductor of 1uH (with Q=50) at a frequency of 20MHz must be utilized in order to achieve 80% efficiency. If 70% efficiency is acceptable, the inductor can be decreased to 0.6uH using a frequency of 90MHz, resulting in a lighter and smaller power converter.

In general, SC offers efficiency benefits for low to medium power applications, especially with high conversion ratios. Also, there is potential for complete on-chip integration for SC designs as capacitors are readily available in standard CMOS. SI converters, on the other hand, shrink in size with increasing output power and are better suited for medium to high power applications. In both cases, integrated passives for output filter capacitors and on-chip inductors will be necessary to achieve a fully integrated system, and given the range of sub-system voltages being considered in MAST, voltage tolerance continues to be a consideration when developing such technologies.

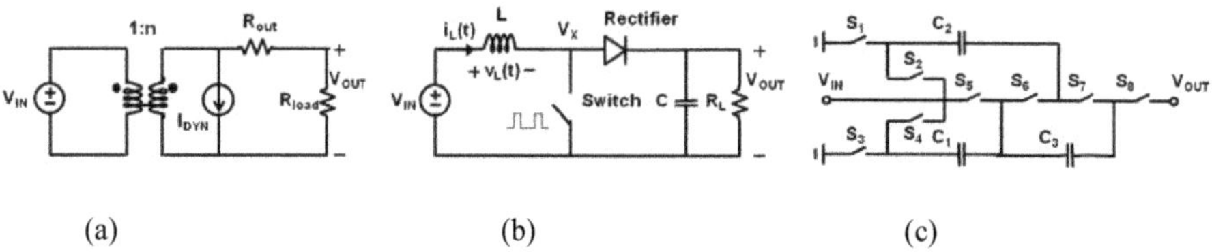

Fig. 6 (a) simplified model for general boost converter (b) basic SI boost converter (c) SC dickson converter

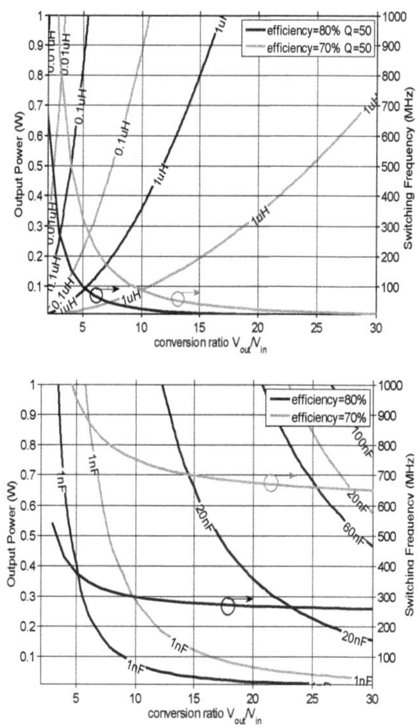

Fig. 7 Simulation results for switched inductor boost (left) and switched capacitor Dickson (right) topologies.

5.2 MEMS magnetic passives

Switched-mode, dc-dc power converters have traditionally relied on high-performance, discrete inductors and transformers. Integration of the magnetic passive components could reduce the size of such power units, but limitations in materials processing and component performance have hampered progress [26]. While high-permeability, high-saturation magnetic alloys can be integrated via electroplating or sputtering, magnetic losses limit their performance in the MHz regime [27,28]. At these frequencies, however, air-core inductor and transformer passives may provide viable performance, thus mitigating the fabrication complexity of integrating magnetic core materials.

Previously, air-core inductors have been studied extensively for GHz operation for RF applications, however the design of integrated passives for power conversion applications in the MHz range requires far greater inductance while maintaining low resistance – a paradox when limited to metal layers commonly available in CMOS processes [29], [30]. We therefore developed a new four-layer, through-mask, electroplated-copper fabrication process, summarized in Fig. 8 [31]. This has enabled stacking of two thick copper spiral layers to take advantage of quadratic gains in inductance through mutual coupling between layers [32]. Both inductors and transformers have been fabricated using the developed process, enabling the process to serve as a platform for device optimization.

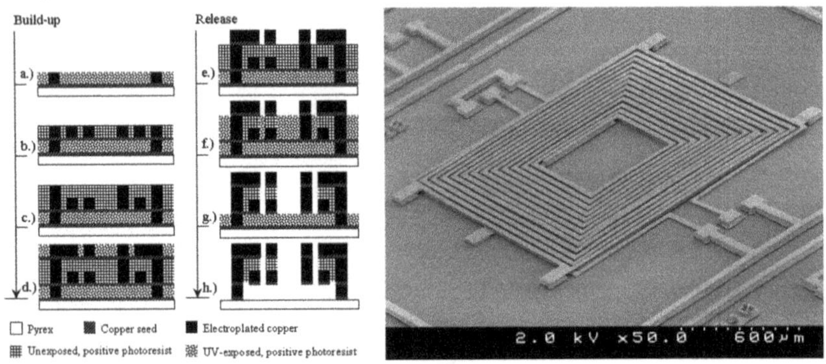

Fig. 8 Fabrication process and SEM perspective image of a completed transformer [31].

High frequency measurements of fabricated inductors have demonstrated inductances exceeding 100 nH/mm^2 with quality factors approaching 20 near 125MHz [31], shown in Fig 8. Step-up transformers with 1:3.3 turns ratios have been tested that exhibit coupling coefficients >0.6 and efficiency >75% around 125 MHz [31] to provide voltage gain with moderate efficiency for state-of-the-art switching converters. As indicated by the plots of primary, secondary, and mutual inductances in Figure 9, the transformers demonstrated inductance densities > 150 nH/mm2 as high inductance density was required in the secondary winding for voltage gain. Simple design changes enable tradeoffs between frequency and inductance density, allowing devices to be tailored for specific applications.

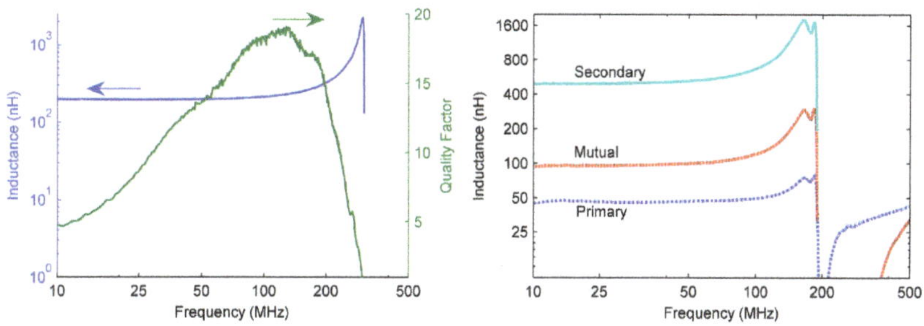

Fig. 9 Example measurements for a (1.5 mm)^2 inductor (left) and a (1.5 mm)^2 transformer (right) [31].

These results clearly demonstrate that air-core integrated inductors and transformers are attractive options for emerging high frequency power conversion applications.

5.3 MEMS piezoelectric transformers

Low frequency bulk piezoelectric resonant transformers (< 5MHz) have been implemented in power conditioning units due to several advantages over magnetic core transformers including high voltage isolation, small size and the absence of induced electromagnetic noise [33]. At the smaller size scales, however, thin film piezo transformer (PT) models show performances comparable with their larger (>4mm^2), state of the art thin film magnetic transformer counterparts at similar switching frequencies which suffer magnetic losses at high frequencies [35].

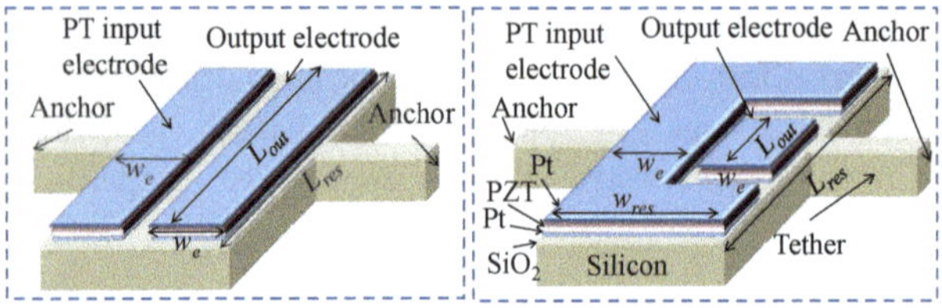

Fig. 10 Illustration of matched input and output coverage. (Right) Illustration of optimizing the electrode coverage for load impedance matching where the L_{out} x w_e area is traded for a larger input electrode area. [35].

An example thin film 21.4 MHz fundamental length extensional piezoelectric resonator using lead zirconium titanate (PZT) is shown in Figure 10 (with L_{res} = 190µm, w_{res} = 40µm and thickness = 20µm [34]). The equivalent series motional resistance, inductance and capacitance of the first length extensional mode are extracted to be 50 Ω, 0.75 mH, and 74 fF, respectively, for the matched input and output coverage case (Figure 10-right).

By tailoring the input and output electrode coverage, it is possible to impedance match to various electrical loads, R_L, which in a power converter directly effects efficiency, voltage buck/boost ratio, and power delivery. The performance is primarily tailored by reducing the output capacitance, C_o, through decreasing the output electrode size. Although this comes at a cost of reduced output electromechanical coupling and higher motional resistance, the tradeoffs in efficiency and boost improve in the higher resistive load / lower power delivery regimes where voltage boosts are desired. The efficiency and voltage gain as a function of R_L are shown in Figure 11 for varying output electrode coverage where the reduced output area is traded with larger input electrode area to maximize η_{in}.

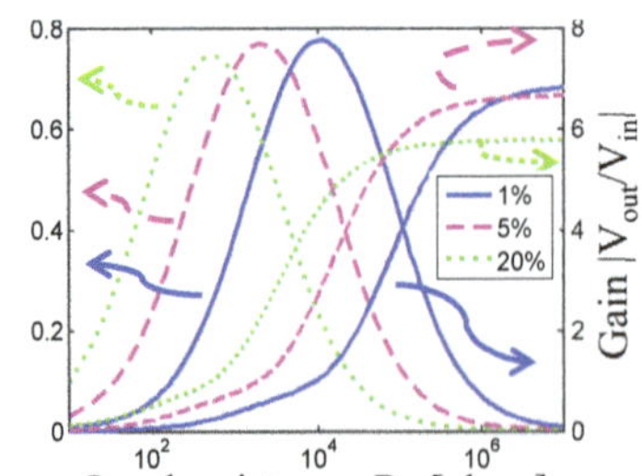

Fig. 11 Piezoelectric transformer voltage gain and efficiency versus R_L for different electrode coverage areas [35].

As evident in the figure, a wide range of efficiencies and voltage gain combinations are possible, where the electrode area should be tailored to match the expected load [35]. For example, if 50% efficiency is targeted, AC/AC voltage boosts > 3.5:1 can be achieved for loads in the 10's of kΩ. If voltage boosts in the low power (high R_L) regime are desired, peak AC/AC boosts can exceed 6:1 (at the cost of low efficiencies). For simple isolation (1:1) transformers, peak efficiencies exceed 75%. If such transformers were inserted into an AC/DC voltage doubler, voltage gains >8:1 at 29% efficiency and 0.4% ripple should be possible (with 1MΩ load). One should note that low efficiency may be acceptable if the system being powered is used infrequently and the tradeoff is a converter of extremely small size.

With each PT occupying a total volume in the hundreds of cubic micrometers, a power unit composed of an array of impedance matched PT is envisioned as a potential ultra-miniature power conditioning unit capable of adapting to a variety of loads.

6. Conclusions & future work

Because future micro-UAV's are expected to perform all manner of ISR, choosing the appropriate power sources for each particular microsystem will depend not only on the primary actuators providing locomotion but also on the number and type of all other UAV subsystems employed. A summary of subsystems currently under development in MAST shows that the power requirements of these subsystems may span several orders of magnitude in both voltage and power. On the other hand, each of the various available power sources has its own ideal operating characteristics including discharge rate and duty cycle.

Highly efficient power converters may provide some of the handshaking that will be needed between power sources and powered subsystems. Increasing converter switching frequencies enables shrinking the size of the passives, which traditionally dominate the overall size of the converter. A simplified model provides a comparison of converter topologies in terms of efficiency at output powers and conversion ratios. Additionally, the development these converters is supported by high-density, high-efficiency inductors and transformers. Air-core devices show strong high-frequency performance (> 100 MHz) with inductance densities greater than 100 nH/mm^2 and quality factors approaching 20. At lower frequencies (<5 MHz) thin film piezo transformers are superior to larger (>4mm^2), state of the art thin film magnetic transformers at similar switching frequencies, which suffer from high frequency magnetic losses [35].

Future work is focused on developing high density capacitors capable of holding off moderate voltages (10's to 100's of V), a challenge posed by the numerous MAST systems shown in the previous section that plan to operate far beyond preferred CMOS voltages of <10V.

References

[1] www.aerovironment.com

[2] Autumn, K., Hsieh, S.T., Dudek, D.M., Chen, J., Chitaphan, C., Full, R.J. 2006. The dynamics of vertical running in geckos. The Journal of Experimental Biology, vol. 209, pp. 260-272. https://doi.org/10.1242/jeb.01980

[3] R. Fearing, "Challenges for Effective Millirobots," Digest of 2006 IEEE. Int. Conf. Robotics and Automation, 2006.

[4] S. Mao and DU Gang, "Lift and power requirements of hovering insect flight," ACTA Mechanica Sinica, Vol. 19, No. 5, October 2003. https://doi.org/10.1007/bf02484580

[5] Steltz, E. Seeman, M. Avadhanula, S. and Fearing, R.S.,``Power Electronics Design Choice for Piezoelectric Microrobots" Int. Conf. On Robotics and Intelligent Systems, Beijing Oct. 9-15, 2006.

[6] Bergbretier S, Pister K S J, "Design of an Autonomous Jumping Microrobot," IEEE Int. Conf. on Robotics and Automation (Rome, Italy) 447-453, 2007. https://doi.org/10.1109/robot.2007.363827

[7] A. Stux and K. Swider-Lyons, "Survey of Commercial Small Lithium Polymer Batteries," NRL/MR/6110–07–9073

[8] Dudney, "Thin Film Micro-Batteries," The Electrochemical Society INTERFACE, Vol 17, No.3, p. 44-48, 2008

[9] Derived from Eagle Picher data sheet at http://www.eaglepicher.com/images/stories/ep_single_page_cfx.pdf

[10] AFRL Public Release Document Number: 88ABW-2009-2300

[11] McConnel, V.P., "Military UAVs claiming the skies with fuel cell power," Fuel Cells Bulletin, Dec 2007, p. 12–15. https://doi.org/10.1016/S1464-2859(07)70438-8

[12] Tiron, R., "The Quest for Better Batteries and Energy Systems: Power-Hungry Vehicles Drive Push for Longer Endurance," Unmanned Systems, Apr 2008, pp. 15 – 17 (www.auvsi.org).

[13] Z. Guo and A Faghri, "Vapor feed direct methanol fuel cells with passive thermal-fluids management system," Journal of Power Sources, Volume 167, Issue 2, 15 May 2007, Pages 378-390. https://doi.org/10.1016/j.jpowsour.2007.02.024

[14] W. Ma, et al, "Thermally Stable, Efficient Polymer Solar Cells with Nanoscale Control of the Interpenetrating Network Morphology," Advanced Functional Materials, Volume 15 Issue 10, Pages 1617 – 1622, 2005. https://doi.org/10.1002/adfm.200500211

[15] A. Barnett, C. Honsberg, D. Kirkpatrick, S. Kurtz, D. Moore, D. Salzman, R. Schwartz, J. Gray, S. Bowden, K. Goossen, M. Haney, D. Aiken, M. Wanlass, K. Emery, "50% efficient solar cell architectures and designs," in: Photovoltaic Energy Conversion, Conf Record of the 2006 IEEE 4th World Conf, V. 2, p 2560-2564, May 2006. https://doi.org/10.1109/wcpec.2006.279768

[16] R. Venkatasubramanian, C. Watkins, C. Caylor, and G. Bulman, "Microscale thermoelectric devices for energy harvesting and thermal management," PowerMEMS 2006, p.1-4, Nov 29-Dec 1, 2006, Berkeley, CA USA.

[17] S. Roundy, "On the Effectiveness of Vibration-based Energy Harvesting," Journal of Intelligent Material Systems and Structures, v. 16. p. 809-823, October 2005. https://doi.org/10.1177/1045389X05054042

[18] http://www.eetasia.com/ART_8800561528_499488_NT_828d37c8.HTM

[19] http://www.ohgizmo.com/2010/01/09/ces2010-rca-airnergy-charger-harvests-electricity-from-wifi/

[20] M. Karpelson, G-Y. Wei, and R.J. Wood, "A review of actuation and power electronics options for flapping-wing robotic insects," 2008 IEEE Int Conf on Robotics & Automation, Pasadena, CA, USA, p.779-786, May 19-23, 2008. https://doi.org/10.1109/robot.2008.4543300

[21] Y. Bar-Cohen, Electroactive Polymer (EAP) Actuators as Artificial Muscles (Second Edition), p. 26.

[22] Aaron M. Hoover, Erik Steltz, Ronald S. Fearing, "RoACH: An autonomous 2.4g crawling hexapod robot," IEEE Int. Conf. on Intelligent Robots and Systems, Nice France Sept. 2008. https://doi.org/10.1109/iros.2008.4651149

[23] R.J. Wood, "Fly, Robot Fly," IEEE Spectrum, March 2008. https://doi.org/10.1109/MSPEC.2008.4457845

[24] R.J. Wood, "Liftoff of a 60mg flapping-wing MAV," IEEE/RSJ Int. Conf. on Intelligent Robots and Systems, San Diego, CA, October 2007. https://doi.org/10.1109/iros.2007.4399502

[25] M. D. Seeman, S.R. Sanders, "Analytical and Practical Analysis of Switched-Capacitor DC-DC Converters," IEEE Trans. Power Electronics, v.23, n.2, pp.841-851, March 2008. https://doi.org/10.1109/TPEL.2007.915182

[26] T. O'Donnell, et al., "Microfabricated Inductors for 20 MHz DC-DC Converters," Proc. APEC, pp. 689-693, 2008. https://doi.org/10.1109/apec.2008.4522796

[27] D. Flynn, et al, "Characterization of Core Materials for Microscale Magnetic Components Operating in the Megahertz Frequency Range," IEEE Trans. Mag., vol. 43, no. 7, pp. 3171-3180, 2007. https://doi.org/10.1109/TMAG.2007.895700

[28] Gardner, et al., "Integrated on-chip inductors using magnetic material (invited)," Journal of Applied Physics, vol. 103, 07E927, pp. 1-6, 2008.

[29] T. Karnik, P. Hazucha, G. Schrom, F. Paillet, and D. Gardner, "High-Frequency DC-DC Conversion: Fact or Fiction," Proc. 2006 IEEE International Symposium on Circuits and Systems, pp. 245-248, 2006. https://doi.org/10.1109/ISCAS.2006.1692568

[30] S. C. O. Mathuna, T. O'Donnell, N. Wang, and K. Rinne, "Magnetics on Silicon: An Enabling Technology for Power Supply on Chip," IEEE Trans. Power Electron., vol. 20, no. 3, pp. 585-592, 2005. https://doi.org/10.1109/TPEL.2005.846537

[31] C. Meyer, S.S. Bedair, B. Morgan, and D. Arnold, "High-inductance-density, air-core, power inductors and transformers designed for operation at 100-500 MHz," INTERMAG, Washington DC USA, January 22, 2010.

[32] A. Zolfaghari, A. Chan, and B. Razavi, "Stacked Inductors and Transformers in CMOS Technology," IEEE Journal of Solid-State Circuits, vol. 36, no.4, pp. 620-628, 2001. https://doi.org/10.1109/4.913740

[33] G. Ivensky, M. Shvartsas, S. Ben-Yaakov, "Analysis and modeling of a voltage doubler rectifier fed by a piezoelectric transformer," IEEE Trans. On Power Electronics 19 542-549, 2004. https://doi.org/10.1109/TPEL.2003.823179

[34] Chandrahalim H, Bhave S, Polcawich R, Pulskamp J, Judy D, Kaul R, Dubey M 2008 Applied Physics Letters, 93, 233504-1. https://doi.org/10.1063/1.3046717

[35] S. Bedair, J. Pulskamp, B. Morgan, and R. Polcawich, "Performance model of electrode tailored thin film piezoelectric transformers for high frequency switched mode power supplies," Proceedings of Power MEMS 2009, p. 435-438, Washington, DC USA, Dec 1-4, 2009.

SECTION III

Power Integration

E. Shaffer, B. Geil
U.S. Army Research Laboratory
Email: Edward.c.shaffer.civ@mail.mil; bruce.r.geil.civ@mail.mil

1. Introduction

The research and engineering work within Power Integration Branch is arranged into three primary interrelated areas. These are:

- Wide bandgap devices – modeling, characterization and analysis for high power/ pulse power applications

- Power conditioning and thermal management – designs for active (power electronics) and passive (packaging) components for high power, pulse power, and military unique applications

- Intelligent energy/ energy networks – architectures, designs, and test beds for advanced control of integrated and distributed energy systems

The research performed in Power Integration Branch is more applied in nature than fundamental. It primarily falls in the category of systematic studies to gain knowledge or understanding of power devices, components, and subsystems necessary to determine the means by which recognized and specific needs may be met. The specific needs addressed by Power Integration include warfighter capabilities for Army Soldiers and platforms, such as advanced mobility/propulsion, protection, lethality, and sustainment. This research is documented in ARL Technical Reports as well as in published reports shared primarily within the engineering community. Introductory comments highlight key aspects of this work.

2. Overview

2.1 Wide Bandgap Devices

ARL has led Army and DoD development of advanced wide band-gap power devices for over twenty years. The motivating need endures to provide robust materials and devices to replace silicon (Si) power electronics in order to realize compact/ reduced footprint, efficient, high-power, high-temperature power electronics for military applications. Wider bandgap power technologies operate with higher efficiency, which leads to fuel savings and less waste heat to manage. This is due to lower conduction and/or switching losses via higher temperature operation, which also allows a smaller cooling system design, greater operating margin, and higher frequency designs to further reduce the size of passive (inductive, capacitive) components. This drives power electronic solutions that are much smaller in weight and volume (critical for military applications). ARL continues to push wide bandgap performance limits through higher voltages, higher power levels, and higher temperatures while reducing costs and improving reliability.

Detailed analyses of selected silicon carbide (SiC) wide bandgap devices illustrate performance characteristics differing from those found in Si devices. Understanding the underlying mechanisms causing these differences is crucial to ensuring reliable device operating regime designs and operation. Issues include stability of device threshold voltages, short-circuit performance, and reliability of both the body diodes and the gate oxides [1].

While significant progress has been made in applying SiC for lower voltage applications, higher voltages needed for continuous and pulse military applications are driving improvements for operating voltages from 15 kV up to 27 kV. ARL physics-based Technology Computer Aided Design (TCAD) device modeling based on doping profiles and structural layers are simulating steady-state and transient characteristics of recently fabricated Ultra High Voltage (UHV) experimental devices [2].

SiC switches for pulse power experience very high instantaneous power for short durations. SiC material properties support very high voltage-blocking across thin epilayers, withstand high instantaneous power and temperature, and enable fast switching speeds. ARL research exploring device turn-on mechanisms of actual high-voltage devices under high current density and fast switching have identified trade-offs for optimizing device chip area and voltage ratings for pulse power applications. In depth analysis of commerical devices include silicon carbide (SiC) “super” gate turn-off thyristors (SGTO); ARL mixed-mode simulation models have shown how switching dependent device parameters affect switching performance. A particularly unique performance feature of wide bandgap devices is voltage protection; ARL studies of SiC avalanche breakdown diodes and use of large chip area SiC PiN diodes for thyristor protection are illustrative [3].

2.2 Power Conditioning and Thermal Management

Efforts in power conditioning have been focused in several areas. Because ARL's mission is oriented toward exploring the innovation space and art of the possible versus final engineering designs, ARL power electronic designs provide opportunities to highlight performance of vertically integrated high power devices and power/ thermal management components. Examples include compact solid-state SiC circuit breakers [4] as an alternative to bulky, slow and vibration susceptible electro-mechanical breakers.

High power devices have provided opportunities to develop advanced high density converters, inverters and pulse power circuits. ARL developed a test bed that incorporated a 28 V to 600 V DC-DC converter for an Army unique pulse power application [5]; this converter provided 2 kW of charging power in a 3.0 L compact footprint without active cooling.

Military systems often demand power solutions beyond what is available from commercial off the shelf products. ARL engineers have provided a number of specialty designs working with Government and industry partners; presented here is the development of a 16-channel, 3-kW, constant-current power distribution unit (PDU) for aircraft applications. Key design considerations included maintenance of electrical power quality, and volume and weight minimization. Despite limited available commercial power-factor correction components for 400-Hz systems, ARL implemented a constant-current (1 A) and 200 V per channel design [6].

Combat vehicles and platforms represent challenging thermal environments for electronic systems. Components must be able to operate from sub-freezing to hundreds of degrees Celsius with power electronics systems requiring some form of cooling to maintain operation below critical operating temperature limits. In most power electronic modules, thermal/electrical/mechanical packaging constraints limit achievable power device power densities. More effective heat removal techniques have been explored through device layout optimization, thermal management design, and improved thermal management techniques. These include single-phase liquid cooling; two-phase liquid-vapor cooling; advanced passive thermal ground plane (TGP) heat spreaders; transient thermal mitigation with solid-liquid phase change materials; and multi-functional packaging elements. These studies have informed advanced integrated packaging power module design concepts [7].

2.3 Intelligent Energy/ Energy Networks

There is significant value in developing integrated energy and power system solutions that take advantage of improvements in efficiency, density, and duration offered by advances in underlying material, device, and component technologies as detailed herein. Although considerable progress has been made by the Army in recent years in advancing power distribution to support expeditionary operations under new Operational Energy imperatives, the future multidomain battlefield will demand more agile, flexible energy solutions. Understanding current and emerging SmartGrid and tactical microgrid architectures has informed the way ahead for future intelligent tactical energy networks.

Focused Army workshops on future tactical microgrids highlighted a number of desirable features posing research and development challenges:

- Intelligent and adaptive—leveraging the Internet of Things, Artificial Intelligence and Machine Learning (AI/ML)
    - Architecture with inherent adaptability (capability enhancing attribute)
    - Enabling energy device and component self-awareness
- Managed services, controls, and safety
    - Automating configuration/reconfiguration (military unique attribute)
    - Distributed agent control and dispatching (military unique attribute)
    - Dynamic turndown of sources, renewables, energy storage
    - Fault tolerance/recovery, stability, security
- Universal connectivity:
    - Interfaces and standards (capability enhancing attribute)
    - Integrating sensing, IT, conditioning, intelligent power management (IPM)
    - Modular, adaptable microgrids that can be combined in any number of sequences/configurations (military unique attribute)
- Physical attributes - Reduced footprint/size, weight and power (SWaP)

Advanced energy architectures for such systems enable interoperability between equipment from different vendors, encourage competition within industry, and facilitate future equipment modernization. In particular, the control methods, communication protocols, and power interfaces should be standardized in a vendor agnostic manner. Evaluating the performance of hybrid energy systems and future microgrids is complex; a recent study at ARL surveyed available modeling tools for such systems. Existing research and software tools related to modeling, simulation, validation, and verification were evaluated against identified enabling capabilities to determine research gaps and develop a prioritized list of modeling and simulation projects with consideration of the requirement to balance technological advancements.

3. Summary

An overview of ARL's Power Integration work is given. Although additional military capability specific work within this area is not included, the power component and electronic efforts represent enabling technologies which underpin the broader Power Integration area.

References (articles in this chapter)

[1] Lelis, A. et al. - Basic Mechanisms of Threshold-Voltage Instability and Implications for Reliability Testing of SiC MOSFETs (2015).

[2] Hinojosa, M. et al. - Electro-thermal TCAD model for 22 kV silicon carbide Insulated-Gate Bipolar Transistors (2013).

[3] Ogunniyi, A. et al. - Simulation Study of Switching-Dependent Device Parameters of High Voltage 4H-SiC GTOs (2017).

[4] Urciuoli, D. et al. - A Compact 100-A, 850-V, Silicon Carbide Solid-State DC Circuit Breaker (2016).

[5] Wood, R. and Urciuoli, D. - Development of a Compact 2-kW, 28- to 600-V DC-DC Battery Charger (2013).

[6] Tipton, W. et al. - Multi-Channel, Constant-Current Power Source for Aircraft Applications (2017).

[7] Boteler, L. et al. - Integrated Thermal Management for Power Electronics Packaging (2018).

Chapter 22

Basic Mechanisms of Threshold-Voltage Instability and Implications for Reliability Testing of SiC MOSFETs

Aivars J. Lelis, Ron Green, Daniel B. Habersat and Mooro El

U.S. Army Research Laboratory, Adelphi, MD 20783 USA

aivars.j.lelis.civ@mail.mil, ronald.green39.civ@mail.mil, daniel.b.habersat.civ@mail.mil, mooro.i.el.civ@mail.mil

Abstract

A review of the basic mechanisms affecting the stability of the threshold voltage in response to a bias-temperature stress is presented in terms of the charging and activation of near-interfacial oxide traps. An activation energy of approximately 1.1 eV was calculated, based on new experimental results. Implications of these factors, including the recovery of some bias-temperature stress-activated defects, for improved device reliability testing are discussed.

Keywords

Oxide Trap, Power MOSFET, Reliability, Silicon Carbide, Threshold Voltage

Contents

1. Introduction

Silicon carbide (SiC) continues to be the most promising near-term wide bandgap semiconductor material to replace silicon (Si) for the development of efficient, high-power, high-temperature electronic devices. This is due in part to its much wider bandgap. Potential benefits include operating with higher efficiency—due to lower conduction and/or switching losses, which leads to fuel savings and less waste heat to manage; operating at higher temperature, which allows for a smaller cooling system and greater operating margin; and operating at higher frequency, which reduces the size of passive components. This in turn can lead to the development of systems that are much smaller in weight and volume. This advantage vis-à-vis Si increases dramatically at higher voltages. But none of this matters if such new-technology devices are not reliable.

However, if successful, then moving from Si to SiC and replacing mechanical functions with power electronics should lead to significant increases in system efficiency, functionality, and reliability. Military systems will be better able to meet the significantly increased requirements for electrical-power generation, distribution, and control. A wide array of commercial systems will greatly benefit as well from the development of SiC MOSFET technology, including hybrid-electric vehicles (HEVs) and renewable-energy production, and the significant improvements in energy efficiency will lead to reduced power costs in every sector of electrical usage.

The SiC MOSFET is the most desirable power switch because it is both voltage-controlled and normally-*OFF*. With the introduction of SiC power MOSFETs into the commercial market place, it is critically important to determine the reliability of these devices. A number of potential issues need to be addressed, including the stability of the device threshold voltage [1-13], and the reliability of both the body diode [14] and the gate oxide [15]. This work focuses on the stability of the threshold voltage of SiC power MOSFETs and the basic mechanisms that affect this stability. A review of these mechanisms will be presented—first the oxide-trap charging mechanisms at room temperature, and then the additional trap-activation mechanisms that are likely present at high temperature. It will be shown how these are all consistent with a near-interfacial oxide-trap model. New experimental results will be offered providing additional insight into these high-temperature mechanisms that affect the response of SiC MOSFETs to a bias-temperature stress (BTS). Finally, the implications of these results for improved reliability testing will be discussed.

2. Background

There may be as many as four different types of interfacial charge that affect the threshold voltage: oxide traps, interface traps, mobile ions, and fixed charge [1]. The closer their proximity is to the interface between the SiC conduction channel and the insulating gate oxide, the greater their effect will be. An increase in the net positive charge (caused, for example, by an increase in the number of positively-charged oxide traps) will result in a negative shift of the threshold voltage, V_T. Likewise, a net negative change in the total interfacial charge (caused, for example, by additional negatively-charged interface traps—or perhaps by electron trapping in the oxide) will result in a positive shift in V_T. Examples of the instability of the device drain-current versus gate-voltage characteristic, I_D-V_{GS}, which results from a shift in V_T, are shown in Fig. 1. These results are for high-temperature stressing, discussed in detail in Section IV below.

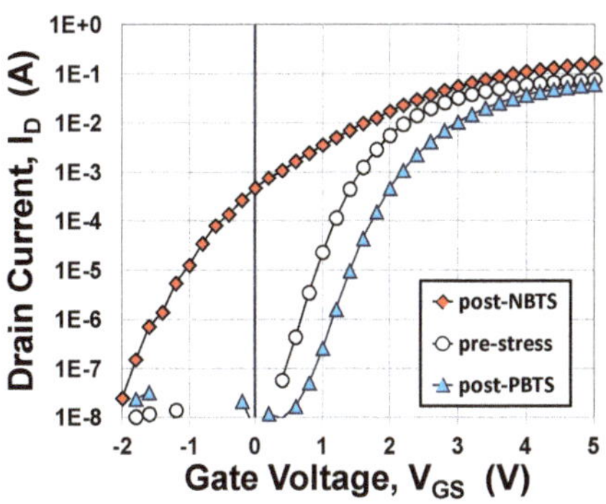

Fig. 1. Subthreshold I_D-V_{GS} characteristics of a commercially-available SiC power MOSFET: typical initial pre-stress measurement at high-temperature (150 °C); following 100 hours NBTS (–15 V, 150 °C); following 600 hours PBTS (+15 V, 150 °C).

Limiting the shift in the threshold voltage can be critically important because an ideal switch has minimal leakage current in the *OFF* state and very low resistance in the *ON* state. However, an excess negative shift of the threshold voltage under high-temperature reverse-bias (HTRB) conditions can lead to a critical increase in *OFF*-state drain-leakage current and potential device failure (for example, the negative shift in Fig. 1 following negative BTS (NBTS) results in a five-order-of-magnitude increase in leakage current). In a similar fashion, a positive shift of the threshold voltage may occur under high-temperature gate-bias (HTGB) conditions, wherein a positive BTS (PBTS) is applied. This can lead to an increase in the *ON*-state resistance and thus a possible decrease in device efficiency. As the temperature increases, the threshold voltage already naturally shifts negatively due to fewer negatively-charged interface traps, as well as an increase in the number of thermal carriers. Oxide-trap charging, discussed below, can make those negative shifts much larger.

3. Basic Mechanisms at Room Temperature

The threshold-voltage instability effect is the shifting of V_T back and forth in response to an applied gate-bias stress. A positive gate bias leads to a positive shift in V_T and a negative gate bias results in a shift back in the negative direction. A typical V_T instability, or hysteresis in the I_D-V_{GS} characteristic, is on the order of 0.25 to 0.50 V [1]. This is when stressed at room temperature with a gate-oxide field in the ±2 to ±3 MV/cm range for a stress time ranging from minutes to hours, and the I_D-V_{GS} characteristic measured with a standard commercially-available parameter analyzer such as an Agilent 4155 that applies a gate-voltage sweep taking 1 to 10 s to complete [1]. Such an instability in V_T is observed in all SiC MOSFETs, including commercially-available power MOSFETs, and is repeatable. Similar instabilities are also observed in MOS capacitors [16]. At room temperature, for the same bias-stress and measurement conditions, the magnitude of the V_T instability typically does not increase or diminish with time. Not only does the difference between V_T measured following a positive-bias stress and a negative-bias stress not change, neither do the individual values of V_T change. In other words, the location of this V_T "envelope" does not drift. This is true whether V_T is calculated by the change in gate voltage for a fixed current level in the sub-threshold ("sub-threshold V_T," 3×10^{-5} A is a typical drain current level used), or by extrapolating on a linear current scale to zero current ("linear V_T") [17]. This implies that no additional defect formation or build up of trapped charge occurs.

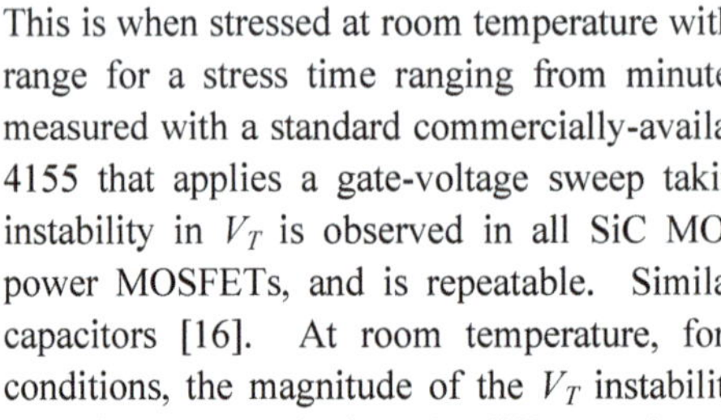

Fig. 2. Stress-and-measure sequence for a back-and-forth bias stress employed to measure the hysteresis in the I_D-V_{GS} characteristics, which can provide a lower bound for the number of switching oxide traps present.

The main mechanism at room temperature is the positive charging (or neutralization) of near-interfacial oxide traps via a direct tunneling mechanism [1]. Under negative gate bias, electrons tunnel out of the oxide, uncovering the positively-charged oxide traps and causing a negative shift of V_T. Under positive bias, electrons can tunnel back into the oxide, neutralizing the positively-charged oxide traps, and causing a positive shift of V_T.

A two-way direct tunneling model was developed that successfully describes this V_T instability as a function of stress time, stress bias, and measurement time [18]. (A two-way tunneling model allows electrons to both tunnel in and out of the oxide at any given moment, based on trap occupancy and local electric field.) The V_T instability typically increases at a linear-with-log-stress-time rate, during a back-and forth stress-and-measure sequence such as that depicted in Fig. 2 [1]. V_T will also shift at a linear-with-log-stress-time rate during a unipolar stress-and-measure sequence—such as that depicted in Fig. 3 for positive-bias stress only [19]. This is the expected result, especially for a single trap energy and uniform spatial distribution of traps in the oxide. The details of this tunneling model reveal that the rate of change in the number of positively-charged oxide traps with respect to time under constant stress conditions is always sharply peaked spatially, with a full-width half-max value of around 2 Å. Furthermore, this peak moves at a linear-with-log-time rate of about 2 Å per decade of time into the oxide, giving rise to the notion of a tunneling front. Traps in the wake of this tunneling front have settled on a new steady-state balance of charged versus neutralized near-interfacial oxide traps, whereas those traps deeper in have yet to be affected by this new bias stress [18].

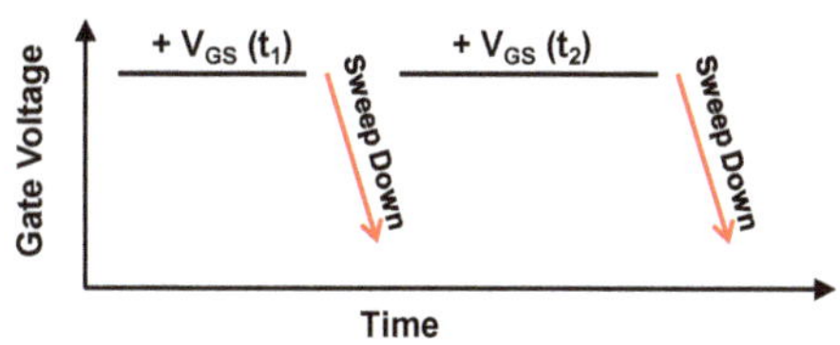

Fig. 3. Stress-and-measure sequence for a unipolar (in this case positive, with increasing stress times) bias stress employed to measure the shift in I_D-V_{GS} characteristics, and whether a device V_T has drifted too far beyond its specified values.

This notion is depicted in Fig. 4, indicating that the magnitude of the stress time determines which oxide traps may change charge state during the stress. The longer the stress, the deeper into the oxide that traps are affected. Shorter stress times leave more oxide traps undisturbed. Traps deeper into the oxide from the SiC interface will not change charge state during either a positive or negative bias stress when stressed in the back-and-forth manner depicted in Fig. 2. Longer stress times are needed to affect those oxide traps located deep into the oxide. Therefore, it is expected that longer stress times will result in larger V_T shifts (for a unipolar stress, as depicted in Fig. 3) or larger hysteresis in V_T (for a back-and-forth stress, as depicted in Fig. 2). This is what has typically been reported in the literature and what the tunneling model predicts [1, 18]. Generally, the only instances where this is not the case is during an elevated temperature stress, where mobile ions may be countering the effects of charge trapping [3, 20, 21].

However, the effect of stress time only tells half the story. Fig. 5 depicts the additional effect of measurement time on what is ultimately observed. This schematic indicates that only those oxide traps within a certain range of the oxide interface will both switch charge state during the stress and be observed following a post-stress measurement. This is because traps very close to the interface will change charge state not only during the

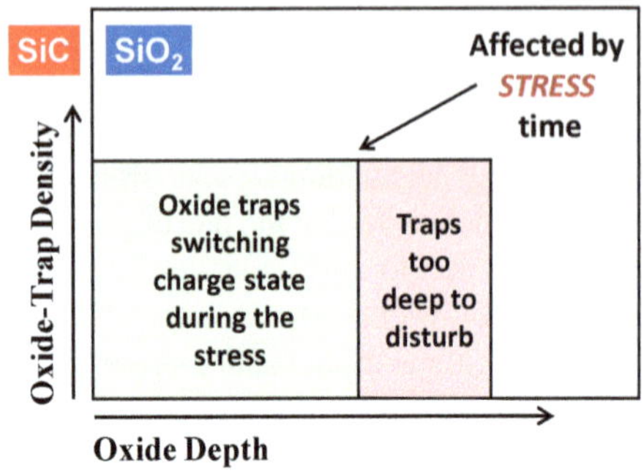

Fig. 4. Schematic indicating that the length of the stress time determines which oxide traps may change charge state during the stress. The longer the stress, the deeper into the oxide (further to the right) traps are affected. For example, a 10^6-s stress would be expected to affect oxide-trap charge states approximately 30 to 40 Å into the oxide.

stress, but also during the measurement. The slower the measurement speed, the deeper into the oxide this region will extend since the gate bias applied during the measurement is in fact a separate stress that has its own effect. So faster measurement speeds are required to observe a greater fraction of the total number of oxide traps switching charge state due to the bias stress; and a slower, more different gate bias during the measurement more strongly counteracts the effect of the previous stress bias [1, 19].

Results in the literature show the effect that measurement speed has, with fast *I-V* measurements taking only 100 μs revealing a V_T hysteresis (ΔV_T) four times greater than measurements taking 1 s to perform [1, 19]. Again, this is because faster measurements expand the number of switching oxide traps observed by reducing the number that change charge state during the measurement, as predicted by a two-way tunneling model [18]. In fact, it is calculated that traps within about 25 Å of the interface may change charge state during a 1-s measurement. Thus, fast *I-V* measurement techniques more accurately reflect the full effect of a gate-bias stress. The measurement speed affects the result, whether a back-and-forth stress or a unipolar stress is applied. This general effect has extremely important implications for proper reliability testing, as discussed in Section V below.

It should be noted that the gate bias should always be swept in the direction away from the previous stress (for example, sweeping from $+V_{GS}$ to $-V_{GS}$ following a positive-bias stress). This is to reduce the effect that the measurement has on the results of the measurement. Thus, sweeping the gate bias in the negative direction following a positive bias stress results in the *I-V* characteristic being further to the right (thus resulting in a larger V_T hysteresis) than if sweeping positively. In the latter case, the oxide traps would be exposed to an opposite-polarity stress sooner, promoting the uncovering of additional positive charge during this measurement, resulting in a smaller shift to the right following the positive-bias stress [19].

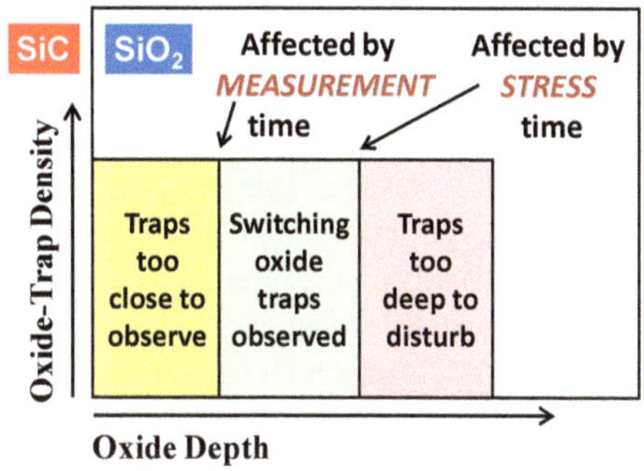

Fig. 5. Schematic indicating that not all oxide traps that switch charge state during the stress will be observed during a subsequent measurement, since traps very close to the interface will change charge state during the measurement itself. These traps are effectively controlled by the measurement bias rather than the stress bias.

It has been typically observed that a larger magnitude gate-bias stress results in a larger V_T instability, and that most of this charging effect is under positive gate bias. The two-way tunneling model can account for this effect as well, due to differences in calculated tunneling probability [18]. It is also typically observed that there is more stretch-out of the sub-threshold slope when sweeping positively following a negative gate-bias stress than when sweeping negatively following a positive bias stress [22]. This effect may be better explained by extending the tunneling model to consider a two-step mechanism involving interface traps as an intermediary since oxide trap states likely lie within the bandgap for at least one of the oxide-trap charge states [18].

Processing conditions can also have a dramatic effect on the magnitude of the V_T instability observed. It has been reported that without the standard post-oxidation NO anneal, the typical room-temperature V_T instability would be three times greater [1] since the N anneal appears to reduce the number of active oxide traps [1, 23, 24]. Other recent results indicate the benefits of a P-based anneal as well [25]. However, this is contrary to reports that the presence of additional N at the interface leads to additional hole trapping [26], although the width of the transition layer at the SiC-gate oxide interface decreases with longer N anneals, which suggests a thinner region of disorder [27].

4. Basic mechanisms at high temperature

Recent results from the literature clearly show that gate-bias stressing at elevated temperature results in much larger V_T instabilities than bias stressing at room temperature [4, 7, 10, 20, 28-35]. This is true whether externally heating the device, or by self-heating when performing an *ON*-state stress and allowing the rated current to flow through the device. Unlike the linear-with-log-stress-time increase in ΔV_T that is observed during room-temperature bias stressing, bias-temperature stressing at high temperature (for example, at 150 °C) generally exhibits a super-linear increase on a logarithmic time scale, especially at longer stress times. The initial cause of this is a large negative shift in V_T following a NBTS, especially when measured in the sub-threshold region, due to a significant stretch-out of the sub-threshold *I-V* characteristic [28-30, 32]. This can lead to an increase in drain-current leakage in the *OFF* or blocking state and potential device failure. In addition, a large positive shift in V_T may be observed in some devices following a PBTS, especially when measured by the linear V_T method. If large enough, this positive V_T shift can in turn lead to an increase in *ON*-state resistance and degradation in device efficiency [30, 32]. Thus a BTS causes a widening of the V_T hysteresis envelope, and may lead to a degradation of both *ON*-state and blocking-state characteristics. It should be noted that no such increase in ΔV_T occurs if no bias stress is applied at elevated temperature [29].

However, it has been observed in some relatively-recent-vintage devices—when performing a back-and-forth stress-and-measure BTS sequence as depicted in Fig. 2, at 150 °C for stress times up to 10^4 s—that a smaller ΔV_T occurs when making a high-temperature measurement than when rapidly (taking less than one minute) cooling the device back to room temperature before measuring. In fact, the high-temperature measurement exhibited a smaller V_T instability than before the BTS [34]. It is not clear whether this diminished high-temperature response is due to the presence of mobile ions [3, 20, 21] in these devices, or to some other mechanism. On the other hand, many recent-vintage devices do in fact exhibit a monotonically increasing ΔV_T with increasing stress temperature, even when measured at high temperature.

Another important aspect of the response is the post-stress recovery in ΔV_T with time back at room temperature. This recovery is especially dramatic regarding the positive shift under positive bias that had previously occurred during bias-temperature stressing [28-30, 32]. Often the entire positive shift recovers, and in some cases V_T measured following a room-temperature positive-bias stress is actually more negative than before the high-temperature bias stress. (On the other hand, very little of the negative shift under negative bias —or the degradation of the sub-threshold slope—recovers [28, 29].)

Thus, depending on the specific key stress and measurement factors—bias, temperature, and time—any of three conclusions might be drawn regarding the effect of a BTS: (1) the BTS actually improved the V_T stability (if measured at high temperature for some sets of devices, for a stress well less than 10^5 s); (2) the BTS definitely degraded V_T stability (if measured at room temperature following rapid cooling); (3) the BTS only had a small effect on V_T stability (if measured at room temperature after some time has passed, especially if the device was unstressed during this time, or had zero or negative bias applied) [34]. It has also been reported that an AC BTS results in a similar post-stress degradation [34]. The implications of all these results for reliability testing are addressed in Section V.

In the previous section, direct tunneling to and from near-interfacial oxide traps was given as the cause of the room-temperature V_T instability. But why should high-temperature bias stressing result in such a big increase in instability when this tunneling mechanism should not be very sensitive to temperature? We have previously suggested that the much larger ΔV_T following a BTS is due to the activation of additional near-interfacial oxide traps, which are then free to participate in the same tunneling mechanisms described above [29, 31]. This is the likely high-temperature mechanism for the following reasons.

First, the main near-interfacial oxide trap is likely some variation of the so-called E'-type oxide defect, which is associated with an oxygen (O) vacancy. This is the only type of oxide defect that has been observed in SiC MOSFETs using electrically detected magnetic resonance (EDMR) techniques [36]. This defect has been previously identified as a hole trap [37], either being positively charged or charge neutral. It should be noted that this is also the same type of oxide defect that was determined to be the cause of the V_T instability in irradiated Si MOSFETs [38], whose instability was successfully explained by a direct tunneling mechanism [38, 39], as well as now being attributed as the cause of bias- temperature instability in Si MOSFETs [40]. The precursor site for this oxide trap is a weak Si-Si bond, owing to the oxygen vacancy, which may be broken by the hole-trapping process. This bond apparently can also be broken by the application of an electric field at elevated temperature. A certain number of these precursor sites have clearly already been activated during device processing, which is why the V_T instability effect is observed at room temperature, before the application of a BTS.

Second, it has been reported using on-the-fly electron spin resonance (ESR) techniques in the study of negative bias temperature instability in Si MOS that the number of E' centers increased dramatically with bias at temperature, but not with temperature stress alone [41]. In addition, they found that the number of oxide defects markedly decreased when the sample was returned to room temperature. These results are very similar to the electrical measurements reported above.

Third, and most telling, are quite recent ESR results showing an increase in the E' signal as a function of stress temperature (during a negative-bias stress), with larger signals—indicating more active oxide traps present—at higher stress temperatures; followed by a decrease in the E' signal back at room temperature [42]. These are consistent with preliminary electrical results from the literature [33], as well as more thorough results presented here.

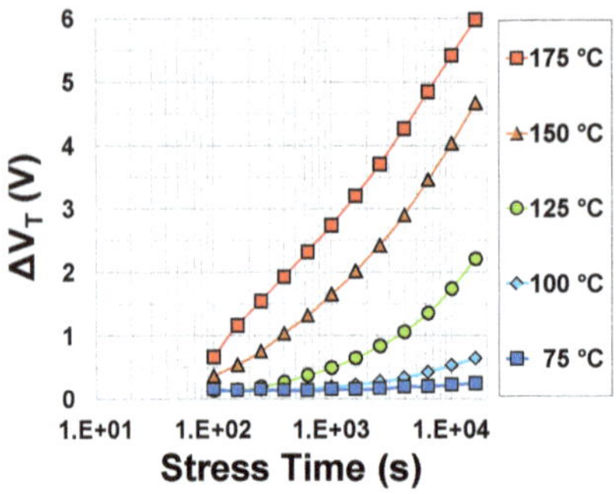

Fig. 6. Temperature dependence of the threshold-voltage instability versus stress time, calculated from the change in linear V_T, for a slightly older vintage SiC power MOSFET ("Gen-1"). Gate-bias stress was ±15 V.

Fig. 6 plots the change in V_T instability, determined by applying the stress-and-measure sequence depicted in Fig. 2, as a function of stress time, for various bias-stress temperatures and a bias stress of ±15 V. The device stressed (and measured) at 75 °C still has a linear-with-log-stress-time response. The devices stressed at 100 °C and above begin to exhibit an accelerated response with temperature, which occurs at earlier and earlier times as temperature is increased. After 10^4 s of first positive and then negative bias stress at 175 °C, a device shows a V_T hysteresis of about 6 V. This large instability is what was observed in state-of-the-art SiC power MOSFETs just a few years back. These older devices were used to better illustrate the effect of bias stress versus temperature.

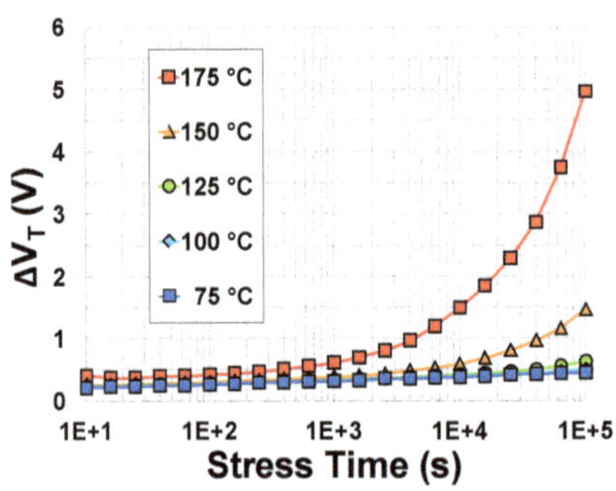

Fig. 7. Temperature dependence of the threshold-voltage instability versus stress time, calculated from the change in linear V_T, for a commercially available SiC power MOSFET of more recent vintage ("Gen-2").

Fig. 7 on the other hand shows results of a similar stress-and-measure sequence as a function of stress temperature for a commercially-available power MOSFET purchased within the past year, which more closely represents the state of the art. (It should be noted that it has very recently been reported that devices with very much improved V_T stability are now commercially available [43].) No super-linear increase in V_T stability is observed at 100 °C and below, out to 10^5 s of stressing, and only the slightest uptick is observed at 125 °C. Evidence of activation appears beyond 10^4 s at 150 °C and beyond 10^3 s at 175 °C. This increase in ΔV_T is particularly significant at 175 °C, reaching 5 V after 10^5 s. However, when comparing the 10^4-s 175 °C ΔV_T

results of Fig. 6 with those of Fig. 7, it is clear that there has been a dramatic (4×) improvement in SiC MOSFET V_T stability.

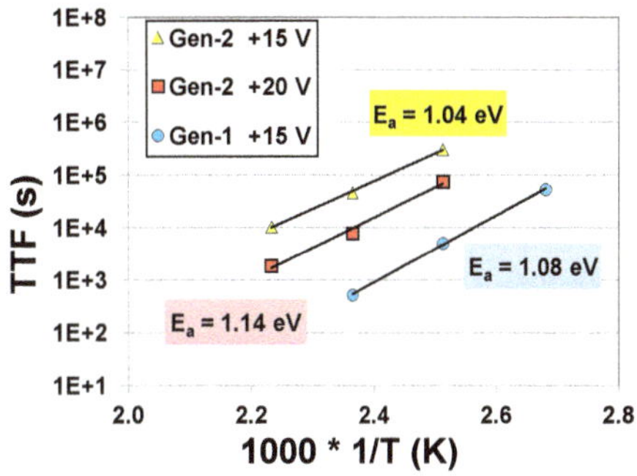

Fig. 8. Arrhenius plot of time to failure (defined as a V_T shift or instability of at least 0.5 V) plotted versus inverse temperature for both older ("Gen-1") and more recent vintage ("Gen-2") SiC power MOSFETs.

Activation energies were calculated for the more recent devices (labeled "Gen-2") stressed at +15 V and +20 V and for slightly older devices (labeled "Gen-1") stressed at +15 V. These results are shown in Fig. 8. Time to failure was defined as the time it took for V_T to shift 0.5 V in the case of unipolar bias stress, or the time it took for the back-and-forth instability of V_T to reach 0.5 V. In all the cases shown, the activation energy, E_a, is around 1.1 eV. These are typical results, although in some cases E_a values as large as 1.5 eV were calculated. This is presumably the energy required to break the weak Si-Si bond of the precursor state, creating an active E'-type oxide defect. In fact, a very recent density functional theory (DFT) analysis of alpha-quartz calculates an activation energy of 1.2 eV to transform from a charge-neutral Si-dimer state to a positively-charged "puckered" configuration in which the Si bonds have been broken and the E'-center complex has relaxed [44]. As expected from the results presented in Figs. 6 and 7, the more-recent-vintage devices take longer to fail. It was also observed that a larger applied gate-bias stress causes failure to occur sooner.

Fig. 9 shows the results of a series of room-temperature back-and-forth stress-and-measure sequences for a SiC power MOSFET of similar vintage as those whose results were shown in Fig. 6. But in this case, a single device was exposed to this series of high-temperature BTS sequences ranging from 75 °C to 175 °C. Each room-temperature bias-stress sequence shown in Fig. 9 was performed after rapidly cooling to room temperature following each BTS step. Rather than plotting the magnitude of the V_T instability, the V_T values following both the positive (closed symbols) and negative (open symbols) bias segments—of the test sequence depicted in Fig. 2—are displayed. First noticeable following the 100 °C stress, the width of the V_T "envelope" is clearly widening with increasing temperature stress. For the first four sequences, V_T remains relatively constant following the positive-bias segment of the room-temperature back-and-forth stress-and-measure sequence, but is drifting more negatively following the negative-bias segment. Following the 150 °C bias stress, the positive side of the V_T envelope begins drifting positively and the negative side does not move that much further than following the 125 °C bias stress. Following the 175 °C bias stress, the room-temperature V_T envelope has grown even wider, but it is all due to a large positive shift following positive bias stress. In fact, V_T following negative bias stress has also now begun to drift in the positive direction. This suggests that not only are additional near-interfacial oxide traps being generated with increasing temperature during a BTS (leading to a widening of the V_T envelope, i.e., an increase in V_T hysteresis), but that it is likely that additional interface traps are being generated as well, causing a general positive drift of the V_T envelope (due to more negatively-charged interface traps). This is consistent with other results in the literature [35, 45]. However, this is not the cause of the increased V_T instability since the same interface-trap charge state should exist when measuring the threshold voltage.

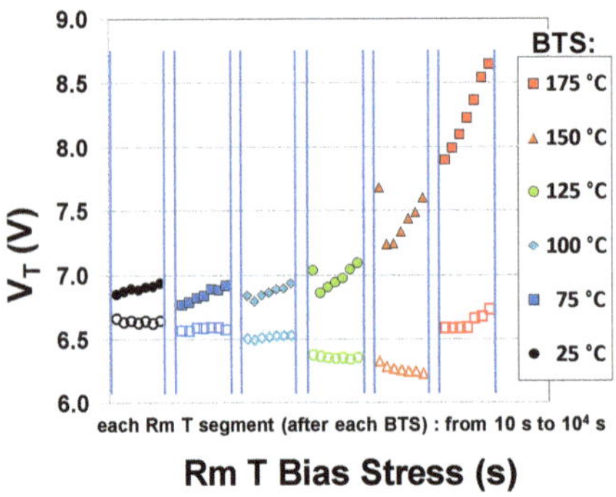

Fig. 9. Results of a series of room-temperature back-and-forth stress-and-measure sequences (as indicated in Fig. 2), each following an ever increasing BTS temperature. (Each BTS itself was performed for one day, for equal times under both +15 V and –15 V.) Closed symbols are V_T measured following the positive bias-stress segment and the open symbols are V_T measured following the negative bias-stress segment.

The negative shift in V_T under negative bias following BTS can be easily explained by additional active E'-type oxide defects, since this type of defect is typically either positively charged or charge neutral [37]. But what about the strong positive shift in V_T that is observed under positive bias following a BTS? Some additional defect must be contributing additional negative charge. We see evidence from Fig. 9 for a build-up of additional interface traps, but what can account for the rapid recovery of this positive shift in V_T with time back at room temperature that is reported in the literature [28-30, 32]? One possibility is that the E' complex is able to trap a second electron and become net negatively charged [46]—and that this only happens at elevated temperature due to a change in energy level of the oxide trap [47]. Such a high-temperature oxide electron-trap may lose its additional charge when cooled back to room temperature—especially if the gate bias is switched from positive to zero or negative applied voltage. Although no such experimental ESR evidence exists at present, theoretical considerations suggest that the E' complex may indeed behave in this fashion [48].

Clearly, this oxide-trap activation, which requires an applied electric field across the gate oxide, is a function of time at temperature. Given the significant super-linear increase of the V_T instability with log stress time that occurs as a result, it is critical to understand and limit the underlying basic mechanism(s); especially since SiC power MOSFETs are expected to operate reliably at elevated temperatures. There are several potential paths to improved device reliability: (1) decrease the number of precursor sites by introducing more O at the interface, eliminating the weak Si-Si bonds; (2) limit the activation process by changing the Si-Si bond strain near the interface [38]; (3) reduce or eliminate potential high-temperature high-electric-field conditions in the oxide during processing; (4) neutralize already active traps by an additional post-stress N or P anneal [11]; (5) balance charge-trapping processes with mobile ions [3, 20, 21] (not recommended!); (6) reduce the number of interface traps instead, if the charging of near-interfacial oxide traps is a two-step process requiring the presence of interface traps.

5. Implications for reliability testing

It is critically important to be able to ensure the reliability of the recently commercialized SiC power MOSFETs. The presence of all these mechanisms described above regarding oxide-trap charging, activation, and recovery result in a rather complex response to bias and temperature stress as a function of stress time. This is especially true if this response is further convoluted by the presence of interface traps and possibly even mobile ions. As a result, the present existing test standards, which are based on the response of Si power devices, are inadequate for SiC.

It has been clearly demonstrated in the literature that the application of existing test standards such as the Automotive Electronics Council Q101 [49]—based on the JEDEC JESD-22 A108C test method [50]—may result in an ambiguous test result [30, 32]. For example, an immediate room-temperature measurement following a PBTS may result in a significant positive shift in V_T. However, delaying that measurement for an hour, especially if unbiased, generally results in a significant recovery of V_T back to near its pre-stress value. The immediate measurement might indicate device failure, depending on the failure criteria for V_T shift, but the delayed measurement would suggest that the device had not failed [30, 32]. The present reliability standard states that "electrical testing shall be completed as soon as possible and no longer than 96 hours after removal of bias from devices" [50]. A 96-hour delay is clearly inadequate! The test method is also generally silent regarding measurement details such as measurement speed. Not only are the commercial test standards inadequate, but so are the military standards. The relevant test methods given in MIL-STD-750—e.g., test method 1042.3 for device burn-in and life-testing—are similarly flawed and also need revising [51].

Recommendations for an improved HTGB test method include minimizing the delay time, reapplying the gate-bias stress if bias was momentarily removed to facilitate testing, and minimizing the measurement time [1, 19]. This is especially important if a gate-bias sweep is performed to calculate V_T—in such a case it is also very important to sweep the gate correctly (negatively following a positive bias stress and positively following a negative bias stress) [19]. It is important to perform an HTGB test under both bias polarities and to make measurements both at high temperature and immediately at room temperature [34].

Beyond improving the standard DC bias test methods, which are relatively simple to perform and provide a universal standard for comparison, it is of great importance to stress under application-specific conditions. This includes testing under AC temperature-stress conditions since typical applications involve constant switching of the power MOSFET from the *ON*-state to the blocking state and back again. (Recent results in the literature suggest that a high-temperature gate-switching (AC) stress is just as detrimental as a DC BTS [34].) Furthermore, it is critically important to be able to identify the worst-case stress and measurement conditions, based on the insight of the controlling mechanisms at both room temperature and high temperature, as a function of bias and time. This will enable a determination of the expected time-to-failure for a given application. The worst case is likely to be when a device is kept under constant bias at temperature, especially the HTRB condition. Even if the device was mostly switching, oxide-trap activation would likely still be occurring so that if the device was put in the blocking mode for just a few minutes, a significant negative shift in V_T could occur (keeping in mind that much larger V_T shifts may be occurring than can be readily measured—compare Fig. 4 with Fig. 5), resulting in a large increase in drain-leakage current and possibly device failure.

Conclusion

The results presented can generally be explained by the oxide hole-trap model that was developed at ARL years ago to explain the V_T instability observed in irradiated Si MOSFETs with similar oxide thicknesses as today's SiC power devices [38]. The precursor site is a weak Si-Si bond owing to an oxygen vacancy in the near-interfacial region. Once this bond is broken (due either to ionizing radiation or BTS mechanisms), an active trap is formed—the E' center. It can either be positively charged, or with the tunneling in of an electron, charge neutral. This process is generally repeatable.

Researchers at ARL were the first to systematically study this V_T instability effect in SiC MOSFETs and suggested that it was likely due to a similar direct-tunneling mechanism to and from near-interfacial oxide traps [1, 18]. Researchers at Penn State later experimentally verified the presence of E'-type oxide defects in SiC MOSFETs [36]. Researchers at ARL then proposed that the significant increase in the V_T instability following BTS at elevated temperature was due to the activation of additional precursor O-vacancy defect sites [29, 31]. This too has now been experimentally verified in the past year [42]. Furthermore, in this work we calculated an activation energy of 1.1 eV, which compares very well with very recent DFT calculations of 1.2 eV [44]. The cause of the rapid room-temperature recovery of a significant positive V_T shift following a positive BTS is still not fully understood [28-30, 32]. One possible explanation is that this same hole trap can also act as a high-temperature electron trap by trapping a second electron [47, 48]. When filled, this can cause a positive V_T shift, but cannot be sustained at room temperature, especially when a zero or negative gate bias is applied and this electron tunnels out of the trap.

Developing a complete understanding of the response of near-interfacial oxide traps with respect to bias, temperature, and time during application-relevant stress conditions is imperative in order to develop an accurate method for estimating time-to-failure and ensuring SiC power MOSFET reliability. Understanding the effect of various processing steps and their variations on the creation and potential passivation of these oxide traps (and perhaps interface traps as well, if they play a role in oxide-trap charging) should result in dramatic improvements in device reliability.

References

[1] A. J. Lelis, D. Habersat, R. Green, A. Ogunniyi, M. Gurfinkel, J. Suehle, and N. Goldsman, "Time Dependence of Bias-Stress-Induced SiC MOSFET Threshold-Voltage Instability Measurements," IEEE Trans. Elec. Dev., 55(8), p. 1835 (2008). https://doi.org/10.1109/TED.2008.926672

[2] M. Gurfinkel, H. D. Xiong, K.P. Cheung, J. S. Suehle, J. B. Bernstein, Y. Shapira, A. J. Lelis, D. Habersat, and N. Goldsman, "Characterization of Transient Gate Oxide Trapping in SiC MOSFETs Using Fast I–V Techniques," IEEE Trans. Elec. Dev., 55(8), p. 2004 (2008). https://doi.org/10.1109/TED.2008.926626

[3] T. Okayama, S. D. Arthur, J. L. Garrett, and M. V. Rao, "Bias-stress induced threshold voltage and drain current instability in 4H-SiC DMOSFETs," Solid-State Electronics, 52, p. 164 (2008). https://doi.org/10.1016/j.sse.2007.07.031

[4] M. J. Tadjer, K. D. Hobart, E. Imhoff, and F. J. Kub, "Temperature and Time Dependent Threshold Voltage Instability in 4H-SiC Power DMOSFET Devices," Mater. Sci. Forum, 600-603, p. 1147 (2009). https://doi.org/10.4028/www.scientific.net/MSF.600-603.1147

[5] M. Grieb, M. Noborio, D. Peters, A. J. Bauer, P. Friedrichs, T. Kimoto, and H. Ryssel, "Comparison of the threshold-voltage stability of SiC MOSFETs with thermally grown and deposited gate oxides," Mater. Sci. Forum, 645-648, p. 681 (2010). https://doi.org/10.4028/www.scientific.net/MSF.645-648.681

[6] H. Yano, Y. Oshiro, D. Okamoto, T. Hatayama, and T. Fuyuki, "Instability of 4H-SiC MOSFET Characteristics due to Interface Traps with Long Time Constants," Mater. Sci. Forum, 679-680, p. 603 (2011). https://doi.org/10.4028/www.scientific.net/MSF.679-680.603

[7] S. DasGupta, R. Brock, R. Kaplar, M. Marinella, M. Smith, and S. Atcitty, "Extraction of trapped charge in 4H-SiC metal oxide semiconductor field effect transistors from subthreshold characteristics," Appl. Phys. Lett., 99, 023503 (2011). https://doi.org/10.1063/1.3611029

[8] M. Okamoto, Y. Makifuchi, M. Iijima, Y. Sakai, N. Iwamuro, H. Kimura, K. Fukuda, and H. Okumura, "Coexistence of Small Threshold Voltage Instability and High Channel Mobility in 4H-SiC(000_1) Metal–Oxide–Semiconductor Field-Effect Transistors," Applied Physics Express, 5, 041302 (2012). https://doi.org/10.1143/APEX.5.041302

[9] S. Noll, D. Scholten, M. Grieb, A. Bauer, and L. Frey, "Electrical Impact of the Aluminum p-implant Annealing on Lateral MOSFET Transistors on 4H-SiC n-epi," Mater. Sci. Forum, 740-742, p. 521 (2013). https://doi.org/10.4028/www.scientific.net/MSF.740-742.521

[10] T. Aichinger, P. Lenahan, and D. Peters, "Interface defects and negative bias temperature instabilities in 4H-SiC PMOSFETs – a combined DCIV/SDR study," Mater. Sci. Forum, 740-742, p. 529 (2013). https://doi.org/10.4028/www.scientific.net/MSF.740-742.529

[11] H. Watanabe, D. Ikeguchi, T. Kirino, S. Mitani, Y. Nakano, T. Nakamura, T. Hosoi, and T. Shimura, "Novel Approach for Improving Interface Quality of 4H-SiC MOS Devices with UV Irradiation and Subsequent Thermal Annealing," Mater. Sci. Forum, 740-742, p. 741 (2013). https://doi.org/10.4028/www.scientific.net/MSF.740-742.741

[12] G. Pobegen and T. Grasser, "Efficient Characterization of Threshold Voltage Instabilities in SiC nMOSFETs Using the Concept of Capture-Emission-Time Maps," Mater. Sci. Forum, 740-742, p. 757 (2013). https://doi.org/10.4028/www.scientific.net/MSF.740-742.757

[13] K. Matocha, K. Chatty, S. Banerjee, and L. B. Rowland, "1700V, 5.5mOhm-cm^2 4H-SiC DMOSFET with Stable 225°C Operation," Mater. Sci. Forum, 778-780, p. 903 (2014). https://doi.org/10.4028/www.scientific.net/MSF.778-780.903

[14] A. Agarwal, H. Fatima, S. Haney, and S.-H. Ryu, "A New Degradation Mechanism in High-Voltage SiC Power MOSFETs," IEEE Elec. Dev. Lett., 28(7), p. 587 (2007). https://doi.org/10.1109/LED.2007.897861

[15] L. C. Yu, G. Dunne, K. Matocha, K. P. Cheung, J. Suehle, and K. Sheng, "Reliability Issues of SiC MOSFETs: A Technology for High-Temperature Environments," IEEE Trans. Dev. and Mater. Rel., 10(4), p. 418 (2010). https://doi.org/10.1109/TDMR.2010.2077295

[16] D. Habersat and A. Lelis, "Improved Observation of SiC/SiO_2 Oxide Charge Trapping Using MOS C-V," Mater. Sci. Forum, 679-680, p. 366 (2011). https://doi.org/10.4028/www.scientific.net/MSF.679-680.366

[17] A. J. Lelis, D. Habersat, F. Olaniran, B. Simons, J. M. McGarrity, F. B. McLean, and N. Goldsman, "Time-Dependent Bias Stress-Induced Instability of SiC MOS Devices" Mater. Res. Soc. Symp. Proc., 911, article 0911-B13-05 (2006).

[18] A. J. Lelis, D. Habersat, R. Green, and N. Goldsman, "Two-way Tunneling Model of Oxide Trap Charging and Discharging in SiC MOSFETs," Mater. Sci. Forum, 717-720, p. 465 (2012). https://doi.org/10.4028/www.scientific.net/MSF.717-720.465

[19] A. Lelis, D. Habersat, R. Green, and M. El, "Key Reliability Issues for SiC Power MOSFETs," ECS Transactions, 58(4), p. 87 (2013). https://doi.org/10.1149/05804.0087ecst

[20] A. J. Lelis, D. Habersat, R. Green, and N. Goldsman, "Temperature-Dependence of SiC MOSFET Threshold-Voltage Instability," Mater. Sci. Forum, 600-603, p. 807 (2009). https://doi.org/10.4028/www.scientific.net/MSF.600-603.807

[21] D.B. Habersat, A. Lelis, and R. Green, "Detection of Mobile Ions in the Presence of Charge Trapping in SiC MOS Devices," Mater. Sci. Forum, 717-720, p. 461 (2012). https://doi.org/10.4028/www.scientific.net/MSF.717-720.461

[22] D. B. Habersat, A. J. Lelis, S. Potbhare, and N. Goldsman, "Improvements in SiC MOS Processing as Revealed by Studies of Fixed and Oxide Trap Charge," Mater. Sci. Forum, 615-617, p. 769 (2009). https://doi.org/10.4028/www.scientific.net/MSF.615-617.769

[23] V. V. Afanas'ev, A. Stesmans, F. Ciobanu, G. Pensl, K. Y. Cheong, and S. Dimitrijev, "Mechanisms responsible for improvement of 4H–SiC/SiO_2 interface properties by nitridation," Appl. Phys. Lett., 82(4), p. 568 (2003). https://doi.org/10.1063/1.1532103

[24] D. B. Habersat, A. J. Lelis, J. M. McGarrity, F. B. McLean and S. Potbhare, "The Effect of Nitridation on SiC MOS Oxides as Evaluated by Charge Pumping," Mater. Sci. Forum, 600-603, p. 743 (2009). https://doi.org/10.4028/www.scientific.net/MSF.600-603.743

[25] H. Yano, T. Araoka, T. Hatayama, and T. Fuyuki, "Improved Stability of 4H-SiC MOS Device Properties by Combination of NO and $POCl_3$ Annealing," Mater. Sci. Forum, 740-742, p. 727 (2013). https://doi.org/10.4028/www.scientific.net/MSF.740-742.727

[26] J. Rozen, S. Dhar, M. E. Zvanut, J. R. Williams, and L. C. Feldman, "Density of interface states, electron traps, and hole traps as a function of the nitrogen density in SiO_2 on SiC," J. Appl. Phys., 105, 124506 (2009). https://doi.org/10.1063/1.3131845

[27] J. A. Taillon, J. H. Yang, C. A. Ahyi, J. Rozen, J. R. Williams, L. C. Feldman, T. S. Zheleva, A. J. Lelis, and L. G. Salamanca-Riba, "Systematic structural and chemical characterization of the transition layer at the interface of NO-annealed 4H-SiC/SiO_2 metal-oxide-semiconductor field-effect transistors," J. Appl. Phys., 113, 044517 (2013). https://doi.org/10.1063/1.4789924

[28] A. Lelis, R. Green, D. Habersat, and N. Goldsman, "Effect of *ON*-state Stress on SiC DMOSFET Subthreshold *I-V* Characteristics," Mater. Sci. Forum, 645-648, p. 983 (2010). https://doi.org/10.4028/www.scientific.net/MSF.645-648.983

[29] A. J. Lelis, R. Green, and D. Habersat, "High-Temperature Reliability of SiC Power MOSFETs," Mater. Sci. Forum, 679-680, p. 599 (2011). https://doi.org/10.4028/www.scientific.net/MSF.679-680.599

[30] R. Green, A. Lelis, and D. Habersat, "Application of Reliability Test Standards to SiC Power MOSFETs," 2011 Intl. Rel. Physics Symposium Proc. 11, p. 756 (2011). https://doi.org/10.1109/IRPS.2011.5784573

[31] A. Lelis, R. Green, and D. Habersat, “Effect of Threshold-Voltage Instability on SiC Power MOSFET High-Temperature Reliability,” ECS Transactions, 41(8), p. 203 (2011). https://doi.org/10.1149/1.3631498

[32] R. Green, A. Lelis, and D. Habersat, “Charge Trapping in SiC Power MOSFETs and its Consequences for Robust Reliability Testing,” Mater. Sci. Forum, 717-720, p. 1085 (2012). https://doi.org/10.4028/www.scientific.net/MSF.717-720.1085

[33] A. Lelis, R. Green, M. El, and D. Habersat, “Effect of Stress and Measurement Conditions in Determining the Reliability of SiC Power MOSFETs,” ECS Transactions, 50(3), p. 251 (2012). https://doi.org/10.1149/05003.0251ecst

[34] R. Green, A. Lelis, M. El, and D. Habersat, “A Study of High Temperature DC and AC Gate Stressing on the Performance and Reliability of Power SiC MOSFETs,” Mater. Sci. Forum, 740-742, p. 549 (2013). https://doi.org/10.4028/www.scientific.net/MSF.740-742.549

[35] D. B. Habersat, A. J. Lelis, R. Green, and M. El, "Evaluation of PBTS and NBTS in SiC MOS Using In Situ Charge Pumping Measurements," Mater. Sci. Forum, 740-742, p. 545 (2013). https://doi.org/10.4028/www.scientific.net/MSF.740-742.545

[36] C. J. Cochrane, P. M. Lenahan, and A. J. Lelis, “An Electrically Detected Magnetic Resonance Study of Performance Limiting Defects in SiC Metal Oxide Semiconductor Field Effect Transistors,” J. Appl. Phys., 109, 014506 (2011). https://doi.org/10.1063/1.3530600

[37] P. M. Lenahan and P. V. Dressendorfer, “Hole traps and trivalent silicon centers in metal/oxide/silicon devices,” J. Appl. Phys., 55(10), p. 3495 (1984). https://doi.org/10.1063/1.332937

[38] A. J. Lelis, T. R. Oldham, H. E. Boesch, Jr., and F. B. McLean, “The Nature of the Trapped Hole Annealing Process,” IEEE Trans. Nucl. Sci., 36(6), p. 1808 (1989). https://doi.org/10.1109/23.45373

[39] T. R. Oldham, A. J. Lelis, and F. B. McLean, “Spatial dependence of trapped holes determined from tunneling analysis and measured annealing,” IEEE Trans. Nucl. Sci., 33(6), p. 1203 (1986). https://doi.org/10.1109/TNS.1986.4334579

[40] T. Grasser, B. Kaczer, W. Goes, H. Reisinger, T. Aichinger, P. Hehenberger, P.-J. Wagner, F. Schanovsky, J. Franco, M. Toledano Luque, and M. Nelhiebel, “The Paradigm Shift in Understanding the Bias Temperature Instability: From

Reaction–Diffusion to Switching Oxide Traps," IEEE Trans. Elec. Dev., 58(11), p. 3652 (2011). https://doi.org/10.1109/TED.2011.2164543

[41] J. T. Ryan, P. M. Lenahan, T. Grasser, and H. Enichlmair, "Observations of negative bias temperature instability defect generation via on the fly electron spin resonance," Appl. Phys. Lett., 96, 223509 (2010). https://doi.org/10.1063/1.3428783

[42] P. M. Lenahan, C. Cochrane, M. Anders, J. Follman, A. Lelis, and S. Arthur, "Defects in 4H SiC MOSFETs," presented at the 8th Annual SiC MOS Workshop, College Park, MD USA, August 2013.

[43] L. Cheng and J. W. Palmour, "Cree's SiC Power MOSFET Technology: Present Status and Future Perspective," presented at the 9th Annual SiC MOS Workshop, College Park, MD USA, August 2014.

[44] X. Shen and S. Pantelides, private communication, August 2014.

[45] C.J. Scozzie and J.M. McGarrity, "Charge Pumping Measurements on SiC MOSFETs," Mater. Sci. Forum, 264-268, p. 985 (1998). https://doi.org/10.4028/www.scientific.net/MSF.264-268.985

[46] M. Walters and A. Reisman, Radiation-Induced Neutral Electron Trap Generation in Electrically Biased Insulated Gate Field Effect Transistor Gate Insulators, J. Electrochem. Soc., 138, p. 2756 (1991). https://doi.org/10.1149/1.2086050

[47] A. Lelis, R. Green, D. Habersat, and M. El, "Basic Mechanisms of SiC MOS Bias-Temperature Instability," presented at the 8th Annual SiC MOS Workshop, College Park, MD USA, August 2013.

[48] X. Shen and S. Pantelides, "Atomic Origin of High-Temperature Electron Trapping in SiO_2," presented at the 9th Annual SiC MOS Workshop, College Park, MD USA, August 2014.

[49] "Stress Test Qualification for Automotive Grade Discrete Semiconductors," AEC-Q101-Rev-C, 2005.

[50] "Temperature, Bias, and Operating Life Std.," JESD22-A108C, 2005.

[51] "Test Methods for Semiconductor Devices," MIL-STD-750E, 2006.

Chapter 23

Integrated Thermal Management for Power Electronics Packaging

Lauren M. Boteler, Darin J. Sharar, and Nicholas R. Jankowski
U.S. Army Research Laboratory, Adelphi, MD, USA

Abstract

The U.S. Army is interested in improving power modules to enable current and future Army platforms. This chapter discusses the unique challenges of developing next-generation cooling solutions and their associated power module implications. The analysis has shown that a combination of approaches is the best method to achieve future Army goals. Understanding the tradeoffs between the various thermal and packaging improvements is critical to optimizing future power modules.

Foreword

This Chapter is a showcase of ARL staff technologists and includes collaborative work with external partners including universities, industry and other government organizations. Please refer to the Acknowledgements section for more details, with reference to PhD and MS advisors, committee members, and collaborators whose guidance, insight, and support, as well as pioneering work, has inspired and informed much of this ongoing research effort.

Keywords

power electronics, thermal management, two-phase cooling, integrated thermal management, electronics packaging, single phase cooling, codesign, multi-functionalization, phase change materials, transient thermal, power module, manifold microchannel cooling.

Contents

1. Introduction

The U. S. Army Research Laboratory (ARL) has made a substantial investment in developing high performance electronic systems to control the distribution of power on soldier, vehicle, and airborne platforms. While power control and conversion elements are ubiquitous in modern technology, appearing in every device that cannot directly operate on the connected wall-plug, battery, or generator electrical output, military electronic products must satisfy a more rigorous set of requirements. Militarized products in general must be able to operate in a wide range of potential environmental conditions, including exposure to high and low ambient temperatures, vibration, and shock as laid out in MIL-STD-810 [1]. In addition to this, performance demands of military electronics components are often much higher than their commercial counterparts. Military platforms are developed under tight size, weight, and power (SWAP) constraints, and subcomponents are required to have higher power and functional density to meet those constraints. These demands result in higher power densities, increased temperatures, and daunting thermal challenges.

The majority of the Army power electronics packaging and thermal work reviewed in this chapter is focused on the vehicle and vehicle payload power electronics cooling challenge, as it is the dominant programmatic element driving the development of high power and high voltage electronics. That said, many of the approaches and solutions described herein are applicable to other thermally challenging applications such as high-performance computing, renewables, and high-power optical and microwave systems, to name a few.

1.1 Power electronics applications

In the most general sense, power electronics are used to convert electrical energy from one form (alternating or direct current, low or high voltage, continuous or pulsed, etc.) to another to enable electrical transfer from a source to a load. Often it is the case that due to transmission or storage characteristics a mismatch exists between the form of electricity available and the form needed for operation. This is especially evident in modern vehicles, where the primary energy sources include a chemical battery and/or a motor-coupled alternator; but the vehicle, passenger, and payload systems may each require different voltages and currents.

Converting from one form of electrical power to another can take many forms: inverting a steady direct current to produce a sinusoidal current (DC-to-AC), rectifying an alternating current to produce a steady output (AC-to-DC), changing voltage or current level (DC-to-DC or AC-to-AC), changing AC frequency, or generating a pulsed output from a non-pulsed input. The exact methods for constructing electrical circuits are far outside the scope of this effort (the interested reader is referred to [2]), but many modern methods of constructing these circuits make use of common elements that define the thermal and packaging challenges that we are trying to address. In particular, the solid-state semiconductor transistor operating in a digital, or switched, mode is one of the primary building blocks of these power conversion circuits. Used in the correct topology with other circuit components (diodes, capacitors, inductors, etc.), these semiconductor switches can perform desired power conversion functions in compact form factors with relatively high efficiency. However, even though the power conversion efficiency of these components can be high (> 95%), their high power density still creates a challenging thermal management problem, especially when being used in a severe vehicle environment.

1.1.1 Vehicle power electronic applications

Just as commercial vehicles have been moving to more electric drivetrain technologies, military vehicle designs have been exploring the same approach. Increasingly electrified vehicles require more power conversion elements to move energy to what were traditionally mechanical components, including electric wheel-motors, drive-by-wire steering elements, and even passenger environmental control systems. Figure 1 depicts a notional military vehicle layout including numerous power conversion elements for drive and payload systems. Moving to an electric or hybrid electric vehicle architecture introduces additional high power electronic components that must operate in a severe under-hood environment.

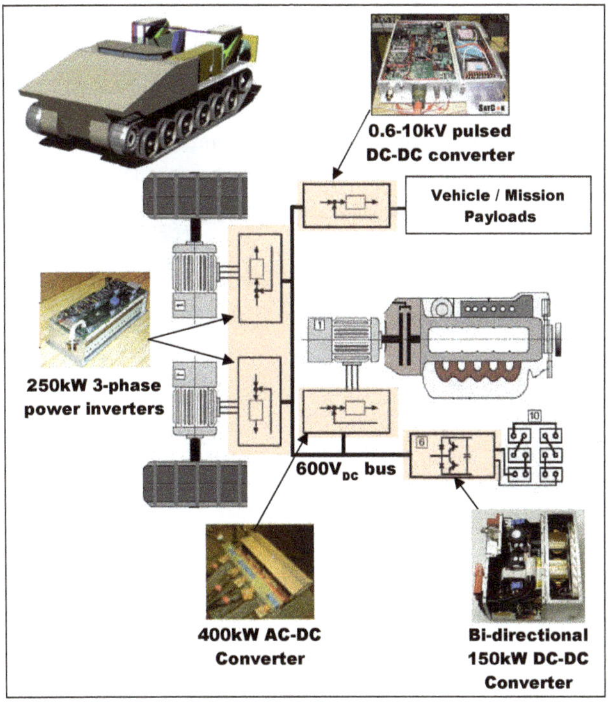

Figure 1 - Conceptual future tactical vehicle hybrid electric drive train and representative high heat flux power electronic components.

In addition to the vehicle systems, many military vehicles are configured with numerous high-power electronic components within the crew cabin, sometimes to the point of being overcrowded as shown in Figure 2, which was adapted from an overview of the Vehicle Electronics and Architecture focus area of the Tank and Automotive Research Development and Engineering Center (TARDEC) [3]. A recent power study of a M2A3 Bradley vehicle detailed several electrical load scenarios with total electrical power ranging from 4 to 10kW [4]. These loads included Command and Control (C2) computers and displays, radio systems, navigation systems and others, and the total equipment electric load far exceeded the designed personnel heat load (approximately 100-150W metabolic thermal load per crew member) [5]. These increasing vehicle cabin and payload components have multiple impacts on the vehicle power and thermal systems: (1) they increase the requirement for non-drive-related electrical supply (more/larger vehicle power electronics), (2) they inefficiently vent heat directly to the cabin air, putting additional load on an already severely overburdened vehicle environmental control system (ECS) [6], and (3) they are generally composed of less robust commercial electronics with lower operating temperature limits. This combination further complicates the overall vehicle thermal management challenge.

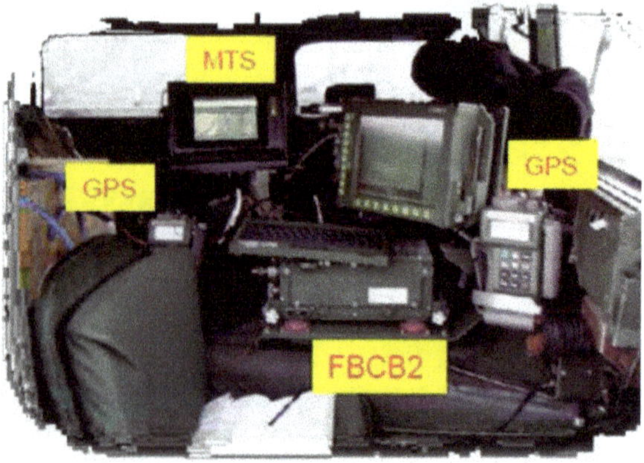

Figure 2 – Crew cabin loaded with an array of militarized electronic equipment contributing to a vehicle climate control overburden, adapted from [3].

1.1.2 Vehicle cooling loop considerations

As described in a recent review by Jankowski and McCluskey, the vehicle presents a challenging thermal environment for electronic systems [7]. These components must be able to operate from sub-freezing temperatures to peak temperatures that could reach hundreds of degrees Celsius depending on the exact location and function. Under-hood engine and transmission components could easily see 150-200°C, while wheel mounted components, and brake- and steer-by-wire components in modern vehicles could see as much as 300°C environments [8]

Modern vehicles handle waste heat using liquid coolant distribution loops and vehicle heat exchangers, eventually rejecting heat to the outside environment using the vehicle radiator. This imposes specific temperature range limits on the electrical components to be cooled. The presence of multiple, disparate components around the vehicle has generally forced a requirement for multiple coolant loops to be used to provide the necessary operating thermal environments. Most power electronics systems require some form of direct cooling, and high power electronics need to be liquid cooled to stay below critical operating temperature limits. Figure 3 provides a multi-loop cooling connection layout for components in a notional military hybrid electric vehicle (HEV), with highlights showing electronic components determined to require connections to the high, midrange, and low temperature loops [9], with the estimated coolant rejection temperatures highlighted in Figure 3(b).

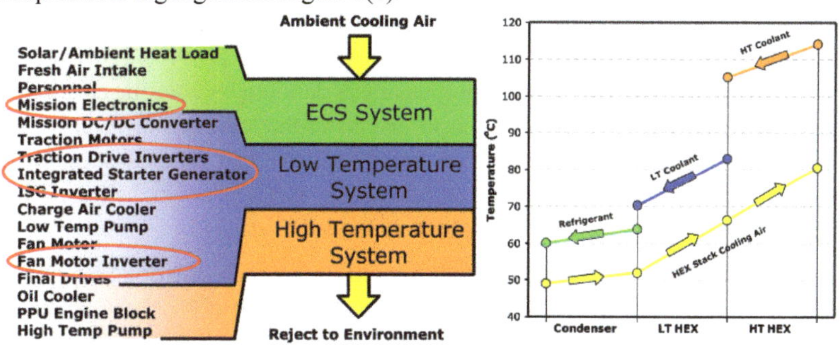

(a) (b)

Figure 3 – (a) Assignment strategy of vehicle loads to cooling loop in a concept multi-loop military HEV cooling architecture, and (b) temperature progression between the heat acquisition and rejection sides of the different loops, adapted from [9].

As can be seen, the electronic components are primarily located in the two lower temperature levels, which is largely a result of the upper temperature limits of the components. Rejecting heat at lower temperatures results in lower heat transfer rates that must be overcome using larger radiators or extra fan power, which comes at a direct cost to system SWAP. Moving the electronics to higher temperature cooling loops has been a vehicle system goal for quite some time in both the civilian and military sectors, but doing so still requires significant thermal improvement in the electrical devices and related components to be able to survive operation at these higher temperatures.

In most vehicles, the lowest temperature loop is the Environmental Control System (ECS) loop, which is a refrigerant loop that generally only provides air conditioning in commercial vehicles. The low and high temperature loops are generally antifreeze coolant loops consisting of an ethylene or propylene glycol and water mix that is pumped to the radiator for heat rejection, in addition to a pumped oil loop for primary engine cooling and lubrication. Temperature ranges of these different loops can vary from system to system, but in both commercial and military vehicles the lower temperature loop will ideally operate in the 65-85°C range, while the upper temperature loops can be anywhere from 80-120°C. Most loops are generally pressurized to maintain the coolant in the liquid phase with the antifreeze boiling temperature being just above 120°C. In fact, the Department of Energy's former FreedomCar program targets were based on a HEV model that had a nominally 65°C lower temperature loop for the vehicle's power conversion electronics and a 105°C higher temperature loop for engine cooling [10] [11]. Similarly, military vehicle development used 80°C lower and 105-120°C upper temperature loop designs [11]. In both cases, there was desire to improve the electronics and associated packaging such that they could operate at the same power level on the higher temperature coolant loops such that overall radiator and fan loads could be reduced, although such high temperatures can cause performance issues in the engine and reliability issues with elements such as the pump, seals, and hoses. [12]

The upper temperature loop has historically been an antifreeze loop consisting of a 50/50 ethylene glycol and water (EGW) solution [13], though there has been a recent commercial movement toward the use of lower toxicity propylene glycol and water (PGW) solutions [14]. The temperature at which the coolant is delivered to the electronic components is highly dependent on operating condition. Most silicon power electronics have a full-power operating temperature limit of about 125 °C [15], and the boiling temperatures for EGW and PGW at ambient pressures are 107 °C and 106 °C, respectively. These high inlet temperatures limit the electronic device temperature increase to only 25–45 °C above inlet temperature, reducing electronic cooling potential and requiring exceptional cooling performance to dissipate high-heat-fluxes and prevent overheating.

In addition to low allowable temperature margins, the EGW coolant limits the electronics thermal budget. EGW is used mainly to provide a wider liquid temperature range for the fluid than a water system would provide (–40 °C–107 °C compared to 0 °C–100 °C at ambient pressure, respectively). However, past research by Sharar et al. at ARL on single-phase cooling has shown that EGW and PGW mixtures can have a 28% to 42% reduction, respectively, in cooling performance compared to a pure water microchannel system [16], mainly due to increased viscosity and decreased thermal conductivity of the

EGW and PGW mixtures compared to pure water. Furthermore, typical cooling systems are not pure 50/50 EGW. Additives that affect fluid properties—borax, sodium Mercaptobenzothiazole (sodium MBT), trisodium phosphate, and antifoaming agents—are often present. It is unclear exactly what effect these additives have on thermal performance or the function of advanced heat exchangers [13].

The single coolant loop architecture also restricts the potential performance of the electronic cooling components. Because the engine is usually the primary cooling concern for this loop, changes to loop parameters, such as fluid type, flowrate, and available pressure, generally cannot be made even if electronics cooling performance could be improved. The loop is generally high flowrate with low pressure drop available for electronics cooling (typically 2–5psi), which complicates or even prevents the use of many cooling solutions that often have driving pressure requirements on the order of 10–100 psi [17] [18] [19]. While these designs may reduce total pumping power, integration to current single loop architectures would require flow conversion.

This single loop architecture is also not designed to handle combined gas and liquid flow, as would occur in boiling heat transfer. Consequently, vapor buildup in any portion of the system could result in coolant venting or reduced coolant flow through the radiator and premature thermal failure due to subsequent reduced heat transfer. Therefore, a margin of safety is required to maintain the system well within single-phase conditions, which can increase flowrate, pumping power, and radiator requirements. Additionally, the specific boiling behavior of EGW and PGW (termed binary fluids) is typically suppressed compared to the onset conditions for boiling of pure components. The resulting suppression, whose severity is dependent on the thermodynamic properties of the specific mixture, presents an additional resistance to heat transfer.

Finally, while the engine coolant loop is a sealed system, it is user-serviced and subject to contamination and particulate accumulation [13]. From a design perspective, contamination either requires a system with large enough fluid lines and cooling geometries to provide immunity to clogging, or stringent filtering requirements to remove contamination from the system. Both of these scenarios have negative consequences; typically larger geometries in cold plates result in reduced thermal performance, and rigorous filtering may cause large system pressure drops and reduced flowrates.

Because of the limitations of a single coolant loop, some consideration has been given to a secondary cooling loop that either exchanges heat with the primary coolant loop or with outside air directly. This approach is most readily apparent in the cabin cooling or air conditioning systems present in almost all modern vehicles. Beyond this, other platforms have examined the use of secondary loops specifically designed for cooling non-engine electronic heat loads [20]. Four potential benefits lie in the ability to design a secondary cooling loop optimized for the specific heat loads in the system:

1. Flexibility of working fluid, including refrigerants in some cases.
2. Possibility to implement a vapor compression refrigeration cycle to deliver coolant to the cold plates below ambient temperature.
3. Pump autonomy to optimize system pressure, flow, and temperature to match specific heat loads.
4. Potential for hermetically sealed systems to reduce contamination and clogging/fouling potential.

There are definite disadvantages involved with implementing a secondary loop; the most obvious of these are the additional size, weight, and power (SWAP) required for the extra loop components such as pumps, hoses/tubing, and heat exchangers/condensers. While the possible drawbacks of such a two-loop system may be significant, the overall potential for cooling performance improvement may be able to offset those negative elements. Implementation of a secondary loop may add system complexity but enables the use of thermal techniques like microchannels and two-phase cooling that would be logistically difficult with a primary engine coolant loop. Please note, this 'Vehicle Cooling Loop Considerations' section is covered in ARL-TR-5323 by Sharar, Jankowski, and Morgan [21]

2. Army power electronics cooling challenges

Electronics devices (Si, SiC, and GaN) have advanced to the point at which the package and thermal dissipation technique is now what is limiting performance. There is a critical need for military electronic systems that are smaller, lower weight, more reliable, cheaper, and better performing. The choice of packaging materials, device and package geometry, interconnection (on and off-chip), and device and package configuration significantly impacts power density. Maintaining appropriate device temperature improves reliability as well as performance. Current power electronics are large and heavy, most of which is the packaging. Current power electronics packaging needs to catch up to device capability.

2.1 Future power module performance targets

There has been a significant investment over the past few decades in developing high power devices from 'wide-bandgap materials', such as silicon carbide and gallium nitride, which promise improved performance and operating temperatures as high as

400°C, but other practical packaging material limits have generally kept operation of those devices under 200°C [22]. Still, while significant progress has been made especially in the commercialization of silicon carbide device technology, most of those devices still remain cost-prohibitive for widespread adoption, and even military circuit demonstrations are still in the prototyping phase [23], [24]. So, because implementation of high temperature semiconductors does not look to be an immediate nor a universal solution to the power electronics thermal problem, it is necessary to examine the heat removal characteristics of the electronics thermal stack and look for improvements there. Figure 4 shows a schematic of the current state-of-the-art of power modules and where the future Army target is for power modules showing a higher heat flux, higher voltage, increased temperature operation, integrated thermal transient suppression and improved reliability.

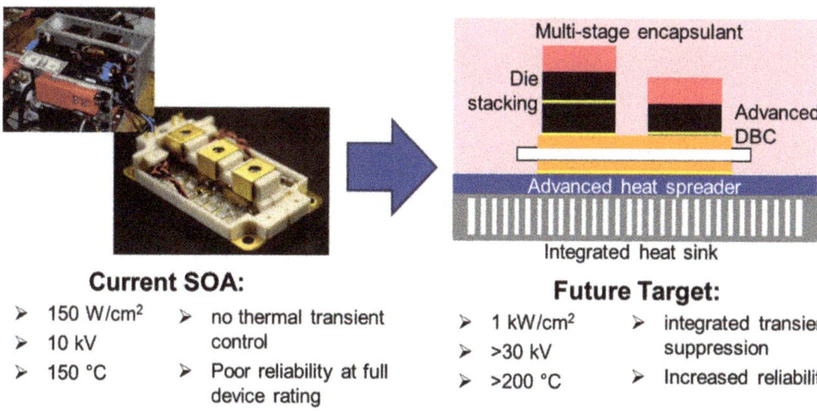

Figure 4. Depiction of the current state of the art of power packaging and the Army's future target.

The goals of reducing the size of power electronic components while operating on higher temperature coolant loops at higher power levels compete directly with each other, primarily because of the realities of dealing with heat generation from the module components. Looking at the composition and structure of a typical high-performance power module shows how module design affects heat removal and provides insight into how we can begin improving the overall thermal performance.

2.2 Standard power module configuration

Figure 5 shows the representative topology of a high-density switch power module made at ARL, containing rows of semiconductor devices (primarily discrete transistors and diodes) wired into an appropriate circuit arrangement in an enclosure or package. The semiconductor devices are the primary heat generating elements in the module, while the package provides structural support, electrical routing, and heat removal pathways.

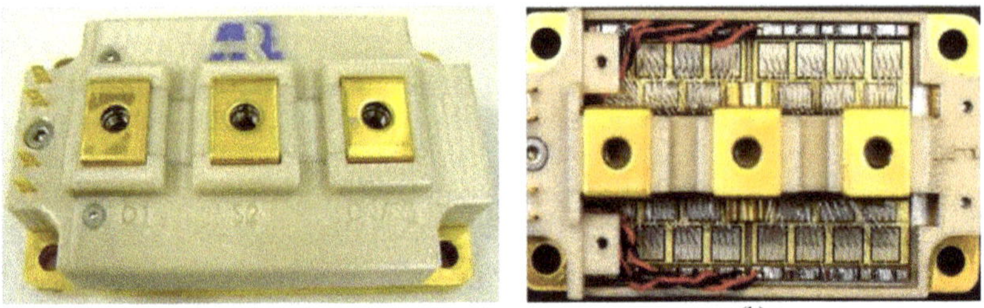

(a) (b)

Figure 5 – (a) ARL 1200V, 400A dual switch module, and (b) internal view of module showing transistor and diode arrays bonded down to substrate and baseplate and wired to external connections, from [25].

A representative cross-sectional model of the package itself can be seen in Figure 6. There the semiconductor switch (device or chip) is approximated as a single block of material with heat generation occurring at the top surface. Most power modules are designed with the primary heat removal path through a heat sink at the base, while the top of the device is typically kept free for wirebonding (typically 10-20 mil aluminum, gold, or copper) or other electrical contact. All elements between the semiconductor junction and the final heat removal mechanism (whether an air cooled heat sink, liquid cold plate, or other structure) comprise what is referred to as the *thermal stack*. Each piece of that stack contributes to the total thermal impedance to heat flow from the device. The sum of these impedances produces a temperature gradient with steady heat flow that can use up the temperature margin available in the system.

The components of the package below the semiconductor chip provide several functions in addition to being a heat removal path. The solder attaches (which are typically gold-tin [AuSn], tin-silver-copper [SAC], or sintered silver) are conductive bonds allowing electrical connection to the metal routing layers while also acting as a mechanical stress buffer between the typically disparate semiconductor and metallic layers. Below that is a ceramic (typically 25 mil aluminum nitride in high power packages) layer metallized on both sides with thick 12 mil copper, the tri-layer stack being called a Direct Bond Copper (DBC) substrate. Alternative substrates include Direct Bond Aluminum (aluminum metal on ceramic alumina) and Active Metal Brazing (AMB). That substrate is bonded down to a typically copper baseplate (CuMo, CuW, AlSiC are also typically used) that helps the heat spread out before it reaches the heat sink or cold plate. As noted in the figure, in a typical power module everything below the spreader or baseplate is a separately manufactured component. Heat sinks or liquid cold plates and attached to the module using some sort of grease, paste, or adhesive, referred to as a Thermal Interface Material. These are typically large area bonds using rather low thermal conductivity materials, and can be a significant thermal impediment. Finally, the actual heat removal element is attached to the base. In lower heat flux systems this may be an air-cooled heatsink, but in most vehicle power electronic applications this would take the form of a pumped liquid cold plate to take advantage of the higher convection rates they provide.

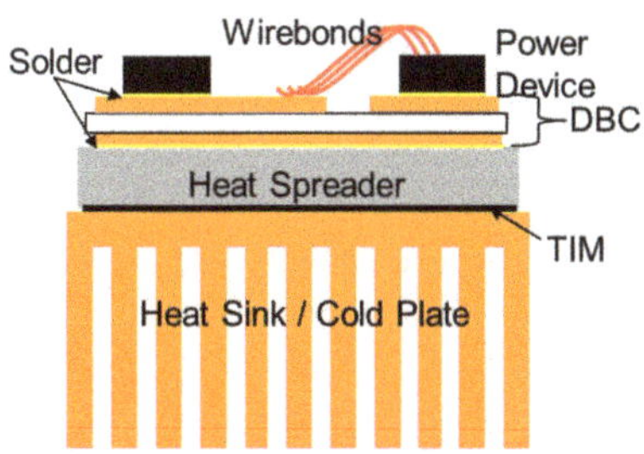

Figure 6 – Power electronics thermal stack depicting multiple thermal resistance components (layers not to scale).

2.3 Current power module challenges

Table 1 lists current challenges of power electronics modules and potential solutions.

Since this is a chapter focused on thermal challenges, we will just briefly touch on parasitic inductance, voltage breakdown and reliability challenges. Although it is important to note that all of these challenges are exacerbated by increased temperature and/or temperature non-uniformity.

Table 1: List of current power packaging challenges and the current approaches to solve each of these challenges

Challenge	Current Solution(s)
Packaging limits performance	Derate devices
Operating temperature (> 200°C)	High temperature materials
Thermal (1 kW/cm^2) – drives the size	Integrated cooling, reduce thermal resistance, PCMs, higher thermal conductivity materials, improved heat sinks
Break-down voltage (up to 25 kV)	Increase package size, thicker/improved dielectrics
Inductance	Snubbers, eliminate wirebonds, intelligent placement of chips on board
Reliability: DBC	Dimple edges, DBA
Reliability: Wirebonds	Eliminate wirebonds
Reliability: Large area contacts	DBC, CTE matched materials (TiW, CuMo), multiple DBC boards, stress-relieving TIMS

Parasitic inductance is primarily caused by the packaging: wirebonds and planar configurations. This parasitic inductance causes voltage overshoot and ringing during switching and is another reason devices modules are derated. Current trends in power modules are to decrease the system voltages but there is also a sharp rise in power requirements from the system. This leads to much higher currents in the modules. This, in combination with an increase in switching speeds, leads to a module where inductance dominates the performance, as is seen in the following equation:

$$V = L\,{}^{dI}\!/_{dt}$$

The standard practice for reducing inductance is to add passives to the system to handle the effects of the parasitic inductance which often have their own additional thermal and packaging challenges causing the system size to increase. Additionally, researchers are looking at low inductance electrical layouts and eliminating the wirebonds [27] [28].

The three primary failure locations in power electronics include DBC cracking, wirebond failures, and large area attach CTE mismatch. Many approaches are being investigated to eliminate these failure mechanisms, as is shown in Table 1. As higher voltage devices are being developed, there is also a challenge of packaging capable of handling these higher voltages [29].

2.4 Thermal challenges in power electronics packaging

Standard silicon and silicon carbide (SiC) power modules, described in Figure 6 and reiterated in Figure 7, have thermal resistivities between 0.6-3.3 cm^2K/W, measured relative to die area heat flux, with the majority of that resistivity in the TIM and heat sink. As was mentioned previously in the chapter, coolant temperatures are moving to be >100°C and a maximum chip operating temperature of 150°C [8], leaving a ΔT of 50°C between the device and the coolant fluid. With the current package, the maximum allowable heat flux is 83 W/cm^2 for only a 50°C rise in chip temperature. There is a desire to move to higher heat fluxes (up to 500 W/cm^2) which required package thermal resistivities (R_{th}) < 0.1 cm^2K/W, a very challenging goal. This work discusses both the challenge of developing next-generation cooling solutions capable of meeting this goal and the complexities of implementation from a module perspective.

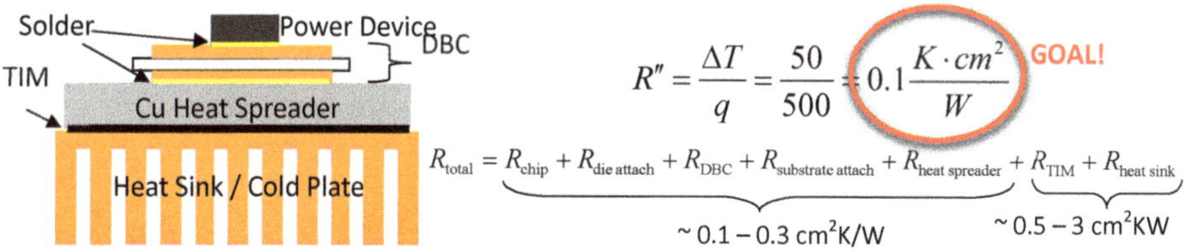

Figure 7. Schematic of a standard power module with the thermal resistance stack equation showing the standard module thermal resistivities as well as the future goal.

Thermal management is crucial in power electronics due to the extremely large amount of heat that is dissipated by the devices. In most power modules, the packaging is limiting the power density, not the devices. If the heat can more effectively be removed, the module can operate at higher power density. In order to achieve the 0.1 $K\text{-}cm^2/W$ goal, there are three primary ways to remove heat that will be explored in this chapter:

1. Improve conduction - reducing thermal stack resistance, thinner materials, higher thermal conductivity materials, spread out devices
2. Improve convection - integrated cooling, improved heat sinks, topside and backside cooling, two-phase heat transfer, pool boiling, microchannels
3. Add latent heat absorption: phase change thermal management

2.4.1 1-dimensional package layer model

Table 2 shows the thermal resistivities of the various layers in a typical power package shown in Figure 7 from the chip through DBC, before the heat spreader and heat sink. It is interesting to note that the solders, which are often times considered negligible, are actually 30% of the total package resistance. The total thermal resistance, calculated by adding the layers together is 0.0915 cm^2K/W. ANSYS was used to approximate the spreading resistance to be 0.0127 cm^2K/W, thus when subtracted indicates a total package resistance of 0.0788 cm^2K/W. Referring back to the initial goal of 0.1 cm^2K/W, 80% of our total available thermal resistance is taken up by just the layers between the device and substrate and only 20% (0.02 cm^2K/W) remains for the TIM, heat spreader and heat sink.

Table 2. Table outlining the various layers in a standard power module from the chip through the DBC showing the typical thicknesses (t) and thermal conductivity (k) of each layer and using them to calculate the thermal resistivity (K-cm²/W).

Layer	t (μm)	k (W/mK)	Rth (cm^2K/W)
SiC Chip	380	380	0.0100
Solder - 2 mil AuSn	50	56	0.0089
DBC Top Cu	305	401	0.0076
DBC AlN	635	170	0.0374
DBC Bottom Cu	305	401	0.0076
Solder 2 - 4 mil SnPb	100	50	0.0200

Therefore, it is interesting to look at how improving the cooling and/or the package will help us achieve our goal. It is important to note that if we move to 1000 W/cm^2, we would need 0.05 cm^2K/W, a significantly harder goal since almost 40% of that is taken up by just the device and solder.

2.4.2 ANSYS Modeling Results

In order to determine the cooling methods required to achieve the 0.1 cm^2K/W package goal, a number of numerical simulations were performed. The baseline module, shown in Figure 8a and b, was previously fabricated at the Army Research Laboratory in Adelphi, MD with an integrated heat sink that achieved an experimental thermal resistivity of 0.2 cm^2K/W using macroscale channels (>1 mm) [25] and the existing engine coolant (50/50 water ethylene glycol (WEG)) [13] [30] [31]. Thermal finite element modeling, shown in Figure 8c, was used to estimate the module effective backside convective coefficient (h) by fitting the parameter to the experimental power and temperature. This was calculated by deleting the existing heat sink from the model, simplifying the geometry, and then varying the h boundary condition on the backside until the model matched the experiment. An effective backside convection coefficient of 50,000 W/cm^2K was found to provide equivalent cooling to the experimental module. This is a summary of the work from [32].

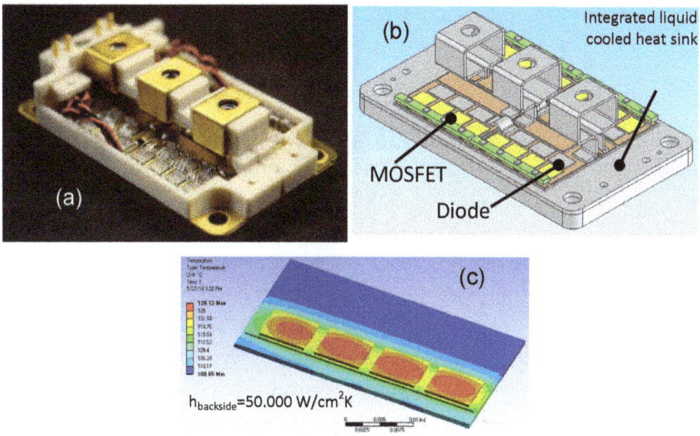

Figure 8. (a) Actual baseline power module with an integrated heat sink used for comparison (b) CAD image of the baseline module indicating device locations (c) ANSYS model used to determine the convective coefficient of the integrated heat sink

The impact of enhancing both topside and backside cooling was investigated to determine impact on the device temperature when compared to the baseline module shown in Figure 8. As was shown in Figure 7, in order to achieve the 0.1 cm^2K/W goal, a chip temperature rise <50°C must be achieved, shown as red dashed lines in each of the plots below.

2.4.2.1 Impact of improving the backside cooling

There is a significant amount of research looking into improving heat sink performance and a number of heat sinks have been proposed including microchannels, jet impingement, and two-phase flow. To determine the benefit of replacing the existing heat sink with a higher performing option, ANSYS models were run with an effective backside convective coefficient varying between 50,000 W/m^2K (baseline) and 1,000,000 W/m^2K. The results are shown in Figure 9 and indicate very little improvement past ~300,000 W/m^{2}K and operation at the baseline of 50,000 W/m^{2}K is not desired due to the fact that it is on the high sensitivity the part of the temperature curve such that small changes in flow rate could cause large fluctuations in temperature.

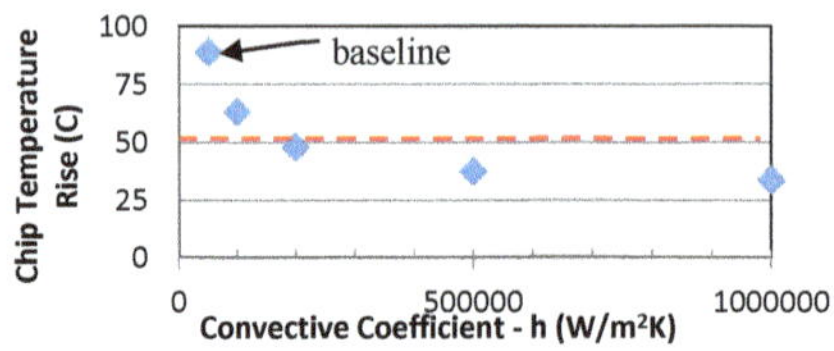

Figure 9: Numerical results showing the impact of improved backside cooling on the chip temperature rise.

Figure 9 also shows that for convective coefficients >200,000 W/m^{2}K, the desired temperature rise can be achieved. This is, however, an extremely challenging goal. Previous work to improve backside cooling of power modules assumed a realistic upper limit of ~80,000 W/m^{2}K [33] Two-phase flow boiling in microchannels is a potential option to achieve these high convective coefficients but have the disadvantages of dry-out failures, and clogging of microchannels. Agonstini has showed that with a two-phase microchannel heat sink, a heat transfer coefficient as high as 250,000 W/m^{2}K can be achieved with R236fa [34]. Lee and Mudawar also has shown a two-phase microchannel heat sink capable of 112,000 W/m^{2}K with HFE-7100 [35]. Jet impingement is also promising but have a number of challenges including significant temperature gradients, clogging of nozzles, and very high pressure drops. Khandlikar summarized some jet impingement results showing heat transfer coefficients as high as 73,000 W/m^{2}K [36].While there have been a few papers that show convective coefficients >200,000 W/m^{2}K, they are measured in controlled laboratory environments and hard to achieve in practice. In addition they typically require large driving pressures and a custom system. The challenges of adding these advanced cooling solutions to a power module system have been discussed previously in the Vehicle Loop Considerations section.

2.4.2.2 Impact of adding topside cooling

Most devices generate heat at the top surface of the chip which makes topside cooling a very attractive option. In order to assess the impact of including topside cooling, a topside boundary convective coefficient was added to the existing baseline model. The existing backside heat sink was assumed to still be used ($h_{backside}$=50,000 W/m^{2}K) and the topside convective coefficient was varied between 10 W/m^{2}K (natural convection) and 100,000 W/m^{2}K. The results are plotted in Figure 10.

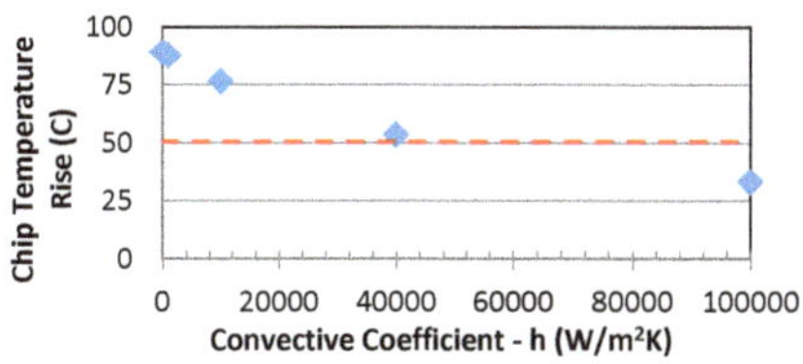

Figure 10: Numerical results showing the impact of adding topside cooling to the baseline module.

Figure 10 shows that for ΔT_{chip}<50°C, the convective coefficient of the topside must be >40,000 W/m^{2}K. Practical topside cooling options could include pool boiling and spray cooling. However, as was the case with backside cooling, these high convective coefficients are very difficult to achieve. Spray cooling of the topside is a potential option but it is very complex and large driving pressures are required. Mudawar has shown a heat transfer coefficient of as high as 40,000 W/m^{2}K using HFE-7100 in boiling mode spray nozzle [37]. In a review paper, the heat transfer coefficient of spray cooling was around 20,000 W/m^{2}K on a flat surface, this can be doubled by using an enhanced surface [38]. As was the case in backside cooling, these results are from a controlled experimental analysis in optimal conditions and would not be easily repeated in the field. Additional concerns arise from the need to apply cooling directly to the top of sensitive electronic components. These concerns include interaction with wirebonds and other electrical contacts, increased complexity, and working with the topside of the devices limits the ability to spread the heat or improve surface area.

2.5 Thermal challenge summary

As described above, the last several decades have witnessed a dramatic increase in the use of electronics on a variety of systems. This trend continues today in hybrid electric vehicle (HEV), electric vehicle (EV), and plug-in hybrid electric vehicle (PHEV) technologies for commercial and military applications. Military use can include both direct vehicle applications, such as propulsion, and indirect applications such as electrically operated arms or interfacing the vehicle electrical system to create a military base microgrid [39]. Power conversion electronics are a ubiquitous and enabling technology for the success of current and future electric vehicle architectures; this is due to the disparate electrical systems common to these programs and the need to draw from a platform's singular electrical bus. This mainly involves the use of power semiconductor switches such as power diodes, metal oxide field effect transistors (MOSFETs), and insulated gate bipolar transistors (IGBTs). Unfortunately, these power electronic devices are not 100% efficient and require aggressive thermal management schemes to allow operation within acceptable temperature limits. This is exacerbated by the fluid and temperature requirements on vehicle applications.

The analysis in the preceding sections has shown that some advanced cooling technologies may be able to achieve the desired 0.1 cm^2K/W resistivity goal. Due to the system complexities, a combination of the investigated approaches (higher thermal conductivity materials, improved heat sink, or adding topside cooling) will be the best solution to the thermal challenge. Understanding the effects and tradeoffs between solutions is critical to optimizing the design. ARL has investigated many ways to improve the standard power electronics module. The remainder of the Chapter focuses on chronicling these efforts at ARL's to reach the 0.1 cm^2K/W resistivity goal.

3. ARL approaches to improving power electronics cooling

ARL is investigating a number of ways to meet the thermal challenges outlined previously. These efforts are broken in to several core competency areas:

1. Single-phase liquid cooling
2. Two-phase liquid-vapor cooling
3. Advanced passive thermal ground plane (TGP) heat spreaders
4. Transient thermal mitigation with solid-liquid phase change materials
5. Multi-functional packaging and thermal management design and
6. Device layout optimization with the ParaPower tool

Each of these core competency areas are discussed in detail below. A combination of these approaches is the most likely solution to the power packaging challenges. This multi-tiered approach, while adding complexity, is necessary to achieve the 0.1 cm^2K/W resistivity goal.

3.1 Advanced single phase cooling

From the discussion on power electronic package thermal analysis it is fairly obvious that thermal improvement requires reducing thermal resistance somewhere in the package thermal stack. The package modeling demonstrated that improvement goals cannot be met simply by improving heat sink performance or finding higher thermal conductivity materials. Instead, ARL pursued a series of investigations into moving the cooling structure closer to the semiconductor devices. This approach generally had two positive effects on cooling. First, by moving into the package certain layers could be completely removed from the thermal stack. Second, because thermal performance is the sum of all thermal resistances, eliminating the thermal resistance contributions of those layers could potentially permit using a less aggressive cooling method with a lower convection rate and still achieve an acceptable total package resistance. That benefits the system in other ways, including things like reducing the coolant flowrates or driving pressure, easing manufacturing complexity/cost, or increasing cooling component reliability.

Considering how to move from standard "module plus heat sink" configurations to "package integrated cooling", we can see by looking at Figure 7 that there are multiple levels into which the cooling function can be moved. The three levels investigated by ARL have been the (1) replacing the heat spreader/baseplate with an integrated heatsink [25], (2) bringing the coolant into the ceramic substrate, and (3) bringing the coolant into direct contact with the device. The following sections will briefly review the work done in these areas

3.1.1 Substrate integrated microchannel substrates

While bringing coolant all the way to the electrical device would provide the lowest potential thermal resistance by completely eliminating any additional material from the thermal stack, it also removes any isolation between the electrical components and the coolant. Vehicle coolants (such as water or ethylene glycol based antifreezes) are generally conductive, creating a potential electrical shorting path. Dielectric coolants (which are primarily fluorocarbons or refrigerants) that could avoid the conductance problem have generally worse cooling performance and have been shown to reduce convection under similar conditions by 50% or more [40]. As such, investigating the cooling integration within the dielectric substrate preserves electrical isolation and enables use of arbitrary coolants.

For the standard high performance power module stack the device substrate is an aluminum nitride direct-bond-copper (DBC) layer. As AlN is a high thermal conductivity dielectric material, providing on the order of 50V/μm of electrical isolation [41], integration into this ceramic layer would remove any lower layers from the thermal stack allowing them to be constructed from lighter or less expensive materials. The challenge, however, lies in determining how to fabricate cooling structures in the ceramic material. While it is a useful packaging material, AlN is also brittle and generally chemically inert. This restricts the use of standard micromachining operations for defining deep microchannels. While there have been a few cases in the literature where deep plasma etching [42] or laser etching [43] have been used on AlN, they have not been particularly scalable for larger power electronic substrates. ARL explored two alternative approaches to fabricating AlN substrate microcoolers: (1) diamond saw cut AlN microchannels, and (2) stereolithographically fabricated microchannel substrates.

3.2 Diamond saw-cut ceramic microchannels

In 2007 Jankowski et al. used a relatively straightforward, high speed fabrication technique to make a series of aluminum nitride microchannel substrates with a Disco DAD320 diamond bladed wafer dicing saw. [44] Partial depth cuts in a 1"x1" AlN substrate, a photograph and SEM micrograph of which are shown in Figure 11. To maintain substrate integrity the channels were about 300 μm deep, or ~1/2 the substrate thickness. They found that with proper zeroing technique the target depth could be achieved to within +/-10μm across a 1-inch (~25.4mm) piece and cut width depended solely on the blade width. The greatest drawback to this technique is that the cut channels must pass all the way across the substrate. This was generally only amenable to straight microchannels which required fluid coupling on the thin substrate edge, which required cumbersome edge-seals for proper manifolding.

Figure 11 – Saw-cut aluminum nitride ceramic substrate. The channels shown were cut 300μm deep with a 250μm blade and a 375μm pitch, adapted from [44].

A series of test substrates fabricated this way were epoxied to a second, flat AlN substrate to cap the channels. They were then metalized and a 4mm SiC Cree pin diode was soldered to the substrate, as shown in Figure 12.

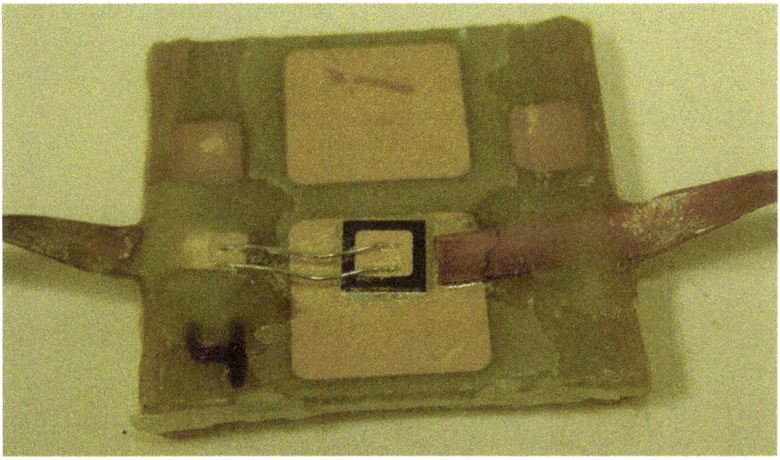

Figure 12 – Representative packaged and sealed AlN substrate with SiC diode heat source, adapted from [44].

Yield problems in fabrication, mounting and testing led to only two substrates being fully testable, but these were able to be placed into a thermo-fluid test rig and characterized for varying flowrates and applied power levels. The resulting flow and thermal performance is shown in Figure 13, where the substrate integrated heat sinks were able to show thermal resistivities approaching 0.15 cm^2K/W with heat fluxes that approached 225 W/cm^2.

3.2.1 Stereolithographically defined ceramic microchannel substrates

Recognizing that one of the primary limitations of the saw cut microchannels was the limited channel depth and lack of geometric design freedom, Jankowski, et al., reported on a 3D printed AlN microchannel substrate effort performed in partnership with Technology Assessment & Transfer, Ltd (TA&T) [45]. TA&T had been developing ceramic materials for use with standard stereolithography equipment. This process used the 3D printing to turn a 3D CAD file into a ceramic green part, which could then be fired into the full ceramic piece. As with the other AlN substrates yield was a primary difficulty, but in this case the issue was largely due to the splitting and warping that would occur during the uncured resin removal and part firing process. Still, a number of parts were successfully made and tested. Figure 14a shows a representative CAD drawing of a microchannel substrate, and Figure 14b shows a photograph of a fully assembled substrate including diode heater. Note that the ability for multi-dimensional channel routing enabled easy fluid coupling to the substrate.

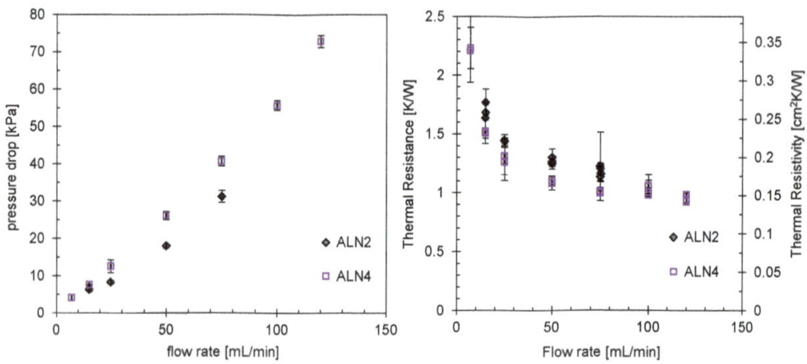

Figure 13 – thermal performance result of tested AlN microchannel substrates, showing die area averaged resistivities approaching 0.15 cm^2K/W, adapted from [44].

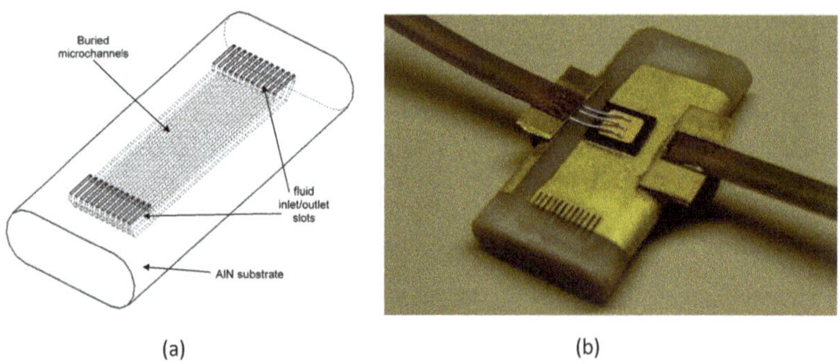

Figure 14 – (a) CAD sketch of microchannel substrate used as the fabrication input file and (b) Fully assembled stereolithographically fabricated AlN microchannel substrate, from [45].

Figure 15 shows the thermal results of testing these substrates at ARL in a manner similar to the saw-cut substrates. These substrates had 2-3mm deep channels with about 200μm channel widths. As can be seen, several of these substrates meet or exceed 0.1 cm^2K/W at die heat flux levels exceeding to 650 W/cm^2. The deeper channels permitted much higher flowrates than the shallower saw cut channels with lower thermal resistances as a result.

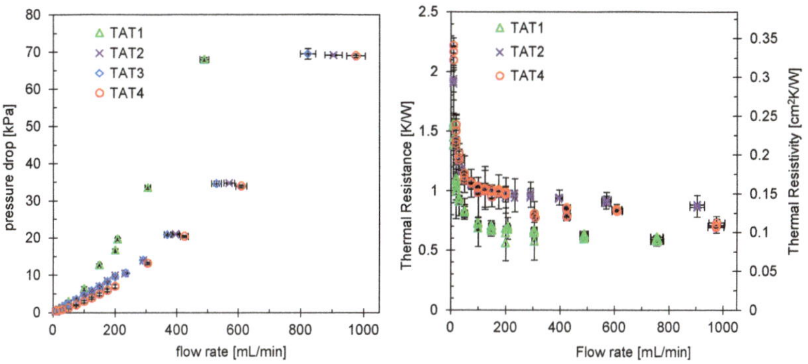

Figure 15 – Thermal resistance of the stereolithographic aluminum nitride microchannel substrates, adapted from [45].

3.2.2 Aluminum nitride manifold-microchannel cooling

As described above, implementation of printed and micro-machined microchannel coolers can lead to a significant reduction in the thermal stack resistivities by bringing the fluid closer to the heated device and improving convective coefficients. However, these benefits are attenuated by the increased fluid flow impedance, temperature non-uniformity, and difficult fluidic coupling. At the expense of increased complexity, Sharar et al. [16] developed a manifold microchannel cooler (MMC) to obtain resistivities comparable to parallel-flow devices with reduced flow restriction. The manifold design distributes fluid evenly over the cooling region, therefore reducing the temperature non-uniformity. By directing the fluid primarily through the large manifolds, and only passing through the constrictive microchannels at the cooling location, the pressure drop can be significantly reduced. An added benefit of this design is the enhanced convection associated with developing flow and liquid impingement. Figure 16 shows the manifold microchannel cooler that was designed and fabricated at ARL. Details on the fabrication can be found in the paper by Sharar et al. [16].

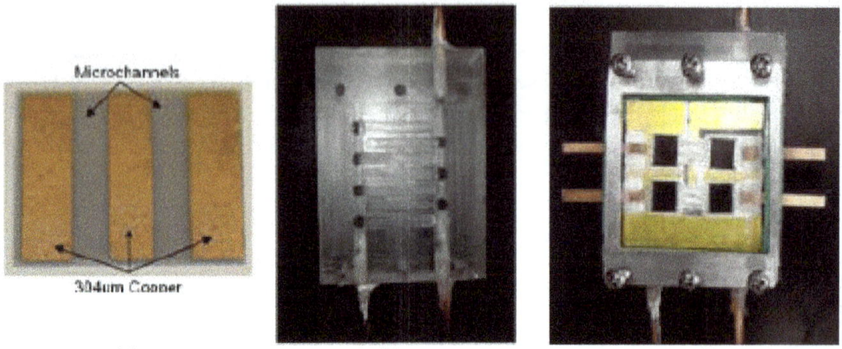

(a) (b) (c)

Figure 16: Fabricated Aluminum Nitride Direct Bond Copper Manifold Microchannel Cooler (a) backside of saw-cut DBC substrate (b) acrylic manifold (c) assembled MMC cooler, from [16].

Sharar et al. [16] tested Army-relevant fluids and fluid temperatures (simulating those that might be found in vehicle applications) including demineralized water at 25°C, water at 80°C, 50% water and ethylene glycol (WEG) at 80°C, 50% propylene glycol and water (PGW) at 80°C, and 0.14% 4-5nm nanodiamond solution in 50% ethylene glycol and water (provided by the International Technology Center) at 80°C. Flowrates ranging from 50 to 650 ccm were tested. As expected, the thermal resistance decreased with increasing flowrates and reached a steady state value at high flowrates (approaching the conduction resistance through the DBC). It was shown that for an Army-relevant pressure drop of 5 psi, the lowest thermal resistance observed was for water at 25°C due to the low viscosity and high thermal conductivity (0.316 Kcm^2/W). All other fluids had less favorable material properties leading to reduced performance. Water at 80°C had a resistivity value of 0.359 Kcm^2/W, EGW at 80°C had a value of 0.486, PGW at 80°C had a value of 0.622, and the nanodiamond solution at 80°C performed worst with a value of 0.628 Kcm^2/W. This demonstrates that pure water is the best option for cooling using a MMC module, with a significant reduction in cooling performance expected at elevated temperatures with the Army-relevant cooling fluids of PGW and EGW. While not equaling the thermal performance of previous Army microchannel designs, 0.1-0.2 Kcm^2/W above, the MMC did provide a vast improvement over the standard package while staying within the Army-relevant 5psi pressure drop requirement.

3.2.3 Package integrated MEMS manifold microchannel cooler

This research presents the analysis and realization of a single phase high performance manifold microchannel cooler for improving the thermal and hydrodynamic performance of multi-chip power electronic module. This heat exchanger, microfabricated directly into the substrate, enables higher power density electronic products by more efficiently removing the high levels of heat generated. Improved heat removal is accomplished by reducing the number of packaging layers between the device and the heat exchanger and by improvement in convective heat transfer. In addition, the efficiency of the device is enhanced by minimizing fluid pressure drop through the use of large manifold channels to transport fluid to the cooling area and smaller crossover microchannels in the active cooling area. This combination of channels also improves the uniformity of the temperature distribution across the device. The improved thermal performance and efficiency of the heat exchanger has been demonstrated using both numerical and experimental techniques. A full description of the work can be seen in the PhD thesis [47].

The manifold microchannel coolers were fabricated and tested both with [48] and without [49] electrical isolation between the chip and the coolant.

Experimentally, the coolers without electrical isolation, shown in Figure 17, demonstrated thermal resistivity values as low as 0.06 $K/(W/cm^2)$, which is up to a 50X improvement over a standard power package with significant size and weight reduction. The coolers with an incorporated aluminum nitride electrical isolation layer experimentally demonstrated up to a 15X improvement.

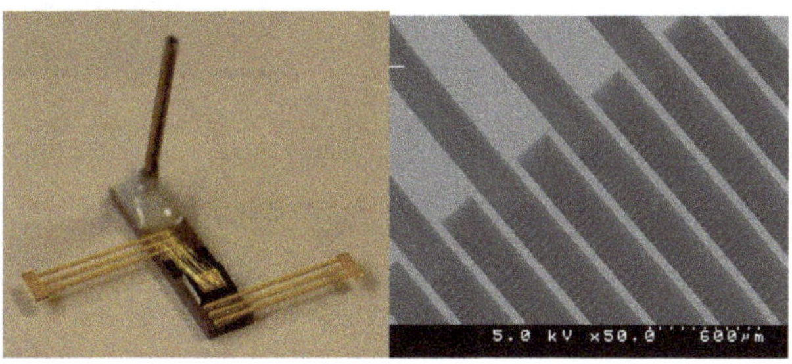

Figure 17: (a) Fabricated manifold microchannel coolers without electrical isolation and (b) image of the relationship between the manifold and microchannels

This research has developed and demonstrated the use of a multi-device MMC cooler fabricated directly into the substrate of an electronics package with an AlN electrical isolation layer that can hold off up to 6 kV and significantly reduce the package size. The fabricated module is shown in Figure 18. Nine different microchannel dimensions were fabricated and tested under a variety of fluid pressures and power densities including a device with no microchannels. The device with no microchannels experimentally showed pressure drops 1.5 to 2.5 times smaller and thermal resistivities on average of 1.33 X larger than the tested devices with microchannels (at 293 W/cm^2 and 200 mL/min). Thermal resistivity values were measured as low as 0.104 $K\text{-}cm^2/W$, which is 6 to 30 X better than standard power packaging. The authors previously published work showed thermal resistivities as low at 0.06 K-cm2/W for a device with no electrical isolation [49]. Comparing to the previous device with no electrical isolation, shown in Figure 17, the current module showed only slightly worse thermal performance that now has an incorporated electrical isolation layer that can hold off up to 6 kV along with the ability to use standard die attach methods, both significant improvements. Additionally, the channels are 3 to 6 times larger to reduce the effects of particulate clogging and improve pressure drop [47].

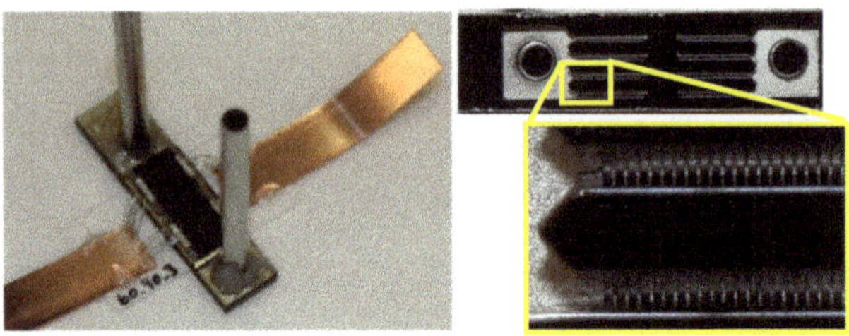

(a) (b)

Figure 18: (a) Completed device with electrical and fluidic connections (b) Image from the backside showing the relationship of the manifold and microchannels.

Figure 19 shows the fabrication process and cross section of the of the fabricated manifold microchannel device with aluminum nitride electrical isolation. Cleanroom microfabrication using deep reactive ion etching (DRIE) was used to fabricate both the microchannel and the manifold wafer and then were then attached using a gold tin bonding process in a wafer bonder.

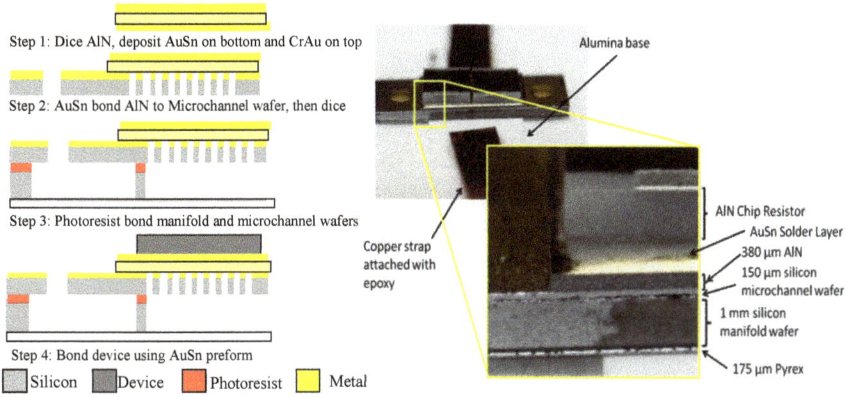

Figure 19: Fabrication process of the manifold microchannel structure and a cross section of the layers of the final assembly.

In addition to experimental results, the interaction between the manifold channels and multiple microchannels was numerically modeled and compared to simpler, one-dimensional approximations based on the Hagen-Poiseuille equation. The comparison shows that the one-dimensional model, while under-predicting total pressure drops, can provide insight into the effect of varying dimensions on system performance. The numerical models were used to identify the impact of varying dimensions across the entire length of the cooler, and a sensitivity analysis was performed with respect to system pressure drop, thermal resistance and uniformity. Additionally, large microchannel velocity gradients, some larger than 10X, were observed along the length of the device which impacts the chip non-uniformity. The simulations showed that when comparing the manifolded design to a comparable straight microchannel cooler, there is a 38X reduction in system pressure drop for similar thermal performance, as is shown in Figure 20 [50]. In addition the models showed that simply modeling a single manifold unit is not sufficient. There is a complex flow structure that varies along the length of the manifold.

In addition the experimental results were compared to CFD simulations and compared favorably. A sample of the results are shown in Figure 21.

Due to the inherent problems with straight microchannels, the manifold microchannel cooler (MMC) design has been shown to be a promising alternative to maintain thermal performance while improving system pressure drop.

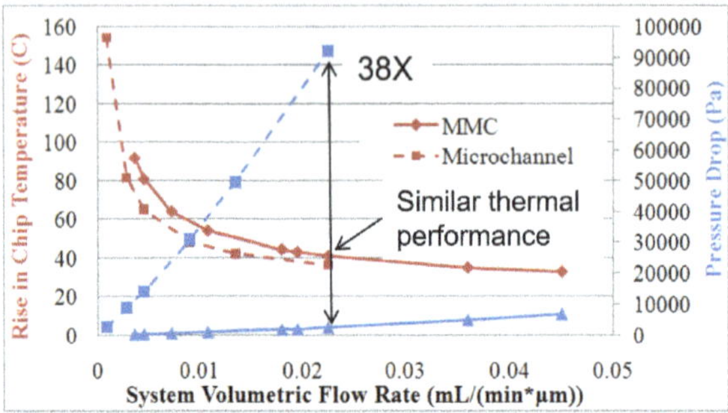

Figure 20: Numerical results showing 38X lower pressure drop for similar thermal performance by moving from straight microchannels into manifold microchannels.

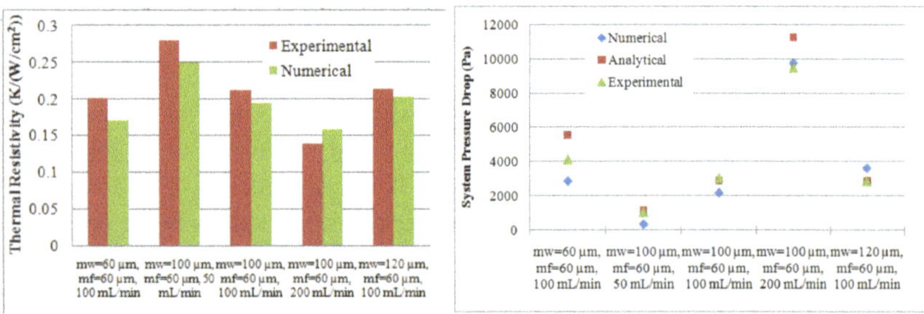

Figure 21: Plots comparing the experimental, numerical and analytical results for both thermal resitivity and pressure drop for various manifold microchannel geometries

3.3 Two-phase cooling

Army programs have focused on increasing the use of power-dense electronic components to improve system weight, fuel usage, design flexibility, and overall functionality. Traditionally, these electronics have relied on air-cooled heat sinks and liquid cooled cold plates to manage waste heat, however, these new power-dense systems are presenting challenges to these conventional cooling approaches. Specifically, the effect of higher heat fluxes for air cooled systems is heavy, large, and expensive fans and heat sinks. This effect is similar for single-phase liquid cooling, whereby higher heat fluxes are requiring larger coolant flowrates [51]. Both of these trends are resulting in bulky (heavy) systems that consume more fuel and directly undermine efforts to reduce system size and cost. Therefore, one goal at ARL is to develop improved cooling components that are smaller and lighter, while also providing increase performance compared to air-cooled and single-phase cold plates.

In 2010, ARL began exploring cooling schemes using liquid-vapor phase change (two-phase cooling) as a cost-conscious and practical next step beyond traditional methods. This concept and several methods (including flow boiling, spray cooling, and jet impingement) were outlined in an internal technical report by Sharar, Jankowski, and Morgan [21]. To summarize, cooling schemes using liquid-vapor phase change have several benefits over standard approaches. First, the latent heat of vaporization can be several orders of magnitude larger than the specific heat for the same fluid. Therefore, boiling allows increased heat absorption per unit mass or volume (flowrate) of fluid. This allows reduced flowrates for two-phase flow and smaller fluid reservoirs, smaller fluid volume, reduced pumping power, smaller pumps, and an overall reduction in system size and weight. Additionally, the latent heat absorption associated with two-phase cooling occurs at isothermal conditions, allowing less temperature deviation across a device and less thermal stress. Finally, two-phase cooling can provide much higher heat transfer coefficients due to the thin film evaporation, nucleate boiling, and mixing associated with this phenomenon. The result of higher heat transfer coefficients is lower total thermal resistance and either a) reduced temperature for a given heat dissipation or b) increased heat dissipation for an allowable temperature rise.

3.3.1 Enhanced cooling with internally-grooved tubes

Based on the above understanding, in 2011 Sharar et al. [52] took an existing ARL single-phase offset fin minichannel IGBT cold plate and operated it under two-phase conditions (and single-phase as a basis for comparison). They demonstrated a reduction in the total thermal resistance by a factor of 4.8, 1.5 to 10 times improvement in cold plate temperature uniformity, and 2.2 to 5.3 times reduction in required pumping power for a fixed heat dissipation compared to single-phase cooling. These were favorable initial results, but ARL understood that specially designed two-phase surface enhancements have been developed to further improve two-phase cooling performance. Specific to flow boiling, enhancement techniques can be classified as either a) nucleate boiling techniques or b) convective vaporization techniques. A summary of these techniques (which can also be found in the PhD Dissertation by Sharar [53]) and are discussed more extensively by Bergles [54] [55], Thome [56], Webb [57], and Kandlikar [58]. Furthermore, a closer look at specific topics including, microporous coatings [59], reentrant cavities [60], nanoparticle fluid additives [61], twisted tape inserts [62] [63], corrugated tubes [62] [64], and internally-grooved tubes can be found in a presentation by Sharar et al. [65].

Based on this survey of available techniques, internally-grooved tubes were identified as the most promising technology for integration into two-phase vehicle power electronics cold plates. Figure 22a shows a photograph of a commercial Wieland Cuprofin internally-grooved tube and Figure 22b shows the typical internally-grooved tube geometry.

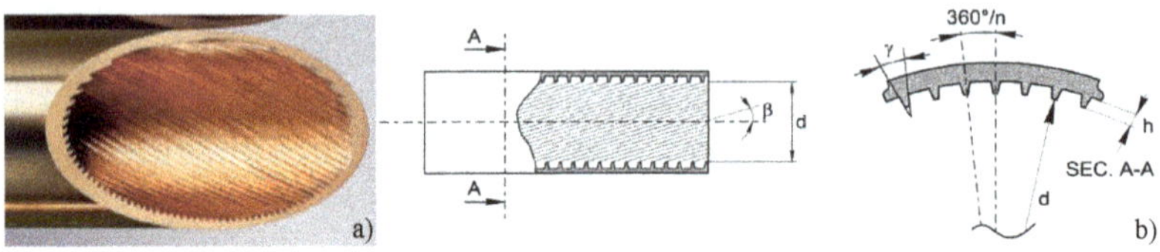

Figure 22: a) Photograph of a grooved tube [adapted from [66]] and b) Schematic of grooved tube, adjusted from [67].

During horizontal operation, internally-grooved tubes provide thermal enhancement ratios up to seven times that of smooth tubes at low mass flux (50 to 100 kg/m²s) and improvement nominally equal to the tube area enhancement, 1.5, at high mass flux [68] [62] [69]. Pressure drop increase is generally minimal, with normal values 1-1.5 times higher than a comparable smooth tube [62]. Due to this superior performance, commercial refrigeration units and air conditioners incorporating internally-grooved tubes are used in industry and, subsequently, in some military equipment. However, until recently electronic cold plates incorporating internally-grooved tubes had not been explored; this provided a unique opportunity to design, fabricate, and test internally-grooved tube two-phase cold plates. Unfortunately, the absence of phenomenological insights and physical models hinders the adoption of internally-grooved tubes for this purpose. Therefore, a more comprehensive understanding of the physical mechanisms leading to enhancement in internally-grooved tubes was necessary.

3.3.1.1 Literature review

Two-phase flow boiling in internally-grooved tubes has been of interest since they were first introduced in 1977. Notable reviews by Thome in 1990 [56] and 2004 [71] set the groundwork for the communities' understanding of heat transfer mechanisms in internally-grooved tubes. Since 2004, there have been a large number of experimental and analytical studies of grooved tubes, but outside of a brief introductory literature review, no researchers made the effort to comprehensively compile recent data and analyze the relationship between flow regime and thermal performance.

In 2014, a full 10 years after Thome's latest review, Sharar and Bar-Cohen published a comprehensive 100+ reference paper in the Journal of Enhanced Heat Transfer [69] titled "Two-phase flow regimes and evaporative heat transfer in internally-grooved tubes." The review paper began with an introduction to smooth tube flow regime maps and heat transfer coefficient correlations and established that this regime-based approach is necessary to accurately predict heat transfer for the full range of diabatic operating conditions. Next, a review of heat transfer and flow regime data in the literature revealed that heat transfer enhancement at low mass flux is a result of early transition to Annular flow in the grooved tubes. Figure 23 shows boroscope view and schematic of Stratified flow and Annular flow in a smooth (left) and internally-grooved (right) tube with HFE-7100 as the working fluid. The most important outcome of the review paper was to establish that the flow regime transition and larger wetted perimeter (and reduced dry angle) observed in the grooved tube, along with enhancement due to the Gregorig effect and turbulence from the fins, is responsible for the observed enhancement.

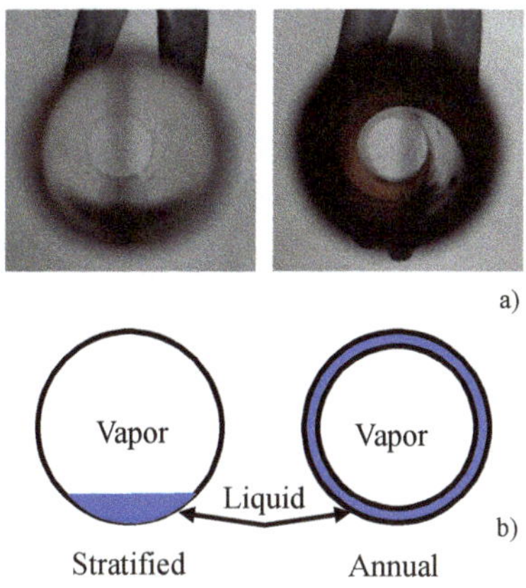

Figure 23: Stratified and Annular flow in a smooth and internally-grooved tube; a) boroscope visualization and b) schematic of flow

3.3.1.2 Experimental setup and flow visualization technique

It was detailed in the aformentioned review paper by Sharar and Bar-Cohen [69] that the flow regime data presented in the literature for internally-grooved tubes is almost exclusively based on visual and verbal descriptions. Unfortunately, due to the complex nature of two-phase flow (and varying definitions from one researcher to the next), this type of subjective determination is unreliable. As a result, in 2014 a non-intrusive optical flow regime determination technique was developed by Sharar et al. [72] to quantify the primary flow regimes of Stratified, Intermittent, Bubbly, Annular, and Dryout. This work was later improved upon to allow identification of the primary flow regimes, as well as the more subtle transitions that exist between dominant regimes. The improvements were published as a paper in the Journal of Heat Transfer Engineering [73], titled "Non-intrusive optical validation of two-phase flow regimes in a small-diameter tube." Figure 24 shows a schematic of the Total Internal Reflection (TIR) technique. Based on the Master's Thesis by Shedd [74], the technique consisted of shining a red diffuse LED on the top of a glass tube, using a camera to capture the total internal reflected light rays from the liquid-vapor interface, and extracting a film thickness profile from the dynamically changed liquid-vapor distribution. It was shown that different flow regimes result in unique temporally-varying film thickness profiles. This method was shown to be a reliable tool for experimentally validating two-phase flow regimes and a significant improvement over subjective visual and verbal descriptions, alone.

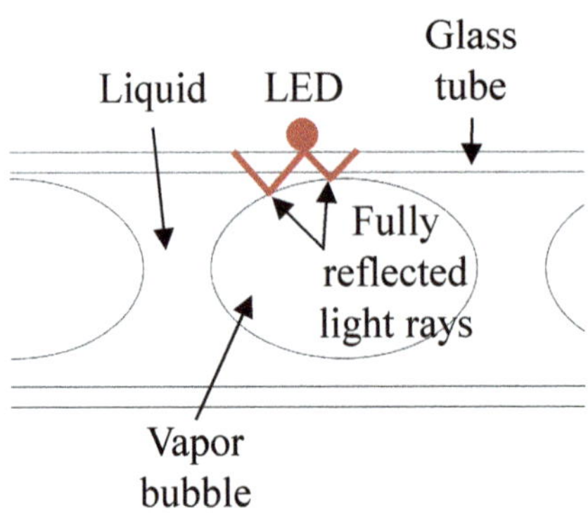

Figure 24: Schematic of the non-intrusive total internal reflection flow regime quantification technique

3.3.1.3 Model development and validation

Flow regime data, obtained with dynamic total-internal-reflection measurements, and heat transfer coefficient data, obtained with infrared thermography and the test setup described by Sharar [53] and Sharar and Bar-Cohen [75], were presented and analyzed for two-phase HFE-7100 flow in horizontal 2.62-8.84 mm diameter smooth and internally-grooved tubes, with mass fluxes from 25-300 kg/m²s, heat fluxes from 4-56

kW/m^2, and vapor qualities approaching 1. Over the course of these studies, an extensive heat transfer coefficient and flow regime database of over 6,500 combined data points for the adiabatic and diabatic smooth and internally-grooved tubes was acquired.

The flow regime results were compared to the Wojtan et al. smooth tube map and used to develop the new Sharar et al. internally-grooved tube map. Details on the model development can be found in references [76] [53]. In summary, the new flow regime map and associated heat transfer coefficient (which is a modification of the Wojtan et al. map) considers the impact of flow regime on enhancement in grooved tubes. The Intermittent to Annular transition and Stratified-Wavy to Annular transition were modified to account for early flow regime transition. In addition, a turbulence factor and empirically derived enhancement factor were introduced to modify the asymptotic heat transfer coefficient correlation. These modification changed the 'dry-angle' calculation and represent the mechanisms responsible for enhancement in internally-grooved tubes.

The smooth tube and internally-grooved tubes heat transfer results were compared at varying mass fluxes and heat fluxes to illustrate the validity of the current regime-based approach to explaining enhancement in internally-grooved tubes. This data was then used to statistically validate the new flow regime based heat transfer coefficient correlation, as well as correlations from the literature. Finally, the new heat transfer model was compared to independent data in the literature. Based on results from the experiments and a reinterpretation of data from the literature, it was firmly established that heat transfer enhancement in internally-grooved tubes at low-to-intermediate mass flux is primarily due to early flow regime transition in internally-grooved tubes. The newly-developed regime-based heat transfer coefficient correlation outperformed empirical correlations from the literature with mean and absolute deviations of 4.0% and 32% for the full range of data collected.

3.3.1.4 Cold plate design and testing

Based on the newfound understanding of flow regimes and heat transfer in internally-grooved tubes, ARL was prepared to accurately and reliably predict performance during two-phase flow boiling in internally-grooved tube cold plates. As shown in Figure 25, serpentine cold plates were designed, fabricated, and tested at ARL. In total, a database of over 280 temperature data points for the smooth and internally-grooved tube cold plates were acquired. It was shown that, the performance in the internally-grooved tube cold plates outperformed the smooth tube results by as much as 400%, and matched well with the newly developed model for the range of operating conditions [77]

Figure 25: Photograph of the cold plate (smooth tube) before applying Boron Nitride, from [77]

3.4 Thermal ground plane (TGP) heat spreaders

In 2008, the Defense Advanced Research Projects Agency (DARPA) launched the Thermal Management Technologies (TMT) program. The portfolio of Thermal Management Technologies, developed as enablers for high-power electronic and military systems, focused on developing and optimizing new heat spreaders, air-cooled heat exchangers, and thermal interface materials. The TMT program was divided into five technical efforts; Thermal Ground Plane (TGP); Microtechnologies for Air-Cooled Exchangers (MACE); NanoThermal Interfaces (NTI); Active Cooling Modules (ACM); and Near Junction Thermal Transport (NJTT). Among these, ARL was instrumental in the success of the Thermal Ground Plane (TGP) program.

The primary goal of the TGP program was to obtain lateral heat spreaders with extremely-high effective thermal conductivities, on the order of 10-20 kW/mK. This goal represents a massive 25-50 times improvement over copper (400 W/mK) and an order of magnitude improvement over diamond substrates (1000 W/mK). Success relied on passive evaporation, condensation, and fluidic wicking (due to micro- and nano-porous structures) of a working fluid (typically water or alcohol) in an evacuated TGP. In addition to metrics regarding hermiticity, coefficient of thermal expansion (CTE), and form factor, TGPs were engineered to maintain performance at elevated inertial loads (up to 20g) in consideration of military and avionic load profiles. Eight teams participated in the TGP program including Raytheon, Northrop Grumman, University of Colorado (Boulder), University of California (Las Angeles), University of California (Santa Barbara), Teledyne, University of California (Berkeley), and General Electric.

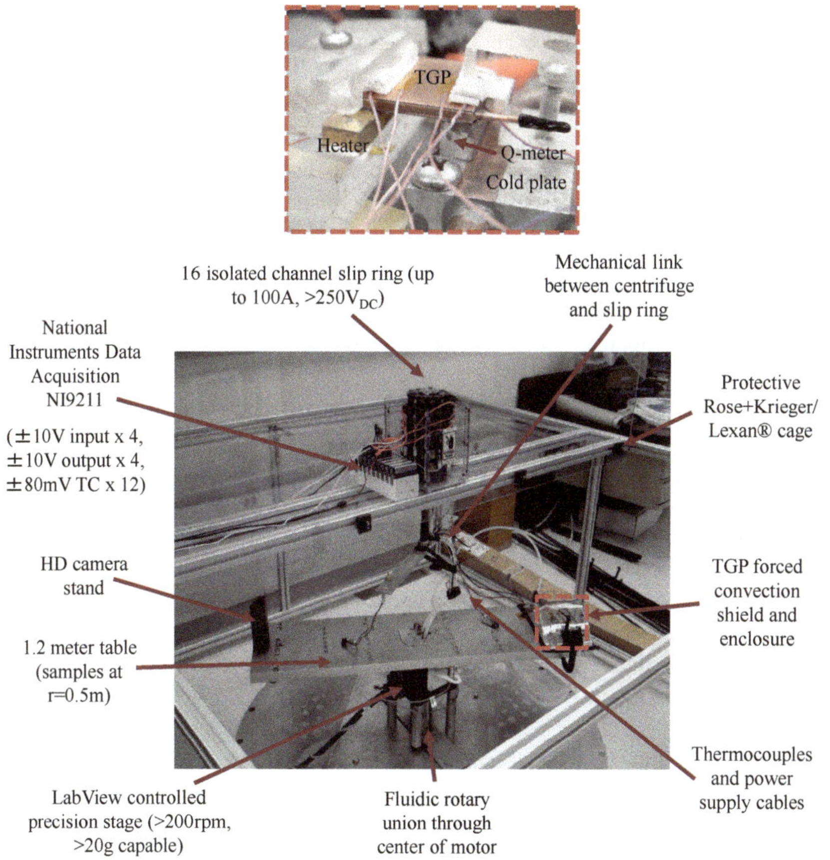

Figure 26: Photograph of the ARL TGP centrifuge and close-up of TGP test section

ARL was commissioned as a third-party testing facility to provide device evaluation through all phases of the TGP program. A photograph of the ARL TGP testing facility, along with a close up of the TGP test section (Phase II) is shown in Figure 26. As described in the DARPA Thermal Ground Plane final report [78] and subsequent journal paper highlighting the TGP development and parametric results [79], ARL was instrumental in testing and analyzing TGP devices for all performers at varying points in the program, as well as testing commercial off-the-shelf (COTS) devices as a benchmark for comparison. Please refer to the above papers, as well as a publication in the Journal of Heat Transfer by Thompson et al. [80] regarding the testing of a flat-plate oscillating heat pipe during high-gravity loading, a publication in the Journal of Micromechanics and Microengineering by Oshman et al. [81] discussing the thermal performance of a flat (flexible) polymer TGP under high acceleration loads, and a conference paper by Altman et al. [82] discussing the design, testing, and characterization of a CTE-matched passive heat spreader in dynamic and high-inertial-load environments.

Please also reference a recent conference paper by Hose et al. [83] discussing the development of a low-CTE, high heat flux vapor chamber that was integrated into the backside of a direct bond copper (DBC) substrate to improve performance in an ARL power electronic module. In this study, modeling results demonstrated thermal conductivities up to 950 W/mK, which represents a 5x improvement over the standard CuMo heat spreader, 2x improvement over pure copper (which has a poor CTE mismatch), and parity with diamond substrates. Experimentally, this resulted in a significant (43°C) reduction in device temperature during operation.

3.5 Transient thermal mitigation and phase change materials

The previous chapters described approaches and efforts to improve the thermal performance of power electronic packages, primarily by addressing the thermal stack and reducing the net thermal resistance of the multiple layers and convective cooling approach. This is often adequate for systems with steady or slowly-changing heat profiles. In practice, however, a steady-state approach to thermal design is inadequate in addressing the cooling requirements of the system. In this chapter we discuss the unique challenge that transient thermal conditions present to power electronics thermal management and review current efforts at the Army Research Laboratory to address this problem using integrated phase change material thermal buffering.

3.5.1 High-rate transient heat transfer and power electronics challenges

Many of the previously described improvements has involved enhancing convective performance, using higher thermal conductivity materials, and reducing the number of layers in the package thermal stack. These improvements have enable significant decrease in steady-state thermal resistance allowing component operation at higher power and/or functional density. However, these steady improvements may not always be fully realized because of the simple fact that in many cases the electronic components are not used in steady thermal systems. Vehicle electronics are a useful example of this where thermal transients can span from milliseconds to hours. For the primary drivetrain power electronics, over any given drive cycle a vehicle can have peak power draws much higher than the average value, with numerous, sudden load changes. Power surges exceeding twice the average power draw, but lasting only a few seconds, occur numerous times throughout the US06 drive cycle as shown in Figure 27 [84] [85]. In addition to vehicle electronics, the U.S. Army has also been investing in pulsed power electronics for a number of offensive and defensive applications, and these pulses can range from microseconds to seconds in duration [86] [87]. Thus, there exists a need to better understand how electronic packages perform under transient load conditions to optimize their development for these applications.

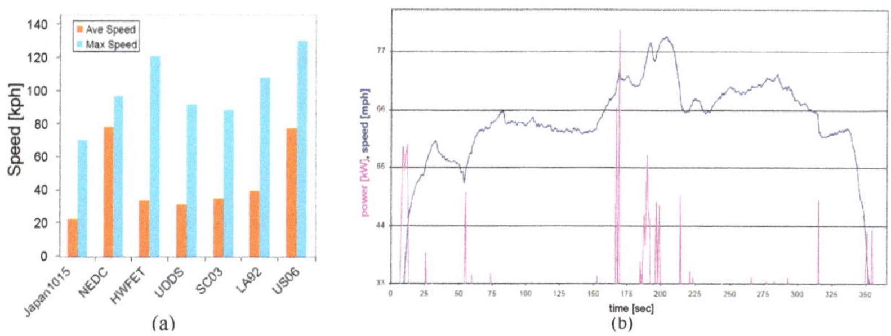

Figure 27 - (a) average and maximum vehicle speeds under different standard driving cycles, adapted from, (b) transient speed and modeled engine power for the highway portion of the US06 drive profile for a hybrid vehicle, adapted from [85] and [84], respectively.

Most electronic packages and cooling systems are designed to accommodate transients by having enough steady capacity to handle the peak expected heat load. If the average power draw is much lower than the peak, however, this creates system level thermal overdesign with associated increased size, weight, and power of the cooling system. An additional concern is that the packages designed for steady-state heat loads may actually have degraded performance for systems with fast transients. Most of the previously described thermal improvement efforts decrease package thermal capacitance (C_{th}) along with thermal resistance (R_{th}). The thermal time constant of the package, which is $R_{th} \cdot C_{th}$, decreases significantly in these cases. As shown in a study by Meysenc on semiconductor transients, a lower RC constant will result in a component that follows transient more closely and could result in higher temperature swings.

Jankowski and McCluskey examined the effect of this on several power electronic package configurations, and confirmed that low thermal-mass packages can indeed exhibit higher temperatures under pulsed loads [88]. Using a finite element equivalent thermal circuit formulation within a SPICE circuit simulation environment, they modeled five power electronics thermal stacks representing progressively increasing levels of cooling integration, shown in Figure 28.

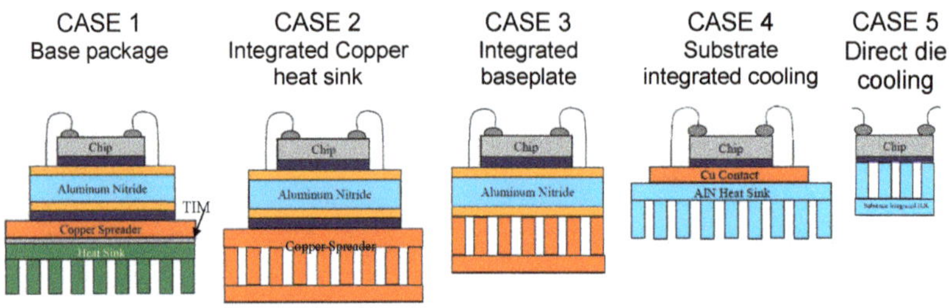

Figure 28 – Power package configurations used to examine transient behavior. Cases 2 through 5 represent increasing levels of cooling integration into the package thermal stack, up to the point of bringing the cooling directly to the backside of the semiconductor device (not to scale), adapted from [88].

The step response of these packages was analyzed for a range in convection coefficients applied to the cooling surface. Two primary results of interest were observed. First, it was noted that for all packages there exists a pulse width below which no amount of increased cooling would provide any significant thermal improvement. A representative case of this is shown in Figure 29, where a 1000x increase in effective convection coefficient, h_{eff}, produces diminishing temperature reduction with smaller pulse widths, eventually showing no thermal improvement below the package's thermal time constant. This suggests.

The second result verified that for certain pulse conditions there exists the possibility that a power package with a higher degree of cooling integration, and supposedly improved thermal performance, will actually perform worse due to the decreased package thermal inertia. This is shown in both the time and frequency domains in Figure 30. In Figure 30(a) a frequency domain plot of the package thermal impedance shows the expected rolloff in junction temperature rise with increasing pulse frequency. However, the decreased thermal time constant for the more highly integrated packages creates a region where the thermal performance is inverted. This should result in higher junction temperatures for the more integrated packages, and a representative example of this is shown in Figure 30(b). Here, the different packages are subjected to a 100ms pulse at a 10% duty cycle. The Case 5 directly cooled package exhibits the largest temperature swings with each pulse, and has three times the temperature rise of the non-integrated Case 1.

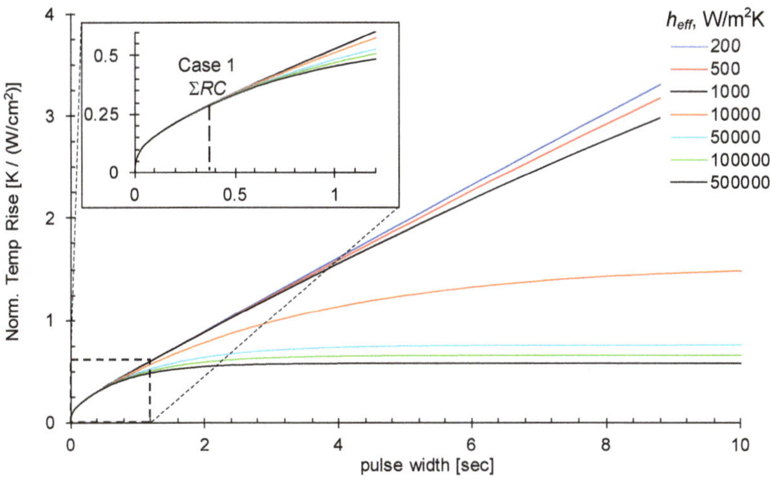

Figure 29 - Junction temperature rise per unit applied heat flux under pulsed excitation for Case 1 with varying convection rates and pulse widths. The inset magnifies the range up to 1 second showing a lack of differentiation between junction heating at small pulse widths. Adapted from [88].

The overall takeaway from this study are the following: (1) there are transient pulse conditions that will show no thermal improvement from improved convection, and (2) better package integration my help steady state thermal response but in some cases can produce degraded performance under transient conditions. Reducing temperature rise in a power electronics package subject to fast thermal transients will require addressing both package thermal resistance and thermal capacitance.

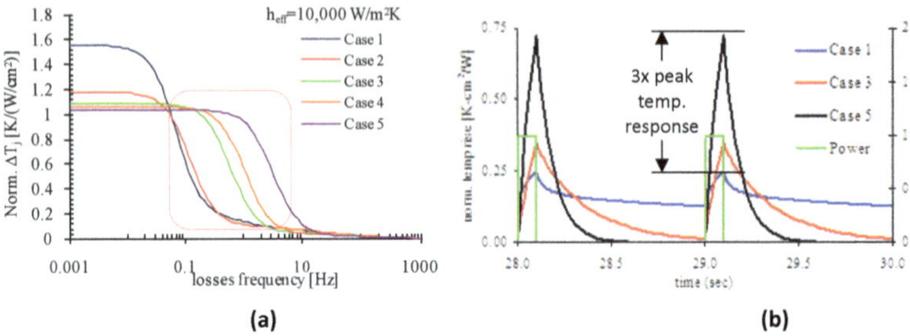

Figure 30 – Inverted temperature response of power packages under certain pulsed loading conditions. (a) The thermal impedance roll-off with increasing pulse frequency showing an inverted response region, and (b) a transient comparison of pulsed temperature rise for several package Cases showing much higher thermal response for the highly integrated package. Adapted from [88].

3.5.2 Engineered thermal capacity – phase change material thermal buffering

The transient challenges described in the previous section create the need to engineer thermal capacity back into the electronics package while retaining as much of the low thermal resistance as possible. One technique that has been investigated for a number of applications is Phase Change Material (PCM) based Thermal Energy Storage (TES), where a material with an appropriate melting temperature is selected to absorb excess thermal energy through solid-liquid phase change, and then release the energy at a more system compatible rate in between pulses or surges. Designed correctly, this would act as a buffer between the high-rate heat source and a slower rate cooling system, permitting it to be designed for the average, rather than peak heat load. Additionally, because phase change involves latent, rather than sensible, energy absorption, the process occurs (ideally) isothermally. These factors all support attempting to utilize PCMs with electronic packages to reduce transient temperature rise.

3.5.2.1 Phase change materials, selection considerations

The use of phase change materials for thermal energy storage is not a new concept, as research into building environmental control and electronic system cooling have been occurring for decades [89] [90]. However, the majority of electronics applications have focused on slower transient systems and have not attempted to integrate directly with power packaging. As a result, they have relied on traditional materials, and results have been sub-par when attempting to obtain faster temporal response.

In an effort to better understand both the available material space and the implications of specific applications, Jankowski and McCluskey published a comprehensive review of phase change materials for vehicle applications, including thermal buffering of high temperature power electronics [7]. This review examined vehicle applications with transient thermal profiles from 0 to 800°C, and included over 700 materials from over a dozen material classes, attempting to identify gaps in the literature and match promising materials to applications. Figure 31 provides a loose categorization of PCMs covered in the study. Historically, the majority of research focus was on waxes and hydrates salts due to their broad availability over the low-medium temperature range below 100°C. For higher temperature ranges there had been some additional work on sugar alcohols, and some limited research had also been performed on metallic PCMs. Figure 32provides an estimate of temperature coverage of the different classes of PCMs.

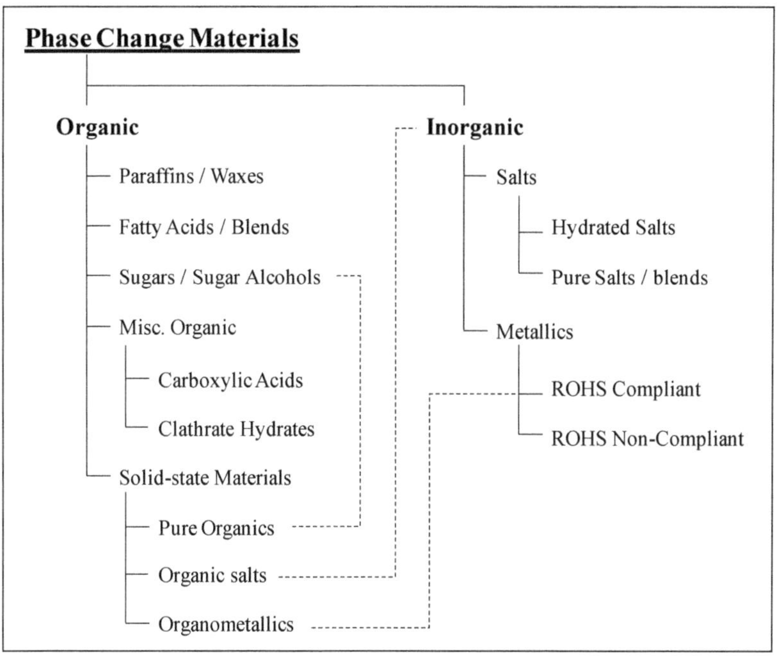

Figure 31 – General phase change material categories with dashed lines showing some categorical overlap. Adapted from [7].

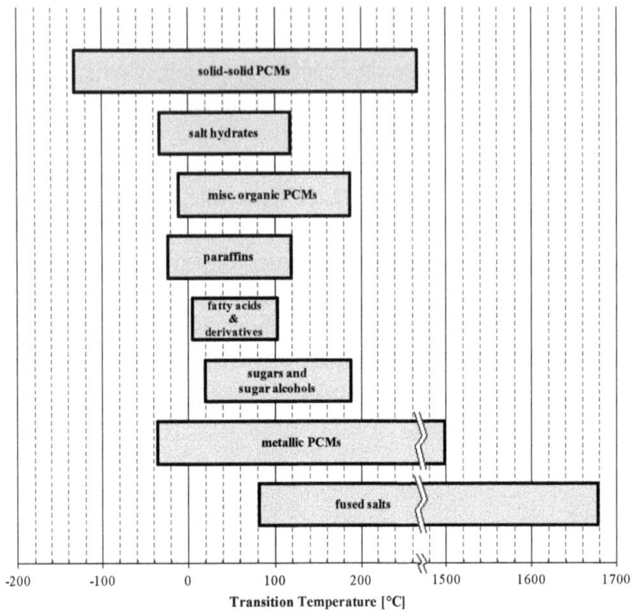

Figure 32 – Transition temperature span of reviewed phase change materials, adapted from [7].

Recommendations were made on further materials research needs, improved PCM property characterization, and PCM usage in specific applications including power electronics. In particular, it was noted that much of the PCM materials work has focused on trying to overcome the low thermal conductivity of the more common PCMs. Most PCMs have been focused on because of their high specific latent heat (i.e., kJ/kg), but their low thermal conductivity and low density make it difficult to drive heat through the material to the melt front. By recasting the materials set on the basis of volumetric latent (i.e., MJ/m^3), many of the traditional materials look far less promising as phase change materials, and underexplored materials like metals and alloys become much more attractive. Additionally, these materials have thermal conductivities one or two orders of magnitude higher than the more common waxes and salt hydrate materials, which can significantly reduce the heat absorption challenge. This is situation is shown graphically in Figure 33, where comparing the same materials using specific and volumetric latent heats produce opposite results in showing the relative merit of paraffin and metallic phase change materials. Finally, materials of interest to power electronics were identified including higher temperature organics like sugar alcohols (e.g., xylitol, erythritol) due to their high specific and volumetric latent heats, and metallic materials like low temperature solders due to their high specific latent heats and thermal conductivities.

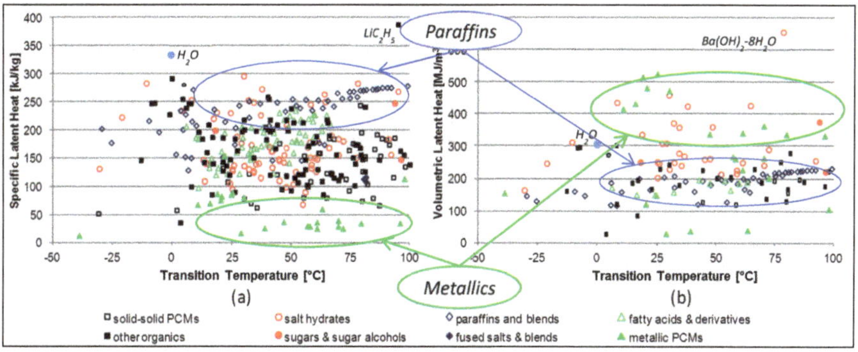

Figure 33 – latent heats of reviewed lower temperature (sub-100°C) PCMs showing the inversion in relative energy absorption potential between paraffins and metallics when comparing materials based on specific and volumetric latent heat.

3.5.2.2 Non-ideal phase change - supercooling

In line with the PCM material investigation, ARL began examining materials material options specifically for use in thermal protection of power electronics. A popular material being investigated for moderate temperature applications was erythritol, previously mentioned as a high latent heat sugar alcohol. Unfortunately one known problem of high latent heat crystalline PCMs is that they have a high tendency to supercool, which means that after melting a material can cool to temperatures below the phase change temperature before resolidifying. This is potentially a critical failure mode in a thermal protection system, as a PCM that has not re-solidified will not absorb heat isothermally on the next thermal transient, and the protected component will likely overheat and fail. Jankowski and McCluskey reviewed successful supercooling mitigation strategies in 2010, including material modification and mechanical agitation, and also presented a preliminary experiment on electrical supercooling mitigation [91]. The review described how using seed crystals and additives have been successful in passively reducing, but not eliminating, supercooling in many materials, while ultrasonic acoustic agitation has been shown to completely eliminate supercooling. Several electrical methods were reviewed, but only a direct current through water with proper electrodes had shown conclusive evidence of effect. ARL's electrical investigation attempted to build on that work with erythritol, applying a moderate voltage across silver wire electrodes suspended in a vial of melted PCM with a separation of about 350μm. Figure 34(a) shows a representative transient thermal profile over an experimental trial where erythritol was allowed to cool until spontaneously nucleating after significant supercooling, in this case reaching supercooling about 50°C below the typical 118°C melting temperature. Two temperatures were recorded, near the electrodes at the center of the melt, and on the edge of the vial. Spontaneous nucleation usually occurs at the coldest spot in the melt, usually the vial edge as shown in (a), as the area with highest degree of supercooling has the highest nucleation potential. Stimulated nucleation is expected to occur at or near the point of stimulation.

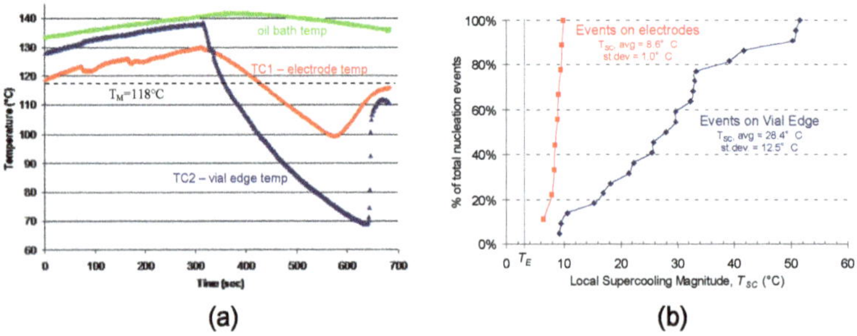

Figure 34 – Erythritol supercooling experimental data, showing (a) a representative cooling curve, and (b) the nucleation event accumulation curves for spontaneous and electrically stimulated nucleation. Adapted from [91].

The results of the experiment are shown in Figure 34(b), depicting accumulation curves for the nucleation events that occurred either on the electrodes or on the edge of the vial. In all cases when $200V_{DC}$ was applied to the electrodes, nucleation occurred on the positive silver electrode at the warmer part of the melt. This nucleation occurred at temperatures much higher than when no voltage was applied and the nucleation occurred on the colder vial edge. The average nucleation temperature at the electrodes was almost 20°C higher than the edge nucleation events, and the variation in nucleation temperature was reduced by more than a factor of ten. Additionally, the fact that nucleation always occurred on the positive electrode, even when switching polarity, indicates the possibility of an electrochemical cause of stimulated nucleation. This work was far from comprehensive in investigating the electrical supercooling mitigation mechanism, but it does support the idea that supercooling can be actively avoided as a critical failure mode in phase change thermal protection systems.

3.5.2.3 PCM substrate integrated package design

At the same time that the Army was investigating architectures and cooling options for Future Combat System vehicles [92], the Department of Energy (DOE) sponsored a project under the FreedomCar program looking at vehicle heat sinks incorporating thermal energy storage [84]. This program examined several PCM-heatsink configurations and materials for a power electronics heat sink, and in 2007 identified materials and a parallel PCM-coolant arrangement called a Thermal Buffer Heat Sink (TBHS) that would maximize transient response while maintaining high cooling performance. Building upon the previous microchannel manifolding and substrate integration work, ARL's first electronics phase change material investigations numerically explored the possibility of implementing a similar THBS design into the electronics substrate in 2009 and 2011 [93] [94].

The DOE computational study found that a parallel arrangement of PCM and coolant channels with a high latent heat PCM (erythritol being the best candidate selected in this case), would provide sufficient steady cooling and enough absorption for the transients encountered over a standard drive cycle. ARL's substrate TBHS design studies focused on the same general structure with sub-millimeter scale channels embedded in an electronics substrate as shown schematically in Figure 35. Here, the substrate channels alternate between cooling channels connected by a manifold and completely sealed PCM channels. In this way, the area devoted to absorption or cooling can be traded off with cooling area for a particular use case.

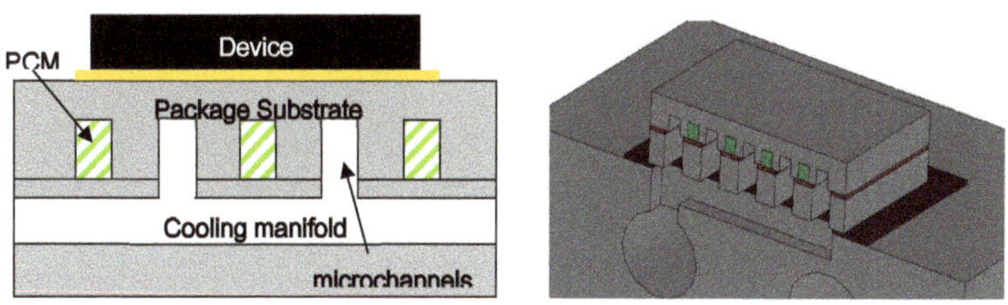

(a) (b)

Figure 35 – (a) Schematic of a substrate integrated thermal buffer heat sinkwith single layer PCM-coolant arrangement over a fluid delivery manifold, and (b) notional 3D rendering. Adapted from [93]. These reports presented a fabrication and assembly method for the substrates as well a parametric study using numerical simulation of thermal suppression to compare the impact of geometric parameters for different coolant and heat input conditions. Figure 36(a) shows a unit cell of the two-dimensional simulation domain, where the widths of the PCM, fin, and coolant regions were the primary variables. Using an implementation of the apparent heat capacity method, a custom polynomial phase change smoothing function, and automatic mesh refinement algorithm (Figure 36 (b) and (c)), maximum temperature rise was calculated over the parametric space.

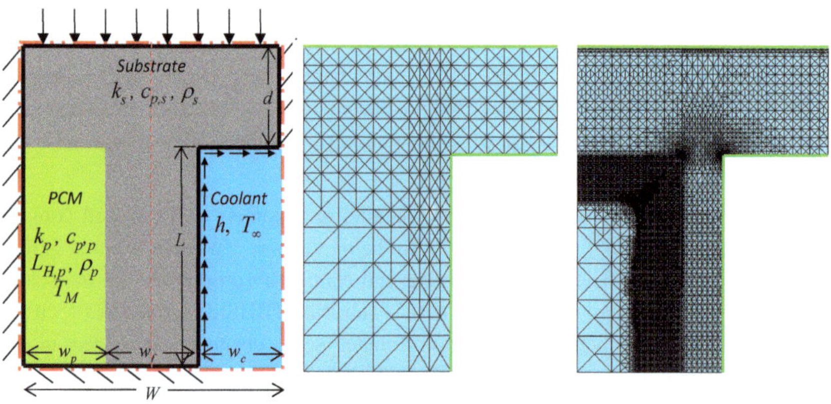

Figure 36 – (a) TBHS simulation domain, representative (b) coarse and (c) refined finite element mesh layout. Adapted from [94].

Sample results of these simulations are shown in Figure 37 and Figure 38. In Figure 37 the transient temperature rise is shown for a 100 W/cm^2, 100 ms, 10% duty cycle pulse train. The TBHS shows significant temperature suppression (4-5°C) relative to an identical heat sink without phase change, although it was recognized that the non-phase change heat sink was perhaps not the best comparison as it had a significant fraction of the potential heat removal area performing no useful function. In Figure 38 the simulated step response of another TBHS is shown, this time in comparison with a substrate that has the PCM volume replaced with additional cooling area. In comparing a PCM substrate with an equivalently manufactured substrate-integrated heatsink it is apparent that there is some potential for thermal suppression for a certain pulse duration, but this benefit diminishes and finally disappears for longer pulse times. Additionally, the suppression was only significantly exhibited for lower heat fluxes and convective cooling capacities.

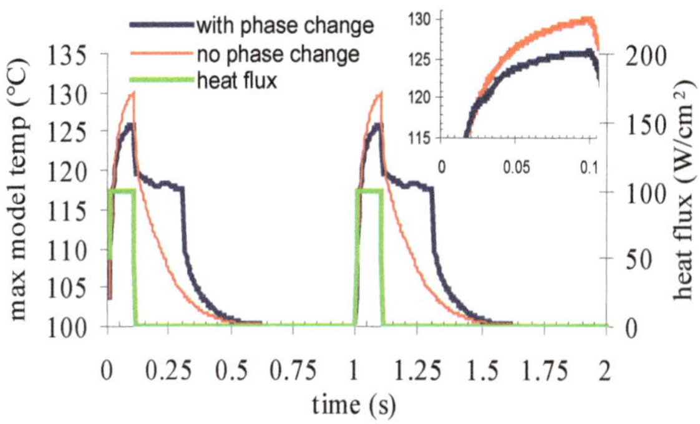

Figure 37 – Simulated transient temperature rise for a heat pulse applied to a silicon TBHS with erythritol and 10kW/m²K cooling, showing 4°C temperature suppression compared to an identical substrate without phase change absorption. Adapted from [93].

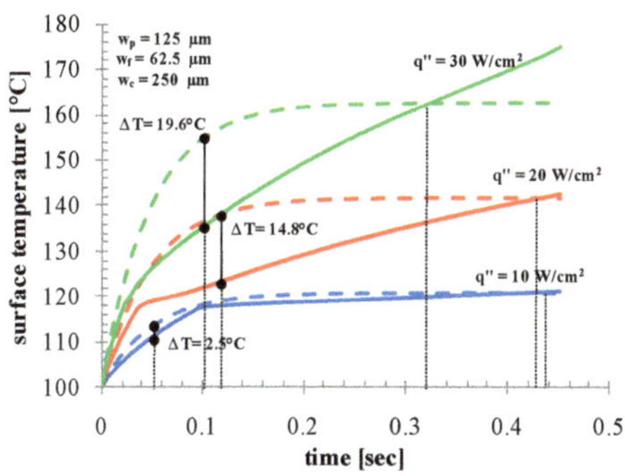

Figure 38 – Simulated transient thermal response to a various step heat loads and 1 kW/m²K cooling for the silicon TBHS (solid lines) and the equivalent dimension microchannel heat sink with coolant replacing the PCM area (dashed lines) showing cooling benefit over a standard heat sink at low heat fluxes and ~100 ms pulse times. Adapted from [94].

The primary takeaway from these substrate integrated TBHS simulations is that while the parallel cooling arrangement does provide for design flexibility in the amount of absorption and cooling in the package, the presence of an equivalent thermal resistance between the heat source and both the PCM and cooling meant that only a small fraction of the heat was being absorbed by the PCM, and it's effectiveness was diminished. This was exacerbated by the fact that the simulated PCM was high latent heat but low thermal conductivity, providing a large thermal resistance to the phase front once melting began. While ARL has continued numerical and experimental evaluation of the TBHS design space and alternative materials, the need for much faster thermal response, a factor of 10 or better faster than those seen above, has led the research group to investigate metallic PCMs and direct PCM contact package designs that will be described in the next section.

3.5.2.4 PCM direct device cooling

Recognizing from the previously described material review and simulation work that achieving higher speed thermal suppression will require removing thermal resistance between the heat source and the phase front, ARL began initial tests on placing a phase change material in direct contact with a power electronic device. Using a low thermal conductivity organic PCM, however, showed no thermal improvement under high-power, 20ms pulses, and devices often overheated and critically failed. This prompted a course of modeling and experiment to understand why there appeared to be no thermal benefit and determine how to design high-speed PCM thermal suppression.

In 2016, Gonzalez-Nino, et al. published the results of a custom phase change model examining heat transfer in the packaged power device and a series of different phase change materials [95]. Using a finite-difference discretization and modified source-sink model of the non-linear phase change problem, results such as that shown in Figure 39 were obtained to explain the device behavior. Figure 39 shows the simulated response of several PCMs to a 20ms, high heat flux thermal pulse. Both of the organic PCMs, octadecane and erythritol, shown almost no thermal suppression compared to a standard non-melting encapsulant, as the PCM heat absorption provides almost no benefit due to their high thermal resistance. Alternatively, same test performed with two metallic PCMs showed over 30°C of thermal suppression in the 5-10ms timescale. Significant thermal benefit is seen even before reaching melting temperatures because of the increased material thermal conductivity and spreading of heat from the pulse. The PCM model is now being applied to the task of exploring the PCM-package parametric design space and optimization for different Army electronic applications.

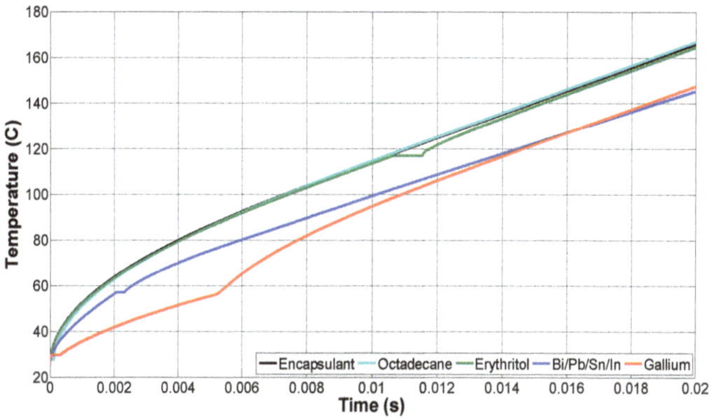

Figure 39 – Comparison of simulated device heating relative to a dielectric encapsulant using different organic and metallic phase change materials under a 1.6 kW/cm², 20ms pulsed load. Adapted from [95].

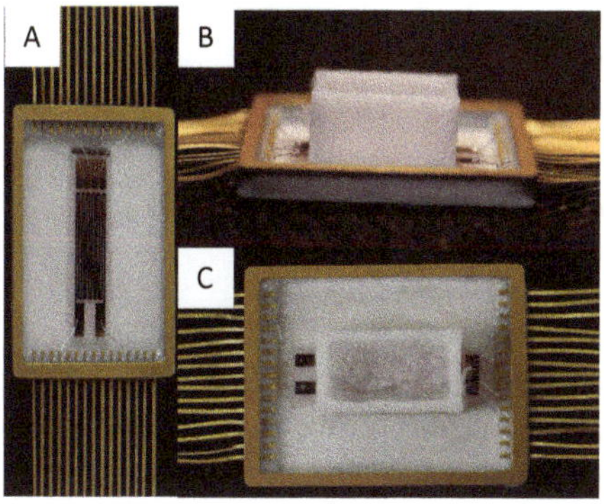

Figure 40 – Experimental device for high-speed PCM testing showing (a) top view of power resistor without PCM, (b) side view of PCM enclosure, (c) top view of filled PCM enclosure. Adapted from [96].

Following on this simulation exercise, an experimental verification of the metallic PCM thermal suppression was published by Gonzalez-Nino, et al. in 2017 [96]. In this work a custom power resistor with integrated micro-temperature sensor was fabricated with a series of PCMs as direct-coupled encapsulants, an example of which is shown in Figure

40. These devices were subjected to a series of high power pulses to measure thermal response.

Subjected to 20ms pulses, the thermal improvement provided by the PCMs is in agreement with the previously described simulation work. As shown in Figure 41, the two PureTemp dielectric phase change materials provide a small reduction in temperature rise, but the two different metallic materials provided 60-80°C of thermal suppression. These results were consistent over a range of powers and pulse times, and suggest that proper design optimization may provide further improvement.

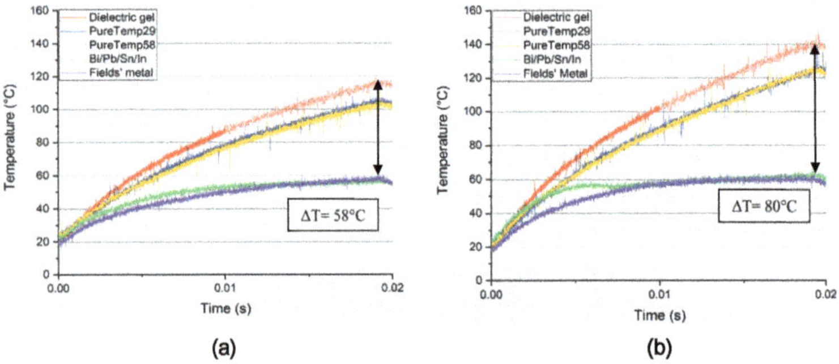

Figure 41 – Experimental results showing thermal improvement from using metallic phase change materials for fast thermal pulses. Adapted from [96].

These experimental results show that high speed thermal suppression can be achieved, but there is still a significant amount of work to be done to identify the best methods of metallic PCM integration. In addition to the more obvious potential issues of electrical shorting and containment presented by a liquefied metal in the package, some other non-ideal issues were identified during the experiment. As described in Gonzalez-Nino, et al., repeated cycling of the devices described above began to show a change in thermal behavior as shown in Figure 42 [97].

As repeatability of thermal response is an important metric of a phase change thermal protection device, the test articles exhibiting increased temperature rise were inspected for faults. The resistor behavior was consistent, but in removing the solidified metallic PCM significant voiding was seen at the heated interface. Follow-on thermal tests were performed showing the development of voids over time even if there were no voids or obvious volatile materials present after initial device assembly, as shown in Figure 43.

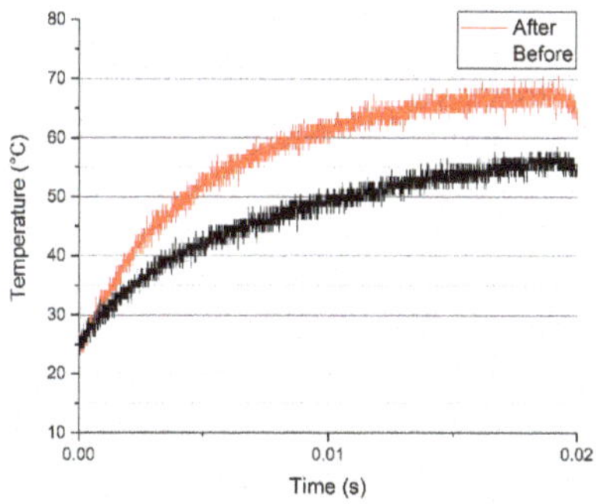

Figure 42 – increase in device thermal response to an 80W before and after thermal cycling, adapted from [97].

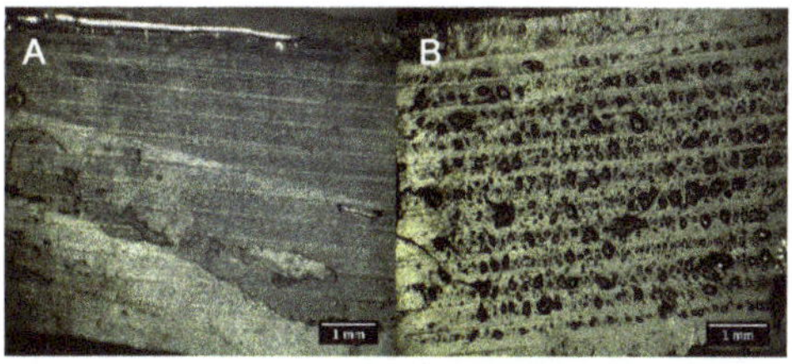

Figure 43 – Photograph of the metallic phase change material through a glass slide before (A) and after (B) thermal cycling showing the development of voids likely responsible for increased temperature rise, adapted from [97].

Metal voiding is a well-known phenomenon in soldering and similar package metallization processes. Efforts are currently underway to examine mitigation strategies for PCM voiding such as metallization or other surface preparation strategies to ensure a low thermal resistance mating surface.

3.5.3 Conclusion –transient thermal management of electronic packages

From the analysis of heat propagation through electronic packages to the investigation of integrated phase change material thermal buffering, it is clear that effective design of electronic packages needs to account for the transient nature of the driving loads and thermal environment. While a number of applications may be successfully treated as steady thermal systems with minimum thermal resistance as the only significant design factor, realistic Army systems will have conditions changing over timescales ranging from days to microseconds. Eliminating thermal overdesign to achieve high power density systems will require developing transient cooling strategies for these truly transient systems, and this will require a significant change in how high power electronic components and systems are designed.

3.6 Multi-functional packaging design

The current electronics design process is linear, as is shown in Figure 44. It starts with an electrical engineer who designs a circuit, then a mechanical engineer designs a reliable package, then a thermal engineer determines the appropriate heat sink for the application. This results in overdesigned, costly, and suboptimal electronic systems. In addition to the overdesign inherent in a linear design process, many components of an electrical package only perform a single function. For example, the only function of a wirebond is to carry electricity and the only function of a heat sink is to remove heat. Most research investigates improving one aspect of electronics packaging: eliminating wirebonds, better thermal interface materials, better heat sinks, etc. ARL aims to holistically optimize the entire package by defining and implementing a new approach to electrical system design through multi-functional components (MFCs). A multi-functional component is one which performs multiple functions (electrical, thermal, and/or mechanical) simultaneously to improve size, weight, cost, reliability, and performance.

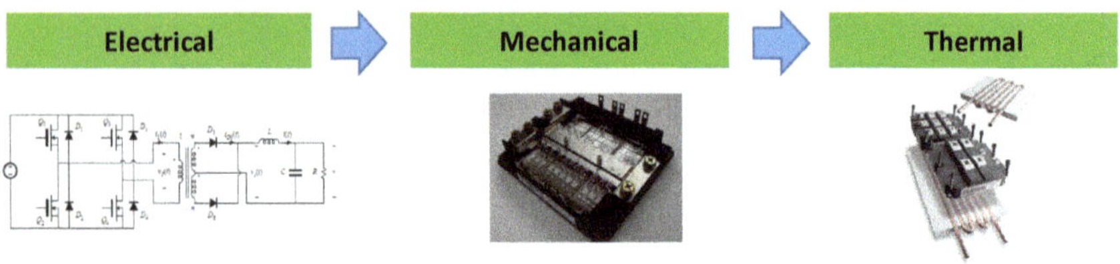

Figure 44: Typical linear design process of a power electronics module.

Multi-functionalized design is a novel approach to design theory and requires a paradigm shift in the typical linear approach and requires electrical, thermal, and mechanical engineers to come together during the initial design phase to understand and iterate through the tradeoffs and requirements across all domains and novel manufacturing methods. It involves using numerical and analytical modeling to understand the trade-off space and perform a parametric study on a wide range of parameters.

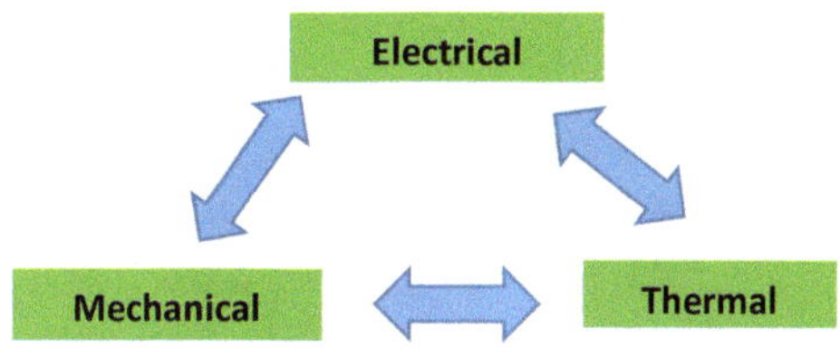

Figure 45: Multi-functionalized design process

An example of a module designed using this multi-functionalized approach is shown in Figure 46 [98]. The module on the left side of the figure shows a commercially available all silicon carbide 1200V/50A power module. The module on the right shows the same module re-designed using a multi-functional design process. The module on the right is 8X smaller and lighter and also includes an integrated heat sink which is not included on the commercial module. In addition, it has 3-10X lower thermal resistivity allowing significantly more power dissipation for the same temperature rise and lower inductance which reduces the need for additional passives.

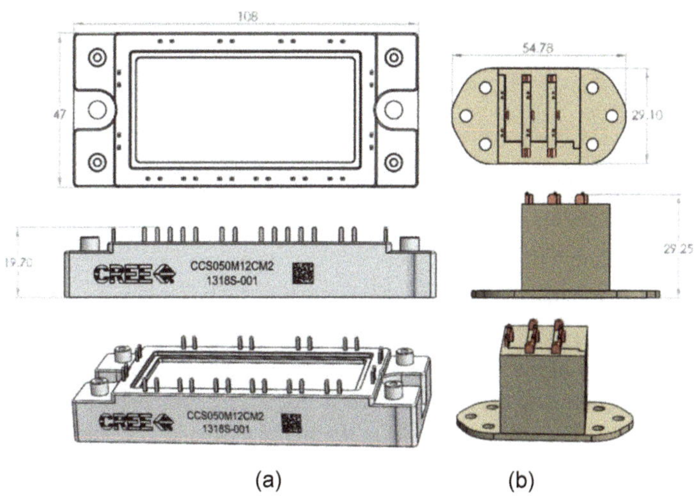

Figure 46: Example of a power module redesigned using a multi-functional approach which shows a >8X reduction in size and weight, 3-10X lower thermal resistivity, lower inductance and improved reliability. The module of the left (a) is a commercially

available 1200V/50A power module and the module on the right (b) is the same modules redesigned that is ~1 in^3.

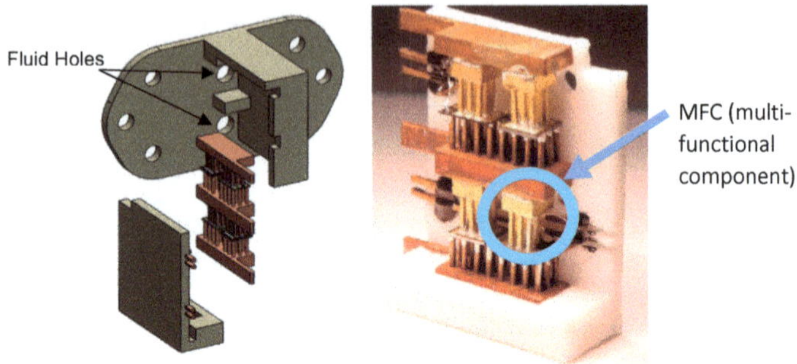

(a) (b)

Figure 47 – Power Tower module packaging approach (a) assembly structure showing the fluid connections (b) fabricated half bridge showing the MFC. Adapted from [98].

The MFC (multi-functional component) is circled in Figure 47. The component simultaneously acts as an electrical, thermal and mechanical interconnect. It has eliminated a number of single function components allowing significant improvement in the module. The advantages of this approach over what is commonly done includes:

1. >8X reduction in weight and volume which increases the power density
2. 3-10X lower thermal resistivity allowing also improves the power density
3. Reduced inductance (faster switching frequencies, reduce passive components)
4. Improved reliability by eliminating 3 primary failure locations (wirebonds, DBC and large area contacts)
5. Improved cost (smaller packaging and eliminating high cost DBC)

The integrated heat sink is a key enabling feature of this power module by allowing double sided cooling of the stacked devices. Stacking devices is very difficult without effectively handing the heat removal. In order to effectively stack two power devices, the ability to effectively remove heat is critical. In addition, it is known that removing heat from both the top and bottom of the device is ideal. Typically heat is only removed through the substrate (backside). A lot of work is looking into topside cooling due to its benefits since the heat is actually generated on the topside. This configuration allows for both top and backside cooling with the heat sink directly in contact with the chip. The locations of the fluidic inlet and outlet are shown in Figure 47. The fluid, which must be a dielectric fluid, flows over both the top and bottom fins of each device, cooling each device as it passes. Figure 48 shows the general path of the fluid using blue arrows. The fluid will heat up as it progresses along this path.

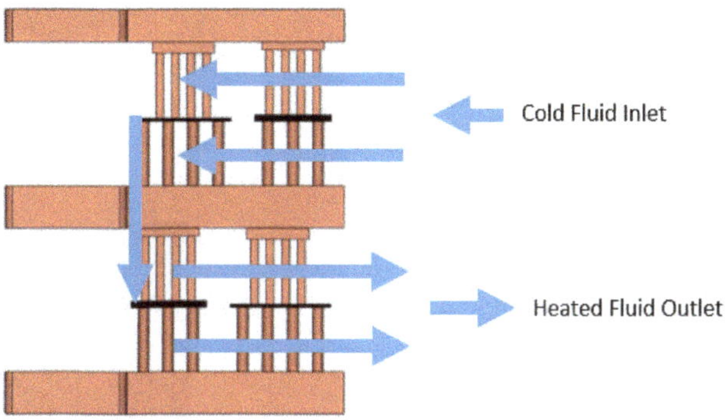

Figure 48 Cross section view of the stacked module showing how the fluid passes over the devices

The basic premise of this design is that by stacking power devices, the power module is no longer constrained by the limitations of planar packaging, and therefore significant improvements can be made to the package. The primary challenge of stacking devices is the significant heat build-up penalty; however, the novel arrangement addresses this challenge by incorporating a heat sink both above and below each of the stacked devices. This is the first power module that both stacks power devices and includes an integrated heat sink. It also eliminates the solid dielectric substrate used in planar packaging and other stacked power device modules.

3.7 ParaPower tool

In addition to the multi-functional design approach, ARL researchers have also developed a tool (ParaPower) which uses a compact 3D thermal resistance network to quickly analyze device temperatures and stresses in a generic power module configuration [99] [100]. A screenshot of the tool is shown in Figure 49. This tool allows quick parametric analysis to help reduce the overdesign and improve the size, weight and performance of the module.

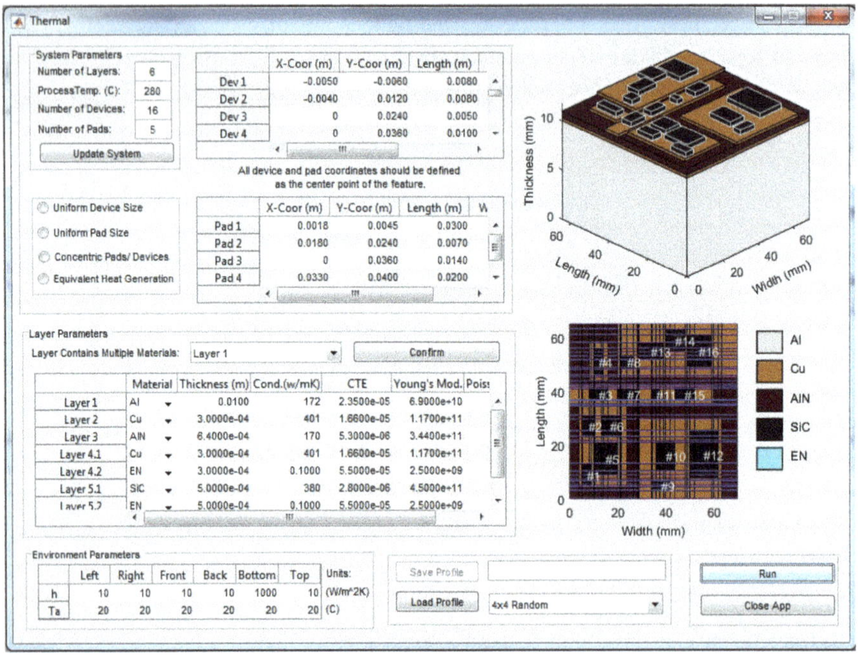

Figure 49: Screenshot of the main ParaPower graphical user interface.

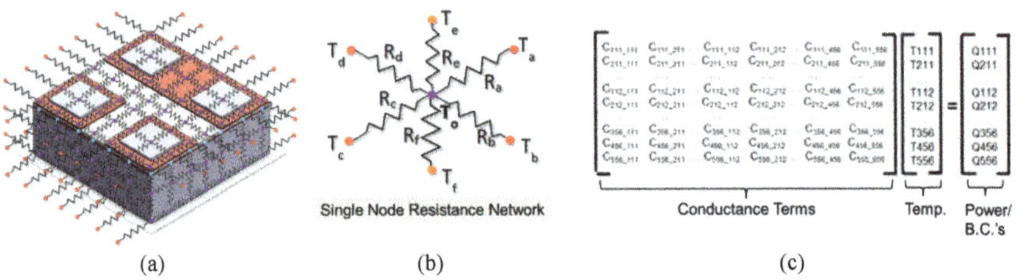

Figure 50: Nodal network analysis used in ParaPower (a) simplified resistor network on a 4 chip power module (b) 3D nodal network around each node (c) matrix used to calculate the temperature at each node.

The method is based on a nodal network, as is shown in Figure 50. A summation is made around each node and a series of equations are developed and solved in MATLAB using matrix inversion. A thermal stress model is also implemented [101] to estimate device stress. When comparing to full FEA using ANSYS, the temperature results agree to <2C and the stress agrees to <20% error while solving >100X faster [100].

Two example parametric studies was performed to assess the impact of varying device spacing and backside cooling. The results are presented in Figure 51a and Figure 51b, respectively. The results were obtained in seconds and can be used to make design decisions about device spacing and heat sink options.

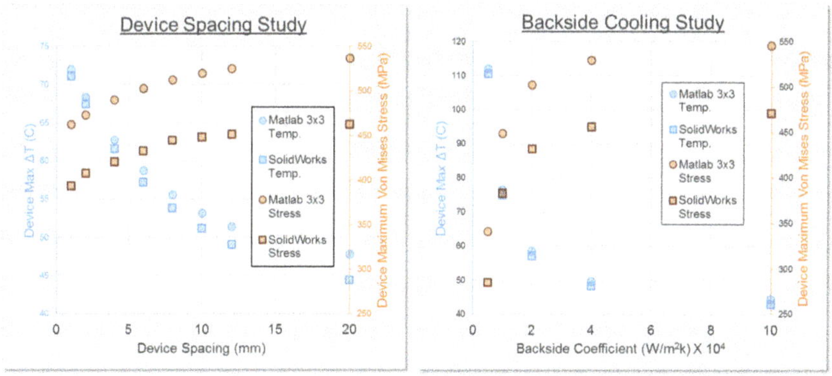

Figure 51: Example parametric study results showing how device spacing (a) and backside cooling (b) effect both the maximum device temperature and module stress (Q_{chip}=64 W/cm^2, $h_{backside}$=1000 W/m^{2}C). The results are also compared to finite element analysis as well showing good agreement.

Codesign tools such as ParaPower create meaningful results at much faster speeds and is significantly more efficient in evaluating the design space as compared to full finite element analysis (FEA) or computational fluid dynamnics (CFD). Computational abilities of standard hardware limits the frequency an FEA solution is generated, and when a typical requirement of 1000's of unique designs are to be considered, ParaPower has the capability to perform the analysis in considerable less time. The key is simple output, each solution just outputs the maximum temperature and stress in the module for comparative purposes. This does not eliminate full FEA or CFD analysis but rather should be used as an initial design tool to narrow the design space and also explore potential configurations which might have been overlooked using a linear design process.

4. Conclusions

In order to realize the full potential of wide bandgap (WBG) devices and to improve the SWaP of power systems, it is critical to address the thermal and packaging challenges. Future Army power electronics systems are moving towards higher power density, higher voltages, higher temperature and increased reliability. Current WBG devices can handle these requirements but the packaging and ability to remove the heat is now limiting their performance. This chapter outlined ARLs efforts to reach the 0.1 cm2K/W thermal resistivity goal. Understanding the effects and tradeoffs between solutions is critical to optimizing the design. ARL has investigated many ways to improve the standard power electronics module. The areas of research discussed in the chapter include: single-phase liquid cooling, two-phase liquid-vapor cooling, advanced passive thermal ground plane (TGP) heat spreaders, transient thermal mitigation with solid-liquid phase change materials, multi-functional packaging and thermal management design, and device layout optimization with the ParaPower tool. Due to the system complexities, a combination of these approaches is the most likely solution to the power packaging challenges. This multi-tiered approach, while adding complexity, is necessary to achieve the 0.1 cm2K/W resistivity goal.

Acknowledgements

Dr. Darin J. Sharar would like to acknowledge his PhD advisor, Dr. Avram Bar-Cohen, for providing the opportunity and motivation to undertake his work on enhanced performance in internally-grooved tubes; please refer to the sections titled ‘Platform Considerations’, ‘Two-Phase Cooling’, and ‘Thermal Ground Plane (TGP) Heat Spreaders’. He would also like to thank his Dissertation defense committee; Professor Michael Ohadi, Professor Bongtae Han, Professor F. Patrick McCluskey, and Professor Mario Dagenais for their advice and guidance through all stages of this work. He would like to give special thanks to Dr. Arthur Bergles, whose guidance has inspired and informed much of this ongoing research effort. He will be sorely missed, both professionally and personally.

Dr. Lauren Boteler would like to acknowledge her PhD advisor, Dr. F. Patrick McCluskey for his expertise and support on her PhD work. She would also like to acknowledge ARL colleagues Dimeji Ibitayo, Bruce Geil, Gail Koebke, Damian Urciuoli, Claude Pullen, and Morris Berman for their support. In addition, Professors Steven Miner and Andrew Smith provided expertise on the codesign work. And colleagues at the University of Puerto Rico supported the metallic phase change work including David Gonzalez-Nino and Professor Pedro Quintero.

Nicholas Jankowski would like to acknowledge his PhD advisor, Dr. F. Patrick McCluskey, along with his dissertation committee, Professor Avram Bar-Cohen, Professor Jungho Kim, Professor Hugh Bruck, and Professor Neil Goldsman for their ongoing work supporting the efforts on electronic transient mitigation and phase change materials.

Collectively, we would like to thank our collaborators at the Army Research Laboratory, University of Maryland, Rochester Institute of Technology, General Technical Services, United States Naval Academy, Defense Advanced Projects Agency, Advanced Cooling Technologies, Raytheon, University of Missouri, Northrop Grumman, University of Colorado (Boulder), Thermacore, Technology Assessment & Transfer. Please see references for a full catalog of those individuals.

References

[1] "Military Specification (MIL)-STD-810G, Test Method Standard for Environmental Engineering Considerations and Laboratory Tests, revision G," United States Department of Defense, 31 Oct 2008.

[2] R. W. Erickson and D. Maksimovic, Fundamentals of Power Electronics, 2nd Edition, New York: Springer Science+Business Media, 2001.

[3] C. Mocnik, "Vehicle Electronics and Architecture," *DTIC Document #ADA543108,* 2011.

[4] J. Conover, H. Husted, J. MacBain and H. McKee, "Logistics and capability implications of a Bradley fighting vehicle with a fuel cell auxiliary power unit," in *Proc. 2004 SAE World Congress*, Detroit, MI, 2004.

[5] K. V. Pahinkar DG, " Analytical model for human thermal comfort in passenger vehicles," *SAE Int J Mater Manuf*, vol. 4, no. 1, pp. 204-215, 2011.

[6] O. H. Middlebrooks and S. E. Heuckeroth, "An Investigation of Combat Vehicle Ventilation Requirements," United States Army Research Laboratory Technical Report, no. ARL-SR-231, 2011.

[7] N. R. Jankowski and F. P. McCluskey, "A review of phase change materials for vehicle component thermal buffering," *Applied Energy,* vol. 113, pp. 1525-1561, 2014.

[8] R. W. Johnson, J. L. Evans, P. Jacobsen, J. R. Thompson and M. Christopher, "The Changing Automotive Environment: High-Temperature Electronics," *IEEE Transactions on Electronics Packaging Manufacturing,* vol. 27, no. 3, p. 164 175, 2004.

[9] S. Jones, J. Mendoza, G. Frazier and G. Khalil, "Integrating environmental control system functionality into a scheme of thermal management system evaluation metrics for military vehicles," *DTIC Document #ADA547740,* 2011.

[10] U.S. Department of Energy, Office of Energy Efficiency and Renewable Energy, "FreedomCAR and Fuels Partnership, Electrical and electronics technical team roadmap," *Technical Report,* 2010.

[11] J. Hitchcock, "Ground vehicle power and mobility overview," *DTIC Document #ADA470330,* 2007.

[12] J. Vetrovec, "Engine Cooling System with a Heat Load Averaging Capability," *SAE Technical Report 2008-01-1168,* 2008.

[13] C. Jordan, "U.S. Army Antifreeze after 50 years," in *American Society for Testing and Materials*, Philadelphia, PA, 1986.

[14] T. Salem and D. Urciuoli, "Power module cooling for future electric vehicle applications: A coolant comparison of oil and PGW," *DTIC Document #ADA481572,* 2006.

[15] J. Balda, F. Barlow, A. Lostetter and A. Mantooth, "Silicon-Carbide (SiC) electronics," University of Arkansas white paper, 2003.

[16] D. Sharar, N. Jankowski and B. Morgan, "Thermal performance of a direct-bond-copper Aluminum Nitride

manifold-microchannel cooler," in *26th IEEE SEMI-THERM Symposium*, Santa Clara, 2010.

[17] D. Tuckerman and R. Pease, "High performance heat sinking for VLSI," *IEEE Electron Device Letters*, vol. 2, no. 5, 1981.

[18] L. Lin and R. Ponnappan, "Critical heat flux of multi-nozzle spray cooling in a closed loop," in *37th Intersociety Energy Conversion Engineering Conference*, 2002.

[19] A. Bhunia, Q. Cali and C. Chen, "Liquid impingement and phase change for high power density electronic cooling," in *41st AIAA Aerospace Science Meeting Exhibit*, 2003, 2003.

[20] M. Kuszewski and M. Zerby, "Next generation Navy thermal management program," Carderock Division Naval Surface Warefare Center Report #CARDIVNSWC-TR-82-2002/12, 2002.

[21] D. Sharar, N. Jankowski and B. Morgan, "Review of two-phase electronics cooling for Army vehicle applications," United States Army Research Laboratory Technical Report, no. ARL-TR-5323, 2010.

[22] F. P. McCluskey, R. Grzybowski and T. Podlesak, High temperature electronics. The electronic packaging series, New York, NY:: CRC Press, 1997.

[23] D. Urciuoli, R. Green, A. Lelis and D. Ibitay, "Performance of a dual, 1200 V, 400 A, silicon-carbide power MOSFET module," in *Proc. IEEE Energy Conversion Congress and Exposition*, Atlanta, GA, 2010 Sep 12-16.

[24] L. D. Stevanovic , K. S. Matocha, P. A. Losee PA, J. S. Glaser , J. J. Nasadoski and S. D. Arthur, "Recent advances in silicon carbide MOSFET power devices," in *Proc. Applied Power Electronics Conference and Exposition,*, Palm Springs, CA, 2010 Feb 21-25.

[25] L. M. Boteler, D. Urciuoli, G. Ovrebo, D. Ibitayo and R. Green, "Thermal performance of a dual 1.2 kV, 400 a silicon-carbide MOSFET power module."," in *26th Annual IEEE SEMI-THERM*, 2010.

[26] [Online]. Available: http://www.irf.com/product-info/directfet/.

[27] R. A. Fillon, B. S. Whitmore, C. S. Korman and A. A. Esser, "Power Overlay Chip Scale Packages for Discrete Power Devices". Patent U. S. Patent 6,306,680, 23 October 2001.

[28] C. Gillot, C. Schaeffer, C. Massit and L. Meysenc, "Double-sided cooling for high power IGBT modules using flip chip technology," *IEEE Transactions on Components and Packaging Technologies*, vol. 24, no. 4, pp. 698-704, Dec 2001.

[29] L. M. Boteler, M. Hinojosa, V. A. Niemann, S. M. Miner and D. Gonzalez-Nino, "High voltage stacked diode package with integrated thermal management," in *IEEE ITherm Proceedings*, Orlando, FL, 2017.

[30] S. T. Jones, J. Mendoza, G. Frazier and G. Khalil, "Developing a Methodology for the Evaluation of Hybrid Vehicle Thermal Management Systems," in *NDIAs Ground Vehicle Systems Engineering and Technology Symposium (GVSETS)*, Troy, Michigan, July 17 2009.

[31] D. Saums, "Vaporizable Dielectric Fluid Cooling of IGBT Power Semiconductors for Vehicle Powertrains.," in *5th IEEE Vehicle Power and Propulsion Conference*, Dearborn MI, Sept. 2009.

[32] L. M. Boteler and D. J. Sharar, "Thermal and Packaging Challenges of Power Electronics Modules," in *GOMACTech*, Orlando, FL, 2016.

[33] M. O. Keefe and K. Bennion, "A comparison of hybrid electric vehicle power electronics cooling options," in *IEEE Vehicle Power and Propulsion Conference*, 2007.

[34] B. Agostini, "High heat flux flow boiling in silicon multi-microchannels–Part I: heat transfer characteristics of refrigerant R236fa," *International Journal of Heat and Mass Transfer,* vol. 51, no. 21, pp. 5400-5414, 2008.

[35] Lee, Jaeseon and Mudawar, "Assessment of the effectiveness of nanofluids for single-phase and two-phase heat transfer in micro-channels," *International Journal of Heat and Mass Transfer,* vol. 50, no. 3, pp. 452-463, 2007.

[36] S. G. Kandlikar and A. V. Bapat, "Evaluation of jet impingement, spray and microchannel chip cooling options for high heat flux removal," *Heat Transfer Engineering,* vol. 28, no. 11, pp. 911-923, 2007.

[37] I. Mudawar, "Two-phase spray cooling of hybrid vehicle electronics.," *IEEE Transactions on Components and Packaging Technologies,* vol. 32, no. 2, pp. 501-512, 2009.

[38] J. Kim, "Spray cooling heat transfer: the state of the art," *International J. of Heat and Fluid Flow,* vol. 28, no. 4, pp. 753-767, 2007.

[39] C. Mi, M. A. Masrur and D. Gao, Hybrid Electric Vehicles: Principles and applications with practical perspectives, West Sussex: John Wiley and Sons Ltd., 2011.

[40] K. Sienski and R. Eden, "3-D Electronic Interconnect Packaging," in *Proceedings of the Aerospace Applications Conference*, 1996.

[41] T. Ruemenapp and D. Peier, "Dielectric Breakdown in Aluminium Nitride," in *Proc. 11th International Symposium on High Voltage Engineering*, 1999.

[42] F. A. Khan, L. Zhou, V. Kumar, I. Adesida and R. Okojie, "High rate etching of AlN using BCl3/Cl2/Ar inductively coupled plasma," *Materials Science and Engineering*, vol. B, no. 95, pp. 51-54, 2002.

[43] J. P. Calame, "Experimental investigation of microchannel coolers for the high heat flux thermal management of GaN-on-SiC semiconductor devices," *Int. J. Heat Mass Transfer*, 2007.

[44] N. R. Jankowski, L. M. Evenhart, B. Morgan, B. Geil and F. P. McCluskey, "Comparing microchannel technologies to minimize the thermal stack and improve thermal performance in hybrid electric vehicles," in *Proc. Vehicle Power and Propulsion Conference*, Arlington, TX, Sept 9-12, 2007.

[45] N. R. Jankowski, L. M. Evenhart, B. Geil, C. W. Tipton, J. Chaney, T. Heil and W. Zimbeck, "Stereolithographically fabricated Aluminum Nitride microchannel substrates for integrated power electronics cooling," in *Proc. Intersociety Conference on Thermal and. Thermomechanical Phenomena in Electronic Systems (ITherm)*, Orlando, FL, 2008.

[46] D. Sharar, N. Jankowski and B. Morgan, "Thermal performance of a direct-bond-copper aluminum nitride manifold-microchannel cooler," in *IEEE SemiTHERM*, Santa Clara, CA, 2010.

[47] L. M. Boteler, Microfabrication and analysis of manifold microchannel coolers for power electronics - PhD Dissertation, University of Maryland, Mechanical Engineering Department, College Park, MD, 2011.

[48] L. Boteler, N. Jankowski, B. Geil and P. McCluskey, "A micromachined manifold microchannel cooler," in *Proc International Mechanical Engineering Congress and Exposition*, Lake Buena Vista, FL, November 13–19, 2009.

[49] L. M. Everhart, N. R. Jankowski, B. Geil, A. Bayba, D. Ibitayo and F. P. McCluskey, "Manifold Microchannel Cooler for Direct Backside Cooling of SiC Power Devices," in *Proceedings of the ASMEICNMM2007*, Puebla, Mexico, 2007.

[50] L. M. Boteler, N. R. Jankowski, F. P. McCluskey and B. Morgan, "Numerical investigation and sensitivity analysis of manifold microchannel coolers," *Internation Journal of Heat and Mass Transfer*, vol. 55, no. 25-26, pp. 7698-7708, 2012.

[51] S. Garimella and V. Singhal, "Single-phase flow and heat transport in microchannel heat sinks," *1st International Conference on Microchannels and Minichannels*, April 2003.

[52] D. Sharar, N. Jankowski, B. Morgan and A. Bar-Cohen, "Two-phase minichannel cold plate for Army vehicle power electronics," *ASME InterPACK*, vol. 2, pp. 133-142, 2011.

[53] D. Sharar, "Flow regime driven thermal enhancement in internally-grooved tubes - PhD Dissertation," University of Maryland, Mechanical Engineering Department, College Park, MD, 2016.

[54] A. Bergles, "Heat transfer enhancement - the encouragement and accommodation of high heat fluxes," *Journal of Heat Transfer*, vol. 119, pp. 8-19, 1997.

[55] A. Bergles, "Techniques to enhance heat transfer," in *Handbook of Heat Transfer*, New York, NY, McGraw-Hill, 1998.

[56] J. R. Thome, Enhanced Boiling Heat Transfer, CRC Press, 1990.

[57] R. Webb, Principles of Enhanced Heat Transfer, New York, NY: John Wiley and Sons, Inc., 1994.

[58] S. Kandlikar and P. Spiesman, "Augmentation of saturated flow boiling heat transfer," in *Chapter 18, Handbook of Phase Change*, Philadelphia, PA, Taylor and Francis, 1999.

[59] G. Moreno, S. Naramanchi, S. Thiagarajan, K. Bennion and T. Venson, "Enhanced surfaces to improve power electronics cooling," in *MEPTEC*, San Jose, 2011.

[60] S. Yilmaz, J. Hwalck and J. Westwater, "Pool boiling heat transfer performance for commercial enhanced tube surfaces," ASME Paper 90-HT-41, 1980.

[61] H. Kim, J. Kim and M. Kim, "Effect of nanoparticles on CHF enhancement in pool boiling of nano-fluids," *International Journal of Heat and Mass Transfer*, vol. 49, pp. 5070-5074, 2006.

[62] J. Thome, "Chapter 11: Boiling heat transfer inside enhanced tubes," in *Wolverine Engineering Data Book III*, 2004.

[63] M. Jensen and H. Bensler, "Saturated forced-convection boiling heat transfer with twisted-tape inserts," *Journal of Heat Transfer,* vol. 108, pp. 93-99, 1986.

[64] S. Laohalertdecha and S. Wongwises, "An experimental study into the evaporation heat transfer and flow characteristics of R-134a refrigerant flowing through corrugated tubes," *International Journal of Refrigeration,* vol. 34, pp. 280-291.

[65] D. Sharar, J. Nicholas, B. Morgan and A. Bar-Cohen, "Enhanced Two-Phase Cooling," in *IMAPS 2012 Chesapeake Chapter*, College Park, MD, 2012.

[66] "Inner-grooved tubes," Wieland, [Online]. Available: http://www.wieland-thermalsolutions.com/internet/en/branchenprodukte/kltetechnik/kaltwasserstze/innenberippterohre/Innenberippte_Rohre.jsp. [Accessed 18 January 2013].

[67] C. Dang, N. Haraguchi and E. Hihara, "Flow boiling heat transfer of carbon dioxide inside a small-sized microfin tube," *International Journal of Refrigeration,* vol. 33, pp. 655-663, 2010.

[68] M. Kim and J. Shin, "Evaporating heat transfer of R22 and R410A in horizontal smooth and microfin tubes," *International Journal of Refrigeration,* vol. 28, pp. 940-948, 2005.

[69] D. Sharar and A. Bar-Cohen, "Two-phase flow regimes and evaporative heat transfer in internally-grooved tubes," *Journal of Enhanced Heat Transfer,* vol. 21, no. 2-3, pp. 195-230, 2014.

[70] J. R. Thome, Enhanced Boiling Heat Transfer, CRC Press, 1990.

[71] J. Thome, "Engineering Data Book III - Chapter 11: Boiling heat transfer inside enhanced tubes," 2004. [Online]. Available: http://www.wlv.com/heat-transfer-databook/. [Accessed 3 February 2015].

[72] D. Sharar, A. Bergles, N. Jankowski and A. Bar-Cohen, "Non-intrusive optical validation of two-phase flow regimes in a small diameter tube," in *ASME FEDSM*, Chicago, 2014.

[73] D. Sharar, A. Bergles, N. Jankowski and A. Bar-Cohen, "Non-intrusive optical validation of two-phase flow regimes in a small diameter tube," *Journal of Heat Transfer Engineering,* vol. 37, no. 11, 2016.

[74] T. Shedd, "An automated optical liquid film thickness measurement method," *Master's thesis, University of Illinois at Urbana-Champaign, Urbana, IL,* 1998.

[75] D. J. Sharar and A. Bar-Cohen, "Flow regimes and evaporative heat transfer coefficients for flow boiling in smooth and internally-grooved tubes," in *ITHERM*, Las Vegas, NV, 2016.

[76] D. Sharar, N. Jankowski and A. Bar-Cohen, "Modified model for improved flow regime prediction in internally-grooved tubes," in *InterPACK*, Burlingame, CA, 2013.

[77] D. Sharar, S. Argenna and A. Bar-Cohen, "Flow boiling performance in smooth and internally-grooved tube u-bend cold plates," in *ITHERM*, Orlando, FL, 2017.

[78] A. Bar-Cohen, K. Matin, K. Bloschock, N. Jankowski and D. and Sharar, "Two-phase ground planes: Technology developement and parametric results," Final Report: DARPA Thermal Ground Plane (TGP) DOC#21479, 2014.

[79] A. Bar-Cohen, K. Matin, N. Jankowski and D. and Sharar, "Two-phase thermal ground planes: technology developement and parametric results," *ASME Journal of Electronic Packaging,* vol. 137, p. 9, 2015.

[80] S. Thompson, A. Hathaway, C. Smoot, C. Wilson, H. Ma, R. Young, L. Greenberg, B. Osick, S. Van Campen, B. Morgan, N. Jankowski and D. and Sharar, "Robust thermal performance of a flat-plate oscillating heat pipe during high-gravity loading," *Journal of Heat Transfer,* vol. 133, no. 10, p. 5, 2011.

[81] C. Oshman, Q. Li, L. Liew, R. Yang, Y. Lee, V. Bright, D. Sharar, N. Jankowski and B. and Morgan, "Thermal performance of a flat polymer heat pipe heat spreader under high acceleration," *Journal of Micromechanics and Microengineering,* vol. 22, p. 12, 2012.

[82] D. Altman, A. Gupta, T. Dubrowski, D. Sharar, N. Jankowski and M. and North, "Analysis and characterization of thermal expansion-matched wick-based multi-chip passive heat spreaders in static and dynamic environments," in *InterPACK*, Burlingam, CA, 2013.

[83] C. Hose, D. Ibitayo, L. Boteler, J. Weyant and B. and Richard, "Integrated vapor chamber heat spreader for power module applications," in *InterPACK*, San Francisco, CA, 2017.

[84] C. W. Ayers, "Thermal Buffer Heat Sink for Time-Averaged Operating Conditions," DoE Annual Progress

Report for the Advanced Power Electronics and Electric Machinery Program, 2007.

[85] J. Kwon, J. Kim, E. Fallas and A. Rousseau, "Impact of Drive Cycles on PHEV Component Requirements," in *2008 World Congress*, Detroit, Michigan, April 14-17, 2008.

[86] L. Kingsley, M. Weiner, A. Kim, R. Pastore and R. &. S. H. C. L. &. F. L. Youmans, "Ultrafast Switching at the U.S. Army Pulse Power Center,," *Ultra-Wideband, Short-Pulse Electromagnetics 2,* pp. 85-96, 1995.

[87] I. McNab, "Developments in pulsed power technology," *IEEE Transactions on Magnetics,* vol. 37, pp. 375-378, 2001.

[88] N. Jankowski and F. P. McCluskey, "Modeling transient thermal response of pulsed power electronic packages," in *IEEE Pulsed Power Conference*, Washington, DC, 2009.

[89] G. A. Lane, Solar heat storage: latent heat materials. Background and scientific principles, vol. I., Boca Raton, FL: CRC Press, 1983.

[90] D. Hale, M. Hoover and M. O'Neill, "Phase change materials handbook. Technical Report no.: NASA-CR-61363," National Aeronautics and Space Administration, Huntsville, AL, Sept 1971.

[91] N. Jankowski and F. P. McCluskey, "Electrical supercooling mitigation in erythritol," in *Proc 14th International Heat Transfer Conference*, Washington, DC, August 7–13, 2010.

[92] A. Feikert, "The Army's Future Combat System (FCS): Background and Issues for Congress (RL32888)," U.S. Congressional Research Service, May 12, 2008.

[93] N. J. Jankowski, B. C. Morgan and F. P. McCluskey, "Design and fabrication of a substrate integrated phase change thermal buffer heat sink," in *Proc International Mechanical Engineering Congress and Exposition,*, Lake Buena Vista, FL, November 13–19, 2009.

[94] N. R. Jankowski and F. P. McCluskey, "Numerical study on the thermal performance of a substrate integrated thermal buffer heat sink," in *Proc International Mechanical Engineering Congress and Exposition*, Denver, CO, November 2011.

[95] D. Gonzalez-Nino, L. M. Boteler, D. Ibitayo, N. R. Jankowski and P. O. & Quintero, "Numerical Evaluation of Multiple Phase Change Materials for Pulsed Electronics Applications," in *ASME 2016 Heat Transfer Summer Conference*, July 2016.

[96] D. Gonzalez-Nino, L. M. Boteler, D. Ibitayo, N. R. Jankowski, D. Urciuoli, I. M. Kierzewski and P. O. Quintero, "Experimental Evaluation of metallic phase change materials for thermal transient mitigation," *International Journal of Heat and Mass Transfer,* vol. 116, pp. 512-519, 2018.

[97] D. Gonzalez-Nino, L. M. Boteler, N. R. Jankowski, D. Ibitayo and P. O. Quintero, "Voiding effects on the thermal response of metallic phase change materials under pulsed power loading," in *ASME Internation Technical Conference and Exhibition on Packaging and Integration of Electronic and Photonic Microsystems*, San Francisco, CA, 2017.

[98] L. M. Boteler, V. A. Niemann, D. P. Urciuoli and S. M. Miner, "Stacked power module with integrated thermal management," in *IEEE International Workshop On Integrated Power Packaging (IWIPP),*, Delft,, 2017.

[99] L. Boteler and A. Smith, "3D thermal resistance network method for the design of highly integrated packages," in *Proc. ASME Summer Heat Transf. Conf.*, 2013.

[100] L. M. Boteler and S. M. Miner, "Power Packaging and Stress Model for Quick Parametric Analyses," in *InterPACK2017 proceedings*, San Francisco, CA, 2017.

[101] C. H. Hsueh, "Thermal Stresses in Elastic Multilayer Systems," *Thin Solid Films,* vol. 418, 2002.

Chapter 24

Multi-Channel, Constant-Current Power Source for Aircraft Applications

C. Wesley Tipton IV[1*], Damian Urciuoli, Donald H. Porschet, and Erik Schroen
U. S. Army Research Laboratory, 2800 Powder Mill Rd, Adelphi, MD 20783 USA
charles.w.tipton6.civ@mail.mil

Abstract

To address the anticipated need of constant-current power sources in aerospace applications, we have developed a 3-kW, multi-channel source powered from a 3-phase, 400-Hz, 115-VAC supply. Special considerations impacting this design were minimizing volume, maintaining system power quality, and providing electrical fault protection. Performance results and a discussion of the design approach are presented.

Keywords

Power-Factor Correction, Constant-Current Source; 400-Hz Power, Aircraft Power, Power Distribution

Contents

1. Introduction

The development of a 16-channel, 3-kW, constant-current power distribution unit (PDU) for aircraft applications is discussed in this section. Electrical loads, such as lighting, de-icing heaters, and actuators may be operated from this compact power conversion unit.

Because of the nature of aircraft systems, two of the most important design considerations are the maintenance of electrical power quality and minimization of weight. To this point, military and commercial standards for power quality are well established in MIL-STD-461F, MIL-STD-704F, Def-Stan 59-41, RTCA DO-160F, EUROCAE/ED-14D. At an operating power of 3 kW, an equipment designer must give special consideration to maintaining power quality according to the selected standard(s) and a power-factor correction (PFC) circuit will be needed to preserve the power quality of the AC source feeding the converter. To our dismay, the majority of PFC product development has been focused on 60-Hz systems as demonstrated by the lack of 400-Hz products. Astrodyne-TDI offers a single-phase, 375-W PFC [1] that operates over the input frequency range of 47 to 880 Hz. VICOR has a 1.4-kW, single-phase PFC [2] operating over the range of 47 to 440 Hz. However, this unit is not recommended for new designs and neither product was an optimal solution for the present application. Fortunately, SynQor produces a 1.5-kW, 3-phase, PFC operating over the range of 45 to 800 Hz [3] which became commercially available in July 2016.

2. PDU topology

Primary power is supplied by the platform's 3-phase, 115-VAC, 400-Hz bus while all other housekeeping power is derived from the 28-VDC bus. The primary power is conditioned by SynQor electro-magnetic interference filter (EMI) and PFC modules. Two EMI modules (MACF-115-3PH-UNV-HT) and two PFC modules (MPFC-115-3PH-270-FP) are used in combination to produce a nominal 275-VDC source with power quality conforming to MIL-STD-704F and MIL-STD-461F specifications. The 275-VDC power is distributed among the 16 constant-current output converters. The outputs of the 2 pairs of filter-PFC modules are not directly interconnected, each PFC module powers 8 channels. Each output channel is comprised of a DC-DC buck controller (Microchip, HV9661), a DC ground-fault (GF) detector, an output voltage monitor, and an isolation relay. Presently, a maximum output current of 1 A may be supplied by each channel. The operation of the PDU is controlled by a 16-bit microcontroller (Microchip, PIC24F32KA304) and a variety of other circuits that support signal processing, platform-PDU communications, and built-in test (BIT) functions.

In one typical operating scenario, the BIT would be performed as part of the power-up sequence. This test verifies the operational status of the analog measurement circuits and other functional blocks. Upon successfully completing the BIT, the user would program operational parameters. As previously stated, the PDU channels are arranged as 2 groups of 8 channels where the output current level may be set on a per-group basis. Otherwise, each channel has independent enable/disable control, ground-fault detection, output voltage monitoring, and galvanic load isolation. Also, that in the case of catastrophic failure, the 275-V supply is fuse-protected at each channel. A few of the salient design features of the PDU are presented below.

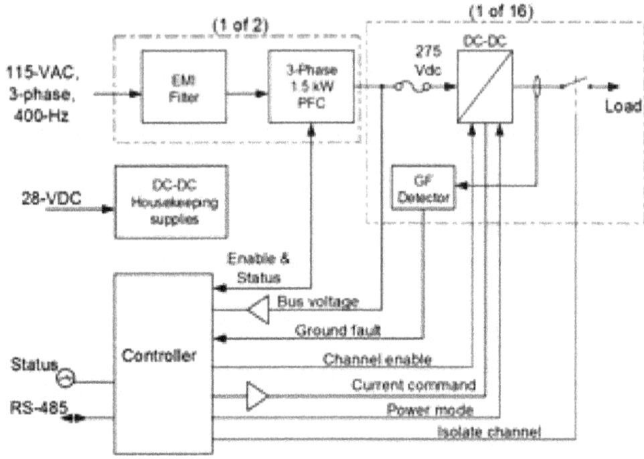

Fig. 1 Power Distribution Unit simplified functional diagram.

Constant-current Converter: The HV9961 average-mode, buck controller is very attractive for this application because it supports high-frequency switching, includes both analog and digital adjustments of the output current, and output over-current protection. A simplified schematic of the buck converter used for each constant-current channel is shown in Figure 2. As documented in its data sheet [4], the current used to energize the buck inductor is monitored as a voltage at the Current Sense (CS) input through a current-sensing resistor and is the control feedback signal. The control scheme holds the Gate output off-time constant and the switching frequency is adjusted to regulate output current. The output current set-point may be adjusted by applying a voltage on the linear dimmer (L_D) input in the range of 0 to 1.5 V. A fixed off-time of 2 μs was chosen for this design and this dictates a buck duty-factor range of 0.33 to 0.75 which translates to an output current range of 0.17 to 1.0 A. To extend the range to lower currents, the current sensing resistor was split into two components, R_{Hi} and R_{Lo}, and a shunt-transistor, under microprocessor control, modulates the resistance value. Lower currents may be obtained by turning off the shunt transistor.

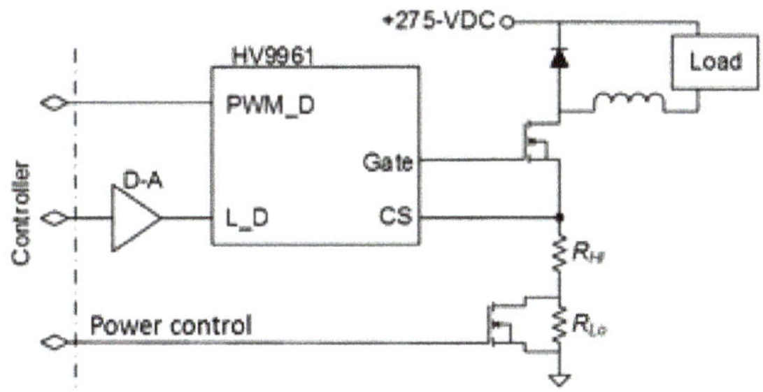

Fig. 2 Simplified schematic of the HV9961 buck converter.

Figure 3 shows the ranges of output currents available in the two modes. In low-power mode, L_D may be varied to give a range of currents from 39 mA (4%) to 140 mA (14%) and in high-power mode, 170 mA (17%) to 1 A (100%).

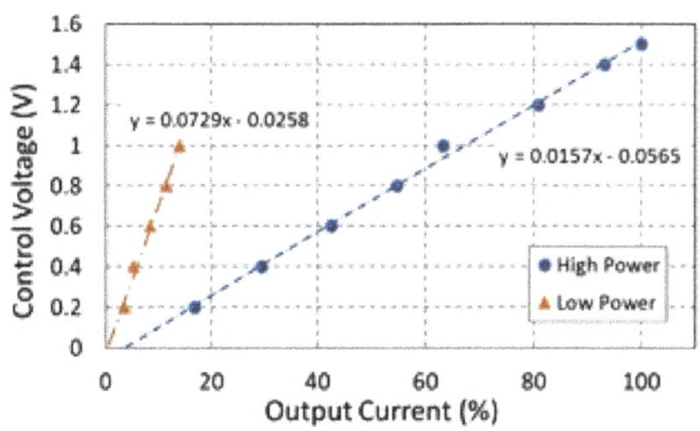

Fig. 3 Output current as a function of the dimmer control voltage for the high- and low-power modes.

One control issue arises from the HV9961's over-current protection scheme. We found that high current set-points could only be achieved by modulating the logical dimming input (PWM_D) during the start-up period. By disabling the gate driver, via PWM_D, the over-current detection circuit was effectively "reset" and the built-in over-current off-time of 400 μs could be significantly reduced thereby allowing the DC output current in the buck inductor to increase. The start-up waveforms for maximum output current are shown in Figure 4. During the time between 10 μs and 81 μs, PWM_D is switched by the microcontroller. Based on this signal, the HV9961 produces a gate drive pulse train at a frequency of 111 kHz and having a duty factor of 0.58. This waveform was empirically optimized for low-current start-up, but is used for all current levels in this prototype system. In Figure 4, the output current saturates at approximately

220 mA at 20 μs. At this level, the converter could have been placed under closed-loop control without a current fault. In the present example, when open-loop control is relinquished at 81 μs, PWM_D is held active and the buck controller attains current regulation, at approximately 106 μs, through closed-loop control without experiencing an over-current condition. At an output current of 1 A, the steady-state gate switching frequency is approximately 130 kHz. Note also in Figure 4 that the initial output voltage (when there is no switching) is approximately 137 V. This is due to the relatively high load resistance and leakage current flowing through the power transistor.

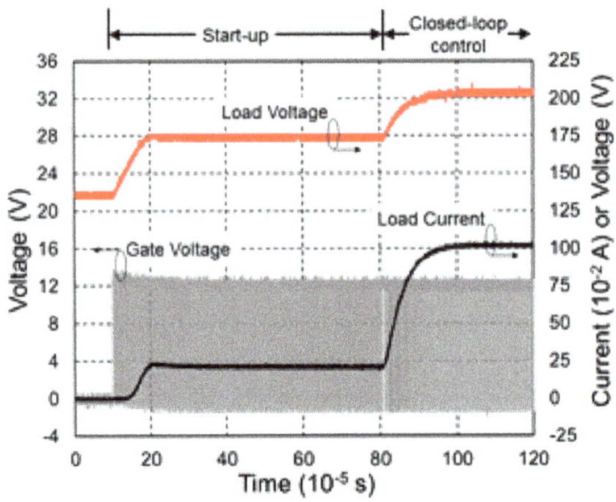

Fig. 4 Load voltage and current waveforms during start-up of the HV9961 buck controller.

Ground-fault Sensor: Because the PFC-rectifier module does not provide galvanic isolation, there exists a possibility that a ground- fault event could damage the PDU. Figure 5 shows a simplified schematic of a drive channel and potential fault scenario. In normal operation, the channel is isolated from the platform chassis and the load current passes to and from the PFC through a bifilar-wound, gapped toroid core (EPCOS, B64290L38X38). With equal (and opposite) winding currents, the instantaneous flux in the toroid core is zero. However, if one of the load's terminals experiences a short-circuit to chassis, for example, or an arcing event occurs, the net current through the windings will not be zero and a flux will be generated. A Hall-effect switch (Allegro Microsystems, A1120), with an operating point of 3.5 mT, is inserted in the toroid gap and, in this design, will respond at a differential current of 200 mA. If any of the GF sensors reports a fault, logic within the control system will identify and isolate the appropriate channel(s) then notify the microcontroller through a hardware interrupt.

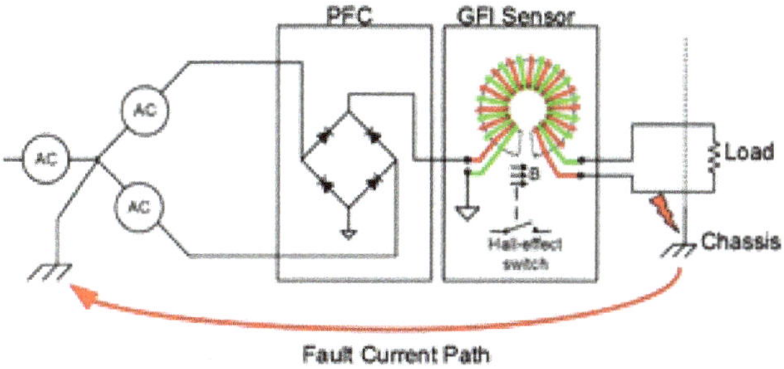

Fig. 5 Load voltage and current waveforms during start-up of the HV9961 buck controller.

Mechanical Configuration: Figure 6 shows the solid model of the constant-current, PDU which has a volume of 540 in3 and a weight of approximately 16 lbs. The PDU is constructed as a motherboard with function-specific daughter-cards. The control processor, complex programmable logic devices, analog interface circuits, and housekeeping power converters are located on the backplane circuit card (motherboard) which is positioned on the bottom of the chassis. Each of the 4 drive cards contains 4 constant-current output channels and has an aluminum plate that transfers waste heat to the chassis through wedge-lock card retainers. The user-interface card is also connected to the motherboard and contains the front-panel connectors, isolated RS-485 communications interface, and status indicators. The PFC is mounted to the chassis' back panel to enhance heat removal.

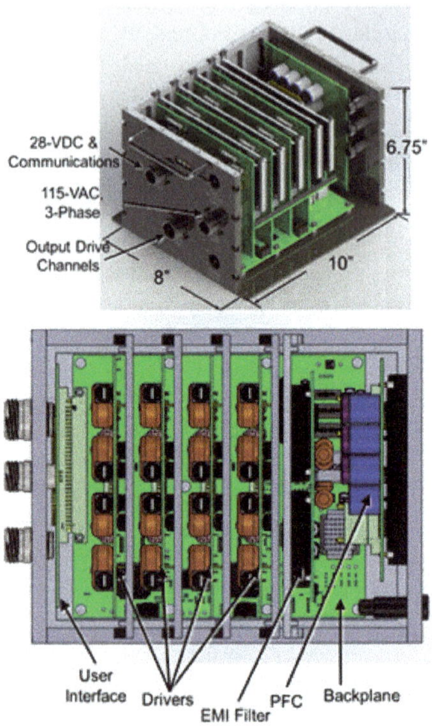

Fig. 6 Rendered solid model of the Power Distribution Unit and identification of major system components.

Summary

This section presented an overview of the design requirements and engineering approach taken to realize a 3-kW, constant-current source compatible with 115-VAC, 400-Hz input power systems. Special considerations addressed in this design are maintaining system power quality and providing electrical fault protection. By manipulating the control parameters of a buck-mode controller, we demonstrate constant-current operation at 1 A and 200 V per channel.

References

[1] "375W AC-DC 1/2 Brick Power Factor Correction Module, 375V," Astrodyne-TDI, http://www.astrodynetdi.com/ecatalog/power-supplies/PFC-375LF accessed on 29 December, 2016.

[2] "MIL-COTS PFC Front End GP-MPFC1H21," VICOR, http://www.vicorpower.com/documents/datasheets/Custom/Granite/ds-MILCOTS-pfc-FrontEnd.pdf accessed on 29 December, 2016.

[3] "Military Grade 3-Phase Power Factor Correction Module," SynQor Inc., http://www.synqor.com/company-press.html?releasename=pr_MPFC-115-3PH-270-FP.html&releasetype=us&releaseyear=2016, accessed on 29 December, 2016.

[4] "LED Driver with Average-Mode Constant Current Control," Supertex Inc., Document number DSFP-HV9961, 2010. ww1.microchip.com/downloads/en/DeviceDoc/hv9961.pdf accessed on 29 December, 2016

Chapter 25

Simulation Study of Switching-Dependent Device Parameters of High Voltage 4H-SiC GTOs

Aderinto Ogunniyi[1, a,*], James Schrock[2], Miguel Hinojosa[1], Heather O'Brien[1], Aivars Lelis[1], Stephen Bayne[2], and Sei-Hyung Ryu[3]

[1]Army Research Laboratory, 2800 Powder Mill Road, Adelphi, MD. USA.

[2]Texas Tech University, 224 Electrical and Computer Engineering, Lubbock, TX. USA.

[3]Wolfspeed, (a Cree company), 3026 E. Cornwallis Rd, Research Triangle Park, NC. USA.

[a]aderinto.a.ogunniyi.civ@mail.mil

Abstract

The silicon carbide (SiC) "Super" gate turn-off thyristor (SGTO) is a viable device for high voltage and fast dI/dt switching applications. These devices are well suited for various pulsed power applications requiring high peak currents in the kilo-amp regime. The turn-on transition speed is determined by the spreading velocity, which depends on applied gate current, applied anode current density, minority carrier lifetime, and both the gate base-width and the drift region of the thyristor. The impact of device parameters on switching performance is discussed in this section.

Keywords

Turn-on Delay, Silicon Carbide, Gate Turn-off Thyristor, Ambipolar Lifetime, Gate Current

Contents

The work has been previously published by Trans Tech Publications Ltd:
A. Ogunniyi, J. Schrock, M. Hinojosa, H. O'Brien, A. Lelis, S. Bayne, and S. H. Ryu, " Simulation Study of Switching-Dependent Device Parameters of High Voltage 4H-SiC GTOs," Material Science Forum, ISSN: 1662-9752, Vol.897, pp. 575-578. doi:10.4028/www.scientific.net/MSF.897.575

1. Introduction

Silicon carbide (SiC) material properties are extremely desirable for power devices because the wide bandgap and high critical field of SiC enables power devices to have a much smaller drift layer for a given blocking voltage than its silicon counterpart [1-3]. Utilizing SiC power devices can simplify power system designs, reduce cooling requirements, and improve efficiency of power electronics and pulsed-power systems. The U.S. Army Research Laboratory (ARL) has collaborated with Wolfspeed in the development of high-voltage SiC "Super" gate turn-off (SGTO) thyristors. This collaboration has led to an increase in the die area and high-voltage blocking capability of SiC SGTOs. Silicon carbide material processing and device fabrication development is continuing to improve through the reduction of material defects in epi-layers and wafers which will lead to much more reliable and stable device performance. A simulation study of the switching-dependent device parameters of SiC SGTOs is presented in this section.

2. SiC SGTO device design

The SGTO is the power device of choice for the high voltages and high current densities required by pulsed Army systems. The two sets of devices utilized in the simulation study are both 1.0 cm^2 SGTOs with blocking hold-off capability of 9-kV and 15-kV. These GTOs are asymmetric devices utilizing a pnpn structure and are grown on an n+ 4H-SiC substrate as shown in Figure. 1. This is essential due to the high resistance of p-type SiC substrates. The low doped p-type blocking layer of the 9-kV SGTO and 15-kV SGTO is 90μm and 120μm, respectively. The high blocking voltage of the SGTO is supported by this thick epi-layer and multi-zone edge junction termination extension (JTE). Details of device development at Wolfspeed and pulse evaluations at ARL were previously published in references [1-2, 4]. The SiC SGTO presented in this research is a current-controlled device that transitions from the OFF-state to the ON-state when a small current pulse-trigger is applied to the anode relative to the gate region.

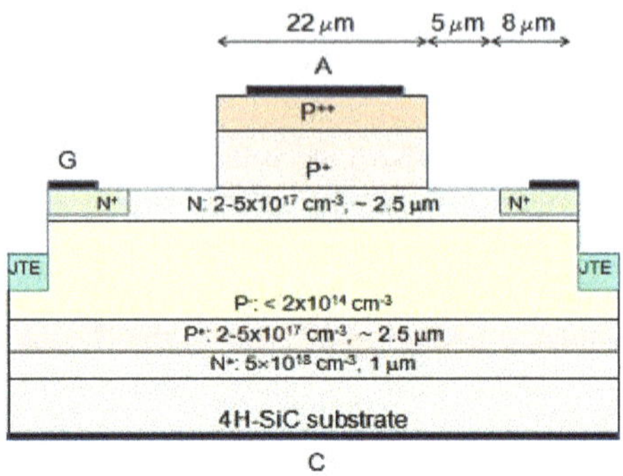

Fig. 1 Simple cross-section view of a 1.0 cm² SiC Super-GTO. The 9-kV SGTO has a 90-μm drift layer thickness and the 15-kV SGTO has a 120-μm drift layer thickness

3. Mixed-mode simulation analysis

High voltage GTOs were simulated in a mixed-mode series resistor-capacitor (RC) circuit as shown in Figure 2. The RC produces a relatively fast-rising high current pulse that is suitable for comparing turn-on delay and switching capabilities of the GTO. It is important to note that the rate-of-rise for the anode current pulse of the device is limited by the discharge rate of the RC circuit. Furthermore, the turn-on delay of the GTO is primarily dependent on the time required to complete the internal feedback loop between the two coupled transistors as shown in Figure. 3 [2, 5]. The simulation study investigated various device parameter effects on turn-on switching performance of the GTO.

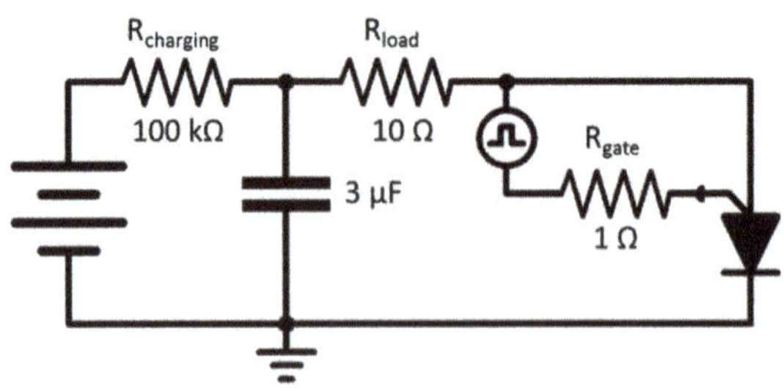

Fig. 2 Mixed-mode series resistor-capcitor (RC) circuit topology for GTO turn-on characterization.

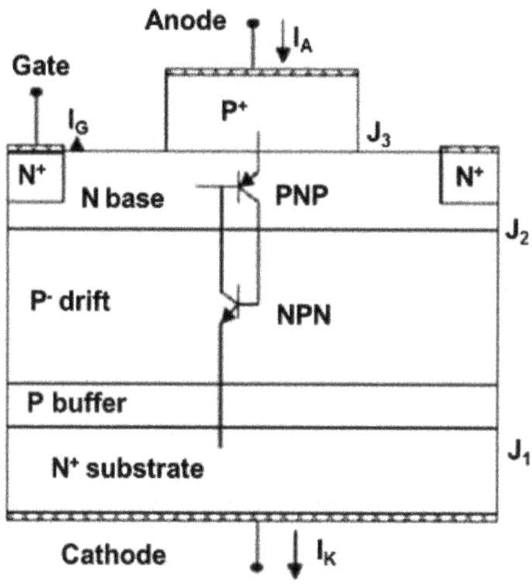

Fig. 3 Generic cross-section with superimposed schematic of two coupled transistors.

The modeling and simulations were performed using the Silvaco Atlas 2-D technology computer aided design (TCAD) software package. The high-voltage SiC SGTO cell structures implemented in Atlas are physics-based and developed using semiconductor equations with physics models that represent the SiC material properties. These parameters include: carrier mobility, generation rate, and recombination rate specific to 4H-SiC properties. A detailed description of the physics models used in this research can be in found in [6]. Figures 4 and 5 compare the turn-on delay dependence of the gate currents on both the 9-kV GTO and 15-kV GTO, respectively. The ambipolar carrier lifetime for both of the high voltage GTOs was 2 μs and the capacitor charge voltage of 1-kV was used to assess the turn-on characteristics of both device. The gate trigger for the simulation of the GTOs were delayed by 1-μs as shown in Figures. 4 and 5. Based on Figures 4 and 5, there is a significant reduction between the time when the gate signal is applied and when the anode current begins to rise, with increasing gate current ranging from 0.5 A to 4 A for both GTO devices. Higher gate current equates to a faster rate-of-current rise applied to the gate terminal, which spreads the charge beyond the gate electrode and creates a more uniform distribution of anode current across the active area of the device. The variation in turn-on delay is much closer in 9-kV GTO due to the 25% reduction in drift region thickness compared to the 15-kV GTO.

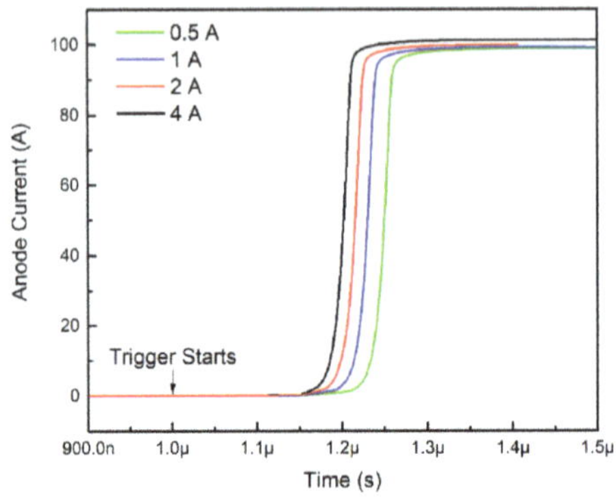

Fig. 4 Simulated turn-on delay dependence vs. applied gate current for a 9-kV SiC SGTO with an ambipolar lifetime of 2 µs and a capacitor charge voltage of 1 kV.

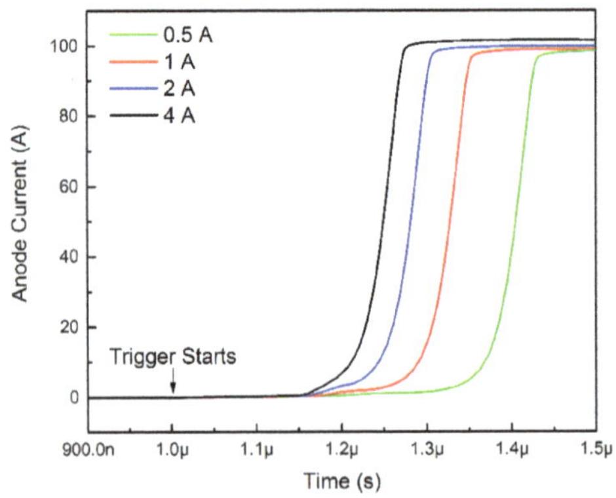

Fig. 5 Simulated turn-on delay dependence vs. applied gate current for a 15-kV SiC SGTO with an ambipolar lifetime of 2 µs and a capacitor charge voltage of 1 kV.

Figures 6 and 7 compare the switching characteristics of 9-kV and 15-kV devices with an applied gate current of 1 A and 2 A, respectively. Since the turn-on delay of the GTO is dependent on the base transit time of the two coupled transistors internal to the device, the 15-kV would naturally have a much longer delay than the 9-kV due to its much larger drift region.

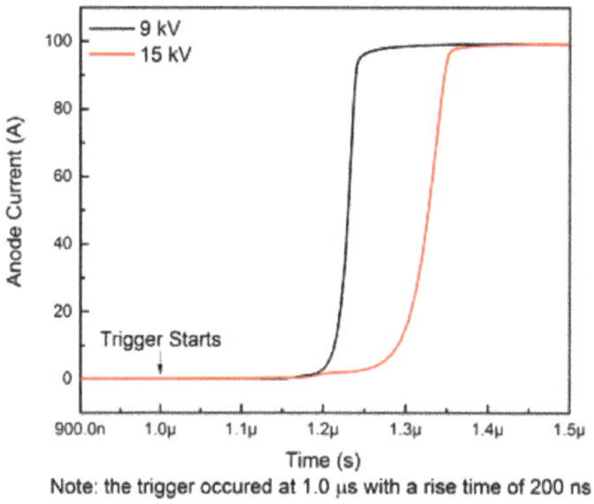

Fig. 6 Simulated results of a 9-kV and a 15-kV SiC SGTO turn-on delay with applied 1 A gate current and 2 μs ambipolar lifetime.

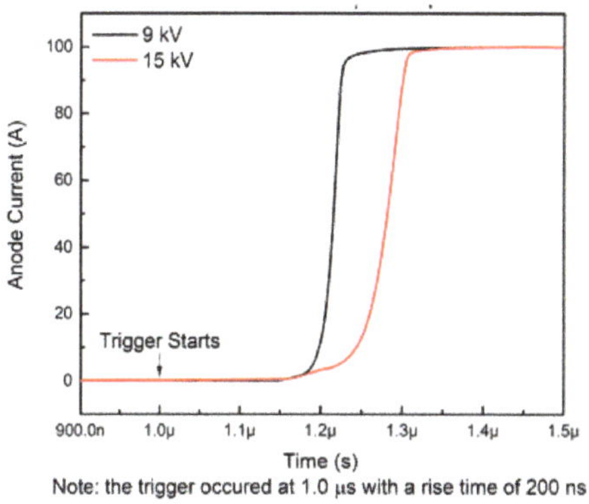

Fig. 7 Simulated results of 9-kV and 15-kV SiC SGTO turn-on delay with applied 2 A gate current and 2 μs ambipolar lifetime.

The increase in anode-gate current causes the turn-on delay difference between the 9-kV and 15-kV GTO to be less than 50 ns (Fig. 7) nearly half the time difference when compared to the 1 A gate current results depicted in Figure 6. Figures. 8 and 9 illustrate the turn-on delay dependency on the gate current and ambipolar lifetime. It is important to note that the ambipolar lifetime has no significant effect on the turn-on delay when a gate current greater than 4 A is applied on both the 9-kV and 15-kV GTO due to the enhanced carrier injection efficiency on the topside of the GTO. Figures. 10 and 11 depict the turn-on delay dependency on the anode voltage and anode current of 9-kV and 15-kV GTOs, respectively.

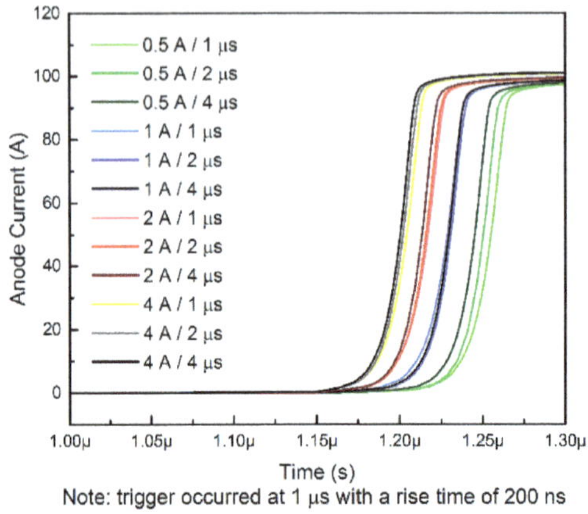

Fig. 8 Simulated results of 9-kV and 15-kV SiC SGTO turn-on delay with applied 2 A gate current and 2 μs ambipolar lifetime.

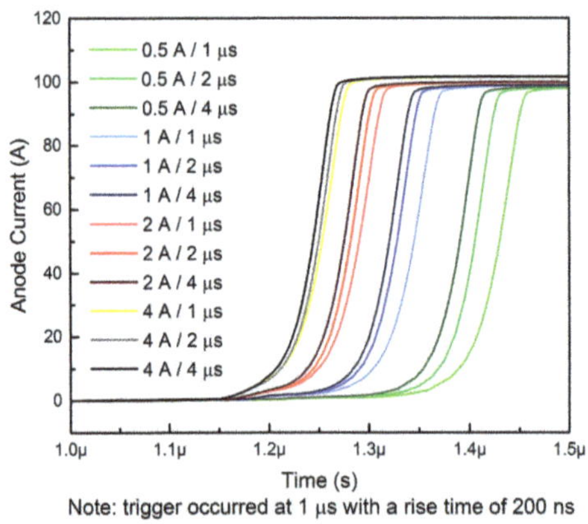

Fig. 9 Simulated results of 9-kV and 15-kV SiC SGTO turn-on delay with applied 2 A gate current and 2 μs ambipolar lifetime.

The results shown in Figures. 10 and 11 imply that an increase in anode current decreases the internal feedback loop in the GTO, which reduces the turn-on delay time [1].

Furthermore, the lateral plasma spreading velocity is increased by the anode current density as shown analytically in [5].

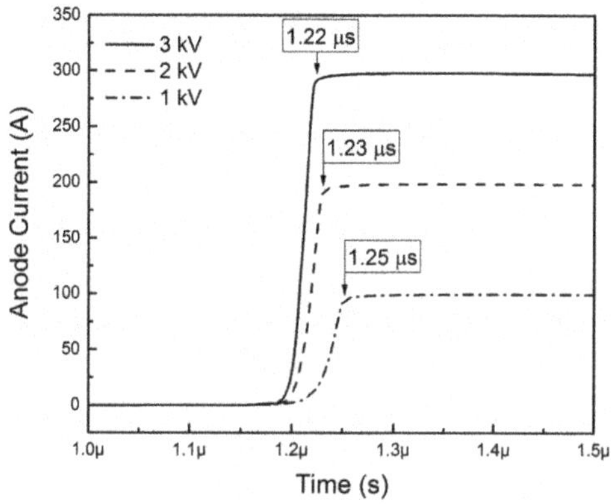

Fig. 10 Simulated results of 9-kV SGTO turn-on delay dependence on anode voltage and anode current with applied gate current of 1A and ambipolar lifetime of 2μs.

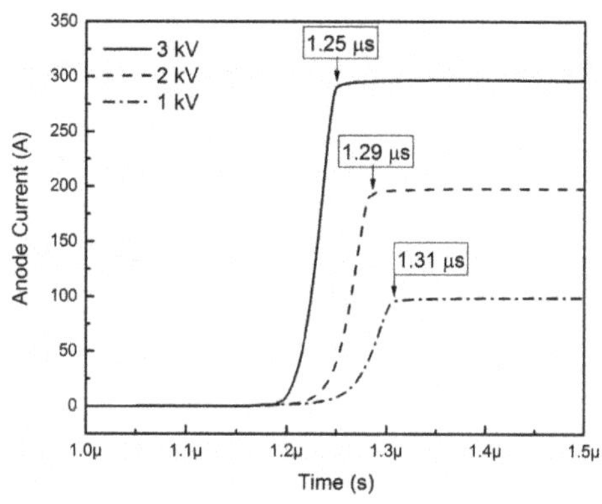

Fig. 11 Simulated results of 15-kV SiC SGTO turn-on delay dependence on anode voltage and anode current with applied gate current of 1A and ambipolar lifetime of 4μs.

Summary

This section focused on analysis of the turn-on transition of high-voltage SiC GTOs through simulation. Factors contributing to the switching transition of the GTO include device geometry, diffusion rates, current density, and the base width of the coupled transistors. The magnitude of the gate current and the gate-current rate-of-rise time (dI_G/dt) also affect the rate at which the thyristor transitions into the ON-state.

References

[1] A. Agarwal, *et al*, "9 kV, 1 cm x 1 cm SiC super GTO technology development for pulse power, " Proc. 17th IEEE PPC, pp. 264-269, 2009. https://doi.org/10.1109/PPC.2009.5386305

[2] A. Ogunniyi, *et al,* "The benefits and current progress of SiC SGTOs for pulsed power applications," Solid-State Electronics, vol. 54, issue 10, pp. 1232-1237, 2010. https://doi.org/10.1016/j.sse.2010.05.018

[3] A. Elasser et al, "Silicon carbide benefits and advantages for power electronics circuits and systems," IEEE, Vol. 90, pp 969, 2002.

[4] L. Cheng, et al "15 kV, Large Area (1 cm2) 4H-SiC p-type Gate Turn-Off Thyristors" Mat. Sci. Forum, Vols 740-742 (2013), 978-981 (2012)

[5] B. Jayant Baliga, *Power Semiconductor Devices.* Boston: PWS Publishing Co., 1996.

[6] Atlas User's Manual Device Simulation Software, Silvaco, Inc., 2015.

Chapter 26

A Compact 100-A, 850-V, Silicon Carbide Solid-State DC Circuit Breaker

D. P. Urciuoli*, D. Ibitayo, G. Koebke, G. Ovrebo and R. Green
U.S. Army Research Laboratory, Adelphi, MD, USA
damian.p.urciuoli.civ@mail.mil*

Abstract

A 100-A, 850-V, solid-state circuit breaker (SSCB) having silicon carbide transistors and diodes was developed. The SSCB conducts 100 A continuously with air cooling in a 32 square-centimeter footprint. It is normally off and unidirectionally blocking, and has a configurable trip response allowing a range of overcurrent transients to be conducted, while maintaining a fast trip response at its fault-current limit. The SSCB also provides remote trip, status output, and reset functions. The time-versus-current trip response of the SSCB was characterized at 25, 50, and 75 degrees Celsius baseplate temperatures using pulsed currents with ramp rates of over 70 amps per microsecond. Saw-tooth-shaped current pulse widths of less than 5 microseconds, with amplitudes of greater than 200 A, were demonstrated.

Keywords

Silicon Carbide, Circuit Breaker, Solid-State, Avalanche Breakdown

Contents

The work has been previously published by IEEE:
D.P. Urciuoli, D. Ibitayo, G. Koebke, G. Ovrebo, and R. Green, "A compact 100-A, 850-V, silicon carbide solid-state DC circuit breaker," Energy Conversion Congress and Exposition (ECCE) 2016, Doi: 10.1109/ECCE.2016.7854656.

1. Introduction

A compact 100-A, 850-V, solid-state circuit breaker (SSCB), having silicon carbide (SiC) transistors and diodes, was developed. The SSCB provides advantages over electromechanical switches in transition speed, reduced degradation, and longer operating life [1] [2]. It can also be more tolerant of shock and vibration. These benefits enable improved performance in a wide range of applications, especially those on mobile platforms. The SSCB is normally- off and unidirectionally blocking, specifically for DC applications. It has low conduction losses that enable it to be used in applications with air cooling. Additionally, the SSCB analog controller was designed with a configurable trip response to allow a range of overcurrent (between the SSCB continuous rating and the fault interrupt rating) transients to be conducted, while maintaining a fast response at its fault-current limit. The SSCB also has remote trip, status output, and reset capabilities. Switched test circuits were built to provide pulsed SSCB currents of over 200-A peak, with slew rates of over 70 A/μs. The trip response of the SSCB was characterized from current pulses ranging from less than 5-μs wide to the SSCB continuous current rating.

2. Device topology

The SSCB was designed for third-generation, 1200-V blocking, 4.04-mm by 6.44-mm, SiC MOSFET engineering samples fabricated by Cree, Inc. Fig. 1 shows typical output characteristics for a packaged MOSFET at 28 °C and 150 °C,

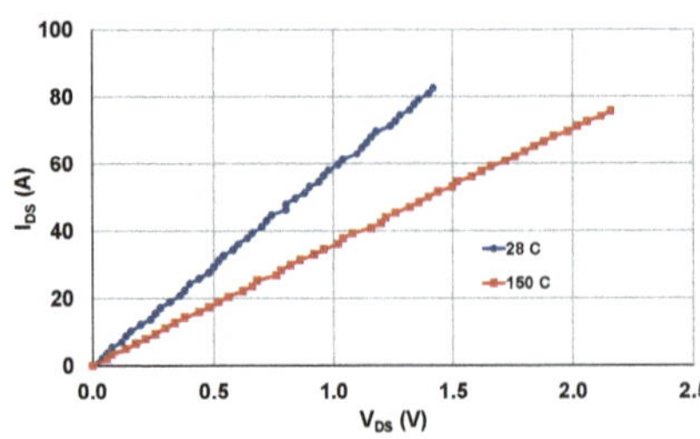

Fig. 1 Typical 4.04 × 6.44-mm third-generation SiC MOSFET forward conduction for VGS = 20 V.

at a 20-V gate bias. At 28 °C, and 59 A, the MOSFET on-state resistance is 17 mΩ. SiC avalanche breakdown diodes (ABDs) were integrated into the module, in parallel with the MOSFETs, for transient voltage suppression during MOSFET turn-off transitions [3]. The ABDs, having a nominal avalanche voltage of 950 V, provide sufficient margin for SSCB blocking-voltages up to 850 V, while also clamping well below the 1200-V rated MOSFET blocking-voltage. SiC 1200-V Schottky diodes (SDs) were connected anti-parallel to the MOSFETs, and additional SDs were connected in series with the MOSFET switch, with their anode at a separate terminal to provide an optional free-wheeling load-side current path. Fig. 2 shows a simplified schematic of the SSCB module in an application circuit with a load (R), and source-side (LS) and load-side (LL) line inductances.

The SSCB topology is very similar to that presented in [4] and requires an isolated power supply for MOSFET gate control. However, it can provide both high-speed fault-current protection and protection from overcurrent. Although self-powered SSCBs, as presented in [5], can interrupt short circuit currents, they can require 4 to 5 volts across the SSCB for operation. This results in either no ability to interrupt overcurrent, or high SSCB conduction losses.

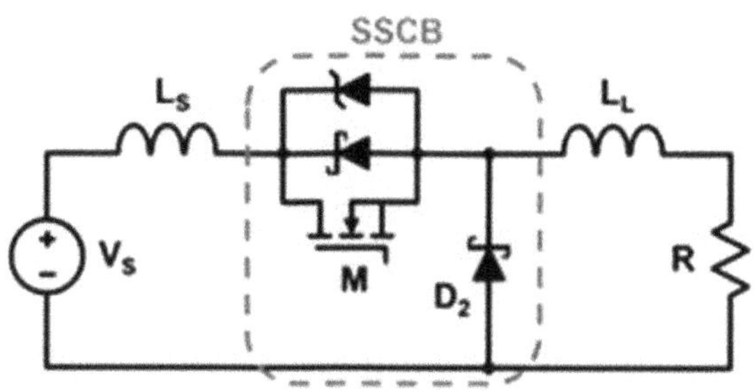

Fig. 2 Simplified schematic of the SSCB module in an application circuit.

3. Power module design

The SSCB footprint area was chosen to approximate that of the GIGAVAC GX14 mechanical contactor. From the manufacturer datasheet, at 200 A, the GX14 is rated for about 25,000 switching cycles at 600 V, and about 800 switching cycles at 800 V. Its maximum release time is 12 ms and its typical operate time is 13 ms, including its maximum bounce time of 7 ms. The footprint of its body (excluding terminals) is 33.4 cm^2. A 32.0 cm^2 (6.4 cm x 5.0 cm) baseplate was designed for the SSCB module to have three MOSFET die, four 0.09-cm^2 SD die, and two 0.12-cm^2 ABD die. A 75% molybdenum, 25% copper, metal composite was selected for the baseplate for stress reduction. Fig. 3. shows the power-module layout. From left to right, the top row of die consists of an SD, MOSFET, ABD, MOSFET, ABD, MOSFET, and SD die. The second row consists of 2 SD die.

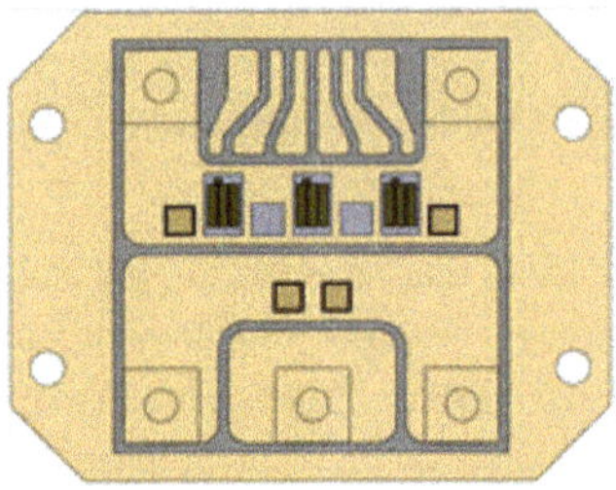

Fig. 3 SiC SSCB module layout with 6.4 × 5.0-cm baseplate.

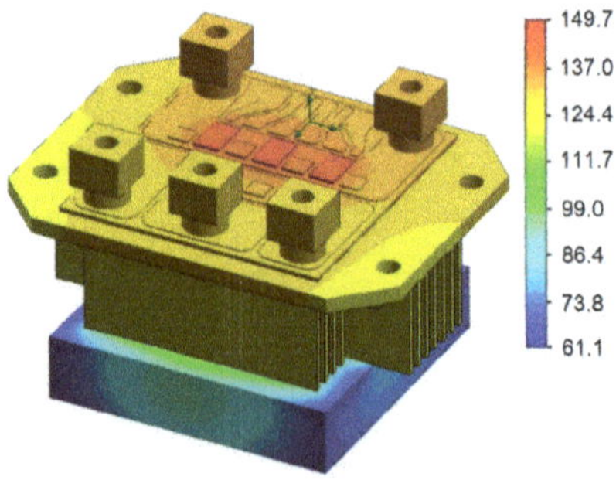

Fig. 4 Simulated thermal image of the SSCB power module, heatsink, and blower, conducting a steady state current of 87 A with 50 °C air at 2.8 CFM with no topside convection.

The module was solder bonded in a vacuum furnace, wire bonded, and packaged. Fig. 5 shows the well-matched on-state curves for each of the 3 MOSFETs at room temperature at a 20-V gate bias. The off-state leakage current of the parallel combination of the 3

MOSFETs, 2 SDs, and 2 ABDs was low at 14 μA at 950 V, and ABD breakdown occurred at 956 V.

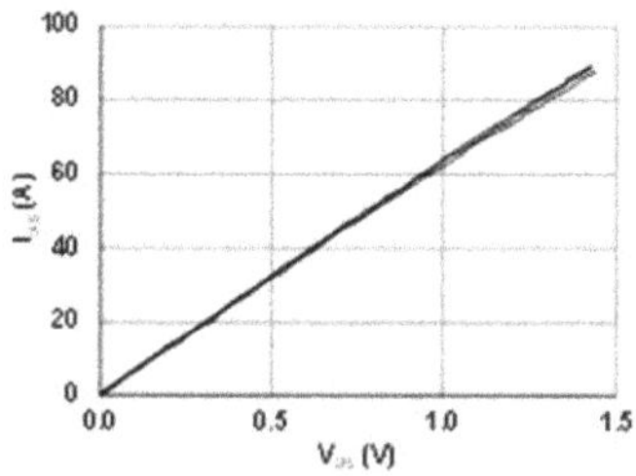

Fig. 5 Individual forward-conduction curves of the 3 SSCB MOSFETs at room temperature and VGS = 20 V.

4. Controller design

For small size, fast response, and noise immunity, the voltage across the MOSFETs was sensed, instead of the conducted current, to determine a trip condition. To sense voltage across the MOSFETs during conduction and to protect the sensing circuitry from high voltage during MOSFET blocking, a low-current-rated SiC 1200-V MOSFET was used as a sense MOSFET in series with a sense resistor. Fig. 6 shows a simplified schematic of the controller. The same gate drive signal (DRV) is used for the power MOSFETs and the sense MOSFET (M1). However, the source of M1 connects to the sense resistor (R3). A sense-resistor value of 49.9 Ω was selected to limit noise at the controller input. The on-state resistance of the sense MOSFET is nominally 280 mΩ at 25 °C. Therefore, approximately 99.5% of the voltage across the power MOSFETs is dropped across the sense resistor when M1 is conducting. The sensed voltage across R3 is input to an inverting amplifier (OPA). The amplifier output charges capacitor C1 through resistors R11 and R3 (with R11 >> R3). Capacitor C1 integrates the amplifier output voltage. The sum of the voltages across C1 and R3 is applied to the inverting input of a comparator (CMP). Therefore, the sensed voltage is provided to the comparator, with the voltage across C1 adding to it during its charging time.

This allows the SSCB to have a configurable high-speed trip point for high sensed voltages, and a configurable trend of trip levels over ranges of sensed voltage and time duration. The comparator has a latching output that drives the MOSFET gate driver to control the state of the SSCB and M1. The status output, reset, and external trigger signals are isolated by optocouplers represented by D1, Q1, and Q2, respectively.

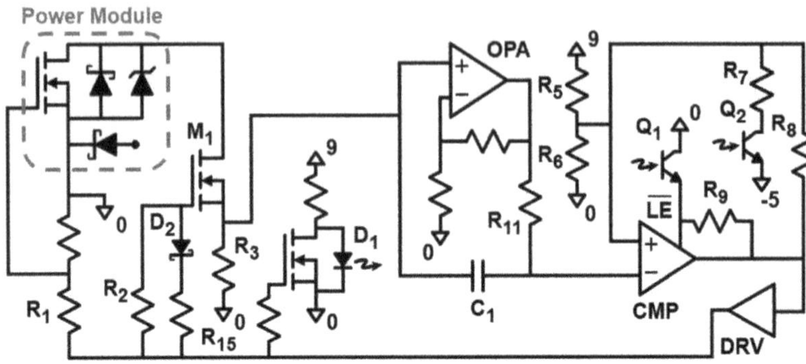

Fig. 6 Simplified schematic of the SSCB control circuit.

During normal operation, with the SSCB conducting, the sensed voltage and amplifier output voltage are below the reference voltage set by resistors R_5 and R_6 at the non-inverting input of the comparator. Thus, the comparator provides a high output to the gate driver to maintain MOSFET conduction. The comparator's active-low latch enable (LE′) pin is pulled up by R_9 to disable the latch and allow the comparator output state to change based on its inputs. When voltage at the inverting input of the comparator exceeds the reference voltage, the comparator output becomes low and pulls LE′ low to latch the output. The gate driver turns off the MOSFETs to trip the circuit breaker. To reset the latch and turn the SSCB on, optocoupler transistor Q_1 is turned on to pull the LE′ up to 0 V from the comparator negative supply voltage of –5 V. Resistor R_8 provides feedback for comparator hysteresis. The SSCB can be tripped using an external signal by turning on optocoupler transistor Q_2 to pull the comparator voltage reference low through R_7. By pulling the reference toward –5 V, the SSCB can be externally tripped even at a sensed voltage of 0 V. To prevent high-current pulses from flowing through the sense MOSFET and sense resistor during SSCB trip and reset transitions, an alternate gate turn-off path is provided by diode D_2 and gate resistor R_{15}. Resistors R_1, R_2, and R_{15} were selected so that sense MOSFET M_1 turns off a short time before the power MOSFETs during an SSCB trip, and M_1 turns on a short time after the power MOSFETs during a reset. A trip current of 210 A was selected for the high-speed trip response, to allow 70 A peak per MOSFET at 28 °C, and to stay close to the repetitive-pulse capability of the two ABDs. From the MOSFET data of Fig. 1, 70 A at 28 °C, corresponds to 1.20 V. The comparator reference was set at 1.19 V. The steady-state trip point was selected to be 93 A based on the 150 °C curve in Fig. 1. This operating point corresponds to 31 A per MOSFET and a drain-to-source voltage of 0.87 V. With the amplifier gain set to 1.37, an output of 1.19 V results for a sensed voltage of 0.87 V. At voltages below 0.87 V, the controller will not trip the SSCB. At voltages between 0.87 V and 1.19 V, the SSCB response time is determined by the amplifier gain and the integrator at the amplifier output.

The plot in Fig. 7 shows simulated controller trip time versus sensed-voltage amplitude. The curve was plotted using voltage amplitudes between 0.88 V and 1.19 V. The curve has vertical asymptotes at about 0.87 V and about 1.19 V, corresponding to the steady-state and high-speed SSCB trip points, respectively. The middle part of the curve can be raised or lowered by changing the integrator and amplifier gain. The upper left and middle parts of the curve look similar to the trip-time versus current characteristics of many mechanical circuit breakers and fuses. However, as seen at the right side of the curve, the controller response does not follow a constant I^2t profile, and can provide very fast trip times near the 1.19-V limit.

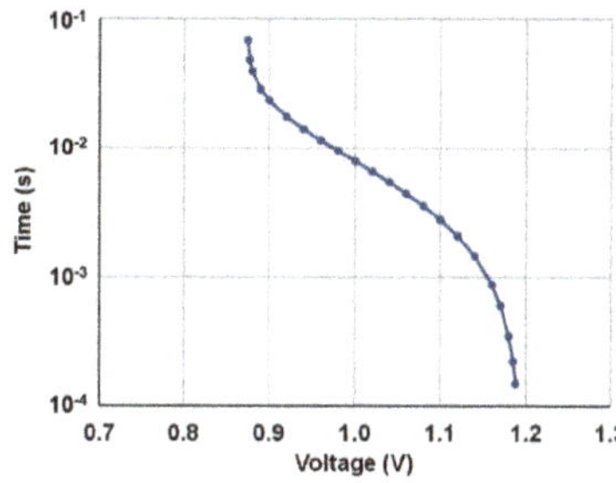

Fig. 7 Simulated controller trip-time versus sensed-voltage amplitude.

Yet, in implementation, as the SSCB controller and power stage reaches it fastest trip response, the bottom of the curve will begin to turn toward the right and become horizontal with a constant minimum trip time for all sensed voltages.

The controller circuit board was mounted on the packaged power module. Two flat copper bus bars provided SSCB drain and source terminals. The connection from the sense-MOSFET drain to the power module was made by a trace on the control board to one of the drain mounting holes. The module gate and source pins were soldered into the controller. Fig. 8 shows the assembled SSCB, heatsink, and forced-air cooler.

Fig. 8 SSCB module with compact forced-air cooler.

5. Pulsed current test circuits

For SSCB response times on the order of 100 ms, the 5-ms rise time of a high-current power supply provided nearly-square current pulses without overshoot. With the SSCB in the on-state, and the supply connected directly to it, the supply output was enabled to produce each long current pulse. For shorter SSCB response times, the output of the power supply was initially enabled and the SSCB was reset from the off-state to initiate a current pulse. However, interactions between SSCB turn-on, sensing, and turn-off provided a different trip response than that from the SSCB on-state. Therefore, pulsed currents were switched-on externally to the SSCB to characterize its trip response from the on-state. The diodes providing an optional free-wheeling load-side current path were not connected during testing, as a load was not present and line inductance was only added to the supply-side of the SSCB to provide worst-case turn-off transients. Two pulsed-current test circuits were used to characterize the SSCB trip response. Sufficient delays between pulses allowed the SSCB to return to its initial temperature before the next pulse. Ideally, it was desired to obtain trip response times as the pulse duration at a constant current amplitude.

However, providing nearly-square current pulses with amplitudes over 200 A for trip response times below 5 μs, required ramp rates of over 100 A/μs. Sourcing such current pulses over a range of amplitudes without overshoot is difficult and can require several multistage pulse-forming networks. To provide pulsed-current ramp rates of approximately 10 A/μs, an insulated-gate bipolar transistor (IGBT) (Q) was added in series with the output of a high-current power supply, as shown in the left schematic of Fig. 9. With *Q* blocking, the supply (V_S) was in voltage-mode with its output capacitance charged. Turning *Q* on allowed the supply output capacitance to discharge through the SSCB and the loop inductance (L). When the supply reached its current limit, it switched to current-mode to maintain the current amplitude for the duration of the pulse.

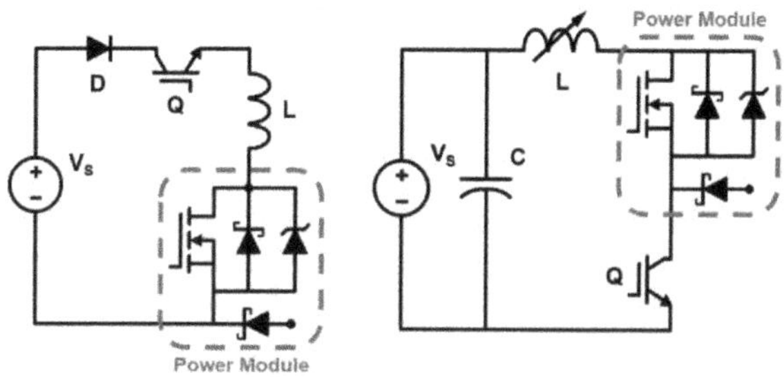

Fig. 9 Switched power-supply pulsed circuit (left) and switched RLC pulsed circuit (right).

To further increase pulsed-current ramp rates to over 70 A/μs, a compact switched RLC circuit was built. Current pulses were provided over a range of ramp rates determined by capacitor voltage and lead inductance. The circuit schematic is shown on the right in Fig. 9. Capacitor (C) was charged by a low-current power supply. IGBT (Q) was gated-on to provide each current pulse. Inductance was changed between tests by adding or removing wire loops to the circuit to function as an air-core inductor. The circuit was made compact enough to form a minimum loop area of approximately 70 cm^2 from the SSCB power terminals. Together, the two pulsed-current circuits facilitated SSCB characterization over 4 orders of magnitude of pulse width. However, current-pulse shapes varied from nearly square for long trip times, to nearly saw-tooth-shaped for short trip times. The trend of the trip response is affected by the shapes of the current pulses used to characterize it. The average value of a square pulse is equal to its peak value. The average value of a saw-tooth pulse divided by its peak value is 0.5, and that same ratio for a sine wave from 0 to $\pi/2$ is 0.637. For current pulses tripping the SSCB and having average values less than their peak values, the pulse amplitude may be below the trip point for a large portion of the pulse. However, the current below the trip point contributes to the trip time. This issue also affects the trip responses of fuses and mechanical circuit breakers. However, because they respond more slowly, the rise times of their current pulses may be neglected. In this evaluation, the trip response was characterized using the total width and peak amplitude of each current pulse. However, because pulse shapes differed across the data set, the average pulsed current divided by the peak current was used as a figure of merit (FOM) to relate pulse shapes. For pulses having lower FOM values, the reported pulse width includes a significantly long sub-interval with current amplitude below the trip point.

6. SSCB trip response

Current pulse widths and amplitudes were recorded with the RLC-pulsed circuit connected to the SSCB with its minimum loop area of about 70 cm^2, and with up to 7 additional loops of the same area. Fig. 10 shows the data for the minimum loop area (0 loops), 1, 2, 3, 5, and 7 loops at baseplate temperatures of 25 °C, 50 °C, and 75 °C. Each series name has a hyphen between the baseplate temperature and the number of loops. Trip currents are lower at the higher baseplate temperatures for the same pulse widths. This result was expected due to higher sensed voltages caused by the positive temperature coefficient of resistance of the MOSFETs.

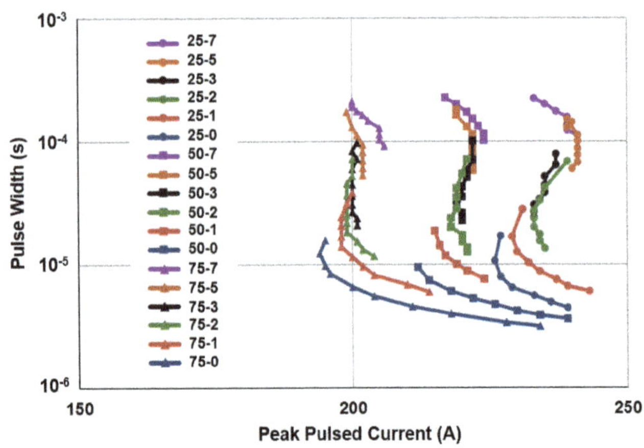

Fig. 10 SSCB trip response with switched RLC circuit for minimum loop area (0), 1, 2, 3, 5, and 7 loops at 25 °C, 50 °C, and 75 °C baseplate temperatures.

For each baseplate temperature set, the data shows an unexpected reverse S-shaped trend. Each 2- and 3-loop series has points with lower peak currents having narrower pulse widths, and points with higher peak currents having wider pulse widths. The trend is also contradicted by the change in current pulse shape within each loop data series. For each loop configuration, the capacitor voltage of the RLC circuit was increased to increase the pulsed-current ramp rate, resulting in narrower pulses with more linear trajectories. Therefore, for each loop configuration, the pulse-shape FOM (average pulsed current divided by peak pulsed current) decreases with decreasing pulse width. Pulses having decreasing FOMs are expected to have increasing peak currents. An explanation for the trend is that the capacitor that functions as an integrator at the output of the controller amplifier stage, functions as a differentiator at the input of the comparator stage. Points in the lower half of the plot bend toward higher trip currents and a lower trip-time limit. This is expected as the SSCB turn-off time becomes significant relative to the current pulse width. Points at the top of the plot show the expected integrator trend of increasing trip current with decreasing pulse width. The IGBT-switched pulsed circuit was used to extend the trip responses to longer pulse widths. Fig. 11 shows the longer pulse-width data with selected data from Fig. 10, to form the trip-response. The data are shown for maximum pulse widths of approximately 100 ms. For pulse widths longer than this, the trends are increasingly affected by baseplate temperature rise, causing trip currents to decrease for increasing pulse widths. The shapes of the responses for the longer pulse-width data look very similar to the simulated trend shown in Fig. 7. The horizontal lines in Fig. 11 show the approximate pulse-shape FOM value. Except for the bottom 2 lines, FOM values generally decrease moving down the plot. It is observed that at each baseplate temperature, trip currents are nearly the same at both FOM values of 0.55. Also, these trip currents are higher than those at the FOM of 0.6. This is expected and suggests that for a constant FOM, the values between about 10-4 and 10-5 s may be more vertical than shown. However, some positive slope is still expected based on the trends of the 2- and 3-loop data series from the pulsed RLC circuit. The SSCB module was attached to the forced-air cooler, as shown in Fig. 8. Current pulses ranging in width from less than 50 ms to over 50 min were supplied by enabling the output of the high-current power supply. Ambient air temperatures ranged from 22 to 25 °C, over the data set. The trip response with forced-air cooling is shown in Fig. 12, with the data from Fig. 11 included.

For pulse widths near 50 ms, the SSCB trips at currents about 6 A lower (about 3.5% lower current) than those with the 25 °C liquid-cooled baseplate. This is a result of additional baseplate temperature rise during each current pulse with air cooling. Moving up the curve, pulse widths increase with decreasing trip current amplitudes. At a pulse width of 1000 s, the SSCB approaches a constant current amplitude of about 100 A. The trend will reach a constant current at the MOSFET thermal-equilibrium point. Following the test having the longest pulse width, at the top of Fig. 12, the SSCB was tested at 102 A for over 3000 s (over 50 min) without tripping. Although the response with 50 °C air-cooling was not tested, a maximum steady-state current can be approximated for 50 °C air-cooling, based on the data in Fig. 12. The offset current between the 50 °C and 25 °C data using liquid cooling is approximately 17 A. Subtracting this current from the tested 102-A steady-state current with 22–25 °C air cooling gives 85 A. This value compares well with the simulated maximum steady-state current of 87 A using 50 °C air-cooling. This also indicates that the maximum MOSFET temperature at the 102-A test point is near the 150 °C operating value.

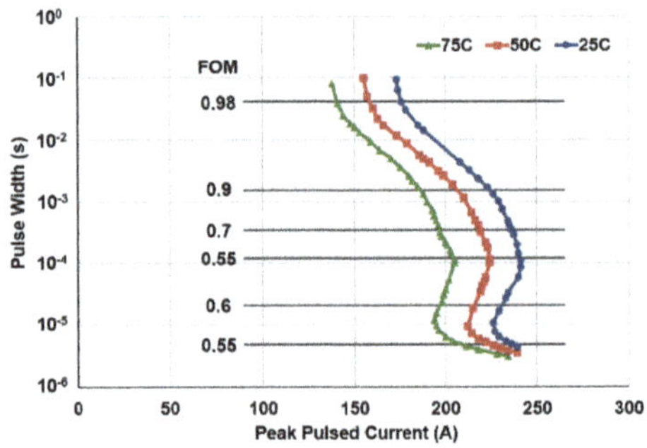

Fig. 11 SSCB trip response including FOM values at 25 °C, 50 °C, and 75 °C baseplate temperatures.

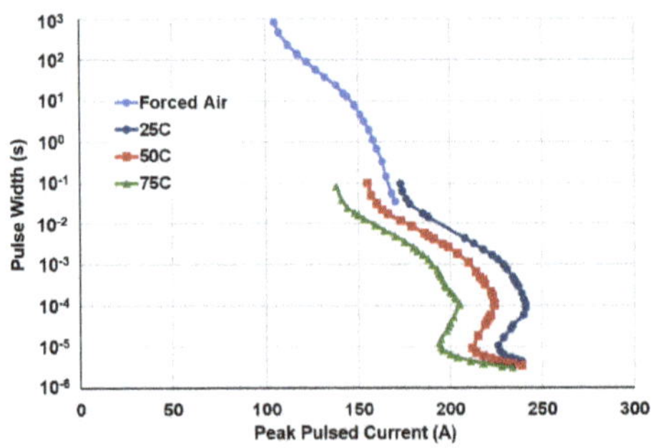

Fig. 12 SSCB trip response including data with 22 to 25 °C forced-air cooling.

Summary

A 100-A, 850-V, SSCB, having SiC MOSFETs, SiC Schottky diodes, and SiC avalanche breakdown diodes was developed. The SSCB has a 32 cm2 footprint and is able to continuously conduct up to 100-A with 25 °C air cooling. It was designed with an analog controller that provides a configurable trip-response characteristic to allow a range of overcurrent transients to be conducted, while maintaining a fast response at its fault-current limit. The controller also provides remote trip, status output, and reset functions. Test-circuits were built to provide current pulses with ramp rates up to over 70 A/µs. A pulse-shape FOM was used to relate pulse shapes provided by the test circuits. The SSCB trip response was characterized at baseplate temperatures of 25 °C, 50 °C, and 75 °C. The trip response showed expected trends associated with the controller integrator and the power-module thermal resistance, and an unexpected differentiator trend. A minimum pulse width of 3.2 µs was demonstrated for a nearly saw-tooth-shaped 234-A current pulse, at a baseplate temperature of 75 °C. With 22–25 °C air cooling, the SSCB conducted 102 A continuously for over 50 min (the test duration) without tripping. In total, the SSCB trip response was characterized over nearly 9 orders of magnitude in current pulse widths.

References

[1] M. Jian-guo, W. Li, and H. Jie, "Research on main circuit topology for a novel DC solid-state circuit breaker," *5th IEEE Industrial Electronics and Applications Conf.*, 2010, pp. 926-930. https://doi.org/10.1109/ICIEA.2010.5515678

[2] D. Urciuoli, V. Veliadis, H. Ha, and V. Lubomirsky, "Demonstration of a 600-V, 60-A, bidirectional silicon carbide solid-state circuit breaker," *Applied Power Electronics Conf. and Expo. (APEC)*, 2011, pp. 354-358. https://doi.org/10.1109/APEC.2011.5744620

[3] D. Urciuoli et al., "Performance of a 1-kV, silicon carbide avalanche breakdown diode," *IEEE Trans. Power Electron. Lett.*, vol. 30, no. 9, pp. 4643-4645, 2015. https://doi.org/10.1109/TPEL.2015.2403199

[4] C. Henze, N. Mohan, C. Henze, and R. Button, "A programmable solid-state active current limiting circuit breaker for space applications," Industrial Electronics Society Annual Conf. (IECON), 2002, vol. 4, pp. 2849-2854. https://doi.org/10.1109/IECON.2002.1182847

[5] Z. Miao et al., "A self-powered bidirectional DC solid-state circuit breaker using two normally-on SiC JFETs," *IEEE Energy Conversion Congress and Expo. (ECCE)*, 2015, pp. 4119-4124. https://doi.org/10.1109/ECCE.2015.7310241

Chapter 27

Performance of a 1-kV, Silicon Carbide Avalanche Breakdown Diode

D. Urciuoli[1*], S. Ryu[2], D. C. Capell[2], D. Ibitayo[1], G. Koebke[1], and C. W. Tipton[1]

[1] U. S. Army Research Laboratory, 2800 Powder Mill Rd, Adelphi, MD 20783 USA

[2] Wolfspeed, a Cree Company, 3026 E. Cornwallis Rd, Research Triangle Park, NC 27709 USA

damian.p.urciuoli.civ@mail.mil*

Abstrac

A SiC avalanche breakdown diode (ABD) having a nominal 1-kV breakdown voltage was fabricated to provide improved suppression of voltage transients induced during hard- switched turn-off of solid-state devices. Three SiC ABDs were pulsed 1000 times in an inductive load circuit at peak currents of over 100 A. Superior performance in peak pulse current, clamping voltage, and peak pulse power was seen, compared to the results of two series-connected commercial TVS devices, collectively having a comparable breakdown voltage. The transient thermal response of the SiC ABDs was calculated using a model for energy dissipation in short pulses. SiC ABD design parameters and test data were used to show that the reported performance of these devices was not related to package thermal impedance.

Keywords

Avalanche Breakdown Diode (ABD), Silicon Carbide (SiC), Solid State Circuit Breaker (SSCB), Solid State Power Controller (SSPC), Snubber, Metal Oxide Varistor (MOV), Transient Voltage Suppression (TVS)

Contents

The work has been previously published by SPIE:
D. Urciuoli, S. Ryu, D. C. Capell, D. Ibitayo, G. Koebke, and C. W. Tipton, "Performance of a 1-kV, Silicon Carbide Avalanche Breakdown Diode," IEEE Transactions on Power Electronics, Vol. 30, Issue 9, Sept. 2015, Doi: 10.1109/TPEL.2015.2403199.

1. Introduction

During hard-switched turn-off, solid-state circuit breakers (SSCBs) or solid-state power controllers (SSPCs) require a means of dissipating energy stored in line inductances. DC and hard-switched AC SSCBs, operating on busses at up to hundreds of volts, can be subjected to transients exceeding their maximum blocking voltages. Although some solid-state switches have sufficient avalanche energy ratings, it is often desirable to limit voltage stress and additional temperature rise by absorbing energy using other components designed to suppress the transient. Crowbar-type devices are undesirable in these applications due to their latching behavior that prevents self-recovery at the circuit operating voltage. Snubbers can be used, but have transient responses that are dependent on voltages and impedances of the power circuits [1], [2]. Snubber capacitors can also be large and have high leakage currents [3]. Another approach is to place a metal-oxide varistor (MOV) in parallel with the switch to suppress the transient [4]. Although an MOV can dissipate high pulsed energies, its dynamic resistance, at even less than 2% of its rated peak current, can result in clamping voltages on the order of twice its rated working stand-off voltage. Because on-state losses generally increase with switch blocking voltage ratings, higher SSCB blocking voltages are undesirable if only needed to provide sufficient margin for such high clamping voltages. Avalanche breakdown diodes (ABDs) or transient voltage suppression (TVS) diodes can have lower dynamic resistances than MOVs over low current ranges, resulting in lower clamping voltages [3], [5]. Like MOVs, ABDs can be connected in series and parallel to provide higher voltage and current operation and higher energy dissipation [6], [7]. In this study, 11.6 mm^2 silicon carbide (SiC) ABDs, having a nominal breakdown voltage of 1 kV, were fabricated and evaluated in a pulsed circuit for a 600-V SSCB application. Performance of the SiC ABDs was compared to that of two series-connected commercial TVS devices, collectively having a similar breakdown voltage.

2. SiC ABD applications and designs

A unidirectional TVS device having a stand-off voltage of greater than 600 V, and the ability to clamp voltage below a 1.2- kV SSCB switch rating, at a peak current of 100 A, was sought. TVS devices with stand-off voltages exceeding 600 V were not found, so devices with lower stand-off voltages were connected in series. After finding that many of these TVS devices are comprised of lower voltage series-stacked diodes, a SiC ABD was developed. A nominal 1-kV breakdown voltage was chosen for the SiC ABD to provide stand-off for applications operating at up to 900 V, and to allow a voltage clamping range of roughly 200 V, up to a 1.2-kV switch rating. The fabrication started on a 5.5-µm n-type epitaxial-layer, with a doping concentration of 1.6×10^{16} cm^{-3}, grown on the Si-face of a 4-degree off-cut n+ 4H-SiC substrate. The wafers received an aluminum ion implantation of 1.3×10^{15} cm^{-2} to form the anode region of the diode, and the guard rings of the edge termination structure [8]. The implants were annealed at approximately 1600 °C. Ni-based ohmic contacts were formed on the anode p+ implants, and on the backside of the n+ substrate, which serves as the cathode of the ABD. A 4-µm Al layer was deposited and patterned for the top anode electrode, and a Ti/Ni/Au layer was deposited on the backside for the cathode electrode. The die size was 3.4 mm × 3.4 mm, with an active area of approximately 0.09 cm^2. Unfortunately, poor adhesion of the gold layer was caused by an equipment malfunction during the backside metallization process. In order to provide an adequate bond to the selected TO-247 package, the backside gold layer was removed and Sn63Pb37 solder paste with flux was used.

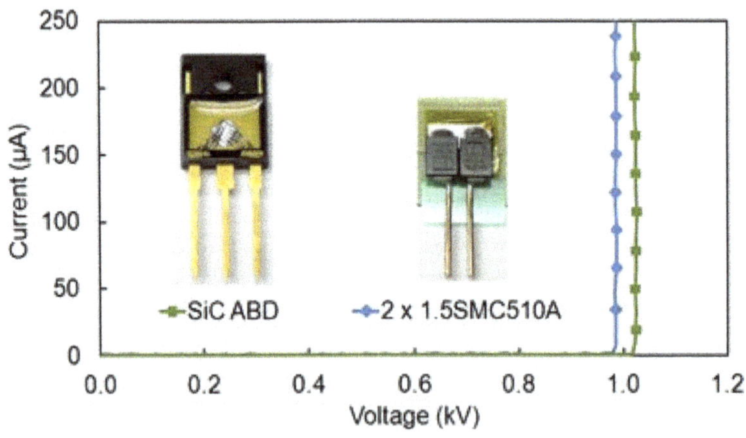

Fig. 1 Typical reverse leakage current versus voltage for a SiC 1-kV ABD and two series-connected 1.5SMC510A TVS devices.

About one-half of the packaged die required a second solder reflow to reattach the die after bond-strength sheer testing. A single SiC ABD was mounted in each package with four 0.25-mm diameter aluminum wire bonds connecting the anode to each of the two outer TO-247 terminals (total of eight wires). A packaged SiC ABD device with gel encapsulant is shown in the inset of Fig. 1.

3. SiC ABD evaluation and results

SiC ABD characteristics were compared to those of two series-connected 1.5SMC510A commercial TVS devices from Littelfuse, Inc., Chicago, IL, USA, each having two 2.9 mm × 2.9 mm series-stacked ABDs. The typical breakdown characteristics of a SiC 1-kV ABD (1.02 kV at 101 μA), and of two series-connected 1.5SMC510A devices (0.99 kV at 101 μA), as measured by a Tektronix 370B curve tracer, are shown in Fig. 1. The ABDs were evaluated in a pulsed inductive load circuit providing the hard-switched turn-off stresses of a 600-V DC SSCB application. Three SiC ABDs were each pulsed separately 1000 times, at a 100-A nominal peak current. Fig. 2. shows the current and voltage responses of pulses 1 and 1000 for each of the SiC ABDs. The schematic of the pulsed circuit is shown in the inset of Fig. 2. Circuit parameter values of 600 V, 2 kΩ, 90 μF, and 24 μH were used for VDC, R, C, and L, respectively. With the capacitor charged by the dc supply, the IGBT is pulsed to the on-state to ramp up inductor current. The IGBT is then turned OFF, and the inductor current ramps down through the device-under-test (DUT). Pulsed currents through the DUT have fast rise time and nearly linear decay, unlike standard ABD impulse wave shapes such as 8/20-μs and 10/1000-μs pulses [7]. A 40-s period was used during repetitive pulsing of the SiC ABDs, to allow die cooling between pulses, while maintaining a 25 °C case temperature. The 1.5SMC510A devices were not repetitively pulsed. Statistical parameters were extracted from captured pulse data using MATLAB. The peak pulse current (I_{PP}) average for the six pulses shown in Fig. 2 is 104 A, and the clamping voltage (V_C) average is 1.11 kV. The resulting peak pulse power (P_{PPM}) average is 115 kW. The current pulses of Fig. 2 have a 0.31-μs average virtual-front-time (t_R , effective rise time) and a 2.64-μs average virtual-time-to-half-peak (t_D , effective pulsewidth). The aver-age V_C is an increase of only 9% over the breakdown voltage measured at 101 μA. All three SiC ABDs showed nearly identical initial pulsed currents and voltages.

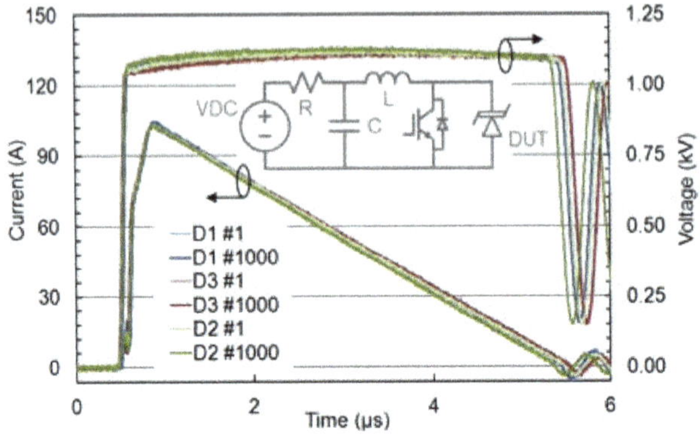

Fig. 2 Avalanche breakdown pulses 1 and 1000 for three SiC ABDs averaging 104-A and 1.11-kV peak at a 25 °C case temperature (pulsed inductive load circuit schematic inset).

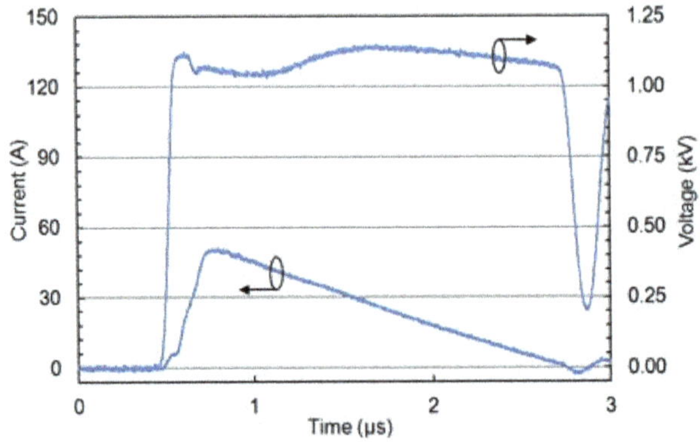

Fig. 3 Avalanche breakdown pulse for two series-connected 1.5SMC510A TVS devices at 51-A and 1.14-kV peak at a 25 °C case temperature.

The V_C value for SiC ABDs D1 and D3 increased by less than 5 V between pulses 1 and 1000, and that of D2 increased by 13 V. Although de-vice degradation cannot be ruled out as a cause, no increases in breakdown voltage were seen during curve tracer measurements taken after testing. The increases in V_C could be attributed to degradation of the solder die attach, seen to have varying mechanical reliability, as previously mentioned. However, the 13V increase of D2 represents only a 1.2% change in V_C. Fig. 3 shows the current and voltage responses of the series-connected 1.5SMC510A devices during evaluation in the pulsed circuit. The inductor charging time was reduced to provide a lower peak current to meet, or slightly exceed, the peak pulse power rating of the devices. For the two series-connected 1.5SMC510As, I_{PP} and V_C are 51 A and 1.14 kV. The P_{PPM} is 58.2 kW, and t_R and t_D are 0.23 μs and 1.22 μs, respectively. The V_C is 15% higher than the breakdown voltage at 101 μA. Assuming equal sharing, each 1.5SMC510A had a P_{PPM} of 29.1 kW, which is above its rating of about 25 kW for a 1.22-μs t_D. The total P_{PPM} of 58.2 kW is about half of the 115-kW average SiC ABD P_{PPM}. Also, the SiC ABD average t_D is over twice that of the 1.5SMC510As. Furthermore, one 11.6 mm^2 SiC ABD die is simpler and smaller than four die having a total area of 33.6 mm^2.

4. Transient thermal response

It was important to determine if the superior performance of the SiC ABD was attributable to its larger TO-247 package, compared to the smaller DO-214 package of the 1.5SMC510A.

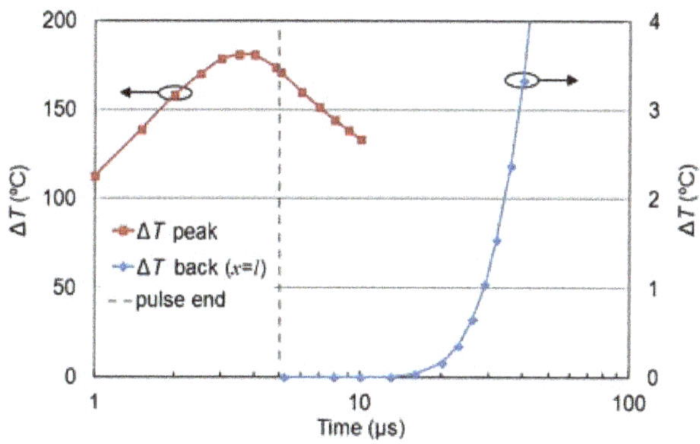

Fig. 4 Calculated SiC ABD backside die temperature rise following an average current pulse shown in Fig. 2 and calculated peak SiC ABD temperature rise through D2 pulse 1.

The transient thermal response of the SiC ABD was modeled using the methods discussed by Clemente for energy dissipated in a semiconductor in a short pulse [9]. Because heat spreads from the junction, beyond the active die thickness during a pulse, calculating a larger effective thickness as a correction factor for the energy dissipating layer improves the accuracy of the thermal model. Using SiC material properties, ABD design parameters, and averaged test data, the effective thickness of the energy dissipating die layer was calculated from the temperature rise based on Equation 1.

$$\Delta T(x,t) = \frac{E}{\rho c l}\left[1 + 2\sum_{n=1}^{\infty} e^{-(\frac{n\pi}{l})^2 \alpha t} \cos(\frac{n\pi x}{l}) \cdot \frac{\sin(\frac{n\pi g}{l})}{(\frac{n\pi g}{l})}\right] \quad (1)$$

where x is the depth from the die surface, t is the pulsewidth (5.0 μs), E is energy per unit area (285 mJ/0.09cm^2), ρ is the density of 4H-SiC (3.21 g/cm^3), c is the specific heat of 4H-SiC (0.69 J/(g·°C)), l is the die thickness (355.5 μm), α is the thermal diffusivity of 4H-SiC (1.7 cm^2/s), and g is the active thickness of the die (5.5 μm) [10]. An effective thickness of 83 μm was calculated by solving for x at 90% of its final temperature, given by the first term E/(ρcl) in (1). In all calculations using (1), the first 100 terms of the summation provided sufficient accuracy. By substituting the effective thickness back into (1) for g, and letting x equal l, backside die temperature can be calculated with time, as plotted in Fig. 4. The curve shows that temperature rise at the backside of the die, closest to the package heat spreader, is negligible at less than 13 μs from the start of the pulse, and takes over 30 μs to rise 1.5 °C. Therefore, the TO-247 package is believed to have no effect on SiC ABD transient thermal response for the pulsed tests reported. Finally, Fig. 4 also shows calculated SiC ABD peak temperature rise from 1 to 10 μs from the start of the pulse. Points were computed using (1) with individually calculated effective thicknesses from incremental time and energy values of ABD D2 pulse 1. The trend shows a peak temperature rise, during the pulse, of approximately 180 °C.

Summary

ABDs are preferred for suppressing overvoltage transients in many applications due to their simplicity, self-recovery, low leakage current, and low dynamic resistance. However, individual ABDs having stand-off voltage ratings above 600 V, for protection of 1.2 kV solid-state switches, were not found during a product search. SiC ABDs having a die area of 11.6 mm^2 and a breakdown voltage of 1.02 kV were fabricated. Three SiC ABDs were pulsed 1000 times in an inductive load circuit. Their performance was compared to that of two series-connected 1.5SMC510A TVS devices having a combined breakdown voltage of 0.99-kV. The SiC ABDs sustained peak pulse power levels of over twice the rating of the two 1.5SMC510As combined, for over twice the effective pulsewidth. For peak pulse currents over twice that of the two 1.5SMC510A devices combined, the SiC ABD clamping voltage was only 9% above its breakdown voltage, compared to an increase of 15% for the 1.5SMC510As. The SiC ABD has a die area of roughly one-third of the total area of the four die packaged in the two 1.5SMC510A devices. Thermal modeling results showed that during pulsed testing, the SiC ABD performance was independent of the thermal impedance of the package. The SiC ABD can provide dramatic performance improvements and reduced complexity compared to commercial TVS devices.

References

[1] J. Park, and J. Candelaria, "Fault detection and isolation in low-voltage DC-bus microgrid system," IEEE Trans. Power Del., vol. 28, no. 2, pp. 779-787, Apr. 2013. https://doi.org/10.1109/TPWRD.2013.2243478

[2] Transient Suppression Devices and Principles, AN9768, Littelfuse, Inc., Chicago, IL, 1998.

[3] Protecting AC Output SSRs against voltage transient phenomena, Crydom Inc., San Diego, CA, Jan. 2012.

[4] J. Magnusson, et al., "On the use of metal oxide varistors as a snubber circuit in solid-state breakers," in Proc. IEEE PES ISGT Europe, Copenhagen, 2013, pp. 1-4. https://doi.org/10.1109/ISGTEurope.2013.6695454

[5] S. N. Friedman, "Robust/low cost solid-state remote power controller (RPC) technology is here," in Proc. Digital Avionics Systems Conference, St. Louis, MO, 1999, pp. 3.B.4-1 – 3.B.4-7. https://doi.org/10.1109/DASC.1999.863710

[6] The ABCs of MOVs, AN9311.6, Littelfuse, Inc., Chicago, IL, 1999.

[7] Avalanche Breakdown Diode (ABD) Transient Voltage Suppressors, JESD210, 2007.

[8] S. Ryu and A. Agarwal, "Multiple floating guard ring edge termination for silicon carbide devices," U.S. Patent 7 026 650, Apr. 11, 2006.

[9] S. Clemente, "Transient thermal response of power semiconductors to short power pulses," IEEE Trans. Power Electron., vol. 8, no. 4, pp. 337-341, Oct. 1993. https://doi.org/10.1109/63.261001

[10] M. E. Levinshtein, S. L. Rumyantsev, and M. S. Shur, Properties of Advanced Semiconductor Materials: GaN, AlN, InN, BN, SiC, SiGe, New York: Wiley & Sons, 2001, pp. 93–95.

Chapter 28

Large Chip Area SiC PiN Diodes Demonstrated for Thyristor Protection in a Pulsed System

Heather K. O'Brien[1*], William Shaheen[2], Aderinto Ogunniyi[1], Charles Scozzie[1], Lin Cheng[3], Miguel Hinojosa[1,4], Kevin Lawson[4], Shelby Lacouture[4], and Stephen Bayne[4]

[1]U.S. Army Research Laboratory, Adelphi, MD 20783, USA

[2]Berkeley Research Associates, Beltsville, MD 20705, USA

[3]Cree, Inc., Durham, NC 27703, USA

[4]P[3]E, Texas Tech University, Lubbock, TX 79409, USA

heather.k.obrien.civ@mail.mil*

Abstract

Asymmetric thyristors require protection from voltage and current reversals in high-inductance capacitor discharge systems. Silicon carbide (SiC) PiN diodes capable of blocking up to 16 kV were demonstrated to have the high-current capability to transmit forward pulse current in a series configuration with a thyristor, and to clamp reverse current in an anti-parallel configuration. In series with a thyristor, diodes were switched 1000 pulses at a single-shot rate at 2000 A peak current (3.8 kA/cm^2 over anode area and 2100 A^{2}s per pulse) without any notable increases in forward voltage or reverse leakage current. In the reverse clamp configuration, a parallel pair of PiN diodes was demonstrated to block 12 kV charge on the capacitor bank, then clamp a total of 4200 A current reversal with good parallel current sharing. These evaluations demonstrate that for high current density pulsing above 10 kV, individual 16 kV PiN diodes yield lower on-state voltage loss (16 V at 2000 A) than series-stacked assemblies of 9 kV SiC PiN diodes or 6 kV Si diodes.

Keywords

Silicon Carbide, Pulse Power Systems, Power Semiconductor Devices, PiN Diode, Pulse Current, High Voltage

The work has been previously published by IEEE:
H. K. O'Brien et al., "Large chip area SiC PiN diodes demonstrated for thyristor protection in a pulsed system," 2014 IEEE International Power Modulator and High Voltage Conference (IPMHVC), Santa Fe, NM, 2014, pp. 538-541. doi: 10.1109/IPMHVC.2014.7287331

Contents

1. Introduction

Mobile Army platforms require reliable, compact, high-performance power electronics to enable future survivability and lethality systems. Thyristors and PiN diodes fabricated in silicon carbide are the choice devices for combining high-voltage blocking in the off-state with high current density conduction in the on-state. The U.S. Army Research Laboratory is developing and thoroughly characterizing high-voltage (>10 kV) silicon carbide components through joint research with Cree, Inc. and Texas Tech University.

The culmination of this research will be robust, large chip area, high-voltage SiC devices that can be packaged in various series and parallel combinations to support a range of high-power systems. Vital progress toward that end includes optimizing device termination and packaging for new high voltage territory, reducing on-state resistance, and achieving uniformity from device-to-device. Beyond the PiN diode characterizations described herein taking place at the Army Research Laboratory, the partnership with the Pulsed Power and Power Electronics laboratory at Texas Tech University adds the ability to study devices for long-term performance and reliability. At Texas Tech, the PiN diode's operation is being evaluated over tens of thousands of pulses at a narrower pulse width for a broader range of applications.

The goals of the investigations reported in this paper are to evaluate the peak pulse current capabilities of Cree's recent 16 kV PiN design and to compare the device's characteristics to those of 9 kV SiC PiN diodes and lower-voltage silicon diodes switched under similar pulsed electrical stresses.

2. High voltage SiC PiN

The PiN diodes are fabricated on 4-inch 4H-SiC wafers by Cree. The footprint of each chip is 1.05 cm^2, with an anode area of 0.52 cm^2. A drift region thickness of 120 μm, combined with a negative bevel edge termination around the anode, enables high voltage blocking up to 16 kV with less than 1 μA of leakage current (Fig. 1) [1]. The surface passivation is a 1-μm thick PECVD oxide. Cree's lifetime enhancement process was utilized to improve conductivity modulation and reduce forward voltage drop in the thick epi [2].

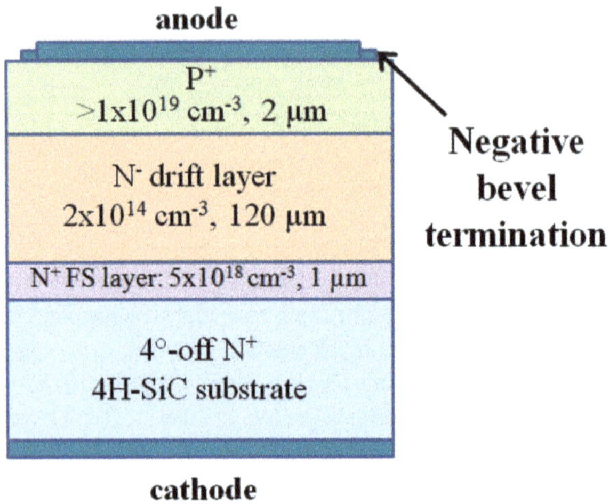

Fig. 1 Cross section of the 16 kV, 1.05 cm², 4H-SiC PiN diode with doping and layer thicknesses. Thicknesses are not drawn to scale. Adapted from [2].

Three PiN diodes were individually packaged in custom high-voltage packages (Fig. 2). The cathode (underside) of the die is attached to the base plate with gold-germanium. Ten-mil aluminum wire bonds connect the anode surface to the tabs of the package. A commercial silicone gel fills the center well to maintain high-voltage blocking and protect the wire bonds. The vertical separation and alignment of the anode tabs and the cathode base plate enable consistent DC voltage blocking through 13 kV before leakage begins to creep along the edge of the insulator connecting the tabs (Fig. 3). During device evaluations in the pulse circuits at the Army Research Laboratory, voltage was kept below 12 kV.

Fig. 2 Custom Cree package for evaluating individual devices. The base plate serves as the cathode. The tabs nearer to the reader serve as the anode connections. The center well is filled with a silicone-based gel.

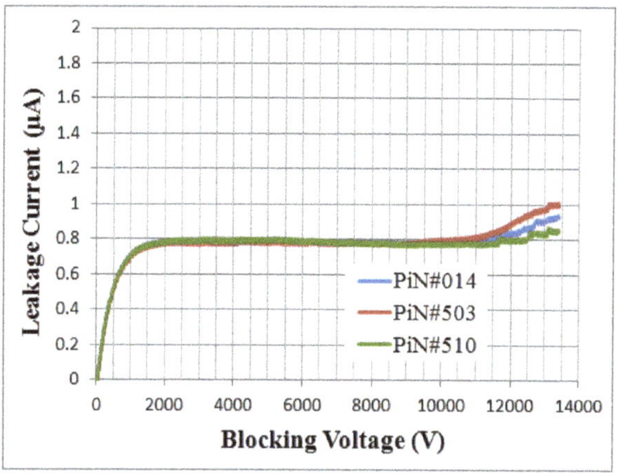

Fig. 3 Breakdown voltage for the three packaged PiN diodes. Leakage along the package's insulator limits the DC high voltage operation to 13 kV.

3. Evaluation

The PiN diodes were evaluated both in series with a thyristor and as an anti-parallel clamp. Pulse evaluations were conducted similarly to previous evaluations of 90-μm thick, 9 kV SiC PiN diodes in order to determine how pulse performance scales with drift region thickness [3].

Series Configuration: The pulse circuit employed in these evaluations generates a sinusoidal current pulse shaped by a large inductor (Fig. 4). Gating the thyristor ON initiates energy discharge from the capacitor; that energy is then dissipated in the load resistor. Configured in series with the thyristor, the PiN diode conducts the same high-action pulse current, then transitions to blocking mode to protect the asymmetric thyristor from the reverse voltage swing. The primary purpose of this evaluation was to determine what current density the PiN diode can conduct under this pulse shape and what the on-state losses are.

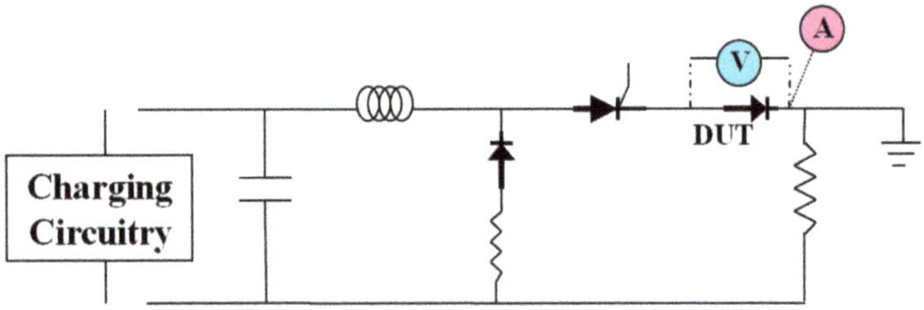

Fig. 4 Circuit topology used to generate sinusoidal current pulse. The Device Under Test (DUT) – the high voltage PiN diode – is the only silicon carbide component.

Each diode was pulsed at increasing current increments up to 2300 A (Fig. 5). At this amplitude, the action (I^2t) under the curve was 2700 A^2s, the current density over the anode was 4.4 kA/cm^2, and the peak instantaneous power was 41 kW at 470 µs from the start of the pulse. At the 2300 A peak, the forward voltage across the diode was 17 V, and remained stable over the course of ten single-shot pulses. The earlier 90-µm PiNs, by comparison, showed stable pulse capability at 3400 A per die (4.7 kA/cm^2 over the anode) with a 14 V drop [3].

Diode #014 had slightly lower on-state resistance than the other two 16 kV PiNs, and it was pulsed to higher current levels, up to 2580 A (5.0 kA/cm^2 over the anode area) (Fig. 6). Over the course of five pulses at 2580 A, the forward voltage across diode #014 increased by 5 V. This may be attributed to current-crowding in lower-resistance areas of this large chip, or to degradation of the ohmic contact causing an overall increase in resistance. The resulting 10 kW increase in instantaneous power dissipation detached the wire bonds and caused visible damage at the anode metallization. The authors intend to use focused ion beam to examine the cross section of the device and determine the depth of the failure.

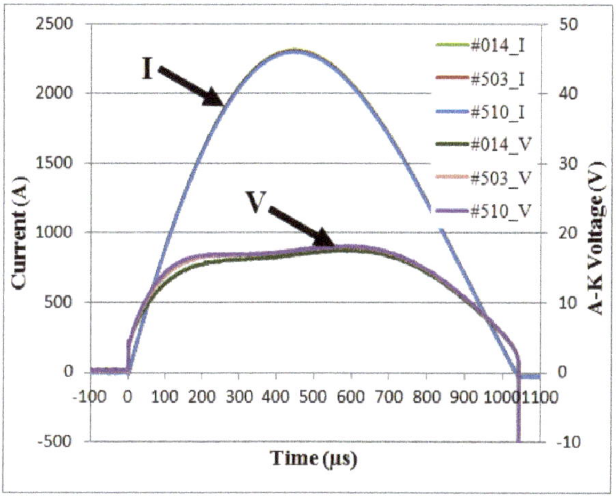

Fig. 5 Single-shot pulse current (2300 A) and on-state voltage (17 V at peak current) for three PiN diodes, as measured at the points marked "V" and "A" in Figure 4.

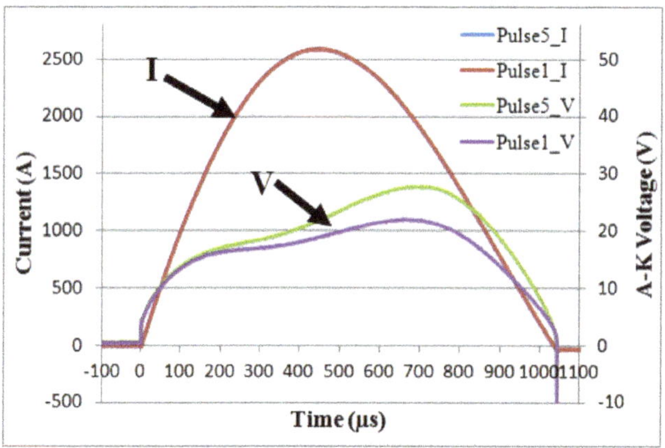

Fig. 6 Diode #014 was pushed beyond 2500 A, at which point the on-state voltage waveform indicated current-crowding effects or contact degradation, increasing by 5 V over the course of 5 pulses.

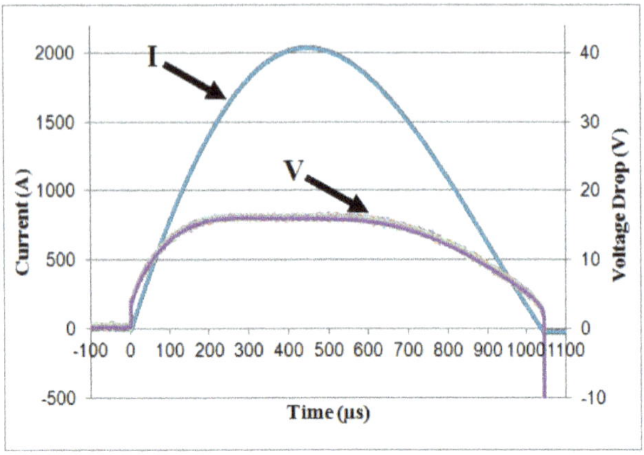

Fig. 7 Overlay of every 100th pulse from #1-1000 for diode #503, showing good repeatability at 2000 A with 16 V on-state drop. Results were identical for diode #510.

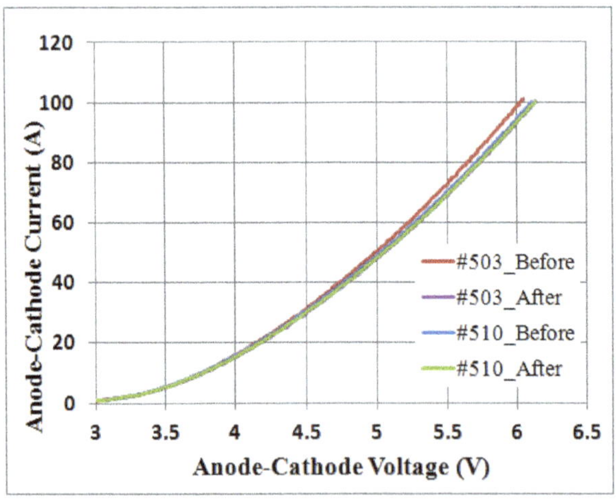

Fig. 8 Low-current curve tracer measurements for PiN diodes #503 and #510 before and after 1000-pulse switching. No notable V_F drift effects were observed after 1000 high-current pulses. The spread for device #503's voltage at 100 A is 90 mV.

Diodes #503 and #510 were individually pulsed 1000 times each at 2000 A (3.8 kA/cm^2 over the anode area or 1.9 kA/cm^2 over the chip footprint) in order to verify reliability and repeatability at that level (Fig. 7). A single-shot rate was maintained, with period no shorter than 15 seconds, so that results would not be affected by accumulating heat in the device. Both PiN diodes completed 1000 pulses without any degradation in high voltage blocking capability or V_F drift (Fig. 8), which was seen in earlier generations of 4H-SiC PiNs in which stacking faults had initiated [4].

Clamp Configuration: Diodes #503 and #510 were next connected in parallel in a position to clamp the reverse current pulse generated by the capacitor and inductor in the circuit (Fig. 9). In this configuration, both diodes were required to block the initial charge voltage on the capacitor, block the transient voltage drop across the thyristor and the load resistor when the thyristor was gated ON, and then conduct the negative current pulse that followed. The goals of this evaluation were to demonstrate PiN operation at full voltage capability (with blocking voltage restricted by the package) and to investigate current-sharing between parallel devices. Current sharing was not controlled by any passive components, only by the PiNs' on-state and package impedances. The diodes demonstrated balanced current sharing at 2100 A per device and continued high voltage blocking capability (Fig. 10).

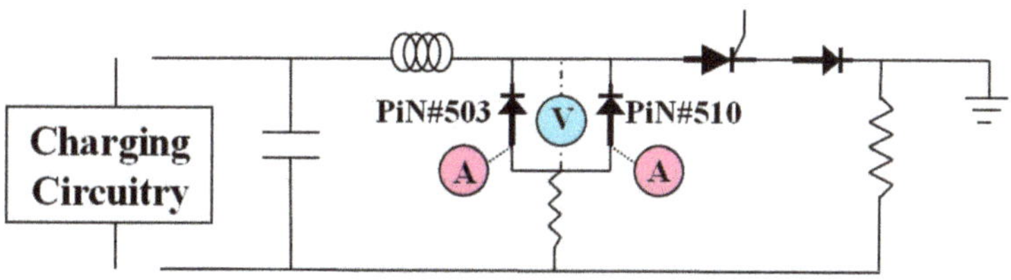

Fig. 9 Diodes #503 and #510 paralleled in the clamping position, where reverse current is steered away from the thyristor. With the diodes in this position, it was possible to obtain a 12 kV blocking waveform followed by 2100 A conduction per device.

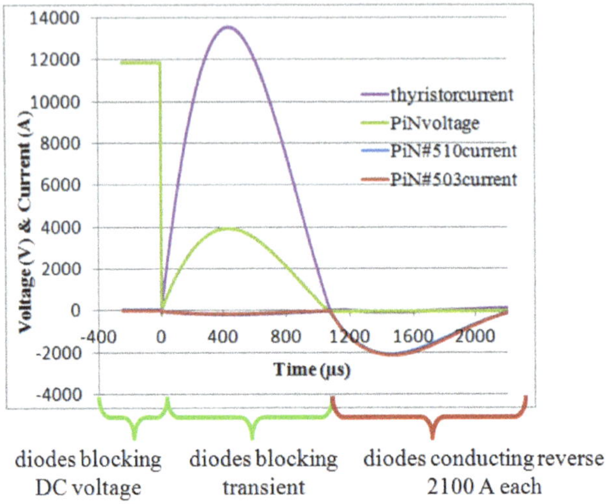

Fig. 10 Switching waveforms for the pair of diodes blocking 12 kV prior to thyristor switching, then clamping 2100 A per device following the thyristor pulse.

Summary

Combining the information from the series and anti-parallel evaluations and comparing the 120-µm SiC PiNs with 90-µm SiC PiNs and 6 kV silicon diodes pulsed in the same circuit [3, 5], it is shown that for high-current pulse operation (>1000 A/cm^2) in the range 10-15 kV, a single 120-µm epi PiN had lower on-state losses than two series 90-µm PiNs or three series silicon diodes would (Fig. 11). Utilizing a single-layer high-voltage diode in this scenario adds benefit by eliminating the need to match series devices at pulse turn-on and turn-off.

The 90-µm, 9 kV PiNs were pulsed to higher current density than the 120-µm thick devices. The packaging of the 90-µm diodes was more favorable for both spreading current evenly and drawing heat away from the semiconductor, utilizing a solder-based anode contact rather than wire bonds [3]. The thicker high-voltage PiNs were likely more limited, though, by carrier lifetime through the large drift layer. While carrier lifetimes of several microseconds have been reported thanks to Cree's lifetime enhancement process, lifetime must be further extended if 120-µm and thicker devices are to meet the pulse current density achievements of 90-µm devices.

The comparison made in Figure 11 is particular to the sinusoidal pulse condition evaluated in this study, and the authors are evaluating the 16 kV PiNs under faster switching conditions as well. The work reported here supports continuing research in material growth and device fabrication with thicker epi and developing new packaging technology for high-voltage, high-current applications.

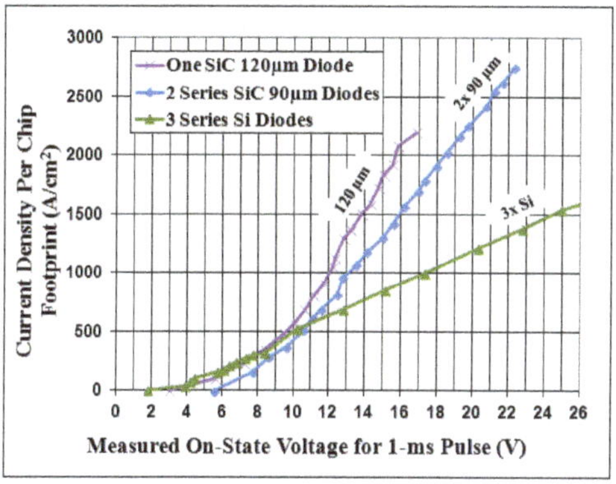

Fig. 11 Pulsed current and voltage measurements (each data point taken at mid-point of pulse, when current reaches peak amplitude). Comparison is made to the sums of two series 1.0 cm^2, 9 kV SiC PiNs and three series 3.5 cm^2, 6 kV Si diodes to show benefit/power savings.

References

[1] L. Cheng, et al, "16 kV, 1 cm^2, 4H-SiC PiN diodes for advanced high-power and high-temperature applications," Materials Science Forum, Vol. 740-742, pp. 895-898, 2013. https://doi.org/10.4028/www.scientific.net/MSF.740-742.895

[2] L. Cheng, et al., "Development of low RON,Diff, 12 kV, 4H-SiC GTOs for high-power and high-temperature applications," Proc. International Conference on High Temperature Electronics (HiTEC 2012), Albuquerque, NM, USA, May 8-10, 2012. https://doi.org/10.4071/HITEC-2012-WA13

[3] H. O'Brien, et al, "1.0 cm2 silicon carbide PiN diodes for pulsed power applications," Proc. 2010 IEEE Power Modulator and High Voltage Conference (IPMHVC), pp. 310-313, 2010. https://doi.org/10.1109/IPMHVC.2010.5958355

[4] N. Thierry-Jebali, et al, "Observation of the generation of stacking faults and active degradation measurements on off-axis and on-axis 4H-SiC PiN diodes," Appl. Phys. Letters, Vol. 101, 222111 (2012). https://doi.org/10.1063/1.4768440

[5] H. O'Brien, A. Ogunniyi, C.J. Scozzie, W. Shaheen, and V. Temple, "Silicon diode evaluated as rectifier for wide-pulse switching applications," Proc. 2009 IEEE Pulsed Power Conference, pp. 627-630, 2009. https://doi.org/10.1109/PPC.2009.5386353

Chapter 29

Study of the Turn-on of Various High-Voltage SiC Thyristors

Heather K. O'Brien[1*], Aderinto Ogunniyi[1], William Shaheen[2], and Sei-Hyung Ryu[3]

[1]U.S. Army Research Laboratory, Adelphi, MD 20783, USA

[2]Berkeley Research Associates, Beltsville, MD 20705, USA

[3]Wolfspeed, a Cree Company, Durham, NC 27703, USA

heather.k.obrien.civ@mail.mil*

Abstract

This research is focused on characterization of the turn-on transition of high voltage SiC thyristors of different epilayer thicknesses and active area sizes to determine their suitability and limitations in high-dI/dt, fast-switching applications. The unique aspects of this study include the very high current density being switched through the thyristors over a short period of time at initial turn-on, resulting in very high instantaneous dissipated power over the small device volume. The devices that were characterized were 6 kV, 0.5 cm^2 super gate turn-off thyristors (SGTOs), 10 kV, 1.05 cm^2 SGTOs, and 15 kV, 1.05 cm^2 SGTOs, all fabricated by Cree, Inc. for the Army Research Laboratory. The highest dI/dt and current density were 13 kA/microsecond and 3.2 kA/cm^2 for a parallel pair of 0.5 cm^2 thyristors, with pulse current peaking 250 ns from initial gate trigger. These evaluations help determine tradeoffs between series-stacking two lower-voltage thyristors versus using a single thicker-epi device, or paralleling two small-area devices versus switching one larger device, for fast-switching applications.

Keywords

Silicon Carbide, Pulse Power Systems, Gate Turn-off Thyristor, Thyristor, Turn-on Delay, Pulse Current, High Voltage.

Contents

The work has been previously published by IEEE:
H. O'Brien, A. Ogunniyi, W. Shaheen and S. H. Ryu, "Study of the turn-on of various high-voltage SiC thyristors," 2015 IEEE 3rd Workshop on Wide Bandgap Power Devices and Applications (WiPDA), Blacksburg, VA, 2015, pp. 5-9. doi: 10.1109/WiPDA.2015.7369283

1. Introduction

The rapid development of high-voltage, large-area silicon carbide (SiC) devices nurtures interest in comparing the switching performance of each generation in order to track advances and determine design tradeoffs for high power applications. Over the past 12 years, through collaboration between the U.S. Army Research Laboratory (ARL) and Wolfspeed (part of Cree, Inc.), SiC thyristors and PiN diodes have increased voltage blocking capability from 1.2 to 20 kV, and chip area from 0.16 to 2.0 cm^2. These SiC devices are of particular interest to pulsed power, when switches experience very high instantaneous power for short durations in order to transfer high voltage and high current to a circuit load. SiC has attractive material properties that support very high voltage-blocking across a relatively thin epilayer, withstand high instantaneous power and temperature, and enable fast switching speeds [1, 2].

The parameters controlling thyristor turn-on vary for the multiple device generations considered in this study. These parameters include epilayer thickness, gate layer thickness, carrier lifetime, and geometric layout. The goal of this study was to investigate the turn-on mechanisms of real high-voltage devices under high current density and fast switching. By comparing turn-on delays, minimum on-resistance, and maximum dI/dt capability, tradeoffs can be identified for selecting optimum device chip area and voltage rating for each pulse power application.

Due to carrier injection limitations and device geometry, there is a finite rate at which the thyristor area can be driven to the minimum-resistance on-state. Established circuit-based means of working around this limit include compensating the passive components (especially the inductance) of the overall circuit in order to limit the rate at which high current can flow, or adding additional components, such as a saturable inductor, to delay full current flow until the thyristor has reached minimum on-resistance. This paper rather focuses on the simplest condition with minimum number of components and minimum volume, where short turn-on delay or high-frequency switching (kHz to MHz) is desired, and the circuit cannot be slowed.

2. Device evaluation

High Voltage GTO Devices: The SiC thyristors that are evaluated and compared are all past generations of research-level super gate turn-off thyristors (SGTOs) fabricated by Cree, Inc., now Wolfspeed. The devices were designed with an attempt to satisfy the widest possible range of high-voltage, high-current needs. The design details of the devices characterized in this study have all been previously published by ARL and Cree [3, 4, 5, 6]. The most recent device included here is the 15 kV, 1.05 cm^2 SGTO, preceded by the 10 kV, 1.05 cm^2 SGTO, and the 6 kV, 0.5 cm^2 SGTO (Fig. 1). A selection of relevant device characteristics is contained in Table I. In order to make fair switching comparisons, two 0.5 cm^2 SGTOs were switched in parallel for comparable total SiC footprint and active area.

Fig. 1 Three thyristors that were compared at turn-on. Clockwise from top: 6 kV, 0.5 cm^2; 15 kV, 1.05 cm^2; and 10 kV, 1.05 cm^2.

Table 1 SiC Device Specifications.

SGTO Blocking Voltage (kV)	Chip Footprint (cm^2)	Active Area (cm^2)	Epi Thickness (μm)	Average Lifetime (μs)
15	1.05	0.52	120	4
10	1.05	0.73	90	1.4
6	0.5	0.36	60	<1

Evaluation Circuitry: The pulse circuit is a simple R-L-C circuit designed and constructed to switch fast and not restrain the current rise time (Fig. 2). Loop inductance was managed by laying out a wide anti-parallel buss between the capacitor and the load. An energy storage capacitor that was designed for rapid, high-current discharge was used to briefly hold the charge before the SGTO was triggered. The large-area bulk ceramic load resistors are components specifically designed to dissipate high energy. A SiC 15 kV, 1.05 cm^2 P-i-N diode was connected in anti-parallel with the SGTO to conduct reverse current. For these evaluations, the focus was on fast-switching time combined with high dI/dt, rather than trying to match the circuit speed to the plasma spread across the device area. The circuit will later be modified to increase the ratios of the capacitance and inductance relative to the load resistance, to determine whether a controlled slower rising current rate would enable higher overall dI/dt by delaying the high current density the device sees. Due to different package dimensions for the smaller 6 kV SGTO, as well as the desire to connect two of these devices in parallel, it was necessary to modify the SGTO connection to the circuit in order to evaluate these devices, resulting in a small increase in overall inductance (Fig. 3).

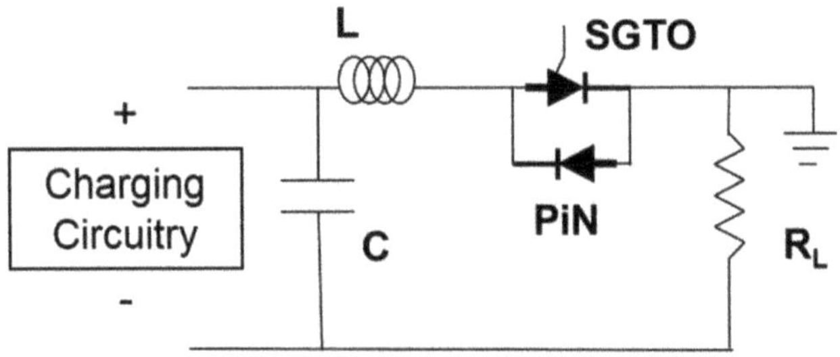

Fig. 2 Circuit schematic for high-current pulse evaluations: R_L= 0.75 ohm, C= 0.15 µF, L~ 100 nH (including component parasitics).

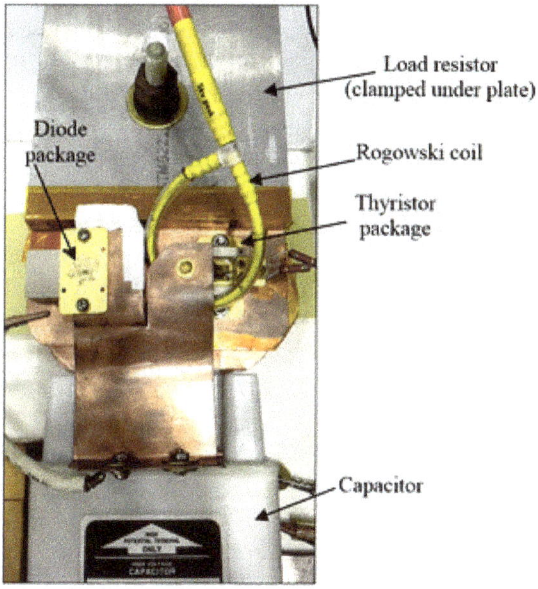

Fig 3. 0.5 cm^2 SGTO in pulse circuit (top) with anti-parallel diode (bottom), energy storage capacitor (right), load resistor (left) and probes.

3. Minimum turn-on delay

The turn-on delay time (t_d) of the SGTO is determined by the base transit times of each of the two transistors that compose the structure of the thyristor. It is dependent on the base widths of each transistor (W_{BP} and W_{BN}) and the diffusion rates of the electrons and holes (D_n and D_p), as shown in (1) [7].

$$t_d = ½(W_{BP}^2/D_n) + ½(W_{BN}^2/D_p) \qquad (1)$$

The subsequent lateral propagation or spreading velocity (υ_s) of the conduction across the full area of the device additionally depends upon the current density (J), charge (q), and carrier lifetime (τ_{HL}), and is affected by temperature and lateral geometry (2) [7].

$$\upsilon_s = (J/q)*(W_N+W_P)^{-1}*(\tau_{HL}/D_a)^{½} \qquad (2)$$

ARL previously published a study of earlier-generation 9 kV, 1.0 cm^2 SiC SGTOs, which found that variation in turn-on delay from device-to-device, for a given design, could be partially overcome by driving the SGTO at higher peak gate current and gate dI/dt slope [8]. In this new evaluation, the SGTOs were driven at even higher gate currents, ranging from 10 to 120 A. The driver circuit consisted of a floating 100 V DC supply maintaining charge on a 110 μF capacitor bank, which was connected across the SGTO's anode-gate junction via a fast MOSFET at trigger time. Series gate resistance at the SGTO was varied from 10 ohm, to 1.5 ohm, to 0.75 ohm. These gate resistors corresponded to peak applied gate currents of 10 A, 64 A, and 128 A, respectively (Fig. 4). Decreasing the gate resistor from 10 ohm to 1.5 ohm increased the front edge of the gate dI/dt from 200 to 400 A/μs, and shortened the delay to cathode current flow from 100 to 75 ns, measured from the time when gate current began to rise (t= 175 ns in all figures in this paper) (Fig. 5). Further reducing the gate resistor to 0.75 ohm, in this case, did not change the gate dI/dt in the first 20 A of rising gate current, and did not shorten the turn-on delay any further. It was decided to standardize all pulse current evaluations to the 0.75 ohm gate drive. For switching the two 0.5 cm^2 SGTOs in parallel, two 1.5 ohm resistors in parallel were used to split the gate current to each device.

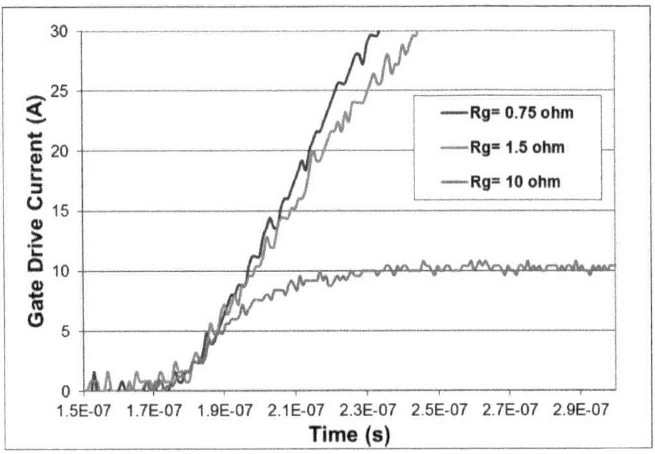

Fig. 4 Focus on the rising edge of different gate drive currents used to determine minimum turn-on delay. The rising slope of the Rg= 10 ohm (10 A) gate current is 200 A/μs, while the rising edge of the Rg= 1.5 ohm and Rg= 0.75 ohm (resulting in peak 64 A and 128 A, respectively) currents are both approximately 400 A/μs.

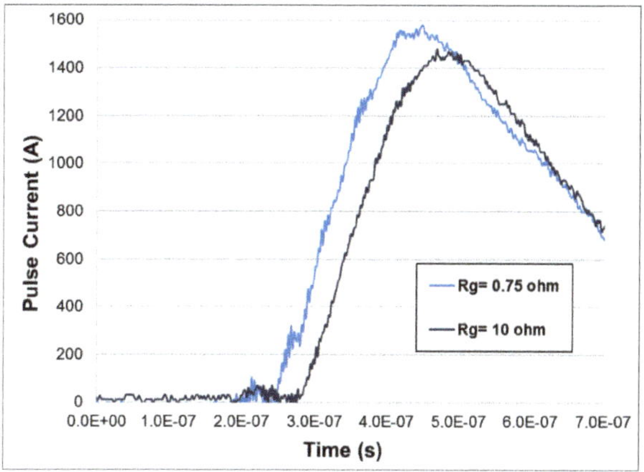

Fig. 5 Increasing the front edge of the gate current to 400 A/μs enabled the SGTOs to turn on 25 ns earlier, starting to conduct pulse current at 75 ns after initial gate current. Example waveforms are shown for the 10 kV, 1.05 cm^2 SGTO with a 3000 V charge on the pulse capacitor.

With sufficient gate drive and pulse current density to minimize the turn-on time, the cathode current delay time was approximately the same for each of the three types of SGTO evaluated (Fig 6). Pulse current began to rise 75 ns after the initial rise in gate current. The time at which the pulse current peaked varied for each device type, as expected, due to varied on-resistance defined by the different epi thicknesses and carrier lifetimes (Fig. 7). The current peaked at 250 ns after the gate signal initiated for the parallel 6 kV, 0.5 cm^2 SGTOs, 260 ns for the 10 kV, 1.05 cm^2 SGTO, and 270 ns for the 15 kV, 1.05 cm^2 SGTO. For many pulse current applications, time difference of 20 ns would not be a hindrance to power delivery to the circuit load, or at least could be accounted for in system-level timing controls. This time spread would be unlikely to force the circuit designer to choose between SGTO designs, but rather allow one to use whichever design is most appropriate for the high operating voltage level.

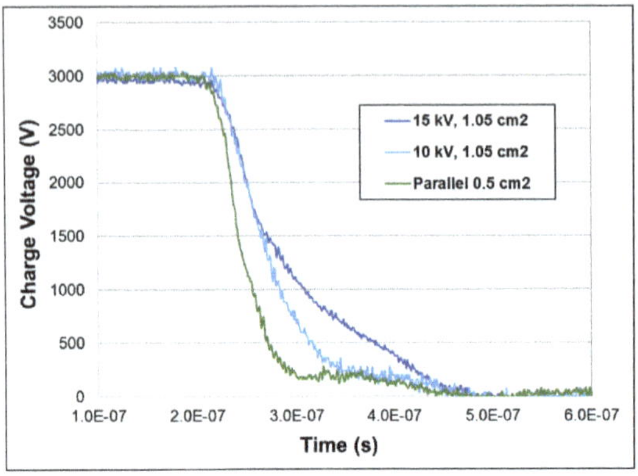

Fig. 6 For all devices evaluated, high voltage began to fall across the device at 40 ns after gate current was initiated. The 3000 V charge level is shown as an example. The slope at which the voltage continued to fall once the SGTO began to conduct high current was affected by each device's epi thickness, carrier lifetime, and instantaneous cathode current density.

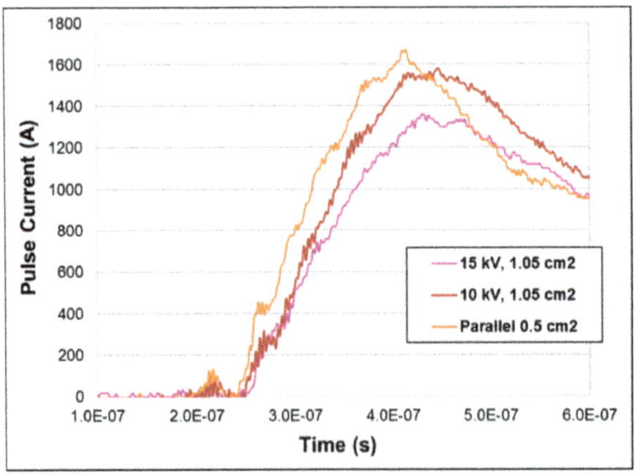

Fig. 7 Variation in peak current amplitude (Δ 200 A), peak current timing (Δ 20 ns), and dI/dt (Δ 4 kA/μs) for the three different generations of devices at the same charge level.

.

4. Turn-on propagation

Spreading velocity is the rate at which the plasma spreads across the area of the thyristor. It affects the anode current rate of rise, which affects the rate at which a given device reaches its minimum on-state resistance during a pulsed switching event. The spread velocity also affects the maximum current density a device can conduct without exceeding the material's catastrophic power or thermal limit [9]. For slow, low-frequency current pulses with rise times on the order of hundreds of amperes per microsecond, even a large 1.05 cm^2 thyristor can be in full conduction by the time pulse current amplitude peaks. For fast switching, though, the percent of device area in conduction will limit peak pulse current density. If too much current is forced through the switch when only the first few cells are conducting, high localized current density will lead to localized heating and to current saturation or failure. Part of the motivation of this study was to determine whether lateral propagation rate limits the chip size one should use in the fastest switching applications. Each device was pushed to the saturation limit under the fast-switching condition. Saturation conditions are detailed in Table 2.

Table 2 Current Saturation Under Fast Turn-on Condition

SGTO Blocking Voltage (kV)	Active Area (cm^2)	Peak Pulse Current (A)	Current Density (kA/cm^2)	Peak Instantaneous Power (kW)
15	0.52	1370	2.6	800
10	0.73	1830	2.5	900
6	0.36	1480	4.1	500

5. Peak dI/dt under fast turn-on condition

The peak dI/dt values achieved with each SGTO are particular to the circuit in which the devices were evaluated. Slower-rising, wider pulse waveforms will yield different results. At capacitor charge levels beyond the ones depicted here, each device entered a saturation condition, where voltage across the device fell at a slower rate once current density through the device began to rise. Peak dI/dt for each device at the fast turn-on condition is depicted in Fig. 8 through 10. As expected, the fastest rise was achieved with the SGTOs with the thinnest epi – the parallel 6 kV, 0.5 cm^2 SGTOs. The dI/dt was 13 kA/μs measured across the 10-90% of the peak amplitude and 14 kA/μs across the fastest portion between 0-20% of peak amplitude.

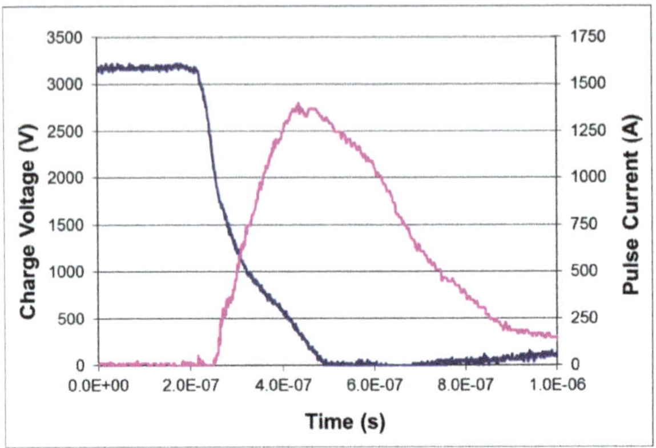

Fig. 8 For the 15 kV, 1.05 cm² SGTO, the peak dI/dt in this circuit was reached at a current amplitude of 1370 A. The dI/dt measured from 10-90% of maximum current level was 8 kA/μs.

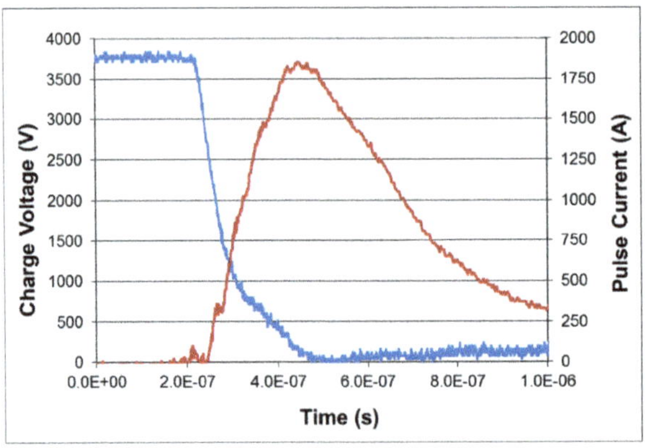

Fig. 9 For the 10 kV, 1.05 cm² SGTO, the peak dI/dt in this circuit was reached at a current amplitude of 1830 A. The dI/dt measured from 10-90% of maximum current level was 11 kA/μs.

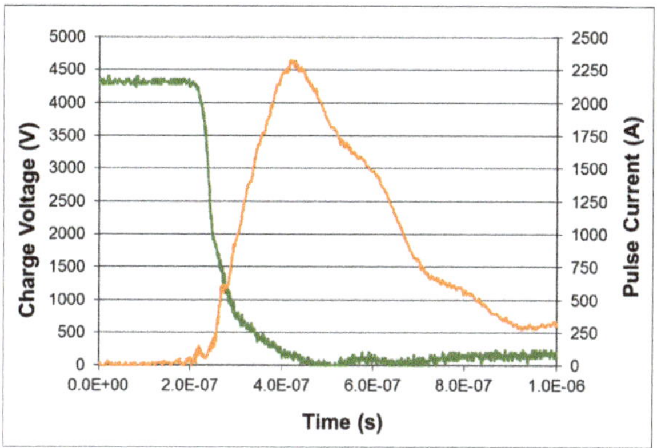

Fig. 10 For the two parallel 6 kV, 0.5 cm^2 SGTOs, the peak dI/dt in this circuit was reached at a total current amplitude of 2300 A. The dI/dt measured from 10-90% of maximum current level was 13 kA/μs

Summary

The results of this study suggest that for practical application in fast-switching pulse circuits, the SGTO design should be selected based upon the required high-voltage blocking and dI/dt of the application. In terms of turn-on delay and total turn-on from gate trigger to pulse current peak, the difference on the order of 10's of nanoseconds would not rule out any of these devices from use. Each device's capability to switch high dI/dt pulse current in relatively short time primes a follow-on investigation of whether they can endure burst-rate switching at frequencies exceeding 1 kHz.

References

[1] R. Singh, "High power SiC PiN rectifiers," International Journal of High Speed Electronics and Systems, vol. 15 no. 4, pp. 867-898, 2005. https://doi.org/10.1142/S0129156405003442

[2] M. A. Rezaei, G. Wang, A. Q. Huang, L. Cheng, and C. Scozzie, "Static and dynamic characterization of a >13 kV SiC p-ETO device," Proc. IEEE ISPSD, pp. 354-357, 2014. L. Cheng, et al, "15 kV, large area (1 cm2), 4H-SiC p-type gate turn-off thyristors," Materials Science Forum, vol. 740-742, pp. 978-981, 2013.

[3] L. Cheng, et al, "Advanced silicon carbide gate turn-off thyristor for energy conversion and power grid applications," Proc. 2012 IEEE ECCE, pp. 2249-2252, 2012. https://doi.org/10.1109/ECCE.2012.6342435

[4] A. Agarwal, et al, "9 kV, 1 cm x 1 cm SiC super gto technology development for pulse power," Proc. 2009 IEEE Pulsed Power Conf., pp. 264-269, 2009. https://doi.org/10.1109/PPC.2009.5386305

[5] H. O'Brien, et al, "Wide-pulse evaluation of 0.5 cm2 silicon carbide SGTO," Proc. 2009 IEEE Pulsed Power Conf., pp. 260-263, 2009.

[6] B.J. Baliga, Fundamentals of Power Semiconductor Devices, PWS Publishing, Boston, 1996.

[7] S.M. Sze and K.K. Ng, Physics of Semiconductor Devices, Wiley, Hoboken,2007.

[8] H. O'Brien, et al, "High dI/dt pulse switching of 1.0 cm2 SiC GTOs," Materials Science Forum, vol. 717-720, pp. 1155-1158, 2012. https://doi.org/10.4028/www.scientific.net/MSF.717-720.1155

Chapter 30

Evaluation of High-Voltage, High-Power 4H-SiC Insulated-Gate Bipolar Transistors

Miguel Hinojosa [1,2], Aderinto Ogunniyi[2], Heather O'Brien[2], Stephen B. Bayne[1] and Charles Scozzie[2]

[1]P^{3}E, Texas Tech University, Lubbock, TX 79409, USA

[2]U.S. Army Research Laboratory, Adelphi, MD 20783, USA

miguel.hinojosa4.civ@mail.mil*

Abstract

This section presents preliminary results on the static and dynamic characterization of 12-kV and 20-kV N-channel silicon carbide Insulated-Gate Bipolar Transistors (IGBTs). These state-of-the-art devices were evaluated for their possible use in pulsed-power and energy conversion applications. The 12-kV IGBTs had a chip area of 0.7 cm^2 and were rated for 10 A. Their active area was 0.32 cm^2, with a drift region of 140 μm and two different field-stop buffers of 5 μm and 2 μm. The 20-kV IGBTs had a chip area of 1 cm^2 and were rated for 12 A. Their active area was 0.37 cm^2, with a drift region of 180 μm, and their FSB was 2 μm. The switching and conduction losses were calculated for both devices with short pulses and low-inductance resistive loads. Both types of IGBTs displayed promising results for possible replacement of gas switches and Si IGBTs in high voltage applications.

Keywords

Silicon Carbide (SiC), High voltage solid state switch, 20-kV 4H-SiC, IGBT, insulated-gate bipolar transistor, pulsed testing, clamped inductive switching.

Contents

The work has been previously published by IEEE:
M. Hinojosa, A. Ogunniyi, H. O'Brien, S. B. Bayne and C. Scozzie, "Evaluation of high-voltage, high-power 4H-SiC insulated-gate bipolar transistors," *2014 IEEE International Power Modulator and High Voltage Conference (IPMHVC)*, Santa Fe, NM, 2014, pp. 101-104. doi: 10.1109/IPMHVC.2014.7287217

1. Introduction

The realm of high voltage and high power applications has been historically reserved for gaseous electronic switches. Vacuum tubes, spark gaps, ignitrons, and thyratrons have been the switches of choice for many years due to their outstanding ability to withstand very high voltages. Solid-state power devices utilizing silicon (Si) have slowly replaced gas switches in many fields with improved reliability, switching speed, space, and cost savings. Yet, there are still many limitations because as the operating voltages increase, a large number of devices must be stacked in series-parallel combinations, which increases system volume and cooling requirements. The U.S. Army Research Laboratory (ARL) is investigating the utilization of solid-state, silicon carbide (SiC) devices as a possible replacement for Si power devices and gas switches. The enhanced properties of this wide-bandgap material include the following benefits: a high critical electric field that allows operation at high voltages and high switching speeds; reduced intrinsic carrier concentration that allows stable operation at high temperatures; high thermal conductivity that provides rapid removal of heat and operation at high current densities; and high radiation resistance. This section presents the evaluation of a small number of experimental N-channel, silicon carbide Insulated-Gate Bipolar Transistors (IGBTs) that were developed under a collaborative agreement between ARL, Cree Inc., and Texas Tech University. In order to test the devices, a series of custom-made high voltage evaluation stations were designed and constructed in-house which provided the necessary infrastructure to test the IGBTs safely. Preliminary laboratory results of the steady-state and transient characteristics for two 12-kV and two 20-kV 4H-SiC IGBTs are included in this work. These devices have some of the highest breakdown voltages of a single MOS-controlled semiconductor reported in literature to this date [1][2][3].

2. High Voltage IGBT Structure

The 12-kV and 20-kV silicon carbide IGBTs were designed and manufactured by Cree Inc. and they were packaged in a high temperature PEEK plastic enclosure as seen in Figure 1. The device structure for both devices was previously introduced in a similar fashion as shown in Figure 2. Both devices had a cell pitch length of 19 µm, a channel length of 1 µm, and an oxide thickness of 500 Å. Their p-well regions were doped with 5 x 10^{17} cm^{-3}, and their N+, P+, and P++ regions with a very high doping concentration (>10^{19} cm^{-3}). Both devices utilized an N-type drift region and an asymmetrical punch-through design. The 12-kV IGBTs had a chip area of 0.7 cm^2 and active area of 0.32 cm^2. The high voltage blocking capability was accomplished with a wide epitaxial drift region of 140 μm at a concentration of 2 x 10^{14} cm^{-3} and with guard ring edge terminations. The 20-kV IGBTs had a chip area of 1 cm^2, active area of 0.37 cm^2, and a wide epitaxial drift region of 180 μm at a concentration of 2 x 10^{14} cm^{-3}. A thin field-stop buffer (FSB) layer was present on both designs, which controlled the injection of holes from the P+ layer during conductivity modulation. The concentration of this layer was 5 x 10^{16} cm^{-3}. The 12-kV device had two different thicknesses of 5 μm and 2 μm, while the 20-kV devices were fixed to 2 μm. The P+ backside injection layer was connected to the collector electrode and it had a doping concentration of 1 x 10^{19} cm^{-3}.

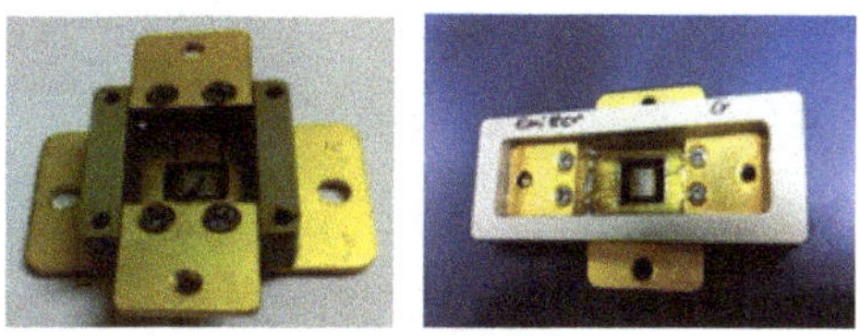

Fig. 1. *Packaged 12 kV IGBT (left) and 20 kV IGBT (right).*

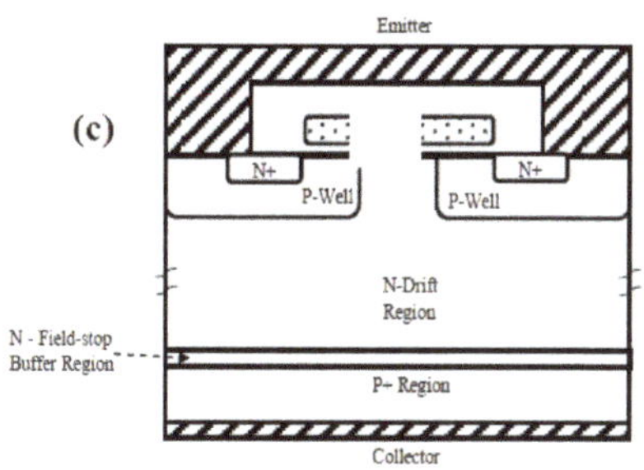

Fig. 2. *Structure of N-type IGBT.*

3. Evaluation Circuits

Hybrid Parametric Device Analyzer: A hybrid parametric device analyzer was developed at ARL to measure the conduction and blocking characteristics of these high voltage IGBTs. Because commercially available parametric curve tracers were limited in the high voltage range, a hybrid approach was implemented to obtain the two-modes of device operation. The forward conduction I-V plots were obtained with a Tektronix 371B high-power curve tracer and the blocking curves with a custom-made FPGA-controlled high voltage curve tracer. This configuration offered up to 3 kW for forward conduction I-V plots, and up to 30 kV and 300 μA for blocking. The HV curve tracer also featured current and voltage compliance guards with active monitoring to prevent accidental destruction during avalanche testing. The overall system diagram can be seen in Figure 3.

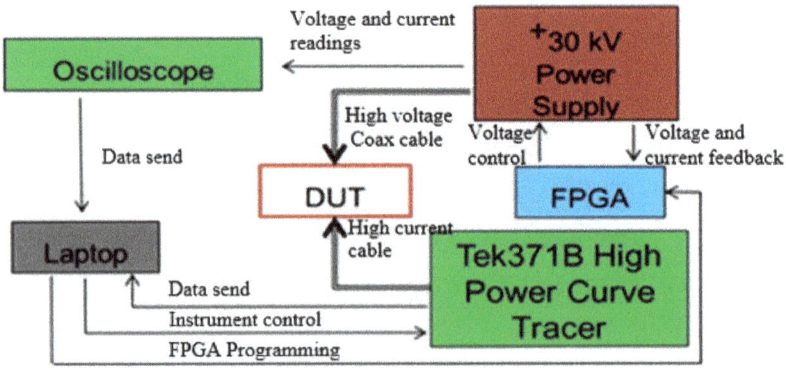

Fig. 3. *Block diagram of the hybrid parametric analyzer.*

IGBT Evaluation Station: The dynamic characteristics of the IGBTs were tested with the 7 kJ high voltage evaluation station shown in Figures 4 and 5. This station consisted of a 15-kV capacitor charger, two series-connected 8 kV protection puck diodes, a 64 μF capacitor bank, four 10 MΩ balancing resistors, a reconfigurable load, a gate driver board, and the device under test (DUT). In this particular instance, the DUT was setup in a common-emitter configuration with a resistive load tied to its collector terminal. The gate driver was designed based on the IXYS IXDD430 IC and it was triggered externally via fiber optic cable. There were two types of resistive loads used to test the IGBTs. For the 12 kV test, nine 86 Ω low-inductive carborundum disk resistors were stacked together to make an equivalent resistance of 774 Ω and obtain a load current of 10.3 A. For the 20 kV IGBTs, two tubular carborundum resistors were used to get an equivalent resistance of 1.2 kΩ and a load of 12.5 A.

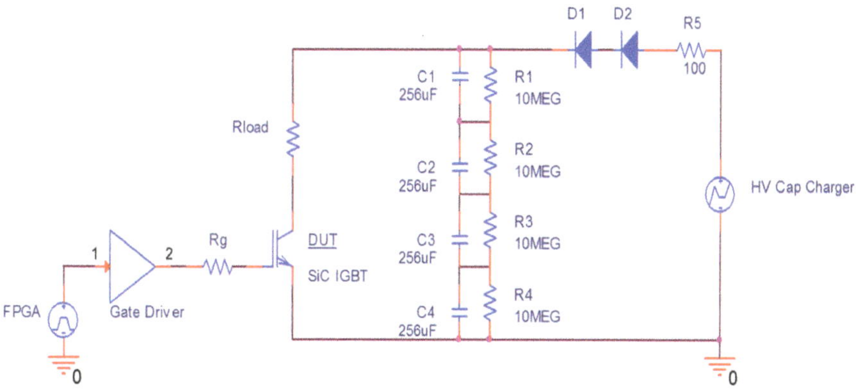

Fig. 4. *Schematic of the high voltage evaluation station.*

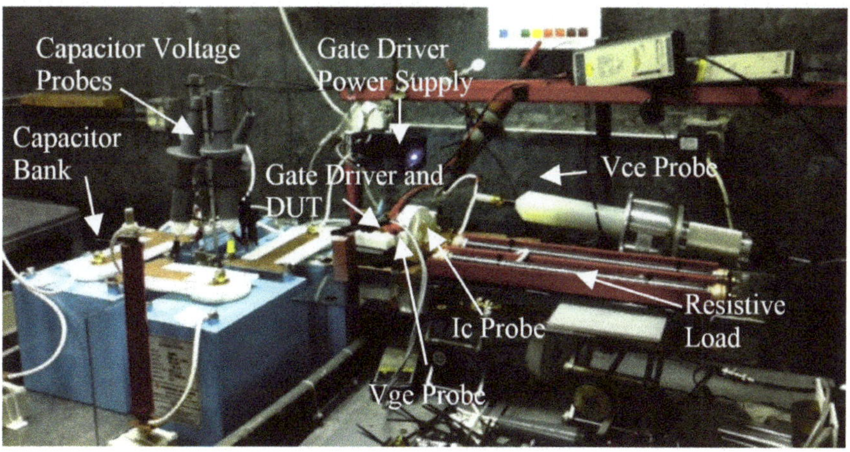

Fig. 5. *Image of the high voltage evaluation test-bed.*

Results for the 12-kV IGBTS: The forward conduction current-voltage characteristics were obtained for the two 12-kV IGBTs as seen in Figures 6 and 7. The device with the 5-µm FSB was labeled dev1_12kv, and the one with the 2-µm FSB was labeled dev2_12kv. The devices had the gate voltage stepped from 0-20 V in increments of 5 V and their collector voltage swept from 0-21 V. The graphs show the knee that is characteristic of IGBTs, and it begins to rise when the collector voltage reaches approximately 2.8 V. This knee is attributed to the built-in potential that must be overcome at the PN junction found between the P+ epitaxial layer and the n-type FSB, before the device turns on. The I-V plots also show some differences between the two devices: The device with the 2-µm FSB, dev2_12kv, was expected to have lower conduction losses because of its thinner buffer layer and higher injection rate as explained in [2]. The field-stop buffer was utilized as a mean to control the injection of holes from the P+ region into the drift region. The thicker FSB delayed the delivery of holes longer than the one with the thinner FSB, and this effectively controlled the conductivity modulation. On the curve where Vge was 20 V and Ic was 50 A, for instance, the dev2_12kv device shows a turn-on voltage of 8.4 V while the dev1_12kv device shows 10.4 V. When Vge was equal to 15 V and 10 V, the current magnitude of the thinner FSB is less, however. In this case, MOS channel effects cause the curves leading to the active region to bend prematurely. When Vge was 15 V and Ic was 50 A, dev2_12kv had a turn-on voltage of 12.4 V whereas dev1_12kv had a turn-on voltage of 11.6 V. When Vge was 10 V and Ic was 10 A, dev2_12kv showed a voltage drop of 7.2 V and dev1_12kV showed 6.4 V.

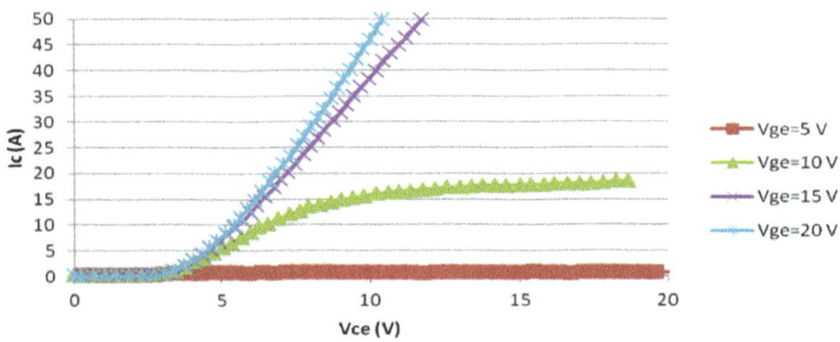

Fig. 6. *I-V curves for the 12-kV, 5-µm FSB IGBT (dev1_12kv).*

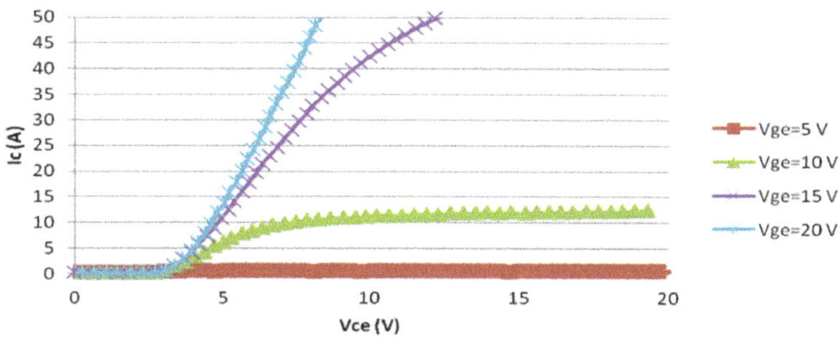

Fig. 7. *I-V curves for the 12-kV, 2-µm FSB IGBT (dev2_12kv).*

The extrapolation technique in the linear region of the transfer curve, seen in Figure 8, was used to extract the threshold voltage. For this measurement the collector voltage was set to 3.2 V, and the gate voltage was swept from 0 to 7 V. The threshold voltage for dev1_12kv was 3.10 V and 3.38 V for dev2_12kv.

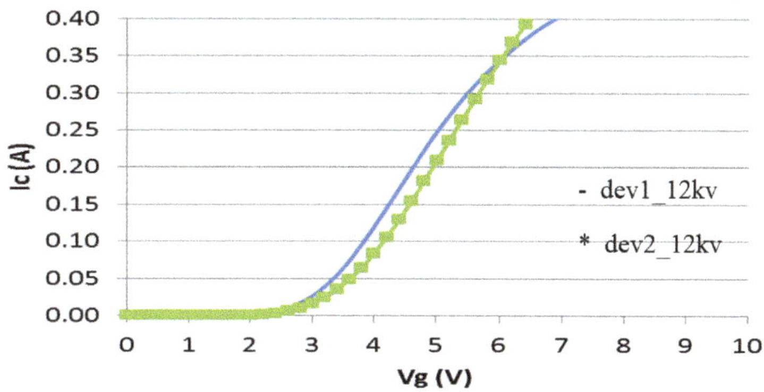

Fig. 8. *Transfer curve for the 12 kV SiC IGBTs.*

The forward blocking characteristics were obtained by applying zero volts to the gate of the DUT and sweeping the collector voltage up to the breakdown voltage. The dev1_12kv IGBT had a leakage current of 38 μA at 12.5 kV and the dev2_12kv IGBT had a leakage current of 8 μA at 12.1 kV as seen in Figure 9.

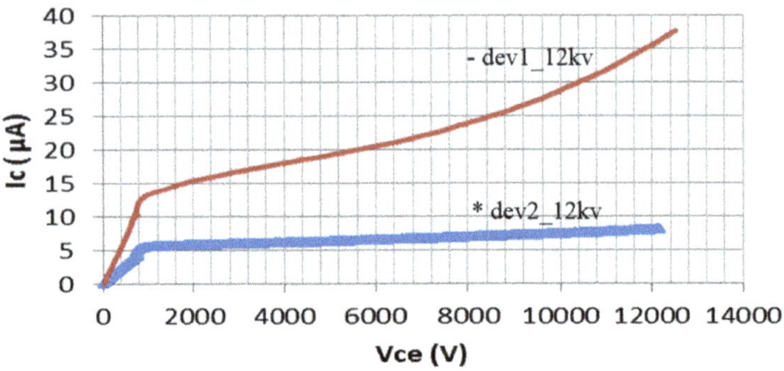

Fig. 9. Forward blocking curves for the 12 kV IGBTs.

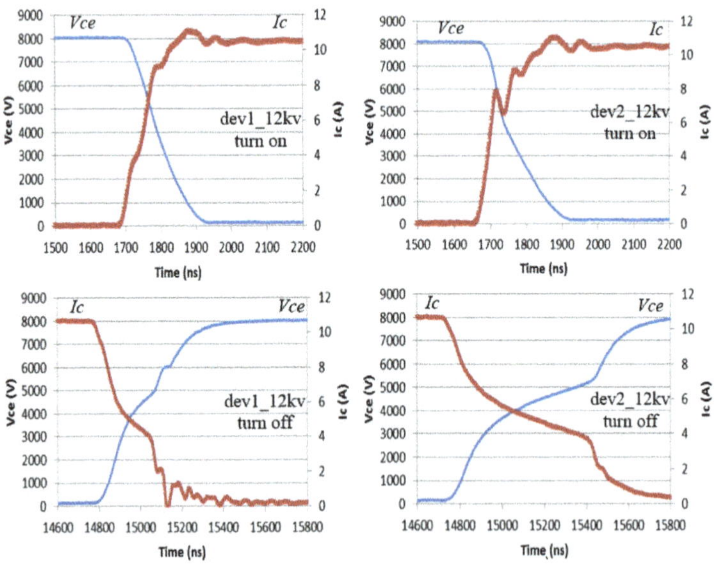

Fig. 10. Switching waveforms for dev1_12kv (L) and dev2_12kv (R).

Various tests were carried out in the lab with various supply voltages, currents, pulse widths, and loads. In this section, the case where the supply voltage was 8 kV, the collector current was approximately 10 A, the pulse width was 13 μs, and the load was 774 Ω is presented. The gate driver voltage was set from -5 V to 15 V as per (Si) industry standard, and the gate resistor was set to 10 Ω. Figure 10 shows the turn-on and turn-off waveforms and Table 1 gives a summary of the results.

Table 1. *Summary of results for 12-kV IGBT resistive switching.*

dev1_12kv (5 μm FSB)	dev2_12kv (2 μm FSB)
▪ Vsupply = 8 kV	▪ Vsupply = 8 kV.
▪ Ic = 10.69 A.	▪ Ic = 10.7 A.
▪ Rload = 774 Ω	▪ Rload = 774 Ω
▪ Rg = 10 Ω	▪ Rg = 10 Ω.
▪ Vsat = 5.82 V	▪ Vsat = 4.9 V
▪ Vfalltime = 150 ns.	▪ Vfalltime = 159 ns.
▪ Irisetime = 120 ns.	▪ Irisetime = 133 ns.
▪ Vrisetime = 390 ns.	▪ Vrisetime = 778 ns.
▪ Ifalltime = 306 ns.	▪ Ifalltime = 765 ns.
▪ Eon = 4.73 mJ.	▪ Eon = 4.17 mJ.
▪ Eoff = 5.91 mJ.	▪ Eoff = 15.55 mJ.
▪ Econd = 0.8 mJ (single pulse)	▪ Econd = 0.69 mJ (single pulse)

The switching results show advantages and disadvantages for both types of devices. The IGBT with the 5-μm FSB showed higher turn-on voltages because its injection of holes was reduced intentionally with the thicker buffer. The thinner 2-μm device shows lower turn-on voltages because of the drop in resistance in its drift region due to conductivity modulation (higher injection). However, another interesting observation was that the turn-on and turn-off energy losses for the dev1_12kv IGBT were very similar, in contrast to dev2_12kv which shows a significant energy loss during turn-off. This energy loss was caused by the lingering current tail which is attributed to the high injection. When the device turns off, the excess carriers in the drift region have to wait to be extracted and recombine.

Results for the 20-kV IGBTS: The forward conduction I-V characteristics were obtained for two 20-kV IGBTs with a fixed 2-μm FSB and they are presented in Figures 11 and 12. The dev1_20kv IGBT showed a slightly better performance than the dev2_20kv device because of its lower resistivity. At Ic = 50 A, dev1_20kv showed a Vce of 15 V while dev2_20kv showed a Vce of 18 V. The difference is attributed to the longer carrier lifetime, which increases the conductivity modulation in the drift region.

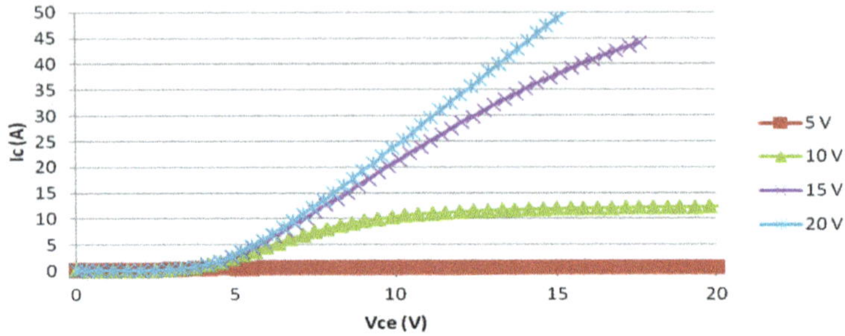

Fig. 11. *I-V curves for dev1_20kv IGBT.*

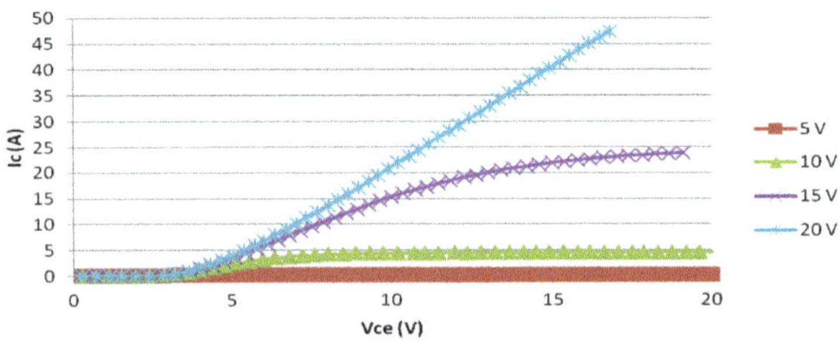

Fig. 12. *I-V curves for dev2_20kv IGBT.*

The threshold voltages for the 20-kV devices were measured with the same process previously described for the 12 kV IGBT. The threshold voltage for the dev1_20kv was 2.6 V and 3.25 V for the dev2_20kv.

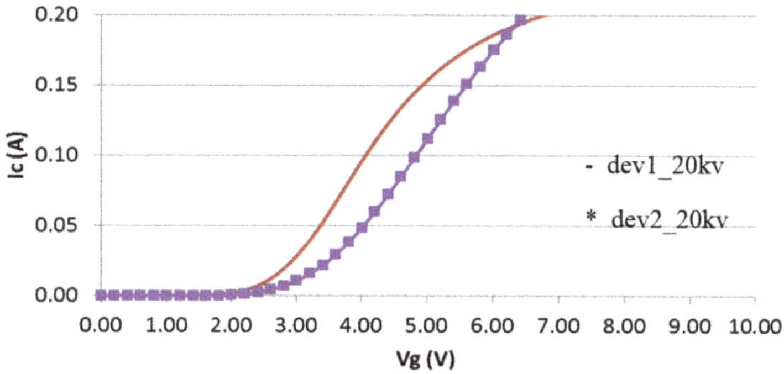

Fig. 13. Transfer curve for 20-kV IGBTs.

The forward blocking characteristics were obtained by applying zero volts to the gate of the DUT and sweeping the collector voltage up to the breakdown voltage. The dev1_20kv IGBT had a leakage current of 5.5 μA at 20.4 kV and the dev2_20kv IGBT had a leakage current of 3.4 μA at 22 kV.

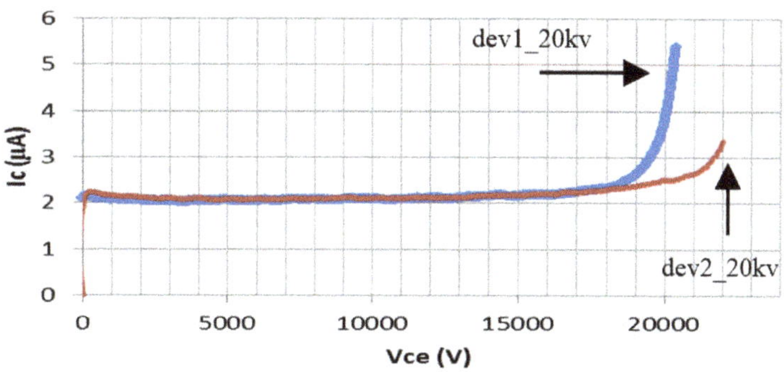

Fig. 14. Forward blocking plots for 20 kV IGBTs.

The dynamic response of the 20-kV IGBTs was tested with a resistive load of 1.2 kΩ. This value was selected to obtain 12 A with a supply voltage set to 15 kV. The turn-on and turn-off waveforms for both devices are shown in Figure 15, and a summary of the results in Table 2. Several tests were carried out in the lab with various supply voltages, currents, pulse widths, and loads. In this section, the case where the supply voltage was 8 kV, the collector current was approximately 10 A, the pulse width was 13 μs, and the load was 774 Ω is presented. The gate driver voltage was set from -5 V to 15 V as per (Si) industry standard, and the gate resistor was set to 10 Ω.

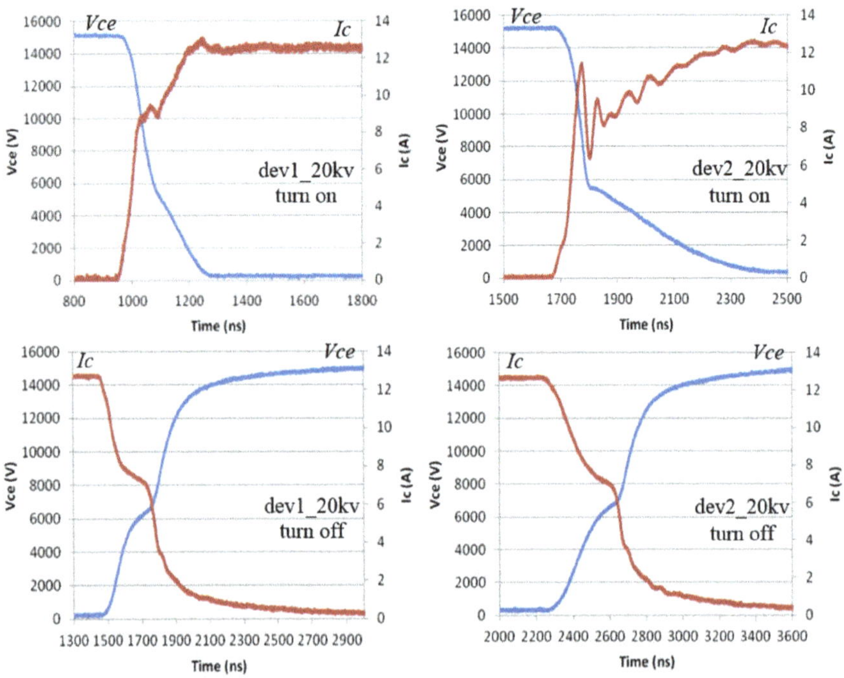

Fig. 15. *Switching waveforms for dev1_20kv (L) and dev2_20kv (R).*

Table 2. *Summary of results for 20 kV IGBT resistive switching.*

dev1 20kv	dev2 20kv
▪ Vsupply = 15 kV	▪ Vsupply = 15 kV.
▪ Ic = 12.5 A.	▪ Ic = 12.5 A.
▪ Rload = 1.2 kΩ	▪ Rload = 1.2 kΩ
▪ Rg = 10 Ω	▪ Rg = 10 Ω.
▪ Vsat = 7.6 V	▪ Vsat = 8.9 V
▪ Vfalltime = 210 ns.	▪ Vfalltime = 450 ns.
▪ Irisetime = 200 ns.	▪ Irisetime = 430 ns.
▪ Vrisetime = 500 ns.	▪ Vrisetime = 540 ns.
▪ Ifalltime = 530 ns.	▪ Ifalltime = 660 ns.
▪ Eon = 13 mJ.	▪ Eon = 20 mJ.
▪ Eoff = 24 mJ.	▪ Eoff = 25 mJ.
▪ Econd = 1.3 mJ (single pulse)	▪ Econd = 1.5 mJ (single pulse)

Summary

This work presented the static and dynamic characterization of state-of-the-art 4H-silicon carbide IGBTs as well as a brief description of the high voltage test beds utilized to obtain the laboratory measurements. The 12-kV IGBT tests showed two different designs with 2- µm and 5-µm field-stop buffers which offered several advantages depending on the application. The thicker FSB layer would be preferred where one might want lower switching losses and faster rise/fall times, at the sacrifice of conduction losses (e.g. pulsed applications). The device with the thinner FSB would be preferred for power conversion and high duty cycle applications, but as with most commercial IGBTs, the current tail losses must be considered. The 20-kV IGBTs showed similar characteristics as the 12-kV devices, but they were a newer generation with a thicker epitaxial drift region and offered operating voltages higher than 15 kV. The two 20-kV sample devices showed very similar characteristics, but the first one had a slightly better performance in terms of switching and conduction losses. Although at this point more reliability tests are needed to determine whether these 12-kV and 20-kV SiC IGBTs devices can actually replace existing commercial Si IGBTs and gas switches, this project showed promising results and it provides a platform to conduct more extensive tests in the future.

References

[1] Sei-Hyung Ryu; Capell, C.; Jonas, C.; Lin Cheng; O'Loughlin, M.; Burk, A.; Agarwal, A.; Palmour, J.; Hefner, A., "Ultra high voltage (>12 kV), high performance 4H-SiC IGBTs," Power Semiconductor Devices and ICs (ISPSD), 2012 24th International Symposium on , vol., no., pp.257,260, 3-7 June 2012

[2] Sei-Hyung Ryu et al., "20 kV 4H-SiC N-IGBTs" Material Science Forum. Materials Science Forum, 778-780, 1030.

[3] Brunt, E.V.; Cheng, L.; O'Loughlin, M.; Capell, C.; Jonas, C.; Lam, K.; Richmond, J.; Pala, V.; Ryu, S.; Allen, S.T.; Burk, A.A.; Palmour, J.W.; Scozzie, C., "22 kV, 1 cm^2, 4H-SiC n-IGBTs with improved conductivity modulation," *Power Semiconductor Devices & IC's (ISPSD), 2014 IEEE 26th International Symposium on* , vol., no., pp.358,361, 15-19 June 2014. https://doi.org/10.1109/ISPSD.2014.6856050

[4] J. Baliga, Fundamentals of Power Semiconductor Devices, Springer, 2008. https://doi.org/10.1007/978-0-387-47314-7

Chapter 31

Electro-Thermal TCAD Model for 22 kV Silicon Carbide IGBTs

Miguel Hinojosa[1, a], Aderinto Ogunniyi[1], Stephen Bayne[2], Edward Van Brunt[3] and Sei-Hyung Ryu[3].

[1]Army Research Laboratory, 2800 Powder Mill Road, Adelphi, MD. USA.

[2]Texas Tech University, 224 Electrical and Computer Engineering, Lubbock, TX. USA.

[3]Wolfspeed, a Cree company, 4600 Silicon Dr., Research Triangle Park, NC. USA.

[a]miguel.hinojosa4.civ@mail.mil

Abstract

This section presents the current progress in the development of an electro-thermal numerical model for 22 kV 4H-silicon carbide IGBTs. This effort involved the creation of a TCAD model based on doping profiles and structural layers to simulate the steady-state and switching characteristics of recently-fabricated experimental devices. The technical challenge of creating this high voltage SiC IGBT model was incorporating semiconductor equations with sub-models representing carrier mobility, generation, recombination, and lattice heat flow effects with parameters conditioned for 4H-silicon carbide material. Simulations of the steady-state and switching characteristics were performed and later verified with laboratory measurements for an N-type SiC IGBT rated for 22 kV with an active area of 0.37 cm^2 and a drift region of 180 μm.

Keywords

4H-SiC, IGBT, Silicon Carbide, High Voltage, TCAD Simulation, Pulsed Power Switch

Contents

The work has been previously published by Trans Tech Publications Ltd:
M. Hinojosa et al., "Electro-Thermal TCAD Model for 22 kV Silicon Carbide IGBTs", Materials Science Forum, Vol. 858, pp. 949-953, 2016

1. Introduction

Significant progress has been made in the development of ultra-high voltage (UHV) silicon carbide IGBTs in the last five years due to improvements in process, device fabrication, and packaging technologies. 4H-SiC IGBTs have been reported with operating voltages from 15 kV up to 27 kV at a rated current of 20 A [1-4]. These next-generation bipolar devices show great potential for use in the energy and medical sectors for pulsed power and grid applications.

This novel research entails the creation of a physics-based TCAD model based on doping profiles and structural layers to simulate the steady-state and transient characteristics of recently fabricated experimental devices [1-2]. While compact SABER and SPICE models for similar devices have been reported in literature [5-6], the intent of the TCAD model is to complement existing simulation tools with the added ability to monitor key areas within the structure during operation. In addition, the numerical physics-based model is intended to be used as an economical way to investigate design tradeoffs and optimizations for specific applications, as well as an eventual predictor and diagnostic resource to pinpoint failure mechanisms under various operating conditions. The modeling and simulations were performed using a commercially available 2-D TCAD software package. In order to validate the results, the simulated data was later plotted against laboratory measurements obtained from an R&D-grade 22 kV N-channel SiC IGBT.

2. Method

The high voltage SiC IGBT model was based on the drift-diffusion transport equations, Poisson's equation, the continuity equations, and the lattice self-heating equation (see Fig. 1), where each node is solved numerically by the 2-D simulator. The semiconductor equations contain independent and dependent variables coupled in areas of carrier mobility, generation, recombination, and lattice heat flow effects.

There were two critical and challenging parts that were necessary for creating a good quantitative model representative of the 20 kV SiC IGBTs: 1) establishment of reliable SiC material properties based on compatible published literature or laboratory measurements and 2) defining SiC device-specific conditions such as scaling factors, layer thicknesses, carrier concentration variability, ambipolar lifetimes in layers and effective channel lengths. Because a comprehensive explanation of each step and detailed description of each sub-model is not possible within the confines of this section, a summary of the various sub-models is presented below on Table 1.

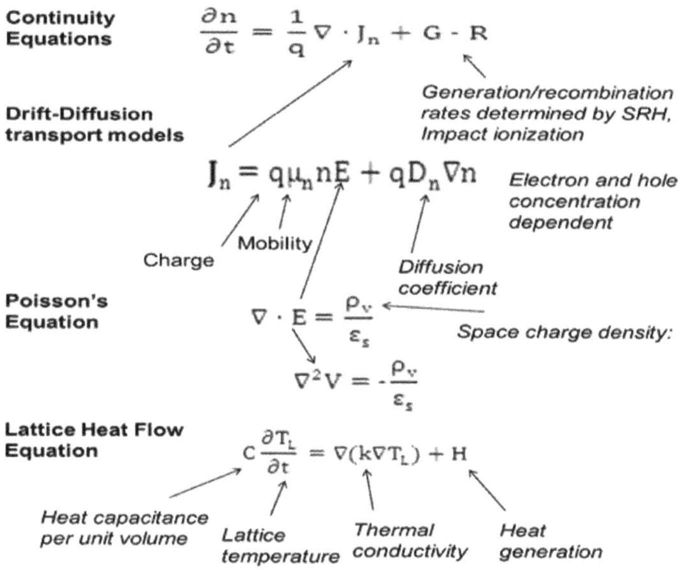

Fig. 1 Simplified 1-D Semiconductor Equations.

A number of 4H-SiC IGBTs with a chip area of 1 cm^2 were designed and fabricated by Wolfspeed with ratings of 20-22 kV and 20 A. The structure was implemented within the simulation tool as seen in Fig. 2, based on details provided in [1-2]. A full cell containing a double-diffused MOS vertical structure was created with a cell pitch of 19 μm, channel length of approximately 1 μm, and with an oxide layer of 500 Å. Each layer was drawn with a Gaussian doping profile in order to create a more realistic non-uniform doping concentration. The IGBT model utilized an N-type epitaxial drift region with a thickness of 180 μm, a concentration of 2 x 10^{14} cm^{-3}, and an asymmetrical punch-through design. The thin field-stop buffer (FSB) was 2 μm, which was used to control the injection of holes from the P+ region, and had a carrier concentration of 5 x 10^{16} cm^{-3}. The P+ substrate, or backside injection layer, had a doping concentration of 1 x 10^{19} cm-3. This layer was located below the FSB and it was connected to the collector electrode, which was also utilized as a heatsink for the thermal model. Because the IGBTs had an active area of 0.37 cm^2, the scaling factor for the two-dimensional structure was 1.95 x 10^6 µm to account for multiple cells and the length across the z-axis.

Table 1 Summary of sub-models used for the 4H-SiC IGBT.

Sub-model:	Area:	Accounts for:
Mobility	Bulk	Low-field effects: Temp. and doping dependent
		High-field effects. Saturation velocity and temp. dependent
		Conductivity modulation carrier-carrier scattering
		Anisotropic effects for perpendicular plane.
	Channel	Phonon scattering effects
		Surface roughness effects
		Charged impurity effects
Generation	Breakdown	Anisotropic impact ionization.
Recombination	Bulk	Two-step Shockley-Read-Hall recombination
		Concentration-dependent lifetime model
	Surface	Surface dangling bonds model
	Auger	High concentration 3-step recombination process
Thermal	Lattice Heating	Thermal conductivity, specific heat
		Heat generation
Carrier Statistics	Fermi	Fermi-Dirac statistics
	BGN	Band-gap narrowing effects
		Incomplete ionization

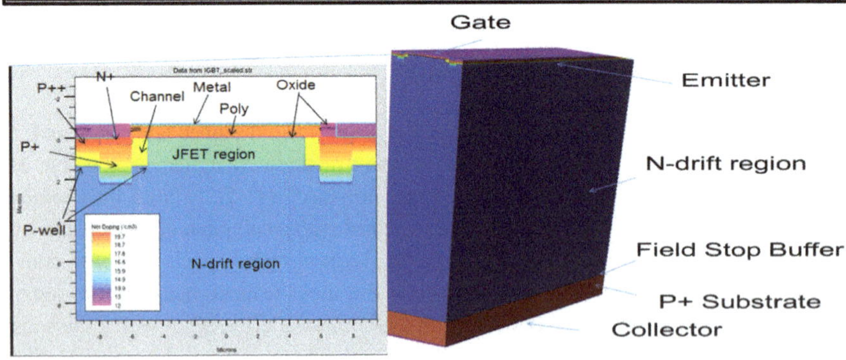

Fig. 2 Proposed cross-section of 22 kV SiC N-IGBT based on [1-2] with 180 µm epitaxial drift region and 2 µm FSB.

3. Steady-state and transient results

The steady-state *Ic-Vce* forward conduction family of curves and the *Ic-Vge* transfer curve were utilized as tuning and optimization tools to adjust the device-specific parameters. The forward-conduction I-V characteristics of the fabricated 4H-SiC IGBTs were first measured with a high-power parametric curve tracer that provided 0-20 V, 300 μs *Vce* pulses. The gate voltage had a range from 5 V to 20 V, and it incremented in 5 V steps every 66 ms. The transfer curve was obtained by setting the collector voltage to a fixed voltage of 3.2 V, which was high enough to forward bias the junction between the P+ substrate and the FSB, and by sweeping the gate voltage *Vge* from 0-15 V. The simulated vs. measured results are illustrated in Fig. 3. At first glance, the matching of the forward conduction I-V plots and the transfer curve seemed like a trivial task, but it turned out to be a challenging exercise because of the interdependence of model parameters.

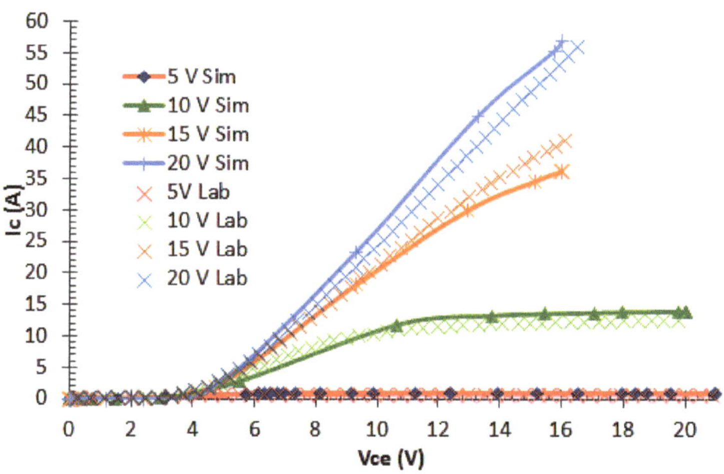

Fig. 3 Simulated vs. measured results of Ic-Vce family of curves

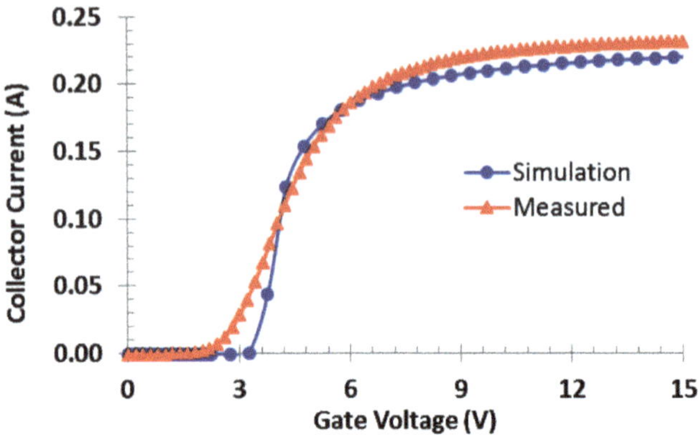

Fig. 4 Simulated vs. measured results of Ic-Vge transfer plot.

Design of experiments (DOE) trials were conducted with the simulator to qualitatively identify the principal components, to observe their effect on the curves, and to quantitatively adjust the coefficients for best fit. Table 2 shows a summary of the optimization factors and their effect.

The dynamic response of SiC IGBT model was also compared with laboratory experiments by switching a resistive load at a bus voltage of 8 kV. A low-inductance 1200 Ω load was selected, which displayed a shunt parasitic capacitance of 34 pF. The bus connection between the load and the IGBT had minimal resistance, but it contained an equivalent series inductance of 38 nH which was included in the simulation. The gate voltage at the negative rail was set to -5.5 V and the overdrive voltage to 15.6 V. The pulse width of this input signal was 13 µs, and its period was 26 µs with a 50 % duty cycle. The gate driver signal was connected to a gate resistor of 14 Ω. Fig. 4 shows the results of the measured vs. simulated waveforms for the gate voltage Vge, the collector voltage Vce, and the collector current Ic from top to bottom, respectively. Although not a perfect match, the waveforms show congruency during turn on and turn off. In order to improve the turn on characteristics, additional parasitic elements need to be considered at the gate and the device terminations. In order to obtain a better match during turn off, the carrier lifetimes and field-stop buffer characteristics need to be further optimized.

Table 2 Summary of optimization factors.

Category	Factors	Levels	Impact
Channel Mobility	Phonon scattering	mref1n, mref1p	Affects Plateau of IV curves.
	Surface roughness	mref2n, mref2p	Most effect on high-field mobility (Plateau of IV curves, Vg = high)
	Coulombic scattering	mref3n, mref3p,al3n, al3p	Affects low-field mobility (Plateau of IV curves, Vg = low)
	Interface charge	Qf	Affects threshold voltage
Recombination	Drift Lifetime	concentration dependent model	Affects slope of IV plots and magnitude significantly. Affects breakdown voltage.
	FSB Lifetime	Concentration dependent model	Minor effect on slope and plateau of IV curves.
Surface Recombination	Interface surface recombination	S.N, S.P	Affects PNP gain.
	Contact surface recombination	Vsurfn, Vsurfp	Affects PNP gain.
Thermal	Thermal conductivity	TC.A, TC.B, TC.C	Affects lattice temperature, severe effects on IV plots
	Specific heat capacitance	HC.A, HC.B, HC.C, HC.D	Affects lattice temperature, severe effects on IV plots
	Boundary conditions	Thermal contact, alpha	Affects lattice temperature, severe effects on IV plots

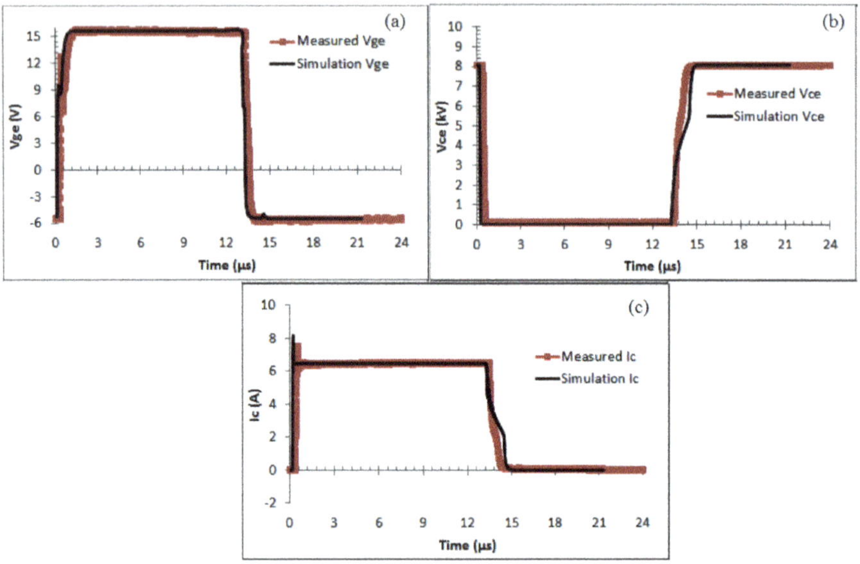

Fig. 4 Measured vs. simulated transient results (a) gate voltage Vge,(b) collector voltage Vce, (c) collector current Ic.

Nevertheless, the implementation of this TCAD model with a commercial numerical simulator shows acceptable results. One of the advantages of the TCAD model is the ability to monitor the structural components during device operation. Fig. 5(a), for example, illustrates the current density at various points in time during the 8 kV, 6.5 A switching event. When the IGBT fully conducts, high current density leads to localized heating in the JFET region which then transitions to the epi-layer region. The JFET region could be improved in future designs by implementing a carrier storage layer (CSL) or similar technique [8]. Fig. 5(b) illustrates an example of resistive switching with similar loading conditions, but this time the IGBT is pulsed for several cycles. Although the temperature rise in this case is minimal, the same method could be used to predict temperature variations at higher voltage and current levels, as presented in Fig. 5(c). This case shows a high current test of 55 A during a short 3 μs, 8 kV pulse. In this condition, a greater temperature rise was observed with hot spots forming near the JFET and channel regions.

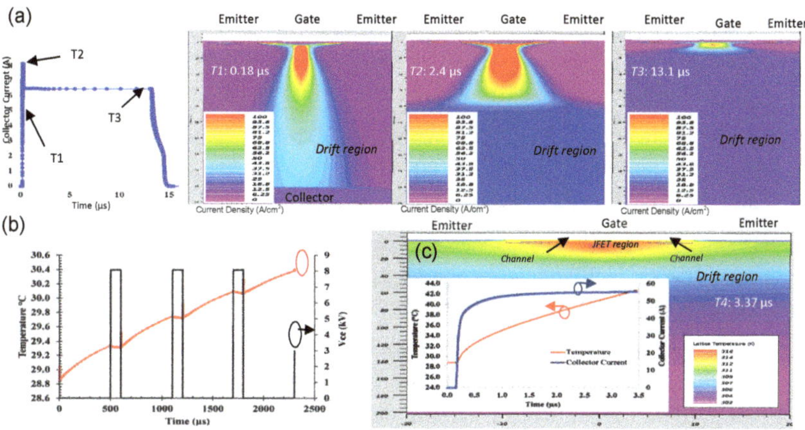

Fig. 5 (a) Current density of the IGBT at various points in time for single pulse resistive switching at 8 kV, 6.5 A. (b) Temperature profile for repetitive switching at 8 kV, 6.5 A with a period of 600 µs and an 83 % duty cycle. (c) Estimated internal temperature heating during turn on with a short 3 µs 8 kV, 55 A pulse.

Summary

This work presented the current progress in developing a physics-based, electro-thermal TCAD model for a 22-kV 4H-SiC IGBT. Simulations of the steady-state and transient characteristics were presented and their results were validated with measurements obtained in the laboratory. Future work includes further optimization of the models, the addition of JTE or guard ring terminations, and the implementation of the IGBT model with more complex circuits and stressing conditions.

References

[1] Brunt, E.V. et al., "22 kV, 1 cm2, 4H-SiC n-IGBTs with improved conductivity modulation," 2014 IEEE 26th International Symposium on Power Semiconductor Devices & IC's (ISPSD). https://doi.org/10.1109/ISPSD.2014.6856050

[2] S. Ryu et al., "20 kV 4H-SiC N-IGBTs", Materials Science Forum, v778-780, p1030, 2014. https://doi.org/10.4028/www.scientific.net/MSF.778-780.1030

[3] S. Ryu et al. "Ultra-high voltage IGBTs in 4H-SiC." In Wide Bandgap Power Devices and Applications (WiPDA), 2013 IEEE Workshop on, pp. 36-39. IEEE, 2013. https://doi.org/10.1109/WiPDA.2013.6695557

[4] E. van Brunt et al, "27 kV, 20 A 4H-SiC n-IGBTs", Materials Science Forum, Vols. 821-823, pp. 847-850, Jun. 2015. https://doi.org/10.4028/www.scientific.net/MSF.821-823.847

[5] Duong, T.H. et al., "Physics-based electro-thermal Saber model and parameter extraction for high-voltage SiC buffer-layer IGBTs," Energy Conversion Congress and Exposition (ECCE), 2014.

[6] Saadeh, M. et al., "A unified silicon/silicon carbide IGBT model," Applied Power Electronics Conference and Exposition (APEC), 2012 Twenty-Seventh Annual IEEE, p.1728,1733, Feb. 2012. https://doi.org/10.1109/APEC.2012.6166055

[7] Silvaco "ATLAS User's Manual", September 11, 2014

[8] Ryu, S., et al. "Ultra high voltage MOS controlled 4H-SiC power switching devices." *Semiconductor Science and Technology* 30.8 (2015): 084001.

Chapter 32

Development of a Compact 2-kW, 28- to 600-V DC-DC Battery Charger

Robert A. Wood, Damian P. Urciuoli
U.S. Army Research Laboratory, 2800 Powder Mill Road, Adelphi, MD 20783, USA
robert.a.wood98.civ@mail.mil, damian.p.urciuoli.civ@mail.mil

Abstract

This section outlines the design and evaluation of a 2-kW, 28- to 600-V DC-DC converter designed to charge a battery of a vehicle pulse power system. An atypical topology was chosen for the converter which reduces the amount of capacitance and volume required. This section also discusses the operation, hardware design, experimental waveforms and operation of the converter.

Keywords

Battery Charger; DC-DC Converter; Electric Vehicles; Current-Fed Half-Bridge; Silicon Carbide.

Contents

The work has been previously published by Sheridan Communications:
Robert A. Wood and Damian P. Urciuoli, "Development of a Compact 2-kW, 28- to 600-V DC-DC Battery Charger," GOMACTech Conference Proceedings, March 31, 2013.

1. Introduction

Pulse power systems require high energy for short durations. However, typical vehicle 28-V busses are incapable of supplying sufficient current to pulsed loads, and would need to be prohibitively cumbersome to do so. One way of overcoming this limitation is to install a high voltage (600-V) battery that can be charged over a longer duration from the 28-V bus. High power pulses are supplied by the 600-V battery, not the vehicle bus, and the battery can be recharged to replenish the energy discharged during the event.

This section focuses on the development of a charger that provides the power conversion from the 28-V low-voltage- side (LVS) bus to the 600-V high-voltage-side (HVS) battery. The nominal design constraints for the charger are shown in Table 1. Note that the input LVS voltage ranges from 25 to 31 V for a nominal 28-V bus, charged by an alternator. The HVS voltage ranges from 500 to 650 V and the max current output of the converter is 3.2 A throughout the LVS and HVS voltage ranges.

There is a requirement for 1200-V of galvanic isolation between the LVS and the HVS. The LVS on the vehicle is typically referenced to the chassis and isolation was added to increase system safety. This isolation was provided by a transformer in the power circuit.

Table 1 Battery Charger Nominal Specifications.

	Min	Nom	Max
Input LVS Voltage (V)	25	28	31
Output HVS Voltage (V)	500	600	650
Output Current (A)	0	3.2	3.2
Output Power (W)	0	1920	2080
Galvanic Isolation (V)		1200	
Case Temperature (°C)	0	25	70

The converter is in contact with the vehicle's chassis which is at a max temperature of 70 °C and is not actively cooled. Typically, each converter operational interval will be short because the 600-V battery will not be deeply discharged. However, the worst-case thermal condition results from a full recharge of the battery at the rated output power and case temperature. This condition guided the converter design.

2. Converter design

Topology: A topology relying more heavily on high temperature rated components was sought to address the cooling limitations. Emphasis was placed on the use of magnetic components such as inductors and transformers, due to typically higher temperature capability and lower effective thermal resistances compared to capacitors. To achieve a high voltage step-up while using a relatively low transformer turns ratio to limit leakage inductance, a current-fed half-bridge (CFHB) topology with a LVS active clamp and HVS voltage doubler was selected [1]-[3]. This CFHB topology is shown in Figure 1. MOSFETs Q1 and Q2 form the CFHB with inductors L1 and L2, the transformer (X) primary, voltage source VS, and capacitor C1. MOSFETs Q3 and Q4 with clamp capacitor CC serve as the active clamp, recovering energy stored in the leakage inductance of the transformer following the turn-off transitions of MOSFETs Q1 and Q2, respectively. MOSFET Q5 and resistor RE are part of a protection circuit that is described in a later section. On the HVS, capacitors C2 and C3 form a voltage doubler with diodes D1 and D2 to boost converter voltage step-up.

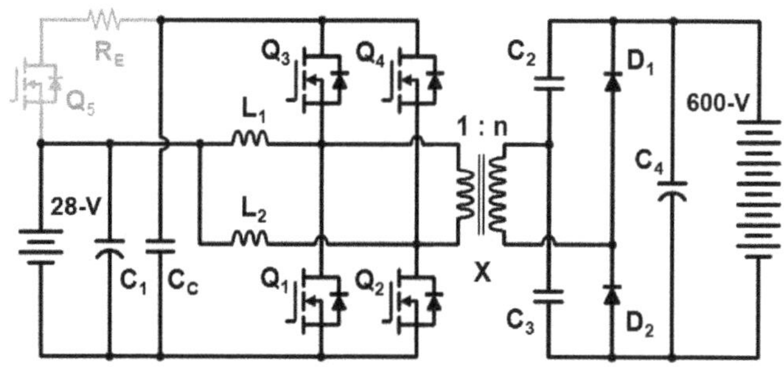

Fig. 1 Schematic of CFHB Topology with Active Clamp and Voltage Doubler.

Circuit Operation: Figure 2 shows selected simulated converter waveforms for operation at a 600-V, 2-kW load. At the operating condition shown in Figure 2, the duty cycle of Q1 and Q2 is greater than 50%, and the inductors operate in continuous conduction mode (CCM). Unlike the converter proposed in [1], this converter has no discrete leakage inductor and operates using only the inherent leakage inductance of the transformer. MOSFETs Q3 and Q4 are gated-on for a small portion of the complementary intervals of Q1 and Q2, respectively, to allow approximately one half-cycle of resonant current flow between the transformer leakage inductance and the clamp capacitor. The low duty cycles of Q3 and Q4 reduce clamp currents and corresponding component ratings by preventing additional current oscillations.

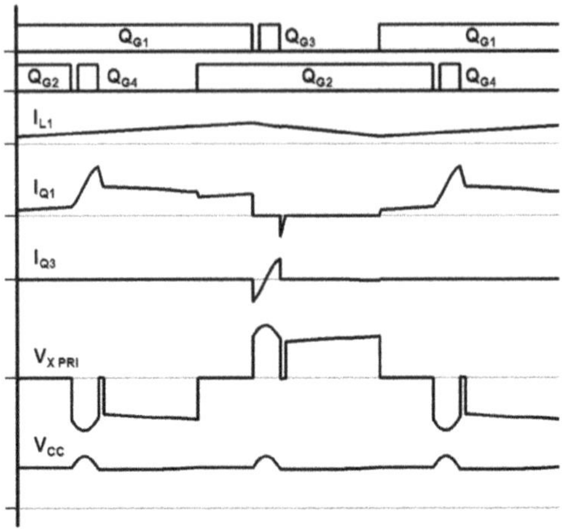

Fig. 2 Simulated Full-Power Converter Waveforms

Mechanical Design: The converter was designed with low passive component values to minimize volume and meet the high temperature constraint. Current de-ratings for active and passive components were also required to meet high operating temperatures. Table 2 lists selected converter component values at the switching frequency of 50 kHz. Two half-bridge modules were used to implement low inductance MOSFET pairs (Q1, Q3) and (Q2, Q4). Two silicon carbide (SiC) junction barrier Schottky (JBS) diodes were paralleled for each of D1 and D2.

Table 2 Converter Component Values

Component	**Value / Rating**
C1	1760 μF / 50 V
CC	1 μF / 250 V
L1, L2	10 μH
C2, C3	0.48 μF / 630 V
C4	14 μF / 900 V
X	1 : 4, L_L = 0.9 μH, L_M = 200 μH
(Q1, Q3), (Q2, Q4)	200 V, 278 A (T_C = 80 °C)
D1, D2	1200 V, 10 A x 2 (T_C = 150 °C)

The converter layout was designed to minimize volume, bus inductance, and internal electromagnetic interference. Figure 3 shows a top view inside the case of the converter with the front and back of the converter on the left and right of the figure, respectively. A footprint 13.3 in. long by 6.0 in. wide and a height of 2.3 in. resulted in a converter volume of 184 in3 (3.0 liters). With the baseplate covering the entire footprint, a large heat transfer area to converter volume ratio was achieved. Connectors for the control / communication interface, the 28-V input, and the 600-V output were placed in a single row on the front panel of the converter to minimize the total effective volume occupied by connectors. A compact 28-V bus, housing capacitance C1, is located between the inductors at the front of the converter. The MOSFET modules extend along the length of the converter behind each inductor, with gate drivers installed above the modules. A four-layer, parallel-plate bus connects to: both inductors, all MOSFET power terminals, the clamp capacitor, and the transformer primary winding. The clamp capacitor and transformer are located on the center line of the converter for symmetric current paths. The voltage doubler, output capacitor C4, current and voltage sensors, and the closed loop controller are located at the back of the converter. The control boards were placed near the back of the converter to distance the controller from the high currents conducted by LVS components during operation.

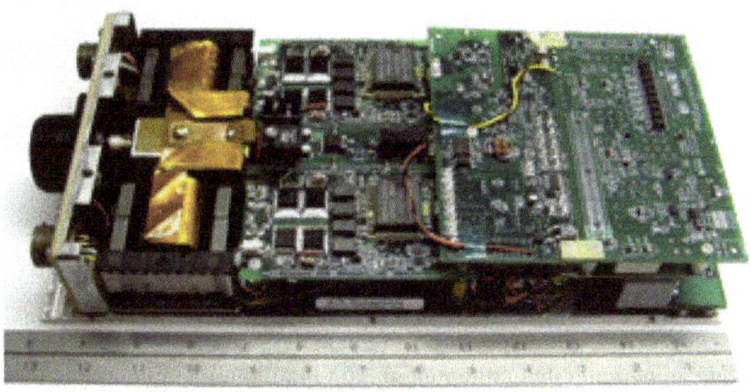

Fig. 3 28- to 600-V, 2-kW CFHB DC-DC Converter.

Software Design: The converter is designed to be a part of a pulse power system, which has specific communication requirements. The communication protocol is controller area network (CAN) bus, which is typically used in vehicles for communications between components. The commands received by the converter include target voltages and currents and the converter operating mode. These commands and operating modes are explained in more detail in the next section. The converter measures the LVS voltage, HVS voltage and HVS current for closed loop control and circuit protection. The baseplate temperature and controller temperatures are also measured by the controller and used to provide thermal shutdown protection. These measured values are reported over the CAN-bus to the rest of the pulse power system.

3. Experimental results

Modes of Operation: The converter has two main modes of operation: precharge and battery charge. Precharge is used to bring the HVS capacitance voltage to the voltage of the battery. Battery voltage is communicated to the converter as a precharge voltage and when enabled, the converter precharges to the battery voltage, minimizing the stress and maximizing the life of the contactor. Battery charge mode is used when the charger is actively charging the battery. The charger operates as a voltage mode power supply with a current limit. The current limit and target voltage are received over CAN-bus by the converter, and the converter charges the battery when enabled.

Precharge Mode: The battery charger HVS was precharged to voltages ranging from 500 to 650 V at input voltages ranging from 25 V to 31 V to document the capability of the converter to precharge the output. Additional capacitance was added to the HVS of the converter to emulate the capacitance of a connected pulse power system. One precharge example is shown in Figure 4, which occurred at a target voltage of 650 V, and an input of 31 V. In this example, the converter reached 90% of the precharge voltage within 0.7 s.

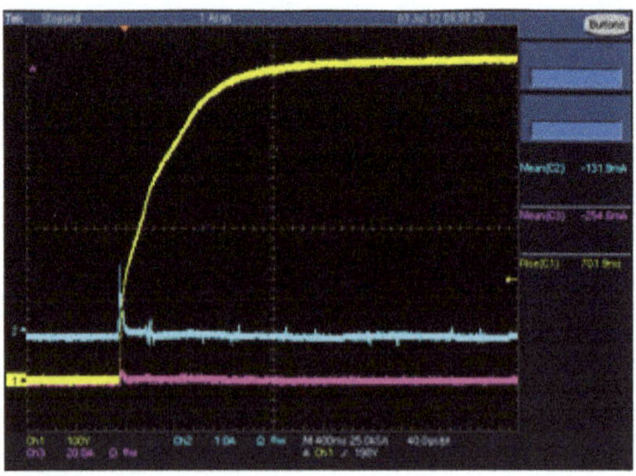

Fig. 4 Precharge Example.

Battery Charge Mode: The second operational mode is battery charge mode where the converter functions like a voltage source with a current limit. First, a target voltage and a max current limit are received on the CAN-bus. If the voltage on the battery is below the voltage limit, the converter will increase the amount of current produced until the max charging current is reached. As the state of charge (SOC) of the battery increases, the amount of current required to reach the target voltage decreases. When the target voltage is reached, the converter regulates the amount of charging current to keep the battery at the target voltage. Eventually, no current will be required to sustain the target voltage and the converter will disable to prevent overcharging the battery. A typical charging waveform is shown in Figure 5, which shows the change from constant current to constant voltage operation.

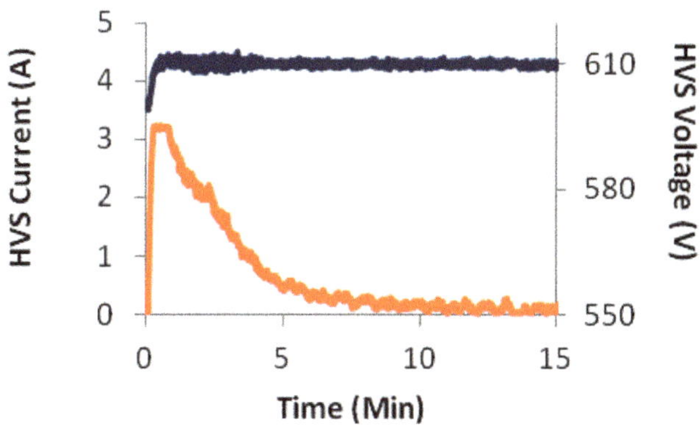

Fig. 5 HVS Voltage and Current During Charging.

The target voltage was set to 610 V and the max current was specified as 3.2 A. While the battery voltage is less than the target voltage of 610 V, the converter outputs the max current of 3.2 A. Once the battery voltage reaches the target voltage, the controller switches to constant voltage operation and reduces the current to regulate voltage, eventually reaching 0.0 A. Waveforms were captured during the charge cycle shown in Figure 5, and those corresponding to the max power of 1952 W (610 V @ 3.2 A) are shown in Figure 6.

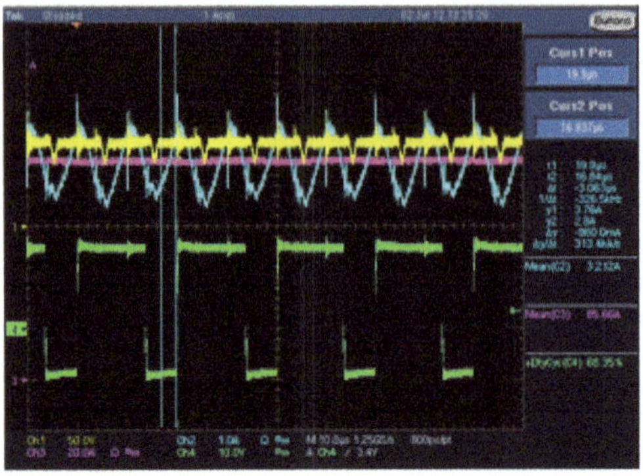

Fig. 6 Converter Waveforms at Full Power.

The MOSFET (Q1) duty cycle is 68% and the converter is operating in CCM with non-zero inductor currents (not shown). The HVS current oscillates at 100 kHz, which is twice the switching frequency. These oscillations are not a concern for the pulse power system as the only converter load is the 600-V battery. However, there are significantly less oscillations on the LVS side, which is shared with additional vehicle electronics. Although the oscillating energy between CC and the leakage inductance of the transformer is evident from the CC voltage ripple, its voltage is a relatively constant 75 V.

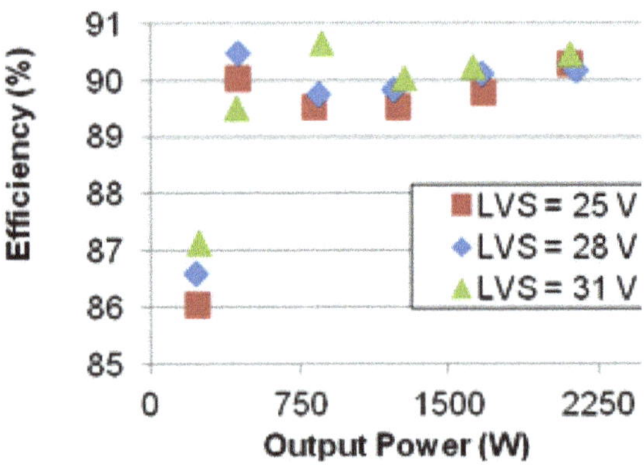

Fig. 7 Converter Efficiency vs. Output Power.

The output power was varied at different LVS voltages to document the variability of the efficiency over the load range. Figure 7 shows the efficiency with respect to output power at a constant HVS output of 650 V for various LVS input voltages. Other than the lightest load point, the average efficiency is approximately 90%. Higher LVS voltages increase the efficiency only slightly over the output power range.

Fault Handling: When the converter is operating at high loads, it enters CCM and the inductors L1 and L2 have stored energy in their magnetic fields. However, if the switches are instantly disabled such as during a fault, the energy in the inductors is transferred to the clamp capacitor (CC). The 200-V rating of the MOSFETs would be exceeded if all the energy in the inductors is transferred to the 1-μF clamp capacitor.

This problem was mitigated by the addition of a MOSFET (Q5) on the LVS connected between the positive 28-V input terminal and the positive terminal of the clamp capacitor, as shown in Figure 1 in grey. A 1.0-Ω resistor (RE) was added in series with Q5 to limit the peak current during a fault. The control was programmed such that if all of the main switches in the circuit are disabled (Q1, Q2, Q3 and Q4), Q5 automatically turns on and limits the voltage on CC. This circuit has the added benefit of returning the energy in the clamp capacitor to the 28-V bus, effectively recycling some of the energy. The protection circuit was tested and the voltage on CC rose from a steady state value of approximately 90 V to a max of 190 V during a full power shutdown as shown in Figure 8. This max voltage is within the system component ratings, and demonstrates the successful limiting of the clamp voltage during a fault.

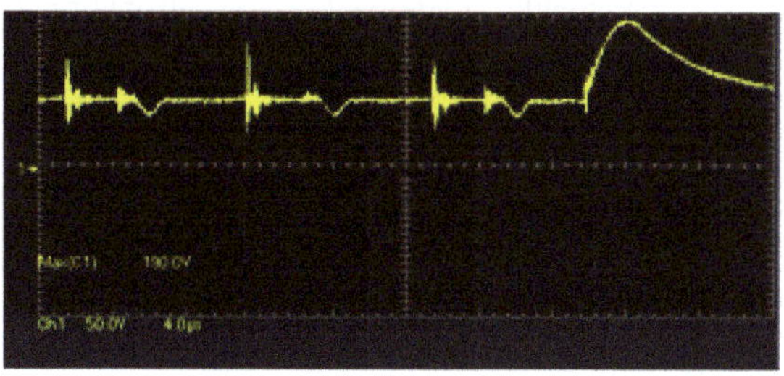

Fig. 8 Clamp Capacitor Voltage During Shutdown.

Summary

This section documents the design and evaluation of a converter that is part of a pulse power system for legacy ground vehicles with a 28-V bus. This converter charges a 600-V battery from the 28-V bus at a nominal power rating of 2-kW without active cooling.

The topology of the converter was chosen to be a current-fed half-bridge with active clamp and voltage doubler due to the temperature constraint on the converter and to minimize volume. The active components of the converter are 200-V MOSFETs and 1200-V SiC JBS diodes. The converter was operated at 50 kHz and has an effective volume of 3.0 liters.

The converter was tested over its full voltage, current, and power ranges. Waveforms were presented documenting the operation of the converter while charging a battery and utilizing target voltage and max current commands received over a CAN-bus network. The efficiency of the converter over the operating space was shown to be approximately 90%. This design meets all of the physical and operational system requirements and is ready to be a part of a successful vehicle based pulse power system.

References

[1] Kim, J., et al, "An Active Clamping Current-Fed Half-Bridge Converter for Fuell-Cell Generation Systems," Proceedings of the 35th Power Electronics Specialists Conference, Aachen, Germany, 2004, Vol. 6, pp.4709-4714.

[2] Kwon, J., and Kwon, B., "High Step-up Active Clamp Converter with Input-current Doubler and Output-voltage Doubler for Fuel Cell Power Systems," IEEE Transactions on Power Electronics, Vol. 24, no. 1, pp-108-115, 2010.

[3] Urciuoli, D., Wood, R A., and Tipton, C.W., "Trade Study and Design of a TRL-4, 100OC, 28-to 600-V Bidirectional DC-DC Converter Power Stage," ARL-TR-5430, Jan 2011. www.arl.army.mil/arlreports/2011/ARL-TR-5430.pdf

Keyword Index

About the Editors

Dr. Edward C. Shaffer received the B.S. degree from the US Military Academy; the M.S. and E.E. degrees from the Massachusetts Institute of Technology; and the Ph.D. in Electrical Engineering from Auburn University. He served in a variety of technical and leadership positions as a US Army officer, including tours in Germany, Korea and the United Kingdom, and as an Associate Professor in the Department of Electrical Engineering & Computer Science at the US Military Academy. Dr. Shaffer was also a Senior Design Engineer with Solectria Corporation in Woburn, MA. As Chief of the Energy & Power Division at ARL, he supervises scientists and engineers developing materials, devices and components for high energy batteries, compact power sources, alternative energy solutions, power electronics, energy networks, and related technologies. He has also served in a number of senior leadership details within ARL, including Acting Deputy Director and Director of the Sensors and Electron Devices Directorate; Acting Chief of the Electronics & RF Division, and Acting Director of the Vehicle Technology Directorate. Dr. Shaffer is the senior Army representative for the DoD Energy & Power Community of Interest; he is also Past Chair of the both the Steering Committee and Electrical Systems Working Group for the Interagency Advanced Power Group (IAPG). His military awards include the Legion of Merit; he is a Senior Member of IEEE and is a licensed Professional Engineer.

Dr. Tsvetanka S. Zheleva received her B.S. and M.S degree in Physics from Sofia University, Bulgaria and her Ph.D. in Materials Science and Engineering from North Carolina State University. She joined ARL with The National Research Council/National Academy of Sciences and The American Association for Engineering Education fellowships. Also, during her fellowship at ARL she was Adjunct Associate Professor at North Carolina State University. Dr. Zheleva's expertise is in the field of condensed matter physics and in particular thin film physics, including electronic materials characterization and synthesis for electronic and optoelectronic devices. This includes: structural characterization by transmission electron microscopy and x-ray diffraction of new materials and interfaces, thin films, and bulk materials; dislocation mechanisms, modeling of epitaxial growth, design of new modes of epitaxy; image processing and simulation, atomistic modeling; other characterization techniques. As Associate Chief of the Energy & Power Division at ARL, she oversees numerous Science and Technology programs in Power and Energy relevant technologies. Dr. Zheleva has published her work in over 150 peer review journals and has 28 US and foreign patentsin the area of wide bandgap semiconductor device structures. Dr. Zheleva has numerous awards from US professional societies and the Department of the Army Commanding General Commander's Award for Civilian Service.

www.ingramcontent.com/pod-product-compliance
Ingram Content Group UK Ltd.
Pitfield, Milton Keynes, MK11 3LW, UK
UKHW020700041026
15590UKWH00001B/372